# Obstacles Facing Hydrogen Green Systems and Green Energy

Jamal Mabrouki
*Faculty of Science, Mohammed V University in Rabat, Morocco*

IGI Global
Scientific Publishing
Publishing Tomorrow's Research Today

| | |
|---|---|
| Vice President of Editorial | Melissa Wagner |
| Director of Acquisitions | Mikaela Felty |
| Director of Book Development | Jocelynn Hessler |
| Production Manager | Mike Brehm |
| Cover Design | Phillip Shickler |

Published in the United States of America by
IGI Global Scientific Publishing
701 East Chocolate Avenue
Hershey, PA, 17033, USA
Tel: 717-533-8845
Fax: 717-533-7115
Website: https://www.igi-global.com E-mail: cust@igi-global.com



Library of Congress Cataloging-in-Publication Data

LCCN: 2024055959 (CIP Data Pending)
ISBN13: 9798369389805
Isbn13Softcover: 9798369389812
EISBN13: 9798369389829

British Cataloguing in Publication Data
A Cataloguing in Publication record for this book is available from the British Library.

# Table of Contents

# Detailed Table of Contents



World wide hydrogen energy is known as the clean energy, which acts as a savior of environment as it will not harm environment and nature.The hydrogen energy is considered as the new source of fuel energy that will bring changes and revolution to the earth. It is expected that all the future vehicles will be dependent on hydrogen energy.Still it is not widely adopted .There are many challenges faced by scientists in declaring hydrogen energy as alternate fuel for future vehicles or energy which can reduce green house effect. Hence in this chapter an effort is taken to find the main barriers of hydrogen energy adoption as per Indian scenario, then SWARA(Stepwise Assessment ratio) and DEMATEL(Decision Making Trial and Evaluation Laboratory) analysis is conducted to find cause and effect relation between the challenges.



The transition towards a sustainable energy future is increasingly focused on hydrogen as a clean, versatile fuel that can decarbonize various sectors, including transportation, industry, and energy storage. Green hydrogen, produced through water electrolysis using renewable energy, is at the forefront of this transition. This chapter provides a comprehensive overview of the various water electrolysis technologies, including alkaline water electrolysis (AWE), proton exchange membrane (PEM) electrolysis, and solid oxide electrolysis (SOE). It discusses their principles, efficiency, advantages, challenges, and the role each technology plays in the context of large-scale green hydrogen production. The chapter also explores recent advancements, ongoing research, and future trends that could shape the development and deployment of these technologies. With a focus on sustainability, the chapter aims to elucidate the critical role of water electrolysis in achieving global carbon reduction targets and fostering a hydrogen-based economy.

This study examines the in vitro antagonistic effect of three native Trichoderma sp. isolates originating from saffron corms, against three Fusarium species plants (F. pseudograminearum, F. moniliforme and F. oxysporum) known to cause rotting in the bulbs and roots of saffron. To introduce Trichoderma into a preventive control program, the Trichoderma isolates were tested by in direct confrontation and volatile substances production. They demonstrated colonization percentages ranging from 50% to 79.6%, significantly inhibiting the mycelial growth and spore germination of pathogenic species, with inhibition rates ranging from 53% to 79% and 51% to 79%, respectively. However, indirect confrontation through the production of diffusible substances showed low inhibition of mycelial growth and spore germination for all pathogenic isolates studied, not exceeding 28.47% and 20.5%, respectively. This study discusses the potential of using Trichoderma as biocontrol agents against a wide range of bioaggressors.



Hydrogen, particularly green hydrogen produced via renewable energy, is a key component in the global shift toward sustainable energy. Its potential to decarbonize hard-to-abate sectors such as heavy industry, transportation, and energy storage positions it as a critical solution in the clean energy transition. However, widespread adoption faces significant challenges. This chapter explores major obstacles to scaling green hydrogen, including inefficiencies in electrolysis, high production costs, and inadequate infrastructure for storage and transport. Additionally, it discusses regulatory hurdles, market limitations, and the need for technological advancements to improve cost-effectiveness. Addressing these challenges is essential for realizing hydrogen's potential in global decarbonization. With innovation, policy support, and international collaboration, hydrogen can become central in achieving carbon neutrality.

Metal oxide nanoparticles (NPs) particularly titanium dioxide have gained significant interest due to their exceptional properties and versatile applications in various sectors such as industrial, medicinal, and environmental areas. TiO2 (NPs) play a crucial role in improving the efficacy and stability of drug formulations. Their high surface area and reactive properties enable them to serve as effective carriers for drug delivery systems, ensuring efficient absorption of therapeutic agents by the body. In addition, TiO2 (NPs) are also pivotal in the development of advanced biosensors that exhibit heightened sensitivity and specificity, enabling precise detection of biological molecules. Furthermore, TiO2 (NPs) have shown remarkable potential in cancer therapy, with their application in photodynamic therapy (PDT) generating reactive oxygen species (ROS) that can selectively induce apoptosis in cancer cells.



This study investigates the gasteroid species within the families Sclerodermataceae and Rhizopogonaceae, focusing on species with significant ecological and economic roles. The species included in the analysis are Rhizopogon luteolus Fr. and Nordhom (1817), Pisolithus microcarpus (Cooke & Massee) G. Cunn. (1931), Scleroderma citrinum Pers. (1801), and Scleroderma meridionale Demoulin & Malençon (1971). These species are part of the larger group of Basidiomycetes, which are classified under the order Boletales. The taxonomic positioning of these species has been discussed in several studies, highlighting their placement within the broader taxonomic framework of the Basidiomycetes.

Genetic and biotechnological developments have made it possible to apply synthetic biology approaches to new classes of microbial hosts, such as photosynthetic microalgae. A biosensor is a device that transforms a biological answer into an electrical signal. These devices have the potential to be used in a wide range of applications, including environmental and water quality monitoring. In biosensing applications, microalgae have been widely used as bioreceptors due to their lower susceptibility to physiochemical changes, as well as being used in medical, environmental, industrial, and commercial fields. To overcome the limitations of conventional analytical techniques, new technologies are being introduced based on integrating different types of microorganisms, including microalgae, to build biosensors. Detecting environmental pollutants is one of the main applications of biosensors. Our study highlights the advantages and disadvantages of using microalgae to produce biosensors as a new technology for detecting wastewater pollutants such as heavy metals.



In this study, an isolate of Trichoderma sp.1, collected from saffron corms, was utilized to evaluate its effect on saffron plant growth and bulb formation.Treating saffron corms with a conidial suspension of the Trichoderma isolate before sowing showed a positive impact on agronomic parameters and bulb formation. After four months of greenhouse cultivation, plants from Trichoderma-treated corms had an average of 31 leaves per plant, compared to 16,5 in the control plants. The average leaf length of treated plants was nearly identical to that of the controls.The average weights of leaves and roots in treated plants were 2,53g and 0,3g, respectively, compared to 1,93g and 0.15g in the controls.Furthermore, plants from Trichoderma-treated corms produced an average of about 10 newly formed bulbs, while control plants produced 5,75. The average bulb weight of treated plants was 16,5g, compared to 10,36g in control plants. These results benefit saffron producers by ensuring planting material availability. Trichoderma stimulates daughter corm formation, boosting production.

Hydrogen is an environmentally friendly energy source that is crucial in addressing global climate change and energy shortages. Microalgae are a highly appealing method for generating biohydrogen, which is a clean and environmentally friendly energy source. This strategy can help reach carbon neutrality and ensure the sustainability of bioenergy. The production of biohydrogen by dark fermentation is seen as a promising way of using clean, renewable energy. Microbial fermentation activities produce the majority of hydrogen in the biosphere. The biological synthesis of hydrogen is an attractive approach to renewable energy production, utilizing both photosynthesis and fermentation processes. Certain types of green algae, blue-green algae (cyanobacteria), fermentative and photosynthetic bacteria, and archaea can produce hydrogen by using light energy and/or various organic substances. Our work aims to study the role and influence of biohydrogen production by dark fermentation using microalgae, as well as improve the production yield, with a view to wider application in the future



Energy transition policy now places a strong emphasis on green hydrogen because of its enormous potential to cut greenhouse gas emissions and its substantial commercial potential. Renewable energy sources like wind and solar power are used to manufacture this eco-friendly substitute. A novel technique uses chlorophyll-rich microalgae, which produce hydrogen as a byproduct of photosynthesis, the process by which sunlight is converted into chemical energy.Converting renewable energy into forms that may be used, like green hydrogen, requires the development of production circuits for renewable PtX (Power-to-X). Water electrolysis, which separates water into oxygen and hydrogen using power from renewable sources, is the technique involved in this procedure. Continuous and dependable manufacturing necessitates a strong infrastructure, which includes effective electrolyzers.Since it is the basis for the production of numerous synthetic fuels and other energy products, green hydrogen is an essential part of PtX.

Technology has significantly accelerated digital entrepreneurship, enabling rapid business establishment. As digital firms increase, understanding the obstacles their founders face and the factors contributing to their success becomes crucial. This study identifies key barriers and highlights the role of human and social capital in overcoming them. Major challenges include difficulty in finding skilled workers, rising competition, limited financial resources, data security concerns, and the need for consumer engagement beyond digital platforms. These issues can hinder business growth and sustainability. By leveraging networks, skills, and relationships, entrepreneurs can better navigate these obstacles. Understanding these barriers and utilizing their resources effectively allows digital entrepreneurs to develop strategies for success. This study offers valuable insights to help them identify challenges and find practical solutions for their ventures.



Hydrogen has a strong potential to accelerate the decarbonization of industries, surpassing other renewable alternatives. However, its overall development remains limited, with analyzes often focused on specific aspects (production, storage, transport). This study provides an overview of the challenges and opportunities of the hydrogen economy. Innovations such as wastewater recycling and electrolysis via microbes provide promising solutions. Technological advances are crucial to secure transport and storage, and an extensive network of refueling stations is necessary for its adoption in transport. Finally, government policies and incentives are essential for the transition to green hydrogen.

The transition to a green hydrogen economy offers a path toward sustainable energy but faces challenges in production efficiency and cost. This chapter explores integrating methanization and biogas technologies within the green hydrogen landscape, focusing on synergies and obstacles. Methanization, which converts organic waste into methane-rich biogas, could supplement hydrogen production via methods like steam methane reforming and methane pyrolysis. The chapter examines pathways for using biogas-derived methane in hydrogen production, highlighting opportunities to enhance yield and reduce reliance on fossil fuels. It also discusses economic implications, policy frameworks, and offers strategic recommendations for advancing the green hydrogen economy and supporting the global energy transition.



These systems are well-established and benefit from relatively low costs and durability, making them suitable for large-scale hydrogen production. However, they may have limitations in terms of response time and efficiency at lower power densities.In contrast, PEM electrolyzers utilize a solid polymer electrolyte membrane to separate the anode and cathode, operating at higher efficiencies and with faster response times. They are more compact and can operate under higher pressure, which can be advantageous for integrating with renewable energy sources and optimizing hydrogen storage. On the one hand, the variability of renewable energy generation, such as intermittent wind or solar power, necessitates advances in electrolyzer technology and energy storage solutions to ensure a stable and reliable hydrogen supply. Additionally, hydrogen can serve as an energy carrier and storage medium, enabling the efficient storage of excess renewable energy and its subsequent use when energy demand exceeds generation capacity.

Drinking water is one of the most important elements on which our everyday life depends, because its importance lies in its multiple uses in different domains. The town of SALE is supplied by the REDAL water towers, which are supplied by the ONEP, which processes the waters of the Bourgreg complex coming from the Sidi Mohamed Ben Abdellah Dam. This work consists in carrying out quality checks of drinking water of the province of SALE, We have carried out sampling, and then physico-chemical analysis in the laboratory, of the distribution points of the castles. According to the standard of the quality of the water intended for the food said that for a water to be potable it must be potable on the bacteriological and physico-chemical side in this report the emphasis is placed solely on the physico-chemical analyses. From this study allowed us to deduce that the physico-chemical quality of the waters of the province of SALE follows the Moroccan standard of potability.



The selection of the T asperellum-based product required several in vitro tests in order to preserve its efficacy during the formulation process, while maintaining the viability of the conidia and safeguarding their antagonistic properties. The addition of talc to all of the compounds added to the formulation of T asperellum's conidia, Moroccan strain isolated from compost, preserved their viability, stability and antagonistic power during the entire duration under storage conditions. P1 and P2's conidia germination percentages while stored for 38 weeks at room temperature were 0% - 12% and at 4°C were 31% - 46.33% respectively. The antagonistic effect of P1 and P2 against V dahliae by direct contact is estimated by the T asperellum colonization percentages, the mycelial growth of V dahliae's inhibition percentages, the mycelial growth of V dahliae's inhibition percentages and the mycelial inhibition percentages by antibiosis are respectively 0 - 79.58%, 0 - 39.24% and 3.25 - 39.84% at room temperature and 80.22 - 81.49%, 39.83 - 44.99% and 42.27 - 43.09% at 4°C.

In recent years, several forms of sustainable or renewable energy like hydroelectric power, solar energy, nuclear fission, and wind energy have been developed. Hydrogen energy is a new form of energy considered to be more useful and sustainable because of its versatility, storage capabilities, potential for energy security, and its role in achieving global climate goals. Currently, more attention is being paid to production of green hydrogen, which can be regarded as the substitute for other widespread methods of hydrogen production. This has created the opportunity for the utilization of renewable energy sources into the energy basket. This chapter captures a detailed discussion on the production of green hydrogen, its potential uses, advantages and disadvantages, inherent challenges, thus comprehensively exploring its importance in the global energy transition process and the potential that the green hydrogen can bring with a view to catalysing positive socio-economic and environmental returns.



Objective: The purpose of this work is to study, to rediscover, and to classify the Amanita mushrooms found in the cork oak forests of Morocco. Methods and Results: Surveys were conducted in various regions of Morocco (Mamora, Northwest, Rif) allowing for the collection of infor-mation on the Amanita species encountered. In the laboratory, macroscopic and microscopic identification criteria for these species were established. In total, twenty-nine species (29) werwe recorded. Among these, five were not found (Amanita codinae, Amanita eliae, Amanita excelsa, Amanita gemmata, and Amanita porphyria). The remaining twenty-four species (24) are con-sistent with the vegetation formations of Morocco. The systematic position, ecology, edibility, and toxicity risks of all the studied species were discussed. Conclusion and Application of the Research: The results of this study contribute to determining the diversity of the Amanita genus in Morocco highlighting that the Moroccan forest ecosystem still holds many mysteries to uncover.



Water constitutes an essential element of the nation's common heritage. Its protection, enhancement, and sustainable development of its usable resources, while respecting natural balances, are of general

Green hydrogen, generated through the electrolysis of water from renewable energies, is seen as an essential energy component in the transition to a low-carbon economy. However, its large-scale manufacture faces a number of significant technical challenges. In this chapter, we examine these challenges, focusing on four key aspects. Firstly, it is necessary to improve the energy efficiency of current processes in order to minimize energy losses throughout the production chain. Secondly, the interruption of renewable energy sources poses a challenge, requiring energy storage and management solutions to ensure continuous hydrogen production. In addition, manufacturing costs remain high compared with grey hydrogen, making it essential to develop more competitive and cost-effective technologies. Finally, the lack of adequate infrastructure for transporting and storing hydrogen represents another brake on its large-scale use.



This chapter will explore the role of green hydrogen as a sustainable energy source for the industrial sector. The will provide an overview of the importance of decarbonizing industry and the challenges associated with conventional energy sources. This will also explain the fundamentals of green hydrogen production through electrolysis, using renewable energy sources like solar, wind, and hydropower. It will compare green hydrogen with grey, blue, and other forms of hydrogen, discussing the environmental impact and energy efficiency. This also focuses on specific industrial applications where green hydrogen has the most potential like Steel Production, Chemical Industry, etc. We also focus on challenges associated with green hydrogen, such as the cost of production, transportation, and storage. At the same we will examine the role of governments in promoting green hydrogen, exploring policies, subsidies, and international agreements. It will also discuss the alignment of green hydrogen initiatives with the UN's Sustainable Development Goals (SDGs).



The aim of this study was to evaluate in vitro and in vivo the effect of four chemical fungicides, Cyprodinil (37,50%) + Fludioxonil (25%), Pyraclostrobin (26,70%) + Boscalid (6,70%), Mepanipyrim (50%) and

Thiram (80%), against Colletotrichum acutatum and C. gloeosporioides, causal agents of anthracnose, one of the important plant diseases in Strawberry in Morocco. Mepanipyrim inhibited mycelial growth from 55.3 to 75% and Thiram from 68.33 to 91.66% and Cyprodinil + Fludioxonil from 93 to 100%. Similarly, Pyraclostrobin + Boscalid and Thiram showed strong inhibition of conidia production in the four Colletotrichum isolates tested ranging from 82.7 to 98.7% and 77 to 99.61%, respectively. Significant inhibition is also observed for Cyprodinil + Fludioxonil 53.72 to 99%. In vivo, the preventive and curative treatments on the aerial parts of Camarosa strawberry plants, with the two fungicides Switch 62.5WG and Signum WG, were shown to be effective against anthracnose, reducing the symptoms of the disease relative to the control.

**Chapter 30**
Transforming Morocco's Road Transport Sector Embracing Circular Economy Practices ............... 531
*Imane Rouichat, Mohammed V University in Rabat, Morocco*
*Souad Nasrdine, Mohammed V University in Rabat, Morocco*
*Jamal Mabrouki, Mohammed V University in Rabat, Morocco*
*Miloudia Slaoui, Mohammed V University in Rabat, Morocco*

The Moroccan automotive sector, a major contributor to the national economy, has experienced rapid growth, accompanied by an increase in carbon emissions that threatens environmental sustainability. This chapter examines the potential of the circular economy (CE) to address these challenges, promoting practices of reuse, recycling, and reduction of materials throughout the product lifecycle. By adopting CE principles, the Moroccan automotive sector can shift away from the traditional "take, make, dispose" linear model, reducing environmental impact and enhancing economic competitiveness. Recognizing the global and national momentum for CE, the Moroccan government has aligned its National Strategy for Sustainable Development with CE principles, aiming to mitigate greenhouse gas emissions and foster sustainable practices.

**Chapter 31**
Trichoderma Use as Seed Treatment for Promoting the Growth of Young Argan Seedlings ........... 547
*Mohamed El Rhoch, Ibn Tofail University, Morocco*
*Soukaina Maazouzi, Laboratory of Plant, Animal, and Agro-Industry Productions, Department of Biology, Faculty of Science, Morocco*
*Najoua Mouden, Laboratory of Molecular Chemistry and Environmental Molecules, Multidisciplinary Faculty of Nador, Morocco*
*Zineb Sellal, Laboratory of Plant, Animal, and Agro-Industry Productions, Department of Biology, Faculty of Science, Morocco*
*Karima Selmaoui, Laboratory of Plant, Animal and Agro-Industry Productions, Department of Biology, Faculty of Science, Morocco*
*Amina Ouazzani Touhami, Laboratory of Plant, Animal, and Agro-Industry Productions, Department of Biology, Faculty of Science, Morocco*
*Allal Douira, Laboratory of Plant, Animal, and Agro-Industry Productions, Department of Biology, Faculty of Science, Morocco*

Seed treatment with Trichoderma spp. has shown to enhance plant growth and alleviate abiotic stress, but its application on endemic species, such as argan (Argania spinosa), remains underexplored. This study investigates the effect of a Trichoderma spp. bioformulation on argan seedling growth under greenhouse conditions over 12 months. Significant improvements were observed in treated seedlings,

# Preface

## INTRODUCTION

For many years, we thought that the world was infinite and that its sources were limitless. We are now discovering that we live in a confined world and that shortages are a threat to our very survival. We discovered some time ago that the climate can interfere with our operations. Now they are discovering that their activity can cause a climate change disaster. Faced with such increasingly oppressive restrictions, we need to make the necessary changes to the world we live in and explore ways of constructing a more prosperous future.

The energy transition that many countries are undergoing aims to change the way we consume energy (fossil, nuclear or renewable), as well as the way we produce it. Green hydrogen can be an energy-efficient solution, given its low-carbon method of production. However, we will also see that the process is not without its difficulties, and that this clean (and theoretically inexhaustible) energy source is not that affordable in practice on the scale of an industry or even a country.

Objective: The point of this work is to explain that, with some qualifications, hydrogen may be one of the alternative sources of energy. More specifically, hydrogen is expected to have a major role to be played in the years to come as a source of propulsion for ground and air vehicles, rockets and spacecraft. The challenges of green hydrogen are many and offer a solution to the energy greed of our countries while at the same time protecting the environment.

Target Audience:
Researchers
Professionals
Phd
Recommended Topics:
-Green hydrogen economy for a renewable energy society
-Prospects and challenges for green hydrogen production
-An overview of water electrolysis technologies for green hydrogen
-Green hydrogen economy
-Green hydrogen renewable
-Process design for green hydrogen
-Building the green hydrogen market
-Hydrogen strategies
-Photovoltaic/photo-electrocatalysis integration
-Green hydrogen for industrial sector

-Desalination-hydrogen system powered
-Green Hydrogen via Desalinated Water
-Green hydrogen cost
-Desalination process
-Techno-economic modelling
-Seawater desalination

## PREFACE

In undertaking this volume, Obstacles Facing Hydrogen Green Systems and Green Energy, we have brought together a panoramic survey of the promises and pitfalls of green hydrogen. For decades, humanity lived under the illusion of boundless resources: infinite reserves of oil, coal, water, land, and an atmosphere ready to swallow our waste. We now understand that our planet is not limitless. In fact, it is alarmingly finite—and the shortages we face threaten our survival.

The urgency of the energy transition is no longer abstract. As temperature records are shattered and once-stable weather patterns turn volatile, we realize that our activities are no longer neutral observers of climate—they are its architects. Countries around the world are restructuring their energy portfolios—shifting from fossil fuels and conventional nuclear plants to renewables and low-carbon solutions. In this context, green hydrogen emerges as a compelling answer: it offers a pathway to store intermittent renewable generation, decarbonize heavy industry, and propel the mobility sector beyond fossil dependence.

Yet, this "clean" energy carrier is not without formidable obstacles. From the capital-intensity of electrolyzer manufacturing to the challenge of sourcing pure water in water-stressed regions, from the intricacies of techno-economic modeling to the creation of a vibrant market infrastructure—each chapter of this book lays bare the technical, economic, and regulatory hurdles we must surmount.

Our objective is clear: to equip researchers, industry professionals, and doctoral candidates with a rigorous analysis of green hydrogen's potential and its practical limitations. We explore water electrolysis technologies, integration with photovoltaics and desalination, process design, cost assessments, and market-building strategies. By confronting these challenges head-on, we aim to hasten hydrogen's transition from promising hypothesis to foundational pillar of a low-carbon future.

## CHAPTER OVERVIEWS

### Chapter 1: Adoption Challenges of Hydrogen Energy as an Alternate Fuel: As Per Indian Perspective

This chapter confronts head-on the disconnect between hydrogen's recognized environmental benefits and its lagging deployment in India. Drawing on SWARA and DEMATEL analyses, it disentangles the causal web of technological gaps, infrastructure deficits, and policy uncertainties that impede hydrogen's widespread uptake in the Indian transport sector. Its insights form a vital foundation for understanding how regional realities shape global ambitions.

### Chapter 2: An Overview of Water Electrolysis Technologies for Green Hydrogen

Surveying the principal electrolysis pathways—alkaline water electrolysis (AWE), proton exchange membrane (PEM) electrolysis, and solid oxide electrolysis (SOE)—this chapter evaluates each technology on efficiency, scalability, and cost. From foundational operating principles to frontier research pushing performance limits, it illuminates water electrolysis as the linchpin of any truly green hydrogen economy.

### Chapter 3: Antifungal Potential of Trichoderma sp. Against Three Species of Fusarium Genus, Responsible for the Root Rot of the Saffron Corms

This study examines the in vitro antagonistic effect of native Trichoderma isolates against three Fusarium species that cause saffron corm rot. Through direct confrontation and volatile-compound assays, the chapter reveals high inhibition rates of mycelial growth and spore germination, while also discussing the limited impact of diffusible substances. It underscores the importance of rigorous, data-driven evaluation in biocontrol strategies.

### Chapter 4: Challenges in Scaling Hydrogen Green Systems for Global Decarbonization

Stepping back to survey macro-scale barriers, this chapter explores electrolyzer inefficiencies, prohibitive production costs, and gaps in storage and transport infrastructure. It further discusses regulatory hurdles, market limitations, and the need for technological advancements—calling for coordinated innovation and policy action to realize hydrogen's decarbonization promise.

### Chapter 5: Comprehensive Insights into the Biomedical Applications of Titanium Dioxide

Focusing on titanium dioxide nanoparticles, this chapter details their roles in drug delivery, advanced biosensors, and photodynamic cancer therapy. It highlights how $TiO_2$'s high surface area and reactive properties enable targeted therapeutic delivery and sensitive biomolecule detection, while calling attention to the techno-economic and safety assessments required for sustainable biomedical applications.

### Chapter 6: Contribution to the Study of Some Gasteroid Macromycetes Species of Morocco

Investigating gasteroid fungi within Sclerodermataceae and Rhizopogonaceae, this chapter offers taxonomic and ecological analyses of species with ecological and economic significance. Its meticulous classification work provides a model for foundational science that underpins applied innovation across disciplines.

## Chapter 7: Current Advances and Future of Microalgae-Based Biosensors for the Detection of Pollutants

Exploring photosynthetic microalgae as living bioreceptors, this chapter reviews biosensor designs that convert biological responses into electrical signals for environmental monitoring. It assesses advantages and limitations—such as robustness versus sensitivity—and highlights emerging integrations of synthetic biology to enhance detection of heavy metals and other wastewater pollutants.

## Chapter 8: Effect of Treatment of Corms with Trichoderma Inoculum on the Development of Crocus sativus Plants and the Formation of Daughter Bulbs

Quantifying the agronomic impact of Trichoderma-treated saffron corms, this chapter reports significant increases in leaf number, bulb formation, and biomass compared to controls. Its results demonstrate how microbial inoculants can boost production, offering practical guidance for saffron cultivators and illustrating the broader theme of harnessing biological partners for enhanced system performance.

## Chapter 9: Environmental Assessment of Microalgal Biomass Fermentation for Green Hydrogen Production

Evaluating dark-fermentation pathways with microalgae, this chapter provides a life-cycle assessment of biohydrogen production. It examines microbial activity, yield optimization strategies, and environmental impacts—positioning this bio-based route as a complementary pathway alongside electrolytic methods in the quest for carbon-neutral hydrogen.

## Chapter 10: Green Hydrogen: The Pillar of the Sustainable Energy Transition

Synthesizing policy and technological trends in Power-to-X circuits, this chapter outlines the conversion of renewable electricity into green hydrogen via water electrolysis and novel microalgal photobioreactors. It emphasizes infrastructure needs for reliable production and situates hydrogen as central to synthesizing fuels and chemicals in a decarbonized economy.

## Chapter 11: Green Hydrogen Sustainable Solution to Combat $CO_2$ Emissions and Mitigate Climate Change

Reviewing the environmental toll of $CO_2$ emissions, this chapter highlights green hydrogen's potential to slash greenhouse gases across transportation, industry, and power generation. It frames hydrogen as both an imperative and an achievable solution, detailing its role in near- and long-term emission reduction strategies.

## Chapter 12: Hydrogen as the Cornerstone of a Renewable Energy Society

Taking a forward-looking perspective, this chapter examines hydrogen's capacity for grid balancing, long-term storage, and sector coupling, particularly where electrification alone falls short. While candid about cost and infrastructure hurdles, it argues that integrating hydrogen with batteries and demand-response systems is essential for a resilient, carbon-neutral energy future.

## Chapter 13: Hydrogen Energy: Adaptation and Challenges

Applying the WASPAS methodology to assess adoption challenges, this chapter reveals that "Formulate Recommendations" ranks as the most critical barrier to hydrogen energy deployment. Economic factors follow closely, with technological readiness and policy frameworks ranking third and fourth, respectively. The analysis underscores the urgent need to develop robust strategic roadmaps—even before scaling technology—to alleviate environmental pressures and enable widespread adoption of hydrogen as a mainstream energy carrier.

## Chapter 14: Initial Tests on Various Formulations and the Recommended Dose of the Moroccan Trichoderma asperellum-Based Product for Stimulating and Protecting Tomato

Through comparative trials of two biocontrol formulations—one pure conidial suspension (P1) and one talc-amended suspension (P2)—this chapter demonstrates significant reductions in verticillium wilt incidence and improvements in tomato yield. Notably, P2 outperforms P1 across dwarfing index, fruit weight, and vegetative growth metrics, while a 40 mL application rate maximizes agronomic benefits. These findings provide practical guidance on optimizing Trichoderma-based products for crop protection.

## Chapter 15: Integration of Renewable Energy in Smart Cities Towards a Clean and Sustainable Future

This chapter examines the multifaceted process of embedding solar, wind, biomass, and hydro power into urban infrastructures. It highlights the environmental and economic dividends of supply diversification, while candidly addressing challenges in storage, grid management, and demand forecasting. By showcasing emerging energy-management platforms and policy incentives, the chapter maps a pathway for smart cities to achieve resilience, lower emissions, and inclusive growth.

## Chapter 16: Navigating the Digital Frontier: Key Challenges and Success Factors for Entrepreneurs in the Age of Technology

Focusing on digital startups, this study identifies five principal obstacles—talent shortages, escalating competition, funding constraints, data-security concerns, and the imperative of multi-channel customer engagement. It then illustrates how founders' human and social capital—networks, specialized skills, and collaborative ties—can mitigate these barriers. The result is a strategic playbook enabling digital entrepreneurs to convert resource limitations into competitive advantages.

## Chapter 17: Opportunities and Challenges for a Green Hydrogen-Based Economy

Offering a panoramic overview of hydrogen's role in decarbonizing heavy industry, this chapter balances ambition with realism. It surveys innovations—from microbial electrolysis to wastewater-recycling loops—that promise to lower production costs, while detailing infrastructure requirements for transport, storage, and refueling networks. Finally, it stresses the indispensability of targeted policy incentives to catalyze an effective transition to a hydrogen economy.

## Chapter 18: Opportunities and Challenges of Integrating Methanization and Biogas Technologies into the Green Hydrogen Economy

This chapter explores synergies between organic-waste methanization and green hydrogen production. It lays out pathways for using biogas-derived methane in steam-methane reforming and pyrolysis processes, identifying opportunities to boost overall yield and reduce feedstock costs. Economic analyses and policy considerations frame the discussion, leading to recommendations for integrated bio-energy strategies that reinforce a sustainable hydrogen value chain.

## Chapter 19: Overview of Water Electrolysis Technologies for Green Hydrogen

Covering both mature alkaline systems and advanced proton-exchange-membrane (PEM) electrolyzers, this chapter contrasts their cost profiles, efficiency curves, and dynamic response characteristics. It emphasizes the need for flexible operation under variable renewable input and examines how system design and power-management solutions can stabilize hydrogen output—thereby positioning electrolysis as the cornerstone of large-scale green hydrogen production.

## Chapter 20: Physico-Chemical Quality Control of Network Water from Different Drinking Water Supply Points in the Province of Salé

Through systematic sampling and laboratory analyses of water distribution nodes fed by the Bourgreg complex, this chapter confirms that Salé's network water meets Moroccan physico-chemical potability standards. Parameters such as pH, turbidity, mineral content, and residual chlorine were assessed, ensuring compliance with national regulations and providing stakeholders with confidence in drinking-water safety across the province.

## Chapter 21: Preliminary In Vitro Tests to Demonstrate Various Formulations' Effect on the Stability of the Moroccan Trichoderma asperellum-Based Product's Biology

Evaluating long-term storage stability of two conidial formulations (with and without talc) at room temperature and 4 °C, this chapter shows that talc-amended preparations maintain higher germination rates and stronger antagonistic activity against Verticillium dahliae. Over 38 weeks, P2 exhibits superior viability and bioefficacy, reinforcing the importance of formulation science in preserving biocontrol potency during storage and distribution.

## Chapter 22: Prospects and Challenges for Green Hydrogen as a Renewable Energy Source

This chapter situates green hydrogen within the broader renewable-energy landscape, contrasting it with hydroelectric, solar, wind, and nuclear options. It delves into production pathways, usage scenarios, and lifecycle emissions, weighing hydrogen's storage versatility against cost and infrastructure challenges. By synthesizing techno-economic data and policy trends, it outlines how green hydrogen can catalyze positive socioeconomic and environmental returns at scale.

## Chapter 23: Rediscovery and Typification of Amanita from the Cork Groves of Morocco

Documenting surveys across Moroccan cork-oak forests, the chapter reports 29 Amanita species encountered, with five species notably absent. It establishes macroscopic and microscopic identification criteria, and discusses each species' systematic position, ecology, edibility, and toxicity. The work enriches our taxonomic understanding of Basidiomycetes in Morocco, underscoring the forest's rich—but still partly unexplored—mycological diversity.

## Chapter 24: Study of Performance and Improvement of the Quality of Treatment of the Purification Station of Wastewater

This chapter analyzes the Agadir wastewater treatment plant, which processes effluents from food industries, domestic sources, and tourism activities. Through systematic monitoring of influent and effluent parameters—such as biochemical oxygen demand (BOD), chemical oxygen demand (COD), suspended solids, and nutrient loads—it identifies process bottlenecks leading to substandard discharge quality. The authors propose targeted upgrades, including optimized aeration control, enhanced primary settling, and constructed wetland polishing, to restore water quality for safe reuse in irrigation and to protect downstream ecosystems.

## Chapter 25: Sustainable and Renewable Strategies for Desalination

Focusing on the integration of renewable energy into desalination, this chapter reviews conventional thermal and membrane technologies alongside emerging hybrid systems. It evaluates staged processes, waste-heat recovery, and photovoltaic- or wind-powered reverse osmosis for coastal and off-grid applications. By comparing levelized cost of water (LCOW) across scenarios, the chapter demonstrates how combining renewables with energy-recovery devices and smart controls can deliver sustainable, cost-effective freshwater—especially critical in arid and resource-constrained regions.

## Chapter 26: Technical Challenges and Innovations in Hydrogen Storage

This chapter presents a comprehensive survey of hydrogen storage options—compressed gas, cryogenic liquid, solid-state materials, and porous media. It assesses each method's volumetric and gravimetric capacity, safety considerations, and energy penalties. Special attention is given to metal-organic frameworks and advanced adsorbents that promise high uptake at moderate pressures. By juxtaposing recent lab-scale breakthroughs with scale-up constraints, the chapter outlines a roadmap for translating novel storage materials into viable industrial solutions.

## Chapter 27: Technical Challenges in Green Hydrogen Production

Examining four core hurdles—energy efficiency losses, renewable intermittency, high production costs, and insufficient transport/storage infrastructure—this chapter delves into strategies to overcome each. It discusses advanced electrolyzer designs to minimize overpotentials, integrated battery-electrolyzer systems for grid smoothing, cost-reduction pathways through giga-scale manufacturing, and modular pipeline or solid-carrier networks for distribution. Together, these innovations form a cohesive plan for scaling green hydrogen production.

## Chapter 28: Techno-Economic Analysis of Green Hydrogen Production for Industrial Use

This chapter benchmarks green hydrogen against grey and blue alternatives through life-cycle costing and levelized cost of hydrogen (LCOH) metrics. It models project economics for steelmaking, chemical synthesis, and ammonia production, incorporating regional renewable tariffs, electrolyzer CAPEX, and carbon pricing. Scenario analyses reveal break-even points under various subsidy and carbon-tax regimes, offering policymakers and industry stakeholders clear guidance on investment strategies aligned with UN Sustainable Development Goals.

## Chapter 29: The Efficacy of Several Fungicides on the Development of Colletotrichum spp. and Strawberry Anthracnose in Morocco

Through in vitro assays, this chapter compares four fungicidal formulations—Cyprodinil + Fludioxonil, Pyraclostrobin + Boscalid, Mepanipyrim, and Thiram—against Colletotrichum acutatum and C. gloeosporioides. It quantifies inhibition of mycelial growth and conidia production, demonstrating near-complete suppression by Cyprodinil + Fludioxonil. Subsequent greenhouse trials with Switch 62.5 WG and Signum WG validate preventive and curative efficacy, providing actionable recommendations for anthracnose management in Moroccan strawberry cultivation.

## Chapter 30: Transforming Morocco's Road Transport Sector by Embracing Circular Economy Practices

This chapter evaluates how circular economy principles—reuse, remanufacture, and material recovery—can decarbonize Morocco's rapidly growing automotive sector. It examines end-of-life vehicle processing, battery second-life applications, and parts remanufacturing to reduce waste and resource extraction. By aligning with national sustainability strategies, the authors demonstrate how circular business models can lower emissions, generate new revenue streams, and enhance competitiveness within the Moroccan transport ecosystem.

## Chapter 31: Trichoderma Use as Seed Treatment for Promoting the Growth of Young Argan Seedlings

Investigating a Trichoderma spp. bioformulation as a seed-treatment for argan, this chapter reports marked improvements in seedling height, biomass, branching, and survival over 12 months under greenhouse conditions. Colonization rates in root tissues reached 95%, correlating with enhanced drought resilience and nutrient uptake. These findings propose a low-cost, biological strategy to bolster argan restoration efforts and support sustainable agroforestry in arid regions.

## Chapter 32: Using Artificial Intelligence to Improve Hydrogen Production Systems

This chapter explores AI-driven optimizations across the green hydrogen value chain. Machine learning models forecast renewable generation and dynamically adjust electrolyzer operating points to maximize efficiency. Neural networks enable predictive maintenance by analyzing vibration and temperature data, reducing downtime and extending equipment life. The authors illustrate case studies where AI integration cuts operating costs and integrates hydrogen production into smart energy grids, paving the way for more resilient, adaptive low-carbon systems.

It is our hope that this work not only informs but also inspires the next generation of innovators, policymakers, and entrepreneurs to turn obstacles into opportunities—and to chart a more prosperous, sustainable future for all.

**Jamal Mabrouki**
*Faculty of Science, Mohammed V University in Rabat, Morocco*

# Chapter 1
# Adoption Challenges of Hydrogen Energy as an Alternate Fuel:
## As per Indian Perspective

**Suchismita Satapathy**
https://orcid.org/0000-0002-4805-1793
*KIIT University, India*

## ABSTRACT

*World wide hydrogen energy is known as the clean energy, which acts as a savior of environment as it will not harm environment and nature.The hydrogen energy is considered as the new source of fuel energy that will bring changes and revolution to the earth. It is expected that all the future vehicles will be dependent on hydrogen energy.Still it is not widely adopted .There are many challenges faced by scientists in declaring hydrogen energy as alternate fuel for future vehicles or energy which can reduce green house effect. Hence in this chapter an effort is taken to find the main barriers of hydrogen energy adoption as per Indian scenario, then SWARA(Stepwise Assessment ratio) and DEMATEL(Decision Making Trial and Evaluation Laboratory) analysis is conducted to find cause and effect relation between the challenges.*

## INTRODUCTION

Energy needs are growing in every industry. Industries rely on power to boost output and run machines. Power plants often burn fossil fuels to make this electricity. Fossil fuels help meet energy needs, but they harm the environment. They pollute the air, making it unhealthy for all. Clean energy is vital to protect our world. Solar and wind power are clean energy sources under study. These systems don't pollute, which is good for the planet. Storing solar and wind power is hard, and costs can be high. So, experts are looking at hydrogen energy. Hydrogen cuts pollution like Sox and NOx, improving air quality. Hydrogen fuel cells only make water when burned, not harmful gases or carbon dioxide. Hydrogen can be made using solar or wind power through electrolysis. This process emits no carbon. Many see hydrogen as

DOI: 10.4018/979-8-3693-8980-5.ch001

a clean energy source. Countries are building hydrogen power plants. However, some nations lack the money, tech, and facilities to use hydrogen.

Hydrogen is hard to transport and store due to its low energy density. Safe hydrogen distribution needs large storage or high-pressure tanks. Many see hydrogen energy as a clean way to cut pollution and fight climate change. Hydrogen may help lower greenhouse gas emissions. Hydrogen fuel cell vehicles use hydrogen to make power. This process creates oxygen and water vapor. This makes them a good energy option. Hydrogen can change how we use power and cars.

Hydrogen can be made in many ways and is easy to find. This makes it a popular green energy source. The public, companies, and governments are very interested. They see its potential as a fuel. Funding is going to hydrogen research and new materials. These materials are for fuel cells, sensors, and storage. Safe hydrogen facilities are key to avoid accidents. Fixing material problems in hydrogen systems is a main goal. Challenges help guide plans for the future. These plans consider money, society, and nature. Hydrogen and new materials can change our energy use for the better. It is important to plan for the benefits of this new technology.

Understanding how materials interact with hydrogen is key. This is especially true for pipelines and hydrogen transport. Embrittlement during storage and transport is also a concern. Hydrogen fuel cells present unique innovations and challenges.

Hydrogen is a good substitute for fossil fuels in many industries. Fuel cells, powered by hydrogen, are coming to the transportation sector. Safe hydrogen handling and storage are crucial in any application. Compressed hydrogen gas is often kept in metal cylinders. Yet, hydrogen embrittlement can weaken these cylinders, creating a safety hazard. High-strength steels are more prone to this embrittlement. Safer hydrogen storage means better materials. Modifying current materials is another option for safer systems. Particularly in India the adoption of hydrogen energy as an ultimate energy has many challenges, hence to find the most important challenges as per Indian perspective SWARA method is used then to find cause and effect relationship between them DEMATEL method is used.

## BACKGROUND

The World Health Organization reports that burning coal, oil, and gas in industries causes air pollution and climate change (U.S. Department of Energy, n.d.; World Health Organization, 2018). These fossil fuels harm human health. Hydrogen energy has become popular as a clean alternative(Endo et al., 2021). Singh et al. (2020) note it can replace fossil fuels. Climate change and global warming result from using oil, coal, and gas. Fossil fuels have a lower density than hydrogen. Yang et al. (2010) predict hydrogen will be a key energy source if production issues are solved. Murray et al. (2009) notes transportation and storage are major challenges. Baykara (2018) explains hydrogen is environmentally friendly. Farrell et al. (2003) state hydrogen will be a main energy source. It could transform transportation by replacing oil. Ewan & Allen (2005) report that gas-steam reform produces 40-45% of hydrogen. Oil reform makes about a third, coal reform a fifth, and water electrolysis about 4%.A EA European Commission group predicts big changes. By 2040, hydrogen fuel may power 35% of European cars, which would replace oil in future (International Renewable Energy Agency, 2019; European Commission, 2003; Chakraborty & Zavadskas, 2014).ropeommission group predicts big changes. By 2040, hydrogen fuel may power 35% of European cars. This would replace oil

## RESEARCH METHODOLOGY

This study used a review of many papers to find issues in using hydrogen as energy. A set of 22 surveys was created for experts. The survey went to teachers, scholars, and energy pros. About 100 yes or no questions were asked (yes=1, no=0). Seventy percent of the surveys were returned. The data was studied with a minitab 17. Around 20 items having Chrobanch's alpha value more than 0.5 are selected and rest discarded. Then SWARA method ranked the top hydrogen energy issues.

his study used a review of many papers to find issues in using hydrogen as energy. A set of 22 surveys was created for experts. The survey went to teachers, scholars, and energy pros. About 100 yes or no questions were asked (yes=1, no=0). Seventy percent of the surveys was studied with a math tool. The WASPAS method ranked the top hydrogen energy issues.

*Table 1. Question*

| Question | Yes=1 | No=0 |
|---|---|---|
| Production of hydrogen by electrolysis method is very costly | | |
| For production, storage, distribution, and dispensing, good infrastructure and facilities are required | | |
| For storing high pressure tank requirements, which are technically difficult | | |
| For transportation of hydrogen is requires different kinds of pipes and transportation systems as highly flammable | | |
| A new innovative durable fuel cell is essential | | |
| Advance technology and more research is required for hydrogen production | | |
| Large-scale energy demands require significant investment and scaling | | |
| Production of green energy is expensive | | |
| Infrastructure development is essential for establishing production facilities, distribution networks, refueling stations, and storage solutions, which requires high investment | | |
| Fuel cell vehicles are in their initial stage and very costly | | |
| Infrastructure hydrogen fuel needs to be developed | | |
| The development of a hydrogen distribution network has challenges like cost, safety etc. | | |
| Lack of refueling station for hydrogen fuel | | |
| Safety standards to be developed for use of hydrogen fuels | | |
| supporting policies for use of hydrogen fuel incentives is required | | |
| Environmental regulations stop fossil fuel use but not support use of hydrogen fuel. | | |
| Production of hydrogen energy from fossil fuels emits carbon, so contradicts the environmental policy | | |
| For production of hydrogen, high use of water resources may create a scarcity of water | | |
| Hydrogen generated from renewable energy also leaves a carbon footprint. | | |
| Hydrogen must compete with well-established energy sources like fossil fuels | | |
| Public awareness and adoption e of hydrogen as fuel is essential | | |
| policies supporting hydrogen infrastructure development, research, and deployment are crucial | | |
| Awareness of hydrogen energy benefits is required. | | |
| Integrating hydrogen with existing energy systems is a challenge | | |
| Public perception and regulatory concerns around safety need to be addressed. | | |

## RESULTS AND DISCUSSION

After collecting data on hydrogen energy adoption hurdles across seven areas, factor analysis was performed. These areas included technical, economic, and infrastructure issues. Environmental, policy, social, and integration problems were also considered. Minitab 17 was used for statistical analysis. Only 20 items had a Cronbach's alpha above 0.5. Items below this value were removed. Factor analysis results are in Table 2. The SWARA method then ranked the challenges by importance.

*Table 2. Factor analysis of challenges*

| Dimensions | Question | Factor |
|---|---|---|
| Technical challenges | Production of hydrogen by electrolysis method is very costly | 0.532 |
| | For production, storage, distribution, and dispensing, good infrastructure and facilities are required | 0.527 |
| | For storing high pressure tank requirements, which are technically difficult | 0.514 |
| | For transportation of hydrogen is requires different kinds of pipes and transportation systems as highly flammable | 0.509 |
| Economic challenge | Green hydrogen production is very costly | 0.542 |
| | Infrastructure development is essential for establishing production facilities, distribution networks, refueling stations, and storage solutions, which requires high investment | 0.503 |
| | Fuel cell vehicles are in their initial stage and very costly | 0.519 |
| Infrastructure challenge | Infrastructure for hydrogen fuel needs to be developed | 0.625 |
| | Development of hydrogen distribution network has challenges like cost, safety etc. | 0.611 |
| | Lack of refueling station for hydrogen fuel | 0.567 |
| Policy challenges | Safety standards to be developed for use of hydrogen fuels | 0.549 |
| | supporting policies for use of hydrogen fuel incentives is required | 0.533 |
| | Environmental regulations stop fossil fuel use but not support use of hydrogen fuel. | 0.510 |
| Environmental challenges | Production of hydrogen energy from fossil fuels emits carbon, so contradicts the environmental policy | 0.713 |
| | For production of hydrogen high use of water resources may create scracity of water | 0.584 |
| | Hydrogen generated from renewable energy also leaves carbon footprint. | 0.557 |
| Sociatal acceptance challenges | Hydrogen must compete with well-established energy sources like fossil fuels | 0.562 |
| | policies supporting hydrogen infrastructure development, research, and deployment are crucial | 0.527 |
| Integration challenges | Integration and acceptance of hydroge energy | 0.634 |
| | Integrating hydrogen with existing energy systems is a challenge | 0.572 |

Then SWARA method is a decision making method used for rank criteria and sub criteria based on their ratio .

Steps of SWARA method.

1. First of all criteria are found, in this chapter technical, economic, and infrastructure issues. Environmental, policy, social, and integration problems are considers as criteria, under every criteria subcriteria's are identified.

2. Experts rank criteria and subcriterias per their relative importance.
3. Coefficient K and weight W are calculated (Table 3 to 10)
4. Lastly their relative weights are determined and ranking is done.(Table3 to 10)

Table.3 to Table.10 shows SWARA ranking of criterias and subcriterias.

*Table 3. Technical challenges by SWARA*

| Parameter | Relative significance of average values ($s_j$) | Coefficients $s_j$=($k_j$+1) | Recalculated Weights $\left(w_j = \frac{x_j - 1}{k_j}\right)$ | Weights $\left(q = \frac{x_i 1}{\sum w_i}\right)$ |
|---|---|---|---|---|
| C1 | | 1 | 1 | 0.272528652 |
| C3 | 0.135 | 1.135 | 0.881057269 | 0.24011335 |
| C4 | 0.25 | 1.25 | 0.908 | 0.247456016 |
| C2 | 0.42 | 1.42 | 0.88028169 | 0.239901982 |

*Table 4. Economic challenge*

| Parameter | Relative significance of average values ($s_j$) | Coefficients $k_j$=($s_j$+1) | Recalculated Weights $\left(w_j = \frac{x_j - 1}{k_j}\right)$ | Weights $\left(q = \frac{x_i 1}{\sum w_i}\right)$ |
|---|---|---|---|---|
| E1 | | 1 | 1 | 0.348629321 |
| E2 | 0.25 | 1.25 | 0.8 | 0.278903456 |
| E3 | 0.17 | 1.17 | 1.068376068 | 0.372467223 |

*Table 5. Infrastructure challenge*

| Parameter | Relative significance of average values ($s_j$) | Coefficients $k_j$=($s_j$+1) | Recalculated Weights $\left(w_j = \frac{x_j - 1}{k_j}\right)$ | Weights $\left(q = \frac{x_i 1}{\sum w_i}\right)$ | |
|---|---|---|---|---|---|
| I1 | | 1 | 1 | 0.37661221 | Ist |
| I2 | 0.2 | 1.2 | 0.833333333 | 0.313843508 | 2nd |
| I3 | 0.46 | 1.46 | 0.821917808 | 0.309544282 | 3rd |

*Table 6. Policy challenge*

| Parameter | Relative significance of average values ($s_j$) | Coefficients $s_j$=($k_j$+1) | Recalculated Weights $\left(w_j = \frac{x_j - 1}{k_j}\right)$ | Weights $\left(q = \frac{x_i 1}{\sum w_i}\right)$ | |
|---|---|---|---|---|---|
| P1 | | 1 | 1 | 0.343093923 | 2nd |
| P2 | 0.35 | 1.35 | 0.740740741 | 0.254143646 | 3rd |
| P3 | 0.15 | 1.15 | 1.173913043 | 0.402762431 | Ist |

*Table 7. Environmental challenge*

| Parameter | Relative significance of average values ($s_j$) | Coefficients $k_j$=($s_j$+1) | Recalculated Weights $\left(w_j = \frac{x_j - 1}{k_j}\right)$ | Weights $\left(q = \frac{x_i 1}{\sum w_i}\right)$ | |
|---|---|---|---|---|---|
| E1 | | 1 | 1 | 0.348837209 | 2nd |
| E2 | 0.5 | 1.5 | 0.666666667 | 0.23255814 | 3rd |
| E3 | 0.25 | 1.25 | 1.2 | 0.418604651 | Ist |

*Table 8. Social challenge*

| Parameter | Relative significance of average values ($s_j$) | Coefficients $s_j$=($k_j$+1) | Recalculated Weights $\left(w_j = \frac{x_j - 1}{k_j}\right)$ | Weights $\left(q = \frac{x_i 1}{\sum w_i}\right)$ | |
|---|---|---|---|---|---|
| S1 | | 1 | 1 | 0.370460048 | Ist |
| S2 | 0.7 | 1.7 | 0.588235294 | 0.217917676 | 2nd |

*Table 9. Integration challenge*

| Parameter | Relative significance of average values ($s_j$) | Coefficients $k_j$=($s_j$+1) | Recalculated Weights $\left(w_j = \frac{x_j - 1}{k_j}\right)$ | Weights $\left(q = \frac{x_i 1}{\sum w_i}\right)$ | |
|---|---|---|---|---|---|
| Int1 | | 1 | 1 | 0.62962963 | Ist |
| Int2 | 0.7 | 1.7 | 0.588235294 | 0.37037037 | 2nd |

*Table 10. Swara ranking of criterias*

| Parameter | Relative significance of average values ($s_j$) | Coefficients $s_j=(k_j+1)$ | Recalculated Weights $\left(w_j = \frac{x_j - 1}{k_j}\right)$ | Weights $\left(q = \frac{x_j 1}{\sum w_i}\right)$ | |
|---|---|---|---|---|---|
| D1 | | 1 | 1 | 0.14885213 | 3rd |
| D2 | 0.23 | 1.23 | 0.81300813 | 0.121017992 | 7th |
| D3 | 0.31 | 1.31 | 0.938931298 | 0.139761924 | 5th |
| D4 | 0.35 | 1.35 | 0.97037037 | 0.144441697 | 4th |
| D5 | 0.29 | 1.29 | 1.046511628 | 0.155775485 | 2nd |
| D6 | 0.46 | 1.46 | 0.883561644 | 0.131520033 | 6th |
| D7 | 0.37 | 1.37 | 1.065693431 | 0.158630738 | Ist |

From Table. 10 it is found that Integrated challenge is the most important challenges ranked 1, Environmental challenge ranked 2nd and technical challenge ranked third as per Indian scenario.Hence steps must be taken to mitigate these challenges .To mitigate these challenges it is necessary to find cause and effect relation between these challenges, So DEMATEL analysis is conducted to find dependent and independent factors.

Steps of Dematel analysis

1. Find decision matrix by using scale (1 to 5) (I.e.,1=no influence, 2=less influence, 3=moderate influence 4=more influence,5=high influence).Table 11 shows Decision matrix
2. Find direct relation ship matrix or normalization matrix (Table.12)
3. Calculate total relation-ship matrix (Table.13)
4. Lastly find cause and effect relationship and rank (table 14), show the relation by Figure.1.

*Table 11. Decision making matrix*

| | H1 | H2 | H3 | H4 | H5 | H6 | H7 | H8 | H9 | H10 | H11 | H12 | H13 | H14 | H15 | H16 | H17 | H18 | H19 | H20 | SUM |
|---|---|---|---|---|---|---|---|---|---|---|---|---|---|---|---|---|---|---|---|---|---|
| H1 | 1 | 3 | 3 | 2 | 4 | 1 | 3 | 2 | 4 | 3 | 3 | 3 | 3 | 4 | 3 | 2 | 1 | 3 | 1 | 4 | 53 |
| H2 | 2 | 2 | 2 | 3 | 2 | 2 | 3 | 1 | 3 | 2 | 1 | 2 | 3 | 4 | 5 | 2 | 1 | 4 | 2 | 5 | 51 |
| H3 | 1 | 2 | 1 | 5 | 3 | 1 | 4 | 3 | 4 | 2 | 1 | 3 | 4 | 2 | 4 | 3 | 2 | 4 | 4 | 1 | 54 |
| H4 | 4 | 1 | 2 | 1 | 1 | 1 | 3 | 2 | 2 | 3 | 1 | 3 | 2 | 3 | 3 | 2 | 2 | 3 | 2 | 4 | 45 |
| H5 | 1 | 1 | 5 | 4 | 1 | 1 | 4 | 2 | 3 | 2 | 1 | 3 | 3 | 4 | 2 | 1 | 2 | 4 | 3 | 1 | 48 |
| H6 | 1 | 1 | 2 | 2 | 3 | 1 | 1 | 2 | 2 | 2 | 1 | 2 | 2 | 3 | 2 | 1 | 3 | 3 | 1 | 1 | 36 |
| H7 | 2 | 1 | 2 | 2 | 2 | 2 | 1 | 3 | 2 | 4 | 1 | 2 | 2 | 3 | 1 | 3 | 3 | 4 | 1 | 2 | 43 |
| H8 | 1 | 2 | 3 | 3 | 1 | 3 | 4 | 1 | 3 | 2 | 1 | 3 | 4 | 5 | 3 | 4 | 4 | 2 | 1 | 1 | 51 |
| H9 | 2 | 2 | 3 | 2 | 4 | 2 | 3 | 3 | 1 | 3 | 3 | 3 | 1 | 4 | 4 | 3 | 3 | 3 | 2 | 1 | 52 |
| H10 | 2 | 1 | 2 | 2 | 1 | 3 | 3 | 5 | 4 | 1 | 2 | 2 | 3 | 1 | 5 | 3 | 3 | 2 | 2 | 1 | 48 |
| H11 | 2 | 2 | 1 | 3 | 2 | 1 | 2 | 3 | 2 | 3 | 1 | 4 | 2 | 3 | 3 | 4 | 2 | 3 | 1 | 2 | 46 |
| H12 | 2 | 4 | 2 | 3 | 3 | 2 | 2 | 3 | 3 | 3 | 1 | 1 | 4 | 3 | 3 | 3 | 3 | 4 | 2 | 0 | 51 |
| H13 | 2 | 2 | 3 | 2 | 2 | 2 | 3 | 2 | 3 | 4 | 1 | 3 | 1 | 4 | 3 | 3 | 3 | 3 | 4 | 3 | 53 |

continued on following page

*Table 11. Continued*

| | H1 | H2 | H3 | H4 | H5 | H6 | H7 | H8 | H9 | H10 | H11 | H12 | H13 | H14 | H15 | H16 | H17 | H18 | H19 | H20 | SUM |
|---|---|---|---|---|---|---|---|---|---|---|---|---|---|---|---|---|---|---|---|---|---|
| H14 | 1 | 3 | 5 | 3 | 4 | 2 | 4 | 4 | 3 | 3 | 2 | 3 | 3 | 1 | 3 | 3 | 3 | 3 | 5 | 3 | 61 |
| H15 | 1 | 4 | 5 | 4 | 3 | 2 | 4 | 3 | 5 | 3 | 4 | 4 | 5 | 3 | 1 | 3 | 3 | 2 | 2 | 1 | 62 |
| H16 | 2 | 3 | 3 | 2 | 3 | 1 | 3 | 2 | 4 | 3 | 5 | 3 | 4 | 3 | 5 | 1 | 3 | 4 | 1 | 1 | 56 |
| H17 | 2 | 3 | 2 | 2 | 3 | 1 | 4 | 3 | 3 | 3 | 4 | 4 | 3 | 3 | 5 | 3 | 1 | 2 | 1 | 1 | 53 |
| H18 | 2 | 3 | 4 | 4 | 5 | 2 | 3 | 2 | 2 | 2 | 3 | 3 | 4 | 4 | 2 | 4 | 2 | 1 | 4 | 3 | 59 |
| H19 | 1 | 1 | 1 | 2 | 4 | 1 | 1 | 1 | 1 | 1 | 1 | 2 | 3 | 5 | 1 | 2 | 2 | 4 | 1 | 4 | 39 |
| H20 | 3 | 3 | 1 | 2 | 1 | 1 | 2 | 1 | 1 | 1 | 4 | 2 | 1 | 4 | 1 | 1 | 1 | 3 | 4 | 0 | 37 |
| SUM | 35 | 44 | 52 | 53 | 52 | 32 | 57 | 48 | 55 | 50 | 41 | 55 | 57 | 66 | 59 | 51 | 47 | 61 | 44 | 39 | |

*Table 12. Normalisation matrix*

| | H1 | H2 | H3 | H4 | H5 | H6 | H7 | H8 | H9 | H10 | H11 | H12 | H13 | H14 | H15 | H16 | H17 | H18 | H19 | H20 |
|---|---|---|---|---|---|---|---|---|---|---|---|---|---|---|---|---|---|---|---|---|
| H1 | 0.015151515 | 0.045454545 | 0.045454545 | 0.03030303 | 0.060606061 | 0.015151515 | 0.045454545 | 0.03030303 | 0.060606061 | 0.045454545 | 0.045454545 | 0.045454545 | 0.045454545 | 0.060606061 | 0.045454545 | 0.03030303 | 0.015151515 | 0.045454545 | 0.015151515 | 0.060606061 |
| H2 | 0.03030303 | 0.03030303 | 0.03030303 | 0.045454545 | 0.03030303 | 0.03030303 | 0.045454545 | 0.015151515 | 0.045454545 | 0.03030303 | 0.015151515 | 0.03030303 | 0.045454545 | 0.060606061 | 0.075757576 | 0.03030303 | 0.015151515 | 0.060606061 | 0.03030303 | 0.075757576 |
| H3 | 0.015151515 | 0.03030303 | 0.015151515 | 0.075757576 | 0.045454545 | 0.015151515 | 0.060606061 | 0.045454545 | 0.060606061 | 0.03030303 | 0.015151515 | 0.045454545 | 0.060606061 | 0.03030303 | 0.060606061 | 0.045454545 | 0.03030303 | 0.060606061 | 0.060606061 | 0.015151515 |
| H4 | 0.060606061 | 0.015151515 | 0.03030303 | 0.015151515 | 0.015151515 | 0.015151515 | 0.045454545 | 0.03030303 | 0.03030303 | 0.045454545 | 0.015151515 | 0.045454545 | 0.03030303 | 0.045454545 | 0.045454545 | 0.03030303 | 0.03030303 | 0.045454545 | 0.03030303 | 0.060606061 |
| H5 | 0.015151515 | 0.015151515 | 0.075757576 | 0.060606061 | 0.015151515 | 0.015151515 | 0.060606061 | 0.03030303 | 0.045454545 | 0.03030303 | 0.015151515 | 0.045454545 | 0.045454545 | 0.060606061 | 0.03030303 | 0.015151515 | 0.03030303 | 0.060606061 | 0.045454545 | 0.015151515 |
| H6 | 0.015151515 | 0.015151515 | 0.03030303 | 0.03030303 | 0.045454545 | 0.015151515 | 0.015151515 | 0.03030303 | 0.03030303 | 0.03030303 | 0.015151515 | 0.03030303 | 0.03030303 | 0.045454545 | 0.03030303 | 0.015151515 | 0.045454545 | 0.045454545 | 0.015151515 | 0.015151515 |
| H7 | 0.03030303 | 0.015151515 | 0.03030303 | 0.03030303 | 0.03030303 | 0.03030303 | 0.015151515 | 0.045454545 | 0.03030303 | 0.060606061 | 0.015151515 | 0.03030303 | 0.03030303 | 0.045454545 | 0.015151515 | 0.045454545 | 0.045454545 | 0.060606061 | 0.015151515 | 0.03030303 |
| H8 | 0.015151515 | 0.03030303 | 0.045454545 | 0.045454545 | 0.015151515 | 0.045454545 | 0.060606061 | 0.015151515 | 0.045454545 | 0.03030303 | 0.015151515 | 0.045454545 | 0.060606061 | 0.075757576 | 0.045454545 | 0.060606061 | 0.060606061 | 0.03030303 | 0.015151515 | 0.015151515 |
| H9 | 0.03030303 | 0.03030303 | 0.045454545 | 0.03030303 | 0.060606061 | 0.03030303 | 0.045454545 | 0.045454545 | 0.015151515 | 0.045454545 | 0.045454545 | 0.045454545 | 0.015151515 | 0.060606061 | 0.060606061 | 0.045454545 | 0.045454545 | 0.045454545 | 0.03030303 | 0.015151515 |
| H10 | 0.03030303 | 0.015151515 | 0.03030303 | 0.03030303 | 0.015151515 | 0.045454545 | 0.045454545 | 0.075757576 | 0.060606061 | 0.015151515 | 0.03030303 | 0.03030303 | 0.045454545 | 0.015151515 | 0.075757576 | 0.045454545 | 0.045454545 | 0.03030303 | 0.03030303 | 0.015151515 |
| H11 | 0.03030303 | 0.03030303 | 0.015151515 | 0.045454545 | 0.03030303 | 0.015151515 | 0.03030303 | 0.045454545 | 0.03030303 | 0.045454545 | 0.015151515 | 0.060606061 | 0.03030303 | 0.045454545 | 0.045454545 | 0.060606061 | 0.03030303 | 0.045454545 | 0.015151515 | 0.03030303 |
| H12 | 0.03030303 | 0.060606061 | 0.03030303 | 0.045454545 | 0.045454545 | 0.03030303 | 0.03030303 | 0.045454545 | 0.045454545 | 0.045454545 | 0.015151515 | 0.015151515 | 0.060606061 | 0.045454545 | 0.045454545 | 0.045454545 | 0.045454545 | 0.060606061 | 0.03030303 | 0 |
| H13 | 0.03030303 | 0.03030303 | 0.045454545 | 0.03030303 | 0.03030303 | 0.03030303 | 0.045454545 | 0.03030303 | 0.045454545 | 0.060606061 | 0.015151515 | 0.045454545 | 0.015151515 | 0.060606061 | 0.045454545 | 0.045454545 | 0.045454545 | 0.045454545 | 0.060606061 | 0.045454545 |
| H14 | 0.015151515 | 0.045454545 | 0.075757576 | 0.045454545 | 0.060606061 | 0.03030303 | 0.060606061 | 0.060606061 | 0.045454545 | 0.045454545 | 0.03030303 | 0.045454545 | 0.045454545 | 0.015151515 | 0.045454545 | 0.045454545 | 0.045454545 | 0.045454545 | 0.075757576 | 0.045454545 |
| H15 | 0.015151515 | 0.060606061 | 0.075757576 | 0.060606061 | 0.045454545 | 0.03030303 | 0.060606061 | 0.045454545 | 0.075757576 | 0.045454545 | 0.060606061 | 0.060606061 | 0.075757576 | 0.045454545 | 0.015151515 | 0.045454545 | 0.045454545 | 0.03030303 | 0.03030303 | 0.015151515 |
| H16 | 0.03030303 | 0.045454545 | 0.045454545 | 0.03030303 | 0.045454545 | 0.015151515 | 0.045454545 | 0.03030303 | 0.060606061 | 0.045454545 | 0.075757576 | 0.045454545 | 0.060606061 | 0.045454545 | 0.075757576 | 0.015151515 | 0.045454545 | 0.060606061 | 0.015151515 | 0.015151515 |
| H17 | 0.03030303 | 0.045454545 | 0.03030303 | 0.03030303 | 0.045454545 | 0.015151515 | 0.060606061 | 0.045454545 | 0.045454545 | 0.045454545 | 0.060606061 | 0.060606061 | 0.045454545 | 0.045454545 | 0.075757576 | 0.045454545 | 0.015151515 | 0.03030303 | 0.015151515 | 0.015151515 |
| H18 | 0.03030303 | 0.045454545 | 0.060606061 | 0.060606061 | 0.075757576 | 0.03030303 | 0.045454545 | 0.03030303 | 0.03030303 | 0.03030303 | 0.045454545 | 0.045454545 | 0.060606061 | 0.060606061 | 0.03030303 | 0.060606061 | 0.03030303 | 0.015151515 | 0.060606061 | 0.045454545 |
| H19 | 0.015151515 | 0.015151515 | 0.015151515 | 0.03030303 | 0.060606061 | 0.015151515 | 0.015151515 | 0.015151515 | 0.015151515 | 0.015151515 | 0.015151515 | 0.03030303 | 0.045454545 | 0.075757576 | 0.015151515 | 0.03030303 | 0.03030303 | 0.060606061 | 0.015151515 | 0.060606061 |
| H20 | 0.045454545 | 0.045454545 | 0.015151515 | 0.03030303 | 0.015151515 | 0.015151515 | 0.03030303 | 0.015151515 | 0.015151515 | 0.015151515 | 0.060606061 | 0.03030303 | 0.015151515 | 0.060606061 | 0.015151515 | 0.015151515 | 0.015151515 | 0.045454545 | 0.060606061 | 0 |

*Table 13. Total relational matrix*

| | H1 | H2 | H3 | H4 | H5 | H6 | H7 | H8 | H9 | H10 | H11 | H12 | H13 | H14 | H15 | H16 | H17 | H18 | H19 | H20 |
|---|---|---|---|---|---|---|---|---|---|---|---|---|---|---|---|---|---|---|---|---|
| H1 | 0.107216132 | 0.162000806 | 0.187432194 | 0.172321584 | 0.196680924 | 0.101572704 | 0.19804036 | 0.159639061 | 0.205101355 | 0.178228562 | 0.152361186 | 0.189542146 | 0.19536903 | 0.229674577 | 0.198940271 | 0.166972063 | 0.140938948 | 0.204353329 | 0.137980699 | 0.160675419 |
| H2 | 0.119123405 | 0.143451637 | 0.167356961 | 0.180698566 | 0.163007458 | 0.112647628 | 0.191060789 | 0.138238973 | 0.184021212 | 0.15778785 | 0.120456765 | 0.168803871 | 0.189662514 | 0.223612737 | 0.220018101 | 0.160448582 | 0.135350047 | 0.212078762 | 0.148680942 | 0.173473391 |
| H3 | 0.109696345 | 0.148018653 | 0.160456985 | 0.217828438 | 0.185540966 | 0.103277194 | 0.215105236 | 0.17549473 | 0.207345804 | 0.167099237 | 0.124061842 | 0.192588295 | 0.214462242 | 0.205284256 | 0.21579892 | 0.18496778 | 0.159705949 | 0.221802088 | 0.181858881 | 0.120307506 |
| H4 | 0.138162205 | 0.115909676 | 0.150004471 | 0.134262301 | 0.132864897 | 0.088406231 | 0.174047151 | 0.139915849 | 0.154062386 | 0.158019437 | 0.10866936 | 0.167364644 | 0.158866493 | 0.189409279 | 0.174816931 | 0.146298125 | 0.136443102 | 0.179201926 | 0.132650672 | 0.14588707 |
| H5 | 0.098935209 | 0.120326834 | 0.203388335 | 0.18927725 | 0.141902809 | 0.093627852 | 0.198961033 | 0.148477931 | 0.176616624 | 0.152085143 | 0.11057749 | 0.176449139 | 0.183148934 | 0.213845273 | 0.169593903 | 0.141374048 | 0.145428796 | 0.204815128 | 0.156946644 | 0.108776834 |
| H6 | 0.078067847 | 0.095783657 | 0.129553061 | 0.128378445 | 0.140503222 | 0.074665706 | 0.122023606 | 0.11992609 | 0.130992015 | 0.122122904 | 0.089211195 | 0.130719629 | 0.135322134 | 0.161945535 | 0.137196332 | 0.110156032 | 0.132226479 | 0.153608654 | 0.099379909 | 0.084464598 |
| H7 | 0.105193515 | 0.110511462 | 0.146258482 | 0.14496185 | 0.142735937 | 0.101081664 | 0.140627217 | 0.151401749 | 0.149632281 | 0.168316852 | 0.104414614 | 0.148406342 | 0.154491614 | 0.18310499 | 0.143273222 | 0.157348711 | 0.14849496 | 0.188080814 | 0.113619949 | 0.11186846 |
| H8 | 0.10465318 | 0.144412366 | 0.183684079 | 0.181463037 | 0.150227133 | 0.12910519 | 0.2090711 | 0.142580789 | 0.188038342 | 0.162558859 | 0.120761088 | 0.186404382 | 0.207455127 | 0.23820553 | 0.197519492 | 0.193586882 | 0.184237814 | 0.184951142 | 0.132371311 | 0.113318944 |
| H9 | 0.119695975 | 0.146023926 | 0.186991397 | 0.1715908 | 0.196448964 | 0.115439811 | 0.197373306 | 0.174197585 | 0.161231087 | 0.177265331 | 0.151544545 | 0.189294465 | 0.167773958 | 0.227158867 | 0.213229759 | 0.181253273 | 0.170477263 | 0.202148438 | 0.14827486 | 0.114454244 |
| H10 | 0.11275468 | 0.121899627 | 0.159483 | 0.158032161 | 0.140628155 | 0.124303097 | 0.184299766 | 0.192157795 | 0.193680651 | 0.137992975 | 0.129368675 | 0.163331843 | 0.183905367 | 0.171479737 | 0.215583403 | 0.17110171 | 0.16213053 | 0.173199956 | 0.135750023 | 0.105239378 |
| H11 | 0.111576811 | 0.134381886 | 0.140799367 | 0.1686703 | 0.150696852 | 0.091582434 | 0.164936333 | 0.158959176 | 0.159559083 | 0.162932544 | 0.111322056 | 0.187211091 | 0.164868011 | 0.194157493 | 0.182261476 | 0.180668777 | 0.141789575 | 0.184122627 | 0.120109552 | 0.118553735 |
| H12 | 0.119450715 | 0.173524057 | 0.170522171 | 0.182967719 | 0.180051256 | 0.115013142 | 0.180728716 | 0.170762294 | 0.188582568 | 0.17589143 | 0.119983408 | 0.157147093 | 0.208860114 | 0.211683453 | 0.198200672 | 0.179140601 | 0.168524639 | 0.214249585 | 0.147485931 | 0.101814094 |
| H13 | 0.121024666 | 0.146365423 | 0.185374616 | 0.170162514 | 0.168346397 | 0.115885573 | 0.196320789 | 0.159262997 | 0.18976833 | 0.191493618 | 0.123404756 | 0.188308037 | 0.166553484 | 0.228506444 | 0.198683675 | 0.180833122 | 0.170454725 | 0.203137163 | 0.179800027 | 0.145604869 |
| H14 | 0.118749462 | 0.175374944 | 0.233044342 | 0.205525605 | 0.214741695 | 0.127216793 | 0.232071133 | 0.204481397 | 0.208954144 | 0.19462053 | 0.150157426 | 0.20828556 | 0.21710841 | 0.209320877 | 0.218976027 | 0.199610737 | 0.187692028 | 0.226081936 | 0.210418762 | 0.159499674 |
| H15 | 0.122855354 | 0.194361104 | 0.237763555 | 0.224234655 | 0.20410163 | 0.130656026 | 0.237782954 | 0.196178901 | 0.244259499 | 0.202119742 | 0.182167419 | 0.228641475 | 0.249905918 | 0.242619877 | 0.198250217 | 0.20573309 | 0.192686299 | 0.216640304 | 0.169674797 | 0.133606254 |
| H16 | 0.128448876 | 0.171396596 | 0.198354702 | 0.183351648 | 0.193288934 | 0.108234775 | 0.210115565 | 0.170516887 | 0.217724374 | 0.190052207 | 0.190398954 | 0.202395203 | 0.223113167 | 0.22701327 | 0.241768684 | 0.164851979 | 0.180615744 | 0.229412189 | 0.144325655 | 0.124038498 |
| H17 | 0.122642116 | 0.164194653 | 0.17479171 | 0.173813453 | 0.183324366 | 0.103426606 | 0.215007002 | 0.177224312 | 0.194632309 | 0.182042887 | 0.168523302 | 0.207209723 | 0.199675405 | 0.216479467 | 0.231929386 | 0.185236333 | 0.144197941 | 0.191056572 | 0.13516646 | 0.116863063 |
| H18 | 0.131081639 | 0.171695085 | 0.2145629 | 0.214868909 | 0.224785874 | 0.122985241 | 0.212036814 | 0.170362718 | 0.189304629 | 0.176305064 | 0.161185124 | 0.203584286 | 0.225458502 | 0.247092744 | 0.199045906 | 0.208108582 | 0.167620249 | 0.192126567 | 0.193129582 | 0.15800537 |
| H19 | 0.083380999 | 0.102152438 | 0.122022828 | 0.134839052 | 0.162284025 | 0.077877171 | 0.1284884 | 0.109506787 | 0.121094989 | 0.112928922 | 0.095405201 | 0.136961523 | 0.156096512 | 0.200901991 | 0.127185289 | 0.129940716 | 0.122075331 | 0.177258934 | 0.108808649 | 0.135672009 |
| H20 | 0.108658784 | 0.12623868 | 0.11280203 | 0.128217017 | 0.11333227 | 0.074332518 | 0.134579756 | 0.104192265 | 0.114887242 | 0.107699708 | 0.133116697 | 0.130709121 | 0.120598698 | 0.179084778 | 0.121512719 | 0.111092996 | 0.101318824 | 0.156354034 | 0.143451261 | 0.076017843 |

*Table 14. Cause and effect relationship matrix*

| | D | R | D+R | D-R | RANK | RELATION |
|---|---|---|---|---|---|---|
| H1 | 3.445041352 | 2.261367916 | 5.706409267 | 1.183673436 | 16 | CAUSE |
| H2 | 3.309980191 | 2.868023512 | 6.178003703 | 0.441956679 | 15 | CAUSE |
| H3 | 3.510701348 | 3.464647186 | 6.975348534 | 0.046054162 | 7 | EFFECT |
| H4 | 2.925262208 | 3.465465303 | 6.390727511 | -0.540203095 | 14 | EFFECT |
| H5 | 3.134555211 | 3.385493766 | 6.520048977 | -0.250938555 | 11 | EFFECT |
| H6 | 2.376247051 | 2.111337355 | 4.487584407 | 0.264909696 | 20 | CAUSE |
| H7 | 2.813824685 | 3.742677026 | 6.556501711 | -0.928852342 | 10 | EFFECT |
| H8 | 3.354605789 | 3.163478289 | 6.518084078 | 0.1911275 | 12 | CAUSE |
| H9 | 3.411867854 | 3.579488927 | 6.991356781 | -0.167621073 | 6 | EFFECT |
| H10 | 3.136322527 | 3.277563802 | 6.413886329 | -0.141241275 | 13 | EFFECT |
| H11 | 3.02915918 | 2.647091103 | 5.676250282 | 0.382068077 | 17 | CAUSE |
| H12 | 3.364583658 | 3.563357868 | 6.927941526 | -0.19877421 | 8 | EFFECT |
| H13 | 3.429291227 | 3.722695633 | 7.15198686 | -0.293404406 | 4 | EFFECT |
| H14 | 3.901931482 | 4.200581175 | 8.102512657 | -0.298649693 | 1 | EFFECT |
| H15 | 4.01423907 | 3.803784387 | 7.818023458 | 0.210454683 | 2 | CAUSE |
| H16 | 3.699417907 | 3.358724137 | 7.058142044 | 0.340693771 | 5 | CAUSE |
| H17 | 3.487437068 | 3.092409245 | 6.579846313 | 0.395027824 | 9 | CAUSE |
| H18 | 3.783345786 | 3.914680149 | 7.698025935 | -0.131334363 | 3 | EFFECT |
| H19 | 2.544881766 | 2.939884565 | 5.484766331 | -0.3950028 | 18 | EFFECT |
| H20 | 2.39819724 | 2.508141256 | 4.906338496 | -0.109944016 | 19 | EFFECT |

*Figure 1. Shows cause and effect relationship*

A standard questionnaire is designed and a study is conducted and factor analysis is done on collected data to find important challenges. Then ranking by SWARA method is done on the criterias and sub criteria of adoption challenges.It is seen that Integrated challenges is the first ranked challenge, followed by environmental challenge which is ranked 2nd and technical challenge ranked third. So, necessary frame work must be developed to mitigate the these challenge on priority basis, such that hydrogen energy can be widely accepted. To find linear relationship between these challenges Dematel analysis is conducted and among these challenges, it is found from figure1 that 8 items (H1, H2, H6, H8 ....etc, those values having high net degree of influence called as influential factors or cause and rest items are called as Effect items. Hence it will help to find most inflential challenges which can be taken care for wide adoption of hydrogen energy .

## CONCLUSIONS

Hydrogen energy needs large investments for future use. New tech is needed for storage and transport. Fueling systems and fuel cells are also vital. Hydrogen energy requires big investments to become a reality. We must have new tech for its storage and movement. Fuel cells and fueling systems are also very important. There are some technical and money issues, but it must be sustainable. New tech must lower the carbon impact when making hydrogen. This step will stop pollution from happening. Solving these issues can help people learn more. Then, hydrogen can work in all parts of the economy. It can take the place of fossil fuels and their damage. Hydrogen energy will make jobs and a cleaner future. It can fix climate change and make air better. Energy security will get better and grow the economy.

Technical and economic issues exist, but sustainability is key. New tech must cut the carbon footprint when making hydrogen. This will prevent pollution. Overcoming issues will raise awareness. Hydrogen can then be used in all sectors. It can replace fossil fuels and their harm. Hydrogen energy will create jobs and a sustainable future. It can solve climate change and boost air quality. Energy security will improve, fostering economic growth.

## REFERENCES


Baykara, S. Z. (2018). Hydrogen: A brief overview of its sources, production, and environmental impact. *International Journal of Hydrogen Energy*, *43*(23), 10605–10614. DOI: 10.1016/j.ijhydene.2018.02.022

Chakraborty, S., & Zavadskas, E. K. (2014). Applications of the WASPAS method in manufacturing decision making. *Informatica (Vilnius)*, *25*(1), 1–20. DOI: 10.15388/Informatica.2014.01

Endo, N., Goshome, K., Matsumoto, T., Segawa, Y., Shimoda, E., & Nozu, T. (2021). Thermal management and power-saving operations for improved energy efficiency within a renewable hydrogen energy system utilizing metal-hydride hydrogen storage. *International Journal of Hydrogen Energy*, *46*(1), 262–271. DOI: 10.1016/j.ijhydene.2020.10.002

European Commission. (2003). *Hydrogen energy and fuel cells: A vision of our future* (Report No. 20719). Directorate-General for Research. https://op.europa.eu/en/publication-detail/-/publication/20719

Ewan, B. C. R., & Allen, R. W. K. (2005). A figure-of-merit assessment of the routes to hydrogen. *International Journal of Hydrogen Energy*, *30*(8), 809–819. DOI: 10.1016/j.ijhydene.2005.02.003

Farrell, A. E., Keith, D. W., & Corbett, J. J. (2003). A strategy for introducing hydrogen into transportation. *Energy Policy*, *31*(13), 1357–1367. DOI: 10.1016/S0301-4215(02)00195-7

International Renewable Energy Agency. (2019). *Global energy transformation: A roadmap to 2050* (2019 Edition). https://www.irena.org/publications/2019/Apr/Global-energy-transformation-A-roadmap-to-2050-2019Edition

Murray, L. J., Dincă, M., & Long, J. R. (2009). Hydrogen storage in metal–organic frameworks. *Chemical Society Reviews*, *38*(5), 1294–1314. DOI: 10.1039/b802256a PMID: 19384439

Singh, R., Altaee, A., & Gautam, S. (2020). Nanomaterials in the advancement of hydrogen energy storage. *Heliyon*, *6*(7), e04231. Advance online publication. DOI: 10.1016/j.heliyon.2020.e04487 PMID: 32743097

U.S. Department of Energy. (n.d.). *Hydrogen: A clean, flexible energy carrier*. U.S. Department of Energy. https://www.energy.gov/eere/articles/hydrogen-clean-flexible-energy-carrier

World Health Organization. (2018). *COP24 special report: Health and climate change*. https://www.who.int/globalchange/publications/COP24-report/en/

Yang, J., Sudik, A., Wolverton, C., & Siegel, D. J. (2010). High-capacity hydrogen storage materials: Attributes for automotive applications and techniques for materials discovery. *Chemical Society Reviews*, *39*(2), 656–675. DOI: 10.1039/B802882F PMID: 20111786

# Chapter 2
# An Overview of Water Electrolysis Technologies for Green Hydrogen

**Dipan Kumar Das**
https://orcid.org/0000-0003-0224-3295
*Centurion University of Technology and Management, India*

**Padmaja Patnaik**
https://orcid.org/0000-0003-4468-7871
*Centurion University of Technology and Management, India*

**Sudip Kumar Das**
https://orcid.org/0009-0003-0814-0884
*Dr. C.V. Raman University, India*

**Nibedita Nayak**
*Centurion University of Technology and Management, India*

**Santanu Kumar Nayak**
https://orcid.org/0009-0004-2143-7655
*Centurion University of Technology and Management, India*

## ABSTRACT

*The transition towards a sustainable energy future is increasingly focused on hydrogen as a clean, versatile fuel that can decarbonize various sectors, including transportation, industry, and energy storage. Green hydrogen, produced through water electrolysis using renewable energy, is at the forefront of this transition. This chapter provides a comprehensive overview of the various water electrolysis technologies, including alkaline water electrolysis (AWE), proton exchange membrane (PEM) electrolysis, and solid oxide electrolysis (SOE). It discusses their principles, efficiency, advantages, challenges, and the role each technology plays in the context of large-scale green hydrogen production. The chapter also explores recent advancements, ongoing research, and future trends that could shape the development and deployment of these technologies. With a focus on sustainability, the chapter aims to elucidate the critical role of water electrolysis in achieving global carbon reduction targets and fostering a hydrogen-*

DOI: 10.4018/979-8-3693-8980-5.ch002

*based economy.*

# 1. INTRODUCTION TO GREEN HYDROGEN AND WATER ELECTROLYSIS

## 1.1. Overview of Hydrogen as a Clean Energy Carrier

Hydrogen is emerging as a promising clean energy carrier to address environmental concerns and fossil fuel depletion (Ishaq et al., 2021). It offers advantages in various applications, including transportation, heating, and power generation, with zero or near-zero emissions at the point of use (Kumar, 2015). Hydrogen production methods encompass both conventional and renewable energy sources, with power-to-hydrogen systems showing potential for storing variable renewable energy (Dawood et al., 2020). However, challenges persist in hydrogen storage, production, and establishing a hydrogen economy. The cleanness of hydrogen as an energy source depends on its production pathway, necessitating the development of a hydrogen cleanness index. Despite these challenges, hydrogen presents opportunities for sustainable development, offering advantages over synthetic carbon-based fuels and potentially serving as both an energy carrier and storage medium (Bicelli, 1986).

## 1.2. Definition and Significance of Green Hydrogen

Green hydrogen is emerging as a crucial element in the transition to low-carbon energy systems. It is commonly defined as hydrogen produced from water electrolysis powered by renewable energy sources (Incer-Valverde et al., 2023). The significance of green hydrogen lies in its potential to decarbonize various sectors of the global economy, offering advantages as both a fuel and energy storage medium. However, challenges remain in production, storage, transportation, and distribution. While green hydrogen is expected to become economically competitive by 2050, blue hydrogen (produced from fossil fuels with carbon capture) is also considered in many climate strategies (Noussan et al., 2020). The development of green hydrogen initiatives, standards, and guarantee of origin schemes varies across regions, with most existing efforts concentrated in Europe (Velazquez Abad & Dodds, 2020). These initiatives differ in their definitions, emission thresholds, and included production technologies, reflecting the complexity of establishing a universal framework for green hydrogen.

## 1.3. Water Electrolysis: Basic Principles and Reaction Mechanisms

Water electrolysis is a promising method for clean hydrogen production, with research focusing on improving catalysts for the oxygen evolution reaction (OER) and hydrogen evolution reaction (HER) (Suen et al., 2017). The process involves thermodynamic considerations, including temperature and pressure effects, and half-cell reaction mechanisms (Millet, 2015). Recent advancements include the development of cobalt oxide-based catalysts and mononuclear ruthenium complexes for homogeneous catalysis. Researchers are exploring nanostructured noble-metal and non-noble metal-based electrocatalysts to enhance activity and stability (Wang et al., 2021). The natural water oxidation process in photosynthesis, involving a manganese-calcium complex, provides insights into potential catalytic mechanisms (Dau et al., 2010). Ongoing challenges include improving catalyst efficiency, stability, and cost-effectiveness for commercial-scale hydrogen production. In situ approaches to studying electrocatalytic behavior during OER are crucial for designing better catalysts.

*Figure 1. Hydrogen production applications and methods (Kumar et al., 2022)*

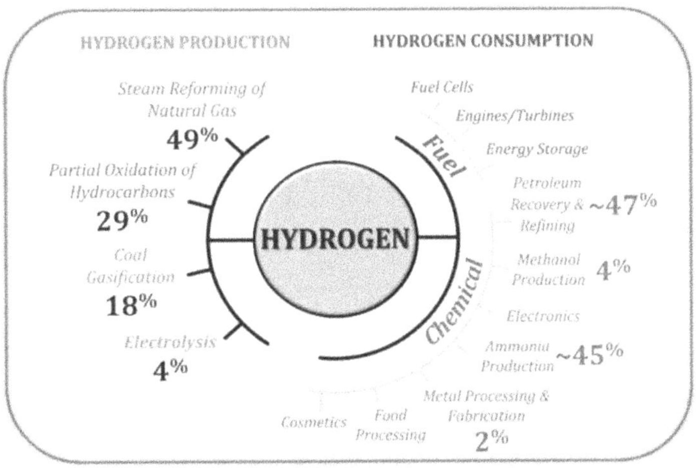


## 1.4. Importance of Renewable Energy in the Production of Green Hydrogen

Renewable energy plays a crucial role in producing green hydrogen, offering a sustainable alternative to fossil fuel-based methods. Solar and wind power are promising renewable sources for hydrogen generation, though their intermittent nature presents challenges. Electrolysis powered by renewable energy is a key technology for green hydrogen production, with potential for cost reduction through continued R&D. Green hydrogen is not only important as a fuel but also as a feedstock for the chemical industry, potentially decarbonizing manufacturing processes (Rambhujun et al., 2020). The transition to renewable hydrogen requires addressing challenges in production, storage, transportation, and distribution. A hydrogen cleanness index has been proposed to ensure the origin and environmental impact of hydrogen production. Overall, renewable energy-based hydrogen production offers significant opportunities for sustainable development and addressing global energy and environmental concerns (Sarker et al., 2023).

# 2. ALKALINE WATER ELECTROLYSIS (AWE)

## 2.1. Historical Development and Basic Working Principles of AWE

Alkaline Water Electrolysis (AWE) has a long history of development and commercialization, spanning over 100 years (Yu et al., 2021). The technology is based on fundamental principles of electrolysis and Faraday's laws. AWE involves the production of hydrogen and oxygen through water splitting in an alkaline electrolyte solution (Stolten & Krieg, 2011). Recent advancements have focused on developing non-precious metal catalysts to enhance efficiency and reduce costs. These catalysts include transition metal oxides, (oxy)hydroxides, and chalcogenides for the oxygen evolution reaction (OER), and alloys, phosphides, and carbides for the hydrogen evolution reaction (HER). Researchers are employing advanced characterization techniques like in situ/operando methods and density functional theory calculations to optimize catalyst performance (Wang et al., 2020). Despite its maturity, AWE continues to evolve, with ongoing research addressing challenges in electrode materials, scaling efficiency, and flexible operation (Emam et al., 2024).

*Figure 2. Alkaline water electrolysis for green hydrogen production (Tüysüz, H., 2024)*

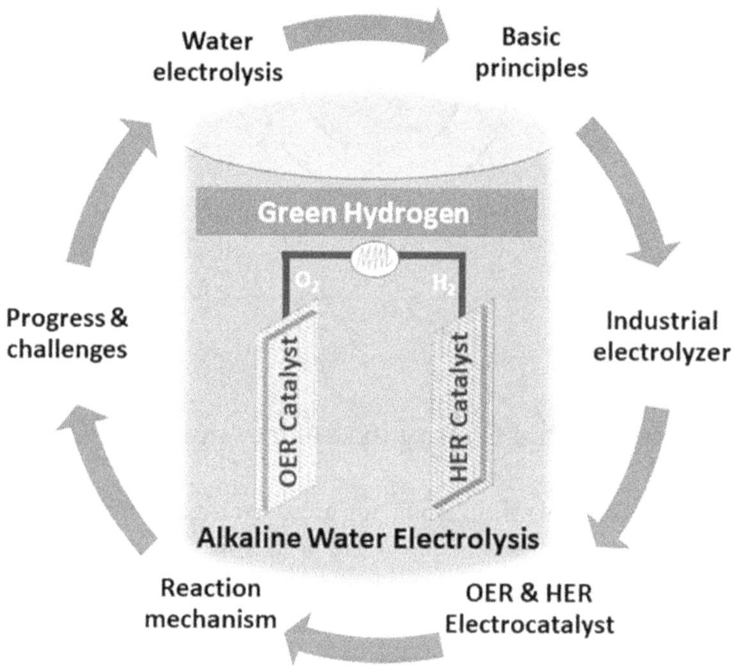


## 2.2. Electrolyte Composition and Electrode Materials

Alkaline water electrolysis (AWE) is a promising technology for hydrogen production, with research focusing on improving electrode materials and electrolyte composition. Studies have shown that NiFe and Ni electrodes perform well in NaOH and KOH electrolytes, with voltage decreasing in the order NiFe<Ni<SUS-316 at higher current densities (Ho-Sang Choi et al., 2012). Advanced electrode materials have demonstrated significant improvements, with some electrodes showing 385 mV lower hydrogen over potential and 50 mV lower oxygen over potential compared to polished nickel. Recent research has explored alternative electrolytes, electrode materials, and catalysts to enhance system performance and economic feasibility. Long-term durability tests have shown promising results, with no significant deterioration in electrode performance over 9000 hours of operation (C. Kjartansdóttir, 2014). Future research directions include improving electrode materials, safety standards, scaling efficiency, and flexible operation to advance AWE technology for sustainable energy applications (Emam et al., 2024).

## 2.3. Operational Conditions and Efficiency Factors

Recent studies on Alkaline Water Electrolysis (AWE) have explored various operational conditions and efficiency factors. Temperature significantly impacts AWE performance, with higher temperatures improving efficiency. Pressure adjustment is crucial for high-temperature operation, balancing efficiency gains with gas crossover concerns (Jang et al., 2023). Electrode spacing, electrolyte concentration, and diaphragm properties also affect AWE efficiency (Ding et al., 2022). Innovative approaches like mechanically induced electrolyte flow and induction heating of electrodes have shown promising results in enhancing AWE performance. Current density and input power mode are important parameters for optimization (Emam et al., 2023). Comprehensive modeling approaches, incorporating thermal balance and machine learning techniques, can help optimize AWE systems for improved efficiency and cost-effectiveness in renewable energy integration (Shangguan et al., 2023).

## 2.4. Advantages and Limitations of AWE

Alkaline water electrolysis (AWE) is a mature technology for hydrogen production, offering cost-effectiveness and potential for large-scale implementation (Kuckshinrichs et al., 2017). However, AWE faces challenges such as lower current densities and slow response to load changes, particularly when integrating with renewable energy sources (Campo Schneider et al., 2024). Recent advancements in materials, cell design, and system integration aim to improve efficiency at high current densities and expand the operational load range. Increasing operating temperature is a promising approach to overcome current density limitations, though it presents challenges in material stability (Lohmann-Richters et al., 2021). Comprehensive modeling efforts, including thermodynamic, electrochemical, thermal, and gas purity models, provide valuable insights for optimizing AWE performance. While electrochemical modeling is well-developed, thermal and gas purity modeling require further research to improve accuracy and develop effective management strategies (Hu et al., 2022).

*Table 1. Case studies of AWE in industrial applications*

| Case Study | Industry | Application | Key Benefits |
|---|---|---|---|
| Hydrogen Production for Ammonia Synthesis | Chemical Industry | Production of hydrogen for ammonia synthesis, a key component of fertilizers and other industrial chemicals. | Reduced carbon footprint, improved energy efficiency. |
| Hydrogen Fuel for Transportation | Transportation Industry | Production of hydrogen fuel for fuel cell vehicles, offering a cleaner alternative to fossil fuels. | Reduced emissions, increased energy efficiency. |
| Hydrogen Storage | Energy Storage | Production of hydrogen for storage in underground caverns or salt domes, providing a means to store excess renewable energy. | Energy security, grid stability. |
| Hydrogen Power Plants | Power Generation | Integration of hydrogen production into power plants, using hydrogen as a fuel or for peaking power generation. | Reduced emissions, increased flexibility. |
| Hydrogen-Based Refinery Processes | Oil and Gas Industry | Use of hydrogen in refinery processes such as hydrocracking and hydrotreating to improve fuel quality and reduce emissions. | Improved fuel quality, reduced emissions. |

# 3. PROTON EXCHANGE MEMBRANE (PEM) ELECTROLYSIS

## 3.1. Introduction to PEM Electrolysis and its Operating Principles

Proton Exchange Membrane (PEM) electrolysis is a promising technology for green hydrogen production, offering high efficiency and rapid response to intermittent renewable energy sources (Corda et al., 2023). PEM electrolysis operates by splitting water into hydrogen and oxygen using a solid thin electrolyte. Key factors affecting PEM electrolyzer performance include temperature, flow rate, cell orientation, and channel patterns. Higher temperatures and lower flow rates generally enhance performance by reducing activation and ohmic losses (Choi et al., 2021). However, gas bubble accumulation, particularly at the anode, can impair cell performance and reduce hydrogen output. While platinum-based catalysts are currently the benchmark for hydrogen evolution reactions, research is ongoing to develop non-noble metal catalysts or reduce noble metal content to improve cost-effectiveness (Perović et al., 2023). Understanding bubble dynamics and optimizing operating conditions are crucial for enhancing PEM electrolyzer efficiency and facilitating large-scale green hydrogen production (Sangtam & Park, 2023).

*Figure 3. PEM water electrolysis (Kumar S. S. and Himabindu, V., 2019)*

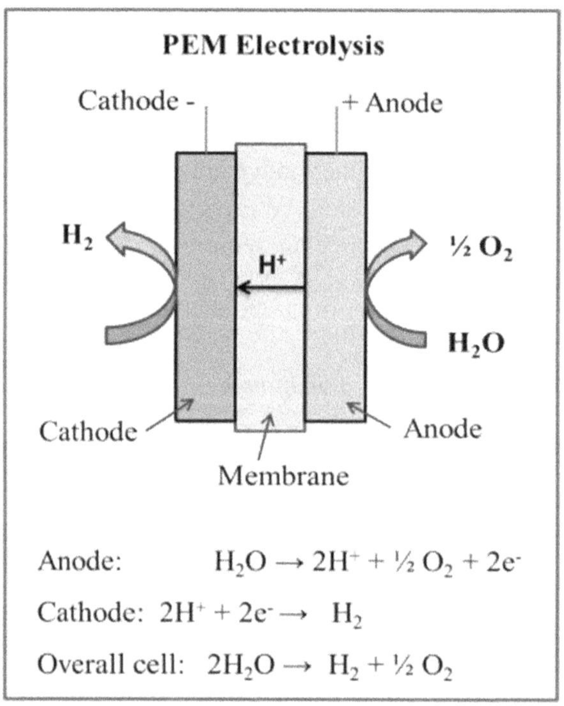

## 3.2. Structure and Function of the PEM Membrane and Catalyst Layers

The structure and function of proton exchange membrane (PEM) fuel cells and electrolyzers are complex, involving multiple layers with specific roles. The catalyst layer (CL) is crucial for electrochemical reactions and mass transfer (Xie et al., 2021). The CL is the core component of PEMFCs where the electrochemical reactions occur, and its composition and structure significantly impact the performance and durability of the fuel cell. The degradation of the Pt catalyst in the CL, including Ostwald ripening, migration, agglomeration, and loss of Pt particles, can lead to a reduction in electrochemical surface area (ECSA) and catalytic efficiency. It can degrade over time, with detachment of catalytic material observed during operation (Panchenko et al., 2020). Synchrotron radiography was used to visualize bubble formation and transport in the porous transport layer of a PEM electrolyzer. Bubbles used preferred pathways with low flow resistance in the PTL, and new transport pathways were activated with increasing current density. Strong particle detachment from the anode catalyst layer was observed, and catalyst diffusion between the anode and cathode was detected. The gas diffusion layer (GDL) connects bipolar plates and catalyst layers, facilitating gas diffusion and water management (Guo et al., 2022). The GDL is a key component in PEMFCs that provides physical and electrical connection between the catalyst layer and bipolar plate, with requirements including high gas permeability, electrical conductivity, thermal conductivity, mechanical strength, and hydrophobicity. The GDL has a bilayer structure consisting of a macroporous substrate (MPS) and a microporous layer (MPL), with the MPL playing an important role in improving the interface and water management. The membrane electrode assembly (MEA) is the core component, consisting of the GDL, CL, and proton exchange membrane (PEM). Modeling and simulation techniques have been employed to optimize these structures, including two-phase calculation models for GDL, microstructure models for CL, and molecular dynamics simulations for PEM. Ongoing research aims to improve performance and durability of PEM fuel cells and electrolyzers through structural optimization and material selection.

## 3.3. Advantages and Limitations of PEM Electrolysis

Proton exchange membrane (PEM) electrolysis is a promising technology for green hydrogen production, offering advantages such as high efficiency, high operating density, fast dynamic response, and the ability to operate at high pressures. PEM electrolyzers produce pure hydrogen without typical catalyst poisons found in reforming processes (Ayers et al., 2010). Compared to alkaline water electrolysis, PEM systems offer greater flexibility and reactivity to network demands (Millet et al., 2020). However, cost and durability remain significant challenges for large-scale implementation. Research efforts focus on developing advanced electrocatalysts to improve efficiency and reduce costs (Wang et al., 2022). Collaborative accelerated stress tests are needed to study degradation and improve durability. Despite these limitations, PEM electrolysis is considered a mature technology for green hydrogen production when integrated with renewable energy sources (Ayers et al., 2021).

## 3.4. Role of PEM Electrolysis in Coupling With Renewable Energy Sources

Proton exchange membrane (PEM) electrolysis coupled with renewable energy sources, particularly photovoltaics (PV), offers a promising pathway for sustainable hydrogen production. Direct coupling of PV arrays to PEM electrolyzers can achieve high energy transfer efficiency, up to 99.52% in optimal configurations (Zhou & Liu, 2013). However, solar radiation variability presents challenges for efficient system sizing and operation. Integrating battery storage can mitigate PV intermittency, ensuring stable hydrogen production and optimizing electrolyzer lifetime, with system efficiencies ranging from 6.6% to 8.81% (Arunachalam & Han, 2024). The seasonal and geographical variations in renewable energy sources necessitate region-specific design considerations for PEM electrolyzers (Ayyaswamy et al., 2023). While PEM electrolysis offers advantages such as high current density and efficiency, the use of expensive materials remains a limitation. Ongoing research focuses on improving key components, including membranes, electrocatalysts, and bipolar plates, to enhance performance and reduce costs (Zhang et al., 2022).

# 4. SOLID OXIDE ELECTROLYSIS (SOE)

## 4.1. Fundamentals of SOE and its Unique High-Temperature Operation

Solid Oxide Electrolysis (SOE) is a high-temperature technology operating at 600-900°C, offering efficient hydrogen production and renewable energy conversion. SOE cells inherit advantages from solid oxide fuel cells but require reconsideration of materials due to changed chemical environments under external bias. The high-temperature operation enables electrothermal catalysis, particularly for converting low-carbon alkanes to ethylene (Ye & Xie, 2021). Recent advancements have achieved remarkable current densities exceeding 6 A/cm$^2$ at 800°C with cell voltages below 1.67 V, surpassing commercial electrolyzer efficiency (Wood et al., 2016). While SOE cells can use materials similar to fuel cells, unique degradation mechanisms necessitate the development of new electrode materials and protective coatings (Elangovan et al., 2011). Modeling plays a crucial role in SOE development, addressing multiscale physical effects and reducing experimental costs (Mueller et al., 2023).

## 4.2. Advantages and Limitations of Solid Oxide Electrolysis (SOE)

Solid oxide electrolysis cells (SOECs) offer significant advantages for renewable energy storage and conversion due to their high operating temperatures, which enable superior thermodynamic and kinetic efficiencies (Hauch et al., 2020). When integrated with natural gas reforming and CO2 capture, SOEC-based power cycles can achieve electrical efficiencies of up to 66% LHV (Campanari & Gazzani, 2014). Combined cycle operations with SOFCs have been explored to further enhance efficiency, utilizing dynamic modeling to evaluate system performance (Colson et al., 2009). A particularly promising configuration involves integrating biomass gasification with SOFCs and gas turbines, potentially achieving electrical efficiencies around 73% (Aravind et al., 2012). This system employs heat pipes between the SOFC and gasifier, using SOFC anode off-gas as a gasifying agent, which significantly reduces exergy losses. These advancements in SOEC and SOFC technologies demonstrate their potential for high-efficiency power generation and energy storage in renewable energy systems.

Solid Oxide Fuel Cells (SOFCs) and Solid Oxide Electrolysis (SOE) systems face significant challenges due to their high operational temperatures (600-1000°C), which impact material stability and system longevity (Li et al., 2024). These elevated temperatures cause degradation of cell components and interfacial reactions, leading to performance deterioration over time (Antoni, 2004). To address these issues, researchers are exploring strategies to lower operating temperatures to the intermediate range (500-700°C), which could enhance materials stability and reduce thermo-mechanical mismatches (Tarancón, 2009). However, lowering temperatures also increases cathode-related electrical losses, necessitating the development of novel cathode materials. Advanced thermal management techniques, including innovative burner and heat exchanger designs, are being investigated to maintain system stability and efficiency. Additionally, nanoengineered cathode materials synthesized through infiltration and exsolution techniques show promise for improving SOE performance and durability (Biswas et al., 2023).

*Figure 4. Solid oxide electrolysis (Kumar S. S. and Himabindu, V., 2019)*

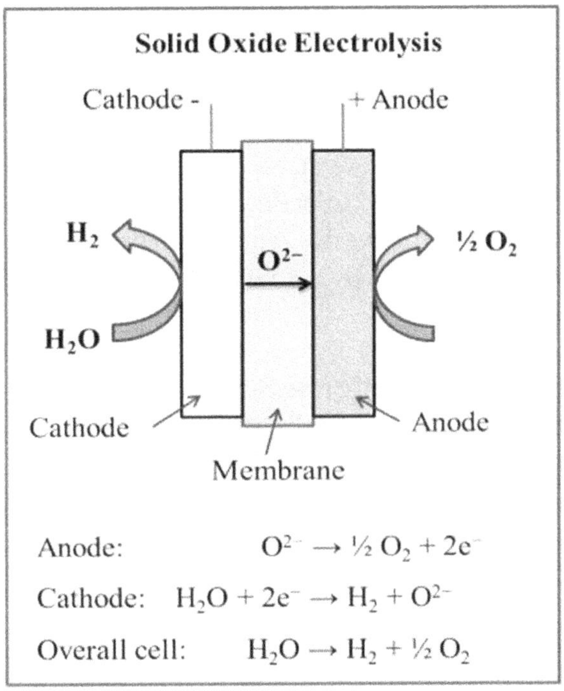


## 5. COMPARATIVE ANALYSIS OF WATER ELECTROLYSIS TECHNOLOGIES

### 5.1. Efficiency Comparison of AWE, PEM, and SOE

This comparative analysis of water electrolysis technologies for hydrogen production examines alkaline water electrolysis (AWE), proton exchange membrane electrolysis (PEM), and solid oxide electrolysis cells (SOEC). PEM offers advantages in current density, efficiency, and compatibility with renewable energy sources (Wang et al., 2022). SOEC demonstrates higher energy and exergy efficiencies in power-to-liquids/power-to-gas processes, while anion exchange membrane electrolysis shows superior carbon efficiency (Gao et al., 2022). A techno-economic analysis reveals that SOEC combined with waste heat sources has the lowest hydrogen production cost at 7.16 \$/kgH2, followed by AWE at 7.60 \$/kgH2 (Jang et al., 2022). Multiphysics modelling highlights differences in I-V characteristics, cell efficiency, and part-load performances among the technologies (Hu et al., 2022). Despite these advancements, challenges remain in reducing capital costs and improving overall efficiency to compete with fossil fuel-based hydrogen production methods.

## 5.2. Economic Analysis: Capital Costs, Operating Costs, and Total Cost of Ownership

Economic analyses of water electrolysis technologies reveal varying costs and future projections. Alkaline water electrolysis (AWE) currently shows lower costs compared to proton exchange membrane electrolysis (PEMEC), with hydrogen production costs of 7.60 \$/kgH2 and 8.55 \$/kgH2 respectively (Jang et al., 2022). Solid oxide electrolysis combined with waste heat (SOEC W.H) appears most competitive at 7.16 \$/kgH2 (Jang et al., 2022). Stack costs for AWE and PEM are expected to decrease significantly by 2030, with PEM showing greater potential for reduction due to its lower maturity (Krishnan et al., 2023). Electricity costs significantly impact overall expenses, as demonstrated in a comparison of AWE sites across Europe (Kuckshinrichs et al., 2017). For PEM electrolyzers, projected high-volume hydrogen production costs range from \$2/kg-H2 to \$7/kg-H2, with uninstalled capital costs between \$230/kW and \$600/kW (Vickers, 2020). Reducing capital costs and subsidizing electricity and tax rates are crucial for competitiveness (Jang et al., 2022).

## 5.3. Suitability for Various Applications: Small-Scale, Decentralized, and Large-Scale Centralized Production

Hydrogen production through water electrolysis is a promising pathway for sustainable energy, with three main technologies: Alkaline Water Electrolysis (AWE), Proton Exchange Membrane (PEM) electrolysis, and Solid Oxide Electrolysis (SOE). AWE is commercially viable for large-scale production, while PEM and SOE offer higher efficiency (Agrawal et al., 2024). PEM electrolysis is particularly suitable for small-scale, decentralized applications due to its high current density, energy efficiency, and compact design (Wang et al., 2022). Small-scale PEM fuel cells have shown potential in transportation, stationary, and portable power generation (Wang et al., 2023). For local hydrogen production, the combination of electrolyzers with renewable power sources is considered viable (Denk et al., 2022). However, challenges remain in developing efficient, durable catalysts and reducing costs for widespread adoption of these technologies.

## 5.4. Environmental Impact Assessment

Water electrolysis technologies for hydrogen production, including alkaline (AWE), proton exchange membrane (PEM), and solid oxide (SOE), have been evaluated for their environmental impacts and integration with renewable energy sources. Life cycle assessments reveal that PEM electrolysis combined with onshore wind power has the lowest global warming potential (Zhang et al., 2023). PEM also offers higher current densities and hydrogen purity compared to AWE . However, PEMEC stack production has higher environmental impacts due to critical material use, particularly platinum and iridium. Water consumption and its impacts on human health and the environment have been analyzed using methods like the Water Scarcity Footprint (Mehmeti et al., 2018). The levelized cost of hydrogen varies among technologies and renewable energy sources, with wind energy-associated PEM ranging from 5.3 to 9.29

$/kg (Ikuerowo et al., 2024). Overall, reducing critical material use in electrolysis stacks is crucial for minimizing environmental impacts (Zhao et al., 2020).

# 6. FUTURE TRENDS AND CHALLENGES IN WATER ELECTROLYSIS FOR GREEN HYDROGEN

## 6.1. Technological Innovations and Emerging Materials in Electrolysis

Recent technological innovations in electrolysis focus on developing advanced materials and techniques for efficient hydrogen production. Seawater electrolysis has emerged as a promising approach, but challenges like electrode corrosion necessitate robust materials and selective catalysts (Jin et al., 2023). Nanostructured electrode materials, including transition metal chalcogenides, carbides, and phosphides, have shown potential for improving water electrolysis performance (Pomerantseva et al., 2017). In alkaline water electrolysis, research has centered on enhancing electrode materials, catalysts, and membrane materials, as well as exploring alternative electrolytes (Emam et al., 2024). The development of new materials for efficient hydrogen and oxygen production in electrolyzers and fuel cells is crucial, with substrate-hydroxide binding energy and double layer interactions serving as key descriptors for catalyst design (Stamenković et al., 2016). These advancements aim to deliver reliable, affordable, and environmentally friendly energy solutions through electrochemical interfaces.

## 6.2. Role of Artificial Intelligence and Machine Learning in Optimizing Electrolysis Processes

Artificial intelligence (AI) and machine learning (ML) are revolutionizing electrochemical processes, including electrolysis for green hydrogen production and battery technologies. These techniques optimize operational conditions, enhance efficiency, and reduce costs. ML models can analyze complex datasets, predict optimal conditions, and provide real-time adjustments, potentially improving energy efficiency by up to 20% in hydrogen production (Bassey & Ibegbulam, 2023). AI algorithms contribute to various aspects of hydrogen and battery technologies, including production, storage, transportation, and safety management (Ramesh et al., 2022). In battery technology, AI aids in material discovery, design optimization, manufacturing improvements, and battery management systems. For electrochemical machining, AI optimizes process parameters, detects defects, predicts maintenance needs, and enhances workpiece and tooling design (Hari Prasadh S et al., 2023). The integration of AI in electrochemistry marks a paradigm shift, bridging theoretical computations with practical applications (Park & Lee, 2024).

## 6.3. Policy, Regulation, and Market Trends Impacting Green Hydrogen Production

Green hydrogen production is gaining momentum as a sustainable energy solution, driven by climate change concerns and the need for renewable energy storage (Wang et al., 2024). Technological advancements in electrolysis, fuel cells, and photocatalysts are accelerating the green hydrogen revolution. However, economic viability, social acceptance, and regulatory challenges remain significant hurdles, particularly in the United States. Policy interventions, such as subsidies and carbon pricing, are crucial for incentivizing green hydrogen adoption (Islam et al., 2024). Research themes in green hydrogen focus on storage, production, electrolysis, and the hydrogen economy. The success of green hydrogen implementation will require substantial investments, public-private collaboration, and regulatory support (Bade et al., 2023). Policy decisions in key countries like the United States, Europe, India, and China are expected to significantly impact global green hydrogen production and storage in the coming years (Raman et al., 2022).

## 7. CONCLUSION

Water electrolysis technologies, including alkaline electrolysis (AEL), proton exchange membrane (PEM) electrolysis, and solid oxide electrolysis (SOE), are pivotal to the production of green hydrogen. As global industries transition toward renewable energy sources, these technologies offer sustainable pathways for generating hydrogen without carbon emissions. Each method has distinct advantages and limitations, with AEL being the most mature, PEM providing flexibility and scalability, and SOE showing promise for high-efficiency hydrogen production at elevated temperatures.

The continued advancement of these technologies is critical for reducing costs, improving efficiencies, and scaling production to meet growing demand. Integrating electrolysis with renewable energy sources like wind and solar will be essential for driving down electricity costs, the primary determinant of green hydrogen viability. Moreover, innovations in material science, catalysis, and system design will further enhance the performance and economic competitiveness of electrolysis systems. Water electrolysis holds immense potential for enabling a hydrogen-based economy, which is key to decarbonizing industries, transportation, and energy storage. However, overcoming challenges related to cost, infrastructure, and energy efficiency will require coordinated efforts across research, industry, and policy to fully realize the promise of green hydrogen.

## REFERENCES


Abad, A. V., & Dodds, P. E. (2020). Green hydrogen characterisation initiatives: Definitions, standards, guarantees of origin, and challenges. *Energy Policy*.

Agrawal, D., Mahajan, N., Singh, S. A., & Sreedhar, I. (2024). Green hydrogen production pathways for sustainable future with net zero emissions. *Fuel*, *359*, 130131. DOI: 10.1016/j.fuel.2023.130131

Antoni, L. (2004). Materials for Solid Oxide Fuel Cells: The Challenge of their Stability. *Materials Science Forum*, *461-464*, 1073–1090. DOI: 10.4028/www.scientific.net/MSF.461-464.1073

Aravind, P., Schilt, C.R., Türker, B., & Woudstra, T. (2012). Thermodynamic Model of a Very High Efficiency Power Plant based on a Biomass Gasifier, SOFCs, and a Gas Turbine. International Journal of Renewable Energy Development.

Arunachalam, M., & Han, D. S. (2024). Efficient solar-powered PEM electrolysis for sustainable hydrogen production: An integrated approach. *Emergent Materials*, *7*(4), 1401–1415. DOI: 10.1007/s42247-024-00697-y

Ayers, K.E., Anderson, E.B., Capuano, C.B., Carter, B.D., Dalton, L.T., Hanlon, G., Manco, J., & Niedzwiecki, M. (2010). Research Advances towards Low Cost, High Efficiency PEM Electrolysis. ECS Transactions.

Ayers, K. E., Danilović, N., Harrison, K. W., & Xu, H. (2021). PEM Electrolysis, a Forerunner for Clean Hydrogen. *The Electrochemical Society Interface*, *30*(4), 67–72. DOI: 10.1149/2.F16214IF

Ayyaswamy, A., Goswami, N., Vishnugopi, B. S., & Mukherjee, P. P. (2023). Examining the Upstream Renewable Energy Source Variability on PEM Water Electrolyzer Behavior. ECS Meeting Abstracts. DOI: 10.1149/MA2023-01361957mtgabs

Bade, S. O., Tomomewo, O. S., Meenakshisundaram, A., Ferron, P., & Oni, B. A. (2023). Economic, social, and regulatory challenges of green hydrogen production and utilization in the US: A review. *International Journal of Hydrogen Energy*.

Bassey, K.E., & Ibegbulam, C. (2023). MACHINE LEARNING FOR GREEN HYDROGEN PRODUCTION. Computer Science & IT Research Journal.

Bicelli, L. P. (1986). Hydrogen: A clean energy source. *International Journal of Hydrogen Energy*, *11*(9), 555–562. DOI: 10.1016/0360-3199(86)90121-7

Biswas, S., Kaur, G., Paul, G., & Giddey, S. (2023). A critical review on cathode materials for steam electrolysis in solid oxide electrolysis. *International Journal of Hydrogen Energy*, *48*(34), 12541–12570. DOI: 10.1016/j.ijhydene.2022.11.307

Campanari, S., & Gazzani, M. (2014). High Efficiency SOFC Power Cycles With Indirect Natural Gas Reforming and CO2 Capture. *Journal of Fuel Cell Science and Technology*, *12*(2), 021008. DOI: 10.1115/1.4029425

Campo Schneider, L. P., Dhrioua, M., Ullmer, D., Egert, F., Wiggenhauser, H. J., Ghotia, K., Kawerau, N., Grilli, D., Razmjooei, F., & Ansar, S. A. (2024). Advancements in Hydrogen Production using Alkaline Electrolysis Systems: A Short Review on Experimental and Simulation Studies. *Current Opinion in Electrochemistry*, *47*, 101552. DOI: 10.1016/j.coelec.2024.101552

Choi, H., Yim, D., Rhyu, C., Kim, J., & Hwang, G. (2012). Study on the Electrode Characteristics for the Alkaline Water Electrolysis.

Choi, Y., Lee, W. K., & Na, Y. (2021). Effect of Gravity and Various Operating Conditions on Proton Exchange Membrane Water Electrolysis Cell Performance. *Membranes*, *11*(11), 11. DOI: 10.3390/membranes11110822 PMID: 34832051

Colson, C., Nehrir, M. H., Deibert, M. C., Amin, M. R., & Wang, C. (2009). Efficiency Evaluation of Solid-Oxide Fuel Cells in Combined Cycle Operations. *Journal of Fuel Cell Science and Technology*, *6*(2), 021006. DOI: 10.1115/1.2971174

Corda, G., Cucurachi, A., Fontanesi, S., & d'Adamo, A. (2023). Three-Dimensional CFD Simulation of a Proton Exchange Membrane Electrolysis Cell. *Energies*, *16*(16), 5968. DOI: 10.3390/en16165968

Dau, H., Limberg, C., Reier, T., Risch, M., Roggan, S., & Strasser, P. (2010). The Mechanism of Water Oxidation: From Electrolysis via Homogeneous to Biological Catalysis. *ChemCatChem*, *2*(7), 2. DOI: 10.1002/cctc.201000126

Dawood, F., Anda, M., & Shafiullah, G. M. (2020). Hydrogen production for energy: An overview. *International Journal of Hydrogen Energy*, *45*(7), 3847–3869. DOI: 10.1016/j.ijhydene.2019.12.059

Denk, K., Paidar, M., Hnát, J., & Bouzek, K. (2022). Potential of Membrane Alkaline Water Electrolysis in Connection with Renewable Power Sources. ECS Meeting Abstracts. DOI: 10.1149/MA2022-01261225mtgabs

Ding, S., Guo, B., Hu, S., Gu, J., Yang, F., Li, Y., Dang, J., Liu, B., & Ma, J. (2022). Analysis of the Effect of Characteristic Parameters and Operating Conditions on Exergy Efficiency of Alkaline Water Electrolyzer. SSRN Electronic Journal.

Elangovan, S., Hartvigsen, J.J., Larsen, D., Bay, I., & Zhao, F. (2011). Materials for Solid Oxide Electrolysis Cells.

Emam, A. S., Hamdan, M. O., Abu-Nabah, B. A., & Elnajjar, E. (2023). Enhancing alkaline water electrolysis through innovative approaches and parametric study. *International Journal of Hydrogen Energy*.

Emam, A. S., Hamdan, M. O., Abu-Nabah, B. A., & Elnajjar, E. (2024). A review on recent trends, challenges, and innovations in alkaline water electrolysis. *International Journal of Hydrogen Energy*, *64*, 599–625. DOI: 10.1016/j.ijhydene.2024.03.238

Gao, R., Zhang, L., Wang, L., Zhang, X., Zhang, C., Jun, K., Ki Kim, S., Park, H., Gao, Y., Zhu, Y., Zhao, T., Wan, H., & Guan, G. (2022). A comparative study on hybrid power-to-liquids/power-to-gas processes coupled with different water electrolysis technologies. *Energy Conversion and Management*, *263*, 115671. DOI: 10.1016/j.enconman.2022.115671

Guo, H., Chen, L., Ismail, S., Jiang, L., Guo, S., Gu, J., Zhang, X., Li, Y., Zhu, Y., Zhang, Z., & Han, D. (2022). Gas Diffusion Layer for Proton Exchange Membrane Fuel Cells: A Review. *Materials (Basel)*, •••, 15. DOI: 10.3390/met13010015 PMID: 36556607

Hari Prasadh, S. (2023). Evaluating the Application of Artificial Intelligence in Electrochemical Manufacturing: A Comprehensive Analysis. *International Journal on Recent and Innovation Trends in Computing and Communication*, *11*(4), 459–463. DOI: 10.17762/ijritcc.v11i4.9859

Hauch, A., Küngas, R., Blennow, P. G., Hansen, A. B., Hansen, J. B., Mathiesen, B. V., & Mogensen, M. B. (2020). Recent advances in solid oxide cell technology for electrolysis. *Science*, *370*(6513), 370. DOI: 10.1126/science.aba6118 PMID: 33033189

Hu, K., Fang, J., Ai, X., Huang, D., Zhong, Z., Yang, X., & Wang, L. (2022). Comparative study of alkaline water electrolysis, proton exchange membrane water electrolysis and solid oxide electrolysis through multiphysics modeling. *Applied Energy*, *312*, 118788. DOI: 10.1016/j.apenergy.2022.118788

Hu, S. H., Guo, B., Ding, S., Yang, F., Dang, J., Liu, B., Gu, J., Ma, J., & Ouyang, M. (2022). A comprehensive review of alkaline water electrolysis mathematical modeling. *Applied Energy*, *327*, 120099. DOI: 10.1016/j.apenergy.2022.120099

Ikuerowo, T., Bade, S. O., Akinmoladun, A., & Oni, B. A. (2024). The integration of wind and solar power to water electrolyzer for green hydrogen production. *International Journal of Hydrogen Energy*, *76*, 75–96. DOI: 10.1016/j.ijhydene.2024.02.139

Incer-Valverde, J., Korayem, A. K., Tsatsaronis, G., & Morosuk, T. (2023). "Colors" of hydrogen: Definitions and carbon intensity. *Energy Conversion and Management*, *291*, 117294. DOI: 10.1016/j.enconman.2023.117294

Ishaq, H., Dincer, I., & Crawford, C. (2021). A review on hydrogen production and utilization: Challenges and opportunities. *International Journal of Hydrogen Energy*.

Islam, A., Islam, T., Mahmud, H., Raihan, O., Islam, M. S., Marwani, H. M., Rahman, M. M., Asiri, A. M., Hasan, M. M., Hasan, M. N., Salman, M. S., Kubra, K. T., Shenashen, M. A., Sheikh, M. C., & Awual, M. R. (2024). Accelerating the green hydrogen revolution: A comprehensive analysis of technological advancements and policy interventions. *International Journal of Hydrogen Energy*, *67*, 458–486. DOI: 10.1016/j.ijhydene.2024.04.142

Jang, D., Cho, H., Lee, S., Park, M., Kim, S., Park, H., & Kang, S. (2023). Investigation of the operation characteristics and optimization of an alkaline water electrolysis system at high temperature and a high current density. *Journal of Cleaner Production*, *424*, 138862. DOI: 10.1016/j.jclepro.2023.138862

Jang, D., Kim, J., Kim, D., Han, W., & Kang, S. (2022). Techno-economic analysis and Monte Carlo simulation of green hydrogen production technology through various water electrolysis technologies. *Energy Conversion and Management*, *258*, 115499. DOI: 10.1016/j.enconman.2022.115499

Jin, H., Xu, J., Liu, H., Shen, H., Yu, H., Jaroniec, M., Zheng, Y., & Qiao, S. (2023). Emerging materials and technologies for electrocatalytic seawater splitting. *Science Advances*, *9*(42), 9. DOI: 10.1126/sciadv.adi7755 PMID: 37851797

Kjartansdóttir, C.K. (2014). Development of Hydrogen Electrodes for Alkaline Water Electrolysis.

Krishnan, S., Koning, V., Theodorus de Groot, M., de Groot, A., Mendoza, P. G., Junginger, M., & Kramer, G. J. (2023). Present and future cost of alkaline and PEM electrolyser stacks. *International Journal of Hydrogen Energy*, *48*(83), 32313–32330. DOI: 10.1016/j.ijhydene.2023.05.031

Kuckshinrichs, W., Ketelaer, T., & Koj, J. C. (2017). Economic Analysis of Improved Alkaline Water Electrolysis. *Frontiers in Energy Research*, *5*, 151. DOI: 10.3389/fenrg.2017.00001

Kumar, S. (2015). Role of Hydrogen in the Energy Sector.

Kumar, S. S., & Himabindu, V. (2019). Hydrogen production by PEM water electrolysis–A review. *Materials Science for Energy Technologies*, *2*(3), 442–454. DOI: 10.1016/j.mset.2019.03.002

Kumar, S. S., & Lim, H. (2022). An overview of water electrolysis technologies for green hydrogen production. *Energy Reports*, *8*, 13793–13813. DOI: 10.1016/j.egyr.2022.10.127

Li, M., Wang, J., Chen, Z., Qian, X., Sun, C., Gan, D., Xiong, K., Rao, M., Chen, C.X., & Li, X. (2024). A Comprehensive Review of Thermal Management in Solid Oxide Fuel Cells: Focus on Burners, Heat Exchangers, and Strategies. Energies.

Lohmann-Richters, F. P., Renz, S., Lehnert, W., Müller, M., & Carmo, M. (2021). Review—Challenges and Opportunities for Increased Current Density in Alkaline Electrolysis by Increasing the Operating Temperature. *Journal of the Electrochemical Society*, *168*(11), 114501. DOI: 10.1149/1945-7111/ac34cc

Mehmeti, A., Angelis-Dimakis, A., Arampatzis, G., McPhail, S.J., & Ulgiati, S. (2018). Life Cycle Assessment and Water Footprint of Hydrogen Production Methods: From Conventional to Emerging Technologies.

Millet, P. (2015). *Fundamentals of water electrolysis. Electrochemical Power Sources: Fundamentals.* Systems, and Applications.

Millet, P., Abdelmotagali, A., Assaud, L., & El Rouby, W.M. (2020). A Comparative Analysis of PEM and Alkaline Water Electrolyzers.

Mueller, M., Klinsmann, M., Sauter, U., Njodzefon, J., & Weber, A. (2023). High Temperature Solid Oxide Electrolysis – Technology and Modeling. *Chemieingenieurtechnik (Weinheim).*

Noussan, M., Raimondi, P. P., Scita, R., & Hafner, M. (2020). The Role of Green and Blue Hydrogen in the Energy Transition—A Technological and Geopolitical Perspective. *Sustainability (Basel)*, *13*(1), 298. DOI: 10.3390/su13010298

Panchenko, U., Arlt, T., Manke, I., Müller, M., Stolten, D., & Lehnert, W. (2020). Synchrotron Radiography for a Proton Exchange Membrane (PEM) Electrolyzer. Fuel Cells, 20.

Park, J., & Lee, J. (2024). Electrochemical energy conversion and storage processes with machine learning. *Trends in Chemistry*, *6*(6), 302–313. DOI: 10.1016/j.trechm.2024.04.007

Perović, K., Morović, S., Jukić, A., & Košutić, K. (2023). Alternative to Conventional Solutions in the Development of Membranes and Hydrogen Evolution Electrocatalysts for Application in Proton Exchange Membrane Water Electrolysis: A Review. *Materials (Basel), 16*(18), 16. DOI: 10.3390/ma16186319 PMID: 37763596

Pomerantseva, E., Resini, C., Kovnir, K., & Kolen'ko, Y. V. (2017). Emerging nanostructured electrode materials for water electrolysis and rechargeable beyond Li-ion batteries. *Advances in Physics: X, 2*(2), 211–253. DOI: 10.1080/23746149.2016.1273796

Raman, R., Nair, V. K., Prakash, V. V., Patwardhan, A., & Nedungadi, P. (2022). Green-hydrogen research: What have we achieved, and where are we going? Bibliometrics analysis. *Energy Reports, 8*, 9242–9260. DOI: 10.1016/j.egyr.2022.07.058

Rambhujun, N., Salman, M. S., Wang, T., Pratthana, C., Sapkota, P., Costalin, M., Lai, Q., & Aguey-Zinsou, K. (2020). Renewable hydrogen for the chemical industry. *MRS Energy & Sustainability : a Review Journal*, •••, 7. PMID: 38624624

Sai Ramesh, A., Vigneshwar, S., Vickram, S., Manikandan, S., Subbaiya, R., Karmegam, N., & Kim, W. (2022). Artificial intelligence driven hydrogen and battery technologies – A review. *Fuel.*

Sangtam, B. T., & Park, H. (2023). Review on Bubble Dynamics in Proton Exchange Membrane Water Electrolysis: Towards Optimal Green Hydrogen Yield. *Micromachines, 14*(12), 14. DOI: 10.3390/mi14122234 PMID: 38138403

Sarker, A. K., Azad, A. K., Rasul, M., & Doppalapudi, A. T. (2023). Prospect of Green Hydrogen Generation from Hybrid Renewable Energy Sources: A Review. *Energies, 16*(3), 1556. DOI: 10.3390/en16031556

Shangguan, Z., Li, H., Yang, B., Zhao, Z., Wang, T., Jin, L., & Zhang, C. (2023). Optimization of alkaline electrolyzer operation in renewable energy power systems: A universal modeling approach for enhanced hydrogen production efficiency and cost-effectiveness. *International Journal of Hydrogen Energy.*

Stamenković, V. R., Strmčnik, D., Lopes, P. P., & Markovic, N. M. (2016). Energy and fuels from electrochemical interfaces. *Nature Materials, 16*(1), 57–69. DOI: 10.1038/nmat4738 PMID: 27994237

Stolten, D., & Krieg, D. (2011). Alkaline Electrolysis - Introduction and Overview. *ChemInform*, •••, 42.

Suen, N., Hung, S., Quan, Q., Zhang, N., Xu, Y., & Chen, H. (2017). Electrocatalysis for the oxygen evolution reaction: Recent development and future perspectives. *Chemical Society Reviews, 46*(2), 337–365. DOI: 10.1039/C6CS00328A PMID: 28083578

Tarancón, A. (2009). Strategies for Lowering Solid Oxide Fuel Cells Operating Temperature. *Energies, 2*(4), 1–21. DOI: 10.3390/en20401130

Tüysüz, H. (2024). Alkaline water electrolysis for green hydrogen production. *Accounts of Chemical Research, 57*(4), 558–567. PMID: 38335244

Vickers, J.C. (2020). DOE Hydrogen and Fuel Cells Program Record 19009: Hydrogen Production Cost from PEM Electrolysis-2019.

Wang, J., Gao, Y., Kong, H., Kim, J., Choi, S., Ciucci, F., Hao, Y., Yang, S., Shao, Z., & Lim, J. (2020). Non-precious-metal catalysts for alkaline water electrolysis: Operando characterizations, theoretical calculations, and recent advances. *Chemical Society Reviews*, *49*(24), 9154–9196. DOI: 10.1039/D0CS00575D PMID: 33140778

Wang, L., Liu, W., Sun, H., Yang, L., & Huang, L. (2024). Advancements and Policy Implications of Green Hydrogen Production from Renewable Sources. *Energies*, *17*(14), 3548. DOI: 10.3390/en17143548

Wang, S., Lu, A., & Zhong, C. (2021). Hydrogen production from water electrolysis: Role of catalysts. *Nano Convergence*, *8*(1), 8. DOI: 10.1186/s40580-021-00254-x PMID: 33575919

Wang, T., Cao, X., & Jiao, L. (2022). PEM water electrolysis for hydrogen production: fundamentals, advances, and prospects. Carbon Neutrality, 1.

Wang, Z., Liu, Z., Fan, L., Du, Q., & Jiao, K. (2023). *Application progress of small-scale proton exchange membrane fuel cell*. Energy Reviews. DOI: 10.1016/j.enrev.2023.100017

Wood, A., He, H., Joia, T., Krivy, M., & Steedman, D. (2016). Communication—Electrolysis at High Efficiency with Remarkable Hydrogen Production Rates. *Journal of the Electrochemical Society*, *163*(5), 163. DOI: 10.1149/2.0341605jes

Xie, M., Chu, T., Wang, T., Wan, K., Yang, D., Li, B., Ming, P., & Zhang, C. (2021). Preparation, Performance and Challenges of Catalyst Layer for Proton Exchange Membrane Fuel Cell. *Membranes*, *11*(11), 11. DOI: 10.3390/membranes11110879 PMID: 34832108

Ye, L., & Xie, K. (2021). High-temperature electrocatalysis and key materials in solid oxide electrolysis cells. *Journal of Energy Chemistry*, *54*, 736–745. DOI: 10.1016/j.jechem.2020.06.050

Yu, Z., Duan, Y., Feng, X., Yu, X., Gao, M., & Yu, S. (2021). Clean and Affordable Hydrogen Fuel from Alkaline Water Splitting: Past, Recent Progress, and Future Prospects. *Advanced Materials*, *33*(31), 33. DOI: 10.1002/adma.202007100 PMID: 34117808

Zhang, J., Wang, Z., He, Y., Li, M., Wang, X., Wang, B., Zhu, Y., & Cen, K. (2023). Comparison of onshore/offshore wind power hydrogen production through water electrolysis by life cycle assessment. *Sustainable Energy Technologies and Assessments*, *60*, 103515. DOI: 10.1016/j.seta.2023.103515

Zhang, K., Liang, X., Wang, L., Sun, K., Wang, Y., Xie, Z., Wu, Q., Bai, X., Hamdy, M. S., Chen, H., & Zou, X. (2022). *Status and perspectives of key materials for PEM electrolyzer*. Nano Research Energy. DOI: 10.26599/NRE.2022.9120032

Zhao, G., Kraglund, M. R., Frandsen, H. L., Wulff, A. C., Jensen, S. H., Chen, M. H., & Graves, C. R. (2020). Life cycle assessment of H2O electrolysis technologies. *International Journal of Hydrogen Energy*, *45*(43), 23765–23781. DOI: 10.1016/j.ijhydene.2020.05.282

Zhou, M., & Liu, Y. (2013). Theoretical study on direct coupling of a PV array to a PEM electrolyser. *2013 International Conference on Materials for Renewable Energy and Environment*, 1, 52-56. DOI: 10.1109/ICMREE.2013.6893613

# Chapter 3
# Antifungal Potential of Trichoderma sp. Against Three Species of Fusarium Genus:
## Responsible for the Root Rot of the Saffron Corms (Crocus sativus L.)


**Samah Ourras**
https://orcid.org/0000-0003-1826-332X
*Ibn Tofail University, Morocco*

**M'hamed El Ouark**
*Ibn Tofail University, Morocco*

**Ismail El Aymani**
*Ibn Tofail University, Morocco*

**Soukaina Msairi**
*Mohammed V University, Morocco*

**Karima Selmaoui**
*Ibn Tofail University, Morocco*

**Moulay Abdelaziz El Alaoui**
*Ibn Tofail University, Morocco*

**Najoua Mouden**
*Mohammed 1st University, Morocco*

**Amina Ouazzani Touhami**
https://orcid.org/0000-0001-9970-7724
*Ibn Tofail University, Morocco*

**Allal Douira**
*Ibn Tofail University, Morocco*



## ABSTRACT

*This study examines the in vitro antagonistic effect of three native Trichoderma sp. isolates originating from saffron corms, against three Fusarium species plants (F. pseudograminearum, F. moniliforme and F. oxysporum) known to cause rotting in the bulbs and roots of saffron. To introduce Trichoderma into a preventive control program, the Trichoderma isolates were tested by in direct confrontation and volatile substances production. They demonstrated colonization percentages ranging from 50% to 79.6%, significantly inhibiting the mycelial growth and spore germination of pathogenic species, with inhibition rates ranging from 53% to 79% and 51% to 79%, respectively. However, indirect confrontation through the production of diffusible substances showed low inhibition of mycelial growth and spore germination for all pathogenic isolates studied, not exceeding 28.47% and 20.5%, respectively. This study discusses*



DOI: 10.4018/979-8-3693-8980-5.ch003

*the potential of using Trichoderma as biocontrol agents against a wide range of bioaggressors.*

## I. INTRODUCTION

Several fungal species carried by corms induce disease in various components of saffron plants (El Aymani et al., 2019; Ourras et al, 2023). Representatives of the genera Fus*arium, Rhizoctonia, Penicillium, Aspergillus, Sclerotium, Phoma, Stromatinia, Cochliobolus and* Rhizopus are associated with saffron diseases (Ahrazem et al., 2010, Rubio-Moraga et al., 2013; El Aymani et al., 2019, Ourras et al., 2023). The most serious diseases reported include *Fusarium oxysporum*-induced rots (Cappelli, 1994; Bentata et al., 2017; Najari et al., 2018), as well as those caused by *Sclerotium rolfsii* (Kalha et al., 2007) and *Rhizoctonia violacea* (Bentata et al., 2017). These saffron diseases cause severe losses across most saffron fields (Palmero et al., 2014). Infected plants exhibit variable symptoms during the flowering period, including yellowing, leaf wilting, and root and corm rot. Typically, pathogens are harbored within corms and contaminated soils (Brayford, 1996). Field plant infections are typically observed shortly after the planting of contaminated corms (Cappelli and Di Minco, 1999).

Chemical control methods against *Fusarium* spp. are not fully effective as the pathogen can penetrate plant tissues (Qostal et al., 2019a). Additionally, crop rotation proves ineffective against these soil fungi due to their ability to survive in crop debris and infect a wide range of host plants (Baha Eddine et al., 2019). Consequently, there is growing interest in alternative control measures, prompting research and development efforts towards sustainable and effective products for managing plant pathogens in Moroccan agriculture (Qostal et al., 2019b). In this context, harnessing fungi belonging to the genus Trichoderma presents an intriguing approach for safeguarding corms and saffron plants against various abiotic and biotic stresses in Morocco.

Various organisms serve as effective biological control agents against soil pathogens. Bacterial and fungal genera commonly utilized and marketed for this purpose comprise Gliocladium, Bacillus, Coniothyrium, Paecilomyces, Phlebiopsis, Pseudomonas, Rhizobium, Serratia, Streptomyces, and Trichoderma (Mazzola and Freilich, 2017; Elbouazaoui et al., 2022; Ait Rahou et al., 2022; Sellami et al., 2023).

Trichoderma are filamentous fungi (Papavizas, 1985) renowned for their antagonistic abilities against numerous phytopathogens (Artigues & Davet, 1984; Camporota, 1985; Mouria et al., 2008; Ouazzani Chahdi et al., 2014; Qostal et al., 2020; Kribel et al., 2020; Mouden et al., 2023; Adnani et al., 2024; Errifi et al., 2024). These fungi are effective soil colonizers (Davet & Camporota, 1986), inhabiting the rhizosphere (Hem & Pang, 2017) and plant roots (Mouria et al., 2008; Qostal et al., 2020; Kribel et al., 2020). They are recognized as biological control agents against a wide range of pathogens, both soil-borne and foliar (Tijerino et al., 2011; Leylaie & Zafari, 2018; Camporota, 1985; Ouazzani-Touhami et al., 1994; Pertot et al., 2015; Qostal et al., 2020; Bastakoti et al., 2017). Several studies have been conducted to evaluate the antagonistic abilities of Trichoderma against various phytopathogens, both in vitro and in vivo. These pathogens include *Fusarium* spp. (Mayo et al., 2015; Moumene et al., 2016; Abbas et al., 2017; Luis Vargas-Inciarte et al., 2019; Adnani et al., 2024), *Botrytis* spp. (De Meyer et al., 1998; Hmouni et al., 1999, 2006; Tronsmo, 1977; Sadek et al., 2020), *Verticillium* spp. (Mouria et al., 2008; Hanson, 2000), *Pythium* spp. (Dumitras & Fratilescu-Sesan, 1980), Colletotrichum spp. (El Kaissoumi et al., 2024), *Sclerotinia rolfsii* (Elad et al., 1980), *Phytophthora infestans* (Moumene et al., 2015), and *Rhizoctonia solani* and *Macrophomina phaseolina* (El-Sebaay & El-Sayed, 2019; Wang & Zhuang, 2019; Ali, 2021; El-Benawy et al., 2020; Errifi et al., 2024).

This study aimed to assess the in vitro antagonistic activity of three Trichoderma strains against three Fusarium species isolated from diseased saffron corms. The goal was to develop a biological control method targeting these highly hazardous pathogens affecting saffron

## II. MATERIAL AND METHODS

### Antagonist

Isolates of *Trichoderma* spp. Tested (*Trichoderma* sp1, Trichoderma sp2 and Trichoderma sp3), originating from saffron corms, are maintained on PSA medium (composed of 200 g of potato, 15 g of Agar-agar and 1000 mL of distilled water) at 28 ° C. and in darkness.

### Pathogen

Isolates of Fusarium species (*F. pseudograminearum, F. moniliforme, and F. oxysporum)* collected from saffron corms were maintained on PSA medium at 28 ° C and in darkness.

### Antifungal Test by Direct Confrontation

- **Inhibition of mycelial growth**

In a Petri dish containing PSA medium, two 5 mm diameter disks, one consisting of the pathogen inoculum and the other of the antagonist inoculum, were placed 40 mm apart symmetrically relative to the center of the dish.

For the control, a mycelial disc of the pathogen was placed in the center of the Petri dish. After incubating for three days at 28 °C in darkness, the percentage inhibition (CI%) of the pathogen's mycelial growth by the antagonists was evaluated according to the method of Sy (1976):

$$IC\ (\%) = (DT - DPA) / DT \times 100$$

**DT:** Diametral growth of control;
**DPA:** Mycelial diametral growth of the pathogen in the presence of the antagonist.

- **Space colonization by *Trichoderma***

In the Petri dishes previously used, the percentage colonization (C%) of the space by *Trichoderma* spp. was measured according to the following formula:

$$C\ (\%) = (DT/DE) \times 100$$

**C%:** Percentage colonization *of Trichoderma*
**DT:** Distance crossed by *Trichoderma*;
**DE:** Distance separating the two mycelial discs

## Inhibition of Germination

Conidial suspensions were prepared from young cultures of the various antagonists and pathogens by washing the cultures with sterile distilled water and filtering through a muslin cloth. The concentrations were adjusted to $10^3$conidia/mL with sterile distilled water using a Malassez cell.

In Petri dishes containing 20 ml of agar water (20 g of Agar-agar, 1000 ml of distilled water), 0.1 ml of the two suspensions were spread successively. After incubation for 20 hours at 28 ° C., the percentage inhibition of germination (IG%) of the conidia of the pathogen is determined using the following formula:

$$IG\,(\%) = (NT - NPA)\,/\,NT \times 100$$

**NT:** Number of germinated conidia in the control;
**NPA:** Number of conidia germinated in the presence of the antagonist.

## Antifungal Test by Production of Volatile Substances

- **Inhibition of mycelial growth**

The production of volatile substances was demonstrated using the method described by Dennis and Webster (1971). A 5 mm diameter mycelial disc of each pathogen and antagonist is placed in the center of the Petri dishes containing the PSA medium. The lids are removed aseptically, then the bottom of each Petri dishes containing the antagonist tested is placed below that containing the pathogen. The two juxtaposed bottoms are then sealed together with layers of parafilm.

A Petri dish bottom containing the medium alone is placed under a Petri dish bottom containing the pathogen for the control. After four days of incubation at 28 °C in the dark, inhibition of mycelial growth is described as above.

- **Inhibition of germination**

Petri dish containing a 48-hour-old Trichoderma culture on PSA was placed beneath a Petri dish containing agar water, on which 0.2 ml of a conidial suspension of the pathogen ($10^3$conidia/mL) was spread. After 20 hours of incubation at 28°C in the dark, the inhibition of germination is assessed as previously described.

## Production of Diffusible Substances

- **Inhibition of mycelial growth**

The secretion of diffusible substances by Trichoderma is detected using the cellophane technique (Dennis and Webster, 1971). The PSA medium is covered with a sheet of sterile cellophane. 24 hours later, a 5mm diameter disc from a seven-day-old Trichoderma crop is placed in the center of the cellophane leaf surface. The dishes are then incubated for 48 hours at 28 °C in the dark. The cellophane membrane and the adherent colony of Trichoderma are then removed and a mycelial disk of the 10-day-old pathogen is placed in the center of the culture medium. After eight days of incubation in darkness and at 28 °C., the percentage inhibition of the mycelial growth of the pathogen is calculated relative to the control.

- **Inhibition de la germination**

After removing the cellophane sheet and the adherent *Trichoderma* colony, 0.2 ml of suspension adjusted to $10^3$conidies/mL was spread over the agar water. After incubation for 20 hours, the inhibition of the germination of the conidia is estimated as above.

## Statistical Analysis

The processing of the data concerned the analysis of the variance by the test ANOVA of order I. A comparison of the means is carried out by the test PPDS (smallest significant difference) if a significant difference is recorded at the probability threshold of 5%.

# III. RESULTS

The antagonistic activity of three *Trichoderma* spp. isolates against three pathogens (*Fusarium pseudograminearum, F. moniliforme, and F. oxysporum*) was studied by assessing different modes of action: competition for space, production of volatile substances, and production of diffusible substances. Additionally, the inhibitory effect of these Trichoderma isolates on the mycelial growth and conidial germination of fungi responsible for saffron root and corm rot was evaluated in vitro.

The space colonization by Trichoderma isolates (C%) (Table 1) yielded different results in the presence of fungal species responsible of saffron roots and corms rot. The highest percentage of colonization of the space was achieved by the isolate *Trichoderma* sp.1, with a percentage of colonization of 76.3% in the presence of *F. pseudograminiarum.* Additionally, the isolate *Trichoderma* sp.3 demonstrated a 65% colonization rate in the presence of *F. oxysporum.*

*Table 1. Space colonization by Trichoderma isolates in the presence of species responsible for root and corms rot of Crocus sativus.*

| **The antagonists** | *Fusarium oxysporum* | *Fusarium pseudograminearum* | *Fusarium moniliforme* |
|---|---|---|---|
| *Trichoderma* sp.1 | $72.6^{a}_{a}$ | $76.3^{a}_{a}$ | $63^{b}_{b}$ |
| *Trichoderma* sp2 | $50^{b}_{cd}$ | $54^{b}_{c}$ | $62.5^{a}_{b}$ |
| *Trichoderma* sp3 | $65^{a}_{b}$ | $62.03^{a}_{bc}$ | $55^{b}_{c}$ |

* Two values read from the same column followed by the same index letter are not significantly different at the 5% threshold.

* Two values read on the same line followed by the same letter while exposing are not significantly different at the 5% threshold

Inhibition of mycelial growth of pathogenic fungal species is a key parameter for measuring Trichoderma antagonism. Most Trichoderma isolates demonstrated significant inhibition of pathogenic isolates, with maximum inhibition percentages ($IC_{CD}$%) *exceeding 78 and 79% for Trichoderm*a sp.1 isolates in the presence *of F. oxysporum, F. pseudograminiarum, and F. moniliforme* (Table 2). For *Trichoderma* sp.2, the inhibition percentages of mycelial growth for the three Fusarium species tested ranged from 76.4% to 72.87%, while for Trichoderma sp.3, the inhibition percentages ranged from 72.54% to 53.50%.

*Table 2. Inhibition of mycelial growth of species responsible for root and corms rot of Crocus sativus plants by Trichoderma isolates in direct confrontation ($ICC_D$%)*

| **The antagonists** | *Fusarium oxysporum* | *Fusarium pseudograminearum* | *Fusarium moniliforme* |
|---|---|---|---|
| *Trichoderma* sp1 | $79.12^{a}_{a}$ | $78.04^{a}_{a}$ | $78.94^{a}_{a}$ |
| *Trichoderma* sp2 | $67.14^{b}_{b}$ | $57.87^{c}_{b}$ | $72.87^{a}_{a}$ |
| *Trichoderma* sp3 | $72.45^{a}_{a}$ | $60.18^{b}_{b}$ | $53.50^{c}_{c}$ |

*Two values read from the same column followed by the same index letter are not significantly different at the 5% threshold.

* Two values read on the same line followed by the same letter while exposing are not significantly different at the 5% threshold

*Figure 1. Antagonistic activity of Trichoderma sp3. (T8) against F. moniliforme*(**a**), *F. oxysporum* (**b**) *and F. pseudograminearum*(**c**) **1**: *Control;* **2**: *Direct confrontation;* **3**: *Indirect confrontation through the production of volatile substances*

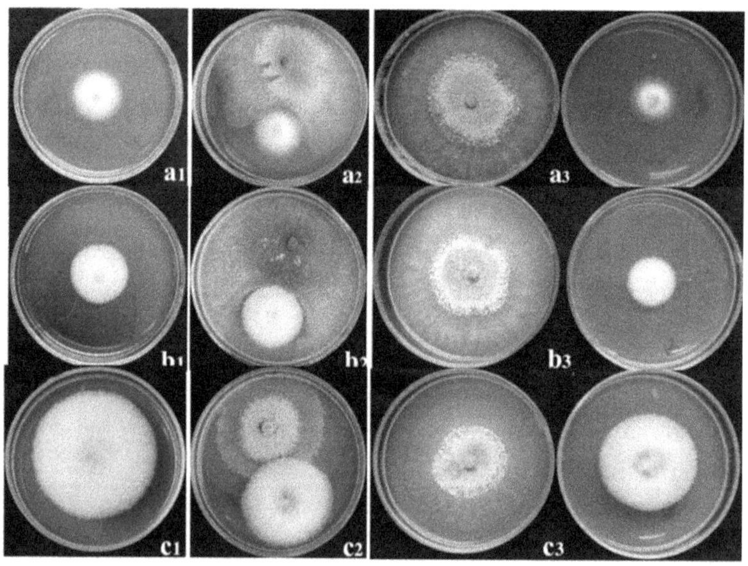


Inhibition of germination of pathogenic fungal species by Trichoderma isolates in direct confrontation ($IGC_D$%) gave satisfactory results (Table 3). The species tested *(F.oxysporum, F. pseudograminiarum, and F. moniliforme)* exhibited high percentages of germination inhibition exceeding 63.5%.Trichoderma sp.1 and sp.2 demonstrated significant inhibition efficacy against these three Fusarium species, with inhibition percentages ranging from 70% to 79.5%.

*Table 3. Inhibition of germination of corm and root rot species of Crocus sativus isolates of Trichoderma in direct confrontation ($IG_{CD}\%$)*

| The antagonists | *Fusarium oxysporum* | *Fusarium pseudograminearum* | *Fusarium moniliforme* |
|---|---|---|---|
| *Trichoderma* sp.1 | $73.5^{a}_{a}$ | $73.25^{a}_{a}$ | $74.25^{a}_{a}$ |
| *Trichoderma* sp2 | $79.5^{a}_{a}$ | $70.5^{a}_{a}$ | $72.25^{a}_{a}$ |
| *Trichoderma* sp3 | $63.5^{a}_{a}$ | $68.5^{a}_{b}$ | $68.5^{a}_{a}$ |

*Two values read from the same column followed by the same index letter are not significantly different at the 5% threshold.
* Two values read on the same line followed by the same letter while exposing are not significantly different at the 5% threshold

The three tested Trichoderma isolates demonstrated high percentages of inhibition of pathogenic fungal growth through the production of volatile substances in indirect confrontation (Table 4). The inhibition percentages ranged from 54.8% to 75.12%.

*Table 4. Inhibition of mycelial growth of species responsible of corms and roots rot of Crocus sativus by Trichoderma isolates in indirect confrontation by production of volatile substances ($IC_{SV}\%$)*

| The antagonists | *Fusarium oxysporum* | *Fusarium pseudograminearum* | *Fusarium moniliforme* |
|---|---|---|---|
| *Trichoderma* sp.1 | $75.12^{a}_{a}$ | $73.61^{a}_{a}$ | $73.1^{a}_{a}$ |
| *Trichoderma* sp2 | $64.88^{a}_{a}$ | $61.75^{a}_{b}$ | $64.3^{a}_{a}$ |
| *Trichoderma* sp3 | $67.11^{b}_{a}$ | $73.14^{a}_{a}$ | $54.8^{c}_{b}$ |

*Two values read from the same column followed by the same index letter are not significantly different at the 5% threshold.
* Two values read on the same line followed by the same letter while exposing are not significantly different at the 5% threshold

Similarly, the studied *Trichoderma* isolates exhibited a high potential for inhibiting germination ($IG_{SV}\%$) by the production of volatile substances (Table 5), the percentages of inhibition fluctuated between 51% and 76%.

*Table 5. Inhibition of germination of species responsible for rot of Crocus sativus roots and corms by Trichoderma isolates in indirect confrontation by volatile substance production ($IGSV_{\%}$)*

| The antagonists | *Fusarium oxysporum* | *Fusarium pseudograminearum* | *Fusarium moniliforme* |
|---|---|---|---|
| *Trichoderma* sp.1 | $72.25^{a}_{a}$ | $72.5^{a}_{a}$ | $76^{a}_{a}$ |
| *Trichoderma* sp2 | $67.5^{a}_{b}$ | $51^{c}_{c}$ | $64^{b}_{b}$ |
| *Trichoderma* sp3 | $62.5^{a}_{b}$ | $56.5^{b}_{c}$ | $66.5^{a}_{b}$ |

*Two values read from the same column followed by the same index letter are not significantly different at the 5% threshold.
* Two values read on the same line followed by the same letter while exposing are not significantly different at the 5% threshold

The Trichoderma sp. isolates examined in indirect test through the production of diffusible substances, as detailed in Tables 6 and 7, did not exhibit significant inhibitory effects on the mycelial growth or germination of any Fusarium species. The observed percentages of inhibition for mycelial growth and germination were both below 28.47% and 20.5%, respectively.

*Table 6. Inhibition of mycelial growth of species responsible for rot of Crocus sativus roots and corms by Trichoderma isolates in indirect confrontation by the production of diffusible substances ($IC_{SD}\%$)*

| **The antagonists** | *Fusarium oxysporum* | *Fusarium pseudograminearum* | *Fusarium moniliforme* |
|---|---|---|---|
| *Trichoderma* sp.1 | $28.47^{a}_{a}$ | $27.12^{a}_{a}$ | $27.54^{a}_{a}$ |
| *Trichoderma* sp2 | $23.05^{a}_{a}$ | $21.66^{a}_{a}$ | $26.47^{a}_{a}$ |
| *Trichoderma* sp3 | $27.1^{a}_{a}$ | $23.05^{a}_{a}$ | $23.14^{a}_{a}$ |

*Two values read from the same column followed by the same index letter are not significantly different at the 5% threshold.

* Two values read on the same line followed by the same letter while exposing are not significantly different at the 5% threshold

*Table 7. Inhibition of germination of root rot species in wheat and barley by Trichoderma isolates in indirect confrontation by the production of diffusible substances ($IC_{SD}\%$)*

| **The antagonists** | *Fusarium oxysporum* | *Fusarium pseudograminearum* | *Fusarium moniliforme* |
|---|---|---|---|
| *Trichoderma* sp.1 | $20.5^{a}_{a}$ | $20.5^{a}_{a}$ | $19^{a}_{a}$ |
| *Trichoderma* sp2 | $13.5^{a}_{a}$ | $19.5^{a}_{a}$ | $16.5^{a}_{a}$ |
| *Trichoderma* sp3 | $17.5^{a}_{a}$ | $16.5^{a}_{a}$ | $19.5^{a}_{a}$ |

*Two values read from the same column followed by the same index letter are not significantly different at the 5% threshold.

* Two values read on the same line followed by the same letter while exposing are not significantly different at the 5% threshold

## IV. DISCUSSION

For many years, researchers have recognized the potential of microorganisms as effective biological control agents against plant diseases (Howell, 2003). Antagonistic interactions have been extensively studied across various plant diseases in the literature (Wells et al., 1972; Elad and Kapat, 1999; Harman, 2000; Hibar et al., 2007; El-Hassni et al., 2007; Kuçuk et al., 2007; Dawidziuk et al., 2016). Numerous organisms have been successfully employed as biological control agents to manage soil pathogens. Among these, bacterial and fungal genera such as Gliocladium, Bacillus, Coniothyrium, Paecilomyces, Phlebiopsis, Pseudomonas, Rhizobium, Serratia, Streptomyces, and Trichoderma are commonly marketed (Mazzola and Freilich, 2017). The fungus *Aspergillus* spp. exhibits antagonistic properties against various phytopathogenic fungi, as demonstrated in studies by Getha et al. (2005) and Hibar et al. (2007). Additionally, Adebola and Amadi (2010) confirmed the antifungal activity of *Aspergillus flavus* in their research focusing on *Phytophthora palmivora*, the causal agent of Cacao black rot.

A biological control agent must occupy an ecological niche similar to that of the plant, the pathogenic microorganism and its mode of action, must interfere specifically and temporally at specific stages in the development of the pathogenic microbe. Several mechanisms of action, acting alone or in combination, have been advanced to explain this biological control power characterizing the antagonism (Figure 15). These methods leverage natural antagonisms between microbial agents to reduce the pressure of the disease while preserving the environment (Moricca et al., 2015). Among the BCA tested and marketed, telluric fungi of the genus Trichoderma are in a good position. Several strains of filamentous fungi *Trich-*

*oderma* spp. (Hypocrea telomorph) have attracted much research as they are an effective biocontrol agent against a wide range of soil and leaf phytopathogenic agents (Deravel et al., 2014; Zhang et al., 2016). Trichoderma are ascomycete fungi with pervasive green spores in the rhizosphere (Hem and Pang 2017). Some strains have the ability to reduce the severity of plant disease by inhibiting phytopathogenic agents, primarily in soil or plant roots (Silva et al., 2017; Hermosa et al., 2012). The main modes of action of *Trichoderma* spp. include nutrient and space competition, production of cell wall-degrading enzymes, production of diffusible and volatile antifungal metabolites, and mycoparasitism (Hermosa et al., 2012).

Direct and indirect confrontation between Trichoderma and pathogens (*Fusarium pseudograminearum, F. moniliforme,* and *F. oxysporum*) responsible for root and corms rot of saffron revealed significant reductions in mycelial growth of these tested pathogens relative to controls with all Trichoderma isolates used. This reduction was represented by the rate of inhibition of mycelial growth. In direct confrontation, this varied rate reached 79%. Indirect confrontation under the effect of volatile substances showed an inhibition rate of 28%. Kribel et al. (2019) reported that isolates of *Trichoderma* spp. (27 isolates), collected from phosphate sludge, were able to effectively inhibit, in direct confrontation and by the production of volatile substances, mycelial growth and conidian germination of *Fusarium culmorum, Fusarium graminearum, Fusarium roseum, Bipolaris sorokiniana, Curvularia spicifera,* pathogens responsible for root rot. In contrast, the production of diffusible substances released by these Trichoderma showed no inhibitory effect on mycelial growth or conidian germination of pathogens. Boughedid and Filali (2015) reported that *Trichoderma harzianum* was able to induce a rate of inhibition of mycelial growth of *Fusarium graminearum,* a barley fusariose agent (*Hordum vulgare* L.), which varies between 50 and 57%. (2018), T*richoderma longibrachiatum* induced a rate of inhibition of mycelial growth of *Fusarium oxysporum f. sp. albedinis*, Bayoud agent *of the date* palm, of the order of 60%. Similarly, Hibar *et al. (*2007) noted that Tr*ichoderma harzianum* was able to show a rate of mycelial growth inhibition of *Fusarium oxysporum f. sp. radici-lycopersici, a*gent of tomato vascular wilt (*Lycopersicum esculentum* L.) which varies between 70 and 76%

This biological control approach study showed the important effect of fungi of the genus Trichoderma tested against Fusarium species, responsible for roots and corms rots of saffron. These promising findings obtained in vitro require validation through in vivo tests conducted in saffron fields contaminated with soilborne pathogens. In this regard, Kribel et al. (2020) demonstrated the beneficial impact of Trichoderma on the growth and yield of wheat and barley plants cultivated in substrates containing escalating concentrations of rock phosphate. Moreover, Trichoderma successfully colonized the studied culture substrates and the roots of wheat and barley plants over time, with colonization percentages surpassing 62.43% eighty days after the initiation of cultivation.

In a study conducted by Qostal et al. (2020), the efficacy of Trichoderma and Thirame application in mitigating the development of root rot symptoms in wheat and barley was evaluated. Seed treatment with either Trichoderma conidia suspension or Thirame solution demonstrated potential in reducing the severity, incidence, and disease index in wheat and barley plants growing on substrates inoculated with root rot agents such as *Curvularia spicifera, Bipolaris sorokiniana,* and *Fusarium roseum.* Additionally, the development of *C. spicifera* and *B. sorokiniana* in the upper levels of wheat and barley plants decreased significantly, and foliar area indices were notably low in seedlings treated with Trichoderma.

## V. CONCLUSION

The evaluation of three Trichoderma isolates collected from Crocus sativus corms based on their antagonistic abilities against agents causing corm and root rot in saffron plants revealed the following findings:

- *Trichoderma* spp. isolates effectively suppressed the mycelial growth and conidial germination of *Fusarium oxysporum*, *Fusarium pseudograminearum*, and *Fusarium moniliforme* through both direct interaction and the production of volatile substances. They also demonstrated strong competitive ability in occupying space when coexisting with these pathogenic species, achieving colonization rates ranging from 55% to 76.3%
- When exposed to diffusible substances produced indirectly, *Trichoderma* spp. isolates demonstrated limited inhibition of mycelial growth (ICCD% < 28.47%) and conidia germination (ICCD% < 20.5%) of the pathogens responsible for corm and root rot in saffron plants.

## ACKNOWLEDGMENT

This study was conducted under the research project: 2nd PMA 2020/11, entitled "Development of a biofertilising and biopesticide product, based on endophytic fungi, adapted to the protection and improvement of the productivity of saffron in Morocco" financed by "Agence Nationale des Plantes Médicinales et Aromatique" and the University Ibn Tofail, Kénitra,

## REFERENCES

Abbas, A., Jiang, D., & Fu, Y. (2017). Trichoderma spp. as antagonist of Rhizoctonia solani. *Journal of Plant Pathology & Microbiology*, *8*(3), 1–9.

Abouamama, S., Noureddine, K., Anis, B., Sadika, H., Bouchra, O., Said, B., & Mebrouk, K. (2018). Pathogenicity and biological control of Bayoud disease by Trichoderma longibrachiatum and Artemisia herba-alba essential oil. *Journal of Applied Pharmaceutical Science*, *8*(4), 161–167.

Adebola, M. O., & Amadi, J. E. (2010). Screening three Aspergillus species for antagonistic activities against the cocoa black pod organism (Phytophthora palmivora). *Agriculture and Biology Journal of North America*, *1*(3), 362–365. DOI: 10.5251/abjna.2010.1.3.362.365

Adnani, M., El Hazzat, N., Msairi, S., El Alaoui, M. A., Mouden, N., Selmaoui, K., Benkirane, R., Ouazzani Touhami, A., & Douira, A. (2024). Exploring the efficacy of a Trichoderma asperellum-based seed treatment for controlling Fusarium equiseti in chickpea. *Egyptian Journal of Biological Pest Control*, *34*(1), 7. DOI: 10.1186/s41938-024-00771-x

Ahrazem, O., Rubio-Moraga, A., Castillo-López, R., Trapero-Mozos, A., & Gómez-Gómez, L. (2010). *Crocus sativus pathogens and defence responses*. Functional Plant Science and Biotechnology. Global Science Book Isleworth.

Ait Rahou, Y., Douira, A., Tahiri, A. I., Cherkaoui, E. M., Benkirane, R., & Meddich, A. (2022). Application of plant growth-promoting rhizobacteria combined with compost as a management strategy against Verticillium dahliae in tomato. *Canadian Journal of Plant Pathology*, *44*(6), 806–827. DOI: 10.1080/07060661.2022.2089235

Ali, A. M. (2021). The Competitive potential of different Trichoderma spp. to control Rhizoctonia root rot disease of pepper (Capsicum annuum L.). *Egyptian Journal of Phytopathology*, *49*(1), 136–150. DOI: 10.21608/ejp.2021.73456.1030

Artigues, M., Davet, P., & Roure, C. (1984). Comparaison des aptitudes parasitaires de clones de Trichoderma vis-à-vis de quelques champignons à sclérotes. *Soil Biology & Biochemistry*, *16*(4), 413–417. DOI: 10.1016/0038-0717(84)90042-7

Azeddine, E., Ouazzani Touhami, A., Selmaoui, K., Benkirane, A. R., & Douira, A. (2024). In Vitro and In Vivo Biological Control by Trichoderma Asperellum Against Rhizoctonia Solani a Causal Agent of Collar and Root Rot in Strawberries. In *Advanced Systems for Environmental Monitoring, IoT and the application of Artificial Intelligence* (pp. 213–235). Springer Nature Switzerland. DOI: 10.1007/978-3-031-50860-8_13

Baha Eddine, S., El-Yousfi, B., & Douira, A. (2019). Interaction of nitrogen fertilizers with wheat growth stage and foliar treatment with urea effects on wheat crown rot induced by Fusarium culmorum. *Plant Archives*, *19*, 2829–2835.

Bastakoti, S., Belbase, S., Manandhar, S., & Arjyal, C. (2017). Trichoderma species as biocontrol agent against soil borne fungal pathogens. *Nepal Journal of Biotechnology*, *5*(1), 39–45. DOI: 10.3126/njb.v5i1.18492

Bentata, F., Lage, M., Bakhy, K., Ibrahimi, M., Jbair, A., El-Aissami, A., Kissayi, K. H., & Labhilili, M. (2017). Sanitary assessment of saffron corms and soil from Great Moroccan production areas: Taliouine and Taznakht. *Acta Horticulturae*, (1184), 263–266. DOI: 10.17660/ActaHortic.2017.1184.37

Boughedid, K., & Filali, M. (2015). *Isolement et identification des champignons antagonistes des champignons phytopathogènes de l'orge*. Mémoire de Master, Université des Frères Mentouri de Constantine.

Brayford, D. (1996). IMI descriptions of fungi and bacteria set 127. *Mycopathologia*, *133*(1), 35–63. DOI: 10.1007/BF00437097

Camporota, P. (1985). Antagonisme in vitro de Trichoderma spp. vis-à-vis de Rhizoctonia solani Kühn. *Agronomie*, *5*(7), 613–620. DOI: 10.1051/agro:19850707

Cappelli, C., & Di Minco, G. (1999). Three-years of trials on saffron diseases in Abruzzo (Central Italy) [Crocus sativus L.]. *Informatore Fitopatologico (Italy), 49*(9).

Chao, W. A. N. G. (2019). Evaluating effective Trichoderma isolates for biocontrol of Rhizoctonia solani causing root rot of Vigna unguiculata. *Journal of Integrative Agriculture*, *18*(9), 2072–2079. DOI: 10.1016/S2095-3119(19)62593-1

Davet, P., Camporota, P., & Roure, C. (1986). Etude comparative de quelques méthodes d'estimation de l'aptitude à la compétition saprophytique dans le sol des Trichoderma. *Agronomie*, *6*(6), 575–581. DOI: 10.1051/agro:19860610

Dawidziuk, A., Popiel, D., Kaczmarek, J., Strakowska, J., & Jedryczka, M. (2016). Optimal Trichoderma strains for control of stem canker of brassicas: Molecular basis of biocontrol properties and azole resistance. *BioControl*, *61*(6), 755–768. DOI: 10.1007/s10526-016-9743-2

De Meyer, G., Bigirimana, J., Elad, Y., & Höfte, M. (1998). Induced systemic resistance in Trichoderma harzianum T39 biocontrol of Botrytis cinerea. *European Journal of Plant Pathology*, *104*(3), 279–286. DOI: 10.1023/A:1008628806616

De Silva, D. D., Crous, P. W., Ades, P. K., Hyde, K. D., & Taylor, P. W. (2017). Life styles of Colletotrichum species and implications for plant biosecurity. *Fungal Biology Reviews*, *31*(3), 155–168. DOI: 10.1016/j.fbr.2017.05.001

Dennis, C., & Webster, J. (1971). Antagonistic properties of species-groups of Trichoderma: I. Production of non-volatile antibiotics. *Transactions of the British Mycological Society*, *57*(1), 25–IN3. DOI: 10.1016/S0007-1536(71)80077-3

Deravel, J., Krier, F., & Jacques, P. (2014). Biopesticides, a complementary and alternative approach to the use of agrochemicals. *RE:view*.

Di Primo, P., Cappelli, C., & Katan, T. (2002). Vegetative compatibility grouping of Fusarium oxysporum f. sp. gladioli from saffron. *European Journal of Plant Pathology*, *108*(9), 869–875. DOI: 10.1023/A:1021204022787

Dumitras, L., & Fratilescu-Sesan, T. (1979). Aspects of the antagonism of Trichoderma viride Pers. ex Fr. to Pythium debaryanum Hesse.

El Aymani, I., Qostal, S., Mouden, N., Selmaoui, K., Touhami, A. O., Benkirane, R., & Douira, A. (2019). Fungi associated with saffron (Crocus sativus) in Morocco. *Plant Cell Biotechnology and Molecular Biology*, *20*, 1180–1188.

El-Benawy, N. M., Abdel-Fattah, G. M., Ghoneem, K. M., & Shabana, Y. M. (2020). Antimicrobial activities of Trichoderma atroviride against common bean seed-borne Macrophomina phaseolina and Rhizoctonia solani. *Egyptian Journal of Basic and Applied Sciences*, *7*(1), 267–280. DOI: 10.1080/2314808X.2020.1809849

El Hassni, M., El Hadrami, A., Daayf, F., Chérif, M., Barka, E. A., & El Hadrami, I. (2007). Biological control of bayoud disease in date palm: Selection of microorganisms inhibiting the causal agent and inducing defense reactions. *Environmental and Experimental Botany*, *59*(2), 224–234. DOI: 10.1016/j.envexpbot.2005.12.008

El Kaissoumi, H., Berber, F., Mouden, N., Ouazzani Chahdi, A., Albatnan, A., Ouazzani Touhami, A., . . . Douira, A. (2022, May). Effect of Trichoderma asperellum on the development of strawberry plants and biocontrol of anthracnose disease caused by colletotrichum gloeosporioides. In International Conference on Advanced Intelligent Systems for Sustainable Development (pp. 609-622). Cham: Springer Nature Switzerland.

El-Sebaay, H. H., & El-Sayed, A. B. (2019). Antagonistic Effect of Rhizobium, Bacillus, Pseudomonas, Trichoderma on Fusarium and Rhizocotonia Compared with Moncut In Vitro. *Journal of Agricultural Chemistry and Biot*

Elad, Y., Chet, I., & Katan, J. (1980). Trichoderma harzianum: A biocontrol agent effective against Sclerotium rolfsii and Rhizoctonia solani. *Phytopathology*, *70*(2), 119–121. DOI: 10.1094/Phyto-70-119

Elad, Y., & Kapat, A. (1999). The role of Trichoderma harzianum protease in the biocontrol of Botrytis cinerea. *European Journal of Plant Pathology*, *105*(2), 177–189. DOI: 10.1023/A:1008753629207

Elbouazaoui, A., Sijilmassi, B., Maafa, I., Allal, D., & Ahmed, S. (2022). Biocontrol activity of Bacillus, Paenibacillus and Pseudomonas against Fusarium wilt of chickpea in Morocco. *Acta Agriculturæ Scandinavica. Section B, Soil and Plant Science*, *72*(1), 847–859. DOI: 10.1080/09064710.2022.2100819

Getha, K., Vikineswary, S., Wong, W. H., Seki, T., Ward, A., & Goodfellow, M. (2005). Evaluation of Streptomyces sp. strain g10 for suppression of Fusarium wilt and rhizosphere colonization in pot-grown banana plantlets. *Journal of Industrial Microbiology & Biotechnology*, *32*(1), 24–32. DOI: 10.1007/s10295-004-0199-5 PMID: 15650871

Hanson, L. E. (2000). Reduction of Verticillium wilt symptoms in cotton following seed treatment with Trichoderma virens.

Harman, G. E. (2000). Myths and dogmas of biocontrol changes in perceptions derived from research on Trichoderma harzinum T-22. *Plant Disease*, *84*(4), 377–393. DOI: 10.1094/PDIS.2000.84.4.377 PMID: 30841158

Hem, R., & Pang, A. (2017). Bio-stimulant effects of Trichoderma spp. on rice (Oryza sativa L.): An initial evaluation using a strongly acidic clay loam soil. *American-Eurasian Journal of Sustainable Agriculture*, *11*(5), 42–48.

Hermosa, R., Viterbo, A., Chet, I., & Monte, E. (2012). Plant-beneficial effects of Trichoderma and of its genes. *Microbiology*, *158*(1), 17–25. DOI: 10.1099/mic.0.052274-0 PMID: 21998166

Hibar, K., Daami-Remadi, M., & El Mahjoub, M. (2007). Effets de certains fongicides de synthèse et biologiques sur la croissance mycélienne et l'agressivité de Fusarium oxysporum f. sp. radicis-lycopersici. *Tropicultura*, *25*(3), 146–152.

Hmouni, A., Douira, A., & Mouria, A. (2006, August). Biological Control of Tomato Grey Mould with Compost Water Extracts," Trichoderma" sp., and" Gliocladium" sp. *Phytopathologia Mediterranea*, 1000–1007.

Hmouni, A., Massoui, M., & Douira, A. M. (1999). Study of antagonistic activity of Trichoderma spp. and Gliocladium spp. against Botrytis cinerea: causal agent of tomato grey mould.

Howell, C. R. (2003). Mechanisms employed by Trichoderma species in the biological control of plant diseases: The history and evolution of current concepts. *Plant Disease*, *87*(1), 4–10. DOI: 10.1094/PDIS.2003.87.1.4 PMID: 30812698

Kaissoumi, H. E., Berbera, F., Mouden, N., Touhami, A. O., Selmaoui, K., Benkirane, R., & Douira, A. (2024). Integrated Control of Strawberry Anthrachnose by Trichoderma Asperellum–Pyraclostrobin/Boscalid Combination. In *Sustainable and Green Technologies for Water and Environmental Management* (pp. 173–187). Springer Nature Switzerland. DOI: 10.1007/978-3-031-52419-6_14

Kalha, C. S., Gupta, V., Gupta, D., & Priya, S. (2007). First report of sclerotial rot of saffron caused by Sclerotium rolfsii in India. *Plant Disease*, *91*(9), 1203–1203. DOI: 10.1094/PDIS-91-9-1203B PMID: 30780677

Kribel, S., Qostal, S., Ouazzani Touhami, A., & Selmaoui, K. (2019). In vitro selection of moroccan phosphate sites' Trichoderma isolates according to their antagonism against the wheat and barley root rot pathogens. *Plant Cell Biotechnology and Molecular Biology*, *20*(15-16), 682–699.

Kribel, S., Qostal, S., Ouazzani Touhami, A., Selmaoui, K., Chliyeh, M., Benkirane, R., & Achbani, E. H. (2020). Effects of Trichoderma on growth and yield of wheat and barley and its survival ability on roots and amended rock phosphate growing substrates. [Journal of Fungal Biology]. *Current Research in Environmental & Applied Mycology*, *10*(1), 400–416. DOI: 10.5943/cream/10/1/32

Küçük, M., Kolaylı, S., Karaoğlu, Ş., Ulusoy, E., Baltacı, C., & Candan, F. (2007). Biological activities and chemical composition of three honeys of different types from Anatolia. *Food Chemistry*, *100*(2), 526–534. DOI: 10.1016/j.foodchem.2005.10.010

Leylaie, S., & Zafari, D. (2018). Antiproliferative and antimicrobial activities of secondary metabolites and phylogenetic study of endophytic Trichoderma species from Vinca plants. *Frontiers in Microbiology*, *9*, 294926. DOI: 10.3389/fmicb.2018.01484 PMID: 30050508

Mayo, S., Gutierrez, S., Malmierca, M. G., Lorenzana, A., Campelo, M. P., Hermosa, R., & Casquero, P. A. (2015). Influence of Rhizoctonia solani and Trichoderma spp. in growth of bean (Phaseolus vulgaris L.) and in the induction of plant defense-related genes. *Frontiers in Plant Science*, *6*, 685. DOI: 10.3389/fpls.2015.00685 PMID: 26442006

Mazzola, M., & Freilich, S. (2017). Prospects for biological soilborne disease control: Application of indigenous versus synthetic microbiomes. *Phytopathology*, *107*(3), 256–263. DOI: 10.1094/PHYTO-09-16-0330-RVW PMID: 27898265

Mazzola, M., & Freilich, S. (2017). Prospects for biological soilborne disease control: Application of indigenous versus synthetic microbiomes. *Phytopathology*, *107*(3), 256–263. DOI: 10.1094/PHYTO-09-16-0330-RVW PMID: 27898265

Messgo-Moumene, S. (2016). Evaluation of the antifungal activity of the protein and non-protein extracts of Trichoderma asperellum and Trichoderma atroviride culture filtrates against Phytophthora infestans (Mont.) de Bary.

Messgo-Moumene, S., Olubunmi, O. F., & Laidani, M. (2015). In Vitro and in Vivo Effects of Aqueous Extract of Rosmarinus officinalis L.(Rosemary) in The Control of Late Blight Disease of Potato Caused by Phytophthora Infestans Mont. De Bary. in Algeria.

Moricca, S., & Ginetti, B. (2015). Saprophytic growth of the alder rust fungus Melampsoridium hiratsukanum on artificial media. *Fungal Biology*, *119*(7), 568–579. DOI: 10.1016/j.funbio.2015.03.002 PMID: 26058533

Mouden, N., Ouazzani Touhami, A., Albatnan, A., Selmaoui, K., Benkirane, R., & Douira, A. (2022, May). Growth Promoting of Tomato Plants by Incorporation of Trichoderma asperellum Enriched Liquid Product via Foliar Spray and the Irrigation System. In International Conference on Advanced Intelligent Systems for Sustainable Development (pp. 599-608). Cham: Springer Nature Switzerland.

Mouria, B., Ouazzani-Touhami, A., & Douira, A. (2008). Effect of various strains of Trichoderma on the growth of a greenhouse tomato crop and their ability to colonize roots and substrate. *Phytoprotection*, *88*(3), 103–110. DOI: 10.7202/018955ar

Najari, G., Nourollahi, K., & Piri, M. (2018). The first report of Fusarium oxysporum causal agent of wild saffron corm rot disease in Iran.

Ouazzani Chahdi, A., Chliyeh, M., Mouria, B., Dahmani, J., Ouazzani Touhami, A., Benkirane, R., & Douira, A. (2014). In vitro and in vivo effect of salinity on the antagonist potential of Trichoderma harzianum and sensitivity of tomato to Verticillium wilt. *International Journal of Recent Scientific Research*, *5*(4), 780–791.

Ouazzani-Touhami, A., Douira, A., Benkirane, R., El Oirdi, M., Bouslim, F., Zidane, L., Gmira, N., & El Haloui, N. E. (1994). Antagonisme in vivo de certaines espèces fongiques vis-à-vis de Verticillium dahliae. Rev. Rés. *APAMA*, *7*, 197–211.

Ourras, S., Aymani, I. E., Mouden, N., Selmaoui, K., Msairi, S., Elouark, M. H., . . . Douira, A. (2022, May). Mycoflora of Dormant Crocus Sativus Corms in Morocco. In *International Conference on Advanced Intelligent Systems for Sustainable Development* (pp. 479-489). Cham: Springer Nature Switzerland.

Palmero, D., Rubio-Moraga, A., Galvez-Patón, L., Nogueras, J., Abato, C., Gómez-Gómez, L., & Ahrazem, O. (2014). Pathogenicity and genetic diversity of Fusarium oxysporum isolates from corms of Crocus sativus. *Industrial Crops and Products*, *61*, 186–192. DOI: 10.1016/j.indcrop.2014.06.051

Papavizas, G. C. (1985). Trichoderma and Gliocladium: Biology, ecology, and potential for biocontrol. *Annual Review of Phytopathology*, *23*(1), 23–54. DOI: 10.1146/annurev.py.23.090185.000323

Pertot, I., Alabouvette, C., Esteve, E. H., & Franca, S. (2015). Mini-paper—The use of microbial biocontrol agents against soil-borne diseases. *Eip-agri focus group soil-borne diseases*, 1-11.

Qostal, S., Kribel, S., Chliyeh, M., Mouden, N., Selmaoui, K., Touhami, A. O., & Douira, A. (2020). Biostimulant effect of Trichoderma on the development of wheat and barley plants and its survival aptitudes on the roots. *Plant Archives*, *20*(2), 7829–7834.

Qostal, S., Kribel, S., Chliyeh, M., Selmaoui, K., Serghat, S., Benkirane, R., Ouazzani Touhami, A., & Douira, A. (2020). Management of wheat and barley root rot through seed treatment with biopestcides and fungicides. *Plant Cell Biotechnology and Molecular Biology*, *21*(35&36), 129–143.

Qostal, S., Kribel, S., Chliyeh, M., Selmaoui, K., Touhami, A. O., Serghat, S., & Douira, A. (2019a). Comparative pathogenesis of 7 Fusarium spp. species and Bipolaris sorokiniana obtained from necrotic lesions of wheat roots and barley plants (North-Western Morocco). *Plant Cell Biotechnology and Molecular Biology*, *20*(5&6), 261–274.

Qostal, S., Kribel, S., Chliyeh, M., Serghat, S., Selmaoui, K., Touhami, A. O., ... & Douira, A. (2019b). Study of the fungal complex responsible for root rot of wheat and barley in the north-west of Morocco.

Rubio-Moraga, Á., Gómez-Gómez, L., Trapero, A., Castro-Díaz, N., & Ahrazem, O. (2013). Saffron corm as a natural source of fungicides: The role of saponins in the underground. *Industrial Crops and Products*, *49*, 915–921. DOI: 10.1016/j.indcrop.2013.06.029

Saddek, D., Messgo-Moumene, S., Chemat-Djenni, Z., Bendifallah, L., & Bencheikh, K. (2020). Antagonism of isolates of Trichoderma spp. at egard of Botrytis cinerea pers. agent of gray rot of tomato (Lycopersicum esculentum Mill.) under greenhouse. *Revue des Sciences Fondamentales et Appliquées*, *12*(2), 583–606. DOI: 10.4314/jfas.v12i2.5

Sy, A. A. (1976). *Contribution à l'étude de Pyricularia oryzae Cav. Recherche in vitro d'antagonistes dans une perspective de lutte biologique* (Doctoral dissertation, Thèse Doct. Ingénieur INP Toulouse).

Sy, J. (1976). A ribosome-independent, soluble stringent factor-like enzyme isolated from a Bacillus brevis. *Biochemistry*, *15*(3), 606–609. DOI: 10.1021/bi00648a024 PMID: 175825

Tijerino, A., Hermosa, R., Cardoza, R. E., Moraga, J., Malmierca, M. G., Aleu, J., Collado, I. G., Monte, E., & Gutierrez, S. (2011). Overexpression of the Trichoderma brevicompactum tri5 gene: Effect on the expression of the trichodermin biosynthetic genes and on tomato seedlings. *Toxins*, *3*(9), 1220–1232. DOI: 10.3390/toxins3091220 PMID: 22069764

Tronsmo, A., & Raa, J. (1977). Antagonistic action of Trichoderma pseudokoningii against the apple pathogen Botrytis cinerea. *Journal of Phytopathology*, *89*(3), 216–220. DOI: 10.1111/j.1439-0434.1977.tb02861.x

Vargas-Inciarte, L., Fuenmayor-Arrieta, Y., Luzardo-Méndez, M., Costa-Jardin, M. D., Vera, A., Carmona, D., & San-Blas, E. (2019). Use of different Trichoderma species in cherry type tomatoes (Solanum lycopersicum L.) against Fusarium oxysporum wilt in tropical greenhouses. *Agronomia Costarricense*, *43*(1), 85–100.

Wells, H. D., Bell, D. K., & Jaworski, C. A. (1972). Efficacy of Trichoderma harzianum as a biocontrol for Sclerotium rolfsii.

Zhang, S., Gan, Y., & Xu, B. (2016). Application of plant-growth-promoting fungi Trichoderma longibrachiatum T6 enhances tolerance of wheat to salt stress through improvement of antioxidative defense system and gene expression. *Frontiers in Plant Science*, *7*, 217856. DOI: 10.3389/fpls.2016.01405 PMID: 27695475

# Chapter 4
# Challenges in Scaling Hydrogen Green Systems for Global Decarbonization

**M'barka El Mouedden**
https://orcid.org/0009-0001-0174-5106
*Faculté des Sciences Rabat, Morocco*

**Jamal Mabrouki**
https://orcid.org/0000-0002-3841-7755
*Mohammed V University in Rabat, Morocco*

**Driss Azdem**
*Mohammed V University in Rabat, Morocco*

**Souad El Hajjaji**
*Mohammed V University in Rabat, Morocco*

## ABSTRACT

*Hydrogen, particularly green hydrogen produced via renewable energy, is a key component in the global shift toward sustainable energy. Its potential to decarbonize hard-to-abate sectors such as heavy industry, transportation, and energy storage positions it as a critical solution in the clean energy transition. However, widespread adoption faces significant challenges. This chapter explores major obstacles to scaling green hydrogen, including inefficiencies in electrolysis, high production costs, and inadequate infrastructure for storage and transport. Additionally, it discusses regulatory hurdles, market limitations, and the need for technological advancements to improve cost-effectiveness. Addressing these challenges is essential for realizing hydrogen's potential in global decarbonization. With innovation, policy support, and international collaboration, hydrogen can become central in achieving carbon neutrality.*

DOI: 10.4018/979-8-3693-8980-5.ch004

# I. INTRODUCTION

The global push for decarbonization and the transition to sustainable energy systems have placed hydrogen, particularly green hydrogen, at the forefront of this movement. Green hydrogen, produced via renewable energy sources such as solar and wind, is unique in that it generates no carbon emissions during its production process (Fahdi, Azdem, et al., 2024) (Azdem, et al 2024). This process primarily involves the electrolysis of water, whereby water molecules are split into hydrogen and oxygen using electricity. The hydrogen produced can then be used as a clean fuel across various industries and applications, positioning it as a cornerstone of future energy systems(Ishaq, et al 2022) (Imane Bencheikh et al., 2021).

Hydrogen can be categorized into different types depending on how it is produced. Green hydrogen is generated entirely through renewable energy(Hassan, et al 2024) (Benchrifa, et al 2023), making it the cleanest option. In contrast, blue hydrogen is derived from fossil fuels but with carbon capture and storage (CCS) technology to mitigate emissions, while purple hydrogen relies on nuclear energy for its production (El-Adawy, et al 2024) (Fatni et al., 2025). Among these, green hydrogen holds the greatest promise for achieving global carbon neutrality, as it offers a pathway to decarbonize sectors that are difficult to electrify(Evro, et al 2024) (I. Bencheikh et al., 2020).

Green hydrogen's role in the global energy transition is multifaceted. It can serve as a key energy carrier, addressing issues such as the intermittency of renewable energy sources and providing a solution for long-term energy storage(Cheekatamarla, 2024) (Azdem, et al 2025). Furthermore, it has the potential to decarbonize multiple sectors, including heavy industry, transportation, heating, and energy storage, where traditional electrification solutions may fall short(Hassan, Algburi, Jaszczur, et al., 2024).

In heavy industries like steel, cement, and chemicals sectors notorious for their carbon-intensive processes green hydrogen offers a viable alternative to fossil fuels, enabling a significant reduction in emissions(Franco, et al 2024) (J. Mabrouki et al., 2020). Similarly, in the transportation sector, hydrogen fuel cells are emerging as a clean solution, particularly for heavy-duty applications such as trucks, buses, and ships, where battery technology may be less effective(Alavi-Borazjani, Adeel, & Chkoniya, 2025) (Fahdi, Mabrouki, et al., 2024). Additionally, hydrogen's ability to store energy from intermittent renewable sources like wind and solar addresses one of the major challenges of renewable energy systems: ensuring a reliable energy supply during periods of low renewable output(Egeland-Eriksen, et al 2021) (El Mouedden, et al 2025). Beyond industrial applications, green hydrogen can also play a critical role in residential and industrial heating, replacing natural gas and other fossil fuels in heating systems(Oliveira, 2021) (Jamal Mabrouki, Benchrifa, et al., 2022).

Despite its clear advantages, the widespread adoption of green hydrogen faces significant challenges(Rasul, et al 2022) (Driss Azdem, et al 2024). The high cost of production, infrastructure limitations, and the lack of supportive regulatory frameworks are major hurdles that need to be addressed (Bade, et al 2024) (Benchrifa & Mabrouki, 2022). Nonetheless, with continued innovation, policy support, and industrial investment, green hydrogen has the potential to become a key player in the global decarbonization effort(Oluleye et al., 2025) (Driss Azdem, et al 2024).

## II. CURRENT STATE OF ELECTROLYSIS TECHNOLOGIES

Electrolysis is the cornerstone of green hydrogen production, and improving its efficiency is key to making green hydrogen more competitive. There are three primary electrolysis methods: **Alkaline Water Electrolysis (AWE)**, **Proton Exchange Membrane (PEM) Electrolysis**, and **Solid Oxide Electrolysis Cells (SOECs)**.

1. Alkaline Water Electrolysis (AWE)

AWE is one of the most promising methods for producing green hydrogen using renewable energy sources like solar or wind power (Tüysüz, 2024) (Loukili et al., 2021). The process involves splitting water into hydrogen and oxygen through electrolysis **(Figure 1).**A basic cell setup with electrodes facilitates the **Hydrogen Evolution Reaction (HER)** and the **Oxygen Evolution Reaction (OER)**. Although it is a mature and widely used technology, AWE suffers from low efficiency in large-scale applications due to the slow kinetics of the OER(Chen et al., 2024) (Driss Azdem, et al 2025).

*Figure 1. Alkaline water electrolysis*

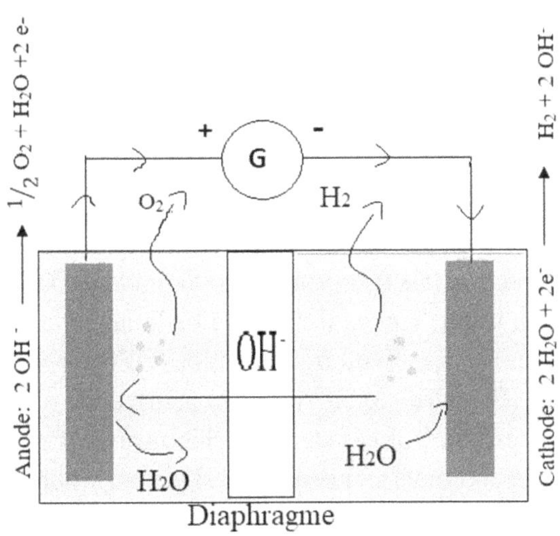


### Oxygen Evolution Reaction (OER)

The Oxygen Evolution Reaction (OER) takes place at the anode during water electrolysis and is a crucial process for generating oxygen from water. However, it is thermodynamically and kinetically more difficult than the Hydrogen Evolution Reaction (HER), requiring more energy and proceeding at a slower rate (Tamilarasi, et al 2024) (Jamal Mabrouki, et al 2020). The OER mechanism involves multiple steps, where hydroxide ions ($OH^-$) are oxidized to form molecular oxygen ($O_2$)(Zhang, Chen, Ye, Xiao, & Liu, 2025) (Driss Azdem, et al 2025) . This process typically follows the Adsorbate Evolution Mechanism (AEM), starting with the adsorption of a hydroxyl ion ($OH^-$) onto the catalyst surface, followed by the formation of intermediates like M–OH and M–O (with "M" representing the catalytic metal), and culminating in the release of electrons and the production of oxygen molecules ($O_2$) (*Figure 2*) (Han et al., 2023) (Jamal Mabrouki, et al 2022).

*Figure 2. Oxygen evolution reaction*

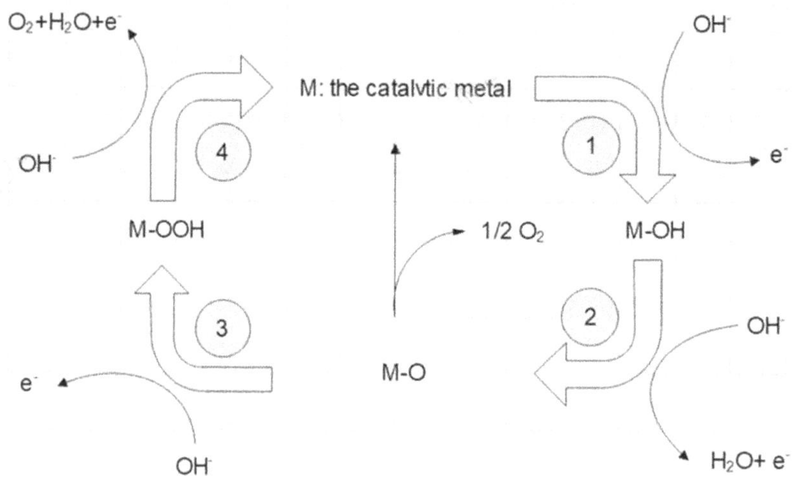


### Hydrogen Evolution Reaction (HER)

The Hydrogen Evolution Reaction (HER) occurs at the cathode during electrolysis, where molecular hydrogen ($H_2$) is produced from water(Yin, et al 2024) (El Alouani et al., 2024). While this reaction is easier to catalyze compared to the Oxygen Evolution Reaction (OER), it still requires a catalyst to enhance efficiency(Rachiq et al., 2021) (Benchrifa, et al 2023). In an alkaline electrolyte, the HER mechanism typically involves two main steps: the Volmer step, in which water is reduced to hydrogen atoms that adsorb onto the catalyst surface, and the Tafel or Heyrovsky step, where the adsorbed hydrogen atoms either combine to form an $H_2$ molecule through the interaction of two hydrogen atoms (Tafel) or react with another water molecule (Heyrovsky) As illustrated in ***Figure 3*** (Al-Jadabi et al., 2023) (Mgalaa et al., 2022).

*Figure 3. Hydrogen evolution reaction*

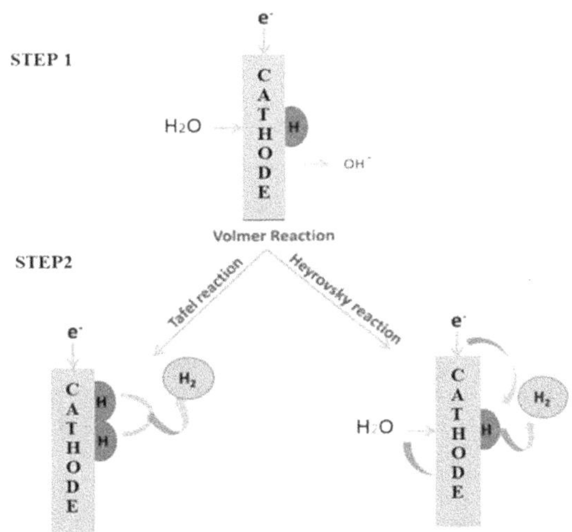


2. Proton Exchange Membrane (PEM) Electrolysis

PEM electrolysis is a highly efficient method for producing green hydrogen, though it is hindered by the high cost of materials used, especially the platinum catalysts in the membrane (Hindson, et al 2024). This method employs a solid polymer electrolyte (PEM) instead of an aqueous solution, which contributes to the system being more compact and efficient compared to traditional alkaline electrolysis(Salehmin, et al 2022) (Ghizlane, et al 2022). The membrane in PEM electrolysis allows only positively charged hydrogen ions (protons) to pass from the anode to the cathode while blocking other ions and gases. At the cathode, these protons react with electrons to produce hydrogen gas, as illustrated in Figure 4, which demonstrates the flow of ions and electrons throughout the electrolysis cell (Hindson & James, 2024).

The key advantage of PEM electrolysis is that it can operate efficiently under varying electrical loads, making it highly suitable for integration with intermittent renewable energy sources, such as solar and wind power. However, the use of precious metals, like platinum, as catalysts substantially increases the cost of production, presenting a significant economic barrier for scaling up to large-scale deployment (Hindson & James, 2024).

Table 1 provides an overview of the main components of PEM devices, including their specific functions, key material requirements, and commonly used materials. This table highlights the important roles of each component in enhancing efficiency and the challenges posed by material selection.

*Figure 4. Proton exchange membrane (PEM) electrolysis*

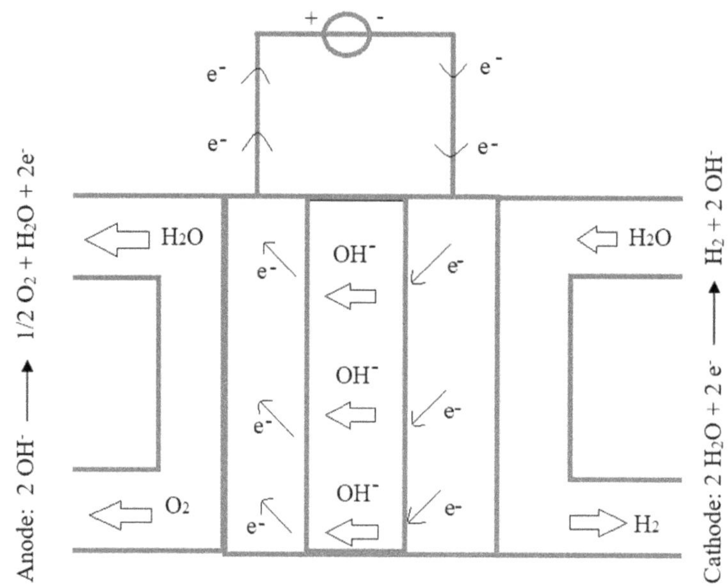


*Table 1. Functions, requirements, and materials of key PEM components*

| Component | Primary Function | Key Requirements | Typical Materials | references |
|---|---|---|---|---|
| BPP (Bipolar Plate) | Facilitates reactant flow to both the anode and cathode | - High thermal and electrical conductivity<br>- Corrosion resistance<br>- Resistance to hydrogen embrittlement<br>- Mechanical durability | - Graphite<br>- Metals (Stainless steel, Aluminum, Titanium) | (Yasin, et al 2024) |
| GDL (Gas Diffusion Layer) | Distributes reactants evenly from BPP to the CL (Catalyst Layer) | - High permeability to water and gases<br>- High electronic conductivity<br>- Thermal conductivity<br>- Mechanical strength<br>- Corrosion resistance | - Carbon paper<br>- Micro-channel metallic mesh | (Tongsh et al., 2024) |
| CL (Catalyst Layer) | Accelerates the anodic and cathodic reactions | - High surface area<br>- Good electronic conductivity<br>- Corrosion resistance | - Carbon-supported platinum (Pt/C) | (Lu et al., 2024) |
| Membrane | Transfers hydrogen ions from anode to cathode | - High ionic conductivity across various temperatures<br>- Gas impermeability<br>- Chemical durability<br>- Resistance to chemical degradation | - Nafion (LT-PEMFC, PEMWE)<br>- PBI (HT-PEMFC) | (Alipour Bonab, et al 2024) |

3. Solid Oxide Electrolysis Cells (SOECs)

SOECs operate at much higher temperatures (700-800°C) and achieve up to 95% efficiency when coupled with external heat sources. Unlike AWE and PEM, SOECs use a solid oxide electrolyte, which allows the system to achieve higher efficiencies by reducing electrical input. This makes SOECs particularly suitable for large-scale hydrogen production where external heat sources are available. However, the materials used in SOECs degrade over time, limiting the durability of these systems (Nechache, et al 2021). Moreover, the high operating temperatures make SOECs less flexible in terms of load variation, restricting their integration with fluctuating renewable energy sources As shown in *Figure 5* (Nechache, et al 2021).

*Figure 5. Solid oxide electrolysis cells*

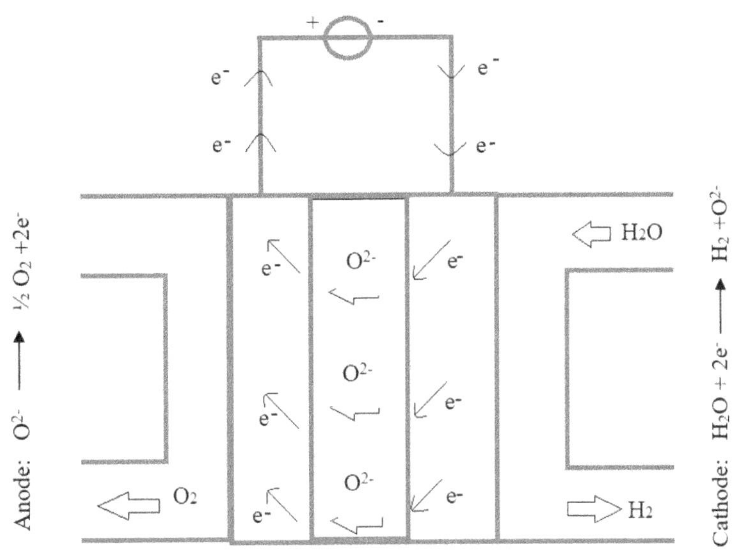


## III. CHALLENGES OF GREEN HYDROGEN PRODUCTION

1. High Production Costs

The production of green hydrogen through electrolysis remains expensive compared to conventional methods like steam methane reforming, which is used for grey hydrogen. Reducing the costs of electrolysis and renewable energy is critical for the widespread adoption of green hydrogen.

2. Infrastructure Limitations

The infrastructure for hydrogen storage and transport is currently underdeveloped. Hydrogen's low volumetric energy density presents challenges for storage, while long-distance transport requires investments in pipelines and other infrastructure. Moreover, refueling stations for hydrogen-powered vehicles are scarce, limiting the use of hydrogen in transport applications.

3. Policy and Regulation

Regulatory frameworks and government support play a crucial role in promoting the development of hydrogen technologies. Policies that include subsidies, tax incentives, and regulatory frameworks are necessary to encourage investments in hydrogen infrastructure and research. Without strong policy support, the hydrogen sector may struggle to gain the momentum required for global decarbonization efforts.

## IV. INFRASTRUCTURE DEVELOPMENT FOR HYDROGEN SYSTEMS

The development of infrastructure for hydrogen production, storage, transport, and distribution is critical for the large-scale adoption of green hydrogen technologies. Currently, hydrogen infrastructure is underdeveloped, and its expansion is a prerequisite for integrating hydrogen into global energy systems(Yazdi, et al 2024).

The transport of hydrogen, due to its low volumetric energy density, presents several challenges. Pipelines, storage tanks, and refueling stations must be specifically designed to handle hydrogen's physical properties. Existing natural gas pipelines, for instance, can be retrofitted for hydrogen transport, but this requires significant investment and technological adaptation. New pipelines specifically designed for hydrogen transport will also be necessary to accommodate long-distance distribution. The development of liquefied hydrogen transport technology, which enables the storage and transport of hydrogen at very low temperatures, is another promising solution that requires further research and investment.

Moreover, refueling infrastructure for hydrogen-powered vehicles is sparse, particularly in regions where hydrogen is not yet a widely used fuel. For the transportation sector to fully benefit from hydrogen fuel cells, investments must be made in constructing hydrogen refueling stations. The establishment of a robust hydrogen infrastructure network is essential to support the widespread use of hydrogen in sectors like heavy-duty transport, where electrification through batteries remains less efficient (Yazdi et al., 2024).

Storage also poses a significant challenge. While hydrogen can be stored in compressed gas or liquid form, both methods require specialized tanks that can withstand high pressures or very low temperatures. Solid-state storage methods, which involve storing hydrogen in metal hydrides or other materials, are still in the early stages of development and are not yet commercially viable. Further research into safe, efficient, and cost-effective storage technologies is needed to advance the use of hydrogen as an energy carrier.

## V. TECHNOLOGICAL ADVANCEMENTS AND FUTURE PROSPECTS

Recent technological advancements in hydrogen production, particularly in electrolysis methods, are driving the growth of the green hydrogen sector. However, to fully realize the potential of green hydrogen, further innovations are needed to overcome existing technological and economic barriers(Islam et al., 2024).

1. Improvements in Electrolysis Technologies

The performance of electrolysis technologies, such as Alkaline Water Electrolysis (AWE), Proton Exchange Membrane (PEM) electrolysis, and Solid Oxide Electrolysis Cells (SOECs), has improved significantly, but challenges remain. Research into enhancing the efficiency and durability of electrolyzers is ongoing, with a focus on developing better electrocatalysts and reducing the reliance on expensive materials like platinum in PEM systems. By optimizing the electrolysis process and improving the efficiency of the oxygen evolution reaction (OER) and hydrogen evolution reaction (HER), the cost of green hydrogen production can be significantly reduced(Islam et al., 2024).

2. Energy Storage and Grid Integration

One of the most significant future prospects for green hydrogen lies in its ability to act as a long-term energy storage solution. As renewable energy sources like wind and solar become more prevalent, energy storage systems are essential for addressing the intermittency of these power sources. Hydrogen can serve as an energy buffer, storing excess energy generated during periods of high renewable output and releasing it when production drops. Advanced hydrogen storage methods, including pressurized and liquefied hydrogen, offer promising solutions to meet future energy needs.

3. Industrial Applications and Scaling

The scalability of hydrogen systems is another area of focus. Scaling up green hydrogen production and integrating it into various sectors, including heavy industry, transport, and heating, will require large-scale industrial plants capable of producing hydrogen at competitive prices. Industrial-scale projects are already underway in several regions. For example, Air Liquide's green hydrogen plant in Quebec produces 8.2 tonnes of green hydrogen per day, showcasing the potential for large-scale production (Ma, et al 2024).

In addition to these developments, international collaborations are vital to supporting the global hydrogen economy. Projects like the Nordic Green Ammonia Powered Ships (NoGAPS), which aims to decarbonize Nordic ports and the shipping industry, are leading the way in demonstrating the feasibility of hydrogen-powered applications on a large scale (Ma et al., 2024).

## VI. STRATEGIES FOR OVERCOMING OBSTACLES IN MOROCCO

Morocco, with its abundant renewable energy resources and strategic geographic position, has the potential to become a leader in green hydrogen production. However, like many other countries, it faces several challenges in its transition to a hydrogen-based economy. To address these obstacles, the following strategies are proposed:

1. Regulatory Framework and Policy Support

Establishing a clear regulatory framework is essential to promote the deployment of hydrogen production technologies in Morocco. This includes streamlining permitting procedures, providing financial incentives, and reducing administrative barriers to entry. Morocco can collaborate with the European Union and other international bodies to align its regulatory standards with global decarbonization goals. Table 2 highlights the capital expenditure needed per decade for Morocco to scale its hydrogen infrastructure, with a particular emphasis on public-private investment in renewable technologies and electrolysis facilities(Bibih, et al 2024).

*Table 2. Capital expenditure (CAPEX) for hydrogen infrastructure development (2020-2050)*

| Category | 2020-2030 (Billion €) | 2030-2040 (Billion €) | 2040-2050 (Billion €) | References |
|---|---|---|---|---|
| **Reference Scenario (Minimum Estimation)** | | | | (Luise, et al 2024) |
| Water Desalination Plant | 0 | 0 | 0.12 | |
| Processing Plant Capacity (PtA) | 0.74 | 2.17 | 3.2 | |
| Conversion Plant Capacity (PtL) | 0.39 | 1.11 | 1.84 | |
| Electrolysis | 1.62 | 5.16 | 6.15 | |
| Renewable Energy Sources (EnR) | 5,57 | 18.02 | 23.65 | |
| **Total CAPEX(Reference Scenario)** | **€8.32 billion** | **€26.46 billion** | **€34.96 billion** | |
| **Optimistic Scenario (Maximum Estimation)** | | | | (Faraldo, et al 2024) |
| Water Desalination | 0 | 0.12 | 0.12 | |
| Processing Plant Capacity (PtA) | 0.84 | 1.94 | 2.67 | |
| Conversion Plant Capacity (PtL) | 0.46 | 1.98 | 3.68 | |
| Electrolysis | 2.38 | 6.642 | 7.23 | |
| Renewable Energy Sources (EnR) | 9.41 | 24.32 | 32.25 | |
| **Total CAPEX (Optimistic Scenario)** | **€13.09 billion** | **€34.78 billion** | **€45.95 billion** | |

2. Water Management and Desalination

A key challenge in green hydrogen production is the significant water requirement for electrolysis. As Morocco is a water-scarce country, developing water management strategies, including desalination plants, is crucial to ensure that hydrogen production does not compromise water availability for agricul-

continued on following page

*Table 3. Continued*

ture and human consumption. Table 3 shows the drastic reduction in per capita water resources between 1960 and 2020, underscoring the importance of this issue. Investment in desalination plants, as shown in Table 2, will be essential to meeting the water needs of hydrogen production.

*Table 3. Per capita water resources in Morocco (1960-2020)*

| Year | Cubic meters /capita |
|---|---|
| 1961 | 2258 |
| 1967 | 1967 |
| 1972 | 1764 |
| 1977 | 1573 |
| 1982 | 1388 |
| 1987 | 1235 |
| 1992 | 1119 |
| 1998 | 1035 |
| 2002 | 969 |
| 2008 | 910 |
| 2013 | 850 |
| 2018 | 785 |
| 2019 | 771 |
| 2021 | 583 |

3. Cost Competitiveness

The production cost of green hydrogen remains a significant barrier to its competitiveness in the global market. Morocco must focus on reducing these costs by enhancing the efficiency of electrolyzers and expanding renewable energy projects. Public-private partnerships (PPPs) are critical to fostering collaboration and investment. By pooling resources, Morocco can accelerate the development of hydrogen infrastructure while maintaining cost competitiveness.

4. Skill Development and Workforce Training

A skilled workforce is essential for supporting the growth of the green hydrogen sector in Morocco. Technical training programs should target the skills needed for managing hydrogen production, infrastructure development, and renewable energy integration. Investing in education and training will ensure that Morocco has the human capital necessary to expand its hydrogen industry in line with its ambitious decarbonization goals.

## CONCLUSION

Green hydrogen holds immense potential as a solution for decarbonizing key sectors like heavy industry, transportation, and energy storage. Its ability to be produced from renewable energy sources without generating carbon emissions makes it a critical tool in the fight against climate change. However, the widespread adoption of green hydrogen faces significant challenges, particularly in reducing production costs, scaling infrastructure, and ensuring consistent regulatory and policy support. Addressing these issues will be crucial for green hydrogen to play a meaningful role in global decarbonization efforts.

Advances in electrolysis technologies, especially in Alkaline Water Electrolysis (AWE), Proton Exchange Membrane (PEM) electrolysis, and Solid Oxide Electrolysis Cells (SOECs), are essential for improving efficiency and reducing costs. Further research and innovation in these technologies are needed to make green hydrogen a cost-competitive alternative to fossil fuels. This includes the development of more durable and affordable catalysts, as well as improvements in energy conversion efficiency. Technological innovation will be a driving force in the future of green hydrogen.

Infrastructure development is another key factor in enabling the growth of the hydrogen economy. The creation of pipelines, storage systems, and refueling stations is necessary to ensure that hydrogen can be produced, transported, and utilized effectively. In particular, countries like Morocco, with its renewable energy resources, stand to benefit from investments in hydrogen infrastructure, but they must also address issues such as water management and the training of a skilled workforce. Public-private partnerships will play an essential role in building the necessary infrastructure and attracting investment.

Ultimately, the future of green hydrogen depends on a collaborative effort among governments, industries, and research institutions. Policy support, international cooperation, and significant investments in technology and infrastructure are required to make green hydrogen a key player in the global energy transition. With the right strategies, green hydrogen can contribute significantly to achieving carbon neutrality and reshaping the energy landscape for a sustainable future.

## REFERENCES


Al-Jadabi, N., Laaouan, M., El Hajjaji, S., Mabrouki, J., Benbouzid, M., & Dhiba, D. (2023). The dual performance of Moringa oleifera seeds as eco-friendly natural coagulant and as an antimicrobial for wastewater treatment: A review. [MDPI.]. *Sustainability (Basel)*, *15*(5), 4280. DOI: 10.3390/su15054280

Alavi-Borazjani, S. A., Adeel, S., & Chkoniya, V. (2025). Hydrogen as a Sustainable Fuel: Transforming Maritime Logistics. [MDPI.]. *Energies*, *18*(5), 1231. DOI: 10.3390/en18051231

Alipour Bonab, S., Waite, T., Song, W., Flynn, D., & Yazdani-Asrami, M. (2024). Machine learning-powered performance monitoring of proton exchange membrane water electrolyzers for enhancing green hydrogen production as a sustainable fuel for aviation industry. *Energy Reports*, *12*, 2270–2282. DOI: 10.1016/j.egyr.2024.08.028

Azdem, D., Fatni, A., Mabrouki, J., & El Hajjaji, S. (2025). The Desalination Strategy and Economic Integration of Morocco: Water Scarcity, Prospects, and Challenges. Circular Economy Applications in Energy Policy, 257-268.

Azdem, D., Mabrouki, J., & El Hajjaji, S. (2024). Methods and Technologies for the Treatment and Disposal of Brine in Seawater Desalination Plants. In Advancements in Climate and Smart Environment Technology (pp. 79-87). IGI Global.

Azdem, D., Mabrouki, J., Fahdi, G., & Hajjaji, S. E. (2025). ENVIRONMENTAL SUSTAINABILITY OF DESALINATION: A CRITICAL REVIEW OF BRINE TREATMENT CHALLENGES AND MITIGATION STRATEGIES. *Water Conservation and Management*, *9*(1), 55–60.

Azdem, D., Mabrouki, J., Moufti, A., El hajjaji, S., & Fatni, A. (2024). Assessment of heavy metal contamination in seawater in Agadir coastline, Morocco. *Desalination and Water Treatment*, *317*, 100129. DOI: 10.1016/j.dwt.2024.100129

Azdem, D., Mabrouki, J., Moufti, A., Hajjaji, S. E., & Fatni, A. (2024). Present Situation and Environmental Impact of Seawater Desalination Plants in Morocco. In *Technical and Technological Solutions Towards a Sustainable Society and Circular Economy* (pp. 581–592). Springer Nature Switzerland.

Bade, S. O., Tomomewo, O. S., Meenakshisundaram, A., Ferron, P., & Oni, B. A. (2024). Economic, social, and regulatory challenges of green hydrogen production and utilization in the US: A review. [Elsevier.]. *International Journal of Hydrogen Energy*, *49*, 314–335. DOI: 10.1016/j.ijhydene.2023.08.157

Bayssi, O., Nabil, N., Azaroual, M., Bousselamti, L., Boutammachte, N., Rachidi, S., & Barberis, S. (2024). Green hydrogen landscape in North African countries: Strengths, challenges, and future prospects. *International Journal of Hydrogen Energy*, *84*, 822–839.

Bencheikh, I., Azoulay, K., Mabrouki, J., El Hajjaji, S., Dahchour, A., Moufti, A., & Dhiba, D. (2020). The adsorptive removal of MB using chemically treated artichoke leaves: Parametric, kinetic, isotherm and thermodynamic study. [Elsevier.]. *Scientific African*, *9*, e00509. DOI: 10.1016/j.sciaf.2020.e00509

Bencheikh, I., Azoulay, K., Mabrouki, J., El Hajjaji, S., Moufti, A., & Labjar, N. (2021). The use and the performance of chemically treated artichoke leaves for textile industrial effluents treatment. *Chemical Data Collections*, *31*, 100597.

Benchrifa, M., Elouardi, M., Fattah, G., Mabrouki, J., & Tadili, R. (2023). *Identification, simulation and modeling of the main power losses of a photovoltaic installation and use of the internet of things to minimize system losses. Advanced technology for smart environment and energy*. Springer.

Benchrifa, M., & Mabrouki, J. (2022). Simulation, sizing, economic evaluation and environmental impact assessment of a photovoltaic power plant for the electrification of an establishment. [Taylor & Francis.]. *Advances in Building Energy Research*, *16*(6), 736–753. DOI: 10.1080/17512549.2022.2096693

Benchrifa, M., Mabrouki, J., Elouardi, M., Azrour, M., & Tadili, R. (2023). Detailed study of dimensioning and simulating a grid-connected PV power station and analysis of its environmental and economic effect, case study. [Springer.]. *Modeling Earth Systems and Environment*, *9*(1), 53–61. DOI: 10.1007/s40808-022-01457-9

Bibih, M., Choukri, K., El Khaili, M., Babkrani, Y., & Chakir, H. (2024). Renewable Energy Integration and Green Hydrogen Prospects in Morocco: A Comprehensive Analysis of Energy Transition Dynamics. *2024 4th International Conference on Innovative Research in Applied Science, Engineering and Technology (IRASET)* (pp. 1–7). Presented at the 2024 4th International Conference on Innovative Research in Applied Science, Engineering and Technology (IRASET). Retrieved September 27, 2024, from https://ieeexplore.ieee.org/abstract/document/10549422

Borges, P. T., Sales, M. B., Guimarães, C. E. C., de França Serpa, J., de Lima, R. K. C., Lopes, A. A. S., & dos Santos, J. C. (2024). Photosynthetic green hydrogen: Advances, challenges, opportunities, and prospects. *International Journal of Hydrogen Energy*, *49*, 433–458.

Cheekatamarla, P. (2024). Hydrogen and the global energy transition—Path to sustainability and adoption across all economic sectors. [MDPI.]. *Energies*, *17*(4), 807. DOI: 10.3390/en17040807

Chen, Y., Mao, J., Zhou, H., Xing, L., Qiao, S., Yuan, J., Mei, B., Wei, Z., Zhao, S., Tang, Y., & Liu, C. (2024). Coordination Shell Dependent Activity of CuCo Diatomic Catalysts for Oxygen Reduction, Oxygen Evolution, and Hydrogen Evolution Reaction. *Advanced Functional Materials*, *34*(10), 2311664. DOI: 10.1002/adfm.202311664

Egeland-Eriksen, T., Hajizadeh, A., & Sartori, S. (2021). Hydrogen-based systems for integration of renewable energy in power systems: Achievements and perspectives. [Elsevier.]. *International Journal of Hydrogen Energy*, *46*(63), 31963–31983. DOI: 10.1016/j.ijhydene.2021.06.218

El-Adawy, M., Dalha, I. B., Ismael, M. A., Al-Absi, Z. A., & Nemitallah, M. A. (2024). Review of Sustainable Hydrogen Energy Processes: Production, Storage, Transportation, and Color-Coded Classifications. [ACS Publications.]. *Energy & Fuels*, *38*(23), 22686–22718. DOI: 10.1021/acs.energyfuels.4c04317

El Alouani, M., Aouan, B., Rachdi, Y., Alehyen, S., El Herradi, E. H., Saufi, H., Mabrouki, J., & Barka, N. (2024). Porous geopolymers as innovative adsorbents for the removal of organic and inorganic hazardous substances: A mini-review. [Taylor & Francis.]. *International Journal of Environmental Analytical Chemistry*, *104*(16), 4784–4796. DOI: 10.1080/03067319.2022.2115891

El Mouedden, M., Azdem, D., Mabrouki, J., & El Hajjaji, S. (2025). *MarocFarm: A digital revolution for agriculture in Morocco*. Circular Economy Applications in Energy Policy. DOI: 10.4018/979-8-3693-6680-6.ch010

Evro, S., Oni, B. A., & Tomomewo, O. S. (2024). Carbon neutrality and hydrogen energy systems. [Elsevier.]. *International Journal of Hydrogen Energy*, *78*, 1449–1467. DOI: 10.1016/j.ijhydene.2024.06.407

Fahdi, G., Azdem, D., Lamchaimech, A., Benrhanem, M., Mabrouki, J., & El Hajjaji, S. (2024). ANALYSIS OF THE IMPACT OF CLIMATE CHANGE ON WATER RESOURCES: CASE OF THE TENSIFT BASIN (MOROCCO). *Water Conservation and Management*, *8*(4), 461–465. DOI: 10.26480/wcm.04.2024.461.465

Fahdi, G., Mabrouki, J., Lamchaimech, A., Azdem, D., Benrhanem, M., & Hajjaji, S. E. (2024). IMPACTS OF CLIMATE CHANGE ON GROUNDWATER QUALITY AND RECHARGE IN THE TENSIFT BASIN, MOROCCO. *Water Conservation and Management*, *8*(4), 466–474. DOI: 10.26480/wcm.04.2024.466.474

Faraldo, F., & Byrne, P. (2024). A Review of Energy-Efficient Technologies and Decarbonating Solutions for Process Heat in the Food Industry. [Multidisciplinary Digital Publishing Institute.]. *Energies*, *17*(12), 3051. DOI: 10.3390/en17123051

Fatni, A., Addich, M., Essekri, A., & Baraka, E. L. (2025). Characterization and mitigation strategies for inorganic scaling in reverse osmosis system treating brackish groundwater. *Sustainable Chemistry One World*, *5*, 100035. DOI: 10.1016/j.scowo.2024.100035

Franco, A., & Rocca, M. (2024). Renewable electricity and green hydrogen integration for decarbonization of "hard-to-abate" industrial sectors. [MDPI.]. *Electricity*, *5*(3), 471–490. DOI: 10.3390/electricity5030024

Ghizlane, F., Mabrouki, J., Ghrissi, F., & Azrour, M. (2022). Proposal for a high-resolution particulate matter (PM10 and PM2. 5) capture system, comparable with hybrid system-based internet of things: Case of quarries in the western rif, Morocco. [University Of Tehran Press.]. *Pollution*, *8*(1), 169–180.

Han, N., Zhang, W., Guo, W., Pan, H., Jiang, B., Xing, L., Tian, H., Wang, G., Zhang, X., & Fransaer, J. (2023). Designing oxide catalysts for oxygen electrocatalysis: Insights from mechanism to application. [Springer.]. *Nano-Micro Letters*, *15*(1), 185. DOI: 10.1007/s40820-023-01152-z PMID: 37515746

Hassan, Q., Algburi, S., Jaszczur, M., Al-Jiboory, A. K., Al Musawi, T. J., Ali, B. M., Viktor, P., Fodor, M., Ahsan, M., Salman, H. M., & Sameen, A. Z. (2024). Hydrogen role in energy transition: A comparative review. *Process Safety and Environmental Protection*, *184*, 1069–1093. DOI: 10.1016/j.psep.2024.02.030

Hassan, Q., Algburi, S., Sameen, A. Z., Salman, H. M., & Al-Jiboory, A. K. (2024). A review of green hydrogen production by renewable resources. [De Gruyter.]. *Energy Harvesting and Systems*, *11*(1), 20220127. DOI: 10.1515/ehs-2022-0127

Hindson, W. A., & James, S. (2024). Effects of surface modification on a proton exchange membrane for improvements in green hydrogen production. *International Journal of Hydrogen Energy*, *49*, 1040–1047. DOI: 10.1016/j.ijhydene.2023.10.322

Ishaq, H., Dincer, I., & Crawford, C. (2022). A review on hydrogen production and utilization: Challenges and opportunities. *International Journal of Hydrogen Energy*, SI: Progress in Hydrogen Production. *Storage and Distribution (Ahmadi)*, *47*(62), 26238–26264.

Islam, A., Islam, T., Mahmud, H., Raihan, O., Islam, M., Marwani, H. M., Rahman, M. M., Asiri, A. M., Hasan, M. M., Hasan, M. N., Salman, M. S., Kubra, K. T., Shenashen, M. A., Sheikh, M. C., & Awual, M. R. (2024). Accelerating the green hydrogen revolution: A comprehensive analysis of technological advancements and policy interventions. *International Journal of Hydrogen Energy*, *67*, 458–486. DOI: 10.1016/j.ijhydene.2024.04.142

Loukili, H., Mabrouki, J., Anouzla, A., Kouzi, Y., Younssi, S. A., Digua, K., & Abrouki, Y. (2021). Pretreated Moroccan natural clays: Application to the wastewater treatment of textile industry. [Elsevier.]. *Desalination and Water Treatment*, *240*, 124–136. DOI: 10.5004/dwt.2021.27644

Lu, M., Liang, J., Zhan, W., Li, Y., Chen, H., Li, M., Liao, S., & Cui, Z. (2024, December 05). Corrosion Mechanism and Mitigation Strategies for Carbon Supports in PEMFCs. [Wiley Online Library.]. *Advanced Sustainable Systems*, *•••*, 2400742. DOI: 10.1002/adsu.202400742

Luise, R., Brisse, A., & Azzaro-Pantel, C. (2024). Paving the way for low-carbon hydrogen supply chain deployment by exploring the potential of renewable energies and multisectoral hydrogen demand: Case study of France. *Renewable Energy Focus*, *50*, 100613. DOI: 10.1016/j.ref.2024.100613

Ma, N., Zhao, W., Wang, W., Li, X., & Zhou, H. (2024). Large scale of green hydrogen storage: Opportunities and challenges. *International Journal of Hydrogen Energy*, *50*, 379–396. DOI: 10.1016/j.ijhydene.2023.09.021

Mabrouki, J. Azrour, M., Farhaoui, Y., & El Hajjaji, S. (2020). Intelligent system for monitoring and detecting water quality. *Big Data and Networks Technologies 3* (pp. 172–182). Springer.

Mabrouki, J. Benchrifa, M., Ennouhi, M., Azoulay, K., Bencheikh, I., Rachiq, T., El-Moustaqim, K., et al. (2022). Geographic information system for the study of water resources in Chaâba El Hamra, Mohammedia (Morocco). *The international conference on artificial intelligence and smart environment* (pp. 469–474). Springer.

Mabrouki, J., Benbouzid, M., Dhiba, D., & El Hajjaji, S. (2022). Simulation of wastewater treatment processes with Bioreactor Membrane Reactor (MBR) treatment versus conventional the adsorbent layer-based filtration system (LAFS). [Taylor & Francis.]. *International Journal of Environmental Analytical Chemistry*, *102*(19), 7458–7468. DOI: 10.1080/03067319.2020.1828394

Mabrouki, J., Moufti, A., Bencheikh, I., Azoulay, K., El Hamdouni, Y., & El Hajjaji, S. (2020). Optimization of the coagulant flocculation process for treatment of leachate of the controlled discharge of the city Mohammedia (Morocco). *Advanced Intelligent Systems for Sustainable Development (AI2SD'2019) Volume 7-Advanced Intelligent Systems for Sustainable Development Applied in Energy and Electrical Engineering* (pp. 200–212). Springer.

Mgalaa, S., Mabrouki, J., Elouardi, M., El Azzouzi, L., Moufti, A., El Hajjaji, S. E. L., & Azzouzi, M.. (2022). Study and evaluation of the degradation of procion blue dye by the ozonation method: Parametric and isothermal study. [Springer.]. *Nanotechnology for Environmental Engineering*, *7*(3), 691–697. DOI: 10.1007/s41204-022-00262-0

Nechache, A., & Hody, S. (2021). Alternative and innovative solid oxide electrolysis cell materials: A short review. *Renewable & Sustainable Energy Reviews*, *149*, 111322. DOI: 10.1016/j.rser.2021.111322

Oliveira, A. M., Beswick, R. R., & Yan, Y. (2021). A green hydrogen economy for a renewable energy society. [Elsevier.]. *Current Opinion in Chemical Engineering*, *33*, 100701. DOI: 10.1016/j.coche.2021.100701

Oluleye, G., McLaughlin, S., Steen, E., Wu, S., Hanna, R., & Heptonstall, P. (2025). Blending interventions to achieve green hydrogen cost competitiveness for industrial decarbonisation. *International Journal of Hydrogen Energy*. Elsevier.

Rachiq, T., Abrouki, Y., Mabrouki, J., Samghouli, N., Fersi, C., Rahal, S., & El Hajjaji, S. (2021). Evaluation of the efficiency of different materials to remove specific pollutants from landfill leachate. [Elsevier.]. *Desalination and Water Treatment*, *238*, 240–250. DOI: 10.5004/dwt.2021.27779

Rasul, M. G., Hazrat, M. A., Sattar, M. A., Jahirul, M. I., & Shearer, M. J. (2022). The future of hydrogen: Challenges on production, storage and applications. [Elsevier.]. *Energy Conversion and Management*, *272*, 116326. DOI: 10.1016/j.enconman.2022.116326

Salehmin, M. N. I., Husaini, T., Goh, J., & Sulong, A. B. (2022). High-pressure PEM water electrolyser: A review on challenges and mitigation strategies towards green and low-cost hydrogen production. [Elsevier.]. *Energy Conversion and Management*, *268*, 115985. DOI: 10.1016/j.enconman.2022.115985

Tamilarasi, B., Jithul, K. P., & Pandey, J. (2024). Non-noble metal-based electro-catalyst for the oxygen evolution reaction (OER): Towards an active & stable electro-catalyst for PEM water electrolysis. *International Journal of Hydrogen Energy*, *58*, 556–582. DOI: 10.1016/j.ijhydene.2024.01.222

Tongsh, C., Wu, S., Jiao, K., Huo, W., Du, Q., Park, J. W., Xuan, J., Wang, H., Brandon, N. P., & Guiver, M. D. (2024). Fuel cell stack redesign and component integration radically increase power density. [Elsevier.]. *Joule*, *8*(1), 175–192. DOI: 10.1016/j.joule.2023.12.003

Tüysüz, H. (2024). Alkaline Water Electrolysis for Green Hydrogen Production. *Accounts of Chemical Research*, acs.accounts.3c00709.

Yasin, M. C., Johar, M., Gupta, A., & Shahgaldi, S. (2024). A comprehensive review of the material innovations and corrosion mitigation strategies for PEMWE bipolar plates. [Elsevier.]. *International Journal of Hydrogen Energy*, *88*, 726–747. DOI: 10.1016/j.ijhydene.2024.09.208

Yazdi, M., Zarei, E., Pirbalouti, R. G., & Li, H. (2024). A comprehensive resilience assessment framework for hydrogen energy infrastructure development. *International Journal of Hydrogen Energy*, *51*, 928–947. DOI: 10.1016/j.ijhydene.2023.06.271

Yin, H., Rong, F., & Xie, Y. (2024). A review of typical transition metal phosphides electrocatalysts for hydrogen evolution reaction. *International Journal of Hydrogen Energy*, *52*, 350–375. DOI: 10.1016/j.ijhydene.2023.08.333

Zhang, H.-J., Chen, Z.-Q., Ye, X.-T., Xiao, K., & Liu, Z.-Q. (2025). Electron Delocalized Ni Active Sites in Spinel Catalysts Enable Efficient Urea Oxidation. [Wiley Online Library.]. *Angewandte Chemie International Edition*, *64*(10), e202421027. DOI: 10.1002/anie.202421027 PMID: 39604019

# Chapter 5
# Comprehensive Insights Into the Biomedical Applications of Titanium Dioxide

**Mariyem Abouri**
*Mohammed V University in Rabat, Morocco*

**Jamal Mabrouki**
https://orcid.org/0000-0002-3841-7755
*Mohammed V University in Rabat, Morocco*

**Mohammed Elouardi**
*Mohammed V University in Rabat, Morocco*

**Adnane El Hamidi**
*Mohammed V University in Rabat, Morocco*

**Mohammed Alaoui El Belghiti**
*Mohammed V University in Rabat, Morocco*

## ABSTRACT

*Metal oxide nanoparticles (NPs) particularly titanium dioxide have gained significant interest due to their exceptional properties and versatile applications in various sectors such as industrial, medicinal, and environmental areas. TiO2 (NPs) play a crucial role in improving the efficacy and stability of drug formulations. Their high surface area and reactive properties enable them to serve as effective carriers for drug delivery systems, ensuring efficient absorption of therapeutic agents by the body. In addition, TiO2 (NPs) are also pivotal in the development of advanced biosensors that exhibit heightened sensitivity and specificity, enabling precise detection of biological molecules. Furthermore, TiO2 (NPs) have shown remarkable potential in cancer therapy, with their application in photodynamic therapy (PDT) generating reactive oxygen species (ROS) that can selectively induce apoptosis in cancer cells.*

DOI: 10.4018/979-8-3693-8980-5.ch005

## INTRODUCTION

In recent decades, nanotechnology has made significant progress, resulting improved performance of different field including electronics, medicine, energy, sensing, and materials science.(Kazemi Shariat Panahi et al. 2024; Sadegh, Dehcheshmeh, et Sadegh 2024; Lambin 2024) These advancements have created new opportunities and enhanced the effectiveness and functionality of many applications. Nanotechnology focuses on using materials at nanometer scale, specifically between 1 and 100 nanometers. Nanoparticles show characteristics in comparison to bulk materials due to their large surface area and volume ratio. (Nishu et Kumar 2023)Researchers had demonstrated significant interest in incorporating metal oxide nanoparticles into the medical field due to their antibacterial, drug delivery, anticancer properties, and ability to identify infections in clinical specimens.(Merugu, Sharma, et Rastogi 2023) Titanium dioxide nanoparticles have been employed in a variety of application and are typically considered environmentally friendly due to their exceptional chemical stability, photocatalytic performance, and low toxicity. These particles serve as a barrier against UV radiation and have been used in many applications such as food colorants, dyes, medicines, cosmetics, and toothpaste.(Shabib Akhtar et al. 2024; Mao et Hao 2024; Dubey et al. 2022; Moon et al. 2024; K, P, et U 2024; L. Zhang et Jin 2024)

In the context of environmental science, $TiO_2$ (NPs) have shown a lot of promise in the purification of air and water. There potent oxidative properties and high photostability make it ideal candidate for degradation pollutants in wastewater and atmospheric systems. This capability has driven extensive research into using $TiO_2$ (NPs) for removing organics contaminants from aquatic environments, highlighting their potential in mitigating pollution and protecting ecosystems. (Mehta et Bhushan 2024) In addition to environmental applications, $TiO_2$ (NPs) have many biological properties, such as antimicrobial, anticancer, and antioxydant effects.(Ali et al. 2024) These properties are particularly advantageous in biomedical applications, where $TiO_2$ (NPs) can be employed in medical imaging, drugs delivery systems, and as components in medical devises and implants. The versability of $TiO_2$ (NPs) in these fields emphasizes their importance and continues demand for research for improving their production and utilization.(Suresh et al. 2022; Hsu et al. 2024) This article focuses on emphasizing role of $TiO_2$ (NPs) in advancing biomedical technologies. Figure 1 shows the different biomedical applications of $TiO_2$.

## BIOMEDICAL APPLICATIONS OF $TIO_2$ (NPS)

### Antimicrobial Applications

Nano-scale materials, including various metallic nanoparticles (NPs) such as titanium dioxide $TiO_2$ (NPs), have appeared as promising antimicrobial agents due to their unique physicochemical properties and high surface area. $TiO_2$ (NPs) have demonstrated notable antibacterial and antifungal activities against a wide range of microorganisms, including bacteria, fungi, and viruses.(Kurutas 2016; Ziental et al. 2020) The efficacy of $TiO_2$ (NPs) in combating microbial infections stems from their photocatalytic action, which generates reactive oxygen species (ROS). These ROS disrupt microbial cell integrity by oxidizing phospholipids in the cell wall, leading to membrane damage and altering cellular functions. (Luksiene 2017) Additionally, $TiO_2$ (NPs) have been shown to inhibit microbial growth through various mechanisms, including interference with respiratory cytosolic enzymes and macromolecule structures. (Kurutas 2016)

Recent advancements in nanotechnology have addressed the limitation of UV light dependency for $TiO_2$ (NPs) activation by developing visible light-activated photocatalysts and doped-$TiO_2$ nanostructures. The antimicrobial efficacy if $TiO_2$ (NPs) extends to a broad spectrum of organisms, including Gram-negative and Gram-positive bacteria, fungi, viruses, algae, protozoa, and bacterial toxins. (Lungu et al. 2013; Evgenidou et al. 2021; Luksiene 2017) In the biomedical field, $TiO_2$ is antimicrobial properties hold significant promise for controlling bacterial colonization on medical devices and implants, addressing the challenge posed by antimicrobial resistance and biofilm formation. Ongoing research efforts continue to explore the diverse application of $TiO_2$ nanostructures, highlighting their potential as effective antimicrobial agents in various biomedical settings.(Bhardwaj et Webster 2017) Figure 2 presents the strategies of $TiO_2$ in executing antimicrobial activities

## Antioxidant Applications

Titanium dioxide nanoparticles exhibit a multifaceted role in oxidative stress responses, displaying not only the potential to induce oxidative stress but also demonstrating effective antioxidant activity. Research has extensively explored the antioxidant properties of $TiO_2$ (NPs), particularly emphasizing their synthesis through plant-mediated methods. These synthesis techniques offer significant advantages, including enhanced biocompatibility, stability, and controlled release of antioxidant functional groups. $TiO_2$ (NPs) have demonstrated rapid scavenging of free radicals, indicating their potential as potent antioxidants.(Suresh et al. 2022) Furthermore, $TiO_2$ (NPs) are reported to possess protective effects against reactive oxygen species (ROS), further highlighting their antioxidant properties. The incorporation of functional groups like phenols and tannins during plant-mediated synthesis aids in stabilizing $TiO_2$ (NPs) and enhancing their antioxidant potential. Studies have shown that $TiO_2$ (NPs) synthesized using plant extracts, such as A.indica and M.indica, exhibit superior antioxidant activity compared to conventional antioxidants like ascorbic acid.(Rana et al. 2024) Also, $TiO_2$ (NPs) synthesized from rhizomes of Curcuma longa linn, Kaempferia galanga and nutmeg in Myristica gragrans have demonstrated antioxidant effects, suggesting their efficacy in combating oxidative stress.(Blessymol et al. 2024)

## Anticancer Applications

Chemotherapy, a cornerstone of cancer treatment, often inflicts collateral damage on healthy cells due to the non-specific targeting of cytotoxic agents. This indiscriminate toxicity can lead to severe side effects and compromise patient well-being. In response to these challenges, nanotechnology has emerged as a revolutionary approach, particularly in the realm of cancer therapeutics. By harnessing the principles of nanoscience and utilizing innovative techniques, nanostructures are engineered to serve as sophisticated carriers for drug delivery.(Wei et al. 2021)

In the context of cancer treatment, the development of nanostructures, such as $TiO_2$ (NPs), represents a change in improving the efficacy and safety of chemotherapy. These nanostructures are synthesized using green synthesis methods, leveraging natural plant extract like A.indica and M. indica. This approach not only ensures the eco-friendliness of the production process but also imbues the resulting nanoparticles with desirable properties such as biocompatibility and non-toxicity.(Rana et al. 2024; Shiva Samhitha et al. 2022)

The multifaceted benefits of $TiO_2$ (NPs) extend beyond their role as drug delivery vehicles. Thes nanoparticles have inherent properties that augment their therapeutic potential, such as their ability to generate reactive oxygen species (ROS) upon illumination by UV light. This property is harnessed on photodynamic therapy (PDT), where $TiO_2$ nanoparticles serve as photosensitizers, including cell death in cancer cells through ROS- mediated mechanisms.(Behnam et al. 2018)

Moreover, surface modifications of $TiO_2$ nanoparticles further enhance their specificity and efficacy in cancer therapy. By functionalizing the nanoparticle surface with targeting moieties, such as folic acid of polyethylene glycol, selective binding to cancer cells is facilitated, further minimizing therapeutic impact.(Zarzzeka et al. 2023) In summary, the integration of nanotechnology into cancer therapy represents a transformative approach towards personalized and precision medicine. $TiO_2$ nanoparticles, synthesized through green and sustainable methods, offer a promising platform for enhancing the efficiency, safety, and selectivity of chemotherapy. As research in this field continues to advance, the potential for nanostructured drug delivery systems to revolutionize cancer treatment paradigms remains immense.(Tiwari et al. 2023)

## Drugs Delivery

Targeted drug delivery is a promising therapeutic strategy that aims to deliver drugs directly to specific sites of action in the body, thereby minimizing unwanted side effects and improving therapeutic efficacy. Titanium dioxide nanoparticles have emerged as promising drug carriers for this application, offering a series of significant advantages. Titanium dioxide nanoparticles have an extremely small size, which allows them to flow freely through the vascular system and penetrate deep into target tissues. Their small size provides precise and controlled drug delivery, thereby improving the distribution and bioavailability of therapeutic agents.(Jafari et al. 2020) The surface of $TiO_2$ (NPs) is highly modifiable, providing significant flexibility in the design of tailored drug delivery systems. Therapeutic molecules can be adsorbed, encapsulated, or chemically bound to the surface of the nanoparticles, allowing targeted and specific delivery to the desired site of action. Additionally, cell-specific targeting ligands can be incorporated onto the surface of nanoparticles to improve their specificity and affinity for target cells, thereby reducing toxicity to surrounding healthy tissues.(Pourmadadi et al. 2023; Yin et al. 2013)

Titanium dioxide nanoparticles function as protective vehicles for drugs, protecting them from enzymatic degradation and early clearance by the immune system. In addition, their structure and reactive surface allow controlled and programmable drug release, ensuring effective therapeutic concentration over a prolonged period. This controlled release can be tailored to meet specific treatment needs, providing personalized and optimized pharmacokinetic profiles.(Ziental et al. 2020; Kawassaki et al. 2021) By delivering drugs directly to specific sites of action, $TiO_2$ (NPs) reduce the need for systemic drug delivery in high doses, thereby minimizing unwanted side effects associated with widespread exposure. This targeted delivery approach also helps reduce drug toxicity to surrounding healthy tissues, thereby improving the safety and tolerability of the treatment. Additionally, nanoparticles can be functionalized to release drugs in response to specific stimuli, such as changes in pH or environmental conditions, allowing for even more precise and controlled delivery. (Jafari et al. 2020; Pulit-Prociak, s. d.) The illustration of drug delivery system using $TiO_2$ is represented in figure 3.

## Biosensing Applications

In recent years, nanotechnology has appeared as a focal point in biosensor research, drawing attention due to its myriad advantages, notably its small scale and high surface-area to volume ratio. This technological approach has facilitated the development of biosensors based on nanoparticles, offering rapid and sensitive detection capabilities.(Song et al. 2023)

Specifically, this article delves into the realm of nano biosensors utilizing titanium dioxide nanoparticles, which have gained substantial interest for their potential applications. Researchers have increasingly explored novel hybrid systems integrating biomolecules with $TiO_2$ nanostructures, resulting in the creation of innovative bio-nano hybrid devices such as biomolecule- sensitized solar cells (BSSCs) and photoelectrochemical cells (PECs). These hybrid systems represent a convergence of biological and nanotechnological principles, offering enhanced functionalities and performance in biosensing applications.(Kusior et al. 2015; Ahmad et al. 2023)

$TiO_2$ nanostructures have been pivotal in the fabricating diverse detecting devices, including humidity and oxygen sensors to hydrogen sensors. Their unique properties, including a porous structure and large specific surface area, make them ideal as electrode materials in biosensors. Additionally, $TiO_2$ serves as an immobilizing matrix for enzymes, preserving their biocatalytic activity and enabling the construction of enzyme-based biosensors with enhanced stability and sensitivity. (J. Zhang et al. 2022)

$TiO_2$-based nanocomposites, like DNA-functionalized structures, showcase $TiO_2$ (NPs) versatility in precise detection of specific biomolecules or genetic sequences, while integration with other materials expands biosensors design possibilities, enabling multifunctional platforms.(Ma, Zhang, et Zhang 2020)

Biosensors employing $TiO_2$ (NPs) often use electrochemical or photoelectrochemical technique for detection, harnessing the distinctive electronic and optical characteristic of $TiO_2$. Particularly, PEC biosensors stand out fir their combined optical and electrochemical detection mechanisms offering synergistic sensitivity and selectivity advantages. Overall, $TiO_2$ nanostructures hold immense promise for advancing biomedical diagnostics and healthcare, offering unprecedented sensitivity, selectivity, and versatility in biosensing applications. (Ahmad et al. 2023) The schematic mechanism of $TiO_2$-based photoelectrochemical cell (PEC) biosensor is shown in figure 4.

## Photodynamic Therapy

Titanium dioxide nanoparticles serve as photosensitizers in photodynamic therapy (PDT), a treatment modality that harnesses light energy to induce localized cytotoxic effects in target cells. Upon exposure to light of specific wavelengths, $TiO_2$ (NPs) absorb photons and undergo photoexcitation, leading to the generation of reactive oxygen species (ROS). These ROS are highly cytotoxic and can induce oxidative damage to cellular components, including lipids, proteins, and nucleic acids, ultimately resulting in cell death via apoptosis or necrosis.(Chen et al. 2019)

One of the key advantages of PDT is its ability to selectively target diseased tissues while sparing surrounding healthy cells. By controlling the parameters of light exposure, such as wavelength, intensity, and duration, PDT can be tailored to selectively activate $TiO_2$ (NPs) in target tissues, thereby minimizing off-target effects and reducing collateral damage to adjacent normal tissues. Moreover, the shallow tissue penetration depth of light allows for precise spatial control over the treatment area, enabling localized therapy without causing systemic toxicity or adverse side effects.(Pourmadadi et al. 2023)

PDT offers several advantages over traditional cancer therapies, including surgery, chemotherapy, and radiation therapy. Unlike surgical resection, which may be invasive and associated with significant morbidity, PDT is minimally invasive and can be performed on an outpatient basis. Additionally, PDT does not induce systemic toxicity or damage to healthy tissues, making it well-tolerated by patients and suitable for repeat treatments if necessary. Furthermore, PDT can be combined synergistically with other therapeutic modalities, such as chemotherapy or immunotherapy, to enhance treatment efficacy and improve patient outcomes.(Bhattacharya et al. 2023) The schematic representation of PDT (photodynamic therapy) is depicted in figure 5

## Biomedical Implants

**Articular prostheses:** Joint prostheses, such as those of the hips and knees, have transformed the treatment of degenerative and traumatic joint diseases. However, these devices are subjected to high mechanical stress and hostile biological environment, which can lead to premature wear and postoperative complications. Titanium dioxide offers a premium solution to these challenges. In the manufacturing of prosthetic joint components $TiO_2$ is preferred due to its corrosion resistance, durability, and biocompatibility.(Abd-Elaziem et al. 2024; Mostakhdemin, Nand, et Ramezani 2021)

Corrosion resistance ensure the structural stability of the implants in a physiological environment, while its durability maintains a long functional life of the prostheses. In addition, its biocompatibility promotes harmonious integration with surrounding tissues, reducing the risk of immune rejection and postoperative inflammation. This strong bone integration, called osseointegration, is essential for the long-term stability of the prothesis and the success of the surgical procedure.(Meng et al. 2023)

**Dental implants:** Dental implants represent a significant advancement in restoring masticatory and aesthetic functions in edentulous or partly edentulous patients. Despite traditional removable dentures, dental implants provide a permanent and stable solution to replace missing teeth. Titanium dioxide is the preferred material for manufacturing these implants due to its unique properties.(Upadhyay et al., s. d.; Ramezani et Ripin 2023) Osseointegration is especially important for dental implants, as it ensures a solid and stable basis for the prosthetic restoration. Titanium dioxide enhances this bone integration by providing a rough surface that stimulates the proliferation and differentiation of bone cells, leading to a strong and durable bone-implant. Additionally, its intrinsic biocompatibility reduces the risk of immunological or inflammatory reactions, ensuring the long-term success of the implant.(Hossain et al. 2023)

**Cardiovascular devices:** Cardiovascular diseases, such as atherosclerosis and heart valve disease, are among the leading causes of morbidity and mortality worldwide. Cardiovascular devices, such as coronary stents and heart valves, are critical for treating these pathologies. Titanium dioxide plays a crucial role in the manufacturing of these devices by offering unique characteristics. In coronary stents, for example, $TiO_2$ is used to coat the metal structure to reduce the risk of thrombosis and restenosis. Its smooth non-reactive surface reduces platelet and endothelial cell adhesion, thereby minimizing the risk of blood clotting and artery constriction. Furthermore, its corrosion resistance guarantees reliable long-term performance of the stent in the dynamic environment of the vascular system.(Ullah et al. 2024; Beshchasna et al. 2019) In Conclusion, Titanium dioxide is an invaluable material in the manufacture of biomedical implants due to its unique properties; including biocompatibility, corrosion resistance, and ability to promote osseointegration. Its use in joint prostheses, dental implants and cardiovascular devices has significantly improved clinical outcomes and quality of life of patient, making it an indispensable element in the field of regenerative medicine and orthopedic, dental, and cardiovascular surgery. (Hashemzadeh et al. 2024; Liu, Wang, et Feng 2023)

## CONCLUSION

To conclude, titanium dioxide nanoparticles present an important advance in biomedical applications due to their unique physicochemical characteristics which include excellent chemical stability, high photocatalytic activity, and low toxicity. Their versatility has enabled them useful in a variety of medical fields, including biosensing and drugs delivery, as well as cancer therapy and antimicrobial treatments. $TiO_2$ (NPs) improve the efficacy and integration of medical implants, give novel approaches for targeted drugs delivery, and have significant antimicrobial properties, making them crucial for combating infections. The promising results in these applications highlight $TiO_2$ NPs potential to improve current medical treatments. Future research should concentrate on improving the biocompatibility and functionalization of $TiO_2$ (NPs) to ensure their safe and effective use in clinical settings.

### Author Contributions

**MA**: Writing – original draft, Conceptualization, Formal analysis, Investigation. **ME**: Writing – review & editing, Methodology. **AE**: Writing – review & editing, Supervision. **JM**: Validation, Writing – review & editing. **MAE**: Writing – review & editing, Supervision

### Conflict of Interest

All authors declare no conflicts of interest.

### Ethics Approval

No applicable in this paper

### Data Availability

No applicable in this paper

## REFERENCES


Abd-Elaziem, W., Darwish, M. A., Hamada, A., & Daoush, W. M. (2024). Titanium-Based alloys and composites for orthopedic implants Applications: A comprehensive review. *Materials & Design*, *241*(mai), 112850. DOI: 10.1016/j.matdes.2024.112850

Ahmad, M., Hasan, M., Tarannum, N., Hasan, M., & Ahmed, S. (2023). Recent advances in optical and photoelectrochemical nanobiosensor technology for cancer biomarker detection. *Biosensors & Bioelectronics: X*, *14*(septembre), 100375. DOI: 10.1016/j.biosx.2023.100375

Akhtar, S., Mohammad, K. C., Saminathan, S., Rajalingam, S. R., Mohsin, N., Ahmed, K. A. A. A., & Alhazmi, Y.. (2024). Nanoengineered Chitosan Functionalized Titanium Dioxide Biohybrids for Bacterial Infections and Cancer Therapy. *Scientific Reports*, *14*(1), 3705. DOI: 10.1038/s41598-024-52847-1 PMID: 38355697

Ali, O. M., Hasanin, M. S., Suleiman, W. B., Helal, E. E.-H., & Hashem, A. H. (2024). Green Biosynthesis of Titanium Dioxide Quantum Dots Using Watermelon Peel Waste: Antimicrobial, Antioxidant, and Anticancer Activities. *Biomass Conversion and Biorefinery*, *14*(5), 6987–6998. DOI: 10.1007/s13399-022-02772-y

Behnam, M. A., Emami, F., Sobhani, Z., & Dehghanian, A. R. (2018). The application of titanium dioxide (TiO2) nanoparticles in the photo-thermal therapy of melanoma cancer model. *Iranian Journal of Basic Medical Sciences.*, *21*(11), 1133–1139. DOI: 10.22038/IJBMS.2018.30284.7304 PMID: 30483386

Beshchasna, N., Au Yeung, K. H., Saqib, M., Kraśkiewicz, H., Wasyluk, Ł., Kuzmin, O., & Duta, O. C.. (2019). Surface Evaluation of Titanium Oxynitride Coatings Used for Developing Layered Cardiovascular Stents. *Materials Science and Engineering C*, *99*(juin), 405–416. DOI: 10.1016/j.msec.2019.01.131 PMID: 30889715

Bhardwaj, G., & Webster, T. (2017). Reduced bacterial growth and increased osteoblast proliferation on titanium with a nanophase TiO2 surface treatment. *International Journal of Nanomedicine*, *12*(janvier), 363–369. DOI: 10.2147/IJN.S116105 PMID: 28123296

Bhattacharya, D., Mukhopadhyay, M., Shivam, K., Tripathy, S., Patra, R., & Pramanik, A. (2023). Recent Developments in Photodynamic Therapy and Its Application against Multidrug Resistant Cancers. *Biomedical Materials (Bristol, England)*, *18*(6), 062005. DOI: 10.1088/1748-605X/ad02d4 PMID: 37827172

Blessymol, B., Yasotha, P., Kalaiselvi, V., & Gopi, S. (2024). An antioxidant study of Titanium dioxide (TiO2) nanoparticles against mace of nutmeg in Myristica fragrans houtt, rhizomes of Curcuma longa linn and Kaempferia galanga extracts. *Results in Chemistry*, *7*(janvier), 101291. DOI: 10.1016/j.rechem.2023.101291

Chen, J.-L., Zhang, H., Huang, X.-Q., Wan, H.-Y., Li, J., Fan, X.-X., Luo, K. Q., Wang, J., Zhu, X.-M., & Wang, J. (2019). Antiangiogenesis-Combined Photothermal Therapy in the Second Near-Infrared Window at Laser Powers Below the Skin Tolerance Threshold. *Nano-Micro Letters*, *11*(1), 93. DOI: 10.1007/s40820-019-0327-4 PMID: 34138046

Dubey, S. K., Dey, A., Singhvi, G., Pandey, M. M., Singh, V., & Kesharwani, P. (2022). Emerging trends of nanotechnology in advanced cosmetics. *Colloids and Surfaces. B, Biointerfaces*, *214*(juin), 112440. DOI: 10.1016/j.colsurfb.2022.112440 PMID: 35344873

El Alouani, M., Aouan, B., Rachdi, Y., Alehyen, S., El Herradi, E. H., Saufi, H., Mabrouki, J., & Barka, N. (2024). Porous geopolymers as innovative adsorbents for the removal of organic and inorganic hazardous substances: A mini-review. *International Journal of Environmental Analytical Chemistry*, *104*(16), 4784–4796. DOI: 10.1080/03067319.2022.2115891

Evgenidou, E., Chatzisalata, Z., Tsevis, A., Bourikas, K., Torounidou, P., Sergelidis, D., Koltsakidou, A., & Lambropoulou, D. A. (2021). Photocatalytic degradation of a mixture of eight antibiotics using Cu-modified TiO2 photocatalysts: Kinetics, mineralization, antimicrobial activity elimination and disinfection. *Journal of Environmental Chemical Engineering*, *9*(4), 105295. DOI: 10.1016/j.jece.2021.105295

Hashemzadeh, S., Bina, F., Khounsari, H. M., & Hashemzadeh, S. (2024). Nanotechnology in the development of cardiac stents. *Journal of Drug Delivery Science and Technology*, *95*(mai), 105596. DOI: 10.1016/j.jddst.2024.105596

Hossain, N., Mobarak, M. H., Islam, M. A., Hossain, A., Al Mahmud, M. Z., Rayhan, M. T., & Chowdhury, M. A. (2023). Recent development of dental implant materials, synthesis process, and failure – A review. *Results in Chemistry*, *6*(décembre), 101136. DOI: 10.1016/j.rechem.2023.101136

Hsu, C.-Y., Mahmoud, Z. H., Abdullaev, S., Ali, F. K., Naeem, Y. A., Mizher, R. M., & Karim, M. M.. (2024). Nano titanium oxide (nano-TiO2): A review of synthesis methods, properties, and applications. *Case Studies in Chemical and Environmental Engineering*, *9*(juin), 100626. DOI: 10.1016/j.cscee.2024.100626

Jafari, S., Mahyad, B., Hashemzadeh, H., Janfaza, S., Gholikhani, T., & Tayebi, L. (2020). Biomedical Applications of TiO2 Nanostructures: Recent Advances. *International Journal of Nanomedicine*, *15*(mai), 3447–3470. DOI: 10.2147/IJN.S249441 PMID: 32523343

Kawassaki, R. K., Romano, M., Dietrich, N., & Araki, K. (2021). Titanium and Iron Oxide Nanoparticles for Cancer Therapy: Surface Chemistry and Biological Implications. *Frontiers in Nanotechnology*, *3*(septembre), 735434. Advance online publication. DOI: 10.3389/fnano.2021.735434

Kumar, S. (2023). Smart and innovative nanotechnology applications for water purification. *Hybrid Advances*, *3*, 100044.

Kurutas, E. (2016). The importance of antioxidants which play the role in cellular response against oxidative/nitrosative stress: Current state. *Nutrition Journal*, *15*(juillet), 71. Advance online publication. DOI: 10.1186/s12937-016-0186-5 PMID: 27456681

Kusior, A., Wnuk, A., Trenczek-Zajac, A., Zakrzewska, K., & Radecka, M. (2015). TiO2 nanostructures for photoelectrochemical cells (PECs). *International Journal of Hydrogen Energy*, *40*(14), 4936–4944. DOI: 10.1016/j.ijhydene.2015.01.103

Lambin, P. (2024). Feature of Review Papers in Nanotechnology and Applied Nanosciences. *Applied Sciences (Basel, Switzerland)*, *14*(5), 1989. DOI: 10.3390/app14051989

Liu, W., Wang, X., & Feng, Y. (2023). Restoring Endothelial Function: Shedding Light on Cardiovascular Stent Development. *Biomaterials Science, 11*(12), 4132–4150. DOI: 10.1039/D3BM00390F PMID: 37161519

Luksiene, Z. 2017. « 16 - Nanoparticles and their potential application as antimicrobials in the food industry ». In Food Preservation, édité par Alexandru Mihai Grumezescu, 567-601. Nanotechnology in the Agri-Food Industry. Academic Press. DOI: 10.1016/B978-0-12-804303-5.00016-X

Lungu, M., Gavriliu, Ş., Enescu, E., Ion, I., Brătulescu, A., Mihăescu, G., Măruţescu, L., & Chifiriuc, M. C. (2013). Silver–Titanium Dioxide Nanocomposites as Effective Antimicrobial and Antibiofilm Agents. *Journal of Nanoparticle Research, 16*(1), 2203. DOI: 10.1007/s11051-013-2203-3

Ma, F., Zhang, Q., & Zhang, C. (2020). Nanomaterial-Based Biosensors for DNA Methyltransferase Assay. *Journal of Materials Chemistry. B, 8*(16), 3488–3501. DOI: 10.1039/C9TB02458A PMID: 32095792

Mabrouki, J., Benbouzid, M., Dhiba, D., & El Hajjaji, S. (2022). Simulation of wastewater treatment processes with Bioreactor Membrane Reactor (MBR) treatment versus conventional the adsorbent layer-based filtration system (LAFS). *International Journal of Environmental Analytical Chemistry, 102*(19), 7458–7468. DOI: 10.1080/03067319.2020.1828394

Mao, X., & Hao, C. (2024). Recent Advances in the Use of Composite Titanium Dioxide Nanomaterials in the Food Industry. *Journal of Food Science, 89*(3), 1310–1323. DOI: 10.1111/1750-3841.16968 PMID: 38343295

Meghavathi, P., & Vijayalakshmi, U. (2024). HAp/TiO2 Composite Coatings and its Effective Use in Biomedical Applications. *Trends in Biomaterials & Artificial Organs, 38*(1), 5–13.

Mehta, M., & Bhushan, I. (2024). Potential of biosynthesized titanium dioxide nanoparticles towards wastewater treatment and antimicrobial activity. 3 Biotech, 14(3), 66.

Meng, M., Wang, J., Huang, H., Liu, X., Zhang, J., & Li, Z. (2023). 3D printing metal implants in orthopedic surgery: Methods, applications and future prospects. *Journal of Orthopaedic Translation, 42*(septembre), 94–112. DOI: 10.1016/j.jot.2023.08.004 PMID: 37675040

Merugu, R., Sharma, S., & Rastogi, M. 2023. « An overview of nanotechnological applications in biomedical engineering ». Materials Today: Proceedings, International Conference on Nanotechnology for Sustainable Living and Environment (ICON NSLE), 76 (janvier):412-17. DOI: 10.1016/j.matpr.2022.12.028

Mgalaa, S., Mabrouki, J., Elouardi, M., El Azzouzi, L., Moufti, A., El Hajjaji, S., EL Azzouzi, M., & El Belghiti, M. A. (2022). Study and evaluation of the degradation of procion blue dye by the ozonation method: Parametric and isothermal study. *Nanotechnology for Environmental Engineering, 7*(3), 691–697. DOI: 10.1007/s41204-022-00262-0

Moon, K.-S., Bae, J.-M., Oh, S., & Choi, E.-J. (2024). Enhanced photocatalytic antibacterial activity of Au-coated Ni–Ti–O nanotubes for biomedical applications. *Materials Chemistry and Physics, 314*(février), 128910. DOI: 10.1016/j.matchemphys.2024.128910

Mostakhdemin, M., Nand, A., & Ramezani, M. (2021). Articular and Artificial Cartilage, Characteristics, Properties and Testing Approaches—A Review. *Polymers, 13*(12), 2000. DOI: 10.3390/polym13122000 PMID: 34207194

Panahi, K. S., Hamed, H. H.-B., Dehhaghi, M., Orooji, Y., Mahian, O., Shahbeik, H., & Kiehbadroudinezhad, M.. (2024). Nanotechnology applications in biodiesel processing and production: A comprehensive review. *Renewable & Sustainable Energy Reviews, 192*(mars), 114219. DOI: 10.1016/j.rser.2023.114219

Pourmadadi, M., Rajabzadeh-Khosroshahi, M., Eshaghi, M. M., Rahmani, E., Motasadizadeh, H., Arshad, R., Rahdar, A., & Pandey, S. (2023). TiO2-based nanocomposites for cancer diagnosis and therapy: A comprehensive review. *Journal of Drug Delivery Science and Technology, 82*(avril), 104370. DOI: 10.1016/j.jddst.2023.104370

Pulit-Prociak, J., Długosz, O., Staroń, A., Radomski, P., Domagała, D., & Banach, M. (2023). In vitro studies of titanium dioxide nanoparticles modified with glutathione as a potential drug delivery system. *Nanotechnology Reviews, 12*(1), 20230126.

Rachiq, T., Abrouki, Y., Mabrouki, J., Samghouli, N., Fersi, C., Rahal, S., & El Hajjaji, S. (2021). Evaluation of the efficiency of different materials to remove specific pollutants from landfill leachate. *Desalination and Water Treatment, 238*, 240–250. DOI: 10.5004/dwt.2021.27779

Ramezani, M., & Ripin, Z. M. (2023). An Overview of Enhancing the Performance of Medical Implants with Nanocomposites. *Journal of Composites Science, 7*(5), 199. DOI: 10.3390/jcs7050199

Rana, A., Pathak, S., Kumar, K., Kumari, A., Chopra, S., Kumar, M., Kamil, D., Srivastava, R., Kim, S.-K., Verma, R., & Sharma, S. N. (2024). Multifaceted Properties of TiO2 Nanoparticles Synthesized Using Mangifera Indica and Azadirachta Indica Plant Extracts: Antimicrobial, Antioxidant, and Non-Linear Optical Activity Investigation for Sustainable Agricultural Applications. *Materials Advances, 5*(7), 2767–2784. DOI: 10.1039/D3MA00414G

Sadegh, N., Dehcheshmeh, I. M., & Sadegh, F. (2024). Review of zeolitic imidazolate framework/graphene oxide: A synergy of synthesis, properties and function for multifaceted applications in nanotechnology. *FlatChem, 44*(mars), 100618. DOI: 10.1016/j.flatc.2024.100618

Samhitha, S. S., Raghavendra, G., Quezada, C., & Bindu, P. H. (2022). Green synthesized TiO2 nanoparticles for anticancer applications: Mini review. *Materials Today: Proceedings, 54*, 765–770.

Song, F. X., Xu, X., Ding, H., Yu, L., Huang, H., Hao, J., & Zhang, S. (2023). Recent progress in nanomaterial-based biosensors and theranostic nanomedicine for bladder cancer. *Biosensors (Basel), 13*(1), 106.

Suresh, S., Imteyaz, S., Murugan, B., Lett, A., Sridewi, N., Kassegn, G., Fatimah, I., & Oh, W.-C. (2022). A comprehensive review on green synthesis of titanium dioxide nanoparticles and their diverse biomedical applications. *Green Processing and Synthesis, 11*(janvier), 44–63. DOI: 10.1515/gps-2022-0005

Tiwari, H., Rai, N., Singh, S., Gupta, P., Verma, A., Singh, A. K., Kajal, , Salvi, P., Singh, S. K., & Gautam, V. (2023). Recent Advances in Nanomaterials-Based Targeted Drug Delivery for Preclinical Cancer Diagnosis and Therapeutics. *Bioengineering (Basel, Switzerland), 10*(7), 760. DOI: 10.3390/bioengineering10070760 PMID: 37508788

Ullah, M., Bibi, A., Wahab, A., Hamayun, S., Rehman, M. U., Khan, S. U., Awan, U. A., Riaz, N., Naeem, M., Saeed, S., & Hussain, T. (2024). Shaping the Future of Cardiovascular Disease by 3D Printing Applications in Stent Technology and its Clinical Outcomes. *Current Problems in Cardiology*, *49*(1, 1, Part A), 102039. DOI: 10.1016/j.cpcardiol.2023.102039 PMID: 37598773

Upadhyay, A., Pradhan, L., Yenurkar, D., Kumar, K., & Mukherjee, S. (2024). Advancement in ceramic biomaterials for dental implants. *International Journal of Applied Ceramic Technology*, *21*(4), 2796–2817.

Wei, G., Wang, Y., Yang, G., Wang, Y., & Ju, R. (2021). Recent progress in nanomedicine for enhanced cancer chemotherapy. *Theranostics*, *11*(13), 6370–6392. DOI: 10.7150/thno.57828 PMID: 33995663

Yin, Z. F., Wu, L., Hua, G. Y., & Yong, H. S. (2013). Recent Progress in Biomedical Applications of Titanium Dioxide. *Physical Chemistry Chemical Physics*, *15*(14), 4844–4858. DOI: 10.1039/c3cp43938k PMID: 23450160

Zarzzeka, C., Goldoni, J., Marafon, F., Sganzerla, W. G., Forster-Carneiro, T., Bagatini, M. D., & Leda, M. S. C. (2023). Use of titanium dioxide nanoparticles for cancertreatment: A comprehensive review and bibliometric analysis. *Biocatalysis and Agricultural Biotechnology*, *50*(juillet), 102710. DOI: 10.1016/j.bcab.2023.102710

Zhang, J., Lei, J., Liu, Z., Chu, Z., & Jin, W. (2022). Nanomaterial-based electrochemical enzymatic biosensors for recognizing phenolic compounds in aqueous effluents. *Environmental Research*, *214*(novembre), 113858. DOI: 10.1016/j.envres.2022.113858 PMID: 35952740

Zhang, L., & Jin, Z. (2024). Antibacterial Activities of Titanium Dioxide (TiO2) Nanotube with Planar Titanium Silver (TiAg) to Prevent Orthopedic Implant Infection. *Journal of Orthopaedic Surgery and Research*, *19*(1), 144. DOI: 10.1186/s13018-024-04596-0 PMID: 38365803

Ziental, D., Czarczynska-Goslinska, B., Mlynarczyk, D. T., Glowacka-Sobotta, A., Stanisz, B., Goslinski, T., & Sobotta, L. (2020). Titanium Dioxide Nanoparticles: Prospects and Applications in Medicine. *Nanomaterials (Basel, Switzerland)*, *10*(2), 387. DOI: 10.3390/nano10020387 PMID: 32102185

## APPENDIX

*Figure 1. Biomedical applications of TiO2.*

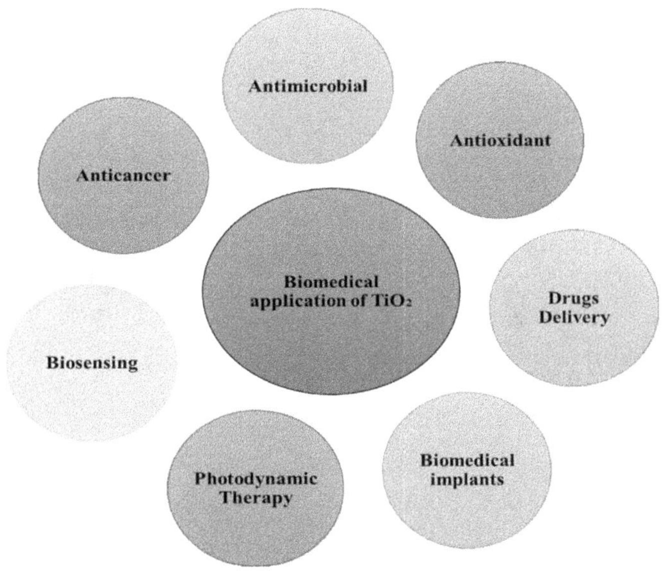


*Figure 2. Strategies of $TiO_2$ in executing antimicrobial activities.(Suresh et al. 2022)*

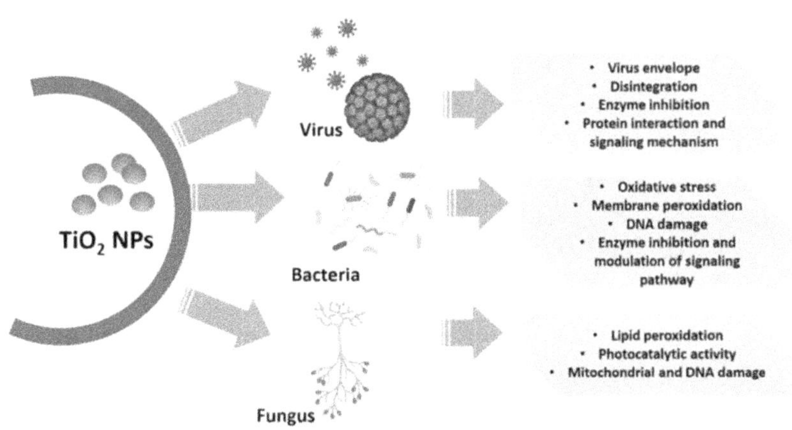

*Figure 3. Scheme of drug delivery system utilizing $TiO_2$ NPs: the nanoparticles are conjugated with desired drug, internalized by cells, inducing DNA damage, and activating the apoptotic pathway.*

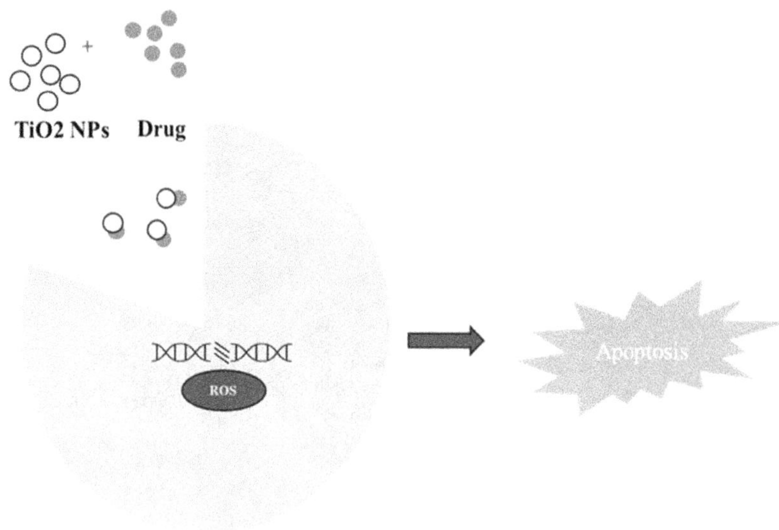


*Figure 4. The schematic representation of the biosensor PEC process.*

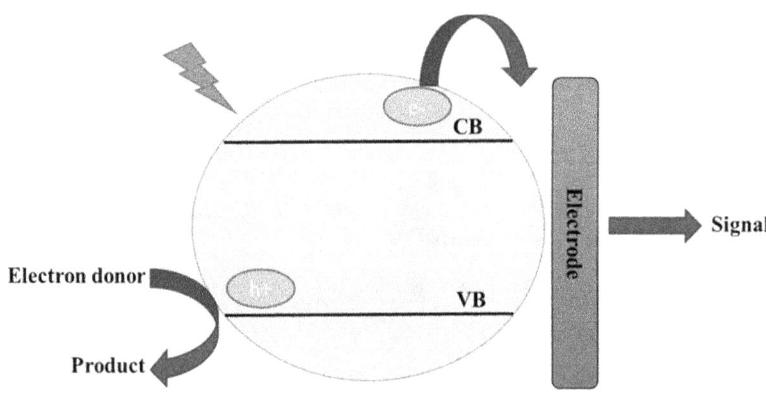

*Figure 5. Representation of PDT protocol for the treatment of tumor.*

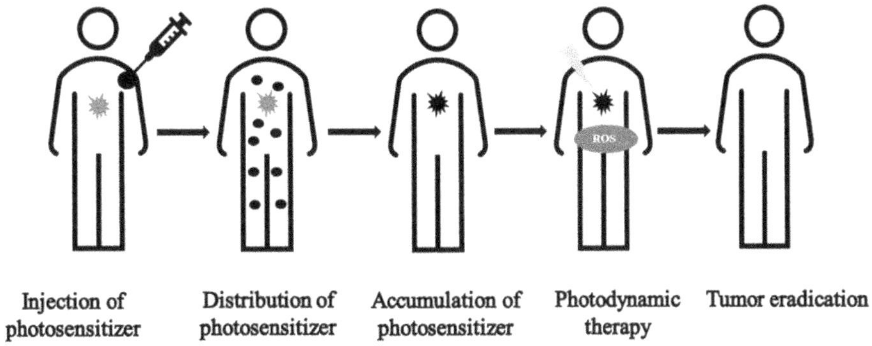

# Chapter 6
# Contribution to the Study of Some Gasteroid Macromycetes Species of Morocco

**Nour Id Elhakkar**
https://orcid.org/0009-0009-8521-5230
*Ibn Tofail University, Morocco*

**Abdesslem Bouchar**
*Faculty of Sciences Rabat, Morocco*

**Ali Outcoumit**
*Faculty of Sciences Rabat, Morocco*

**Soumaya El Gabardi**
*Faculty of Sciences Rabat, Morocco*

**Abdelkrim El Assfouri**
*Faculty of Sciences Rabat, Morocco*

**Karima Selmaoui**
*Faculty of Sciences Rabat, Morocco*

**Moulay Abdelaziz El Alaoui**
*Faculty of Sciences Rabat, Morocco*

**Amina Ouazzani Touhami**
https://orcid.org/0000-0001-9970-7724
*Faculty of Sciences Rabat, Morocco*

**Allal Douira**
*Faculty of Sciences Rabat, Morocco*

## ABSTRACT

*This study investigates the gasteroid species within the families Sclerodermataceae and Rhizopogonaceae, focusing on species with significant ecological and economic roles. The species included in the analysis are Rhizopogon luteolus Fr. and Nordhom (1817), Pisolithus microcarpus (Cooke & Massee) G. Cunn. (1931), Scleroderma citrinum Pers. (1801), and Scleroderma meridionale Demoulin & Malençon (1971). These species are part of the larger group of Basidiomycetes, which are classified under the order Boletales. The taxonomic positioning of these species has been discussed in several studies, highlighting their placement within the broader taxonomic framework of the Basidiomycetes.*

DOI: 10.4018/979-8-3693-8980-5.ch006

## INTRODUCTION

The Order of Boletales, defined by E.-J. Gilbert in 1931, includes 19 families. These include the Amylocorticiaceae, Boletaceae, Calostomataceae, Coniophoraceae, Diplocystidiaceae, Gastrosporiaceae, Gasterellaceae, Gomphidiaceae, Gyroporaceae, and These genera are Coniophoropsis Hjortstam and Ryvarden 1986, Corditubera Henn. In 1897, Corneomyces was described by Ginns and Jaapia was described by Bres. In 1911, Marthaella States and Fogel 1999 published their findings, which were later corroborated by Phaerodulum Pat.1900 (Kirk et al., 2008).

In this study, the gasteroid species belonging to the families Sclerodermataceae and Rhizopogonaceae were investigated. The following gasteroid species were also included in the study: *Rhizopogon luteolus* Fr. and Nordhom 1817, *Pisolithus microcarpus* (Cooke and Massee) G. Cunn. In 1931, *Scleroderma citrinum* Pers. 1801, and Scler*oderma meridionale* Demoulin and Malençon 1971.

## METHODOLOGY

Our prospections, carried out between 2003 and 2020 in various forests in Morocco, have enabled us to harvest basidiocarps from gasteroids belonging to Basidiomycetes fungi. The basidiocarps were photographed and some macroscopic details were noted. The basidiocarps were then taken to the laboratory to be fully described, including their shape, colour and other characteristics. This study was further enhanced by a microscopic description of the spores and cuts at the hymenium, cuticle, flesh and stem levels. Microscopic observations were made with tap water (Ajana et al. (2020a, 2020b); Lotfi et al., (2019, 2020); Nmichi et al. (2021) El Kholfy et al. (2021a and b); El Assfouri et al., 2024 ; Bouchar et al. 2023). The species was determined using the keys of determination (Smith 1964; Demoulin, 1968 and 1974; Guzmán, 1970; Hosford and Trappe, 1980; Miller, 1986; Agerer, 1990; Pulcinelli and Migliozzi, 1997; Mornand, 1988 and 1999; Guzmán et al., 1997; Pulcinelli and Migliozzi, 1998). The following sources were consulted: Diez *et al.,* 2001; Coccia *et al.,* 1990; Calonge, 1998; Jeppson, 1998; Poumarat, 1999 and 2003; Jeppson and Piatek, 2005; Yousaf *et al.,* 2012; Rusevska e*t al.*, 2014.

## RESULTS

### Rhizopogon luteolus Fr. and Nordhom 1817, harvest of 23 January 2018 in Sidi Bouknadel (Mamora forest) (Fig. 1)

The fertile portion initially exhibits a white coloration, which subsequently transitions to pinkish yellow, reddish, and finally, brownish hues. The olfactory perception is that of a fruity aroma, while the gustatory experience is that of a sweet taste. The application of potassium (10%) and ammonia had no effect on the staining of sporiferous tissues and the dyeing of spores, which remained ochraceous brown. The spores (5-7 x 2 to 2.5 µm) are smooth and oblong to elliptical in shape.The fructifications, which measure between 1.5 and 5 cm, initially exhibit hypogynous growth, subsequently undergoing partial semi-epigynous to epigynous development, and frequently displaying bosses or deep folds. The shape of these fructifications is irregular, typically oval, reniform, or globular. The peridium initially exhibits a whitish coloration, ranging from pale yellow to ochraceous, and displays a network of white rhizomorphs that are appressed. With age, these undergo a change in colour, becoming reddish or reddish brown. Subsequently, the coloration of the fructification undergoes a transformation, shifting from a yellowish hue to a chamois appearance, then to a yellow-brown tone, and finally to a pinkish brown. The fertile part is first white and becomes pinkish yellow, reddish and in late brownish. The smell is fruity and the flavor is sweet. Potash (10%) and ammonia had no effect on the staining of sporiferous tissues and dyed spores in olive to ochraceous brown. Spores (5-7 x 2 to 2.5 µm) are smooth and oblong to elliptical.

*Figure 1. Rhizopogon luteolus Fr. and Nordhom: A yellowish white reniform basidioma, in situ(a); Basidioma with long rhizomorph (b); sporophores of different shapes and colours (greenish yellow, yellowish or pinkish) and fertile brown parts after maturation (c and d); pleated and bumped peridium surface; longitudinal cut before ripening, whitish fertile part and vinous red coloring under exoperidium (f); boletoid spores in water (x 400).*

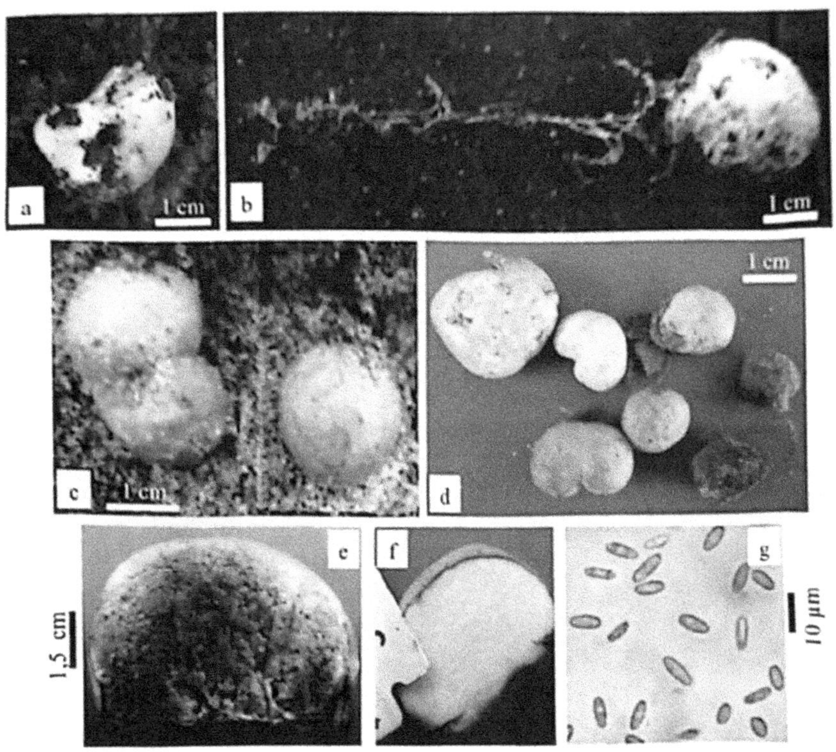


## Sclerodermataceae Corda 1842

### Pisolithus Alb. and Schwein. 1805

*Pisolithus microcarpus* (Cooke and Massee) G. Cunn. In 1931, the harvest was conducted on 15 February 2009 and 2019 under the Eucalyptus sp. in the Lalla Mimouna region between December and April (FIG. 2). The basidiomata, which measure between 5 and 12 cm on average (occasionally exceeding this range), are semi-hypogent, globular, oval or irregular in shape. The basidiomata are characterised by a grey-brown, chocolate brown or blackish colouration. The peridium is thin. The gleba is initially formed of shiny logettes. Upon reaching maturity, the peridium tears irregularly at the top, releasing the spores. The spore is olive green in colour. The interior of the hyphae of the gleba, including the basides and ornaments of the spores, is amyloid. The spores (7-7.5 x 7.5-7.75 µm) are globular in shape and exhibit characteristics of both ovoids and wiggles, with taste buds surrounding them. In aqueous solutions, the spores are brown, while in iodinated solutions, their amyloid ridges appear green.

*Figure 2. Pisolithus microcarpus (Cooke and Massee) G. Cunn. Basidiomas on land in situ, at the foot of a tree of Eucalyptus sp. (a); Two extracted basidiomas showing the variation in pseudostipe length as a function of substrate depth (b and c); Longitudinal section of a specimen: Spore maturation occurs from the top to the base (d); spores in water and iodinated water (x 400), respectively, the spore ornaments are amyloid. (The specimens presented were encountered near Lalla Mimouna, the Gharb).*

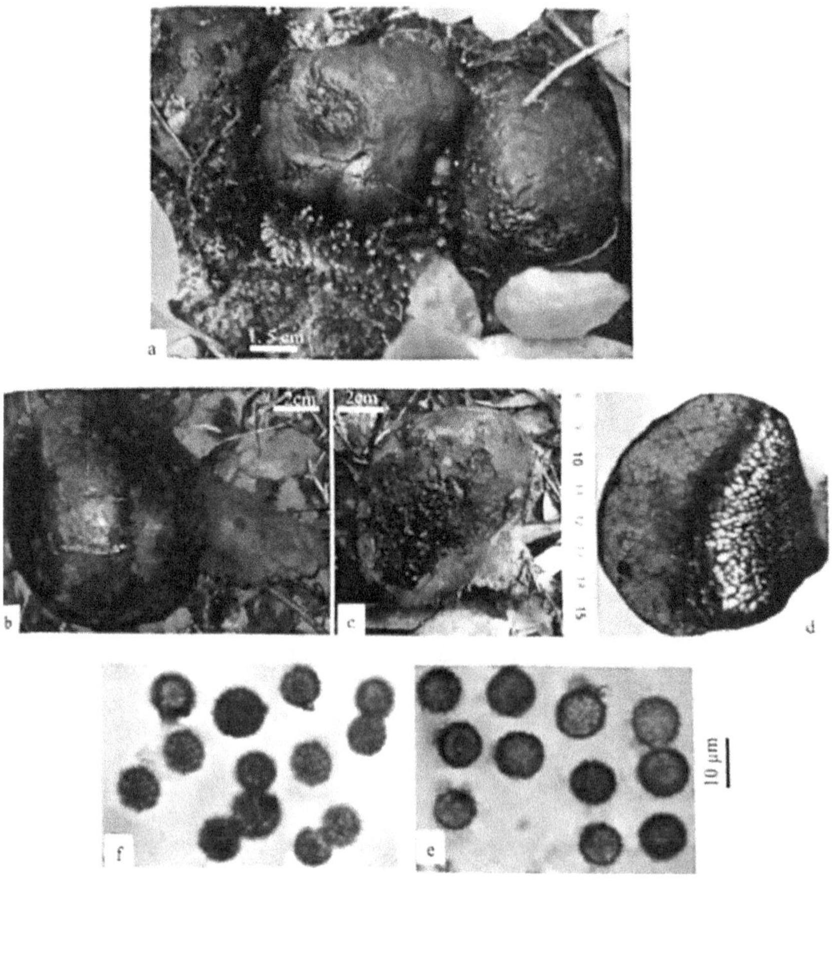

## Scleroderma Pers.

### Scleroderma bovista Fr. 1829

Harvest of November 25, 2010 and December 03, 2018, under *Quercus suber* in the Mamora (Fig. 3). Basidiomas of Scleroderma bovista (3 to 5 cm) are globular and generally without foot. They are grey-beige, orange grey yellowish to brownish red. The surface is more or less smooth or scaly-flaky. The thickness of the peridium is approximately 1 mm thick. The surface of the basidioma turns reddish under the effect of KOH (10%). The peridium also turns red at the cup. The spores (8-12 µ m) are provided with thorns (1 µ m) and roughly crosslinked.

*Figure 3. Scleroderma bovista Fr.: basidioma with pseudostipe and white rhizomorphs in situ (a); brownish yellow and reddening basidioma surface under the effect of KOH concentrate (b); Perimeter thickness that reddens at cutting and glèba masture (c); spores with ridges (d) and crosslinked spores (e).*

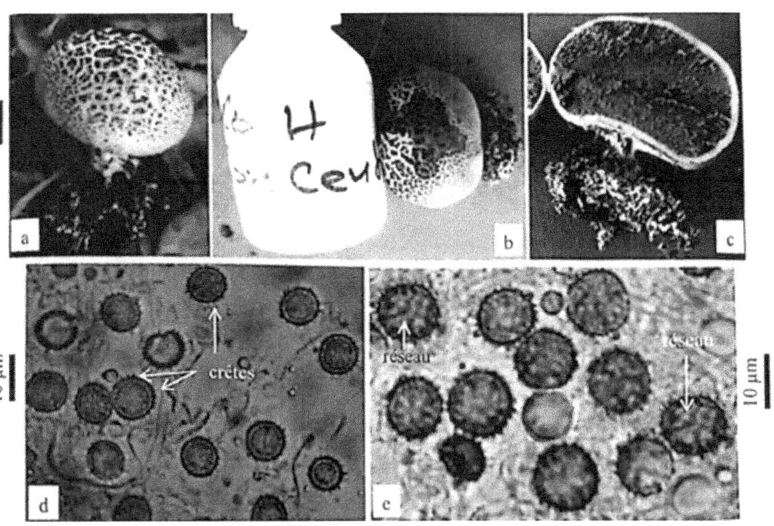


## Scleroderma Pers.

The harvests of 22 November 2006 and 28 December 2020 were obtained from sandy terrain in the Mamora. The basidiomata of *Scleroderma citrinum* (510 (20 cm)) are globose or ovoid in shape, yellow in colour, and have an egg yolk-like appearance, ranging from lemon yellow to brownish ochre. It is possible that the stipulations ma vary in degree. The peridium is thick, measuring between 2 and 3 millimeters. The surface of the peridium may be smooth, scaly, or roughly partitioned.The spores (8-13 µm) are wart-like and irregularly cross-linked.

## Southern Scleroderma Demoulin anD Malençon 1971

The harvests were conducted on the 13th of February, 2010, and the 24th of January, 2018, under Pinus pinea in the Mamora (**Figure 4**).

The basidiomata (3 (8) 15 cm) are ovoid in shape and initially whitish in colour, subsequently becoming dirty beige, yellowish to yellowish ochre, and occasionally exhibiting brownish spots. The pseudo stipe (5-6 cm) can be of varying length, dependent on the depth of the substrate. It is typically well-defined. It can be united or composed of elements that are grouped and prolonged in the soil by whitish rhizomorphs. The peridium is thick (3–6 mm) and its surface is smooth. When ripe, it opens irregularly. The surface of the peridium exhibits a dark brownish red colouration under the influence of 10% potassium hydroxide (KOH). The mycelial filaments of the glebe are divided into distinct segments. Upon reaching maturity, the glebe assumes a purple-brown hue, which subsequently darkens to a uniform purple-brown colour. The mass spore is initially purple brown in colour, but subsequently turns black. The spores (10-12.5 µm), which are coarsely crosslinked, are globular to ovoid in shape and have peaks of 0.5-2 µm. In aqueous solution, which are coarsely crosslinked, are globular to ovoid in shape and have peaks of 0.5-2 µm. In aqueous solution, they exhibit a dark brown colour.

*Figure 4. Scleroderma meridionale Demoulin and Malençon: Spherical basidiome stipulated on sandy soil, pseudostipe with white rhizomorphs (a); stipulated knots (b); segregated gléba (c); thick peridium (2 to 3 mm) (d); spiny and crosslinked spores (x 400) in water and pattern of a spore (f).*

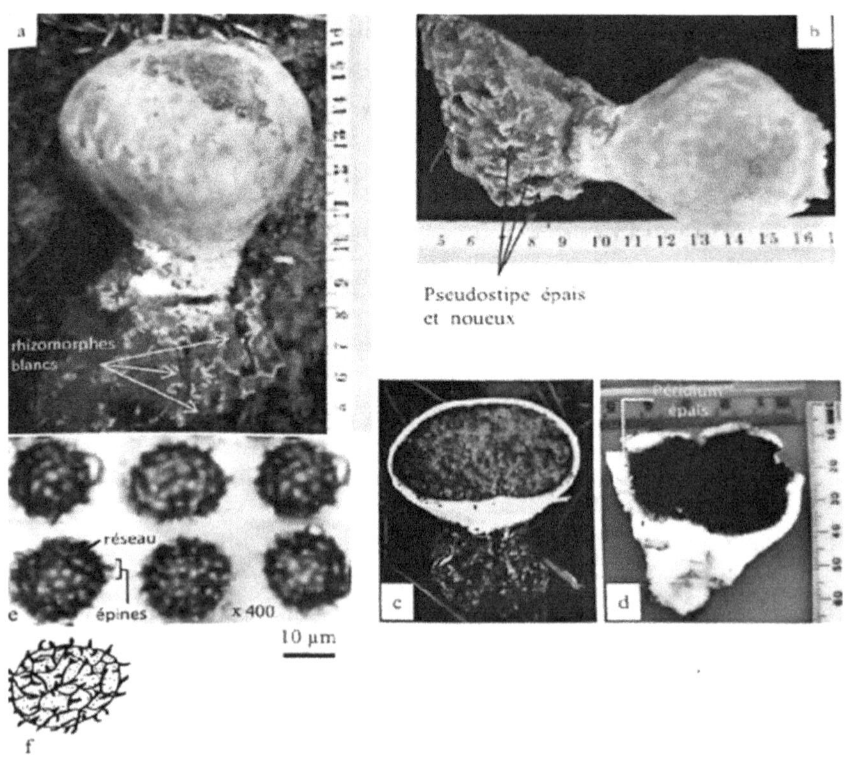

## DISCUSSION

***Scleroderma bovista****(S. bovista* Fr. (= *S. fuscum* (Corda) E. Fisch S. texense Berk. S. verrucosum subsp. Bovista (Fr.) Šebek *S. verrucosum* var. *bovista* (Fr.) Sebek) (Sclerodermataceae, Boletales) (Binder and Bresinsky 2002) is a species that occurs on sandy soils in open environments and under *Q. Suber*. *Scleroderma verrucosum*, is a close species that also occurs in the Mamora. It is distinguished from *S. bovista* by its perimeter, which is brownish-coloured under the effect of KOH (10%) and its thorny but non crosslinked spores. However, both taxa are sometimes considered synonymous (Kirk *et al.* 2008) *Scleroderma bovista* and *S. verrucosum* were described by Ait Aguil (2005). Scleroderma bovista is considered a mycorrhiziene species (Rinaldi *et al.,* 2008; Walbert, 2008).

The genus Pisolithus Alb. and Schwein. has been classified within the family Lycoperdaceae (Lycoperdales, Gasterromycetes, as defined by Clements, 1954). The new system, based on the results of molecular biology, has transferred it to the Pisolithaceae family and then to that of the Sclerodermataceae (Boletales, Agaricomycetes) (Binder and Bresinsky, 2002). In the 10th edition of the Dictionary of the Fungi (Kirk et al., 2008), Pisolithus is classified within the Sclerodermataceae Corda (Boletales). The latter family, Sclerodermataceae, consists of the genera Chloragaster Læssøe and Jalink 2004, Favillea Fr. 1848, Horakiella Castellano and Trappe 1992, and Pisolithus Alb. & Schwein. 1805 and Scleroderma Pers. (Cannon and Kirk, 2007). The 23 species of the genus Pisolithus are associated with tree roots by mycorrhizal symbiosis. *Pisolithus microcarpus* is considered to be an ectomycorrhizal species of Eucalyptus and Pinus (Aouadj et al., 1997; Hitchcock, 2010; Nounsi et al., 2014*). Pisolithus microcarpus* is a species that occurs in abundance under the Eucalyptus Mountains in the Mamora Forest, Sidi Yahia El- Gharb region, near Moulay Boussalham, Lalla Mimouna, Khmiss-es-Sahel (Larache region). Pisolithus microcarpus is also known as P. tinctorius (Mont.). (P. arenarius

Alb. and Schwein = *P. arhizus* (Scop.) Rauschert) are two ectomycorrhizal species with the former species associated with Eucalyptus and the latter with cork oak. In the Mamora, *P. tinctorius* is found under Quercus suber, and it was also observed in April 2007 in Dardara (Chefchaouenne region) and Khmiss-es-Sahel. *P. tinctorius* was described by Ait Aguil et al. (2002) and El-Assfouri et al. (2006).

*Scleroderma bovista* (S. bovista Fr. (= S. fuscum (Corda) E. Fisch S. texense Berk. S. verrucosum subsp. Bovista (Fr.) Šebek S. verrucosum var. bovista (Fr.) Sebek) (Sclerodermataceae, Boletales) (Binder and Bresinsky 2002) is a species that occurs on sandy soils in open environments and under Q. Suber. Scleroderma verrucosum, is a close species that also occurs in the Mamora. It is distinguished from S. bovista by its perimeter, which is brownish-coloured under the effect of KOH (10%) and its thorny but non crosslinked spores. However, both taxa are sometimes considered synonymous (Kirk et al. 2008) Scleroderma bovista and S. verrucosum were described by Ait Aguil (2005). Scleroderma bovista is considered a mycorrhiziene species (Rinaldi et al., 2008; Walbert, 2008).

*Scleroderma citrinum* Pers1801 is considered to be synonymous with S. *vulgare* Hornem. According to Kirk et al. (2008), 1829 is a common species in the Mamora. It is found in sandy terrain and clearings of Quercus suber. It is a mycorrhizal species of oak (Garbaye et al., 2004). Additionally, Sclerderma citrinum was described in the Mamora by Ait Aguil et al. (2002).

*Scleroderma meridionale* Demoulin and Malençon was first described in Portugal in 1970 by Demoulin and Malençon (Demoulin, 1974). It is a relatively common occurrence in Europe and North Africa. It is a southern species (Richter 2004).In Morocco, *S. meridionale* is known to occur in the Lalla Mimouna region (Bertalt 1976) and in the Mamora (Ait Aguil et al. 2002). In the latter southern region, S. meridionale is relatively common from autumn to spring. Indeed, it can be encountered in bare terrain, under *Quercus suber*, and is particularly abundant under *Pinus pinea*. Northern Scleroderma, which is found in the United States (in Florida and Arizona), is similar in appearance to *S. meridionale*. It differs in its western geographic distribution, the colour of its rhizomorphs, and the length of its spore spines, which are slightly shorter at 1-2 µm (Guzman and Ovrebo 2000; Richter, 2000). It is likely that Southern Scleroderma is mycorrhizal with pines (Phosri et al. 2009). The condition known as Southern scleroderma Demoulin and Malençon was first described in Portugal in 1970 by Demoulin and Malençon (Demoulin, 1974).

## Observations on the Family Sclerodermataceae and on the Genus Scleroderma

Basidiomas of Sclerodermataceae are rarely hypoginated, often more or less spherical without a stipe or extended by a root foot. The peridium is typically uncomplicated and rarely comprises two distinct layers. The tissue is firm, membranous, and rarely thin. The peridium opens by an irregular rupture. The glebe is typically well-defined and subdivided into sectors, which areisolated by sterile veins. In its juvenile stage, the fungus is white in colour, whereas in its mature state it assumes a brown, dark purple or purple hue. Basidia are formed in a dispersed manner within the sectors. The spores are released in the form of a powder mass. The spores are approximately spherical in shape, brown in colour, and possess thick walls. They are characterised by the presence of spines or warts with a crosslinked appearance (Hawksworth et al., 1995; Guzmán et al., 2000; Phosri et al., 2009). The family Sclerodermataceae is comprised of four genera: Chlorogaster Læssøe and Jalink 2004, Favillea Fr. 1848, Horakiella Castellano and Trappe 1992 and Pisolithus Alb. and Schwein. 1805. The type genus is Scleroderma Pers. (Kirk et al., 2008). The majority of Sclerodermataceae species are mycorrhizal (Phosri et al., 2009).

The present study focused on the examination of one species of the genus Pisolithus and three species of the genus Scleroderma. The genus Scleroderma, as defined by Persoon in 1801, is a very large one, with a number of species that is far from being precisely defined. The authors posit that this number is in the range of 25 to 30 (Guzmán, 1970; Baseia and Milanez, 2000;

Gurgel et al., 2008; cited by Phosri et al., 2009). In the tenth edition of Index Fungorum (Kirk et al. 2008), following the removal of species that have been transferred to other families, the

number of taxa in this genus is approximately 164. The species of the genus Scleroderma are found in a wide range of habitats, including temperate and tropical regions (Guzmán 2000; Phosri et al., 2009). The majority of species in this genus are considered ectomycorrhizal, with a wide range of forest trees (Nounsi et al., 2014). To date, ten species of this genus have been reported in Morocco. Four of these species have been described. S. aurantium (L.) Pers., S. sud Demoulin and Malençon, and S. verrucosum (Bull.) Ait Aguil et al. (2002) described Pers. and S. citrinum Pers. and S. verrucosum (Bull.). Pers. have been described by El-Assfouri et al. (2006). All of these species were discovered in the Mamora. Other species, including S. aurantium var. spadiceum (Schaeff.) Šebek, S. bovista Fr., S. cepa Pers., S. geaster (J.F. Gmel.) Pers., S. vulgare and Scleroderma sp., were only reported by Bertault (1979). This work represents the first description of S. bovista in Morocco. Prior to the advent of molecular biology in systematics, the criteria that have been widely adopted to distinguish species of the genus Scleroderma are basidioma shape, perimeter thickness, perimetry, the presence or absence of a pseudo stipe, and the anatomy of the spore ornamentation (Gibson 2003 In light of the aforementioned criteria, Guzmán (1970), as cited by Phosri et al. (2009), proposed the subdivision of the genus Scleroderma into three distinct "groups" or subgenera.

The species in question are characterized by spores that are spinous in nature, with spines that are not connected by a network. The genus is represented by Scleroderma albidum Pat. and

Trab. In 1899, S. areolatum Ehrenb. was described. In 1818, S. cepa Pers. was first described.

In 1801, S. laeve Lloyd 1916 and S. verrucosum (Bull.) were also identified. Pers.

- Sub-crosslinked species: an incomplete network exists between thorns. This group is represented by *S. floridanum* Guzmán 1967 and S. polyrhizum (J.F. Gmel.) Pers. 1801.

The species with crosslinked spores, which have a complete network connecting the thorns, are as follows: This section is represented by S. bovista (Fr.) Šebek 1953 and S. fuscum (Corda) E. Fisch. In 1900, S. hypogaeum Zeller 1922 was described, followed by S. meridionale Demoulin and Malençon in 1923.

According to Phosri et al. (2009) the results of molecular phylogeny corroborate the hypothesis that the genus Scleroderma is comprised of three subgenera. Indeed, molecular analysis has demonstrated that the genus Scleroderma is represented by three clades, which correspond to the subgenera defined by Guzman (1970). One of these subgenera is represented by the genus

Sclerangium (Lev.), with the species type being Scleroderma polyrhizum (J.F. Gmel.) Pers. The genus Sclerangium was established by Leveillé in 1948, based on the observation that the opening of the star-shaped peridium resembles that of the geaster peridium and that the spores are retained in a mass rather than dispersed in powder form, as is the case in the species of the genus Scleroderma. Phosri et al. (2009) cite Leveillé's work in support of this generic classification. The aforementioned author posits that the results of molecular biology have demonstrated that Scleroderma polyrhizum (J.F. Gmel.) Pers. belongs to the clade with sub- crosslinked spores and should be placed in the genus Scleroderma. In the absence of further evidence, Sclerangium should be returned to its original taxonomic position as a subgenus of Scleroderma

## CONCLUSION

It is important to note three key points. The gasteroid species mentioned by Mayor and Werner (1937) and Malençon and Bertault (1961) in the lists of accounts rendered by the Society of Natural and Physical Sciences of Morocco between 1922 and 1975 were only reported. In essence, Ait Aguil et al. (2002) conducted the inaugural study of this group of fungi at a time when boletoid gasteroid species were still included within the Gasteromycetes class. Following the work of Ait Aguil et al. (2002), there have been significant developments in the field of macromycete fungal systematics, necessitating an update to this study.

In this study, four species, namely *Rhizopogon luteolus, Pisolithus microcarpus*, *Scleroderma bovista*, *S. citrinum* and *S. meridionale*, which were previously classified within the Gasteromycetes, were subjected to investigation among the gasteroid Boletales. The species *Rhizopogon luteolus* represents a new addition to the flora of Mamora. All species presented in this work arc cctomycorrhizal (Phosri et al., 2009; Vama et al., 1998; Baseia and Milanez, 2000).

## REFERENCES


Agerer, R. (1990). Studies on ectomycorrhizae. XXIV. Ectomycorrhizae of Chroogomphus kelveticus and C. rutilus (Gomphidiaceae, Bastdiomycetes) and their relationship to those of Suillus and Rhizopogon. *Nova Hedwigia*, *50*(1-2), 1–63. DOI: 10.1127/nova.hedwigia/50/1990/1

Ait Aguil, F. (2005). Contribution à l'étude de la biodivérsité de la flore fongique des forêts marocaines: cas des Basidiomycètes du Rif Oriental et du Rif Occidental et des Gastéromycètes de la Mamora (Doctoral dissertation, Thèse de Doctorat, Université Ibn Tofail, Faculté des Sciences, Kénitra, Maroc, 154p). [Contribution to the study of the biodiversity of the fungal flora of Moroccan forests: case of Basidiomycetes of the Eastern Rif and the Western Rif and of the Gastromycetes of the Mamora (Doctoral dissertation, Doctoral Thesis, Ibn Tofail University, Faculty of Sciences, Kenitra, Morocco)].

Ait Aguil, F., El Houssni, A., Ouazzani Touhami, A., & Douira, A. (2002). Contribution à la connaissance des Gastéromycètes de la forêt de la Mamora. Annales de la Recherch Forestière, [Contribution to the knowledge of the Gastromycetes of the Mamora forest. Annals of Forest Research], Maroc, 35: 77-87.

Aouadj, R., Es-Sgaouri, A., & Abourouh, M. (1997). Croissance et assimilation des nitrates chez le champignon ectomycorhizien *Pisolithus tinctorius.Revue Marocaine des Sciences Agronomiques et Vétérinaires*, *17*(4), 209–216.

Baseia, I. G., & Milanez, A. I. (2000). – First record of *Scleroderma polyrhizum* Pers.(Gasteromycetes) from Brazil. *Acta Botanica Brasílica*, *14*(2), 181–184. DOI: 10.1590/S0102-33062000000200006

Bertault, R. (1979). Boletus species from Morocco [Fungi, new taxa]. *Bulletin Trimestriel de la Societe Mycologique de France*.

Binder, M., & Hibbett, D. S. (2006). Molecular systematics and biological diversification of Boletales. *Mycologia*, *98*(6), 971–981. DOI: 10.1080/15572536.2006.11832626 PMID: 17486973

Bonjean, A. P., & Fortin, J. A. (2023). Les mycorhizes, interactions plantes-champignons encore sous-exploitées. In Le Déméter 2023 (pp. 251-264). IRIS éditions.

Bouchar, A., Sellal, Z., Maazouzi, S., Msairi, S., El Kholfy, S., Benkirane, R., Ouazzani Touhami, A., & Douira, A. (2023). Inventory of Corticolous Lichens and Argan Wood Fungi. In: Kacprzyk, J., Ezziyyani, M., Balas, V.E. (eds) International Conference on Advanced Intelligent Systems for Sustainable Development. AI2SD 2022. Lecture Notes in Networks and Systems, vol 713. Springer, Cham. DOI: 10.1007/978-3-031-35248-5_47

Calonge, F. D. (1998). Gasteromycetes, I. Lycoperdales, Nidulariales, Phallales, Sclerodermatales, Tulostomales. In *Flora Mycologica Iberica* (Vol. 3). CSIC Real Jardin Botanico, Madrid et J. Cramer.

Cannon, P. F., & Kirk, P. M. (2007). *Fungal families of the world*. Cabi. DOI: 10.1079/9780851998275.0000

Clements. (1954). The genera of fumgi hafner published Co. New York 496 p.

Coccia, M., Migliozzi, V., & Lavorato, C. (1990). Studio sul genere Scleroderma Persoon. *AMER*, *20-21*, 3–59.

Courtecuisse, R., & Duhem, B. (2000). Guide des champignons de France et d'Europe. Ed. Delachaux et Niestlé, S. A. Lausane (Switzerlande), Paris, 476 p.

Demoulin, V. (1968). Gastéromycètes de Belgique: Sclerodermatales, Tulostomatales. *Lycoperdales. Bull. Jard. Bot. Nat. Belge*, *38*(1), 1–101. DOI: 10.2307/3667475

Demoulin, V. (1974). Scleroderma meridionale Demoulin & Malençon, the correct name for the larger Scleroderma of Great Lakes sand dunes. *Michigan Botanist*, *13*, 68–71.

Diez J., Anta B., Monjón J. L. & Honrubia M. (2001). Genetic variability of *Pisolithus* isolates associated with native hots and exotic eucalyptus in the western mediterranean region. New Phytologist, 149: 577-587.p

El-Assfouri, A., Ouazzani-Touhami, A., & Douira, A. (2006). Etudes de quelques Russules de la forêt de la Mamora. *Bull. Soc. Mycol. Strasbourg.*, *95*, 10–16.

El-Assfouri, A., Touhami, A. O., & Douira, A. (2024). A New Mycorrhizal Association of *Quercus Suber* (Morocco): *Lactarius glaucescens* Crossl., 1900. In: Ezziyyani, M., Kacprzyk, J., Balas, V.E. (eds) International Conference on Advanced Intelligent Systems for Sustainable Development (AI2SD'2023). AI2SD 2023. Lecture Notes in Networks and Systems, vol 930. Springer, Cham. DOI: 10.1007/978-3-031-54318-0_1

El Kholfy, S., Nmichi, A., Bouchar, A., Belahbib, N., Ooazzani Touhami, A., & Douira, A. (2021a). First report of *Mycena haematopus* (Pers.) P. Kumm., Führ. Pilzk. (Zerbst) for the fungal flora of Mamora (Morocco). *Plant Cell Biotechnology and Molecular Biology*, *22*(21-22), 34–37.

El Kholfy, S., Nmichi, A., Bouchar, A., Ouabbou, A., Belahbib, N., Ouazzani Touhami, A., & Douira, A. (2021b). Study of three species of Basidiomycetes of the genus *Bolbitius*, two news for the fungal diversity of Morocco: *Bolbitius grandiusculus* and *Bolbitius lacteus.Plant Cell Biotechnology and Molecular Biology*, *22*(43-44), 62–68.

Guzmán, G. (1970). Monografia del genero Scleroderma Pers. emend. *Fr. Darwiniana*, *16*(1-2), 233–407.

Guzmán, G., & Ovrebo, C. L. (2000). New observations on Sclerodermataceous fungi. *Mycologia*, *92*(1), 174–179. DOI: 10.1080/00275514.2000.12061141

Guzmán, G., Tapia, F., Navarro, P., & Ramírez-Guillén, F. J. (1997). Sclerodermataceous fungi (Basidiomycota, Gasteromycetideae) from Mexico and new reports. *Documents Mycologiques*, *26*(104), 53–58.

Hosford, D. R., & Trappe, J. M. (1980). Taxonomic studies on the genus *Rhizopogon*. II. Notes and new records of species from Mexico and Caribbean countries. *Boletin de la Sociedad Mexicana de Miculogia*, *14*, 3–5.

Jeppson, M. (1998). *Scleroderma meridionale*, a new gasteromycete from North-European sand dunes. *Karstenia*, *38*(2), 37–43. DOI: 10.29203/ka.1998.328

Jeppson, M., & Piatek, M. (2005). *Scleroderma meridionale* (Fungi, Badidiomycetes), first records from central Europe. *Polish Botanical Journal*, *50*(1), 15–17.

Kirk, P. M., Cannon, P. F., Minter, D. W., & Stalpers, J. A. (2008). *Ainsworth & Bisby's Dictionary of the Fungi* (10th ed.). CAB International. DOI: 10.1079/9780851998268.0000

Lotfi, N., El Kholfy, S., Ajana, M., Nmichi, A., El Assfouri, A., Ouabbou, A., Ouazzani Touhami, A., Benkirane, R., & Douira, A. (2019). Some data on the almost permanent presence of *Volvariella bombycina* (schaeff.) singer (1951) in the forest of Mamora (Morocco). *Plant Archives*, *19*(2), 3050–3052.

Lotfi, N., El Kholfy, S., Bouchar, A., Nmichi, A., Outcoumit, A., Selmaoui, K., Ouazzani Touhami, A., Benkirane, R., & Douira, A. (2020). *Lepiota pseudofelina* J.E. Lange ex J.E. Lange, A new fungal taxon in Morocco. *Plant Cell Biotechnology and Molecular Biology*, *21*(7&8), 1–5.

Maire R. et Werner R.G., (1937). Fungi Maroccani. Catalogue raisonée des champignons connus jusqu'ici au Maroc. Alem. Soc. Sci. Nat. Maroc., Rabat, 45: 1-147.

Malençon, G. et Bertault R., (1961). Champignons de la région de Tanger (champignons de la région de Tanger: 8ème liste de Tanger). Comptes rendus mensuels des séances de la Société des Sciences Naturelles et Physiques du Maroc, 27(3), 65-67.

Miller, S. L. (1986). Hypogeous fungi from the southeastern United States. I. The genus Rhizopogon. *Mycotaxon*, *27*, 193–218.

Mohamed, A., El-Assfouri, A., El Kholfy, S., Nmichi, A., Touhami, A. O., Benkirane, R., & Douira, A. (2020). A new decomposer of Quercus suber (Morocco): Gymnopilus liquiritiae (Pers.) P. Karsten (1879).

Mornand, J. (1988). Les Gastéromycètes de France. (4. Sclerodermatales). *Documents Mycologiques*, *18*(72), 1–11.

Mousain, D., Matumoto-Pintro, P., & Quiquampoix, H. (1997). Le rôle des mycorhizes dans la nutrition phosphatée des arbres forestiers. *Revue Forestière Française*, *XLIX*(sp), 67–81. DOI: 10.4267/2042/5676

Nmichi, A., El Kholfy, S., Imrani, N., Belahbib, N., Oouazzani Touhami, A., & Douira, A. (2021). Two new species of basidiomycetes for fungal diversity in Morocco: *Cortinarius saturninus* and *Cortinarius leucopus.Plant Cell Biotechnology and Molecular Biology*, *22*(25-26), 72–78. https://ikprress.org/index.php/PCBMB/article/view/6166

Nounsi, A., Outcoumit, A., Selmaoui, K., Ouazzani Touhami, A., Benkirane, R., & Douira, A. (2014). Inventaire des champignons ectomycorhiziens du Maroc. *Journal of Applied Biosciences*, *79*(1), 6826–6854. DOI: 10.4314/jab.v79i1.10

Phosri, C., Martín, M. P., Watling, R., Jeppson, M., & Sihanonth, P. (2009). – Molecular phylogeny and reassessment of some *Scleroderma* spp. (Gasteromycetes). *Anales del Jardin Botanico de Madrid*, *166*(S1), 83–91. DOI: 10.3989/ajbm.2199

Poumarat S., (1999). Clés des espèces européennes du genre *Scleroderm*a Persoon. Bull. FAMM, N. S., 15: 40-46.

Poumarat, S. (2003). Clé des Gastéromycètes épigés d'Europe. Phallales: Geastraceae, Hysterangiaceae, Phallaceae; Agaricales: Lycoperdaceae, Mycenastraceae, Nidulariaceae, Phelloriniaceae, Tulostomataceae; Boletales: Sclerodermataceae (genres sécotioïdes exclus). Monographies Mycologiques de la FAMM, n° 2, 2ème édit. revue et augmentée, Edit. FAMM, Nice, 2003, 100 p

Richter, C., Wittstein, K., Kirk, P. M., & Stadler, M. (2015). An assessment of the taxonomy and chemotaxonomy of Ganoderma. *Fungal Diversity*, *71*(1), 1–15. DOI: 10.1007/s13225-014-0313-6

Richter, D. L. (2004). *Rare and unusual fungi (Basidiomycota) of the Huron mountains, Marquette County, Michigan*. Michigan Publishing, University of Michigan Library.

Rinaldi, A. C., Comandini, O., & Kuyper, T. W. (2008). Ectomycorrhizal fungal diversity: Separating the wheat from the chaff. *Fungal Diversity*, *33*, 1–45.

Rusevska, K., Karadelev, M., Phosri, C., Dueñas, M., Watling, R., & Martín, M. P. (2014). – Rechecking of the genus *Scleroderma* (Gasteromycetes) from Macedonia using barcoding approach. *Turkish Journal of Botany*, *38*, 375–385. DOI: 10.3906/bot-1301-36

Smith, A. H. (1964). Rhizopogon, a curious genus of false truffle. *Michigan Botanist*, *3*, 13–19.

Smith, A. H. (1971). *The boletes of Michigan*. Monograph.

Walbert, K. (2008). Ectomycorrhizal communities associated with a Pinus radiata plantation in the North Island, New Zealand (Doctoral dissertation, Lincoln University).

Yousaf, N., Khalid, A. N., & Niazi, A. R. (2013). New records of Scleroderma species (Sclerodermataceae, Agaricomycetes) from Pakistan. *Mycotaxon*, *122*(1), 43–50.

Chapter 7

# Current Advances and Future of Microalgae-Based Biosensors for the Detection of Pollutants

**Khadija El-Moustaqim**
https://orcid.org/0009-0007-9193-6638
*Ibn Tofail University, Morocco*

**Aziza Lamchaimech**
https://orcid.org/0009-0002-8393-7167
*Mohammed V University in Rabat, Morocco*

**Karima Azoulay**
*Mohammed V University in Rabat, Morocco*

**Jamal Mabrouki**
https://orcid.org/0000-0002-3841-7755
*Faculty of Science Rabat, Morocco*

**Driss Hmouni**
*Faculty of Science Kenitra, Morocco*

## ABSTRACT

*Genetic and biotechnological developments have made it possible to apply synthetic biology approaches to new classes of microbial hosts, such as photosynthetic microalgae. A biosensor is a device that transforms a biological answer into an electrical signal. These devices have the potential to be used in a wide range of applications, including environmental and water quality monitoring. In biosensing applications, microalgae have been widely used as bioreceptors due to their lower susceptibility to physiochemical changes, as well as being used in medical, environmental, industrial, and commercial fields. To overcome the limitations of conventional analytical techniques, new technologies are being introduced based on integrating different types of microorganisms, including microalgae, to build biosensors. Detecting environmental pollutants is one of the main applications of biosensors. Our study highlights the advantages and disadvantages of using microalgae to produce biosensors as a new technology for detecting wastewater pollutants such as heavy metals.*

DOI: 10.4018/979-8-3693-8980-5.ch007

## 1. INTRODUCTION

A biosensor is a device for research that uses an amplifier, a signal processor, and a bioreceptor to identify, communicate, and record any changes in the analytes' physiology or biochemistry (Low et al. 2021; Malhotra et Ali 2018). One crucial component of environmental problems is the health of the aquatic ecosystem. Water is, in fact, a necessary resource for the majority of human endeavors (El-Moustaqim, Benchrifa, et al. 2024; El-Moustaqim, Sbai, El Yousfi, Mabrouki, et al. 2024). Early warning systems for online and in-person pollution monitoring are necessary to protect the quality of the water. Such devices were also employed to find harmful substances in aquatic environments, specifically cadmium ions (Inda et Lu 2020). Applies for biosensors include monitoring the quality of the environment and water, in addition to therapeutic uses. Biosensors are employed because they can be quickly analyzed, are small, inexpensive, simple to use, and offer fast analysis. Biosensors have applications in the food control, medicinal, industrial process control, and environmental monitoring fields (Wasito et al. 2019; Gupta et al. 2019). An optical transducer integrated with a light source can transform a biological event into an electrical signal. In optical biosensors, the transduction mechanism responds to the physical/ chemical changes brought about by the biorecognition events by altering the absorption (El-Moustaqim, Tallou, Mabrouki, Slaoui, et al. 2024; Tallou et al. 2024), transmission, reflection, refraction, phase, amplitude, frequency, or polarization of light (Naresh et Lee 2021; J. Wang et al. 2013). The sensing element of a whole-cell biosensor might be able to recognize certain species of interest and convert that identification into meaningful data. Modern cell-based biosensors primarily rely on immobilizing cells onto a transducer under cellular conditions. Whole-cell biosensors hold a lot of potential as environmental detection tools (Patwari, Pruckner, et Fabris 2023; Boron, Juárez, et Battaglini 2020; Allouzi et al. 2022). We discover that microalgae are frequently employed in a wide range of applications, including the treatment of wastewater (El-Moustaqim et al. 2023) and the integration of IoT to address environmental issues (El-Moustaqim, Mabrouki, et Azrour 2023). Numerous recent investigations have demonstrated the broad usage of biosensors in a range of microalgae applications. Microalgae are used as sensor elements because of their rapid response, short regeneration time, low cost, and simplicity in maintenance and storage (Kumar et al. 2020; Xie et al. 2022; Chouler et al. 2019). The most preferred biological method for determining how contaminants affect microalgae is to measure photosynthetic activity, which is indicated through the creation of oxygen or chlorophyll fluorescence as a sign within the photosystem (Chaudhary et al. 2020). In addition, suspended microalgae are typically more effective than free cells and can reduce the chance of these cells escaping throughout the surrounding environment (Kashem et al. 2019). Due to their excellent susceptibility to alterations in metabolic activity after contact with toxins (Jacob et al. 2021; Miazek et al. 2015) and their consistency in these tests, microalgae have found application in biosensors (Khan et al. 2021). The culture composition of microalgae is significantly impacted by toxic substances (i.e., home, commercial, and agricultural wastewater) or food waste sources (R. H. Mahmoud et Hassan 2023; Vargas-Estrada et al. 2020). These modifications can be translated through electrical or optical data inside microalgae-based biosensors (Sharma M. et al. 2023). Chemical quantities present in water can be detected by biosensors; nevertheless, more study is needed to understand the various substance types and concentrations to create a microalgal biosensor that works in a range of environments (Nandimandalam et Gude 2019; H. Ali, Khan, et Ilahi 2019).

Thus, the purpose of this work is to examine current developments and methods for incorporating microalgae into analytical plans for pollutant detection using microalgae-based biosensors. Furthermore, the benefits of every type of microalgae biosensor are examined, along with the opportunities and difficulties facing this technology.

## 2. BIOSENSOR BASED ON MICROALGAE

When compared to other types of microorganisms, microalgae are less susceptible to physiochemical changes because of their high degree of environmental adaptability (Amina Antonacci et Scognamiglio 2020). Furthermore, microalgae are highly sensitive to toxins, less hazardous, and have minimal effects on the environment (Eom et al. 2021; Perullini et al. 2014; Turemis et al. 2018). In addition to eliminating heavy metals and microorganisms like coliform, the production of microalgal nanoparticles has also been utilized to lower wastewater-related levels of nitrogen, phosphorus (El-Moustaqim, Mohammed, Mabrouki, et Hmouni 2024; El-Moustaqim, Rachiq, Mabrouki, Slaoui, et al. 2024; El-Moustaqim, Mabrouki, et Hmouni 2024), and biological and chemical oxygen demand (Palacios et al. 2023; Déniel et al. 2019; Chai et al. 2021; Ahmed et al. 2022). The use of microalgae in biological wastewater treatment has become essential. Its potential as a cost-effective and sustainable wastewater treatment approach has been demonstrated by its capacity to biologically purify wastewaters from various sources while using wastewater as the growth substrate (Gonçalves, Pires, et Simões 2017; Cuellar-Bermudez et al. 2017). Microalgae's utilization for wastewater remediation is a novel field that focuses on developing an economical and ecologically friendly technique. Recently, the potential of several microalgae and bacterial consortiums to cleanse wastewater from different sources has been assessed (A. Ali et al. 2023).

*Figure 1. A schematic layout of a biosensors based on microalgae*

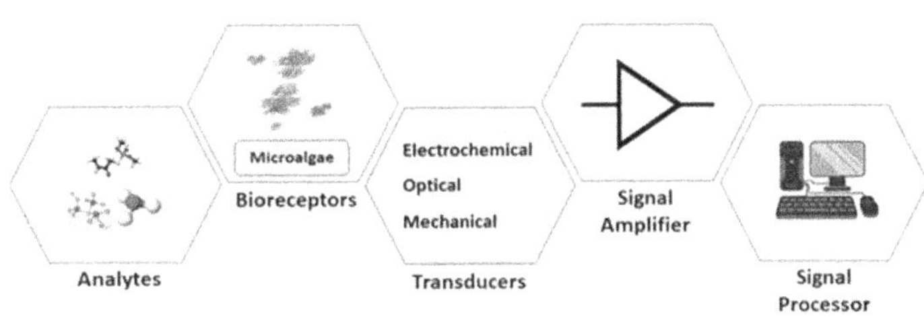


As a result, there has been a sharp rise in interest in microalgae biosensors for the monitoring of heavy metals, marine pollution, herbicides in food and water, etc (El-Moustaqim, Sbai, El Yousfi, Mabrouki, et al. 2024; El-Moustaqim, Sbai, Mabrouki, et Hmouni 2024; El-Moustaqim, Mabrouki, Azrour, Hadine, et al. 2024; El-Moustaqim, Mabrouki, Azrour, et Hmouni 2024). Microalgal-based biosensors provide a number of benefits, including ease of genetic alteration, affordability, high sensitivity, sustainability, and adaptability (Agarwal, Gupta, et Agarwal 2019; Chalupowicz et al. 2020). Microalgal biosensors typically use a dose-effective detection strategy that depends on the metabolic activity of the desired molecule

(**Fig.1**). Numerous parameters related to the analytes inhibitory effects are assessed electrochemically or optically, including growth rate, photosynthetic activity, fluorescence induction, and metabolism (Han et al. 2019; Rinken et Kivirand 2019; Berberich, Li, et Sahle-Demessie 2019; Ejeian et al. 2018). The best settings for microalgal biosensors might be directly linked to the detection of these biosensors based on changes in growth-related parameters. Generally, low saturation but high-intensity light, neutral pH, and temperatures between 16 and 35 °C are the ideal settings for microalgae growth (Gao et al. 2019; Otondo et al. 2018; Gatamaneni, Orsat, et Lefsrud 2018).

## 3. PHOTOSYNTHETIC EFFICIENCY CHANGE DETECTION

Abiotic variables including temperature, light, pH, oxygen, carbon dioxide, and salinity, as well as biotic factors like the presence of pathogens, can all have an impact on how effective microalgal biosensors are. Efficiency can also be affected by operational variables such as dilution rate, aeration and mixing, and harvesting frequency, as well as by the presence of other inhibitory compounds (Kholssi, Lougraimzi, et Moreno-Garrido 2023; Olguín et al. 2022). Consequently, monitoring variations in the growth-related metrics of microalgal biosensors can be directly associated with their ideal circumstances. Microalgae typically thrive best in a pH-neutral environment with low saturation and high intensity light, and at temperatures between 16 and 35 °C (R. H. Mahmoud et Hassan 2023; Prudkin-Silva et al. 2021; Xu et al. 2019). In addition, the growth of microalgae for protein synthesis and as a source of micronutrients is significantly affected by the presence of inorganic nutrients such as carbon, nitrogen, and phosphorus, as well as heavy metals (R. Mahmoud et Hassan 2023; Plyusnina et al. 2019). Regarding microalgal biosensors based on growth-related parameters, they are primarily used for conducting toxicity tests in wastewater. Various microalgal species that have been used as biosensors include Chlorella (**Fig.2**), Chlamydomonas, Arthrospira, Nannochloropsis and Scenedesmus sp. because they are tolerant and adaptable to harsh environmental conditions.

*Figure 2. Example of electrochemical biosensor detecting based-green microalgal*

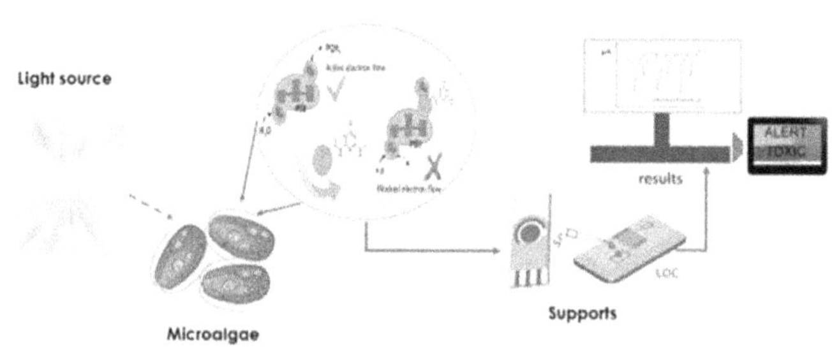


Chlorella sp. has been the most widely utilized for the detection of heavy metals in water due to its capacity to respond quickly and its appropriateness for detecting a wide range of metals (Zhuang et al. 2023; Perin et al. 2019). Microalgae biosensors function similarly to other microbial biosensors. A decrease in sensor sensitivity to inhibitors or toxicants implies possible pollution of the surrounding environment (Qian et al. 2024; Li et al. 2024). Through protein chelation and disruption of numerous physiological and biochemical functions, heavy metals are hazardous to microalgae. subsequently heavy metals enter algal cells via active transport or endocytosis. They bind to sulfhydryl groups of proteins, altering their structure and displacing crucial components (Mathanmohun et al. 2024; Hassan, Khurshid, et Bhat 2023; Tonelli et al. 2023).

## 4. MICROALGAL BIOSENSOR SUPPORT MATERIALS

The critical process of manufacturing biosensors affects the device's essential functioning characteristics, such as sensitivity, stability, and repeatability. As a result, selecting a support material for biosensor manufacture is a critical step (Abdul Manaf et al. 2021). The material that provides support should be beneficial to living cells, insoluble, lightweight, stable, non-biodegradable, and affordable (Giese et al. 2020). Although nanomaterials pose toxicity risks to cellular biological systems, their utilization is limited. Nevertheless, nanotechnology has been applied in various fields. Nanomaterials have proven to be excellent substrates for immobilizing microalgae, primarily because of their extensive surface area, non-porous composition, heightened surface reactivity, super-paramagnetic attributes, and exceptional adsorption capacity for various ions (Vasilieva et al. 2021; Valdivia-Rivera et al. 2021).

Carbon nanotubes, graphene, metallic and noble metal nanoparticles, nanospheres, magnetic nanoparticles, nano-silica, mesoporous silica nanoparticles, paper-based substrates, and fluorescent nanocrystals are examples of nanomaterials used (Nurrohman et al. 2017). The species Chlamydomonas reinhardtii was used to detect pollutants such as Atrazine on various substrates, using a variety of support materials and biosensor modalities (A. Antonacci et al. 2021). These comprise paper-based screen-printed electrodes nanomodified with carbon black, screen-printed electrodes modified with carbon black, and hydrophilic quantum dots nanoparticles used in electro-optical, amperometry, and optical biosensors, respectively (Finney et al. 2022; Rea et al. 2009).

## 5. MICROALGAL BIOSENSOR APPLICATION FOR DETECTION OF HEAVY METALS

Pollution and habitat destruction have caused greater water shortages and a decline in aquatic biodiversity. Wastewater is made up of harmful elements that are produced and released during home, industrial, commercial, or agricultural activities (R. Mahmoud et Hassan 2023). In the past few years, there has been a notable increase in focus on evaluating water quality, driven by the pressing requirement to safeguard vital freshwater reservoirs including lakes, rivers, reservoirs, and groundwater. Deterioration in water quality can stem from multiple sources, including the introduction of heavy metals and organic pollutants. Biosensors, including optical, electrochemical, and thermal sensors, are frequently employed for monitoring wastewater (Chakravorty et al. 2023; Zhao et al. 2023). Microalgae biosensors are increasingly being used to detect heavy metals. These biosensors use the unique features of microalgae to detect and measure heavy metal concentrations in a variety of settings (Ling Shing et al. 2017). These biosensors, which use microalgae's sensitivity to heavy metal toxicity, provide a low-cost and environmentally benign option for monitoring heavy metal pollution in water bodies, soil, and the air (Mahboob et al. 2022). Furthermore, microalgae biosensors can be customized to detect specific heavy metals of concern, enabling precise and dependable detection capabilities for environmental monitoring and remediation activities (Velusamy et al. 2022).

*Figure 3. Working of optical biosensors for detecting heavy metals*

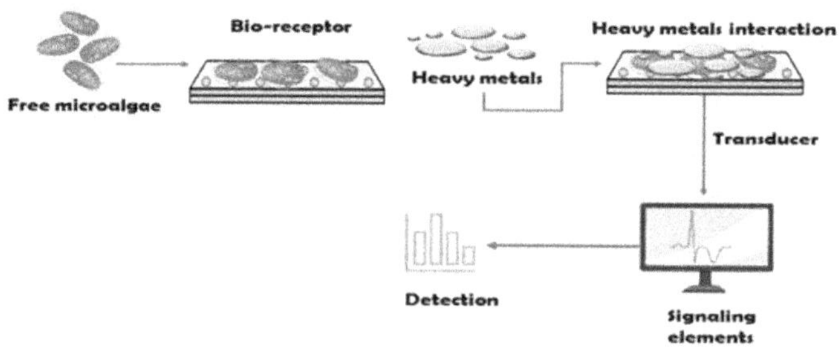


For monitoring, a heavy metal microalgae-based electrochemical biosensor is used. **Fig. 3** illustrates the design and components of an electrochemical biosensor used for detecting heavy metals (Song et al. 2022). Many electrochemical biosensors have been created to detect heavy metals, including silt. The literature mentions three types of fuel cells: sediment microbial fuel cells, single chamber batch microbial fuel cells, and direct methanol fuel cells (DMFC). The aforementioned biosensor monitors voltage changes caused by bacteria's immobilized electrogenic activity (Singh et al. 2023; Mahajan 2024). This biosensor has a low detection threshold. As a result, it is being used as a sensor to check wastewater quality (Hara et Singh 2021). This technique involves immobilizing the biological component on the current electrode and adsorbing the nanomaterial on the working electrode. The target can be detected using the reference electrode. When the analyte interacts with the biological component, the transducer generates a signal and transforms it to a legible format (L. Wang et al. 2020; Meng et al. 2023).

## 6. CONCLUSION

Developing algal-based biosensors with enhanced analytical performance is being aided by interdisciplinary technologies including nanotechnology, microfluidics, materials science, and rational design. Thus, biosensor technology can be advanced for use in useful agri-environmental and safety applications through precisely calibrated algal-based devices. Biosensor technology has improved significantly over the previous ten years, particularly in the detection of environmental toxins such as heavy metals. Across the globe, researchers are continually investigating new designs and manufacturing methods for biosensors capable of detecting specific pollutants or biomolecules. Biosensors offer various advantages, including durability, cost-effectiveness, speed, selectivity, and increased sensitivity. While they may have some limitations, such as being sensitive to pH and temperature swings and requiring sample preparation, these sensors function comparably to other sensor types. Microalgae-based biosensors are widely employed in a variety of industries due to their sensitivity, durability, and adaptability. Nanomaterials can serve as support materials for biosensors, improving their sensitivity and efficacy. Microalgae can be immobilized on biosensors in a variety of ways, used in microfluidics, and even undergo genome engineering and editing to improve their performance and attributes and tailor them to specific requirements.

## REFERENCES


Abdul Manaf, S. A., Mohamad Fuzi, S. F. Z., Abdul Manas, N. H., Md Illias, R., Low, K. O., Hegde, G., Che Man, R., Wan Azelee, N. I., & Matias-Peralta, H. M. (2021). Emergence of Nanomaterials as Potential Immobilization Supports for Whole Cell Biocatalysts and Cell Toxicity Effects. *Biotechnology and Applied Biochemistry*, *68*(6), 1128–1138. DOI: 10.1002/bab.2034 PMID: 32969042

Agarwal, P., Gupta, R., & Agarwal, N. (2019). Advances in Synthesis and Applications of Microalgal Nanoparticles for Wastewater Treatment. *Journal of Nanotechnology*, *2019*(janvier), e7392713. DOI: 10.1155/2019/7392713

Ahmed, S. F., Mofijur, M., Parisa, T. A., Islam, N., Kusumo, F., Inayat, A., Le, V. G., Badruddin, I. A., Khan, T. M. Y., & Ong, H. C. (2022). Progress and challenges of contaminate removal from wastewater using microalgae biomass. *Chemosphere*, *286*(janvier), 131656. DOI: 10.1016/j.chemosphere.2021.131656 PMID: 34325255

Ali, A., Khalid, Z., Allam, A. A., & Ajarem, J. S. (2023). Wastewater treatment by using microalgae: Insights into fate, transport, and associated challenges. *Chemosphere*, *338*(octobre), 139501. DOI: 10.1016/j.chemosphere.2023.139501 PMID: 37453525

Ali, H., Khan, E., & Ilahi, I. (2019). Environmental Chemistry and Ecotoxicology of Hazardous Heavy Metals: Environmental Persistence, Toxicity, and Bioaccumulation. *Journal of Chemistry*, *2019*, 1–14. Advance online publication. DOI: 10.1155/2019/6730305

Allouzi, M. M. A., Allouzi, S., Al-Salaheen, B., Khoo, K. S., Rajendran, S., Sankaran, R., & Show, P. L. (2022). Current advances and future trend of nanotechnology as microalgae-based biosensor. *Biochemical Engineering Journal*, *187*, 108653.

Antonacci, A., Attaallah, R., Arduini, F., Amine, A., Giardi, M. T., & Scognamiglio, V. (2021). A Dual Electro-Optical Biosensor Based on Chlamydomonas Reinhardtii Immobilised on Paper-Based Nano-modified Screen-Printed Electrodes for Herbicide Monitoring. *Journal of Nanobiotechnology*, *19*(1), 145. Advance online publication. DOI: 10.1186/s12951-021-00887-4 PMID: 34001124

Antonacci, A., & Scognamiglio, V. (2020). Biotechnological Advances in the Design of Algae-Based Biosensors. *Trends in Biotechnology*, *38*(3), 334–347. DOI: 10.1016/j.tibtech.2019.10.005 PMID: 31706693

Berberich, J., Li, T., & Sahle-Demessie, E. 2019. « Chapter 11 - Biosensors for Monitoring Water Pollutants: A Case Study With Arsenic in Groundwater ». In *Separation Science and Technology*, édité par Satinder Ahuja, 11:285-328. Evaluating Water Quality to Prevent Future Disasters. Academic Press. DOI: 10.1016/B978-0-12-815730-5.00011-9

Boron, I., Juárez, A., & Battaglini, F. (2020). Portable Microalgal Biosensor for Herbicide Monitoring. *ChemElectroChem*, *7*(7), 1623–1630. DOI: 10.1002/celc.202000210

Chai, W. S., Tan, W. G., Heli, S. H. M., Gupta, V. K., Ho, S.-H., & Show, P. L. (2021). Multifaceted roles of microalgae in the application of wastewater biotreatment: A review. *Environmental Pollution*, *269*(janvier), 116236. DOI: 10.1016/j.envpol.2020.116236 PMID: 33333449

Chakravorty, M., Nanda, M., Bisht, B., Sharma, R., Kumar, S., Mishra, A., Vlaskin, M., Chauhan, P. K., & Kumar, V. (2023). Heavy metal tolerance in microalgae: Detoxification mechanisms and applications. *Aquatic Toxicology (Amsterdam, Netherlands)*, *260*(mai), 106555. DOI: 10.1016/j.aquatox.2023.106555 PMID: 37196506

Chalupowicz, D., Veltman, B., Droby, S., & Eltzov, E. (2020). Evaluating the use of biosensors for monitoring of Penicillium digitatum infection in citrus fruit. *Sensors and Actuators. B, Chemical*, *311*(mai), 127896. DOI: 10.1016/j.snb.2020.127896

Chaudhary, R., Nawaz, K., Khan, A. K., Hano, C., Abbasi, B. H., & Anjum, S. (2020). An Overview of the Algae-Mediated Biosynthesis of Nanoparticles and Their Biomedical Applications. *Biomolecules*, *10*(11), 1498. DOI: 10.3390/biom10111498 PMID: 33143289

Chouler, J., Monti, M. D., Morgan, W. J., Cameron, P. J., & Di Lorenzo, M. (2019). A photosynthetic toxicity biosensor for water. *Electrochimica Acta*, *309*(juin), 392–401. DOI: 10.1016/j.electacta.2019.04.061

Cuellar-Bermudez, S. P., Aleman-Nava, G. S., Chandra, R., Garcia-Perez, J. S., Contreras-Angulo, J. R., Markou, G., Muylaert, K., Rittmann, B. E., & Parra-Saldivar, R. (2017). Nutrients utilization and contaminants removal. A review of two approaches of algae and cyanobacteria in wastewater. *Algal Research*, *24*, 438–449. DOI: 10.1016/j.algal.2016.08.018

Déniel, M., Errien, N., Daniel, P., Caruso, A., & Lagarde, F. (2019). Current methods to monitor microalgae-nanoparticle interaction and associated effects. *Aquatic Toxicology (Amsterdam, Netherlands)*, *217*, 105311. DOI: 10.1016/j.aquatox.2019.105311 PMID: 31730931

Ejeian, F., Etedali, P., Mansouri-Tehrani, H.-A., Soozanipour, A., Low, Z.-X., Asadnia, M., Taheri-Kafrani, A., & Razmjou, A. (2018). Biosensors for wastewater monitoring: A review. *Biosensors & Bioelectronics*, *118*(octobre), 66–79. DOI: 10.1016/j.bios.2018.07.019 PMID: 30056302

El-Moustaqim, K., Benchrifa, M., Mabrouki, J., & Hmouni, D. (2024). *« Interaction between the Microalgae Cultivation System and Its Effects on Nutrient Removal from Wastewater ». In, 70-79.*, DOI: 10.1201/9781003436218-8

El-Moustaqim, K., El Bakraoui, H., Mabrouki, J., Fouad, Y., Slaoui, M., Hmouni, D., Benyeogor, M. S., & Igbigbi, T. L. (2023). Combination of Microalgae Method, Decantation, and Filtration for Domestic Wastewater Treatment. *Sustainability (Basel)*, *15*(22), 16110. DOI: 10.3390/su152216110

El-Moustaqim, K., Mabrouki, J., & Azrour, M. (2023). *« Monitoring of Water Toxicity Through the Internet of Things to Protect the Health of the Population ». In, 85-92.*, DOI: 10.1201/9781003430735-7

El-Moustaqim, K., Mabrouki, J., Azrour, M., Hadine, M., & Hmouni, D. (2024). Enabling Smart Agriculture Through Integrating the Internet of Things in Microalgae Farming for Sustainability. In Azrour, M., Mabrouki, J., Alabdulatif, A., Guezzaz, A., & Amounas, F. (Eds.), *Smart Internet of Things for Environment and Healthcare* (pp. 209–222). Springer Nature Switzerland., DOI: 10.1007/978-3-031-70102-3_15

El-Moustaqim, K., Mabrouki, J., Azrour, M., & Hmouni, D. 2024. « How Can Green Energy Be Improved by Integrating IoT into the Microalgae Process? » In, 239-52. DOI: 10.1007/978-3-031-52419-6_17

El-Moustaqim, K., Mabrouki, J., & Hmouni, D. 2024. « Applied Sciences Green Microalgae for Future Biomass Development ». In, 203-11. DOI: 10.1007/978-3-031-50860-8_12

El-Moustaqim, K., Mohammed, B., Mabrouki, J., & Hmouni, D. 2024. « Application of IoT-Based Intelligent Systems in Microalgae Biofuel Production ». In *Advanced Technology for Smart Environment and Energy*, édité par Jamal Mabrouki et Azrour Mourade, 293-305. Environmental Science and Engineering. Cham: Springer Nature Switzerland. DOI: 10.1007/978-3-031-50871-4_20

El-Moustaqim, K., Rachiq, T., Mabrouki, J., Slaoui, M., & Hmouni, D. 2024. « Application of Microalgae in Green Hydrogen Production: A Diverse Biomass for Multiple Applications ». In *Advanced Technology for Smart Environment and Energy*, édité par Jamal Mabrouki et Azrour Mourade, 231-40. Environmental Science and Engineering. Cham: Springer Nature Switzerland. DOI: 10.1007/978-3-031-50871-4_15

El-Moustaqim, K., Sbai, S. E., El Yousfi, Y., Mabrouki, J., & Hmouni, D. (2024). New Strategy for the Advancement of Modern Agriculture through the Use of Microalgae as Biofertilizers. *Euro-Mediterranean Journal for Environmental Integration*, (octobre). Advance online publication. DOI: 10.1007/s41207-024-00678-x

El-Moustaqim, K., Sbai, S. E., Mabrouki, J., & Hmouni, D. (2024). Critical Review of Green Microalgal Biomass Development through CO2 Sequestration. *Euro-Mediterranean Journal for Environmental Integration*, (mai). Advance online publication. DOI: 10.1007/s41207-024-00548-6

El-Moustaqim, K., Tallou, A., Mabrouki, J., Slaoui, M., & Hmouni, D. 2024. « Phytoremediation Processes for the Removal of Heavy Metals ». In, 33-46. DOI: 10.1007/978-3-031-55787-3_3

Eom, H., Park, M., Jang, A., Kim, S., & Oh, S.-E. (2021). A simple and rapid algal assay kit to assess toxicity of heavy metal-contaminated water. *Environmental Pollution*, *269*(janvier), 116135. DOI: 10.1016/j.envpol.2020.116135 PMID: 33290954

Finney, L., Skrodzki, P., Peskosky, N., Burger, M., Nees, J., Krushelnick, K., & Jovanovic, I. (2022). Ultrafast laser filament-induced fluorescence for detecting uranium stress in Chlamydomonas reinhardtii. *Scientific Reports*, *12*(octobre), 17205. Advance online publication. DOI: 10.1038/s41598-022-21404-z PMID: 36229516

Gao, F., Yang, H.-L., Li, C., Peng, Y.-Y., Lu, M.-M., Jin, W.-H., Bao, J.-J., & Guo, Y.-M. (2019). Effect of organic carbon to nitrogen ratio in wastewater on growth, nutrient uptake and lipid accumulation of a mixotrophic microalgae Chlorella sp. *Bioresource Technology*, *282*(juin), 118–124. DOI: 10.1016/j.biortech.2019.03.011 PMID: 30852331

Gatamaneni, B. L., Orsat, V., & Lefsrud, M. (2018). Factors Affecting Growth of Various Microalgal Species. *Environmental Engineering Science*, *35*(10), 1037–1048. DOI: 10.1089/ees.2017.0521

Giese, E. C., Silva, D. D. V., Costa, A. F. M., Almeida, S. G. C., & Dussán, K. J. (2020). Immobilized microbial nanoparticles for biosorption. *Critical Reviews in Biotechnology*, *40*(5), 653–666. DOI: 10.1080/07388551.2020.1751583 PMID: 32299253

Gonçalves, A. L., Pires, J. C., & Simões, M. (2017). A review on the use of microalgal consortia for wastewater treatment. *Algal Research*, *24*, 403–415.

Gupta, N., Renugopalakrishnan, V., Liepmann, D., Paulmurugan, R., & Malhotra, B. D. (2019). Cell-based biosensors: Recent trends, challenges and future perspectives. *Biosensors & Bioelectronics*, *141*(septembre), 111435. DOI: 10.1016/j.bios.2019.111435 PMID: 31238280

Han, S., Zhang, Q., Zhang, X., Liu, X., Lu, L., Wei, J., Li, Y., Wang, Y., & Zheng, G. (2019). A digital microfluidic diluter-based microalgal motion biosensor for marine pollution monitoring. *Biosensors & Bioelectronics*, *143*, 111597. DOI: 10.1016/j.bios.2019.111597 PMID: 31442755

Hara, T., & Singh, B. (2021). Electrochemical Biosensors for Detection of Pesticides and Heavy Metal Toxicants in Water: Recent Trends and Progress. *ACS ES&T Water*, *1*(février), 462–478. Advance online publication. DOI: 10.1021/acsestwater.0c00125

Hassan, S., Khurshid, Z., & Bhat, R. A. (2023). Green synthesized nanomaterials for biosensors. In *Synthesis of Bionanomaterials for Biomedical Applications* (pp. 339–355). Elsevier. DOI: 10.1016/B978-0-323-91195-5.00024-6

Inda, M. E., & Lu, T. K. (2020). Microbes as Biosensors. *Annual Review of Microbiology*, *74*(1), 337–359. DOI: 10.1146/annurev-micro-022620-081059 PMID: 32660390

Jacob, J. M., Ravindran, R., Narayanan, M., Samuel, S. M., Pugazhendhi, A., & Kumar, G. (2021). Microalgae: A prospective low cost green alternative for nanoparticle synthesis. *Current Opinion in Environmental Science & Health*, *20*(avril), 100163. DOI: 10.1016/j.coesh.2019.12.005

Kashem, M. A., Kimoto, K., Iribe, Y., & Suzuki, M. (2019). Development of Microalgae Biosensor Chip by Incorporating Microarray Oxygen Sensor for Pesticides Sensing. *Biosensors (Basel)*, *9*(4), 133. DOI: 10.3390/bios9040133 PMID: 31726653

Khan, M. J., Rai, A., Ahirwar, A., Sirotiya, V., Mourya, M., Mishra, S., Schoefs, B., Marchand, J., Bhatia, S. K., Varjani, S., & Vinayak, V. (2021). Diatom microalgae as smart nanocontainers for biosensing wastewater pollutants: Recent trends and innovations. *Bioengineered*, *12*(2), 9531–9549. DOI: 10.1080/21655979.2021.1996748 PMID: 34709977

Kholssi, R., Lougraimzi, H., & Moreno-Garrido, I. (2023). Effects of global environmental change on microalgal photosynthesis, growth and their distribution. *Marine Environmental Research*, *184*(février), 105877. DOI: 10.1016/j.marenvres.2023.105877 PMID: 36640723

Kumar, G., Shekh, A., Jakhu, S., Sharma, Y., Kapoor, R., & Sharma, T. R. (2020). Bioengineering of microalgae: Recent advances, perspectives, and regulatory challenges for industrial application. *Frontiers in Bioengineering and Biotechnology*, *8*, 914. DOI: 10.3389/fbioe.2020.00914 PMID: 33014997

Li, Z., Li, S., Wu, Q., Gao, X., & Zhu, L. (2024). Physiological responses and removal mechanisms of ciprofloxacin in freshwater microalgae. *Journal of Hazardous Materials*, *466*, 133519. DOI: 10.1016/j.jhazmat.2024.133519 PMID: 38278073

Low, S. S., Chen, Z., Li, Y., Lu, Y., & Liu, Q. (2021). Design principle in biosensing: Critical analysis based on graphitic carbon nitride (G-C3N4) photoelectrochemical biosensor. *Trends in Analytical Chemistry*, *145*(décembre), 116454. DOI: 10.1016/j.trac.2021.116454

Mahajan, P. (2024). Review—Fundamentals to Applications of MXenes for Biosensing. *Journal of the Electrochemical Society*, *171*(février), 027516. Advance online publication. DOI: 10.1149/1945-7111/ad26e2

Mahboob, S. M. Lakshmi, Madhav Anand, S. Vijayalakshmi, Ranjitha Lingam, et Junaid Malik. 2022. « Microalgae—A Promising Tool for Heavy Metal Remediation ». In, 277-95. DOI: 10.1007/978-3-031-08830-8_13

Mahmoud, R., & Hassan, R. (2023). *« Algae-derived biosensor materials and their applications ». In, 203-18.*, DOI: 10.1016/B978-0-443-18816-9.00016-2

Mahmoud, R. H., & Hassan, R. Y. A. 2023. « Chapter 12 - Algae-derived biosensor materials and their applications ». In *Algae Materials*, édité par K. Arunkumar, A. Arun, Rathinam Raja, et Ramasamy Palaniappan, 203-18. Developments in Applied Microbiology and Biotechnology. Academic Press. DOI: 10.1016/B978-0-443-18816-9.00016-2

Malhotra, B. D., & Ali, M. A. (2017). Nanomaterials in biosensors: Fundamentals and applications. Nanomaterials for biosensors, 1.

Mathanmohun, M., Sagadevan, S., Rahman, M. Z., Lett, J. A., Fatimah, I., Moharana, S., Garg, S., & Al-Anber, M. A. (2024). Unveiling Sustainable, Greener Synthesis Strategies and Multifaceted Applications of Copper Oxide Nanoparticles. *Journal of Molecular Structure*, *1305*, 137788. DOI: 10.1016/j.molstruc.2024.137788

Meng, R., Zhu, Q., Long, T., He, X., Luo, Z., Gu, R., Wang, W., & Xiang, P. (2023). The innovative and accurate detection of heavy metals in foods: A critical review on electrochemical sensors. *Food Control*, *150*, 109743. DOI: 10.1016/j.foodcont.2023.109743

Miazek, K., Iwanek, W., Remacle, C., Richel, A., & Goffin, D. (2015). Effect of Metals, Metalloids and Metallic Nanoparticles on Microalgae Growth and Industrial Product Biosynthesis: A Review. *International Journal of Molecular Sciences*, *16*(10), 23929–23969. DOI: 10.3390/ijms161023929 PMID: 26473834

Nandimandalam, H., & Gude, V. G. (2019). Indigenous biosensors for in situ hydrocarbon detection in aquatic environments. *Marine Pollution Bulletin*, *149*(décembre), 110643. DOI: 10.1016/j.marpolbul.2019.110643

Naresh, V., & Lee, N. (2021). A Review on Biosensors and Recent Development of Nanostructured Materials-Enabled Biosensors. *Sensors (Basel)*, *21*(4), 1109. DOI: 10.3390/s21041109 PMID: 33562639

Nurrohman, D. T., Oktivina, M., Suharyadi, E., Suyono, E. A., & Abraha, K. 2017. « Monitoring Microalgae Population Growth by Using Fe3O4 Nanoparticles-Based Surface Plasmon Resonance (SPR) Biosensor ». In. Vol. 202. DOI: 10.1088/1757-899X/202/1/012077

Olguín, E. J., Sánchez-Galván, G., Arias-Olguín, I. I., Melo, F. J., González-Portela, R. E., Cruz, L., De Philippis, R., & Adessi, A. (2022). Microalgae-Based Biorefineries: Challenges and Future Trends to Produce Carbohydrate Enriched Biomass, High-Added Value Products and Bioactive Compounds. *Biology (Basel)*, *11*(8), 1146. DOI: 10.3390/biology11081146 PMID: 36009773

Otondo, A., Kokabian, B., Stuart-Dahl, S., & Gude, V. G. (2018). Energetic evaluation of wastewater treatment using microalgae, Chlorella vulgaris. *Journal of Environmental Chemical Engineering*, *6*(2), 3213–3222. DOI: 10.1016/j.jece.2018.04.064

Palacios, O. A., León-Vega, R. A., López, B. R., de-Bashan, L. E., Choix, F. J., & Cuevas-Rodríguez, G. (2023). Microalga-bacteria interaction mitigates adverse effects on microalga produced by ZnO nanoparticles. *Algal Research*, *74*(juillet), 103198. DOI: 10.1016/j.algal.2023.103198

Patwari, P., Pruckner, F., & Fabris, M. (2023). Biosensors in microalgae: A roadmap for new opportunities in synthetic biology and biotechnology. *Biotechnology Advances*, *68*, 108221. DOI: 10.1016/j.biotechadv.2023.108221 PMID: 37495181

Perin, G., Bellan, A., Bernardi, A., Bezzo, F., & Morosinotto, T. (2019). The potential of quantitative models to improve microalgae photosynthetic efficiency. *Physiologia Plantarum*, *166*(1), 380–391. DOI: 10.1111/ppl.12915 PMID: 30578540

Perullini, M., Ferro, Y., Durrieu, C., Jobbágy, M., & Bilmes, S. A. (2014). Sol–gel silica platforms for microalgae-based optical biosensors. *Journal of Biotechnology*, *179*(juin), 65–70. DOI: 10.1016/j.jbiotec.2014.02.007 PMID: 24637376

Plyusnina, T. Yu., Khruschev, S. S., Frolov, A. E., Degtereva, N. S., Konyukhov, I. V., Pogosyan, S. I., Riznichenko, G. Yu., & Rubin, A. B. (2019). Monitoring of the Photosynthetic Activity of the Microalgae Chlorella under Nitrogen Depletion Conditions. *Biophysics*, *64*(3), 358–366. DOI: 10.1134/S0006350919030175

Prudkin-Silva, C., Lanzarotti, E., Álvarez, L., Vallerga, M. B., Factorovich, M., Morzan, U. N., Gómez, M. P., González, N. P., Acosta, Y. M., Carrizo, F., Carrizo, E., Galeano, S., Lagorio, M. G., Juárez, Á. B., Ithuralde, R. E., Romero, J. M., & Urdampilleta, C. M. (2021). A cost-effective algae-based biosensor for water quality analysis: Development and testing in collaboration with peasant communities. *Environmental Technology & Innovation*, *22*(mai), 101479. DOI: 10.1016/j.eti.2021.101479

Qian, W., Yang, Y., Chou, S., Ge, S., Li, P., Wang, X., Zhuang, L.-L., & Zhang, J. (2024). Effect of N/P ratio on attached microalgae growth and the differentiated metabolism along the depth of biofilm. *Environmental Research*, *240*, 117428. DOI: 10.1016/j.envres.2023.117428 PMID: 37875171

Rea, G., Polticelli, F., Antonacci, A., Scognamiglio, V., Katiyar, P., Kulkarni, S., Johanningmeier, U., & Giardi, M. (2009). Structure-based design of novel Chlamydomonas reinhardtii D1-D2 photosynthetic proteins for herbicide monitoring. *Protein Science : a Publication of the Protein Society*, *18*(octobre), 2139–2151. DOI: 10.1002/pro.228 PMID: 19693932

Rinken, T., & Kivirand, K. (2019). *Biosensors for Environmental Monitoring*. BoD – Books on Demand. DOI: 10.5772/intechopen.73763

Sharma, M., Sujata, S., Bansal, D., & Kaushik, P. (2023). Mini Review on the Potential of Algal Biosensors in Wastewater Monitoring. *Nanoscience & Nanotechnology-Asia*, *13*(4), 24–27. DOI: 10.2174/2210681213666230517123150

Shing, L., Wong, S. J., Voon, B., Tee, L., Tan, K., Murti, M., & Chai, M. (2017). Bioluminescent Microalgae-Based Biosensor for Metal Detection in Water. *IEEE Sensors Journal PP*, (décembre), 1–1. DOI: 10.1109/JSEN.2017.2787786

Singh, E., Kumar, A., Mishra, R., Singh, A., Madhav, S., Shukla, S., & Kumar, S. (2023). *« Measurement techniques for detection of metals in water resources ». In, 1-20.*, DOI: 10.1016/B978-0-323-95919-3.00009-4

Song, H., Huo, M., Zhou, M., Chang, H., Li, J., Zhang, Q., & Zhang, D. (2024). Carbon nanomaterials-based electrochemical sensors for heavy metal detection. *Critical Reviews in Analytical Chemistry, 54*(7), 1987–2006.

Tallou, A., Erraji, H., Al Masmoudi, Y., Abbatantuono, F., Ali, S. A., Mabrouki, J., ... & Vivaldi, G. A. (2024). Phytoremediation of Contaminated Environments Using Halophytes: General Overview. Integrated Solutions for Smart and Sustainable Environmental Conservation, 143-163.

Tonelli, F. M. P., Silva, C. S., Vinicius, M. S. D., & Flávia, C. P. T. (2023). Algae-based green AgNPs, AuNPs, and FeNPs as potential nanoremediators. *Green Processing and Synthesis, 12*(1), 20230008. DOI: 10.1515/gps-2023-0008

Turemis, M., Silletti, S., Pezzotti, G., Sanchís, J., Farré, M., & Giardi, M. T. (2018). Optical biosensor based on the microalga-paramecium symbiosis for improved marine monitoring. *Sensors and Actuators. B, Chemical, 270*(octobre), 424–432. DOI: 10.1016/j.snb.2018.04.111

Valdivia-Rivera, S., Ayora-Talavera, T., Lizardi-Jiménez, M. A., García-Cruz, U., Cuevas-Bernardino, J. C., & Pacheco, N. (2021). Encapsulation of Microorganisms for Bioremediation: Techniques and Carriers. *Reviews in Environmental Science and Biotechnology, 20*(3), 815–838. DOI: 10.1007/s11157-021-09577-x

Vargas-Estrada, L., Torres-Arellano, S., Longoria, A., Arias, D. M., Okoye, P. U., & Sebastian, P. J. (2020). Role of nanoparticles on microalgal cultivation: A review. *Fuel, 280*, 118598. DOI: 10.1016/j.fuel.2020.118598

Vasilieva, S., Lobakova, E., Grigoriev, T., Selyakh, I., Semenova, L., Chivkunova, O., Gotovtsev, P., Antipova, C., Zagoskin, Y., Scherbakov, P., Lukyanov, A., Lukanina, K., & Solovchenko, A. (2021). Bio-inspired materials for nutrient biocapture from wastewater: Microalgal cells immobilized on chitosan-based carriers. *Journal of Water Process Engineering, 40*(avril), 101774. DOI: 10.1016/j.jwpe.2020.101774

Velusamy, K., Periyasamy, S., Kumar, P. S., Rangasamy, G., Nisha Pauline, J. M., Ramaraju, P., Mohanasundaram, S., & Nguyen Vo, D.-V. (2022). Biosensor for heavy metals detection in wastewater: A review. *Food and Chemical Toxicology, 168*(octobre), 113307. DOI: 10.1016/j.fct.2022.113307 PMID: 35917955

Wang, J., Sun, J., Song, Y., Xu, Y., Pan, X., Sun, Y., & Li, D. (2013). A label-free microfluidic biosensor for activity detection of single microalgae cells based on chlorophyll fluorescence. *Sensors (Basel), 13*(12), 16075–16089. DOI: 10.3390/s131216075 PMID: 24287532

Wang, L., Peng, X., Fu, H., Huang, C., Li, Y., & Liu, Z. (2020). Recent advances in the development of electrochemical aptasensors for detection of heavy metals in food. *Biosensors & Bioelectronics, 147*, 111777. DOI: 10.1016/j.bios.2019.111777 PMID: 31634804

Wasito, H., Fatoni, A., Hermawan, D., & Susilowati, S. S. (2019). Immobilized bacterial biosensor for rapid and effective monitoring of acute toxicity in water. *Ecotoxicology and Environmental Safety*, *170*(avril), 205–209. DOI: 10.1016/j.ecoenv.2018.11.141 PMID: 30529914

Xie, Y., Khoo, K. S., Chew, K. W., Devadas, V. V., Phang, S. J., Lim, H. R., Rajendran, S., & Show, P. L. (2022). Advancement of renewable energy technologies via artificial and microalgae photosynthesis. *Bioresource Technology*, *363*(novembre), 127830. DOI: 10.1016/j.biortech.2022.127830 PMID: 36029982

Xu, X., Gu, X., Wang, Z., Shatner, W., & Wang, Z. (2019). Progress, challenges and solutions of research on photosynthetic carbon sequestration efficiency of microalgae. *Renewable & Sustainable Energy Reviews*, *110*(août), 65–82. DOI: 10.1016/j.rser.2019.04.050

Zhao, D., Lai, S., Ng, E.-P., Khoo, K. S., Show, P.-L., & Ling, T. (2023). Symbiosis of microalgae and bacteria consortium for heavy metal remediation in wastewater. *Journal of Environmental Chemical Engineering*, *11*(avril), 109943. DOI: 10.1016/j.jece.2023.109943

Zhuang, L.-L., Tian, W., Yang, Y., Ge, S., Li, P., Sun, S., Zhang, J., & Liang, S. (2023). Quantified trend of photosynthetic rate along the depth of microalgae biofilm. *The Science of the Total Environment*, *876*, 162801. DOI: 10.1016/j.scitotenv.2023.162801 PMID: 36907420

# Chapter 8
# Effect of Treatment of Corms With Trichoderma Inoculum on the Development of Crocus sativus Plants and the Formation of Daughter Bulbs

**Samah Ourras**
https://orcid.org/0000-0003-1826-332X
*Ibn Tofail University, Morocco*

**M'hamed El Ouark**
*Faculty of Sciences Rabat, Morocco*

**Ismail El Aymani**
*Faculty of Sciences Rabat, Morocco*

**Soukaina Msairi**
*Faculty of Sciences Rabat, Morocco*

**Karima Selmaoui**
*Faculty of Sciences Rabat, Morocco*

**Moulay Abdelaziz El Alaoui**
*Faculty of Sciences Rabat, Morocco*

**Najoua Mouden**
*Nador-Mohammed 1st University, Morocco*

**Amina Ouazzani Touhami**
https://orcid.org/0000-0001-9970-7724
*Faculty of Sciences Rabat, Morocco*

**Allal Douira**
*Faculty of Sciences Rabat, Morocco*

## ABSTRACT

*In this study, an isolate of Trichoderma sp.1, collected from saffron corms, was utilized to evaluate its effect on saffron plant growth and bulb formation.Treating saffron corms with a conidial suspension of the Trichoderma isolate before sowing showed a positive impact on agronomic parameters and bulb formation. After four months of greenhouse cultivation, plants from Trichoderma-treated corms had an average of 31 leaves per plant, compared to 16,5 in the control plants. The average leaf length of treated plants was nearly identical to that of the controls.The average weights of leaves and roots in treated plants were 2,53g and 0,3g, respectively, compared to 1,93g and 0.15g in the controls.Furthermore, plants from Trichoderma-treated corms produced an average of about 10 newly formed bulbs, while*

DOI: 10.4018/979-8-3693-8980-5.ch008

*control plants produced 5,75. The average bulb weight of treated plants was 16,5g, compared to 10,36g in control plants. These results benefit saffron producers by ensuring planting material availability. Trichoderma stimulates daughter corm formation, boosting production.*

## INTRODUCTION

Saffron (*Crocus sativus* L.) is a geophytic, sterile, hermaphrodite-like species that propagates vegetatively, forming female corms from the mother corms (Tahiri et al., 2022). These authors reported that several factors influence the multiplication of corms under natural conditions and thus limit the availability of planting material. Among these factors, both biotic and abiotic stresses impede bulb multiplication (Parray et al., 2012; Menia et al., *2018,*El Aymani et al., 2023). The spread of saffron also depends on human labor, as bulbs or corms must be manually removed, separated and replanted (Tahiri et al., 2022).

Research has reported that organic fertilizers used in the cultivation of *C. sativus* L. are more beneficial than chemical fertilizers. Organic fertilizers have been shown to increase the size of mother bulbs and improve flower yield (Koocheki and Seyyedi, 2015). Work by Tahiri et al. (2022) indicated that humic substances can improve saffron growth and development, but they have no significant effect on the formation of daughter corms.

In recent years, several microorganisms, such as *Trichoderma* spp., *Pseudomonas fluorescens* and *Bacillus* spp., have been identified as potential candidates to stimulate growth and protect saffron against telluric diseases. Sharaf-Eldin et al. (2007) reported that the application of the PGPR bacterium *Bacillus subtilis* offers significant advantages for saffron producers by accelerating bulb growth, increasing biomass output, and enhancing the contents of chemical components (crocine, safranal, picrocrocine, and crocétine). Additionally, according to Gupta and Vakhlu (2015), PGPR can also be used as a biofungicide for bulb rot diseases in *C. sativus.*

The use of microbial inocula composed of PGPR bacteria (Pérez-Montaño et al., 2014) and arbuscular mycorrhizal fungi as biofertilizers represents a potential means of reducing the negative environmental impacts resulting from the continued use of chemical fertilizers, contributing to sustainable agriculture. In this context, El Aymani et al. (2019a) and Ourras et al. (2021) reported that the rhizosphere of saffron plants is rich in arbuscular mycorrhizal fungi. Inocula based on these endomycorrhizal fungi have shown biostimulating effects on root growth, leaf number and length, and the number of female corms. Some of these inocula promoted the sporulation of arbuscular mycorrhizal fungi at the root and rhizosphere levels of saffron plants (El Aymani et al., 2019b and 2023; Ourras et al., 2022).

*Trichoderma* are fungi commonly used in agriculture and horticulture for its various beneficial properties. They are known for their ability to promote plant growth (Hmouni et al. 2006, Mouria et al. 2008, Sghir et al. 2014; Sellal et al., 2020 and 2024; Khirallal et al. 2017; Kribel et al., 2020; El Kaissoumi et al., 2024; Mouden et al., 2023) and their protection from land-based bioaggressors, as biological control agents acting primarily through the production of antimicrobial compounds or by parasitism of plant pathogens (Mouria et al., 2013, 2015; Qostal et al., 2020; Kotba et al., 2022; El Kaissoumi et al., 2023; Errifi et al., 2024; Adnani et al., 2024). Inoculation with *Trichoderma* is therefore an interesting approach to develop in order to reduce the use of chemical inputs and pesticides, especially when it comes to the cultivation of local products and organic farming. This is precisely the case for the cultivation of saffron in the Talouine-Taznakht region, where the concern to improve the productivity of the crop should go with a better valorization of the product, saffron, Morocco, as organic and high quality product.

In this context, the main objective of this study is to develop a formulation based on *Trichoderma* spores to stimulate the growth and development of plants as well as the formation of female corms in saffron. Indeed, in this sense, it is important to recall that the production of corms occurs only once a year, which causes difficulties for the production of new saffron corms (Negbi et al., 1989; Renau-Morata et al., 2013; Sampathu et al., *1984*). Therefore, the extension of the growing area needs corms in quantity and quality. According to Devis et al. (2006), low corm multiplication rates and infection with fungal pathogens (El Aymani et al., 2019b; Gupta et al., 2021; Belfiori et al., 2021; Ourras et al., 2023) reduce productivity and quality, limiting the availability of planting material.

## MATERIALS AND METHODS

### Isolation and Incubation of *Trichoderma* sp.

Isolation of *Trichoderma* sp. from saffron corms, originating in the Taliouine and Taznakht regions, were disinfected by dipping these corms in 96% ethanol solution for 5 minutes, followed by three rinsing operations with sterile distilled water and then drying on filter paper (El Aymani et al., 2019; Ourras et al., 2023).

*The Trichoderma* isolate tested was cultured on a potato sucrose agar medium (PSA: 200g potato, 20g sucrose, 15g agar, 1000ml distilled water) and incubated at 28 ° C. for 7 days in darkness. This isolate develops rapidly on the PSA medium and exhibits a whitish-coloured mycelium at the beginning which becomes, with age, green and with a flourish appearance (Figure1). The conidiophore is branched, and produces phialides which carry at their apex conidia, unicellular, globular or ovoid. Colonies also develop chlamydospores, resistance spores.

*Figure 1. Culture of Trichoderma sp. on Potato Sucrose Agar (PSA), after 7 days incubation at 25°C;* **A:** *Macroscopic appearance in rect;* **B:** *Phialides (Ph);* **C:** *Conidia (cd);* **D:** *Chlamydospores (Chd)*

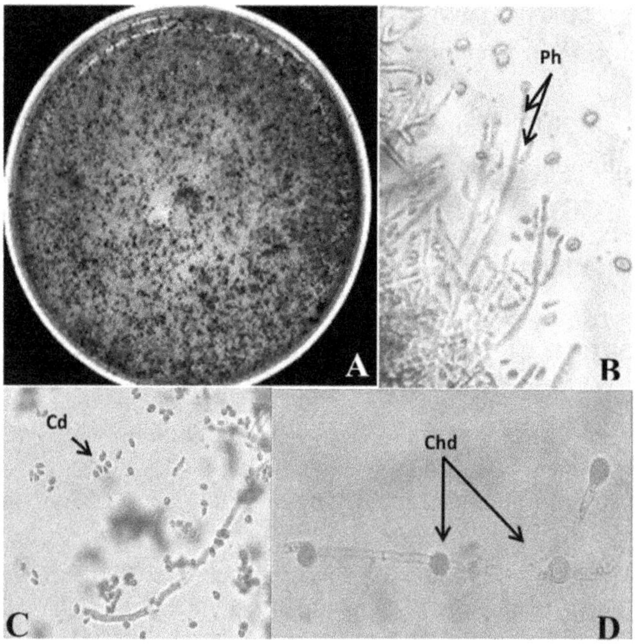


## Inoculation of Bulbs with *Trichoderma* sp.

After incubation, the culture surface, charged with *Trichoderma* sp. conidia, was scraped and then filtered to separate the mycelium from the conidia. Subsequently, the concentration of the conidian suspension was adjusted to $10^7$conidia/ml (Qostal et al., 2020).

The solution used for the treatment of the bulbs consists of 10% gelatin, 0.1% sucrose and the conidian suspension of *Trichoderma* sp.

Prior to inoculation of saffron corms, they were disinfected and then immersed in the *Trichoderma* sp-based inoculum. For 4 hours at 28 ° C. These corms were rolled on the clay until a uniform and fine layer was formed. The treated corms were dried in the air under a hood for one night. The control corms underwent the same steps but without the presence *of Trichoder*ma sp.

## Greenhouse Tests

The treated or untreated corms are cultured in a greenhouse on a substrate consisting of Mamora sand, low in phosphorus (sandy soil, pH (extracted with water), 7.5, organic matter 0.70%, total nitrogen: 0.05%, total phosphorus: 2.239%, total potassium: 0.15 (meq/100 mg, Magnesium: 0.20 (meq/100 g), Calcium: 6.30 (meq/100 g)). The substrate is worked in the form of polyethylene-coated lines, with a drip irrigation system. 20 corms are grown in each line.

## Measurement of Agronomic Parameters

After 4 months of cultivation, the plants are brought back from the greenhouse, the length and number of leaves and their weights are measured. The number of bulbs formed (bulbs) and their weights are also recorded.

## Re-Isolation of *Trichoderma* sp.

Re-isolation of *Trichoderma* sp. from saffron roots of treated corms is carried out after 4 months of cultivation in greenhouse. The roots were cut into a 1 cm fragment, disinfected with 96° ethanol for 1 minute, then rinsed three times with sterile distilled water, dried on sterile filter paper and deposited on a PSA medium and incubated in darkness at 28 ° C. The developing colonies were then observed for species determination.

The percentage of re-isolation **(Pr%)** was obtained by applying the following formula:

$$Pr = Ns\ Px / NT\ x\ 100$$

**Ns Px:** Number of segments containing *Trichoderma***sp**.
**NT:** Total number of segments.

## Statistical Analysis

The analysis of the variance and of the comparisons of means by the LSD test (p = 5%) was carried out using the software STATISTICA.

# RESULTS AND DISCUSSION

In this study, *Trichoderma* sp. improved various saffron corm growth parameters compared with the control after a 4-month inoculation period. All corms inoculated by *Trichoderma* sp. showed a more pronounced development compared to uninoculated corms. The number and weight of corms and leaves, as well as the weight of the roots of treated corms, showed an increase over those of untreated corms, recording respectively (10/5.75); (16.5g / 10.36g); (31/16.5) ; (2.53g / 1.93g); (0.3g / 0.15g). However, the treatment had no impact on the length of the leaves of the inoculated corms, which was 21 cm, compared with that of the non-inoculated corms, which was 23 cm (Table 1).

*Table 1. Effect of Trichoderma sp. on saffron growth parameters and bulb multiplication*

| Agronomic parameters | Control | Treatment with *Trichoderma* sp. |
|---|---|---|
| Average number of corms | $5.75^{a}$ | $10^{b}$ |
| Weight of corms (g) | $10.36^{a}$ | $16.5^{b}$ |
| Number of leaves | $16.5^{a}$ | $31^{b}$ |

continued on following page

*Table 1. Continued*

| Agronomic parameters | Control | Treatment with *Trichoderma* sp. |
|---|---|---|
| Length of leaves (cm) | 23[a] | 21[a] |
| Leaf weight (g) | 1.93[a] | 2.53[b] |
| Weight of roots (g) | 0.15[a] | 0.3[b] |

Two results, read on the same line, followed by two different letters differ significantly at the 5% threshold

Treatment with *Trichoderma* sp. applied to saffron corms showed a significant improvement in the production of new corms (Bubbles), roots and leaves, with the exception of their length, compared to the control (Figure 2)

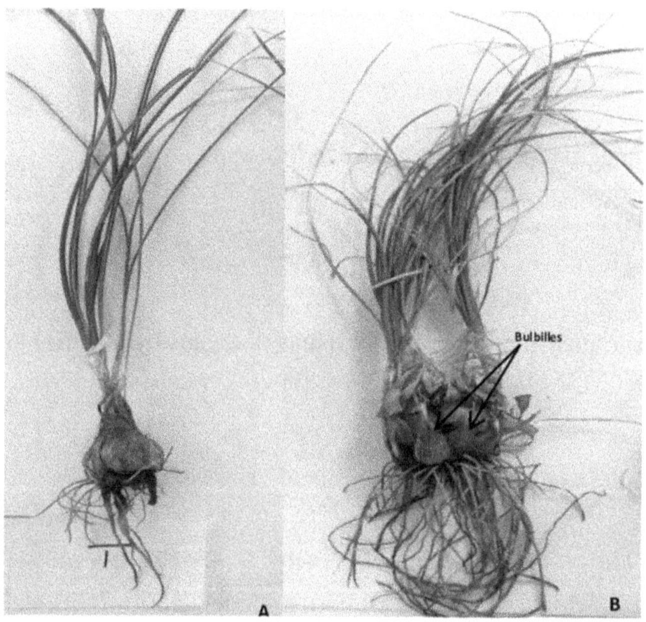


Based on these results, *Trichoderma* sp. had a positive effect on both the aerial part and the root system of the saffron. This is attributed to the colonizing power of the saffron rhizosphere by *Trichoderma* sp. and establishing an interaction with its roots. Root fixation Makes available various nutrients such as copper, iron, phosphorus, manganese, and sodium present in the soil, which facilitates their uptake by the plant (Yedidia et al., 2001; Vinale et al., 2016). Therefore, this is reflected in the plant's vigour and resistance to abiotic stress.

Previous research has reported the influence of Trichoderma on root development and biomass production (Yao et al., 2023; andrzejak et al., 2022; Tančić-Živanov et al., 2020; Nieto-Jacobo et al., 2017), this is in part consistent with the results obtained by Kribel et al., (2020) in wheat and barley plants, Chliyeh et al., (2014) in tomato plants, in particular on the length of the northern part, and Tančić-Živanov et al., (2020) in root pepper plants, acting on.

Trichoderma species have promising plant growth properties through various mechanisms such as nutrient mobilization in soil and water retention in case of water stress (Hmouni et al., 2006; Mouria et al. 2007), phosphorus solubilization (kribel et al. 2019) and the synthesis of plant growth regulators (Reghmit et al., 2022; Nieto-Jacobo et al., 2017). However, not all Trichoderma isolates have the ability to stimulate plant growth. In addition, the interaction between the plant and Trichoderma sp. depends on the genotype of the plant studied. This may be the cause of the lack of significant impact on leaf length which is a specific aspect of saffron plant growth. In addition to these two factors, plant species, Trichoderma isolate, and also environmental conditions may influence these growth parameters.

## Re-Isolation of *Trichoderma* sp.

*Trichoderma* sp. was successfully re-isolated from saffron roots, with a rate of re-isolation of 20% using the PSA medium (Figure 3). This suggests that *Trichoderma* sp. can colonize saffron roots, potentially protecting the plant from biotic stress (Agostini et al., 2023), and plays a crucial role in promoting plant growth and health (Schalamun et al., 2023; Taylor et al., 2022).

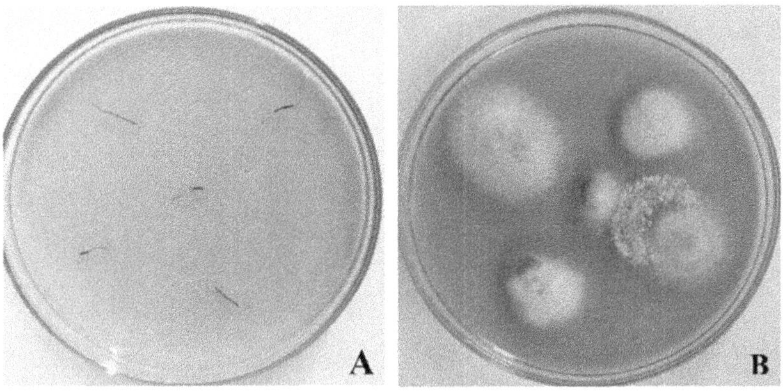


The treatment of corms with an inoculum based on Trichoderma conidia makes it possible to improve the growth and development of saffron plants as well as the multiplication of bulbs. This treatment, which offers important results, could be a very effective method for improving crop production. It is a simple and inexpensive technique that avoids the use of chemicals that can be harmful to the environment and human health.

## CONCLUSION

The results suggest that *Trichoderma* sp. had a positive influence, notably on the formation of bulbils, the improvement of the aerial part, as well as on the development of the root system of plants from treated corms.

Treatment of corms with *Trichoderma* sp. emerges as a promising approach, still in the development phase, aimed at promoting the growth parameters of plants derived from these corms as well as the improvement of the yield of this crop.

Further research is needed to better understand the mechanisms underlying the beneficial effects of this species on *Crocus sativus* and to optimize its use in stimulating the formation of female corms, responding to the availability of planting material, and in the management of cryptogamic diseases associated with saffron.

## ACKNOWLEDGMENT

This study was conducted under the research project of the 2nd PMA 2020/11, entitled 'Elaboration d'un produit biofertilisant et biopesticide, à base de champignons endophytes, adapté à la protection et l'amélioration de la productivité du safran au Maroc' financed by 'Agence Nationale des Plantes Médicinales et Aromatique' et l'Université Ibn Tofail, Kénitra, Morocco'

## REFERENCES

Abdenaceur, R., Farida, B. T., Mourad, D., Rima, H., Zahia, O., & Fatma, S. H. (2022). Effective biofertilizer Trichoderma spp. isolates with enzymatic activity and metabolites enhancing plant growth. *International Microbiology*, *25*(4), 817–829. DOI: 10.1007/s10123-022-00263-8 PMID: 35854164

Adnani, M., El Hazzat, N., Msairi, S., El Alaoui, M. A., Mouden, N., Selmaoui, K., Benkirane, R., Ouazzani Touhami, A., & Douira, A. (2024). Exploring the efficacy of a Trichoderma asperellum-based seed treatment for controlling Fusarium equiseti in chickpea. *Egyptian Journal of Biological Pest Control*, *34*(1), 7. DOI: 10.1186/s41938-024-00771-x

Agostini, R. B., Ariel, F., Rius, S. P., Vargas, W. A., & Campos-Bermudez, V. A. (2023). Trichoderma root colonization in maize triggers epigenetic changes in genes related to the jasmonic and salicylic acid pathways that prime defenses against Colletotrichum graminicola leaf infection. *Journal of Experimental Botany*, *74*(6), 2016–2028. DOI: 10.1093/jxb/erac518 PMID: 36575905

Andrzejak, R., & Janowska, B. (2022). Trichoderma spp. improves flowering, quality, and nutritional status of ornamental plants. *International Journal of Molecular Sciences*, *23*(24), 15662. DOI: 10.3390/ijms232415662 PMID: 36555304

Azeddine, E., Ouazzani Touhami, A., Selmaoui, K., Benkirane, A. R., & Douira, A. (2024). In Vitro and In Vivo Biological Control by Trichoderma Asperellum Against Rhizoctonia Solani a Causal Agent of Collar and Root Rot in Strawberries. In *Advanced Systems for Environmental Monitoring, IoT and the application of Artificial Intelligence* (pp. 213–235). Springer Nature Switzerland. DOI: 10.1007/978-3-031-50860-8_13

Belfiori, B., Rubini, A., & Riccioni, C. (2021). Diversity of endophytic and pathogenic fungi of saffron (Crocus sativus) plants from cultivation sites in Italy. *Diversity*, *13*(11), 535. DOI: 10.3390/d13110535

Chliyeh, M., Chahdi, A. O., Selmaoui, K., Touhami, O. A., Filali Maltouf, A., El Modafar, C., & Douira, A. (2014). Effect of Trichoderma harzianum and arbuscular mycorrhizal fungi against Verticillium wilt of tomato. *International Journal of Recent Scientific Research*, *5*(2), 449–459.

El Aymani, I., El Gabardi, S., Artib, M., Chliyeh, M., Selmaoui, K., Ouazzani Touhami, A., Benkirane, R., & Douira, A. (2019a). Effect of the number of years of soil exploitation by saffron cultivation in Morocco on the diversity of endomycorrhizal fungi. *Acta Phytopathologica et Entomologica Hungarica*, *54*(1), 9–24. DOI: 10.1556/038.54.2019.002

El Aymani, I., Ourras, S., Mouden, N., Chliyeh, M., Selmaoui, K., Msairi, S., Benkirane, R., El Modafar, C., Touhami, A. O., & Douira, A. (2023). Effect of endomycorrhizal fungi inoculum on agro morphological behavior and productivity of saffron (crocus sativus l.) under water and salinity stress. *Acta Universitatis Agriculturae et Silviculturae Mendelianae Brunensis*, *71*(4), 183–192. DOI: 10.11118/actaun.2023.013

El Aymani, I., Qostal, S., Mouden, N., Selmaoui, K., Touhami, A. O., Benkirane, R., & Douira, A. (2019b). Fungi associated with saffron (*Crocus sativus*) in Morocco. *Plant Cell Biotechnology and Molecular Biology*, *20*, 1180–1188.

El Kaissoumi, H., Berber, F., Mouden, N., Ouazzani Chahdi, A., Albatnan, A., Ouazzani Touhami, A., . . . Douira, A. (2022, May). Effect of Trichoderma asperellum on the development of strawberry plants and biocontrol of anthracnose disease caused by colletotrichum gloeosporioides. In *International Conference on Advanced Intelligent Systems for Sustainable Development* (pp. 609-622). Cham: Springer Nature Switzerland.

Gupta, R., & Vakhlu, J. (2015). Native Bacillus amyloliquefaciens W2 as a potential biocontrol for *Fusarium oxysporum* R1 causing corm rot of *Crocus sativus.European Journal of Plant Pathology*, *143*(1), 123–131. DOI: 10.1007/s10658-015-0670-3

Gupta, V., Sharma, A., Rai, P. K., Gupta, S. K., Singh, B., Sharma, S. K., Singh, S. K., Hussain, R., Razdan, V. K., Kumar, D., Paswal, S., Pandit, V., & Sharma, R. (2021). Corm Rot of Saffron: Epidemiology and Management. *Agronomy (Basel)*, *11*(2), 339. DOI: 10.3390/agronomy11020339

Hmouni, A., Douira, A., & Mouria, A. (2006, August). Biological Control of Tomato Grey Mould with Compost Water Extracts," Trichoderma" sp., and" Gliocladium" sp. *Phytopathologia Mediterranea*, 1000–1007.

Kaissoumi, H. E., Berbera, F., Mouden, N., Ouazzani Chahdi, A., Touhami, A. O., Selmaoui, K., & Douira, A. (2024). Tomato Growth Promotion by Trichoderma Asperellum Laboratory-Made Bioproduct. In *Sustainable and Green Technologies for Water and Environmental Management* (pp. 161–171). Springer Nature Switzerland. DOI: 10.1007/978-3-031-52419-6_13

Khirallah, W., Touhami, A. O., Diria, G., Gaboun, F., Benkirane, R., & Douira, A. (2017). Variability and Genetic Structure of a Natural Population of Trichoderma spp. Isolated from Different Substrates in Morocco. *Annual Research & Review in Biology*, *17*(1), 1–11. DOI: 10.9734/ARRB/2017/35389

Koocheki, A., & Seyyedi, S. M. (2015). Relationship between nitrogen and phosphorus use efficiency in saffron (Crocus sativus L.) as affected by mother corm size and fertilization. *Industrial Crops and Products*, *71*, 128–137. DOI: 10.1016/j.indcrop.2015.03.085

Kotba, I., El Kinany, S., Bouaichi, A., Achbani, E. H., Ouazzani Touhami, A., & Douira, A. (2022). Pathogenicity study of Rhizoctonia solani and its integrated management on different crop species in Morocco. *Archiv für Phytopathologie und Pflanzenschutz*, *55*(20), 2301–2316. DOI: 10.1080/03235408.2022.2164160

Kribel, S., Qostal, S., Ouazzani Touhami, A., Selmaoui, K., Chliyeh, M., Benkirane, R., & Achbani, E. H. (2020). Effects of Trichoderma on growth and yield of wheat and barley and its survival ability on roots and amended rock phosphate growing substrates. [Journal of Fungal Biology]. *Current Research in Environmental & Applied Mycology*, *10*(1), 400–416. DOI: 10.5943/cream/10/1/32

Kribel, S., Qostal, S., Ouazzani Touhami, A., Selmaoui, K., Mouria, A., Benkirane, R., Achbani, E. H., & Douira, A. (2019). Quantitative and qualitative estimation of Moroccan Trichoderma isolates capacity to solubilize rock phosphate. *Acta Phytopathologica et Entomologica Hungarica, 54*(2), 157–171. DOI: 10.1556/038.54.2019.016

Menia, M., Iqbal, S., Zahida, R., Tahir, S., Kanth, R. H., Saad, A. A., & Hussian, A. (2018). Production technology of saffron for enhancing productivity. *Journal of Pharmacognosy and Phytochemistry, 7*(1), 1033–1039.

Mouden, N., Ouazzani Touhami, A., Albatnan, A., Selmaoui, K., Benkirane, R., & Douira, A. (2022, May). Growth Promoting of Tomato Plants by Incorporation of Trichoderma asperellum Enriched Liquid Product via Foliar Spray and the Irrigation System. In *International Conference on Advanced Intelligent Systems for Sustainable Development* (pp. 599-608). Cham: Springer Nature Switzerland.

Mouria, B., Ouazzani-Touhami, A., & Douira, A. (2008). Effect of various strains of Trichoderma on the growth of a greenhouse tomato crop and their ability to colonize roots and substrate. *Phytoprotection, 88*(3), 103–110. DOI: 10.7202/018955ar

Mouria, B., Ouazzani Touhami, A., & Douira, A. (2013). Effet du compost et de Trichoderma harzianum sur la suppression de la verticilliose de la tomate. *Journal of Applied Biosciences, 70*(1), 5531–5543. DOI: 10.4314/jab.v70i1.98751

Mouria, B., Ouazzani-Touhami, A., Mouria, A., Benkirane, R., & Douira, A. (2015). Effect of compost and antagonistic fungi on suppression of Tomato Grey Mold. *Biolife : An International Quarterly Journal of Biology and Life Sciences, 3*(2), 378–390. DOI: 10.17812/blj2015.32.5a

Negbi, M., Dagan, B., Dror, A., & Basker, D. (1989). Growth, flowering, vegetative reproduction, and dormancy in the saffron crocus (Crocus sativus L.). *Israel Journal of Plant Sciences, 38*(2-3), 95–113.

Nieto-Jacobo, M. F., Steyaert, J. M., Salazar-Badillo, F. B., Nguyen, D. V., Rostás, M., Braithwaite, M., De Souza, J. T., Jimenez-Bremont, J. F., Ohkura, M., Stewart, A., & Mendoza-Mendoza, A. (2017). Environmental growth conditions of Trichoderma spp. affects indole acetic acid derivatives, volatile organic compounds, and plant growth promotion. *Frontiers in Plant Science, 8*, 102. DOI: 10.3389/fpls.2017.00102 PMID: 28232840

Ourras, S., Aymani, I. E., Mouden, N., Selmaoui, K., Msairi, S., Elouark, M. H., . . . Douira, A. (2022, May). Mycoflora of Dormant Crocus Sativus Corms in Morocco. In *International Conference on Advanced Intelligent Systems for Sustainable Development* (pp. 479-489). Cham: Springer Nature Switzerland.

Ourras, S., Gabardi, S. E., El Aymani, I., Mouden, N., Chliyeh, M., Selmaoui, K., & Douira, A. (2021). Diversity of arbuscular mycorrhizal fungi in the rhizosphere of saffron (Crocus sativus) plants along with age of plantation in Taliouine region in Morocco. *Acta Biologica Szegediensis, 65*(2), 199–209.

Parray, J. A., Kamili, A. N., Hamid, R., & Husaini, A. M. (2012). In vitro cormlet production of saffron (Crocus sativus L. Kashmirianus) and their flowering response under greenhouse. *GM Crops and Food: Biotechnology in Agriculture and the Food Chain, 3*(4), 289–295. DOI: 10.4161/gmcr.21365 PMID: 22892688

Pérez-Montaño, F., Alías-Villegas, C., Bellogín, R. A., Del Cerro, P., Espuny, M. R., Jiménez-Guerrero, I., López-Baena, F. J., Ollero, F. J., & Cubo, T. (2014). Plant growth promotion in cereal and leguminous agricultural important plants: From microorganism capacities to crop production. *Microbiological Research, 169*(5-6), 325–336. DOI: 10.1016/j.micres.2013.09.011 PMID: 24144612

Qostal, S., Kribel, S., Chliyeh, M., Selmaoui, K., Serghat, S., Benkirane, R., & Douira, A. (2020). Management of wheat and barley root rot through seed treatment with biopestcides and fungicides. *Plant Cell Biotechnology and Molecular Biology, 21*(35&36), 129–143.

Renau-Morata, B., Moyá, L., Nebauer, S. G., Seguí-Simarro, J. M., Parra-Vega, V., Gómez, M. D., & Molina, R. V. (2013). The use of corms produced under storage at low temperatures as a source of explants for the in vitro propagation of saffron reduces contamination levels and increases multiplication rates. *Industrial Crops and Products, 46*, 97–104. DOI: 10.1016/j.indcrop.2013.01.013

Sampathu, S. R., Shivashankar, S., Lewis, Y. S., & Wood, A. B. (1984). Saffron (Crocus sativus Linn.)—Cultivation, processing, chemistry and standardization. *CRC Critical Reviews in Food Science and Nutrition, 20*(2), 123–157. DOI: 10.1080/10408398409527386

Schalamun, M., Li, G., Hinterdobler, W., Grosskinsky, D. K., Compant, S., Dreux-Zigha, A., . . . Schmoll, M. (2023). Plant recognition by Trichoderma harzianum elicits upregulation of a novel secondary metabolite cluster required for colonization. BioRxiv, 2023-04.

Sellal, Z., Touhami, A. O., Mouden, N., Selmaoui, K., Dahmani, J., Benkirane, R., & Douira, A. (2023, October). Effect of the Combined Treatment of Endomycorrhizal Inoculum and Trichoderma Harzianum on the Growth of Argan Plants. In *International Conference on Advanced Intelligent Systems for Sustainable Development* (pp. 312-319). Cham: Springer Nature Switzerland.

Sellal, Z., Touhami, O. A., Chliyeh, M., Mouden, N., Selmaoui, K., Dahmani, J., & Douira, A. (2020). Effect of seeds treatment with Trichoderma harzianum on argan plants growth. *Plant Cell Biotechnology and Molecular Biology, 21*(11&12), 69–77.

Sghir, F., Chliyeh, M., Touati, J., Mouria, B., Touhami, O. A., Filali-Maltouf, A., & Douira, A. (2014). Effect of a dual inoculation with endomycorrhizae and Trichoderma harzianum on the growth of date palm seedlings. *Int. J. Pure App. Biosci, 2*(6), 12–26.

Sharaf-Eldin, M. A., Elkholy, S., Fernández, J. A., Junge, H., Cheetham, R. D., Guardiola, J. L., & Weathers, P. J. (2007). The effect of Bacillus subtilis FZB24® on flowers quantity and quality of saffron (Crocus sativus L.). *Planta Medica, 73*(09), P_607.

Tahiri, A., Karra, Y., Aabd, N. A., Qessaoui, R., Bouharroud, R., & Mimouni, A. (2022). Effet des substances humiques de la Leonardite sur la croissance et la multiplication végétative du safran (Crocus sativus L.). *African and Mediterranean Agricultural Journal-Al Awamia*, (137), 163–177.

Tančić-Živanov, S., Medić-Pap, S., Danojević, D., & Prvulović, D. (2020). Effect of Trichoderma spp. on growth promotion and antioxidative activity of pepper seedlings. *Brazilian Archives of Biology and Technology, 63*, e20180659. DOI: 10.1590/1678-4324-2020180659

Taylor, J. T., Harting, R., Shalaby, S., Kenerley, C. M., Braus, G. H., & Horwitz, B. A. (2022). Adhesion as a Focus in Trichoderma–Root Interactions. *Journal of Fungi (Basel, Switzerland)*, *8*(4), 372. DOI: 10.3390/jof8040372 PMID: 35448603

Vinale, F., Strakowska, J., Mazzei, P., Piccolo, A., Marra, R., Lombardi, N., Manganiello, G., Pascale, A., Woo, S. L., & Lorito, M. (2016). Cremenolide, a new antifungal, 10-member lactone from Trichoderma cremeum with plant growth promotion activity. *Natural Product Research*, *30*(22), 2575–2581. DOI: 10.1080/14786419.2015.1131985 PMID: 26728227

Yao, X., Guo, H., Zhang, K., Zhao, M., Ruan, J., & Chen, J. (2023). Trichoderma and its role in biological control of plant fungal and nematode disease. *Frontiers in Microbiology*, *14*, 1160551. DOI: 10.3389/fmicb.2023.1160551 PMID: 37206337

Yedidia, I., Srivastva, A. K., Kapulnik, Y., & Chet, I. (2001). Effect of Trichoderma harzianum on microelement concentrations and increased growth of cucumber plants. *Plant and Soil*, *235*(2), 235–242. DOI: 10.1023/A:1011990013955

Chapter 9

# Environmental Assessment of Microalgal Biomass Fermentation for Green Hydrogen Production

**Khadija El-Moustaqim**
https://orcid.org/0009-0007-9193-6638
*Ibn Tofail University, Morocco*

**Toufik Rachiq**
https://orcid.org/0009-0001-8930-3328
*Mohammed V University in Rabat, Morocco*

**Imane Bencheikh**
*Mohammed V University in Rabat, Morocco*

**Jamal Mabrouki**
https://orcid.org/0000-0002-3841-7755
*Faculty of Science Rabat, Morocco*

**Driss Hmouni**
*Faculty of Science Kenitra, Morocco*

## ABSTRACT

*Hydrogen is an environmentally friendly energy source that is crucial in addressing global climate change and energy shortages. Microalgae are a highly appealing method for generating biohydrogen, which is a clean and environmentally friendly energy source. This strategy can help reach carbon neutrality and ensure the sustainability of bioenergy. The production of biohydrogen by dark fermentation is seen as a promising way of using clean, renewable energy. Microbial fermentation activities produce the majority of hydrogen in the biosphere. The biological synthesis of hydrogen is an attractive approach to renewable energy production, utilizing both photosynthesis and fermentation processes. Certain types of green algae, blue-green algae (cyanobacteria), fermentative and photosynthetic bacteria, and archaea can produce hydrogen by using light energy and/or various organic substances. Our work aims to study*

DOI: 10.4018/979-8-3693-8980-5.ch009

*the role and influence of biohydrogen production by dark fermentation using microalgae, as well as improve the production yield, with a view to wider application in the future*

## 1. INTRODUCTION

The conventional techniques used to produce hydrogen, however, are linked to environmental pollution around the world (Benchrifa et al. 2024). For this reason, the energy and environment sectors are working to develop more sustainable ways of producing hydrogen from renewable resources (Abanades et Rodat 2022; Qureshi et al. 2023). The production of green hydrogen is widely acknowledged to be environmentally friendly and sustainable, as it utilizes a renewable source (Machhirake et al. 2024). Provided that the electricity utilized is derived from a renewable energy source such as hydroelectricity, wind or solar power, or biomass, all aspects of the manufacturing cycle have a minimal carbon impact (El-Moustaqim, Mabrouki, Azrour, et al. 2024). The hydrogen generated is referred to as green hydrogen (El-Moustaqim, Mabrouki, et Hmouni 2024b). This manufacturing process produces minute amounts of greenhouse gas emissions (El-Moustaqim, Mohammed, Mabrouki, et al. 2024; El-Moustaqim, Rachiq, Mabrouki, et al. 2024). Green hydrogen constitutes around 2% of the total worldwide output. The present production costs of this substance exceed those of grey hydrogen (Jarosz et al. 2024). Policymakers and researchers are paying more attention to biohydrogen since it is seen as an environmentally benign energy source that may play a major part in global sustainability efforts (El-Moustaqim et al. 2024; El-Moustaqim, Mabrouki, et Hmouni 2024a; Leong et al. 2024). It is also regarded as a green carbon-free energy source because it emits no products that include carbon dioxide (Sankir et Sankir 2024; Sarkar, Katakojwala, et Mohan 2021). The potential of biomass, including substrates such as food waste, wastewater, sludge, algae, and lignocellulosic materials, for biohydrogen production with reduced carbon emissions is considerable (Buffi, Prussi, et Scarlat 2022). From the circular economy perspective, using microalgae in wastewater treatment offers an ecologically friendly and potentially economically viable substitute for conventional treatment techniques (Udaypal, Goswami, et Verma 2024). However, the general adoption of this technology still needs more study and advancement to surmount the obstacles of enhancing prices and adjusting to various environmental situations (Akram et al. 2024). Most hydrogen supply is now derived from the reforming or gasifying of hydrocarbons (Jain et al. 2024). Water electrolysis is expected to emerge as the predominant technique for hydrogen generation due to its clean nature (eliminating carbon emissions when combined with a renewable energy source) and ability to provide hydrogen of exceptional purity (El-Moustaqim et al. 2024; Omer et al. 2024; Oraby et Shawqi 2024). The production of biohydrogen by dark fermentation is seen as a promising way of using clean, renewable energy (Suastes-Rivas et al. 2024; Sun et al. 2022; Yarkent et al. 2024). Dark fermentation is a widely recognized technique that functions without light (Montiel-Corona et Buitrón 2024). As one of the metabolic byproducts, hydrogen is produced here when anaerobic microbes break down a range of organic substances, including biomass and organic waste (Jung et al. 2013; Kucharska et al. 2021; Xia et al. 2016). This technique is very adaptable and may make use of a broad spectrum of raw materials (Haruna et al. 2024).

Our work plays a role in analyzing methods for producing biohydrogen by dark fermentation using microalgae, as well as improving production yields, highlighting factors affecting its productivity and key strategies to enhance its productivity, and also emphasizing the economic assessment of microalgal biohydrogen production with present challenges. with a view to wider application in the future.

## 2. PRINCIPLE OF DARK FERMENTATION PRODUCTION

Biofuels derived from microalgae are not a novel concept. Indeed, the cultivation of microalgae began to be motivated by a period of food shortage in the 1940s-50s, when they were seen as an alternative source of proteins, lipids, and sugars (Fei-Yue, Yu, et Gao 2022). Microalgae are particularly attractive for the production of biosource diesel, which has attracted the interest of numerous research programs (Dębowski et al. 2020; Loomba et Khan 2023). Through dark fermentation (**Fig. 1**), bacterial activity transforms organic waste into tiny organic acids, solvents, and gases (Jain et al. 2024). The accumulation of reduced biochemical intermediates, such as formate, nicotinamide adenine dinucleotide (NADH), and ferredoxin (FdH2), can lead to the production of hydrogen to regulate pH or keep the internal redox potential constant (NADH and FdH2) (Kucharska et al. 2021). Micro-organisms with specific hydrogen-producing enzymes, known as hydrogenases, can release excess reduced elements in the form of hydrogen gas by reducing two protons in the following reaction:

$$2H^{+} + 2e^{-} \leftrightarrow H_2$$

Dark fermentation is the transformation of substrates by anaerobic bacteria that flourish in the absence of oxygen. Hydrogen serves as the primary substrate in the metabolic processes of several anaerobic bacteria. Depending on the availability of energy-rich hydrogen molecules, these microorganisms can utilize the electrons obtained from hydrogen oxidation to generate energy. When organisms lack external electron acceptors, they accumulate an excess of electrons during metabolic activities that result in the reduction of protons to form hydrogen molecules. Hydrogenases are the primary enzymes to control hydrogen metabolism.

*Figure 1. Descriptive diagram of the dark fermentation process.*

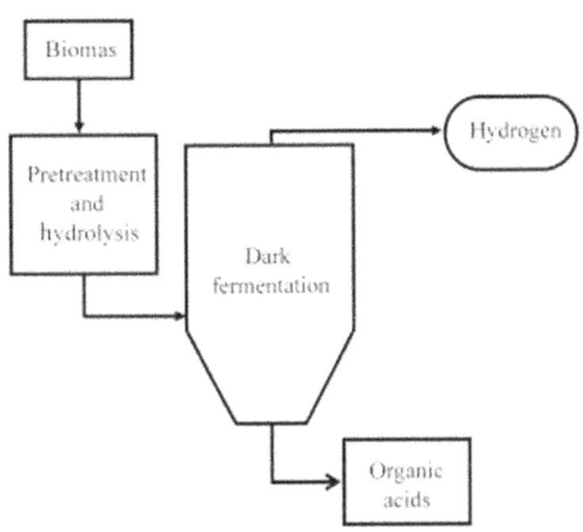


During the first step of this fermentation process, glycolysis, glucose is transformed into pyruvate, the crucial intermediate in the production of reduced nicotinamide adenine dinucleotide (NADH). Pyruvate can be converted into acetyl-CoA under anaerobic circumstances via a mechanism facilitated by pyruvate ferredoxin oxidoreductase (PFOR). Besides acetyl-CoA and $CO_2$, the reaction products also consist of reduced ferredoxin, which participates in the reduction of [FeFe]-hydrogenases. These enzymes subsequently reduce protons to generate hydrogen. An alternative route for transforming pyruvate into acetyl-CoA while producing formate is by the catalytic action of pyruvate formate lyase (PFL).

## 3. SEVERAL APPROACHES TO IMPROVING THE FERMENTATION PROCESS

### 3.1 The Origin and Choice of Microorganisms That Produce Biohydrogen

Pure cultures produce biohydrogen using a single microbial species capable of degrading and consuming an organic substrate (Oraby et Shawqi 2024). This makes it possible to generate biohydrogen through various fermentation pathways such as acetate, butyrate, formate, lactate, etc. Depending on the temperature range at which their development is optimal, the microorganisms used can be strict anaerobes (*Clostridium sp., Ruminococcus sp.*) or facultative anaerobes (*Escherichia coli sp., Enterobacter sp., Citrobacter sp.*), mesophiles (25-40°C), thermophiles (40-80°C) or hyperthermophiles (> 80°C) (Fei-Yue et al. 2022; Suastes-Rivas et al. 2024). Overall, the yields and biohydrogen rates obtained from pure cultures are extremely unstable (Machhirake et al. 2024). The results are largely influenced by the substrate to be degraded, the culture conditions, and the bacterial strain used (Akram et al. 2024). It should be emphasized that the process requires the use of strict anaerobic strains, which require a total absence of oxygen in the culture medium. Finally, the major disadvantage of pure culture fermentations is the need to remain in axenic conditions to avoid contamination by other microorganisms during inoculation and culture (Buffi et al. 2022).

### 3.2 Organic Materials Used in the Manufacture of Biohydrogen

The effect of organic components on yields, bio-hydrogen generation, and the economics of the dark fermentation process is substantial. The primary determinants of this phenomenon are the carbohydrate composition of the substrate and the accessibility and rate of breakdown of the organic residue (Qureshi et al. 2023). Typically, basic substrates like monosaccharides and disaccharides and more intricate substrates like agricultural and food waste are employed. The primary goal of research with basic substrates is to rapidly assess the process's efficiency (Sankir et Sankir 2024). Nevertheless, biohydrogen synthesis by dark fermentation necessitates using substrates derived from sustainable sources to preserve energy and the environment. The breakdown of these substrates is effectively facilitated by optimizing the temperature at which the enzymes exhibit their highest level of activity (Udaypal et al. 2024).

## 4. UTILIZATION OF MICROALGAE AS A PRIMARY SOURCE FOR HYDROGEN GENERATION BY FERMENTATION

The release of hydrogen as a by-product by fermentative bacteria, which use their carbon sources in the absence of oxygen, has been harnessed for many years in the synthesis of biohydrogen (Palumbo et al. 2024). In order to generate ATP, these bacteria employ protons as terminal electron acceptors to release hydrogen, therefore ensuring the maintenance of electrical neutrality within the cells (Ren et al.

2011). Fermentation-based hydrogen generation has great potential for enhancing the value of waste materials, and certain industrial effluents can serve as suitable substrates for dark fermentation.

*Figure 2. Various pathways from microalgae to biofuel energy production*

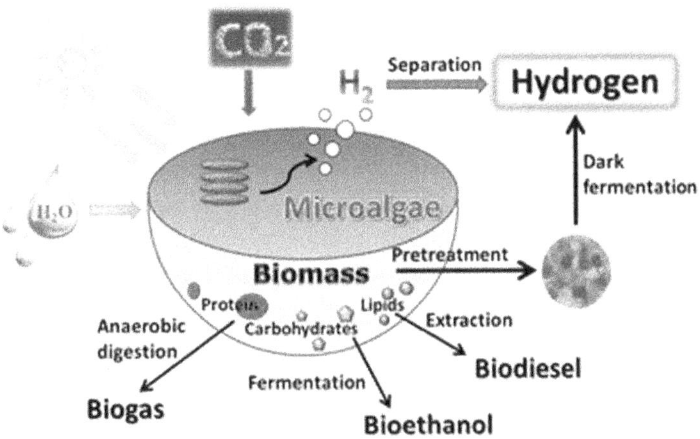


During fermentation, the release of inhibitors pre-treatment (**Fig. 2**) (Wang et Yin 2018), such as furan aldehydes, hinders the successful use of this biomass for hydrogen production (Jung et al. 2013; Sun et al. 2011). Compared with bio-photolysis, hydrogen production by fermentation is free from all the disadvantages of an algae culture system (Haruna et al. 2024), including sufficient light for algae culture and maintaining the particular levels of nutrients and stress required for hydrogen production (Kucharska et al. 2021; Qiu et al. 2023). Microalgae lack lignin and their relatively simple cell walls contain cellulose and hemicellulose, making them easy to pre-treat. By nature, microalgae can accumulate lipids or carbohydrates, varying from strain to strain (El-Moustaqim et al. 2024; Varras et Chalaris 2023).

## 5. EFFICIENCY OF MICROALGAE IN THE PRODUCTION OF GREEN HYDROGEN BY FERMENTATION

The degradation of microalgae cell walls and the liberation of carbohydrates for utilization by hydrogen-generating bacteria need the implementation of very effective pre-treatments (Abessa Rahma 2014; Dev, Bhatt, et Gupta 2024). Although whole cells were used as substrates without hydrogen production, the generation of hydrogen was much increased when the cell wall ruptured by ultrasonic or methanol treatment. Currently, the process of hydrolyzing microalgal biomass involves the use of mechanical, thermal, chemical, and enzymatic pre-treatments (Ghosh et Basu 2024; Oneby et al. 2010; Xia et al. 2016). Bead beating and grinding are mechanical techniques that are effective for microalgae strain. They may be used to open the microalgae cell wall by applying shear stresses, therefore releasing the cell contents for hydrolysis (Algapani et al. 2016; Chum et Overend 2001; Jariyaboon, O-Thong, et Kongjan 2015). Due to the great selectivity and specificity of the enzymes utilized and the comparatively moderate reaction conditions compared to other pre-treatment procedures, enzymatic hydrolysis is the most effective mechanism (Hall 2007; Li et Fang 2007). Employing hydrolytic enzymes is not advisable for achieving competitive hydrogen generation due to their high cost. Certain researchers employ partially purified or crude enzymes that are readily accessible in their laboratory for the purpose of enzymatic hydrolysis, therefore rendering the procedure economically viable (Hasnaoui 2021; Kothari et al. 2012; Li et Fang 2007).

*Table 1. Production of green hydrogen by fermentation.*

| Fermenting bacterial cell | Microalgal raw material | Pre-treatment process | Quantity of hydrogen produced | Refs |
|---|---|---|---|---|
| Anaerobic sludge | *Arthrospira platensis, Laminaria digitata* | Heating to 135°C with 2.5% $H_2SO_4$ | 85.0 ml/g VS | (Xia et al. 2016) |
| Heat-treated anaerobic sludge | *Chlorella sp.* | – | 7.13 ml/g VS | (Sun et al. 2011) |
| Mixed microbial consortia | *Chlorella pyrenoidosa* | Chauffage à la vapeur avec de l'acide dilué | 83.3 ml/g VS | (Xia et al. 2013) |
| *Thermotoga neopolitana* | *Chlamydomonas reinhardtii* | Hydrolyse enzymatique | 311.1 ml $H_2$/g monosaccharides | (Tam et al. 2010) |
| Heat-treated anaerobic seed sludge | *Chlorella Vulgaris* | 0,79% HCl, 49 600 kJ/kg dcw sonication | 42,1 ml $H_2$/g de poids cellulaire sec | (Jung et al. 2013) |
| *Enterobacter aerogenes* ATCC 13048 | *Nannochloropsis sp.* | Biomasse microbienne déshuilée | 60,6 ml $H_2$/g de biomasse sèche | (Nobre et al. 2012) |

The synthesis of biohydrogen by anaerobic fermentation is a subject of significant research and publication efforts globally. An abridged documentation of hydrogen generation with microalgal biomass as the raw material is presented in **Table 1**. Microalgal biomass has apparent potential as a viable feedstock for biohydrogen generation (Ren et al. 2011). Waste oil from the biofuel sector may also be reclaimed using this method, therefore enhancing the efficiency of energy extraction from biomass (El-Moustaqim, Rachiq, Mabrouki, et al. 2024). Considerable effort has been dedicated to diminishing chlorophyll antennas in microalgae. The pertinent question is whether this endeavor will provide positive results in a system where the primary goals are to decrease the efficiency of photosynthesis and the production of hydrogen (El-Moustaqim, Mabrouki, et Hmouni 2024b).

## 6. CHALLENGES IN PRODUCING HYDROGEN FROM MICROALGAE

Utilizing biological methods for hydrogen generation is the optimal choice for achieving long-term sustainability and is an ecologically benign approach (Obiora et al. 2024). Hydrogen, when used as a biofuel, encounters similar obstacles to other renewable energy systems: comparatively limited output, efficiency, and prohibitive manufacturing expenses, mostly associated with raw materials (Goria et al. 2024). It has been extensively recorded that renewable feedstocks, such as de-oiled microalgal biomass, a byproduct of microalgal biodiesel synthesis, are used in the fermentation-based generation of hydrogen (Khosravitabar et Spetea 2024). Optimizing yield and productivity may be mainly accomplished by employing engineering techniques and developing a highly efficient microbial strain through genetic engineering and metabolic engineering methods, applicable to both microalgae and anaerobic bacteria (Wei et Ma 2024). Typically, microalgae develop large chlorophyll antennas in their photosystem to maximize light absorption in natural settings with restricted incident light (Rady, Ali, et El-Sheekh 2024). Under efficient artificial lighting, the result is that more light is absorbed than is needed for photosynthesis and is dissipated in the form of heat (Worku et al. 2024). This phenomenon takes place on the uppermost layer of the crop, and the resulting shadowing hinders the effective absorption of light by the cells located at a considerable depth within the culture vessel. Therefore, the suppression of light absorption in the uppermost layer and the reduction of photosynthesis in the dark regions significantly decrease the effectiveness of photosynthesis. (Encinar et al. 1996; Kandasamy et al. 2024; Saidi, Chaouchi-Abada, et Chergui-Bouafia 2007).

## 7. CONCLUSION

This work has contributed to the investigation of bioprocesses for biohydrogen generation by dark fermentation of various types of organic waste, therefore enhancing production yields for potential large-scale applications. This work demonstrates a shift from a basic methodology (including mixed cultures and methods for regulating and sustaining hydrogen generation using various substrates) to a more practical methodology. The main obstacles to producing hydrogen from algae by biophotolysis are the low photon conversion efficiency and oxygen sensitivity of the hydrogenases, while dark fermentation is hampered by inefficient substrate conversion and thermodynamic limitations. Recent studies have documented the development of efficient phototrophic systems for the synthesis of $H_2$ under aerobic circumstances and the fermentation of $H_2$ in the absence of light using algal substrates. The potential for hydrogen generation with these microorganisms and techniques is quite encouraging. Subsequent investigations in this area may contribute to the realization of a carbon-neutral hydrogen economy.

## REFERENCES


Abanades, S., & Rodat, S. (2022). Solar Fuel Production by Thermochemical Dissociation of Water and Carbon Dioxide. Concentrating Solar Thermal Energy: Fundamentals and Applications, 303-328.

Abessa, R., & Tabet, S. (2014). Étude microbiologique et génotoxique des boues des eaux usées de la ville de Guelma. Mémoire de master. Université, 8.

Akram, F., Fatima, T., Ibrar, R., & Haq, I. (2024). Biohydrogen: Production, promising progressions and challenges of a green carbon-free energy. *Sustainable Energy Technologies and Assessments*, *69*, 103893. DOI: 10.1016/j.seta.2024.103893

Algapani, D. E., Qiao, W., Su, M., di Pumpo, F., Wandera, S. M., Adani, F., & Dong, R. (2016). Bio-Hydrolysis and Bio-Hydrogen Production from Food Waste by Thermophilic and Hyperthermophilic Anaerobic Process. *Bioresource Technology*, *216*, 768–777. DOI: 10.1016/j.biortech.2016.06.016 PMID: 27295255

Benchrifa, M., Bencheikh, I., Azoulay, K., Mabrouki, J., Elouardi, M., Azaz, N., Ouaamou, W. A., Barka, B. D., & El Moustaqim, K. (2024). Hydrogen Production Through Gasification of Organic Matter. In Mabrouki, J., & Mourade, A. (Eds.), *Advanced Technology for Smart Environment and Energy, Environmental Science and Engineering* (pp. 197–207). Springer Nature Switzerland. DOI: 10.1007/978-3-031-50871-4_12

Buffi, M., Prussi, M., & Scarlat, N. (2022). Energy and environmental assessment of hydrogen from biomass sources: Challenges and perspectives. *Biomass and Bioenergy*, *165*, 106556. DOI: 10.1016/j.biombioe.2022.106556

Chum, H. L., & Overend, R. P. (2001). Biomass and renewable fuels. *Fuel Processing Technology*, *71*(1-3), 187–195. DOI: 10.1016/S0378-3820(01)00146-1

Dębowski, M., Zieliński, M., Kisielewska, M., Kazimierowicz, J., Dudek, M., Świca, I., & Rudnicka, A. (2020). The Cultivation of Lipid-Rich Microalgae Biomass as Anaerobic Digestate Valorization Technology—A Pilot-Scale Study. *Processes (Basel, Switzerland)*, *8*(5), 517. DOI: 10.3390/pr8050517

Dev, R., Bhatt, S., & Gupta, M. (2024). Effect of lactic acid fermentation on the physico-chemical, functional, and antioxidant properties, and in vitro protein digestibility of malted ragi (Eleusine coracana L.). *Sustainable Food Technology*, *2*(4), 1128–1138. Advance online publication. DOI: 10.1039/D4FB00134F

EL-Moustaqim. K., Mabrouki, J., Azrour, M., & Hmouni, D. (2024). How Can Green Energy Be Improved by Integrating IoT into the Microalgae Process? In Sustainable and Green Technologies for Water and Environmental Management (pp. 239-252). Cham: Springer Nature Switzerland.

El-Moustaqim, K., Benchrifa, M., Mabrouki, J., & Hmouni, D. (2024). Interaction between the microalgae cultivation system and its effects on nutrient removal from wastewater. In *Artificial Intelligence Systems in Environmental Engineering* (pp. 70–79). CRC Press.

El-Moustaqim, K., Mabrouki, J., & Hmouni, D. (2024a). Applied sciences green microalgae for future biomass development. In *Advanced Systems for Environmental Monitoring, IoT and the application of Artificial Intelligence* (pp. 203–211). Springer Nature Switzerland.

El-Moustaqim, K., Mabrouki, J., & Hmouni, D. (2024b). From Microalgal Biomass to Biohydrogen: A Novel Approach to Environmentally Sustainable Technology. Advancements in Climate and Smart Environment Technology, 29-38.

El-Moustaqim, K., Mohammed, B., Mabrouki, J., & Hmouni, D. (2024). Application of IoT-Based Intelligent Systems in Microalgae Biofuel Production. In Mabrouki, J., & Mourade, A. (Eds.), *Advanced Technology for Smart Environment and Energy, Environmental Science and Engineering* (pp. 293–305). Springer Nature Switzerland. DOI: 10.1007/978-3-031-50871-4_20

El-Moustaqim, K., Rachiq, T., Mabrouki, J., Slaoui, M., & Hmouni, D. (2024). Application of Microalgae in Green Hydrogen Production: A Diverse Biomass for Multiple Applications. In Mabrouki, J., & Mourade, A. (Eds.), *Advanced Technology for Smart Environment and Energy, Environmental Science and Engineering* (pp. 231–240). Springer Nature Switzerland. DOI: 10.1007/978-3-031-50871-4_15

El-Moustaqim, K., Sbai, S. E., Mabrouki, J., & Hmouni, D. (2024). Critical Review of Green Microalgal Biomass Development through CO2 Sequestration. *Euro-Mediterranean Journal for Environmental Integration*. Advance online publication. DOI: 10.1007/s41207-024-00548-6

Encinar, J. M., Beltrán, F. J., Bernalte, A., Ramiro, A., & González, J. F. (1996). Pyrolysis of Two Agricultural Residues: Olive and Grape Bagasse. Influence of Particle Size and Temperature. *Biomass and Bioenergy*, *11*(5), 397–409. DOI: 10.1016/S0961-9534(96)00029-3

Fei-Yue, G., Yu, P.-C., & Gao, M.-R. (2022). Seawater electrolysis technologies for green hydrogen production: Challenges and opportunities. *Current Opinion in Chemical Engineering*, *36*, 100827. DOI: 10.1016/j.coche.2022.100827

Ghosh, S., & Basu, S. (2024). Solid Oxide Electrolysis Cell for Hydrogen Generation: General Perspective and Mechanism. In *Climate Action and Hydrogen Economy: Technologies Shaping the Energy Transition* (pp. 231–260). Springer Nature Singapore.

Goria, K., Singh, H. M., Singh, A., Kothari, R., & Tyagi, V. V. (2024). Insights into biohydrogen production from algal biomass: Challenges, recent advancements and future directions. *International Journal of Hydrogen Energy*, *52*, 127–151. DOI: 10.1016/j.ijhydene.2023.03.174

Hall, K. I. 2007. *Final Technical Report: Hydrogen Codes and Standards Outreach*. DOE/SF22341-1. National Hydrogen Association - Washington, DC. DOI: 10.2172/903333

Haruna, L., Abano, E., Teye, E., Tukwarlba, I., Yeboah, W., Agyei, K., & Lukeman, M. (2024). Effects of Predrying and Spontaneous Fermentation Treatments on Nib Acidification, Fermentation Quality, and Flavour Attributes of Ghanaian Cocoa (Theobroma cacao) Beans. *International Journal of Food Sciences*, *2024*(1), 5198607. Advance online publication. DOI: 10.1155/2024/5198607 PMID: 39145148

Hasnaoui, S. (2021). Optimisation de la production du bio-hydrogène par Spirulina sp via un photobioréacteur électrochimique (EPBR): application au traitement des eaux de lac de Réghaia.

Jain, R., Panwar, N. L., Jain, S. K., Gupta, T., Agarwal, C., & Meena, S. S. (2024). Bio-Hydrogen Production through Dark Fermentation: An Overview. *Biomass Conversion and Biorefinery*, *14*(12), 12699–12724. DOI: 10.1007/s13399-022-03282-7

Jariyaboon, R., O-Thong, S., & Kongjan, P. (2015). Bio-Hydrogen and Bio-Methane Potentials of Skim Latex Serum in Batch Thermophilic Two-Stage Anaerobic Digestion. *Bioresource Technology*, *198*, 198–206. DOI: 10.1016/j.biortech.2015.09.006 PMID: 26386423

Jarosz, Z., Kapłan, M., Klimek, K., Anders, D., Dybek, B., Herkowiak, M., Hołaj-Krzak, J. T., Syrotyuk, S., Korobka, S., Syrotyuk, H., & Wałowski, G. (2024). Evaluation of Biohydrogen Production Depending on the Substrate Used—Examples for the Development of Green Energy. *Energies*, *17*(11), 2524. DOI: 10.3390/en17112524

Jung, K.-W., Kim, D.-H., Oh, Y.-K., Cho, S.-K., & Shin, H.-S. (2013). Optimization of dark fermentative H2 production from microalgal biomass by combined (acid+ultrasonic) pretreatment. *Bioresource Technology*, *141*, 220–226. Advance online publication. DOI: 10.1016/j.biortech.2013.02.054 PMID: 23561952

Kandasamy, S., Jayabalan, J., Manickam, N., Dhandapani, B., & Myneni, V. R. (2024). Developments, challenges, and future perspectives in renewable bioenergy production using microalgae. In Sangeetha, J., & Thangadurai, D. (Eds.), *Microalgal Biomass for Bioenergy Applications, Woodhead Series in Bioenergy* (pp. 341–354). Woodhead Publishing. DOI: 10.1016/B978-0-443-13927-7.00017-7

Khosravitabar, F., & Spetea, C. (2024). Evaluating technical parameters for microalgae immobilization to optimize green hydrogen photoproduction: A comparative review. *International Journal of Hydrogen Energy*, *67*, 925–932. DOI: 10.1016/j.ijhydene.2024.01.299

Kothari, R., Singh, D. P., Tyagi, V. V., & Tyagi, S. K. (2012). Fermentative hydrogen production – An alternative clean energy source. *Renewable & Sustainable Energy Reviews*, *16*(4), 2337–2346. DOI: 10.1016/j.rser.2012.01.002

Kucharska, K., Makoś-Chełstowska, P., Słupek, E., & Gebicki, J. (2021). Management of Dark Fermentation Broth via Bio Refining and Photo Fermentation. *Energies*, *14*(19), 6268. DOI: 10.3390/en14196268

Leong, Y. K., Su, R.-H., Varjani, S., Lee, D.-J., & Chang, J.-S. (2024). Integrated biohydrogen production and dairy manure wastewater treatment via a microalgae platform. *International Journal of Hydrogen Energy*, *52*, 404–417. DOI: 10.1016/j.ijhydene.2023.06.308

Li, C., & Fang, H. H. P. (2007). Fermentative Hydrogen Production From Wastewater and Solid Wastes by Mixed Cultures. *Critical Reviews in Environmental Science and Technology*, *37*(1), 1–39. DOI: 10.1080/10643380600729071

Loomba, S., Khan, M. W., & Mahmood, N. (2023). Seawater to Green Hydrogen: Future of Green Energy. *ChemElectroChem*, *10*(24), e202300471. Advance online publication. DOI: 10.1002/celc.202300471

Machhirake, N., Vanapalli, K. R., Kumar, S., & Mohanty, B. (2024). Biohydrogen from waste feedstocks: An energy opportunity for decarbonization in developing countries. *Environmental Research*, *119028*, 119028. Advance online publication. DOI: 10.1016/j.envres.2024.119028 PMID: 38685297

Montiel-Corona, V., & Buitrón, G. (2024). Practical Applications of Dark Fermentation for Hydrogen Production. In *Wastewater Exploitation: From Microbiological Activity to Energy* (pp. 47–66). Springer Nature Switzerland.

Nguyen, T. A. D., Kim, K. R., Nguyen, M. T., Kim, M. S., Kim, D., & Sim, S. J. (2010). Enhancement of fermentative hydrogen production from green algal biomass of Thermotoga neapolitana by various pretreatment methods. *International Journal of Hydrogen Energy, 35*(23), 13035–13040.

Nobre, B., Villalobos, F., Barragan, B., Oliveira, A., Batista, A., Marques, P., Mendes, R. L., Sovová, H., Palavra, A. F., & Gouveia, L. (2012). A biorefinery from Nannochloropsis sp microalga - Extraction of oils and pigments. Production of biohydrogen from the leftover biomass. *Bioresource Technology, 135*, 128–136. Advance online publication. DOI: 10.1016/j.biortech.2012.11.084 PMID: 23265815

Obiora, N. K., Ujah, C. O., Asadu, C. O., Kolawole, F. O., & Ekwueme, B. N. (2024). Production of hydrogen energy from biomass: Prospects and challenges. *Green Technologies and Sustainability, 2*(3), 100100. DOI: 10.1016/j.grets.2024.100100

Omer, S. N., Saravanan, P., Kumar, P., Kannan, R. R., Rajasimman, M., & Shanmugam, V. (2025). Insights into renewable biohydrogen production from algal biomass: Technical hurdles and economic analysis. *Biomass Conversion and Biorefinery, 15*(3), 3281–3300.

Oneby, M. A., Bromley, C. O., Borchardt, J. H., & Harrison, D. S. (2010). Ozone Treatment of Secondary Effluent at U.S. Municipal Wastewater Treatment Plants. *Ozone Science and Engineering, 32*(1), 43–55. DOI: 10.1080/01919510903482780

Oraby, M., & Shawqi, A. (2024). Green Hydrogen Production Directly from Seawater with No Corrosion Using a Nonmetallic Electrode: A Novel Solution and a Proof of Concept. *International Journal of Energy Research, 2024*(1), 1–11. DOI: 10.1155/2024/5576626

Palumbo, J., Batista, N., Schwan, R., Martins, P., Borém, F., Salvio, L., Dias, D., & Ribeiro, L. (2024). Impact of different fermentation times on the microbiological, chemical, and sensorial profile of coffees processed by self-induced anaerobiosis fermentation. *Brazilian Journal of Microbiology : Publication of the Brazilian Society for Microbiology, 55*(3), 1–14. DOI: 10.1007/s42770-024-01370-6 PMID: 38775907

Qiu, P., Li, C., Liu, B., Yan, D., Li, J., & Jia, L. (2023). Materials of solid oxide electrolysis cells for H2O and CO2 electrolysis: A review. *Journal of Advanced Ceramics, 12*(8), 1463–1510. Advance online publication. DOI: 10.26599/JAC.2023.9220767

Qureshi, F., Yusuf, M., Tahir, M., Haq, M. U., Ibrahim, M., Kamyab, H., Nguyen, H.-H., Vo, D.-V., & Ibrahim, H. (2023). Renewable hydrogen production via biological and thermochemical routes: Nanomaterials, economic analysis and challenges. *Process Safety and Environmental Protection, 179*, 68–88. Advance online publication. DOI: 10.1016/j.psep.2023.07.075

Rady, H. A., Ali, S. S., & El-Sheekh, M. M. (2024). Strategies to enhance biohydrogen production from microalgae: A comprehensive review. *Journal of Environmental Management, 356*, 120611. DOI: 10.1016/j.jenvman.2024.120611 PMID: 38508014

Ren, N.-Q., Guo, W.-Q., Liu, B.-F., Cao, G., & Ding, J. (2011). Biological hydrogen production by dark fermentation: Challenges and prospects towards scaled-up production. *Current Opinion in Biotechnology, 22*(3), 365–370. DOI: 10.1016/j.copbio.2011.04.022 PMID: 21612910

Saidi, A., Chaouchi-Abada, B., & Chergui-Bouafia, S. (2007). Produire de l'hydrogene à partir d'eau et de lumiere grace aux microorganismes.

Sankir, M., & Sankir, N. D. (2024). *Towards Green Hydrogen Generation*. John Wiley & Sons. DOI: 10.1002/9781394234110

Sarkar, O., Katakojwala, R., & Venkata Mohan, S. (2021). Low Carbon Hydrogen Production from a Waste-Based Biorefinery System and Environmental Sustainability Assessment. *Green Chemistry*, *23*(1), 561–574. DOI: 10.1039/D0GC03063E

Suastes-Rivas, J. K., Romero-Pineda, M. J., Monje-Ramírez, I., Velasquez-Orta, S. B., Velasco, A., & Orta-Ledesma, M. T. (2024). Direct biohydrogen production from chitosan harvested microalgae biomass and an isolated yeast. *Energy Conversion and Management*, *302*, 118081. DOI: 10.1016/j.enconman.2024.118081

Sun, F., He, D., Yang, K., Qiu, J., & Wang, Z. (2022). Hydrogen Production and Water Desalination with On-demand Electricity Output Enabled by Electrochemical Neutralization Chemistry. *Angewandte Chemie International Edition*, *61*(31), e202203929. Advance online publication. DOI: 10.1002/anie.202203929 PMID: 35452186

Sun, J., Yuan, X., Shi, X., Chu, C., Guo, R., & Kong, H. (2011). Fermentation of Chlorella sp. for anaerobic bio-hydrogen production: Influences of inoculum–substrate ratio, volatile fatty acids and NADH. *Bioresource Technology*, *102*(22), 10480–10485. DOI: 10.1016/j.biortech.2011.09.016 PMID: 21967710

Udaypal, U., Goswami, R. K., & Verma, P. (2024). « Decarbonization of Biomass Feedstocks with Green Hydrogen Production: Important with Processes and End-Products ». P. 199-218 in *Green Hydrogen Economy for Environmental Sustainability*. Volume 1*: Fundamentals and Feedstocks*. Vol. 1473, *ACS Symposium Series*. American Chemical Society. DOI: 10.1021/bk-2024-1473.ch009

Varras, G., & Chalaris, M. (2023). Critical Review of Hydrogen Production via Seawater Electrolysis and Desalination: Evaluating Current Practices. *Journal of Electrochemical Energy Conversion and Storage*, *21*(4), 1–26. DOI: 10.1115/1.4064381

Wang, J., & Yin, Y. (2018). Fermentative hydrogen production using pretreated microalgal biomass as feedstock. *Microbial Cell Factories*, *17*(1), 22. DOI: 10.1186/s12934-018-0871-5 PMID: 29444681

Wei, L., & Ma, W. (2024). Photosynthetic H2 Production: Lessons from the Regulation of Electron Transfer in Microalgae. *Global Change Biology. Bioenergy*, *16*(1), e13118. DOI: 10.1111/gcbb.13118

Worku, A. K., Ayele, D. W., Deepak, D. B., Gebreyohannes, A. Y., Agegnehu, S. D., & Kolhe, M. L. (2024). Recent Advances and Challenges of Hydrogen Production Technologies via Renewable Energy Sources. *Advanced Energy and Sustainability Research*, *5*(5), 2300273. DOI: 10.1002/aesr.202300273

Xia, A., Cheng, J., Ding, L., Lin, R., Huang, R., Zhou, J., & Cen, K. (2013). Improvement of the energy conversion efficiency of Chlorella pyrenoidosa biomass by a three-stage process comprising dark fermentation, photofermentation, and methanogenesis. *Bioresource Technology*, *146C*, 436–443. DOI: 10.1016/j.biortech.2013.07.077 PMID: 23955091

Xia, A., Jacob, A., Tabassum, M., Herrmann, C., & Murphy, J. (2016). Production of hydrogen, ethanol and volatile fatty acids through co-fermentation of macro- and micro-algae. *Bioresource Technology*, *205*, 118–125. Advance online publication. DOI: 10.1016/j.biortech.2016.01.025 PMID: 26820925

Yarkent, Ç., Oral, İ., Öncel, D. Ş., & Oncel, S. S. (2024). Focusing on a new biohydrogen production strategy (HS-HS) using *Chlamydomonas reinhardtii* mutants: The effects of mutation points on biohydrogen production capacity. *International Journal of Hydrogen Energy*, *52*, 88–102. DOI: 10.1016/j.ijhydene.2023.02.034

# Chapter 10
# Green Hydrogen:
## The Pillar of the Sustainable Energy Transition

**Hajar Khabouche**
https://orcid.org/0009-0003-4466-5784
*Higher School of Technology, Morocco*

**Jamal Mabrouki**
*Faculty of Science Rabat, Morocco*

**Ayoub Ghailane**
*Mohammed V University in Rabat, Morocco*

**Miloudia Slaoui**
*Higher School of Technology, Morocco*

**Imane Rouichat**
*Higher School of Technology, Morocco*

**Houria Al Bakraoui**
*Higher School of Technology, Morocco*

## ABSTRACT

*Energy transition policy now places a strong emphasis on green hydrogen because of its enormous potential to cut greenhouse gas emissions and its substantial commercial potential. Renewable energy sources like wind and solar power are used to manufacture this eco-friendly substitute. A novel technique uses chlorophyll-rich microalgae, which produce hydrogen as a byproduct of photosynthesis, the process by which sunlight is converted into chemical energy.Converting renewable energy into forms that may be used, like green hydrogen, requires the development of production circuits for renewable PtX (Power-to-X). Water electrolysis, which separates water into oxygen and hydrogen using power from renewable sources, is the technique involved in this procedure. Continuous and dependable manufacturing necessitates a strong infrastructure, which includes effective electrolyzers.Since it is the basis for the production of numerous synthetic fuels and other energy products, green hydrogen is an essential part of PtX.*

## 1. INTRODUCTION

The demand for energy is continual due to population expansion, technological advancements, and ongoing economic issues. Actually, a measure of a nation's political, social, and economic growth may be found in its energy sector (Brodny et al., 2020). the generation of energy is still largely dependent on fossil fuels and that annual growth in energy consumption is 2% (Nam et al., 2020 ; Khan et al., 2022). Policies and actions to reduce reliance on fossil fuels, greenhouse gas emissions, and carbon emissions

DOI: 10.4018/979-8-3693-8980-5.ch010

are required. One such approach is energy management via the use of sustainable and renewable energy systems (Yu et al., 2019; Li et al., 2019).

Integrating renewable energy sources such as biomass, solar, wind, hydroelectric power, and hydroelectricity into current energy systems is known as renewable energy integration (Papadis & Tsatsaronis, 2020). By providing a greener, more sustainable fuel substitute, these sources help to lower greenhouse gas emissions and slow down the effects of climate change (Mahapatra et al., 2021). Conversely, climate resilience pertains to the ability of individuals, communities, and systems to predict, adjust, and recuperate from the consequences of climate-related risks (de Graaf-van Dinther & Ovink, 2021). Through diversifying their energy sources, lowering their susceptibility to energy disruptions, and advancing sustainable development, societies can strengthen their resilience to climate change by utilizing renewable energy resources (Ghorbani et al., 2023). Building resilience against the effects of climate change and addressing it urgently have grown more and more important in recent decades. Integration of renewable energy has become an important tactic in this effort, providing a mechanism to reduce the negative consequences of climate change and decarbonize energy systems (Arent et al., 2022). Examining the opportunities, tactics, and difficulties related to this crucial component of sustainable development, this thorough review aims to investigate the complex relationship between climate resilience and the integration of renewable energy (Ajayi & Udeh, 2024 ; Familoni et al., 2024).

In order to achieve sustainable growth, the majority of countries in the world today have looked into the development of renewable energy sources. Solar and wind power generating is one of the most technologically advanced forms of renewable energy. The electrical grid's balance is one of the major obstacles to rising renewable resource penetration, however there are other obstacles as well (Shabani & Kalantar, 2021a ; Shabani et al., 2021b). Potential energy storage solutions to this problem are "Power-to-X (PtX)" technology, where "X" can stand for heat, power, gas, or liquid fuels. With the help of this technology, excess power can be transformed into other types of energy or energy carriers (Cui et al., 2022). Because water is the only feedstock used in the electrolysis process to make hydrogen, this fuel may be deployed almost anywhere. When employed as fuel in properly engineered fuel cell or combustion systems, hydrogen can be used directly as fuel for stationary energy generation or transportation, leaving only water as a byproduct. In this light, hydrogen as a renewable energy carrier will be essential in numerous ways to decarbonize energy systems throughout the entire energy value chain (Sorrenti et al., 2022).

This chapter's goal is to draw attention to how crucial green hydrogen is to the energy transition as a way to lower greenhouse gas emissions and promote the advancement of Power-to-X (PtX) technologies. It also seeks to address the difficulties associated with green hydrogen generation, transportation, storage, and sustainability, highlighting the necessity of appropriate infrastructure and stringent environmental regulations.

## 2. THE CONCEPT OF POWER-TO-X

The technology known as "Power to X," or "PtX," works by converting electricity into a different kind of energy. In order to feed heating networks or meet industrial needs, this "X-vector" could be heat (Power to Heat). It might also be a synthesis gas (Power to Gas), such methane, which can be injected into the gas network for industrial, heating, or mobility demands, or hydrogen for mobility (Ait Almouh, 2019). We need to quickly transition our economy to operate emission-free in order to achieve climate neutrality by 2050. Even though this transition requires only renewable energy, some industries and the transportation sector cannot yet directly electrify to achieve climate neutrality. Power-to-X is the piece that's missing till they can (Fraunhofer – ISI, 2019). Gases (such as hydrogen and methane) have given way to liquid fuels, chemicals (such as ammonia and methanol), heat, desalination, and materials (such as silicon carbide, steel, and carbon fibers) in the Power-to-X concept (Breyer et al., 2022 ; Sterner & Specht, 2021 ; Palys & Daoutidis, 2022). The development of robust energy system models that describe the entire energy-industry system and cover all relevant Power-to-X routes, together with the anticipated abundance of low-cost renewable electricity and the insights on Power-to-X routes and the concept of sector coupling (Brown et al., 2018 ; Bogdanov et al., 2021 ; Mathiesen et al., 2015), allowed for new structural insights on transition pathways towards the future energy system (Breyer et al., 2022). High levels of electrification are used in these pathways for reverse osmosis desalination, battery-electric road mobility, industrial process heat, and space heating. However, around 40% of the final energy and feedstock needs, which includes steelmaking, chemicals, and fuels for long-distance transportation and aviation, cannot be directly electrified. These needs can be sustainably provided by hydrogen produced from renewable power. While most of the hydrogen is transformed into liquids for aviation, ammonia and methanol for marine fuels, and specifically chemicals, hydrogen-to-X pathways summarize the hydrogen (Breyer et al., 2022 ; Ram et al., 2022). Figure 1 shows the Power-to-X concept .

*Figure 1. Power-to-X: The concept beyond hydrogen*

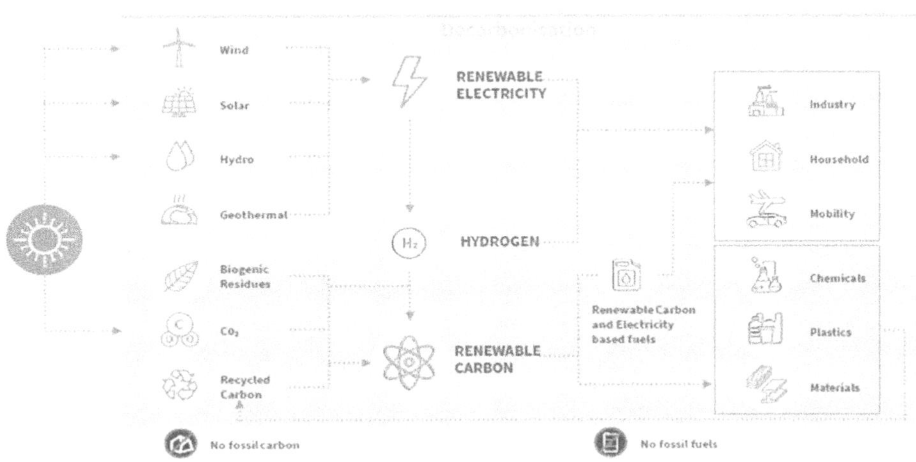


## 3. OPTIONS FOR HYDROGEN PRODUCTION FROM RENEWABLE SOURCES

### 3.1 Renewable Energy-Powered Electrolysis

Nowadays, electrolysis operations account for less than 4% of hydrogen generation, with hydrogen serving as the primary by-product of chlorine production. Due to high electricity costs and conflicting legislation, the production costs of electrolysis processes are higher than those of conventional fossil sources, which is what is causing this predicament (Dincer & Acar, 2015).Through the electrolysis process, which employs only renewable energy sources to produce pure oxygen and hydrogen, green hydrogen is created from 100 percent renewable sources (ESMAP, 2020 ; Luo et al., 2018). The terms "gray hydrogen" and "remaining hydrogen" from chemical reactions in chlor-alkali plants together describe hydrogen produced by the steam methane reforming (SMR) process. Blue hydrogen is defined as gray hydrogen that has gone through the carbon capture and utilization (CCU) process (Luo et al., 2018 ; Natural Resources Canada, 2020).

Three technologies are available for the electrolysis of water: Solid oxide electrolysis (SOEL), alkaline water electrolysis (AEL), and proton exchange membrane (also known as polymer electrolyte membrane) electrolysis (PEMEL) (Figure 2) (Cui et al., 2022 ; Sorrenti et al., 2022 ; Ait Almouh, 2019 ; Fraunhofer – ISI, 2019). The high-temperature SOEL technology is still in the research and development stage, whereas the low-temperature technologies, AEL and PEMEL, both offer high degrees of technological readiness (Carmo et al., 2013). High-temperature electrolysis (SOEL) or steam electrolysis is another name for it. This process converts gaseous water into hydrogen and oxygen at temperatures ranging from 700 to 900 °C. Theoretically, favorable thermodynamic effects on power consumption at higher temperatures allow for stack efficiencies close to 100%.Concentrated lye is used as the electrolyte in alkaline water electrolysis, and a gas-impermeable separator is needed to keep the resultant gases from mixing. Electrocatalytically coated non-noble metals, such as nickel, make up the electrodes. Noble metals like platinum and iridium oxide are used as the electrocatalysts in PEMEL, with a humidified polymer membrane serving as the electrolyte. Both systems can function at pressures of up to 30 bar and at temperatures between 50 and 80 oC. Both approaches have a nominal stack efficiency of roughly 70% (David et al., 2019 ; Buttler & Spliethoff, 2018). Even though SOEL is effective, large-scale applications are not a good fit for it because it needs high temperatures and sources of industrial waste heat. AEL and PEMEL, on the other hand, have larger capacity; AEL is more affordable (800–1500 €/kW) and has a longer lifespan. PEMEL systems are recommended for dynamic operation because of their quick start-up time and load flexibility, even though they cost more (1400–2100 €/kW).

*Figure 2. Schematic illustration of different technologies used in water electrolysis*

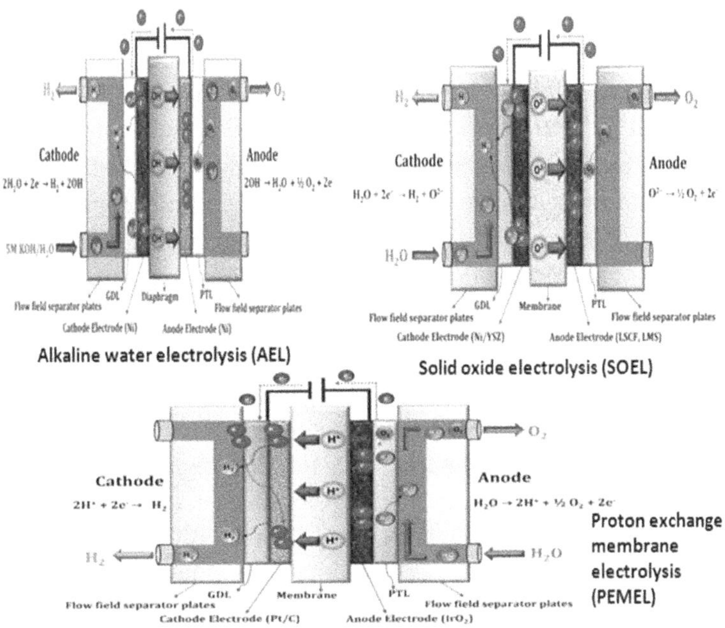


## 3.2 Hydrogen Production From Biomass

As a plentiful renewable resource, biomass has important benefits for the environment and the economy. Utilizing biomass reduces carbon emissions and permits carbon absorption within the cycle, in contrast to using fossil fuels (Huang et al., 2021). Through gasification, biomass which includes organic materials like forest and agricultural leftovers as well as specially grown energy crops can be transformed into useful syngas (Saravanakumar et al., 2023 ; Hoang et al., 2022). Using biomass to produce hydrogen is a viable source of sustainable energy.The advantages and difficulties associated with each strategy are distinct, which adds to the variety of hydrogen generation systems. Hydrogen production from biomass involves both thermochemical methods like pyrolysis, gasification, liquefaction and combustion employ heat and chemical reactions to transform biomass into hydrogen-rich syngas or liquids, offering diverse pathways for sustainable hydrogen generation. In contrast, biological methods, such as dark fermentation, bio-photolysis and photo fermentation, which utilize microorganisms to convert organic materials into hydrogen gas through metabolic processes. Using biomass to produce hydrogen is a viable source of sustainable energy.The advantages and difficulties associated with each strategy are distinct, which adds to the variety of hydrogen generation systems.

The technique of producing hydrogen by biological methods, which can be classed as either light-dependent or light-independent, involves employing microorganisms to produce hydrogen through metabolic activities like photosynthesis or fermentation (Kossalbayev et al., 2024). The synthesis of hydrogen from bacteria, microalgae, and algae has the advantages of being less expensive than other production techniques and yielding high yields at low temperatures (often between 20 and 45 oC) (Trchounian & Trchounian, 2015). While dark fermentation occurs through a light-independent process utilizing bacteria, light-dependent techniques include photo-fermentation and bio-photolysis (direct or indirect) utilizing algae or microalgae (Copa et al., 2020). The primary source of hydrogen produced through bio-photolysis is photosynthetic oxygenic microorganisms, such as cyanobacteria (such as Nostoc sp. and Cyanothece sp.) and green microalgae (such as Chlamydomonas reinhardtii, Chlorella vulgaris, Scenedesmus obliquus, and Chlorococcum minutum) (Márquez-Reyes et al., 2015 ; Torzillo et al., 2009 ; Rashid et al., 2011 ; Paramesh & Chandrasekhar, 2020 ; Mona et al., 2011 ; Zhang et al., 2014).In dark fermentation, anaerobic microorganisms are mainly used in the absence of light. These microorganisms can be found in facultative anaerobes such as Enterobacter cloacae, Escherichia coli, Klebsiella pneumoniae, and Bacillus subtillis, or in strict anaerobes such as Clostridium butyricum, Clostridium acetobutylicum, Clostridium thermocellum, and Thermoanaerobacterium thermosaccharolyticum. Occasionally, they can be used at extreme thermophilic temperatures (>80 oC). Under these circumstances, in addition to hydrogen, additional gases such as carbon dioxide, carbon monoxide, and hydrogen sulfide may be emitted. The glucose molecules found in carbohydrates and other feedstocks are a major source of hydrogen. Hydrogen is produced during photo-fermentation when organic compounds are broken down by photosynthetic bacteria, particularly purple non-sulfur bacteria (PNS bacteria) such as Rhodobacter sphaeroides, Rhodobacter capsulatus, Rhodobacter sulfidophilus, Rhodopseudomonas palustris, and Rhodospirillum rubrum. Nitrogenase uses light energy to catalyze this reaction (Eroglu & Melis, 2011). The tricarboxylic acid (TCA) cycle, sometimes referred to as the citric acid cycle, metabolizes the organic substrates and supplies the carbon intermediates and reducing power required to produce hydrogen (Mishra et al., 2019).

Thermochemical processes employ heat and chemical transformation of biomass to produce hydrogen in order to convert it to renewable H2. By depolymerizing biomass, these processes can produce hydrogen directly from biomass or from intermediary products like methanol and ethanol. Hydrothermal reactions, pyrolysis, and gasification are examples of thermochemical techniques. The process of biomass gasification involves using thermochemistry to transform biomass into syngas, a combustible gas combination that mostly consists of carbon dioxide, methane, hydrogen, and carbon monoxide (Koku et al., 2002). There are multiple steps in the gasification process, including Moisture in biomass needs to be eliminated before gasification can occur. The biomass is heated during this phase typically to 200°C in order to remove the moisture.This happens in a regulated setting where biomass is heated to high temperatures (usually between 800 and 1000°C) with the help of an oxygen, steam, or air gasifying agent. Gasification partially oxidizes biomass to produce high-energy syngas, as opposed to combustion, which completely oxidizes biomass to produce carbon dioxide and water. The biomass proceeds through pyrolysis at temperatures between 200 and 500°C, breaking down into tar, char, and volatile gases (Maitlo et al., 2022).

Pyrolysis is a thermochemical process in which heat is applied in an inert atmosphere to break down biomass into solid, liquid, and gaseous fractions (Lewandowski et al., 2020). The big hydrocarbon molecules found in biomass are broken down into smaller molecules via a variety of chemical processes during the pyrolysis process.The first part of biomass to decompose is hemicellulose, which does so between 220 and 315 °C. Acetic acid, other organic acids, sugars, and furans are the products of this process. The ensuing process of cellulose degradation takes place between 315 and 400 oC, during which levoglucosan and other anhydrocelluloses are formed (Zhang et al., 2024).

*Figure 3. Hydrogen production from biomass*

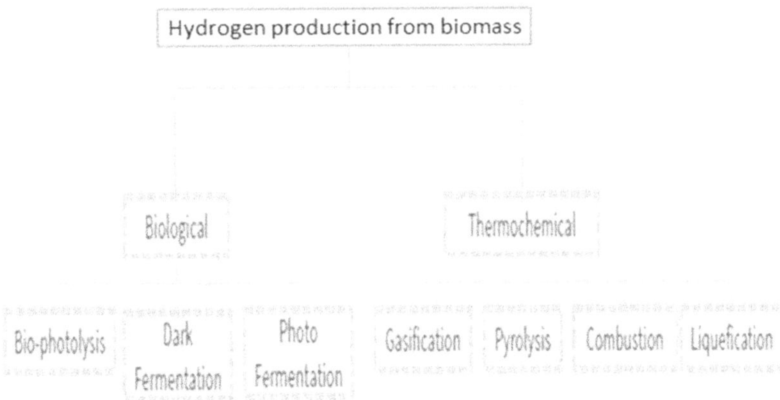


## 4. HYDROGEN TRANSPORT AND STORAGE

A major obstacle to the widespread use of hydrogen as a clean energy source is the logistical difficulties associated with its storage and transportation. Owing to the distinct physical characteristics of hydrogen, such as its high reactivity and low density, specific infrastructure is needed to guarantee secure transportation and storage. Safety issues must also be taken into account, as hydrogen transfer occurs in less stable conditions than storage. Selecting the right transportation techniques that fit these criteria is crucial because hydrogen has varying properties depending on its state.A critical component of the whole hydrogen supply chain is hydrogen transportation, in addition to storage. Since regional differences exist in hydrogen demand, cost is the main factor to be taken into account while transporting hydrogen. A procedure using hydrogen also needs to be handled carefully because of its complex physicochemical characteristics. Large vessels, pipelines, and trucks and trailers on the road are the three main methods that hydrogen is carried. Road transportation also includes rail transit. It won't be covered individually because it shares many of the same economic benefits and hydrogen transportation principles as truck trailer transport. Figure 4 displays the main hydrogen transport mechanisms and process models (Dhyani & Bhaskar, 2018 ; Hunt et al., 2024).

*Figure 4. Schematic of hydrogen transportation*

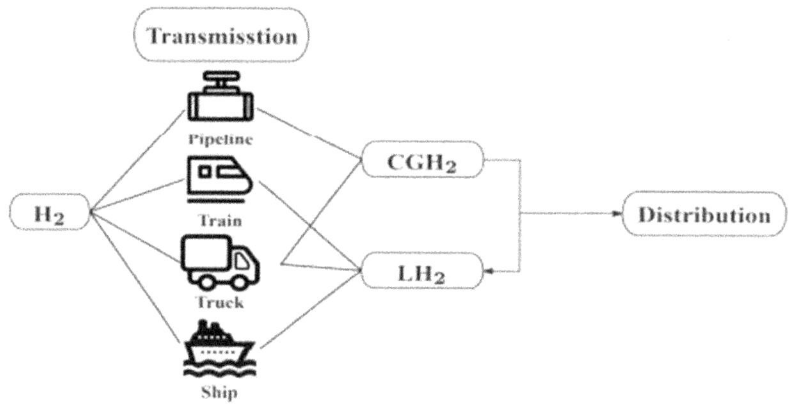


Two parameters have a major influence on the cost and energy consumption of hydrogen transportation: the energy density of the vehicle and the percentage of hydrogen in the overall cargo. The current priorities are to minimize the weight of storage tanks and maximize the amount of hydrogen that can be stored per unit volume. In contrast, lowering total transportation costs becomes a more pressing concern in commercial applications, necessitating a compromise between maximizing storage capacity and cutting prices. Hydrogen is delivered under high pressure utilizing gas cylinders inside trucks or trailers, with a pressure range of 20 to 50 MPa (Genovese et al., 2024 ; Lan et al., 2023).

LH2 (liquid hydrogen) is a clean and effective fuel. It is widely employed in many different industries, such as energy storage, transportation for industrial operations like semiconductor fabrication and metal refining, and, more and more, developing technologies like fuel cells and hydrogen-powered cars. It is also used in space applications as rocket fuel. One important avenue for the development of new energy technology is LH2. Good insulating measures are essential to reduce energy loss during storage. The low heat flux density of vacuum insulation makes it a popular choice for cryogenic gas storage tanks.The most popular type of hydrogen storage used in hydrogen fuel cell or hydrogen-powered cars is compressed gaseous hydrogen storage (Mao et al., 2011). Vehicle hydrogen tank-related regulations and guidelines from various groups state that the technology pertaining to these tanks is evolving quickly and can withstand pressures of up to 70 MPa (Wang et al., 2019). At the moment, ammonia borane, sodium borohydride, and magnesium hydride are the three most widely used solid hydrogen storage compounds.Metal hydride materials have been developed since the 1960s and now include elements including titanium, zirconium, magnesium, and several rare earth elements.Acquisition prices, hydrogen storage capacity, and hydrogen release temperature are the main selection criteria. Because of their low molecular weight, which results in a high gravimetric hydrogen storage capacity, and their abundance in the earth's crust, which may result in lower material costs, magnesium-based materials are particularly noteworthy.Magnesium-based materials' high hydrogen release temperature, which usually exceeds 300°C, is a major drawback (Du et al., 2018 ; Li et al., 2021 ; Landi et al., 2020 ; The Elec, 2024).

## 5. FUTURE PERSPECTIVES AND CHALLENGES

PtX technologies have great potential to decarbonize difficult-to-electrify industries by converting renewable power into hydrogen and other energy carriers. Further developments in electrolyzer technology, better storage options, and regulatory incentives for infrastructure investment are necessary for hydrogen to realize its full potential in the global energy revolution.However, a significant infrastructure for hydrogen production and a vast amount of renewable energy are need for PtX to succeed. Achieving a carbon-neutral economy requires the use of hydrogen, which can be produced in conjunction with carbon capture and storage (CCS) technologies to help lower overall carbon footprints. The main obstacles to hydrogen transport, storage, biomass production, electrolysis powered by renewable energy sources, and Power-to-X (PtX) technologies include those related to cost reduction, infrastructure development, and technological innovation. The infrastructure required to move hydrogen, whether via trucks, ships, or pipelines, is also underdeveloped, which makes large-scale commercialization of the gas more difficult. Because of the peculiar physical characteristics of hydrogen, including its high reactivity, low energy density, and capacity to produce material embrittlement, new materials and storage technologies that can safely manage compressed or liquefied hydrogen are needed. Scalability and cost-effectiveness issues are present in the production of hydrogen from biomass. Although the present biological and thermochemical approaches provide a carbon-neutral and sustainable pathway, they still require efficiency improvements to be economically viable. The foundation of producing green hydrogen, electrolysis fueled by renewable energy sources, is another crucial area that needs innovation. Continuous hydrogen generation is made more difficult by the intermittent nature of renewable energy sources like solar and wind, as well as the high cost and low efficiency of current electrolyzers. To guarantee a steady supply of hydrogen, it will be imperative to increase the capacity of renewable energy sources and enhance the integration of energy storage systems.

## 6. CONCLUSION

As environmental worries about greenhouse gas emissions grow, so does the transition to a future free of fossil fuels, one in which renewable energy sources will inevitably play a major role. Hydrogen accounts up a mere 2% of the energy mix in the EU, even though fossil fuels still account for 95% of the region's energy production. Since hydrogen has the ability to meet 20–50% of transportation energy needs and 5–20% of industrial energy needs, the EU is promoting hydrogen as an enabler. Since they employ water electrolysis to transform renewable energy into green hydrogen, technologies like Power-to-Hydrogen and Hydrogen-to-X have attracted a lot of attention as ways to aid in this transition. There is growing recognition of hydrogen's significance as a fuel of the future. Biomass-derived hydrogen is a very attractive substitute for fossil fuels in the production of clean energy.In order to reduce emissions of CO2 and other greenhouse gases and move toward sustainable energy sources, the production of hydrogen from biomass wastes is essential.Carbon dioxide is naturally captured and stored by biomass, which makes it a useful tool for removing the gas from the atmosphere. Through biological and/or thermal techniques, biomass leftovers can be valorized to produce economically useful compounds and clean, renewable energy with zero carbon dioxide emissions. Combining biological and thermochemical technologies to manufacture H2 seems like a promising way to take use of the production potential associated with various biomass waste streams and the unique features of each locality in order to achieve a 60% reduction in $CO_2$ emissions by 2050.

## REFERENCES


Ait Almouh, H. (2019). “Power To X: Quel intérêt pour le Maroc?”, *lavieeco.com,* March 12, available at: https://www.lavieeco.com/economie/energie/power-to-x-pour-le-maroc-quel-interet-pour-le-maroc/

Ajayi, F.A., & Udeh, C.A. (2024). Review of crew resilience and mental health practices in the marine industry: Pathways to improvement. Magna Scientia Advanced Biology and Pharmacy, 11(02), 033–049. https://doi.org/DOI: 10.30574/msabp.2024.11.2.0021

Arent, D. J., Green, P., Abdullah, Z., Barnes, T., Bauer, S., Bernstein, A., Berry, D., Berry, J., Burrell, T., Carpenter, B., Cochran, J., Cortright, R., Curry-Nkansah, M., Denholm, P., Gevorian, V., Himmel, M., Livingood, B., Keyser, M., King, J., & Turchi, C. (2022). Challenges and opportunities in decarbonizing the US energy system. *Renewable & Sustainable Energy Reviews*, *169*, 112939. DOI: 10.1016/j.rser.2022.112939

Bogdanov, D., Gulagi, A., Fasihi, M., & Breyer, C. (2021). Full energy sector transition towards 100% renewable energy supply: Integrating power, heat, transport and industry sectors including desalination. *Applied Energy*, *283*, 116273. Advance online publication. DOI: 10.1016/j.apenergy.2020.116273

Breyer, C., Bogdanov, D., Ram, M., Khalili, S., Vartiainen, E., & Moser, D.. (2022). Reflecting the energy transition from a European perspective and in the global context – relevance of solar photovoltaics benchmarking two ambitious scenarios. *Progress in Photovoltaics: Research and Applications*, •••, 1–27. DOI: 10.1002/pip.3659

Breyer, C., Khalili, S., Bogdanov, D., Ram, M., Oyewo, A. S., Aghahosseini, A., Gulagi, A., Solomon, A. A., Keiner, D., Lopez, G., Ostergaard, P. A., Lund, H., Mathiesen, B. V., Jacobson, M. Z., Victoria, M., Teske, S., Pregger, T., Fthenakis, V., Raugei, M., & Sovacool, B. K. (2022). On the history and future of 100% renewable energy systems research. *IEEE Access : Practical Innovations, Open Solutions*, *10*, 78176–78218. DOI: 10.1109/ACCESS.2022.3193402

Brodny, J., Tutak, M., & Saki, S. A. (2020). Forecasting the Structure of Energy Production from Renewable Energy Sources and Biofuels in Poland. *Energies*, *13*(10), 2539. DOI: 10.3390/en13102539

Brown, T., Schlachtberger, D., Kies, A., Schramm, S., & Greiner, M. (2018). Synergies of sector coupling and transmission reinforcement in a cost-optimised, highly renewable European energy system. *Energy*, *160*, 720–739. DOI: 10.1016/j.energy.2018.06.222

Buttler, A., & Spliethoff, H. (2018). Current Status of Water Electrolysis for Energy Storage, Grid Balancing and Sector Coupling via Power-to-Gas and Power-to-Liquids: A Review. *Renewable & Sustainable Energy Reviews*, *82*, 2440–2454. DOI: 10.1016/j.rser.2017.09.003

Carmo, M., Fritz, D. L., Mergel, J., & Stolten, D. (2013). A Comprehensive Review on PEM Water Electrolysis. *International Journal of Hydrogen Energy*, *38*(12), 4901–4934. DOI: 10.1016/j.ijhydene.2013.01.151

Copa, J. R., Tuna, C. E., Silveira, J. L., Boloy, R. A. M., Brito, P., Silva, V., Cardoso, J., & Eusébio, D. (2020). Techno-Economic Assessment of the Use of Syngas Generated from Biomass to Feed an Internal Combustion Engine. *Energies*, *13*(12), 3097. DOI: 10.3390/en13123097

Cui, X., & Knudsen Kær, S. (2022). M.P. Nielsen Energy analysis and surrogate modeling for the green methanol production under dynamic operating conditions. *Fuel*, *307*, 121924. DOI: 10.1016/j.fuel.2021.121924

David, M., Ocampo-Martínez, C., & Sánchez-Peña, R. (2019). Advances in Alkaline Water Electrolyzers: A Review. *Journal of Energy Storage*, *23*, 392–403. DOI: 10.1016/j.est.2019.03.001

de Graaf-van Dinther, R., & Ovink, H. (2021). The five pillars of climate resilience. Climate Resilient Urban Areas: Governance, design and development in coastal delta cities, 1-19.

Dhyani, V., & Bhaskar, T. (2018). A Comprehensive Review on the Pyrolysis of Lignocellulosic Biomass. *Renewable Energy*, *129*, 695–716. DOI: 10.1016/j.renene.2017.04.035

Dincer, I., & Acar, C. (2015). Review and Evaluation of Hydrogen Production Methods for Better Sustainability. *International Journal of Hydrogen Energy*, *40*(34), 11094–11111. DOI: 10.1016/j.ijhydene.2014.12.035

Du, Y.-N., Wang, Z.-B., Tang, X.-Y., Yuan, Y.-W., Xue, X.-L., & Jiang, C.-J. Investigation on Standards on Hydrogen Cycle of Composite Tanks for Storage of High Pressure Hydrogen. In *Proceedings of the Pressure Vessels and Piping Conference*, Prague, Czech Republic, 15–20 July 2018; American Society of Mechanical Engineers: New York, NY, USA, 2018; Volume 51593 DOI: 10.1115/PVP2018-84694

Eroglu, E., & Melis, A. (2011). Photobiological Hydrogen Production: Recent Advances and State of the Art. *Bioresource Technology*, *102*(18), 8403–8413. DOI: 10.1016/j.biortech.2011.03.026 PMID: 21463932

ESMAP. (2020). *Green Hydrogen in Amsterdam*. ESMAP.

Familoni, B. T., Abaku, E. A., & Odimarha, A. C.Babajide Tolulope FamiloniEmmanuel Adeyemi AbakuAgnes Clare Odimarha. (2024). Blockchain for enhancing small business security: A theoretical and practical exploration. *Open Access Research Journal of Multidisciplinary Studies*, *7*(1), 149–162. DOI: 10.53022/oarjms.2024.7.1.0020

Fraunhofer – ISI. (2019). "Carbon-neutral energy from power-to-X: Economic opportunity and ecological limitations for Morocco", isi.fraunhofer.de, September 2019, available at: https://www.isi.fraunhofer.de/en/presse/2019/presseinfo-24-klimaneutrale-energie-aus-power-to-x-marokko.html

Genovese, M., Pagnotta, L., Piraino, F., & Fragiacomo, P. (2024). Fluid-dynamics analyses and economic investigation of offshore hydrogen transport via steel and composite pipelines. *Cell Reports. Physical Science*, *5*(4), 101907. DOI: 10.1016/j.xcrp.2024.101907

Ghorbani, Y., Zhang, S. E., Nwaila, G. T., Bourdeau, J. E., & Rose, D. H. (2023). Embracing a diverse approach to a globally inclusive green energy transition: Moving beyond decarbonisation and recognising realistic carbon reduction strategies. *Journal of Cleaner Production*, •••, 140414.

Hoang AT, Huang ZH, Ni˘zeti´c S, et al. Characteristics of hydrogen production from steam gasification of plant-originated lignocellulosic biomass and its prospects in Vietnam[J]. Int J Hydrogen Energy 2022;47(7):4394–425. . [8] Xu H, Li H. Alcohol-assisted hydrodeoxygenatDOI: 10.1016/j.ijhydene.2021.11.091

Huang, B., Liu, Y., & Xie, Z. (2021). Two dimensional nanocarbons from biomass and biological molecules: Synthetic strategies and energy related applications [J]. *Journal of Energy Chemistry*, *54*, 795–814. DOI: 10.1016/j.jechem.2020.06.033

Hunt, J. D., Zakeri, B., Nascimento, A., de Freitas, M. A. V., do Carmo Amorim, F., Guo, F., Witkamp, G. J., van Ruijven, B., & Wada, Y. (2024). Hydrogen balloon transportation: A cheap and efficient mode to transport hydrogen. *International Journal of Hydrogen Energy*, *53*, 875–884. DOI: 10.1016/j.ijhydene.2023.11.305

Khan, Z. A., Hussain, T., Haq, I. U., Ullah, F. U. M., & Baik, S. W. (2022). Towards efficient and effective renewable energy prediction via deep learning. *Energy Reports*, *8*, 10230–10243. DOI: 10.1016/j.egyr.2022.08.009

Koku, H., Erolu, I., Gündüz, U., Yücel, M., & Türker, L. (2002). Aspects of the Metabolism of Hydrogen Production by Rhodobacter sphaeroides. *International Journal of Hydrogen Energy*, *27*(11-12), 1315–1329. DOI: 10.1016/S0360-3199(02)00127-1

Kossalbayev, B. D., Yilmaz, G., Sadvakasova, A. K., Zayadan, B. K., Belkozhayev, A. M., Kamshybayeva, G. K., Sainova, G. A., Bozieva, A. M., Alharby, H. F., Tomo, T., & Allakhverdiev, S. I. (2024). Biotechnological Production of Hydrogen: Design Features of Photobioreactors and Improvement of Conditions for Cultivating Cyanobacteria. [CrossRef]. *International Journal of Hydrogen Energy*, *49*, 413–432. DOI: 10.1016/j.ijhydene.2023.09.001

Lan, Z., Hong, F., Shi, W., Zhao, R., Li, R., Fan, Y., Liu, H., Zhou, W., Ning, H., & Guo, J. (2023). Effect of MOF-derived carbon–nitrogen nanosheets co-doped with nickel and titanium dioxide nanoparticles on hydrogen storage performance of MgH2. *Chemical Engineering Journal*, *468*, 143692. DOI: 10.1016/j.cej.2023.143692

Landi, D., Vita, A., Borriello, S., Scafà, M., & Germani, M. (2020). A methodological approach for the design of composite tanks produced by filament winding. *Computer-Aided Design and Applications*, *17*(6), 1229–1240. DOI: 10.14733/cadaps.2020.1229-1240

Lewandowski, W. M., Ryms, M., & Kosakowski, W. (2020). Thermal biomass conversion: A review. *Processes (Basel, Switzerland)*, *8*(5), 516. DOI: 10.3390/pr8050516

Li, J.; Zhang, L.; Li, R.; Yang, X.; Zhang, T. High-pressure gaseous hydrogen storage vessels: Current status and prospects. Energy Storage Sci. Technol. 2021, 10, 1835. Energies 2024, 17, 4070 27 of 30

Li, L. L., Wen, S. Y., Tseng, M. L., & Wang, C. S. (2019). Renewable energy prediction: A novel short-term prediction model of photovoltaic output power. *Journal of Cleaner Production*, *228*, 359–375. DOI: 10.1016/j.jclepro.2019.04.331

Luo, M., Yi, Y., Wang, S., Wang, Z., Du, M., Pan, J., & Wang, Q. (2018). Review of hydrogen production using chemical-looping technology. *Renewable & Sustainable Energy Reviews*, *81*, 3186–3214. DOI: 10.1016/j.rser.2017.07.007

Mahapatra, S., Kumar, D., Singh, B., & Sachan, P. K. (2021). Biofuels and their sources of production: A review on cleaner sustainable alternative against conventional fuel, in the framework of the food and energy nexus. *Energy Nexus*, *4*, 100036. DOI: 10.1016/j.nexus.2021.100036

Maitlo, G., Ali, I., Mangi, K. H., Ali, S., Maitlo, H. A., Unar, I. N., & Pirzada, A. M. (2022). Thermochemical conversion of biomass for syngas production: Current status and future trends. *Sustainability (Basel), 14*(5), 2596. DOI: 10.3390/su14052596

Mao, J., Guo, Z., Yu, X., & Liu, H. (2011). Improved hydrogen storage properties of NaBH4 destabilized by CaH2 and Ca(BH4)2. *The Journal of Physical Chemistry. C, Nanomaterials and Interfaces, 115*(18), 9283–9290. DOI: 10.1021/jp2020319

Márquez-Reyes, L. A., del Pilar Sánchez-Saavedra, M., & Valdez-Vazquez, I. (2009). Improvement of Hydrogen Production by Reduction of the Photosynthetic Oxygen in Microalgae Cultures of Chlamydomonas gloeopara and Scenedesmus obliquus. Int. J. Hydrogen Energy 2015, 40, 7291–7300. 41. Torzillo, G.; Scoma, A.; Faraloni, C.; Ena, A.; Johanningmeier, U. Increased Hydrogen Photoproduction by Means of a SulfurDeprived Chlamydomonas reinhardtii D1 Protein Mutant. *International Journal of Hydrogen Energy, 34*, 4529–4536.

Mathiesen, B. V., Lund, H., Connolly, D., Wenzel, H., Østergaard, P. A., Möller, B., Nielsen, S., Ridjan, I., Karnøe, P., Sperling, K., & Hvelplund, F. K. (2015). Smart Energy Systems for coherent 100% renewable energy and transport solutions. *Applied Energy, 145*, 139–154. DOI: 10.1016/j.apenergy.2015.01.075

Mishra, P., Krishnan, S., Rana, S., Singh, L., Sakinah, M., & Ab Wahid, Z. (2019). Outlook of Fermentative Hydrogen Production Techniques: An Overview of Dark, Photo and Integrated Dark-Photo Fermentative Approach to Biomass. *Energy Strateg. Rev., 24*, 27–37. DOI: 10.1016/j.esr.2019.01.001

Mona, S., Kaushik, A., & Kaushik, C. P. (2014). Hydrogen Production and Metal-Dye Bioremoval by a Nostoc linckia Strain Isolated from Textile Mill Oxidation Pond. Bioresour. Technol. 2011, 102, 3200–3205. 45 Zhang, X.; Sherman, D.M.; Shermana, L.A. The Uptake Hydrogenase in the Unicellular Diazotrophic Cyanobacterium Cyanothece sp. Strain PCC 7822 Protects Nitrogenase from Oxygen Toxicity. *Journal of Bacteriology, 196*, 840–849.

Nam, K., Hwangbo, S., & Yoo, C. (2020). A deep learning-based forecasting model for renewable energy scenarios to guide sustainable energy policy: A case study of Korea. *Renewable & Sustainable Energy Reviews, 122*, 109725. DOI: 10.1016/j.rser.2020.109725

Natural Resources Canada (NRCan). *Seizing the Opportunities for Hydrogen.* Natural Resources Canada (NRCan); Hamilton, ON, Canada: 2020

Palys, M. J., & Daoutidis, P. (2022). P. Daoutidis Power-to-X: A review and perspective. *Computers & Chemical Engineering, 165*, 107948. Advance online publication. DOI: 10.1016/j.compchemeng.2022.107948

Papadis, E., & Tsatsaronis, G. (2020). Challenges in the decarbonization of the energy sector. *Energy, 205*, 118025. DOI: 10.1016/j.energy.2020.118025

Paramesh, K., & Chandrasekhar, T. (2020). Improvement of Photobiological Hydrogen Production in Chlorococcum minutum Using Various Oxygen Scavengers. *International Journal of Hydrogen Energy, 45*(13), 7641–7646. DOI: 10.1016/j.ijhydene.2019.05.216

Ram, M., Bogdanov, D., Satymov, R., Lopez, G., Mensah, T. N. O., & Sadovskaia, K.. (2022). *Accelerating the European renewable energy transition Lappeenranta.*

Rashid, N., Lee, K., & Mahmood, Q. (2011). Bio-Hydrogen Production by Chlorella vulgaris under Diverse Photoperiods. *Bioresource Technology*, *102*(2), 2101–2104. DOI: 10.1016/j.biortech.2010.08.032 PMID: 20826084

Saravanakumar A, Vijayakumar P, Hoang AT, et al. Thermochemical conversion of large-size woody biomass for carbon neutrality: Principles, applications, and issues[J]. Bioresour Technol 2023;370:128562. https://doi.org/. biortech.2022.128562.DOI: 10.1016/j

Shabani, H. R., & Kalantar, M. (2021). M. KalantarReal-time transient stability detection in the power system with high penetration of dfig-based wind farms using transient energy function. *International Journal of Electrical Power & Energy Systems*, *133*, 107319. DOI: 10.1016/j.ijepes.2021.107319

Shabani, H. R., Kalantar, M., & Hajizadeh, A. (2021). A. HajizadehReal-time transient instability detection in the power system with high dfig-wind turbine penetration via transient energy. *IEEE Systems Journal*, *16*(2), 3013–3024. DOI: 10.1109/JSYST.2021.3079253

Sorrenti, I., Rasmussen, T. B. H., You, Q., & Wu, S. (2022). The role of Power-to-X in hybrid renewable energy systems: A comprehensive review. *Renewable & Sustainable Energy Reviews*, *165*, 112380. DOI: 10.1016/j.rser.2022.112380

Sterner, M., & Specht, M. (2021). M. Specht Power-to-Gas and power-to-X—The history and results of developing a new storage concept. *Energies*, *14*(20), 6594. DOI: 10.3390/en14206594

The Elec, Korea Electronics Industry Media. Hanwha Solutions Developing Type 5 Hydrogen Tank. Available online: https: //www.thelec.net/news/articleView.html?idxno=3450 (accessed on 6 July 2024).

Trchounian, K., & Trchounian, A. (2015). Hydrogen Production from Glycerol by Escherichia coli and Other Bacteria: An Overview and Perspectives. *Applied Energy*, *156*, 174–184. DOI: 10.1016/j.apenergy.2015.07.009

Wang, D., Liao, B., Zheng, J., Huang, G., Hua, Z., Gu, C., & Xu, P. (2019). Development of regulations, codes and standards on composite tanks for on-board gaseous hydrogen storage. *International Journal of Hydrogen Energy*, *44*(40), 22643–22653. DOI: 10.1016/j.ijhydene.2019.04.133

Yu, R., Gao, J., Yu, M., Lu, W., Xu, T., Zhao, M., Zhang, J., Zhang, R., & Zhang, Z. (2019). LSTM-EFG for wind power forecasting based on sequential correlation features. *Future Generation Computer Systems*, *93*, 33–42. DOI: 10.1016/j.future.2018.09.054

Zhang, M., Zhang, C., Zhu, B., Wang, C., Jia, X., Guan, G., Zeng, X., Hu, E., Han, Z., & Xu, G. (2024). *Pyrolysis of Biomass to Produce H-Rich Gas Facilitated by Simultaneously Occurring Magnesite Decomposition*. Carbon Resour. Convers. DOI: 10.1016/j.crcon.2024.100265

Chapter 11

# Green Hydrogen Sustainable Solution to Combat CO2 Emissions and Mitigate Climate Change

**Youssef El Mourabit**
*University of Ibn Zohr, Morocco*

**Mustapha Agnaou**
*University of Ibn Zohr, Morocco*

**Jamal Mabrouki**
https://orcid.org/0000-0002-3841-7755
*Mohammed V University in Rabat, Morocco*

**Maryam Nadir**
*University of Ibn Zohr, Morocco*

**Jaouad Abou Oualid**
*University of Ibn Zohr, Morocco*

**Aicha Ait Alla**
*University of Ibn Zohr, Morocco*

**Abdellatif Moukrim**
*University of Ibn Zohr, Morocco*

## ABSTRACT

*The persistent increase in carbon dioxide (CO2) emissions from traditional fossil fuel-based energy production methods has led to significant environmental challenges, including climate change, ocean acidification, and biodiversity loss. This review examines the detrimental effects of CO2 on the environment and human health, highlighting the urgent need for a transition to cleaner energy sources. It explores the limitations of conventional energy systems and the ongoing efforts to adopt more sustainable practices. In this context, green hydrogen emerges as a promising alternative. Produced through water electrolysis using renewable energy, green hydrogen offers a clean and sustainable solution to reduce CO2 emissions across various sectors, including transportation, industry, and energy.*

DOI: 10.4018/979-8-3693-8980-5.ch011

## 1. HARMFUL EFFECTS OF $CO_2$

The production of carbon dioxide ($CO_2$) is primarily driven by the combustion of fossil fuels such as coal, oil, and natural gas, which are essential to many modern industries including energy production, transportation, and various manufacturing sectors (Paraschiv and Paraschiv, 2020; Ameyaw et al., 2019). However, this widespread reliance on fossil fuels has led to massive emissions of $CO_2$ into the atmosphere, disrupting the Earth's natural carbon cycle. The result is a significant rise in atmospheric concentrations of $CO_2$, surpassing historical norms, and leading to alarming changes in the global climate. This increase in $CO_2$ levels is closely linked to global warming, characterized by a steady rise in the planet's average temperature. While the effects of $CO_2$ on climate change are well-documented, its harmful impacts are not limited to environmental concerns. Elevated levels of $CO_2$ and other pollutants in the atmosphere have direct consequences on human health. Research shows that increased exposure to these pollutants is associated with higher rates of respiratory and cardiovascular diseases, particularly in densely populated urban areas (Verma et al., 2022; Zaman et al., 2021). Air pollution, exacerbated by high CO2 concentrations, reduces air quality, leading to a rise in chronic diseases and a reduction in the overall quality of life. The urgency of these health risks has prompted global recognition of the need to mitigate $CO_2$ emissions.

## 2. NATURAL AND ANTHROPOGENIC SOURCES OF $CO_2$

$CO_2$ is naturally released into the atmosphere through processes such as respiration, combustion, and the decay of organic matter as showed in fig 1. These natural processes contribute approximately 58 gigatonnes of $CO_2$ annually to the atmosphere, playing a crucial role in the carbon cycle of terrestrial and oceanic ecosystems (Yurak and Fedorov, 2024; Zaman et al., 2021). The Earth's natural carbon sinks, including forests and oceans, help regulate atmospheric $CO_2$ levels, but human activities have overwhelmed these systems. But anthropogenic sources of $CO_2$, such as fossil fuel combustion, transportation, and cement production (Fig. 1), have significantly accelerated the rise of atmospheric carbon dioxide on the planet. Currently, the burning of fossil fuels, release about 63 gigatonnes of $CO_2$ annually, far exceeding the capacity of natural sinks to absorb this excess carbon (Yurak and Fedorov, 2024). This imbalance between natural and anthropogenic $CO_2$ emissions is a primary driver of the current climate crisis. As $CO_2$ concentrations in the atmosphere continue to rise, the effects are being felt not only in terms of global warming but also in the health of ecosystems. For example, increased $CO_2$ levels in surface waters have led to a more than 25% rise in $CO_2$ concentrations, which can potentially contribute to the global expansion of harmful algal blooms (HABs) (Hallegraeff, 2010; Fu et al., 2012; Flynn et al., 2015). Such changes have significant implications for marine life and the broader ecological balance.

*Figure 1. Primary contributors to $CO_2$ emissions worldwide*

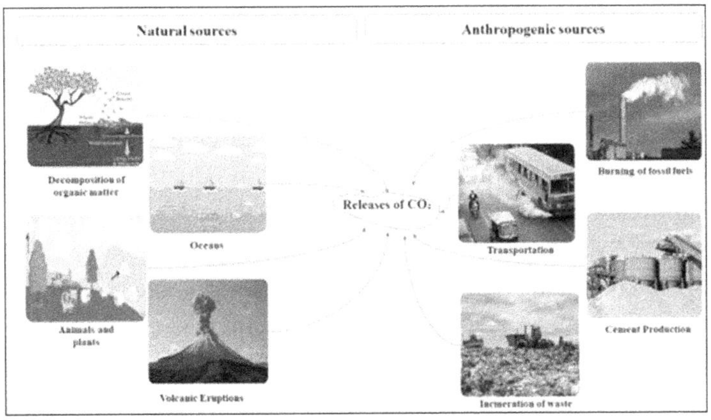


## 3. THE COMPLEX EFFECTS OF RISING $CO_2$ ON ECOSYSTEMS

The rise in atmospheric $CO_2$ has complex and varied effects on ecosystems, particularly in aquatic environments (Table 1). Higher concentrations of dissolved inorganic carbon in surface waters, driven by increasing $CO_2$ levels, can impact the ability of phytoplankton to fix carbon through photosynthesis (Badger et al., 1998; Giordano et al., 2005; Raven et al., 2017). The extent to which different species of phytoplankton may benefit from these elevated $CO_2$ levels depends on a wide array of physiological, ecological, and biogeochemical factors (Fu et al., 2010).

In addition, the increase in $CO_2$ levels in the atmosphere has direct and profound consequences on biodiversity and terrestrial ecosystems. In forests, for example, climate change alters the growth cycles of trees and plants, affecting species composition and ecosystem resilience (Addison et al., 2024). Higher temperatures and prolonged drought periods exacerbate stress on forests, leading to an increased risk of devastating wildfires (Mildrexler et al., 2016). The disruption of natural cycles in grasslands also impacts animal species dependent on these ecosystems for food and habitat.

Climate warming and drying have resulted in decreased tree growth, indicating physiological drought stress, and are expected to cause productivity declines across large regions of the boreal forest (Walker et al., 2015). The combination of drought and high temperatures is directly linked to increasing forest stress and mortality. Forest restoration thinning has been shown to mitigate some of these effects by reducing drought impacts and increasing soil moisture, which benefits forest health during extreme drought periods (Mildrexler et al., 2016; Sankey and Tatum, 2022; Xu et al., 2020). Furthermore, climate extremes, such as heat and drought stress, can significantly affect forest carbon fluxes, with the magnitude and direction of these changes being highly dependent on seasonal dynamics and timing (Mildrexler et al., 2016; Sankey and Tatum, 2022).

Physiological responses of plants to climatic stress, including temperature, are crucial for evaluating forest vulnerability and resilience to climate change. Forest sensitivity to drought and low temperatures, especially under high radiation, are key stressors driving climate-induced forest range shifts globally (Hirata et al., 2022). Effective forest management strategies, including sustainable, community, and urban forest management, are essential to mitigating forest fire incidences. However, rising temperatures increase soil-water evaporation, leading to drier soils and water loss in forest flora, which negatively impacts forest resilience (Mansoor et al., 2022).

In aquatic ecosystems, rising dissolved $CO_2$ concentrations associated with climate change may benefit phytoplankton, though their response is influenced by a complex interplay of physiological, ecological, and biogeochemical factors. Elevated $CO_2$ levels can enhance photosynthetic carbon fixation in some phytoplankton groups, improving their growth and productivity (Riebesell, 2024). However, this response varies significantly among species and strains (Raven et al., 2020). Additionally, resource use efficiency (RUE) is closely linked to rising dissolved $CO_2$ levels. Elevated CO2 enhances carbon use efficiency in organisms like phytoplankton, allowing them to optimize their photosynthetic processes, reduce oxidative stress, and promote better growth, thereby increasing overall ecosystem productivity (Sobrino et al., 2014). Ecologically, rising $CO_2$ levels can alter species composition within phytoplankton communities. Some harmful algal blooms (HABs) may become more dominant under higher CO2 conditions (Raven et al., 2020; Van de Waal et al., 2011). Zooplankton grazing and competition among phytoplankton species also play crucial roles, with elevated CO2 interacting with these factors to influence overall biomass and community structure (Paquette and Beisner, 2018).

From a biogeochemical perspective, increased $CO_2$ significantly affects nutrient dynamics by promoting stratification and altering nutrient availability, which in turn influences phytoplankton growth and species composition (Wang et al., 2022; Häder and Gao, 2015). Changes in $CO_2$ levels also impact the biological carbon pump, affecting the vertical flux of biogenic material and the overall carbon storage capacity of oceans (Riebesell, 2024; Asselot et al., 2024).

*Table 1. Negative impacts of $CO_2$ on the planet*

| Environment | Harmful effects of $CO_2$ |
|---|---|
| **Atmosphere** | - **Global Warming**: $CO_2$ is a major greenhouse gas, trapping heat and contributing to the overall increase in Earth's temperature, leading to climate change. |
| | - **Extreme Weather Events**: Increased $CO_2$ levels intensify weather patterns, causing more frequent heatwaves, storms, and droughts. |
| | - **Air Quality Degradation**: Higher $CO_2$ concentrations can indirectly affect air quality by increasing ground-level ozone and other pollutants. |
| **Surface Water** | - **Acidification**: Elevated $CO_2$ levels in the atmosphere can lead to acid rain, which lowers the pH of lakes, rivers, and streams, harming aquatic life. |
| | - **Algal Blooms**: Higher temperatures and changes in nutrient cycling caused by CO2-induced warming promote harmful algal blooms, depleting oxygen levels in the water (hypoxia). |
| | - **Impact on Aquatic Organisms**: Acidified water affects the health of fish, amphibians, and invertebrates, disrupting freshwater ecosystems. |
| **Ocean** | - **Ocean Acidification**: $CO_2$ dissolves in seawater and forms carbonic acid, lowering the pH and harming marine life, particularly organisms with calcium carbonate shells, like corals, mollusks, and some plankton species. |
| | - **Coral Bleaching**: Increased water temperatures due to higher $CO_2$ levels cause corals to expel the algae (zooxanthellae) living in their tissues, leading to widespread coral bleaching and reef damage. |
| | - **Disruption of Marine Ecosystems**: Changes in ocean chemistry can alter marine food chains, threatening species that depend on calcifying organisms. |

continued on following page

*Table 1. Continued*

| Environment | Harmful effects of $CO_2$ |
|---|---|
| Ground | - **Soil Acidification**: Increased $CO_2$ can cause acid rain, which lowers soil pH, affecting soil fertility and microbial activity, leading to reduced crop productivity. |
| | - **Changes in Vegetation Patterns**: Higher $CO_2$ concentrations can accelerate plant growth in some cases, but excessive levels of $CO_2$ can also disrupt natural vegetation, reducing biodiversity in certain ecosystems. |
| | - **Permafrost Thawing**: Increased atmospheric $CO_2$ raises global temperatures, accelerating permafrost thawing, which releases stored greenhouse gases ($CO_2$ and methane) into the atmosphere, creating a feedback loop. |

## 4. TRADITIONAL AND POLLUTING ENERGY PRODUCTION METHODS

Traditional energy production methods primarily rely on the combustion of fossil fuels, which are major sources of $CO_2$ as well as other pollutants such as sulfur oxides (SOx) and nitrogen oxides (NOx) (Table 2). These pollutants contribute to the formation of acid rain, which damages both terrestrial and aquatic ecosystems (Tontu and Tontu, 2022). Coal-fired power plants, for instance, are particularly notorious for emitting large quantities of CO2, NOx, SO2, and particulate matter (PM), making them significant contributors to air pollution with severe environmental and health impacts (Dincer, 2018; Wong and Amador, 2010).

In addition to their direct emissions, these energy production methods have considerable environmental impacts throughout their entire value chain. The extraction of fossil fuels often leads to the destruction of vast tracts of land, groundwater pollution, and incidents such as oil spills that cause substantial ecological damage (Saravanan and Ramaswamy, 2022). Furthermore, the transportation of these fuels globally exposes the environment to increased risks of industrial accidents (Chen et al., 2019).

The impacts of fossil fuels are not limited to greenhouse gases alone. The combustion of fossil fuels also releases fine particles, which can have a detrimental effect on human health. When inhaled, these particles penetrate deep into the lungs and can cause severe respiratory diseases such as asthma and lung cancer (Tontu and Tontu, 2022; Batmunkh et al., 2008). Additionally, fine particles contribute to global warming by absorbing heat in the atmosphere (Haseli, 2020). Thus, transitioning to cleaner energy sources is crucial not only for reducing CO2 emissions but also for decreasing air pollution and improving public health (Dincer, 2018; Vikhrova et al., 2020).

Thermal power plants, for example, often discharge heated water and pollutants into water bodies, affecting aquatic ecosystems (Vikhrova et al., 2020). Moreover, the deposition of airborne pollutants can lead to soil contamination, impacting agriculture and natural vegetation (Vikhrova et al., 2020; Batmunkh et al., 2008).

*Table 2. Comparative overview of energy types on the planet*

| Energy Type | Definition | Source | Efficiency | Environmental Impact | Examples |
|---|---|---|---|---|---|
| **Coal Power** | Energy produced by burning coal to generate electricity. | Fossil fuel (coal) | Medium (30-40% efficiency) | - High CO2 emissions. | Coal-fired power plants, steam turbines. |
| | | | | - Air pollution (SOx, NOx, particulate matter). | |
| | | | | - Significant mining impact. | |
| **Natural Gas Power** | Energy produced by burning natural gas for electricity or heating. | Fossil fuel (natural gas) | High (45-60% efficiency) | - Lower CO2 emissions than coal, but still significant. | Combined-cycle gas plants. |
| | | | | - Methane leakage from extraction and transport. | |
| | | | | - Water contamination risk (fracking). | |
| **Oil Power** | Energy produced by burning petroleum (oil) for electricity, heating, or transportation. | Fossil fuel (oil) | Medium (30-40% efficiency) | - High CO2 emissions. | Oil-fired power plants, internal combustion engines. |
| | | | | - Air pollution (SOx, NOx, particulates). | |
| | | | | - Oil spills and environmental degradation from extraction. | |
| **Oil Shale and Tar Sands** | Energy produced from unconventional oil resources extracted from oil shale or tar sands. | Shale oil, tar sands | Low (10-20% efficiency) | - Very high CO2 emissions. | Tar sands mining, in-situ processing. |
| | | | | - Water contamination and air pollution. | |
| | | | | - Extensive land destruction for extraction. | |
| **Peat Power** | Energy produced by burning peat, an organic material formed from decayed vegetation. | Peat (partially decayed organic matter) | Low (10-20% efficiency) | - High CO2 emissions. | Peat-fired power plants. |
| | | | | - Destruction of peatlands, which are important carbon sinks. | |
| | | | | - Air pollution. | |

## 5. THE GLOBAL RESPONSE TO $CO_2$ EMISSIONS

Acknowledging the severity of the climate crisis, scientists and policymakers have emphasized the importance of limiting global warming to below 2°C, ideally to 1.5°C, compared to pre-industrial levels, to avoid the most devastating impacts of climate change. Achieving this goal requires substantial transformations across multiple sectors, including energy, industry, and transportation, which are major contributors to $CO_2$ emissions (Jeon et al., 2016; Zaman et al., 2021). These sectors rely heavily on the burning of fossil fuels, which is responsible for the majority of $CO_2$ emissions globally. As such, efforts to mitigate $CO_2$ levels include strategies such as carbon capture and storage, the adoption of renewable energy, and improvements in energy efficiency… (Table 3). These measures aim to reduce emissions in a cost-effective and environmentally sustainable way.

Furthermore, deforestation and land clearance for agricultural purposes also contribute significantly to $CO_2$ emissions, releasing stored carbon into the atmosphere (Zaman et al., 2021). Agricultural practices, including livestock production and land cultivation, add another layer of complexity to the issue, as they further increase the volume of $CO_2$ released annually. The interconnectedness of industrial, energy, and agricultural activities underscores the multifaceted nature of $CO_2$ emissions and highlights the need for a holistic approach to mitigation.

*Table 3. Approaches to decreasing global $CO_2$ levels*

| Field | Solutions to mitigate $CO_2$ |
|---|---|
| **Energy** | - **Shift to Renewable Energy**: Increase the use of solar, wind, geothermal, and hydropower. |
| | - **Carbon Capture and Storage (CCS)**: Capture and store CO2 from power plants. |
| | - **Energy Efficiency**: Improve efficiency in power generation. |
| | - **Nuclear Energy**: Invest in nuclear as a low-carbon source. |
| | - **Distributed Energy Systems**: Use local renewable energy like rooftop solar panels. |
| **Industry** | - **Carbon Capture and Utilization (CCU)**: Capture and reuse CO2 for industrial products. |
| | - **Electrification of Industrial Processes**: Shift from fossil fuels to electricity powered by renewables. |
| | - **Process Optimization**: Improve efficiency in manufacturing. |
| | - **Green Hydrogen**: Use renewable hydrogen for industrial processes. |
| | - **Circular Economy**: Promote recycling and reuse to minimize raw material processing. |
| **Transportation** | - **Electric Vehicles (EVs)**: Increase adoption of electric cars, buses, and trucks. |
| | - **Public Transportation**: Invest in trains, buses, subways to reduce vehicles on the road. |
| | - **Biofuels**: Use sustainable biofuels in aviation and heavy transport. |
| | - **Fuel Efficiency Standards**: Implement stricter standards for vehicles. |
| | - **Urban Planning**: Design walkable cities, bike lanes, and reduce car dependency. |
| **Agriculture** | - **Sustainable Farming Practices**: Implement no-till farming, crop rotation, and organic farming to improve soil health and carbon sequestration. |
| | - **Agroforestry**: Integrate trees into agricultural systems to absorb CO2. |
| | - **Reduced Fertilizer Use**: Use precision farming and organic fertilizers to reduce nitrous oxide emissions. |
| | - **Methane Reduction**: Improve livestock diets and manure management to lower methane emissions. |
| | - **Biochar**: Use biochar to store carbon in soils and improve crop yields. |
| **Waste Management** | - **Waste-to-Energy**: Convert waste into energy through incineration or biogas production, reducing landfill methane emissions. |
| | - **Recycling and Composting**: Reduce landfill waste by recycling materials and composting organic waste. |
| | - **Landfill Gas Capture**: Collect methane from landfills and use it for energy generation. |
| | - **Reduce, Reuse, Recycle**: Promote waste reduction strategies at all levels to lower carbon emissions associated with waste production. |

continued on following page

*Table 3. Continued*

| Field | Solutions to mitigate $CO_2$ |
|---|---|
| **Buildings** | - **Energy-Efficient Buildings**: Implement insulation, energy-efficient windows, and smart building technologies to reduce energy use. |
| | - **Green Building Materials**: Use sustainable materials (e.g., recycled steel, low-carbon cement) to reduce CO2 emissions during construction. |
| | - **Heating and Cooling Efficiency**: Use heat pumps, solar water heaters, and energy-efficient HVAC systems. |
| | - **Rooftop Solar Panels**: Install solar panels on buildings to generate clean energy. |
| | - **Retrofitting Existing Buildings**: Upgrade old buildings with energy-efficient technologies to reduce emissions. |

## 6. ATTEMPTS TO TRANSITION TO CLEANER METHODS

Efforts to transition towards cleaner energy production and industrial practices are essential for achieving long-term sustainability. Key to this transition are new technological and management approaches, such as the adoption of renewable energy sources, replacement of raw materials, and policy reforms. These shifts are vital for guiding industries, governments, and societies towards more sustainable practices (Almeida et al., 2017; Almeida et al., 2015). Technologies like post-combustion CO2 capture and the development of cleaner energy sources—such as renewables and nuclear power—are being actively explored to reduce the environmental impact of energy production (Ayhan et al., 2024; Haseli, 2020; Wong and Amador, 2010). Renewable energy sources like wind, solar, and hydroelectric power are particularly appealing as they produce little to no greenhouse gas emissions, though nuclear energy also presents a low-emission alternative, albeit with its own set of challenges (Ayhan et al., 2024; Yalamati et al., 2023; Wong and Amador, 2010).

One key approach in this transition is the Cleaner Production (CP) principle, which advocates for maximizing output while minimizing resource consumption and environmental impact. Although CP has gained traction in large energy-intensive industries, several barriers—such as cost and infrastructure limitations—hinder its widespread adoption (Nagesha, 2018; Day and Enterprises, 2005). On a smaller scale, innovations in cost-effective energy designs, particularly in regions with limited resources, have shown potential in providing high efficiency at a lower cost (Sleiti and Mehrabian, 2015). At the industrial level, technologies like Combined Heat and Power (CHP) plants have demonstrated substantial energy savings, converting up to 80% of fuel into both electricity and heat. This efficiency makes CHP plants a strong candidate for reducing carbon emissions and promoting energy self-sufficiency (Pierce, 2007).

Beyond individual technologies, energy efficiency plays a critical role in ensuring that energy systems operate optimally. While renewable energy technologies—such as solar, wind, biomass, and biofuels—continue to advance and become more integrated into sustainable systems, it is important not to overlook the importance of energy-saving measures. There is often a disproportionate focus on deploying renewables at the expense of broader energy efficiency improvements, which can undermine the overall effectiveness of sustainable energy transitions (Baldini and Jacobsen, 2016).

Equally important in this shift are supportive policies and investments. Legislative changes that favor low-emission industries and increased funding for innovative energy technologies are critical to making cleaner energy options viable on a larger scale (Pierce, 2007; Sáez-Martínez et al., 2016). However, significant challenges remain, including the high costs associated with transitioning to new technologies and the need for substantial investment in infrastructure upgrades, particularly in developing countries. Existing infrastructures, often designed around traditional energy systems, are not always compatible with newer, cleaner technologies, necessitating additional investments to retrofit or replace these systems.

In summary, while there have been significant advancements in cleaner technologies—such as renewable energy and carbon capture—the successful transition to sustainable energy systems will require a balanced approach. This includes technological innovation, a strong focus on energy efficiency, and robust policy and financial support to overcome existing barriers. Only by addressing these interconnected challenges can we achieve the scale of transformation needed to mitigate the environmental impacts of energy production and consumption.

*Table 4. Comparison of energy types: definitions, subtypes, advantages, and inconveniences*

| Energy Type | Definition | Subtypes | Advantages | Inconvenients |
|---|---|---|---|---|
| **Solar Energy** | Energy harnessed from sunlight using photovoltaic cells or concentrated solar power systems. | - Photovoltaic (PV) | - Renewable and abundant. | - Weather dependent. |
| | | - Concentrated Solar Power (CSP) | - Low operational costs. | - High initial installation cost. |
| | | | - Scalable for residential or industrial use. | - Requires large areas for CSP systems. |
| **Wind Energy** | Energy generated by converting wind flow into electricity using wind turbines. | - Onshore Wind | - Clean and renewable. | - Intermittent power generation. |
| | | - Offshore Wind | - Low operational costs. | - Aesthetic and noise pollution. |
| | | | - No emissions during operation. | - Can impact bird and bat populations. |
| **Hydropower (Hydroelectricity)** | Energy generated by capturing the kinetic energy of flowing or falling water. | - Run-of-River | - Reliable and efficient. | - Can disrupt aquatic ecosystems. |
| | | - Reservoir-Based | - Can store energy. | - High initial construction cost. |
| | | - Pumped-Storage | - Low greenhouse gas emissions. | - Risk of flooding and displacing communities. |
| **Geothermal Energy** | Energy derived from the Earth's internal heat. | - Dry Steam | - Consistent and reliable. | - Limited to geothermal-active areas. |
| | | - Flash Steam | - Not weather-dependent. | - High upfront costs for exploration. |
| | | - Binary Cycle | - Low environmental impact. | - May cause minor earthquakes. |
| **Biomass Energy** | Energy generated from organic materials such as plants, agricultural residues, and waste. | - Direct Combustion | - Reduces waste. | - Can lead to deforestation. |
| | | - Anaerobic Digestion | - Can be carbon-neutral if managed sustainably. | - Air pollution if not properly managed. |
| | | - Biofuels | - Can provide continuous energy. | - Land use competition with food production. |

continued on following page

*Table 4. Continued*

| Energy Type | Definition | Subtypes | Advantages | Inconvenients |
|---|---|---|---|---|
| **Tidal Energy** | Energy generated by harnessing the power of tidal movements using underwater turbines. | - Tidal Stream | - Predictable and reliable. | - High construction and maintenance costs. |
| | | - Tidal Barrage | - Zero emissions during operation. | - Limited to suitable coastal areas. |
| | | | | - Potential impact on marine life. |
| **Wave Energy** | Energy generated from the motion of surface ocean waves. | - Oscillating Water Columns | - Renewable and consistent in coastal areas. | - Expensive to develop. |
| | | - Point Absorbers | - Can operate in diverse weather conditions. | - Difficult to maintain in rough seas. |
| | | | | - Potential interference with marine ecosystems. |
| **Ocean Thermal Energy** | Energy derived from the temperature difference between warmer surface waters and cooler deep waters. | - Ocean Thermal Energy Conversion (OTEC) | - Available in tropical regions. | - Limited to tropical areas. |
| | | | - Constant energy source. | - High upfront costs and technology still in development. |
| | | | - Can desalinate water as a byproduct. | |
| **Hydrogen Energy** | Energy produced by converting hydrogen into electricity, typically through fuel cells. | - Green Hydrogen | - High energy density. | - Expensive to produce (green hydrogen). |
| | | - Blue Hydrogen | - Can store renewable energy. | - Infrastructure for storage and transport not widely available. |
| | | | - Produces only water vapor as a byproduct (green hydrogen). | |
| **Nuclear Energy (Fusion)** | Energy generated through nuclear fusion reactions (not yet commercially available). | - Fusion | - Virtually limitless fuel supply. | - Technology still in experimental phase. |
| | | | - No long-lived radioactive waste. | - Extremely high cost of research and development. |
| | | | - Clean energy potential. | - Safety concerns. |

## 7. EMERGENCE OF HYDROGEN ENERGIES AS AN INNOVATIVE SOLUTION IN THE WORLD

The emergence of hydrogen energy, particularly in the forms of grey, blue, pink, and green hydrogen, represents an innovative and evolving solution to meet global energy demands while addressing varying degrees of environmental impact. Among these, grey hydrogen currently dominates the market, accounting for 96-98% of global hydrogen production. This form of hydrogen is primarily produced through steam methane reforming (SMR), a process heavily reliant on natural gas, which results in significant CO2 emissions (Kannimuthu et al., 2022; Mio et al., 2023; Freund and Sánchez, 2022). However, despite its cost-effectiveness, largely due to the well-established SMR infrastructure and low natural gas prices, grey hydrogen lacks carbon capture technologies, contributing to its high carbon footprint (Linh Le et al., 2024; Howarth and Jacobson, 2021; Nishi et al., 2021). Over time, rising carbon prices may diminish its economic advantage, potentially bringing the levelized cost of grey hydrogen closer to that of green hydrogen (Mio et al., 2023; Freund and Sánchez, 2022).

In contrast, blue hydrogen presents a transitional solution aimed at mitigating some of the environmental drawbacks of grey hydrogen. Blue hydrogen utilizes CCS (carbon capture and storage) technologies to reduce carbon emissions, employing processes such as SMR, Partial Oxidation of Methane (POM), and Auto-Thermal Reforming (ATR) (Hamed et al., 2023; Longden et al., 2020). Nonetheless, while blue hydrogen does reduce carbon output compared to grey hydrogen, it is not without its challenges. Methane leakage during natural gas extraction and processing, coupled with inefficiencies in the carbon capture process, still results in significant greenhouse gas emissions (Howarth and Jacobson, 2021; Pettersen et al., 2022). Furthermore, Life Cycle Assessments (LCA) show that blue hydrogen has a larger greenhouse gas footprint than green hydrogen, and its success in reducing emissions hinges on improving CCS efficiency and minimizing methane leaks (Bauer et al., 2022; Riemer & Duscha, 2023; Hamed et al., 2023; Alghool et al., 2024). Despite these issues, blue hydrogen is often considered a temporary "bridge" technology, providing a lower-emission alternative until green hydrogen becomes more affordable and widely adopted (Tetteh and Salehi, 2023; Longden et al., 2020; George et al., 2022; Nishi et al., 2021).

Moving forward, pink hydrogen—produced using nuclear energy—offers another pathway toward decarbonization. By leveraging nuclear power, pink hydrogen can significantly reduce emissions compared to fossil fuel-based methods (Fernández-Arias et al., 2024; Venizelou and Poullikkas, 2024). Nevertheless, this form of hydrogen faces technical uncertainties, particularly regarding its cost-effectiveness, which necessitates further research and development (Venizelou and Poullikkas, 2024). However, the increasing interest and research in pink hydrogen highlight its growing importance in the global energy transition (Fernández-Arias et al., 2024).

In contrast to all these options, green hydrogen stands out as the most environmentally sustainable and promising solution. It is produced through the electrolysis of water, using renewable energy sources such as solar, wind, and hydroelectric power, which results in zero net greenhouse gas emissions (Majewski et al., 2023; Marouani et al., 2023; Jumah, 2024). Although green hydrogen is currently more expensive than both grey and blue hydrogen, the decreasing costs of renewable energy, combined with potential carbon pricing mechanisms, could soon make it a competitive option (Linh Le et al., 2024; Kim et al., 2023; Diab et al., 2022).

Green hydrogen is arguably the best long-term solution for several reasons. First and foremost, it eliminates the carbon footprint entirely by avoiding the use of fossil fuels, thus bypassing issues like methane leakage and the need for carbon capture technologies (Bauer et al., 2022; Hamed et al., 2023). Moreover, as renewable energy sources become more abundant and affordable, the production costs of green hydrogen will continue to decline, enhancing its economic viability (Majewski et al., 2023; Linh Le et al., 2024). Additionally, green hydrogen offers versatility across various sectors, from energy storage to industrial processes, providing a scalable and sustainable energy solution (Majewski et al., 2023). Finally, with global initiatives focused on reducing carbon emissions, green hydrogen is poised to benefit from increased government incentives and investments, further accelerating its adoption on a global scale (Kim et al., 2023).

*Figure 2. Type of hydrogen production: (A) Grey hydrogen; (B) Blue hydrogen; (C) Green hydrogen; and (D) Pink hydrogen.*

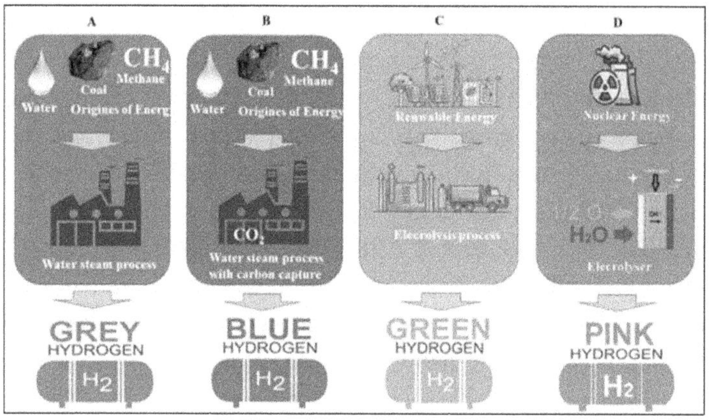


*Table 5. Hydrogen categories: grey, blue, green, and pink - characteristics and comparisons*

| Category | Grey Hydrogen | Blue Hydrogen | Green Hydrogen | Pink Hydrogen |
|---|---|---|---|---|
| **Environmental Benefits** | High carbon dioxide ($CO_2$) emissions due to fossil fuel use, contributing to climate change. | Reduced $CO_2$ emissions with carbon capture and storage (CCS), but still involves fossil fuels. | Zero greenhouse gas emissions at the point of use; water is the only byproduct. | Low to zero carbon emissions, depending on nuclear energy source. |
| **Renewable and Sustainable** | Relies on finite fossil fuel resources (natural gas, coal). | Still relies on finite fossil fuels, but CCS helps mitigate emissions. | Derived from renewable energy sources like wind and solar, which are inexhaustible. | Relies on nuclear energy, which is a low-carbon but finite source. |
| **Energy Security** | Dependent on fossil fuel availability and prices, which can fluctuate due to geopolitical factors. | Slightly more secure than grey hydrogen due to CCS but still tied to natural gas supply. | Enhances energy security by reducing dependence on imported fossil fuels through the use of domestic renewable sources. | Increases energy security through domestic nuclear energy, reducing reliance on fossil fuel imports. |
| **Versatility** | Widely used, but continues to perpetuate fossil fuel dependence and associated environmental impacts. | Versatile like grey hydrogen but with reduced carbon footprint due to CCS. | Can be used in transportation, power generation, industry, and as energy storage, offering flexibility in energy systems. | Similar versatility to green hydrogen; useful in power, transport, and energy storage applications. |
| **Economic opportunities** | Limited innovation, tied to established fossil fuel industries with fewer opportunities for new economic expansion. | Potential for growth in CCS technologies, but largely remains within the traditional fossil fuel sector. | Creates new industries and jobs in renewable energy sectors, driving economic growth and diversification. | Opportunities in nuclear energy and related technologies, although more capital intensive. |

continued on following page

*Table 5. Continued*

| Category | Grey Hydrogen | Blue Hydrogen | Green Hydrogen | Pink Hydrogen |
|---|---|---|---|---|
| **Health and air Quality** | Releases pollutants such as nitrogen oxides (NOx), contributing to air pollution and health problems. | Reduces air pollutants like $CO_2$, but still generates some emissions through natural gas extraction and processing. | Does not produce air pollutants like NOx, $SO_2$, or particulate matter, improving health and air quality. | Low emissions; nuclear energy generates no direct air pollutants, but nuclear waste must be managed carefully. |
| **Technological innovation** | Mature technology with little room for significant innovation. | Encourages advancements in CCS technology, though still fossil fuel-dependent. | Promotes innovation in renewable energy technologies, electrolysis systems, and energy storage solutions. | Supports technological development in nuclear power and hydrogen production but is more capital-intensive. |

## 8. IMPACT OF USE OF GREEN HYDROGEN ON MITIGATE CLIMATE CHANGE

The use of green hydrogen plays a transformative role in mitigating climate change by offering a clean alternative to fossil fuels across various sectors. One of the most critical benefits of green hydrogen is its ability to decarbonize hard-to-abate sectors, such as heavy industry, aviation, and shipping, which are among the largest contributors to greenhouse gas emissions and difficult to electrify directly using renewable energy (Sharma et al., 2024; Saleemi et al., 2024; Shaari et al., 2020). Specifically, it can replace fossil fuels in processes like steel manufacturing and cement production, significantly reducing $CO_2$ emissions (Bouzas et al., 2024; Simões and Santos, 2024). Furthermore, in the transportation sector, green hydrogen has the potential to revolutionize clean transportation by serving as a fuel for hydrogen-powered vehicles, including trucks, buses, and trains, thereby lowering emissions from road transport, aviation, and shipping (Rose and Neumann, 2020; Wallington et al., 2024). As it is produced through the electrolysis of water using renewable energy sources, green hydrogen allows for a drastic reduction in reliance on fossil fuels, which leads to lower emissions of carbon dioxide and methane (de Kleijne et al., 2022; Braga, 2022). Moreover, it acts as an energy storage medium that stabilizes the power grid by storing excess renewable energy for later use. Additionally, green hydrogen can reduce methane emissions by replacing natural gas, which poses significant risks for climate change due to its potency as a greenhouse gas (Ocko and Hamburg, 2022). The combustion of green hydrogen does not release harmful pollutants, leading to improved air quality and reduced incidence of respiratory diseases. Furthermore, by decreasing $CO_2$ emissions, green hydrogen can help limit ocean acidification, thereby protecting marine ecosystems (Hammi et al., 2024). However, the widespread adoption of green hydrogen faces challenges, such as the significant water and renewable energy requirements for its production and the need for substantial investment in hydrogen infrastructure (Shen et al., 2024). Overall, while green hydrogen presents a promising pathway for achieving climate goals, careful management of its challenges is essential for successful implementation.

*Figure 3. Use of green hydrogen on mitigate climate change*

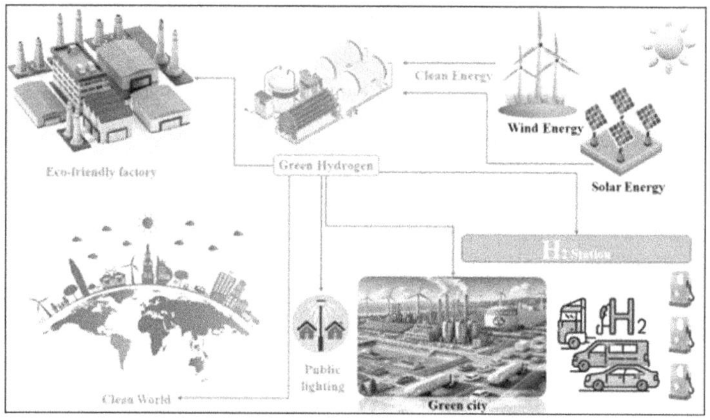


## 9. IMPACT OF USE OF GREEN HYDROGEN ON OCEAN HEALTH

The transition to green hydrogen as an energy source has significant implications for ocean health, encompassing various interconnected aspects of the marine environment. By replacing fossil fuels with renewable energy sources such as green hydrogen, we can substantially reduce greenhouse gas emissions, which is critical in combating climate change (Jayachandran et al., 2024). This shift is essential in mitigating sea level rise—a direct consequence of global warming characterized by melting glaciers and ice caps (Ferreira et al., 2022). Consequently, slowing the rate of emissions can help preserve coastal ecosystems and marine habitats that are increasingly threatened by rising sea levels (de Kleijne et al., 2022; Li et al., 2023).

Moreover, the biodiversity of oceanic flora and fauna is likely to benefit from the adoption of green hydrogen. More stable ocean temperatures and acidity levels foster healthier marine ecosystems (Wallace et al., 2023; Upton and Folger, 2014; Xu et al., 2024). By gradually substituting oil with green hydrogen, global demand for oil would decline, thereby reducing extraction, refining, and transportation activities associated with fossil fuels (Azadnia et al., 2023; Noussan et al., 2020). This transition is vital because oil production poses significant environmental risks, including oil spills that severely impact marine life and habitats (Midilli, 2016; Rahimpour et al., 2023). The reduction in oil transportation diminishes the likelihood of shipwrecks and accidental spills, which are major environmental hazards. Consequently, oil spills contaminate the water column and seabed, adversely affecting marine organisms, from phytoplankton—the foundation of ocean food chains—to fish and marine mammals (Midilli, 2016; Kumar et al., 2023).

As a result, the transition to green hydrogen helps limit these ecological disasters, allowing ocean ecosystems the opportunity to regenerate. With fewer pollutants in the environment, natural food chains can be restored, enabling vital species such as phytoplankton to regain their critical role in marine ecosystems. This restoration benefits all species, from the smallest organisms to the largest predators, contributing to a more stable ecological balance in the oceans.

Furthermore, a decrease in offshore oil extraction facilities significantly lowers the risk of marine pollution, thereby promoting the resilience of fish populations and other marine organisms. Chronic exposure to pollutants has been shown to lead to genetic bottlenecks and reduced reproductive capacity in fish (Santos et al., 2013; Martínez-Gómez and Vethaak, 2019). By reducing pollution levels, fish populations can recover, maintaining genetic diversity that is essential for resilience in changing environmental conditions (Santos et al., 2013; Martínez-Gómez and Vethaak, 2019). This recovery supports both the ecological health of marine environments and the sustainability of fisheries vital for human food security and livelihoods.

In addition, coral reefs, known for their biological diversity, are particularly susceptible to pollution and the effects of warming waters driven by fossil fuel activities (Knowlton and Jackson, 2013; Schulz et al., 2019). By reducing offshore oil extraction, we can lower the risks of coral bleaching and other stressors threatening these vital habitats (Knowlton and Jackson, 2013; Schulz et al., 2019). The health of coral reefs is critical not only for the corals themselves but also for the diverse ecosystems that rely on them for habitat and sustenance.

The concentration of CO2 in the oceans is closely linked to greenhouse gas emissions. Therefore, adopting green hydrogen can help lower atmospheric CO2 levels, leading to reduced concentrations of dissolved CO2 in ocean waters. This reduction is crucial for mitigating ocean acidification, which poses significant threats to marine ecosystems (Upton and Folger, 2014; Gopika et al., 2022; Noor and Das, 2019). Increased atmospheric CO2 leads to higher levels of CO2 dissolving in oceans, resulting in acidification that decreases seawater pH and disrupts marine life dependent on calcium carbonate, such as corals, mollusks, and certain plankton species (Upton and Folger, 2014; Form and Riebesell, 2012).

The long-term impacts of ocean acidification could include reduced growth rates and increased mortality of shell-forming organisms, which would have profound ecological and economic consequences (Upton and Folger, 2014; Noor and Das, 2019). The decline of these organisms not only threatens biodiversity but also jeopardizes fisheries and marine resources critical for human communities.

By transitioning to green hydrogen, we can play a significant role in reducing CO2 emissions, thereby helping to stabilize ocean chemistry. This transition not only addresses the immediate challenges posed by ocean acidification but also fosters the resilience of marine ecosystems. Ultimately, healthy marine ecosystems are essential for maintaining biodiversity, supporting fisheries, and ensuring the overall health of our oceans.

*Figure 4. Comparison between rising $CO_2$ on ocean and use of green hydrogen consequences*

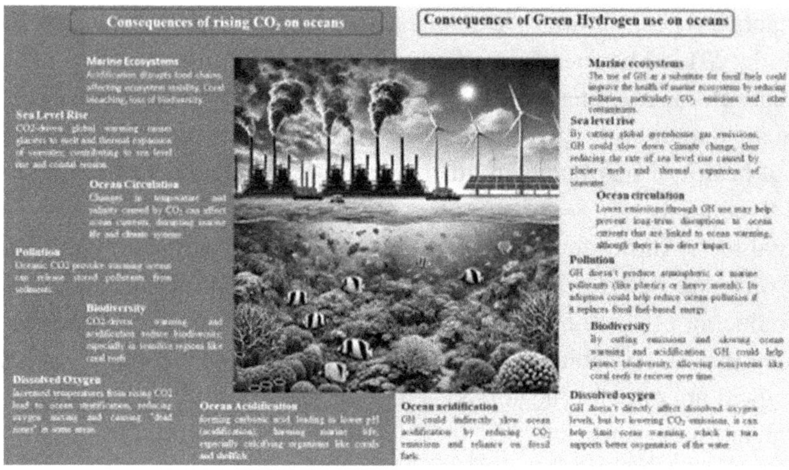


## 10. CONCLUSION

This study underscores the urgent need to address $CO_2$ emissions in order to mitigate environmental degradation and pollution. Indeed, anthropogenic sources, particularly fossil fuel combustion, have contributed to a variety of environmental crises, including climate change, ocean acidification, and biodiversity loss. However, despite global efforts to curb emissions, the continued reliance on traditional, polluting energy production methods significantly hampers progress. Therefore, green hydrogen, produced from renewable energy sources, emerges as a promising solution, as it has the potential to decarbonize sectors that are difficult to electrify, while simultaneously reducing pollution and safeguarding ecosystem health. Moreover, its ability to mitigate $CO_2$ emissions and reverse environmental damage is especially critical for marine ecosystems, where lowering carbon levels can help counteract ocean acidification. Nonetheless, for green hydrogen to achieve its full potential, substantial investments, infrastructure development, and policy support are required. In this context, the paper provides valuable insights into the role of green hydrogen in tackling climate change and protecting ecosystem health, highlighting its essential role in the transition toward sustainable energy systems that safeguard both terrestrial and marine environments.

## AKNOWLEDGMENT

Gratitude is extended to Ibn Zohr University for its vital support and encouragement of this research. The university's commitment to academic excellence and provision of resources have played a key role in advancing the study. The opportunities and guidance offered by the faculty have been instrumental in facilitating the successful completion of the project.

## REFERENCES

Addison, S. L., Rúa, M. A., Smaill, S. J., Singh, B. K., & Wakelin, S. A. (2024). Partner or perish: Tree microbiomes and climate change. *Trends in Plant Science*, *29*(9), 1029–1040. DOI: 10.1016/j.tplants.2024.03.008 PMID: 38641475

Alghool, D., Haouari, M., & Trucco, P. (2024). It is not the same blue: A comparative LCA study of blue hydrogen supply network pathways. *International Journal of Hydrogen Energy*, *81*, 214–224. DOI: 10.1016/j.ijhydene.2024.07.165

Ameyaw, B., Yao, L., Oppong, A., & Agyeman, J. K. (2019). Investigating, forecasting and proposing emission mitigation pathways for CO2 emissions from fossil fuel combustion only: A case study of selected countries. *Energy Policy*, *130*, 7–21. DOI: 10.1016/j.enpol.2019.03.056

Asselot, R., Holden, P. B., Lunkeit, F., & Hense, I. (2024). A missing link in the carbon cycle: Phytoplankton light absorption under RCP emission scenarios. *Earth System Dynamics: ESD*, *15*(4), 875–891. DOI: 10.5194/esd-15-875-2024

Ayhan, F., Yenilmez, M. I., Elal, O., & Dursun, S. (2024). Can technological progress, renewable and nuclear energy consumption be the remedy for global climate crises? An examination of leading OECD countries. *Environmental Science and Pollution Research International*, *31*(1), 228–248. DOI: 10.1007/s11356-023-30627-0 PMID: 37919508

Azadnia, A. H., McDaid, C., Andwari, A. M., & Hosseini, S. E. (2023). Green hydrogen supply chain risk analysis: A european hard-to-abate sectors perspective. *Renewable & Sustainable Energy Reviews*, *182*, 113371. DOI: 10.1016/j.rser.2023.113371

Badger, M. R., Andrews, T. J., Whitney, S. M., Ludwig, M., Yellowlees, D. C., Leggat, W., & Price, G. D. (1998). The diversity and coevolution of Rubisco, plastids, pyrenoids and chloroplast-based CO2-concentrating mechanisms in algae. *Canadian Journal of Botany*, *76*(6), 1052–1071. DOI: 10.1139/b98-074

Baldini, M., & Jacobsen, H. K. (2016, June). Optimal trade-offs between energy efficiency improvements and additional renewable energy supply: a review of international experiences. In 2016 13th International Conference on the European Energy Market (EEM) (pp. 1-5). IEEE. DOI: 10.1109/EEM.2016.7521245

Batmunkh, S., Tseyen-Oidov, T. S., & Battogtokh, Z. (2008, June). Survey to develop standards on air polluting emissions from the power plants. In *2008 Third International Forum on Strategic Technologies* (pp. 615-619). IEEE. DOI: 10.1109/IFOST.2008.4602992

Bauer, C., Treyer, K., Antonini, C., Bergerson, J., Gazzani, M., Gencer, E., Gibbins, J., Mazzotti, M., McCoy, S. T., McKenna, R., Pietzcker, R., Ravikumar, A. P., Romano, M. C., Ueckerdt, F., Vente, J., & Van der Spek, M. (2022). On the climate impacts of blue hydrogen production. *Sustainable Energy & Fuels*, *6*(1), 66–75. DOI: 10.1039/D1SE01508G

Biabani, H., Aminlou, A., Hayati, M. M., Majidi-Gharehnaz, H., & Abapour, M. (2024). Green Hydrogen Research and Development Projects in the European Union. In *Green Hydrogen in Power Systems* (pp. 301–320). Springer International Publishing. DOI: 10.1007/978-3-031-52429-5_12

Bouzas, B., Teliz, E., & Díaz, V. (2024). Green hydrogen production in Uruguay: A techno-economic approach. *International Journal of Chemical Reactor Engineering*, *22*(7), 783–795. DOI: 10.1515/ijcre-2024-0066

Braga, D. (2022). *The hydrogen role in the transition towards 100% energy from renewable sources*. Energy. Environment. Efficiency. Resources. Globalization.

Byrne, R. H. (2014). Measuring ocean acidification: new technology for a new era of ocean chemistry.

Camp, E. F. (2022). Contingency planning for coral reefs in the Anthropocene; The potential of reef safe havens. *Emerging Topics in Life Sciences*, *6*(1), 107–124. DOI: 10.1042/ETLS20210232 PMID: 35225326

Campos Rodrigues, L., Freire-González, J., González Puig, A., & Puig-Ventosa, I. (2018). Climate change adaptation of Alpine ski tourism in Spain. *Climate (Basel)*, *6*(2), 29. DOI: 10.3390/cli6020029

Chen, H., He, L., Chen, J., Yuan, B., Huang, T., & Cui, Q. (2019). Impacts of clean energy substitution for polluting fossil-fuels in terminal energy consumption on the economy and environment in China. *Sustainability (Basel)*, *11*(22), 6419. DOI: 10.3390/su11226419

Day, F., & Enterprises, G. F. GHG Reduction Through Industrial Energy Efficiency in the Developing World: The GERBI and CBLA Projects–Variations on a Theme.

de Kleijne, K., de Coninck, H., van Zelm, R., Huijbregts, M. A., & Hanssen, S. V. (2022). The many greenhouse gas footprints of green hydrogen. *Sustainable Energy & Fuels*, *6*(19), 4383–4387. DOI: 10.1039/D2SE00444E PMID: 36276614

Diab, J., Fulcheri, L., Hessel, V., Rohani, V., & Frenklach, M. (2022). Why turquoise hydrogen will Be a game changer for the energy transition. *International Journal of Hydrogen Energy*, *47*(61), 25831–25848. DOI: 10.1016/j.ijhydene.2022.05.299

Dincer, I. (Ed.). (2018). *Comprehensive energy systems*. Elsevier.

Doney, S. C., Ruckelshaus, M., Duffy, J. E., Barry, J. P., Chan, F., English, C. V. A., Galindo, H. M., Grebmeier, J. M., Hollowed, A. B., Knowlton, N., Polovina, J., Rabalais, N. N., Sydeman, W. J., & Talley, L. D. (2012). Climate change impacts on marine ecosystems. *Annual Review of Marine Science*, *4*(1), 11–37. DOI: 10.1146/annurev-marine-041911-111611 PMID: 22457967

Ejike, C. (2022, August). The advancement of green hydrogen and prospects for the future: a brief overview. In SPE Nigeria Annual International Conference and Exhibition (p. D032S003R005). SPE. DOI: 10.2118/211901-MS

Fernández-Arias, P., Antón-Sancho, Á., Lampropoulos, G., & Vergara, D. (2024). Emerging Trends and Challenges in Pink Hydrogen Research. *Energies*, *17*(10), 2291. DOI: 10.3390/en17102291

Flynn, K., Clark, D., Mitra, A., Heiner, F., Hansen, P. J., Glibert, P. M., Wheeler, G., Stoecker, D., Blackford, J., & Brownlee, C. (2015). Ocean acidification with (de) eutrophication alters phytoplankton growth and succession. *Proceedings. Biological Sciences*, *282*(1804), 20142604. DOI: 10.1098/rspb.2014.2604 PMID: 25716793

Form, A. U., & Riebesell, U. (2012). Acclimation to ocean acidification during long-term CO2 exposure in the cold-water coral L ophelia pertusa. *Global Change Biology*, *18*(3), 843–853. DOI: 10.1111/j.1365-2486.2011.02583.x PMID: 23504841

Freund, S., & Sánchez, D. (2022). Green hydrogen market and growth. In *Machinery and Energy Systems for the Hydrogen Economy* (pp. 605–635). Elsevier. DOI: 10.1016/B978-0-323-90394-3.00001-1

Fu, F.-X., Place, A. R., Garcia, N. S., & Hutchins, D. A. (2010). CO2 and phosphate availability control the toxicity of the harmful bloom dinoflagellate Karlodinium verficium. *Aquatic Microbial Ecology*, *59*, 55–65. DOI: 10.3354/ame01396

Fu, F.-X., Tatters, A. O., & Hutchins, D. A. (2012). Global change and the future of harmful algal blooms in the ocean. *Marine Ecology Progress Series*, *470*, 207–233. DOI: 10.3354/meps10047

George, J. F., Müller, V. P., Winkler, J., & Ragwitz, M. (2022). Is blue hydrogen a bridging technology?-The limits of a CO2 price and the role of state-induced price components for green hydrogen production in Germany. *Energy Policy*, *167*, 113072. DOI: 10.1016/j.enpol.2022.113072

Giordano, M., Beardall, J., & Raven, J. A. (2005). CO2 concentrating mechanisms in algae: Mechanisms, environmental modulation, and evolution. *Annual Review of Plant Biology*, *68*(1), 3701–3716. DOI: 10.1146/annurev.arplant.56.032604.144052 PMID: 15862091

Gómez-Vadillo, M., Mingarro, M., Ursul, G., & Wilson, R. J. (2022). Assessing climate change exposure for the adaptation of conservation management: The importance of scale in mountain landscapes. *Land (Basel)*, *11*(11), 2052. DOI: 10.3390/land11112052

Gopika, K. V., Kumar, N., & Ramesh, M. V. (2022, September). Iot based ocean acidification monitoring system with ml based edge analytics. In 2022 4th International Conference on Inventive Research in Computing Applications (ICIRCA) (pp. 345-353). IEEE.

Häder, D. P., & Gao, K. (2015). Interactions of anthropogenic stress factors on marine phytoplankton. *Frontiers in Environmental Science*, *3*, 14. DOI: 10.3389/fenvs.2015.00014

Hallegraeff, G. (2010). Ocean climate change, phytoplankjton community responses, and harmful algal blooms: A formidable predictive challenge. *Journal of Phycology*, *46*(2), 220–235. DOI: 10.1111/j.1529-8817.2010.00815.x

Hamed, A. M., Kamaruddin, T. N. A. T., Ramli, N., & Wahab, M. F. A. (2023, December). A review on blue and green hydrogen production process and their life cycle assessments. [). IOP Publishing.]. *IOP Conference Series. Earth and Environmental Science*, *1281*(1), 012034. DOI: 10.1088/1755-1315/1281/1/012034

Hammi, Z., Labjar, N., Dalimi, M., El Hamdouni, Y., & El Hajjaji, S. (2024). Green hydrogen: A holistic review covering life cycle assessment, environmental impacts, and color analysis. *International Journal of Hydrogen Energy*, *80*, 1030–1045. DOI: 10.1016/j.ijhydene.2024.07.008

Haseli, Y. (2020, September). Analytical formulation of the performance of the Allam power cycle. In *Turbo Expo: Power for Land, Sea, and Air* (Vol. 84157, p. V006T09A003). American Society of Mechanical Engineers. DOI: 10.1115/GT2020-15070

Hirata, A., Kominami, Y., Ohashi, H., Tsuyama, I., Tanaka, N., Nakao, K., Hijioka, Y., & Matsui, T. (2022). Global estimates of stress-reflecting indices reveal key climatic drivers of climate-induced forest range shifts. *The Science of the Total Environment*, *824*, 153697. DOI: 10.1016/j.scitotenv.2022.153697 PMID: 35143798

Howarth, R. W., & Jacobson, M. Z. (2021). How green is blue hydrogen? *Energy Science & Engineering*, *9*(10), 1676–1687. DOI: 10.1002/ese3.956

Iyer, S. B., Salim, S. N. S., & Jais, M. (2023, October). Cost to Decarbonise. In *Abu Dhabi International Petroleum Exhibition and Conference* (p. D031S092R004). SPE. DOI: 10.2118/216536-MS

Jayachandran, M., Gatla, R. K., Flah, A., Milyani, A. H., Milyani, H. M., Blazek, V., Prokop, L., & Kraiem, H. (2024). Challenges and opportunities in green hydrogen adoption for decarbonizing hard-to-abate industries: A comprehensive review. *IEEE Access : Practical Innovations, Open Solutions*, *12*, 23363–23388. DOI: 10.1109/ACCESS.2024.3363869

Jeon, D. H., Bae, S. T., & Park, S. J. (2016). Preparation and characterization of chemically activated carbon materials for CO 2 capture. *Carbon Letters*, *17*(1), 85–89. DOI: 10.5714/CL.2016.17.1.085

Jumah, A. (2024). A comprehensive review of production, applications, and the path to a sustainable energy future with hydrogen. *RSC Advances*, *14*(36), 26400–26423. DOI: 10.1039/D4RA04559A PMID: 39175671

Kannimuthu, K., Kibria, M. G., & Kundu, S. (2022). Role of hydrogen generation technologies for renewable hydrogen production. In *Sulfide and Selenide Based Materials for Emerging Applications* (pp. 377–407). Elsevier. DOI: 10.1016/B978-0-323-99860-4.00023-X

Kibria, M. G., Tjoelker, M. G., Marchin, R. M., Arndt, S. K., & Rymer, P. D. (2024). Can species climate niche predict canopy growth, functional traits and phenotypic plasticity in urban trees? *Urban Forestry & Urban Greening*, *98*, 128417. DOI: 10.1016/j.ufug.2024.128417

Kim, J. H., Han, S. M., & Yoo, S. H. (2023). Price premium for green hydrogen in South Korea: Evidence from a stated preference study. *Renewable Energy*, *211*, 647–655. DOI: 10.1016/j.renene.2023.05.035

Klemeš, J. J., Varbanov, P. S., Pierucci, S., & Huisingh, D. (2010). Minimising emissions and energy wastage by improved industrial processes and integration of renewable energy. *Journal of Cleaner Production*, *18*(9), 843–847. DOI: 10.1016/j.jclepro.2010.02.028

Knowlton, N., & Jackson, J. (2013). Corals and coral reefs. In *Encyclopedia of biodiversity* (2nd ed., pp. 330–346). Elsevier., DOI: 10.1016/B978-0-12-384719-5.00237-9

Knowlton, N., & Jackson, J. (2024). Corals and coral reefs. In *Encyclopedia of biodiversity* (3rd ed., Vol. 1, pp. V1-764–V1-782). Elsevier. DOI: 10.1016/B978-0-12-822562-2.00210-3

Kumar, P., & Brewster, C. (2022). Co-production of climate change vulnerability assessment: A case study of the Indian Lesser Himalayan region, Darjeeling. *Journal of Integrative Environmental Sciences, 19*(1), 39–64. DOI: 10.1080/1943815X.2022.2033792

Le, L., Viet, N., & Liu, S.-H. (2024). Challenges and solutions in hydrogen energy. In *Circular Economy and Sustainable Energy Materials: A Net-Zero Emissions Approach* (pp. 119–133). Routledge., DOI: 10.1201/9781003463009-9

Li, X., Raorane, C. J., Xia, C., Wu, Y., Tran, T. K. N., & Khademi, T. (2023). Latest approaches on green hydrogen as a potential source of renewable energy towards sustainable energy: Spotlighting of recent innovations, challenges, and future insights. *Fuel, 334*, 126684. DOI: 10.1016/j.fuel.2022.126684

Lloret, F. (2017). Trade-offs in high mountain conservation. High mountain conservation in a changing world, 37-59.

Longden, T., Beck, F., & Jotzo, F. (2020). Assessing the long term costs of blue and green hydrogen under zero-emissions objectives. In Applied Energy Symposium: MIT A+B, AEAB 2020 (p. 6858). DOI: 10.46855/energy-proceedings-6858

Majewski, P., Salehi, F., & Xing, K. (2023). Green hydrogen. *AIMS Energy, 11*(5), 878–895. DOI: 10.3934/energy.2023042

Mansoor, S., Farooq, I., Kachroo, M. M., Mahmoud, A. E. D., Fawzy, M., Popescu, S. M., Alyemeni, M. N., Sonne, C., Rinklebe, J., & Ahmad, P. (2022). Elevation in wildfire frequencies with respect to the climate change. *Journal of Environmental Management, 301*, 113769. DOI: 10.1016/j.jenvman.2021.113769 PMID: 34600426

Mansoor, S., Farooq, I., Kachroo, M. M., Mahmoud, A. E. D., Fawzy, M., Popescu, S. M., Alyemeni, M. N., Sonne, C., Rinklebe, J., & Ahmad, P. (2022). Elevation in wildfire frequencies with respect to the climate change. *Journal of Environmental Management, 301*, 113769. DOI: 10.1016/j.jenvman.2021.113769 PMID: 34600426

Marcovecchio, J. E., Buzzi, N. S., Tartara, M. N., Spetter, C. V., & Simonetti, P. (2013). Potential Effects of Climate Changes on the Marine Ecosystem Stability. *Marine Ecology in a Changing World*, 1.

Marouani, I., Guesmi, T., Alshammari, B. M., Alqunun, K., Alzamil, A., Alturki, M., & Hadj Abdallah, H. (2023). Integration of renewable-energy-based green hydrogen into the energy future. *Processes (Basel, Switzerland), 11*(9), 2685. DOI: 10.3390/pr11092685

Martínez-Gómez, C., & Vethaak, A. D. (2019). Understanding the impact of chemicals on marine fish populations: The need for an integrative approach involving population and disease ecology. *Current Opinion in Environmental Science & Health, 11*, 71–77. DOI: 10.1016/j.coesh.2019.08.001

Mazo, A. F. (2022). Overlapping legal regimes in international law to address ocean acidification [La superposición de regímenes jurídicos en el derecho internacional para abordar la acidificación de los océanos]. *Revista Electrónica de Estudios Internacionales, 44*. Advance online publication. DOI: 10.17103/reei.44.05

Midilli, A. (2016). Green hydrogen energy system: A policy on reducing petroleum-based global unrest. *International Journal of Global Warming, 10*(1-3), 354–370. DOI: 10.1504/IJGW.2016.077906

Mildrexler, D., Yang, Z., Cohen, W. B., & Bell, D. M. (2016). A forest vulnerability index based on drought and high temperatures. *Remote Sensing of Environment, 173*, 314–325. DOI: 10.1016/j.rse.2015.11.024

Mio, A., Barbera, E., Pavan, A. M., Danielis, R., Bertucco, A., & Fermeglia, M. (2023). Analysis of the energetic, economic, and environmental performance of hydrogen utilization for port logistic activities. *Applied Energy*, *347*, 121431. Advance online publication. DOI: 10.1016/j.apenergy.2023.121431

Nagesha, N. (2018). Cleaner Production: A brief literature review. *Materials Today: Proceedings*, *5*(9), 17944–17951. DOI: 10.1016/j.matpr.2018.06.124

Nishi, M., Yamamoto, H., & Takei, K. (2021). Cost comparisons of hydrogen produced from fossil fuels and renewable energy in 2030. In The Proceedings of the International Conference on Power Engineering (ICOPE) 2021.15. The Japan Society of Mechanical Engineers.

Noor, N. M., & Das, S. K. (2019). Effects of elevated carbon dioxide on marine ecosystem and associated fishes. *Thalassas*, *35*(2), 421–429. DOI: 10.1007/s41208-019-00161-3

Noussan, M., Raimondi, P. P., Scita, R., & Hafner, M. (2020). The role of green and blue hydrogen in the energy transition—A technological and geopolitical perspective. *Sustainability (Basel)*, *13*(1), 298. DOI: 10.3390/su13010298

Ocko, I. B., & Hamburg, S. P. (2022). Climate consequences of hydrogen emissions. *Atmospheric Chemistry and Physics*, *22*(14), 9349–9368. DOI: 10.5194/acp-22-9349-2022

Palombo, C., Marchetti, M., & Tognetti, R. (2014). Mountain vegetation at risk: Current perspectives and research reeds. *Plant Biosystems*, *148*(1), 35–41. DOI: 10.1080/11263504.2013.878410

Paquette, C., & Beisner, B. E. (2018). Interaction effects of zooplankton and CO 2 on phytoplankton communities and the deep chlorophyll maximum. *Freshwater Biology*, *63*(3), 278–292. DOI: 10.1111/fwb.13063

Paraschiv, S., & Paraschiv, L. S. (2020). Trends of carbon dioxide (CO2) emissions from fossil fuels combustion (coal, gas and oil) in the EU member states from 1960 to 2018. *Energy Reports*, *6*, 237–242. DOI: 10.1016/j.egyr.2020.11.116

Pettersen, J., Steeneveldt, R., Grainger, D., Scott, T., Holst, L. M., & Hamborg, E. S. (2022). Blue hydrogen must be done properly. *Energy Science & Engineering*, *10*(9), 3220–3236. DOI: 10.1002/ese3.1232

Pierce, J. (2007). Greener choices. *Engineer*, *293*(7726), 33–36.

Pinto, D. D. D., Limpach, J. M., & Knuutila, H. K. (2024). Simulation-based assessment of the potential of offshore blue hydrogen production with high CO2 capture rates with optimised heat recovery. *Gas Science and Engineering*, *121*, 205177. DOI: 10.1016/j.jgsce.2023.205177

Rahimpour, M. R., Omidvar, B., Shirazi, N. A., & Makarem, M. A. (2023). *Crises in oil, gas and petrochemical industries: Disasters and environmental challenges*. Elsevier., DOI: 10.1016/C2021-0-01864-2

Raven, J. A., Beardall, J., & S'anchez-Baracaldo, P. (2017). The possible evolution, and future, of CO2 concentrating mechanisms. *Journal of Experimental Botany*, *68*(14), 3701–3716. DOI: 10.1093/jxb/erx110 PMID: 28505361

Raven, J. A., Gobler, C. J., & Hansen, P. J. (2020). Dynamic CO2 and pH levels in coastal, estuarine, and inland waters: Theoretical and observed effects on harmful algal blooms. *Harmful Algae*, *91*, 101594. DOI: 10.1016/j.hal.2019.03.012 PMID: 32057340

Riebesell, U. (2004). Effects of CO 2 enrichment on marine phytoplankton. *Journal of Oceanography*, *60*(4), 719–729. DOI: 10.1007/s10872-004-5764-z

Riebesell, U., Schulz, K. G., Bellerby, R. G. J., Botros, M., Fritsche, P., Meyerhöfer, M., Neill, C., Nondal, G., Oschlies, A., Wohlers, J., & Zöllner, E. (2007). Enhanced biological carbon consumption in a high CO2 ocean. *Nature*, *450*(7169), 545–548. DOI: 10.1038/nature06267 PMID: 17994008

Riemer, M., & Duscha, V. (2023). Carbon capture in blue hydrogen production is not where it is supposed to be-Evaluating the gap between practical experience and literature estimates. *Applied Energy*, *349*, 121622. DOI: 10.1016/j.apenergy.2023.121622

Rose, P. K., & Neumann, F. (2020). Hydrogen refueling station networks for heavy-duty vehicles in future power systems. *Transportation Research Part D, Transport and Environment*, *83*, 102358. DOI: 10.1016/j.trd.2020.102358

Sáez-Martínez, F. J., Lefebvre, G., Hernández, J. J., & Clark, J. H. (2016). Drivers of sustainable cleaner production and sustainable energy options. *Journal of Cleaner Production*, *138*, 1–7. DOI: 10.1016/j.jclepro.2016.08.094

Saleemi, S., Noor, N., Khan, S. M., Riaz, T., Mannan, H. A., & Gull, N. (2024). Nanomaterial for carbon dioxide capturing and carbon emission control. In *Nanomaterials in Biomass Conversion* (pp. 397–410). Woodhead Publishing. DOI: 10.1016/B978-0-443-13500-2.00019-5

Sankey, T., & Tatum, J. (2022). Thinning increases forest resiliency during unprecedented drought. *Scientific Reports*, *12*(1), 9041. DOI: 10.1038/s41598-022-12982-z PMID: 35641556

Santos, E. M., Hamilton, P. B., Coe, T. S., Ball, J. S., Cook, A. C., Katsiadaki, I., & Tyler, C. R. (2013). Population bottlenecks, genetic diversity and breeding ability of the three-spined stickleback (Gasterosteus aculeatus) from three polluted English Rivers. *Aquatic Toxicology (Amsterdam, Netherlands)*, *142*, 264–271. DOI: 10.1016/j.aquatox.2013.08.008 PMID: 24071362

Saravanan, K., & Ramaswamy, A. (2022, December). Feasibility Analysis for Sustainable Energy Generation from Medium and Large-Scale Industries. In *2022 International Conference on Computer, Power and Communications (ICCPC)* (pp. 468-473). IEEE. DOI: 10.1109/ICCPC55978.2022.10072116

Schickhoff, U., Bobrowski, M., Mal, S., Schwab, N., & Singh, R. B. (2022). The world's mountains in the Anthropocene. *Mountain landscapes in transition: effects of land use and climate change*, 1-144.

Schulz, M., Los, A., Skowronek, P., & Strachecka, A. (2019). Influence of environmental and anthropogenic agents on the occurrence of coral reef bleaching syndrome and coral diseases. *MEDYCYNA WETERYNARYJNA-VETERINARY MEDICINE-SCIENCE AND PRACTICE*, *75*(4), 221–231.

Shaari, N. F., Abdul-Rahim, A. S., & Afandi, S. H. M. (2020). Are Malaysian airline passengers willing to pay to offset carbon emissions? *Environmental Science and Pollution Research International*, *27*(19), 24242–24252. DOI: 10.1007/s11356-020-08662-y PMID: 32306253

Sharma, K., Tyagi, S., Bhardwaj, V., Tyagi, D., Gautam, Y. K., & Singh, B. P. (2024). Greenhouse gas emissions from the industries. In *Advances and technology development in greenhouse gases: Emission, capture and conversion: Greenhouse gases emissions and climate change* (pp. 165–181). Elsevier., DOI: 10.1016/B978-0-443-19231-9.00014-4

Shen, H., del Granado, P. C., Jorge, R. S., & Löffler, K. (2024). Environmental and climate impacts of a large-scale deployment of green hydrogen in Europe. *Energy and Climate Change*, *5*, 100133. DOI: 10.1016/j.egycc.2024.100133

Simões, F. L., & Santos, D. M. (2024). A SWOT Analysis of the Green Hydrogen Market. *Energies*, *17*(13), 3114. DOI: 10.3390/en17133114

Sleiti, A. K., & Mehrabian, A. (2015). Case study of a cost-effective small wind turbine. *Energy Sources. Part B, Economics, Planning, and Policy*, *10*(2), 132–138. DOI: 10.1080/15567240903419385

Sobrino, C., Segovia, M., Neale, P. J., Mercado, J. M., García-Gómez, C., Kulk, G., Lorenzo, M. R., Camarena, T., van de Poll, W. H., Spilling, K., & Ruan, Z. (2014). Effect of CO2, nutrients and light on coastal plankton. IV. Physiological responses. *Aquatic Biology*, *22*, 77–93. DOI: 10.3354/ab00590

Tetteh, D. A., & Salehi, S. (2023). The blue hydrogen economy: A promising option for the near-to-mid-term energy transition. *Journal of Energy Resources Technology*, *145*(4), 042701. DOI: 10.1115/1.4055205

Tontu, M., & Tontu, A. S. (2022). Energy and environment nexus: Environmental impact mitigation strategies. *International Journal of Global Warming*, *27*(3), 213–236. DOI: 10.1504/IJGW.2022.124199

Upton, H. F., & Folger, P. (2014). Ocean acidification. In *Climate, energy, and environment: Issues, analyses, and developments* (Vol. 2, pp. 21–40). Nova Science Publishers.

Upton, H. F., & Folger, P. F. (2018). *Ocean acidification*. Congressional Research Service.

Van de Waal, D. B., Verspagen, J. M., Finke, J. F., Vournazou, V., Immers, A. K., Kardinaal, W. E. A., Tonk, L., Becker, S., Van Donk, E., Visser, P. M., & Huisman, J. (2011). Reversal in competitive dominance of a toxic versus non-toxic cyanobacterium in response to rising CO2. *The ISME Journal*, *5*(9), 1438–1450. DOI: 10.1038/ismej.2011.28 PMID: 21390081

Velez, P. C. O., Blekhman, D., & Ariza, O. (2024). Bibliometric Analysis of Non-Renewable Thermal, Renewable Thermal, and Solar Energy in the Era of VUCA Phenomenon. In *Organizational Management Sustainability in VUCA Contexts* (pp. 204–218). IGI Global. DOI: 10.4018/979-8-3693-0720-5.ch011

Venizelou, V., & Poullikkas, A. (2024). Comprehensive Overview of Recent Research and Industrial Advancements in Nuclear Hydrogen Production. *Energies*, *17*(12), 2836. DOI: 10.3390/en17122836

Verma, S. K., Prajapati, A. K., Tripathi, M., & Bhatnagar, A. (2023). Waste and biomass-based nanomaterials for CO2 capture. In *Nanomaterials for Carbon Dioxide Capture and Conversion Technologies* (pp. 137–151). Elsevier. DOI: 10.1016/B978-0-323-89851-5.00013-5

Walker, X. J., Mack, M. C., & Johnstone, J. F. (2015). Stable carbon isotope analysis reveals widespread drought stress in boreal black spruce forests. *Global Change Biology*, *21*(8), 3102–3113. DOI: 10.1111/gcb.12893 PMID: 25683740

Wallace, D. W., Jutras, M., Nesbitt, W. A., Donaldson, A., & Tanhua, T. (2023). Can green hydrogen production be used to mitigate ocean deoxygenation? A scenario from the Gulf of St. Lawrence. *Mitigation and Adaptation Strategies for Global Change*, *28*(8), 56. DOI: 10.1007/s11027-023-10094-1

Wallington, T. J., Woody, M., Lewis, G. M., Keoleian, G. A., Adler, E. J., Martins, J. R., & Collette, M. D. (2024). Green hydrogen pathways, energy efficiencies, and intensities for ground, air, and marine transportation. *Joule*, *8*(8), 2190–2207. DOI: 10.1016/j.joule.2024.07.012

Wang, J., Song, Y., & Yan, X. (2022). Research progress on water eutrophication induced by climate change. *Journal of Hohai University*, *38*(4), 145–155.

Wilson, R. W., Reardon, E. E., & Perry, C. T. (2011). A fishy tale: A missing part of the inorganic ocean carbon cycle. *The Biochemist*, *33*(3), 30–34. DOI: 10.1042/BIO03303030

Wong, K.-F. V., & Amador, G. (2010). **Power generation for the near future**. In *ASME 2010 International Mechanical Engineering Congress and Exposition, IMECE 2010* (Vol. 5, Parts A and B, pp. 449-460). ASME. https://doi.org/DOI: 10.1115/IMECE2010-37714

Xu, B., Arain, M. A., Black, T. A., Law, B. E., Pastorello, G. Z., & Chu, H. (2020). Seasonal variability of forest sensitivity to heat and drought stresses: A synthesis based on carbon fluxes from North American forest ecosystems. *Global Change Biology*, *26*(2), 901–918. DOI: 10.1111/gcb.14843 PMID: 31529736

Xu, Y., Teng, Y., & Luo, Y. (2024). Recent advances in hydrogen metabolism and its environmental bioremediation function [ ]. *Turang Xuebao*, *61*(4), 916–928. DOI: 10.11766/trxb-202301180025Yalamati, H. P. S., Vij, R. K., & Srivastava, R. (2023). Hydrogen production driven by nuclear energy. In *Solar-Driven Green Hydrogen Generation and Storage* (pp. 347–362). Elsevier. DOI: 10.1016/B978-0-323-99580-1.00017-0

Yong, J. Y., Klemeš, J. J., Varbanov, P. S., & Huisingh, D. (2016). Cleaner energy for cleaner production: Modelling, simulation, optimisation and waste management. *Journal of Cleaner Production*, *111*, 1–16. DOI: 10.1016/j.jclepro.2015.10.062

Yurak, V. V., & Fedorov, S. A. (2024). Review of natural and anthropogenic emissions of carbon dioxide into the earth's atmosphere. *International Journal of Environmental Science and Technology*, •••, 1–18.

Zaman, M., Kleineidam, K., Bakken, L., Berendt, J., Bracken, C., Butterbach-Bahl, K., & Müller, C. (2021). *Greenhouse gases from agriculture*. Springer International Publishing. DOI: 10.1007/978-3-030-55396-8_1

Chapter 12
# Hydrogen as the Cornerstone of a Renewable Energy Society

**J. Mangaiyarkkarasi**
https://orcid.org/0000-0003-1431-9584
*N.M.S.S. Vellaichamy Nadar College, India*

**J. Shanthalakshmi Revathy**
https://orcid.org/0000-0003-1724-7117
*Velammal College of Engineering and Technology, India*

**H. Bhoomeeswaran**
https://orcid.org/0009-0005-8997-4470
*SRM Madurai College for Engineering and Technology, India*

## ABSTRACT

*The global transition to renewable energy has positioned green hydrogen as a pivotal element in reshaping energy production, storage and consumption. This chapter explores hydrogen's critical role in creating a renewable energy society, particularly its potential to decarbonize industries, enhance energy security and provide long-term energy storage. Unlike batteries, hydrogen can store excess renewable energy and deliver it during peak demand, offering a versatile solution for grid balancing. Key sectors such as transportation, heavy industry, and residential applications benefit from hydrogen's flexibility, especially where electrification alone is insufficient. Despite its promise, the hydrogen economy faces challenges, including high costs, infrastructure needs and public acceptance. The chapter highlights that while hydrogen is not a standalone solution, its integration with other clean technologies is crucial to achieving a sustainable, carbon-neutral future.*

## 1. INTRODUCTION

An essential component of the shift to a civilization powered by renewable energy is hydrogen. It is a flexible energy source that can be created by water electrolysis and other renewable energy sources like solar, wind, and hydroelectric power. This makes hydrogen an essential element in storing and distributing clean energy Hydrogen produces just water in fuel cells, as opposed to fossil fuels, offering a

DOI: 10.4018/979-8-3693-8980-5.ch012

zero-emission substitute for industry, transportation, and power production. Hydrogen was recognized as a distinct matter (element), initially (1713-1810) found by Henry Cavendish, an English Chemist. The aspect is very well known for its unique features. In the periodic table, it is the first element with atomic number Z = 1. Compared to other materials, it is the lightest of all because the hydrogen atom is made of 1 electron and 1 proton, thus the total mass remains very small. Because of its extreme reactivity, it may also be blended with other elements with ease (Juangsa et al., 2021). As we know, a hydrogen molecule contains two atoms and therefore it can be easily combined with other chemicals, to form new compounds in the periodic table. Apart, it is the essential constituent of all living matters present in the universe. When it comes to chemistry, the atoms and molecules of Hydrogen play an influential role and that is the reason why it is cited as an example for rendering several theories. For instance, water also known as $H_2O$, is an inevitable compound that covers the earth by 3/4th (75%) approximately. $H_2O$ has a 12% weight of Hydrogen that covers the earth's surface in terms of oceans, rivers, lakes, and other bodies of $H_2O$. Furthermore, it is the element that is more abundant on Earth that is present in compounds. Additionally, there is water in the atmosphere, and planets and animals have a large Hydrogen content due to the water they contain(Lamb & Webb, 2022). Apart from those, a further amount of combined hydrogen is present in plants and creatures in the form of 3 main classes of biological substances proteins, fats, and carbohydrates. It is also a part of the fossil fuels that the sun and other stars have and which are used in nuclear fusion. It is nothing but nuclei of hydrogen that produce Helium and an enormous quantity of energy. This process occurs in the sun provides the all necessary energies required to live on the Earth. Also, it is to be noted that hydrogen is continuously lost from the atmosphere by escaping from the gravity of the earth(Zhu et al., 2023). It is easily replaced by the decay of organic matter, formation of volcanic matter, and photochemical dissociation of water. Hydrogen is an element in the periodic table that is unique in that it does not fall under any particular group. Like Hydrogen, the bonds formed by it known as Hydrogen bonds have unique features for their life-sustaining quality as well as responsible for many biological molecular functions. In the absence of a hydrogen bond, $H_2O$ would not exist as a liquid. The concentration of its ion H+ (in the PH Scale) is used to express the acidity of aqueous solutions. Hydrogen is set as a benchmark or standard for expressing the properties of elements in the modern chemical industry. Valency is one such to be noted(Tarhan & Çil, 2021). Other than valency, the formation of electrochemical series is another one, where hydrogen is assigned arbitrarily to an EMF value of 0 and the EMF values of other elements are expressed related to hydrogen. Metals above Hydrogen in the EMF series will displace it from an acid solution.

Long-term storage and transportation of hydrogen improves the flexibility of renewable energy systems by resolving problems such as the intermittent nature of solar and wind power. Moreover, challenging industries that are challenging to electrify, such as heavy manufacturing, aviation, and shipping, can be decarbonized with green hydrogen. Hydrogen is emerging as a key component for creating a sustainable, renewable energy economy, promoting energy independence, and lowering greenhouse gas emissions as countries and companies strive to achieve carbon neutrality.

## 2. PROPERTIES OF HYDROGEN

In the previous section we have listed certain salient features and properties of Hydrogen and it is more important to know about its physical properties. It can be classified into molecular and atomic properties, along with its values which are listed separately. The chemical characteristics of hydrogen are shown in Table 1.

*Table 1. Molecular properties of hydrogen*

| Property | Value |
|---|---|
| Melting point | -259.2 (°C) |
| Boiling point | -252.8 (°C) |
| Density | -0.899 ($gl^{-1}$) |
| Bond Length | 0.0742 (nm) |
| Bond dissociation energy | 436 ($KJmol^{-1}$) |

Table 2 explains the atomic properties of Hydrogen. The values in the above tables are very important and one must be aware of it to prepare the Hydrogen in an industrial process called steam refining as well by the laboratory method.

*Table 2. Atomic properties of hydrogen*

| Property | Value |
|---|---|
| Ionization Energy | 1310 ($KJmol^{-1}$) |
| Electron Affinity | -72.8 ($KJmol^{-1}$) |
| Electronegativity | 2.1 (no units because it is a property) |
| Atomic radius | 0.037 (nm) |
| Ionic radius | 0.21 (nm) |

Finally, without Hydrogen it is difficult to produce methanol, hydrogenation, etc. Hydrogenation including oil and coal is also improbable without Hydrogen. Thanks to that invention, with the aid of it we can use it as fuel. In addition to that, along with Hydrogen its isotopes called protium, deuterium, and tritium plays equal importance in the modern revolution of chemical plants and its subordinate discipline called Hydrides, which is a binary compound of an element with it. A well-known example is Lithium+ Hydrogen as LiH, which is highly electropositive than Hydrogen. It is denoted as Li+ and H. However, as we discussed in the previous section, Hydrogen itself prefers to combine with several elements with more electronegative. The typical examples are HF, HCl, and $H_2O$. This is because F, Cl, and O are more electronegative than Hydrogen.

## 3. LITERATURE REVIEW

Many studies have been conducted in academic, business, and policy-focused literature on the role of hydrogen as a crucial energy carrier in the shift to a society reliant on renewable energy sources. This review surveys key findings on hydrogen production methods, their applications across different sectors, and the associated challenges of storage, distribution, and economic viability. The generation of hydrogen is essential to its function in a society powered by renewable energy sources, with an increasing emphasis on green hydrogen—produced by water electrolysis driven by renewable energy (Bhandari et al., 2014). It is emphasized that although fossil fuel-based processes such as steam methane reforming (SMR) dominate the current hydrogen generation process, green hydrogen is gaining popularity because of its potential for zero emissions. To increase the cost-competitiveness of green hydrogen, the cost of renewable energy is essential. Proton Exchange Membrane (PEM) and Alkaline Electrolyzers, in particular, are thought to be the most sustainable and scalable forms of electrolysis for producing hydrogen. Notwithstanding the benefits to the environment, the high cost of producing green hydrogen continues to be a major obstacle to its widespread use. Technological advancements and economies of scale will be key to reducing these costs, potentially making hydrogen competitive with traditional fuels in the next few decades.

The long-term energy storage capabilities of hydrogen make it a perfect match for intermittent renewable energy sources like solar and wind. The ability of hydrogen to store excess electricity during times of high generation and subsequently convert it back to electricity when demand exceeds supply helps mitigate the intermittent nature of renewable energy sources. The research supports the idea that hydrogen can serve as a long-term energy storage medium, providing seasonal storage options that are not possible with batteries (Juangsa et al., 2021). Further research by researchers has shown that integrating hydrogen storage into a renewable energy grid could significantly reduce the need for fossil fuel backup power plants. These findings support the idea that hydrogen can provide grid stability and energy security in a fully renewable society(Tarhan & Çil, 2021). Hydrogen is considered an essential remedy for decarbonizing sectors and businesses that are challenging to electrify, like long-distance transportation, aviation, steel manufacturing, and chemical industries (Tashie-Lewis & Nnabuife, 2021). Hydrogen can replace natural gas in industrial processes like steel manufacturing through methods such as hydrogen-based direct reduction, significantly reducing $CO_2$ emissions. In applications where batteries are less practical due to weight and range constraints, such as heavy-duty trucks, ships, and aviation, hydrogen has the potential to be a clean fuel. The difficulties with hydrogen fuel cells in the transportation industry are also mentioned, with a focus on how inefficient they are at converting energy when compared to battery-electric cars that utilize direct power. Despite this, fuel cells remain a viable option for sectors where energy density and fast refueling are critical(Abe et al., 2019). Hydrogen storage and distribution are well-documented challenges in the literature. Because of its low energy density, hydrogen is difficult to store as a cryogenic liquid or as a compressed gas. They stress that although promising, hydrogen storage solutions like chemical carriers and metal hydrides are still costly and high-tech (Raza et al., 2022).

According to the literature, hydrogen is a revolutionary energy carrier that could be essential to a civilization reliant on renewable energy sources. From balancing intermittent renewable energy to decarbonizing heavy industry and transport, hydrogen offers solutions to many of the challenges posed by a green energy transition. However, its large-scale integration will depend on overcoming significant technical, economic, and infrastructural barriers. It will need sustained scientific advancement, cost-cutting, and supporting legislative environments to establish hydrogen as the cornerstone of a global energy system that is sustainable (Yao et al., 2020).

## 4. THE ROLE OF HYDROGEN IN DECARBONIZING THE ENERGY SECTOR

Hydrogen is a key component of the global energy sector decarbonization movement because it provides a flexible way to cut greenhouse gas emissions in a variety of businesses. Hydrogen has emerged as a crucial element in accomplishing comprehensive decarbonization as the globe shifts from fossil fuels to renewable energy, particularly in sectors where electrification is either impracticable or inefficient. Hydrogen's potential as a clean energy carrier is one of its main benefits. It produces no direct carbon emissions when it is made using sustainable techniques, like electrolysis using solar or wind energy. This type of hydrogen, sometimes referred to as "green hydrogen," can replace carbon-intensive fuels like coal, oil, and natural gas in fuel cells, industrial operations, and even transportation networks (Ferrada et al., 2023). Hydrogen presents a competitive option to battery electric vehicles (BEVs) in the transportation industry, especially for heavy-duty vehicles, long-haul trucks, and ships, where batteries would not be able to deliver the required power or range. Hydrogen fuel cells are an eco-friendly way to cut emissions from the transportation industry because they convert hydrogen into energy with water as their only byproduct.

Hydrogen is also essential for decarbonizing some of the hardest sectors to decarbonize, like chemicals, steel, and cement. Fossil fuels and high-temperature procedures are essential to these sectors. These businesses have the potential to drastically lower their carbon footprint by switching to hydrogen in place of conventional fuel. For example, the use of hydrogen in steelmaking techniques such as direct reduction may be able to completely remove carbon emissions from the steelmaking process (Babonneau et al., 2021). Hydrogen also provides a solution for intermittency, which is one of the main problems with renewable energy. Although renewable and plentiful, the production of solar and wind energy is not continuous. As a long-term energy storage method, hydrogen can be created when the supply of renewable energy sources exceeds the demand, stored, and then used to produce electricity when needed. In sectors where direct electrification is either not practical or not adequate, hydrogen is anticipated to play a crucial role as nations and companies strive to meet ambitious climate targets. Hydrogen is a key component of future energy systems because of its ability to decarbonize these challenging industries, opening the door to a low-carbon, sustainable energy civilization.

## 5. PRODUCTION METHODS: GREEN HYDROGEN AND BEYOND

There are various ways to manufacture hydrogen, and each has a different influence on the environment. To differentiate between the various approaches' carbon footprints, color coding is frequently used. Here's an overview of key hydrogen production methods, with a focus on sustainable processes like electrolysis using renewable energy.

## 5.1. Electrolysis (Green Hydrogen)

When using renewable energy, electrolysis is a sustainable and widely used technique for creating green hydrogen—hydrogen produced without any accompanying carbon emissions. Utilizing electricity, water (H2O) is split into hydrogen ($H_3$) and oxygen ($O_2$). The hydrogen generated is regarded as "green" when the electricity comes from renewable energy sources like solar, wind, or hydropower because it doesn't add to greenhouse gas emissions.

### 5.1.1 The Process of Electrolysis

An electrolyzer, which has two primary components—a cathode, or negative electrode, and an anode, or positive electrode—that are separated by an electrolyte is used in electrolysis to break down water molecules by an electrochemical reaction (Awad et al., 2024). Ions can move more easily between the two electrodes thanks to the electrolyte. The electrolysis process is depicted in Figure 1.

*Figure 1. Process of electrolysis*

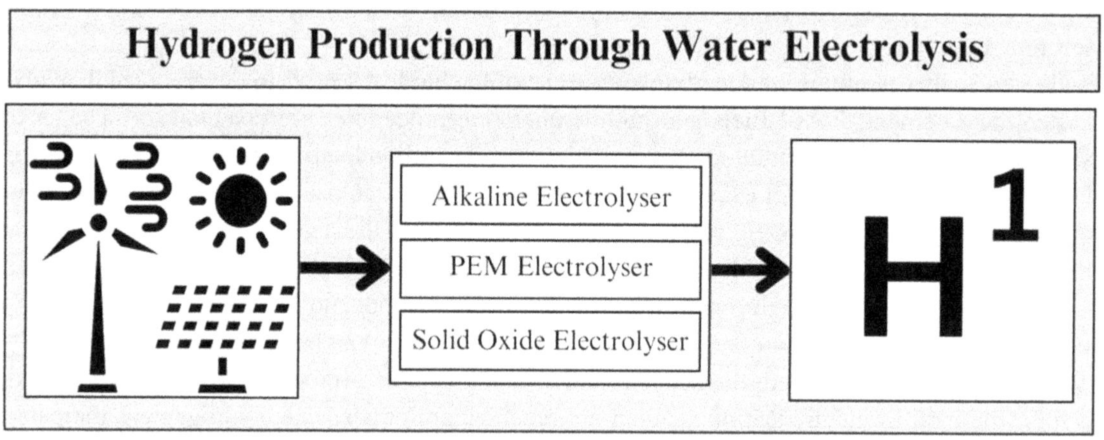


Initially, water ($H_2O$) is fed into the electrolyzer. Then a direct current (DC) of electricity is applied across the anode and cathode. The electricity can come from renewable sources such as solar panels or wind turbines, ensuring the process is environmentally friendly. At the anode, water molecules ($H_2O$) lose electrons (oxidation) and split into oxygen ($O_2$) molecules and positively charged hydrogen ions ($H^+$ or protons). The oxygen ($O_2$) is released as a byproduct, usually into the atmosphere. At the cathode, the hydrogen ions ($H^+$) migrate through the electrolyte to the cathode, where they gain electrons (reduction) to form hydrogen gas ($H_2$). The hydrogen gas ($H_2$) is collected, compressed, and can be stored or transported for use in various energy applications, such as powering fuel cells, industrial processes, or transportation.

### 5.1.2 Types of Electrolyzers

Hydrogen generation uses a variety of electrolyzer types, each with a unique operating environment and efficiency:

The alkaline electrolyzer is a highly developed and extensively utilized technology. An alkaline liquid electrolyte is used in it, usually potassium hydroxide, or KOH. Comparatively inexpensive, but it's not as efficient as other kinds. Second, an electrolyzer called a proton exchange membrane (PEM) is employed. The electrolyte is a solid polymer membrane. Works more efficiently and reacts faster to changes in the electricity supply, which makes it a good fit for integration with renewable energy sources. More costly as a result of the catalysts made of precious metals like iridium and platinum. Additionally, a solid oxide electrolyzer was used, which can increase overall efficiency because it runs at very high temperatures (about 700–1,000°C). Compared to alkaline and PEM electrolyzers, which are more widely used in commerce, these devices are still under development (Reza et al., 2022).

The advantages of Electrolysis for Green Hydrogen Production are,

i. *Zero Carbon Emissions*: When powered by renewable energy, electrolysis produces no direct emissions of carbon dioxide or other pollutants, making it an ideal solution for decarbonizing sectors like transport, industry, and power generation.
ii. *Energy Storage*: Excess renewable energy can be stored in the hydrogen created by electrolysis. Hydrogen can be created using renewable energy when there is more supply than demand (such as on windy or sunny days). Hydrogen can then be converted back into electricity when needed.
iii. *Versatility:* Green hydrogen has a wide range of uses, including fuel cells for automobiles, industrial operations (such as the manufacturing of steel and ammonia), and even as a natural gas substitute for heating.
iv. *Scalability*: Electrolysis is a scalable technology that can be deployed from small decentralized systems (e.g., at a wind farm or solar plant) to large industrial-scale plants capable of producing hydrogen for entire regions.

Challenges of Electrolysis are,

i. *Cost*: At the moment, steam methane reforming (SMR) and other conventional ways of producing hydrogen are less expensive than electrolysis. The primary cost drivers are the electrolyzers themselves and the electricity needed for the process. As renewable energy becomes more affordable and electrolyzer technologies improve, costs are expected to decrease.
ii. *Energy Intensity:* A substantial amount of electricity is needed for electrolysis. Green hydrogen needs to be produced responsibly using renewable energy sources. Green hydrogen has no environmental benefits if clean electricity isn't available.
iii. *Infrastructure:* The current infrastructure for producing, storing, and distributing hydrogen is inadequate, and significant expenditures are required to establish a sustainable hydrogen economy.

Even with the obstacles that still need to be overcome, electrolysis has enormous promise for sustainable energy in the future. Green hydrogen generation is anticipated to become more competitive with fossil fuel-based hydrogen as the price of renewable power continues to decline and technology developments make electrolyzers more effective and economical. To reach net-zero carbon emissions, governments and businesses are investing more money in green hydrogen technologies (Ahmad Kamaroddin et al., 2021). Electrolysis is viewed as a critical route to decarbonize industries that are challenging to electrify, such as heavy manufacturing, aviation, and long-distance transportation. This is especially true when electrolysis is fueled by wind or solar energy. Large-scale production of green hydrogen can help nations move away from fossil fuels and toward a world powered mostly by renewable energy.

## 5.2 Steam Methane Reforming (SMR)

Steam methane reforming (SMR) is currently the most widely used technique for producing hydrogen. This technique creates hydrogen and carbon dioxide by heating steam and natural gas, mainly methane (CH3), to high temperatures. This hydrogen is called "gray hydrogen" because it is produced with significant carbon emissions (Katebah et al., 2022). This method is widely used and well-established technology. Lower cost compared to green hydrogen. Despite these advantages, large amounts of $CO_2$ are released during the process, contributing to climate change and unsustainable due to reliance on fossil fuels. Figure 2 shows the Steam Methane Reforming method.

*Figure 2. Steam methane reforming method*

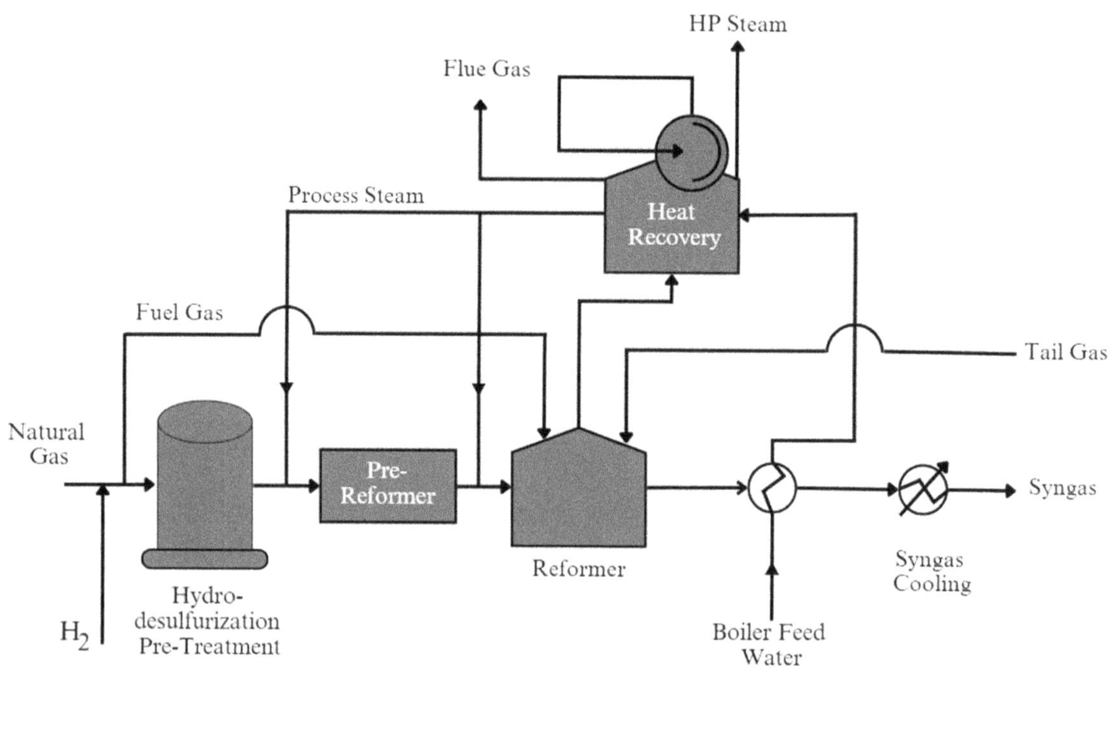


The same steam methane reforming method is used to create blue hydrogen, but carbon capture, utilization, and storage (CCUS) technology is also added. This technique lowers the overall carbon footprint by capturing the CO2 emissions produced during the manufacture of hydrogen and using them in other industrial operations or storing them underground (Boretti & Banik, 2021). Reduces $CO_2$ emissions by capturing up to 90% of the carbon produced. More sustainable than gray hydrogen. Still reliant on fossil fuels (natural gas). Carbon capture technologies are expensive and energy-intensive.

### 5.3. Biomass Gasification (Renewable Hydrogen)

Carbon monoxide (CO), carbon dioxide (CO2), and hydrogen are produced by heating organic materials like wood, agricultural waste, or other biomass at high temperatures without oxygen. This process is known as biomass gasification. Because the biomass used is replaced throughout time, the hydrogen produced is seen as renewable (Cao et al., 2020). Figure 3. Shows Biomass Gasification method.

*Figure 3. Biomass gasification method*

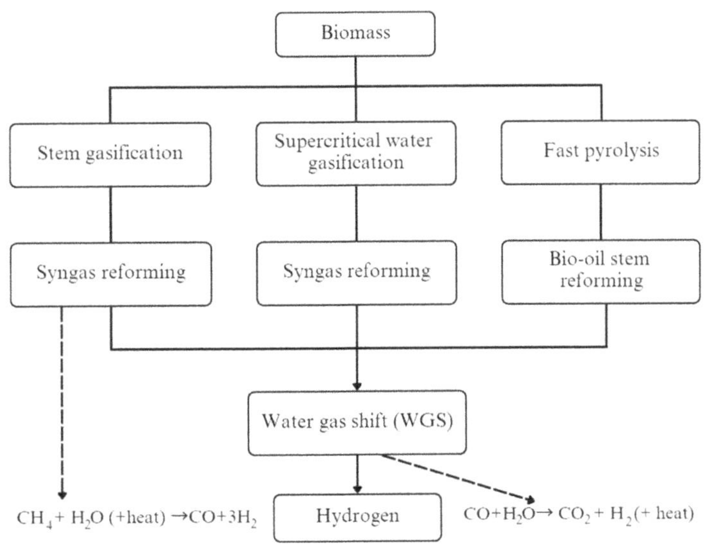


This method utilizes waste products from agriculture or forestry. It can be carbon-neutral if managed sustainably, as the $CO_2$ released during production can be offset by the $CO_2$ absorbed during the growth of biomass. It has lower efficiency compared to electrolysis. The availability of sustainable biomass is limited.

### 5.4. Thermochemical Water Splitting

Thermochemical water splitting is an experimental technique that separates water into hydrogen and oxygen using high-temperature heat, frequently obtained from nuclear power or concentrated solar power (CSP). Although it is still in the research and development phase, it may provide an extremely effective means of producing hydrogen without the need for fossil fuels (Oudejans et al., 2022). Figure 4 shows the Thermochemical water-splitting method

*Figure 4. Thermochemical water splitting*

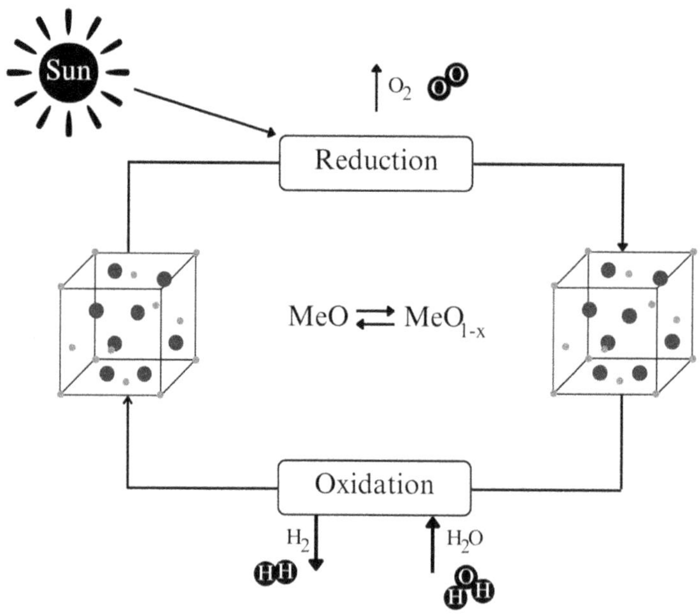


This method has the potential for high efficiency if integrated with nuclear or solar thermal technologies. No carbon emissions during hydrogen production. Still in development and not commercially viable at scale. Requires extremely high temperatures, which present engineering challenges.

## 5.5. Photoelectrochemical (PEC) Water Splitting

This process splits water straight into hydrogen and oxygen using light and specific materials. PEC water splitting mimics natural photosynthesis and is seen as a potential pathway for producing green hydrogen, though it is still in the experimental phase. Figure 5. Shows Photoelectrochemical (PEC) Water Splitting method.

*Figure 5. Photoelectrochemical (PEC) water splitting method.*

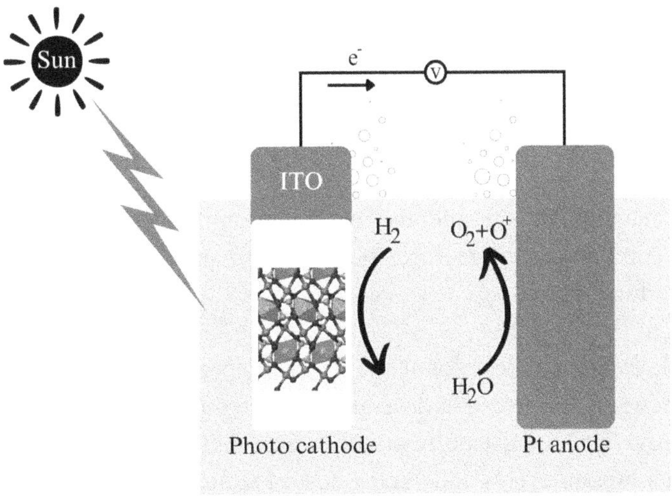


The photoelectrochemical (PEC) water-splitting method directly uses sunlight, a renewable resource, to produce hydrogen. Could be an efficient and low-cost method in the future. Low efficiency and high cost of the materials used in current technologies. Still in the early stages of development.

Green hydrogen, which is created through electrolysis using renewable energy, has the most potential for sustainability among these production techniques. It depicts a time when large-scale hydrogen production will be possible without producing carbon emissions. Green hydrogen has the potential to be a key component of decarbonized energy systems as long as renewable energy costs keep down and electrolysis technology advances (El ouardi et al., 2024). We can lessen our dependency on fossil fuels, decarbonize industries that are difficult to electrify and build a cleaner, more sustainable energy future by making investments in renewable hydrogen generation techniques, particularly electrolysis powered by solar and wind

## 6. HYDROGEN STORAGE AND DISTRIBUTION CHALLENGES

Hydrogen, while offering immense potential as a clean energy carrier, faces several significant challenges when it comes to storage and distribution. Unlike conventional fuels such as gasoline, hydrogen is a lightweight, low-density gas, which makes it difficult and costly to store and transport efficiently. The widespread use of hydrogen as the cornerstone of a civilization reliant on renewable energy depends on overcoming these obstacles.

## 6.1. Hydrogen Storage Challenges

Hydrogen has unique physical and chemical characteristics that make storage difficult. Since hydrogen is the smallest and lightest molecule, effectively holding and storing it involves special challenges (Meda et al., 2023). There are three primary ways to store hydrogen, and each has unique difficulties:

### 6.1.1. Compressed Gas Storage

To lower its volume, hydrogen is frequently kept under high pressure as a gas. In specialized tanks, hydrogen is normally compressed to pressures between 350 and 700 bar (5,000 and 10,000 psi). The difficulties with this storage are

*High energy requirement*: It takes a lot of energy to compress hydrogen to high pressures; typically, 10-15% of the energy in the stored hydrogen is used in this process.
*Material durability*: Tanks must be made of strong, lightweight materials like carbon fiber to withstand the high pressure. This adds to the cost of storage.
*Low energy density*: Compared to traditional fuels like gasoline or diesel, hydrogen has a poor energy density even when compressed, requiring huge amounts to store a comparable amount of energy.

### 6.1.2. Liquid Hydrogen Storage

When hydrogen is cooled to cryogenic temperatures (-253°C or -423°F), it can also be kept in a liquid state. Liquid hydrogen is more volumetrically efficient for storage than compressed hydrogen gas since it occupies a lot less space. Among the difficulties are

*Cryogenic temperatures:* Maintaining hydrogen in a liquid state requires extremely low temperatures, which necessitates expensive, highly insulated storage tanks.
*Energy-intensive process*: Hydrogen liquefying is less efficient since it uses between 30 and 40 percent of the hydrogen's energy content.
*Boil-off losses*: Some hydrogen evaporates over time, especially during transportation or in storage tanks, leading to potential losses, especially over long periods.

### 6.1.3. Hydrogen Carriers (Chemical and Metal Hydrides)

Liquid organic hydrogen carriers (LOHCs) and metal hydrides are two examples of solid or liquid carriers that can hold hydrogen and release it when needed (Sazali, 2020). These carriers offer a more stable and potentially safer way to store hydrogen, especially at lower pressures. Challenges are,

*Weight and efficiency*: Metal hydrides can be heavy and may store relatively low amounts of hydrogen per unit mass, reducing the overall energy density.
*Complexity*: The process of loading and releasing hydrogen from chemical carriers or hydrides can be energy-intensive and slow, and it may require high temperatures or specific catalysts.

## 6.2 Hydrogen Distribution Challenges

Distributing hydrogen efficiently and safely is another critical hurdle that must be addressed for a hydrogen economy to become viable. Hydrogen's physical properties make it more difficult to transport than traditional fossil fuels, and the infrastructure for hydrogen distribution is still in its infancy.

### 6.2.1 Pipelines

Hydrogen can be distributed through pipelines, much like natural gas, but this presents a set of challenges:

*Materials compatibility*: Hydrogen can cause embrittlement of steel and other metals, leading to cracks or leaks in pipelines. Specialized materials or coatings are needed for pipelines to safely transport hydrogen, which increases costs.
*Retrofit vs. new infrastructure*: Existing natural gas pipelines may be adapted to carry hydrogen, but this requires significant retrofitting. Building new hydrogen-specific pipelines from scratch involves enormous capital investment.
*Leakage risk:* Because hydrogen is a small molecule, it can easily leak through tiny cracks or seals in pipelines, creating safety risks and leading to energy losses(Abe et al., 2019).

### 6.2.2 Trucking and Rail Transport

In cases where pipelines are not available, hydrogen can be transported via trucks or trains in either compressed gas or liquid form. Specially designed trailers or tankers are required for this. Challenges are,

*Low capacity*: Hydrogen trucks or tankers can carry less energy compared to fossil fuel equivalents, requiring more frequent trips and increasing transport costs.
*Safety risks:* The high-pressure tanks required for compressed gas or cryogenic tanks for liquid hydrogen present additional safety risks, such as the potential for explosions in the event of an accident.
*Energy loss:* Transporting hydrogen, especially as a liquid, involves energy losses due to boil-off or inefficiencies in the transport process.

### 6.2.3 Hydrogen Distribution for Fueling Stations

The distribution of hydrogen is essential for the widespread use of hydrogen fuel cell vehicles (FCEVs). Compressed hydrogen is required for FCEVs, but putting up this infrastructure is costly and difficult. Hydrogen fuelling stations are necessary (Pareek et al., 2020). Challenges are,

*Cost of fueling infrastructure*: Building hydrogen refueling stations, which must handle compressed hydrogen at very high pressures, is significantly more expensive than building traditional gasoline or electric vehicle charging stations.
*Lack of widespread coverage:* There are currently very few hydrogen refueling stations globally, creating a "chicken-and-egg" problem where vehicle manufacturers are hesitant to scale up production without infrastructure, and infrastructure developers are slow to invest without sufficient demand.

## 6.3. Energy Losses in Storage and Distribution

One of the major overarching challenges in both storage and distribution is the loss of energy. Energy inputs for hydrogen storage (particularly in liquid and compressed forms) and transportation (by truck, pipeline, or rail) are substantial and frequently account for 10–40% of the energy content of the hydrogen molecule. These inefficiencies make hydrogen more expensive and less competitive compared to other energy carriers like electricity, which can be transported and stored more easily (Orfila et al., 2021).

## 6.4. Safety Concerns

Hydrogen's properties also present unique safety concerns in storage and distribution:

Hydrogen is highly flammable and has a wide flammability range, making it more dangerous than some conventional fuels if leaks occur. Hydrogen burns with an invisible flame, making it difficult to detect fires. As mentioned earlier, hydrogen's small molecular size makes it prone to leakage, and even small leaks can create explosive environments. Despite these hurdles, research and innovation are underway to improve hydrogen storage and distribution technologies: New materials, such as carbon nanostructures, are being explored for hydrogen storage at lower pressures and higher densities. Countries like Germany and the Netherlands are experimenting with hydrogen pipelines, while others are investigating blending hydrogen with natural gas in existing networks to reduce costs. Small-scale, modular hydrogen storage and distribution systems are being developed to reduce infrastructure costs and enable decentralized hydrogen production, reducing reliance on large-scale transport networks. Overcoming the storage and distribution challenges of hydrogen is essential for realizing its full potential as a clean energy carrier. While technical and economic hurdles remain, continued investment in research, infrastructure, and safety solutions can help hydrogen play a transformative role in decarbonizing sectors like transportation, industry, and energy. The future of hydrogen depends on solving these logistical challenges, which will ultimately enable a smooth transition to a renewable energy society (Warren et al., 2022).

# 7. FUTURE PROSPECTS: INTEGRATING HYDROGEN INTO A GLOBAL RENEWABLE ENERGY GRID

In the future, hydrogen could play a significant role in the global renewable energy system, balancing out unreliable sources like solar and wind power. Hydrogen can be used as a fuel, storage medium, and energy carrier as the world works toward decarbonization. This can assist address issues like intermittent energy, long-term storage, and decarbonization of difficult-to-electrify sectors.

## 7.1. Balancing Intermittent Renewable Energy Supply

One of the biggest challenges in a renewable energy-dominated grid is the intermittency of power generation from wind and solar. These energy sources are weather-dependent and variable, meaning their output fluctuates throughout the day and across seasons. Hydrogen can address this issue by acting as an energy buffer, providing a solution for energy storage and load balancing(Marouani et al., 2023). Excess electricity generated by wind or solar during periods of high production can be used to produce hydrogen through electrolysis. This green hydrogen can be stored in large quantities over long periods and later converted back into electricity using fuel cells or gas turbines when renewable energy generation is low or demand is high. This process ensures a stable and reliable energy supply, even when renewable sources are not generating power. Unlike batteries, which are suitable for short-term energy storage, hydrogen can store energy for weeks, months, or even longer. This makes hydrogen ideal for seasonal energy storage, addressing long-term supply-demand imbalances in a renewable grid.

## 7.2. Decarbonizing Hard-to-Electrify Sectors

Many industries can benefit from electrification, but it can be challenging to decarbonize particular businesses—like heavy manufacturing, long-distance transportation, and aviation—through direct electrification alone. Hydrogen offers a flexible alternative for these hard-to-electrify sectors by providing a zero-emission energy source that can replace fossil fuels (Di Lullo et al., 2021). Industries such as steel, cement, and chemicals, which require high temperatures and use carbon-intensive processes, can switch to hydrogen as a clean fuel. Hydrogen can be burned in industrial furnaces, reducing reliance on coal and natural gas, or used in direct reduction of iron for steelmaking, significantly reducing emissions. Hydrogen is a promising solution for decarbonizing transportation, especially for heavy-duty vehicles, long-haul trucks, shipping, and aviation, where battery-electric solutions may be less practical due to range, weight, and power density limitations. Hydrogen fuel cells, which convert hydrogen into electricity, can provide zero-emission power with fast refueling and extended range. Hydrogen or hydrogen-based fuels (such as ammonia or synthetic fuels) are considered viable options for decarbonizing sectors like shipping and aviation, which face significant challenges in using batteries due to weight and energy density constraints.

## 7.3. Hydrogen for Power Generation

Hydrogen not only stores and balances energy from renewable sources but also directly contributes to power production:

*Hydrogen Power Plants:* Burning hydrogen produces just water vapor as a byproduct when it burns in combined cycle gas turbines or modified natural gas turbines to create power. Particularly when the output of renewable energy is low, these hydrogen power plants can supply dispatchable power to supplement intermittent renewable energy sources (Oliveira et al., 2022).

*Hydrogen Blending:* To gradually decarbonize power plants, hydrogen can be mixed with natural gas in the current gas infrastructure soon. Some areas have already begun testing the blend of up to 20% hydrogen into natural gas pipelines, which will speed up the switch to a completely hydrogen-based energy system while also assisting in lowering emissions from conventional gas-fired power plants.

## 7.4. Hydrogen as a Global Energy Carrier

A global energy commodity, hydrogen has the potential to connect renewable energy sources across geographical boundaries. Nations with copious amounts of renewable energy resources might manufacture green hydrogen and export it to areas with lower potential for renewable energy sources or where direct electrification is less practical. This could reshape global energy markets, similar to how fossil fuels are traded today. Countries with high solar or wind potential (e.g., Australia, Chile, Saudi Arabia) are already investing in large-scale green hydrogen projects to become global hydrogen exporters. Hydrogen could be transported internationally via pipelines or in liquefied form, or converted into derivatives like ammonia for easier shipping. Countries with high energy demand and limited renewable resources (e.g., Japan, South Korea, Germany) are looking to hydrogen imports to meet their future clean energy needs, especially for sectors that cannot easily electrify.

### 7.5. Technological Advancements and Cost Reductions

The widespread use of hydrogen in the world energy system will rely on continued breakthroughs in technology and cost-cutting. While there are technologies for producing, storing, and distributing hydrogen, their current costs surpass those of traditional fossil fuels or certain renewable energy sources (Di Lullo et al., 2021). However, several factors are expected to drive down costs in the future: Continued improvements in electrolyzer technologies, such as Proton Exchange Membrane (PEM) and Solid Oxide Electrolyzers (SOE), will enhance hydrogen production efficiency, reduce capital costs, and lower overall cost of green hydrogen. As demand for hydrogen increases and large-scale hydrogen projects come online, economies of scale will further reduce production costs, making hydrogen more competitive with fossil fuels. Investments in hydrogen pipelines, storage facilities, refueling stations, and international transport systems (such as hydrogen shipping) will be critical to building a global hydrogen economy. Public-private partnerships and government support are likely to accelerate this process.

## 8. POLICY AND REGULATORY SUPPORT

To encourage investment and innovation, robust legislative frameworks and regulatory support will be necessary for the eventual integration of hydrogen into the global renewable energy grid. Governments around the world are increasingly recognizing the potential of hydrogen and are developing hydrogen strategies and roadmaps to support its growth.

i. *Subsidies and Incentives*: Governments can provide subsidies for green hydrogen production and infrastructure, reduce taxes on hydrogen technologies, and set up incentive programs to encourage industries to adopt hydrogen solutions.
ii. *Carbon Pricing*: Hydrogen will become more economically appealing through the implementation of carbon pricing mechanisms, such as carbon taxes or cap-and-trade systems, which penalize high-carbon energy sources and businesses.
iii. *International Cooperation:* Global cooperation on hydrogen standards, infrastructure development, and trade agreements will be essential to creating a seamless hydrogen market and integrating hydrogen into the global energy system(Hermesmann & Müller, 2022).

Hydrogen integration offers an exciting chance to tackle some of the major obstacles in the clean energy transition when it comes to a worldwide renewable energy grid. In addition to being a worldwide energy transporter, hydrogen can be used to store excess renewable energy and decarbonize difficult-to-electrify areas. Hydrogen has the potential to be a key component of a low-carbon, sustainable energy future if policy support and technical breakthroughs continue to coincide. As we move toward a world powered entirely by renewable energy sources, hydrogen has the potential to play a major role in advancing both the clean energy economy and the global climate goals.

## 9. CONCLUSION

Hydrogen is emerging as a pivotal element in the global transition to a renewable energy society. It is well suited to support and improve renewable energy systems due to its versatility as a fuel, energy carrier, and storage medium. When generation is limited, hydrogen can act as a dependable backup by storing excess energy from sporadic sources like solar and wind power. By doing this, it tackles variability, one of the major problems with renewable energy. Hydrogen has a wider role in decarbonizing nations than just producing electricity; it can also solve problems in hard-to-electrify industries including heavy industrial, long-haul transportation, and aviation. Hydrogen can be a genuinely clean energy source because to the production of green hydrogen via electrolysis that is fueled by renewable energy. While current production costs are higher than fossil-based hydrogen, technological advancements, economies of scale, and increased investments in hydrogen infrastructure are expected to drive down costs, making hydrogen more competitive shortly. Moreover, hydrogen has the potential to become a globally traded energy commodity, linking regions with abundant renewable energy resources to those with high energy demands, thereby fostering international cooperation in the renewable energy transition. Deploying hydrogen as the cornerstone of a civilization powered by renewable energy will need large expenditures in infrastructure, regulatory support, and technological innovation. Governments and businesses must work together to ensure the development of safe and effective production techniques while constructing the infrastructure required for hydrogen storage, distribution, and refueling. Hydrogen will be essential in building a low-carbon, sustainable global energy system when these obstacles are removed, guaranteeing future generations a healthier and more resilient future.

## REFERENCES


Abe, J. O., Popoola, A. P. I., Ajenifuja, E., & Popoola, O. M. (2019). Hydrogen energy, economy and storage: Review and recommendation. *International Journal of Hydrogen Energy, 44*(29), 15072–15086. DOI: 10.1016/j.ijhydene.2019.04.068

Ahmad Kamaroddin, M. F., Sabli, N., Tuan Abdullah, T. A., Siajam, S. I., Abdullah, L. C., Abdul Jalil, A., & Ahmad, A. (2021). Membrane-Based Electrolysis for Hydrogen Production: A Review. *Membranes, 11*(11), 810. DOI: 10.3390/membranes11110810 PMID: 34832039

Awad, M., Said, A., Saad, M. H., Farouk, A., Mahmoud, M. M., Alshammari, M. S., Alghaythi, M. L., Abdel Aleem, S. H. E., Abdelaziz, A. Y., & Omar, A. I. (2024). A review of water electrolysis for green hydrogen generation considering PV/wind/hybrid/hydropower/geothermal/tidal and wave/biogas energy systems, economic analysis, and its application. *Alexandria Engineering Journal, 87*, 213–239. DOI: 10.1016/j.aej.2023.12.032

Babonneau, F., Barrera, J., & Toledo, J. (2021). Decarbonizing the Chilean Electric Power System: A Prospective Analysis of Alternative Carbon Emissions Policies. *Energies, 14*(16), 4768. DOI: 10.3390/en14164768

Bhandari, R., Trudewind, C. A., & Zapp, P. (2014). Life cycle assessment of hydrogen production via electrolysis – a review. *Journal of Cleaner Production, 85*, 151–163. DOI: 10.1016/j.jclepro.2013.07.048

Boretti, A., & Banik, B. K. (2021). Advances in Hydrogen Production from Natural Gas Reforming. *Advanced Energy and Sustainability Research, 2*(11), 2100097. Advance online publication. DOI: 10.1002/aesr.202100097

Cao, L., Yu, I. K. M., Xiong, X., Tsang, D. C. W., Zhang, S., Clark, J. H., Hu, C., Ng, Y. H., Shang, J., & Ok, Y. S. (2020). Biorenewable hydrogen production through biomass gasification: A review and future prospects. *Environmental Research, 186*, 109547. DOI: 10.1016/j.envres.2020.109547 PMID: 32335432

de Oliveira, A. C. L., Tótola, L. A., Lorentz, J. F., Amaral e Silva, A., de Assis, L. R., dos Santos, V. J., & Calijuri, M. L. (2022). Spatial analysis of energy indicators and proposition of alternative generation sources for the Brazilian territory. *Journal of Cleaner Production, 356*, 131894. DOI: 10.1016/j.jclepro.2022.131894

Di Lullo, G., Oni, A. O., & Kumar, A. (2021). Blending blue hydrogen with natural gas for direct consumption: Examining the effect of hydrogen concentration on transportation and well-to-combustion greenhouse gas emissions. *International Journal of Hydrogen Energy, 46*(36), 19202–19216. DOI: 10.1016/j.ijhydene.2021.03.062

El Idrissi, A., Arab, M., Zbair, M., Haspel, H., Saadi, M., & Ahsaine, H. A. (2024). Review of photoelectrochemical water splitting: From quantitative approaches to effect of sacrificial agents, oxygen vacancies, thermal and magnetic field on (photo) electrolysis. *International Journal of Hydrogen Energy, 51*, 1044–1067.

Ferrada, F., Babonneau, F., Homem-de-Mello, T., & Jalil-Vega, F. (2023). The role of hydrogen for deep decarbonization of energy systems: A Chilean case study. *Energy Policy*, *177*, 113536. DOI: 10.1016/j.enpol.2023.113536

Hermesmann, M., & Müller, T. E. (2022). Green, Turquoise, Blue, or Grey? Environmentally friendly Hydrogen Production in Transforming Energy Systems. *Progress in Energy and Combustion Science*, *90*, 100996. DOI: 10.1016/j.pecs.2022.100996

Juangsa, F. B., Irhamna, A. R., & Aziz, M. (2021). Production of ammonia as potential hydrogen carrier: Review on thermochemical and electrochemical processes. *International Journal of Hydrogen Energy*, *46*(27), 14455–14477. DOI: 10.1016/j.ijhydene.2021.01.214

Katebah, M., Al-Rawashdeh, M., & Linke, P. (2022). Analysis of hydrogen production costs in Steam-Methane Reforming considering integration with electrolysis and CO2 capture. *Cleaner Engineering and Technology*, *10*, 100552. DOI: 10.1016/j.clet.2022.100552

Lamb, K. E., & Webb, C. J. (2022). A quantitative review of slurries for hydrogen storage – Slush hydrogen, and metal and chemical hydrides in carrier liquids. *Journal of Alloys and Compounds*, *906*, 164235. DOI: 10.1016/j.jallcom.2022.164235

Marouani, I., Guesmi, T., Alshammari, B. M., Alqunun, K., Alzamil, A., Alturki, M., & Hadj Abdallah, H. (2023). Integration of Renewable-Energy-Based Green Hydrogen into the Energy Future. *Processes (Basel, Switzerland)*, *11*(9), 2685. DOI: 10.3390/pr11092685

Meda, U. S., Bhat, N., Pandey, A., Subramanya, K. N., & Lourdu Antony Raj, M. A. (2023). Challenges associated with hydrogen storage systems due to the hydrogen embrittlement of high strength steels. *International Journal of Hydrogen Energy*, *48*(47), 17894–17913. DOI: 10.1016/j.ijhydene.2023.01.292

Orfila, M., Sanz, D., Linares, M., Molina, R., Sanz, R., Marugán, J., & Botas, J. Á. (2021). H2 production by thermochemical water splitting with reticulated porous structures of ceria-based mixed oxide materials. *International Journal of Hydrogen Energy*, *46*(33), 17458–17471. DOI: 10.1016/j.ijhydene.2020.04.222

Oudejans, D., Offidani, M., Constantinou, A., Albonetti, S., Dimitratos, N., & Bansode, A. (2022). A Comprehensive Review on Two-Step Thermochemical Water Splitting for Hydrogen Production in a Redox Cycle. *Energies*, *15*(9), 3044. DOI: 10.3390/en15093044

Pareek, A., Dom, R., Gupta, J., Chandran, J., Adepu, V., & Borse, P. H. (2020). Insights into renewable hydrogen energy: Recent advances and prospects. *Materials Science for Energy Technologies*, *3*, 319–327. DOI: 10.1016/j.mset.2019.12.002

Raza, A., Arif, M., Glatz, G., Mahmoud, M., Al Kobaisi, M., Alafnan, S., & Iglauer, S. (2022). A holistic overview of underground hydrogen storage: Influencing factors, current understanding, and outlook. *Fuel*, *330*, 125636. DOI: 10.1016/j.fuel.2022.125636

Reza, M. S., Rahman, N., Wali, S. B., Hannan, M. A., Ker, P. J., Rahman, S. A., & Muttaqi, K. M. (2022). Optimal Algorithms for Energy Storage Systems in Microgrid Applications: An Analytical Evaluation Towards Future Directions. *IEEE Access : Practical Innovations, Open Solutions*, *10*, 10105–10123. DOI: 10.1109/ACCESS.2022.3144930

Sazali, N. (2020). Emerging technologies by hydrogen: A review. *International Journal of Hydrogen Energy*, *45*(38), 18753–18771. DOI: 10.1016/j.ijhydene.2020.05.021

Tarhan, C., & Çil, M. A. (2021). A study on hydrogen, the clean energy of the future: Hydrogen storage methods. *Journal of Energy Storage*, *40*, 102676. DOI: 10.1016/j.est.2021.102676

Tashie-Lewis, B. C., & Nnabuife, S. G. (2021). Hydrogen Production, Distribution, Storage and Power Conversion in a Hydrogen Economy - A Technology Review. *Chemical Engineering Journal Advances*, *8*, 100172. DOI: 10.1016/j.ceja.2021.100172

Warren, K. J., Tran, J. T., & Weimer, A. W. (2022). A thermochemical study of iron aluminate-based materials: A preferred class for isothermal water splitting. *Energy & Environmental Science*, *15*(2), 806–821. DOI: 10.1039/D1EE02679H

Yao, L., Lyu, X., Zhang, J., Liu, Y., Zhu, Y., Lin, H., Zhang, Y., & Li, L. (2020). Remarkable synergistic effects of Mg2NiH4 and transition metal carbides (TiC, ZrC, WC) on enhancing the hydrogen storage properties of MgH2. *International Journal of Hydrogen Energy*, *45*(11), 6765–6779. DOI: 10.1016/j.ijhydene.2019.12.139

Zhu, Y., Li, J., Yang, L., Huang, Z., Yang, X.-S., Zhou, Q., Tang, R., Shen, S., & Ouyang, L. (2023). Closed loops for hydrogen storage: Hydrolysis and regeneration of metal borohydrides. *Journal of Power Sources*, *563*, 232833. DOI: 10.1016/j.jpowsour.2023.232833

# Chapter 13
# Hydrogen Energy:
## Adaptation and Challenges

**Kutubuddin Sayyad Liyakat Kazi**
https://orcid.org/0000-0001-5623-9211
*Brahmdevdada Mane Institute of Technology, Solapur, India*

## ABSTRACT

*There will be a direct correlation between the success of the adoption and this particular element. Following the application of the WASPAS methodology to the investigation of the challenges, it has become abundantly evident that the issue of Formulate Recommendations is the individual that occurs in the highest position. Since the WASPAS approach was utilized, this is the situation that has arisen. The issue that is rated second is the Economic Factors challenge, which is next followed by the challenge which is ranked third, that is the Technological Readiness challenge, and finally, the challenge which is ranked fourth is the Policy Frameworks problem: the order in which the challenges are ranked. It is for this reason that the important building needs to be constructed as quickly as feasible to ease the environmental problem as quickly as possible. This is a key step that must be taken with the goal of hydrogen energy to fulfill the conditions for widespread adoption.*

## INTRODUCTION

It is said by Sultabanu (2023) that electric energy is an essential component of modern living. Electric energy serves as the source of power for a broad variety of devices, ranging from home appliances to industrial equipment. According to Needa (2024), as society gets progressively nearer towards a future that will be characterized by increasing levels of electrification, it is becoming more important than it ever has been before to have a comprehensive understanding of the complexity of electric energy. Within the scope of this investigation, we will investigate a wide range of aspects pertaining to electric energy, including its generation, distribution, consumption, and the significance of energy management that is focused on efficiency.

According to Prashant (2023), the generation of electric energy is accomplished through the utilization of a variety of distinct modes of operation. Utilizing fossil fuels (like coal, natural gas, oil), nuclear power, and renewable sources (like solar, wind, hydropower, and geothermal) are some of the ways that fall under this category. Every single generational strategy comes with its own unique set of advantages

DOI: 10.4018/979-8-3693-8980-5.ch013

and challenges that individuals must take into consideration. However, despite the fact that power plants that run on fossil fuels are typically dependable and have the ability to provide substantial quantities of electricity, these also serve as a source of pollution as well as emissions of greenhouse gases. The use of renewable energy sources is lauded for its long-term viability and their minimal impact on the environment. Nevertheless, contingent upon the weather conditions, these sources may be subject to intermittency issues.

In recent years, that has evolved a distinct trend towards renewable energy, that has been fueled by technological developments as well as an increased awareness towards the implications of climate change. This trend has been pushed by both of these factors concurrently. The intermittent nature of renewable energy sources is causing a number of problems, but recent advancements in energy storage technologies, like as batteries, are helping to remedy some of these problems. These new developments are helping to facilitate this transformation.

Therefore, Kutubuddin (2024) asserts that Due to the fact that it is an indispensable component in our day-to-day activities, as well as in the operation of industry and the economy on a global scale, electric energy is perhaps the most significant aspect of modern life. It would be impossible to overstate the significance of this energy source due to the fact that it is both environmentally friendly and adaptable. This is especially true when taking into consideration the present drive towards energy solutions that are both more environmentally friendly and sustainable than those that are currently under development.

Advantages of making use of electric power:

- • Electric energy is incredibly efficient in terms of the transfer of energy, which it's a task of the explanations for why technology so convenient to use. Not only is it easy to generate, but it also has the capability of being sent over long distances with just a little amount of loss experienced. A high population density in metropolitan regions needs an abundance of energy sources which are both trustworthy and immediate. This efficiency is especially crucial in metropolitan areas because of the high population density.
- Repercussions for the Natural World: Electric power, particularly when it is produced from renewable sources like wind, solar, and hydroelectric power, produces a far smaller impact on the environment in comparison to fossil fuels. This is especially true when the electricity is created from renewable sources. This is accomplished by reducing emissions of greenhouse gases, which not only serves to improve air quality but also contributes to the fight against climate change.
- Progresses in the Field of Technology An extensive range of modern technology, like electric vehicles, smart houses, and mobile devices like laptops and smartphones, all get their electricity from electric energy. Electric energy constitutes the source of electricity for all of these technologies. As battery technology and energy management systems continue to progress, electric energy becomes an increasingly viable option for a wide range of applications. This is due to the fact that electric energy can be stored in batteries. The accessibility of electric energy is also being improved as a result of these improvements, which are also increasing its efficiency.

In the field of electric energy, particularly in the field of renewable energy, there is a substantial amount of employment creation taking place. This industry also makes a substantial contribution to the expansion of the economy. Several regions are seeing economic growth as a consequence of the transition towards a greener economy, particularly promotes investments and the development of new technologies. This is helping to drive economic growth.

Electricity's Drawbacks include the following:

- Concerns Regarding the Existing Infrastructure The current electrical infrastructure in many areas has aged past its best and requires a significant amount of funding to be rebuilt. This is why upgrading the grid is necessary. A lack of reliability in the infrastructure can result in a number of undesirable results, including power outages and inefficient energy consumption. This is a problem that is especially prevalent in regions that have been underdeveloped or rural.
- Dependence on Resources: Although electric energy might be generated from a broad variety of sources, specific techniques, like hydroelectric power, might be dependent upon particular geographical conditions. This is because those methods are dependent on the resources that are available. Furthermore, the mining and processing of minerals for batteries may provide rise to issues concerning the environment as well as ethical questions involving the extraction of natural resources. These issues might be caused by the processing of minerals.
- Initial Costs Involved: In many cases, a substantial initial expenditure is required in order to transition to electric energy systems. This is especially true in the framework of renewable energy installations or electric automobiles. In spite of the fact that there is the possibility of long-term cost savings, it is likely that individuals and organizations will feel dissuaded from making the transition due to the cost barrier.
- The stabilization and storage of energy: One of the most major problems that are related with this kind of energy is the fact that renewable electric energy is characterized by its intermittent nature. In addition, the production of energy from solar and wind sources has the potential to be unpredictable. It is vital to have energy storage systems that are efficient in order to stay in a state of equilibrium between supply and demand; however, despite the fact that there is progress being made, that continued to be a significant challenge.

Inference: According to Liyakat (2024), electric energy is a vital component of contemporary civilization, and it has a bright future that is directed towards the advancement of technological innovation and more environmentally friendly practices. Therefore, electric energy represents a vital element of modern civilization. Electric energy is connected with a multitude of enticing advantages, some of which include its efficiency, its impact on the environment, and the possible economic rewards that it may offer. Therefore, eliminating its limits, including the demands for infrastructure and the dependence on resources, shall be of crucial importance as society progresses towards a paradigm of energy that is more sustainable. This is despite the fact that it is present. It would appear that the future of electric energy is bright, and it will provide prospects for a society that is cleaner and more efficient as we continue to learn new things and adapt to changing conditions.

Over the past few years, the topic of the generation of electric energy has turned into one that is becoming increasingly significant. This development is a direct outcome of the expanding expectations of contemporary society as well as the urgent need for energy solutions that are favorable to the environment. As we move into an era that is becoming increasingly characterized by climate change and environmental consciousness, it is impossible to overstate the significance of making the transition from conventional fossil fuels to energy sources that are more ecologically friendly. This is because it is impossible to overestimate the significance of this transition. In this chapter, a comprehensive investigation is conducted into the existing landscape of electric energy generation technologies, including their efficiencies, issues, and probable future orientations.

According to Ewan (2005), one of the most important characteristics of hydrogen is its ability to be adapted to different situations. It can be used in fuel cells, which are used to generate energy, power automobiles, and even power vehicles. In addition to providing a feedstock for a variety of industrial processes, it can also be used in fuel cells. Because the combustion of hydrogen only leads in the formation of water vapor as a byproduct, hydrogen is a naturally pure energy source. This is in contrast to the combustion of fossil fuels, which results in the generation of carbon dioxide.

Furthermore, hydrogen may be produced from a wide range of resources, including water, natural gas, and biomass, which makes it feasible to diversify energy sources in a more complete manner. Hydrogen can be made from a wide variety of resources. The idea of green hydrogen, that's hydrogen which is produced through electrolysis using renewable energy sources like the sun and wind, is even more exciting than the possibility of green hydrogen. Green hydrogen is hydrogen that is manufactured through sustainable energy sources. The implementation of this method not only reduces the amount of carbon emissions, however it additionally offers the capability to take advantage of excessive renewable electricity, which contributes to the maintenance of grid stability.

When it comes to the hunt for energy sources that are less harmful to the environment, hydrogen energy is being increasingly recognized as a major actor. Hydrogen is being explored by the world at a time when it is battling with climate change and trying to transition away from fossil fuels. Hydrogen is a feasible option that is both adaptable and clean, and it is being considered by the world. The purpose of this study is to present an overview of the most significant features of hydrogen energy, including its benefits and drawbacks, as well as the potential role that it could play in the future of our energy system.

The element that can be found in the highest abundance across the entirety of the cosmos is one that is hydrogen. Hydrogen can be extracted through a number of different techniques, the most prominent of which include water electrolysis, steam methane reforming, and biomass gasification. Hydrogen can also be extracted through other processes. During the process of being utilized as a source of energy, hydrogen has the capability of creating electricity through fuel cells, with the emission of water vapor being the sole waste that is produced. In terms of reducing emissions of greenhouse gases, the fact that it possesses this property makes it particularly interesting as a potential solution.

According to Endo(2021), in the process of moving to a more sustainable energy future, hydrogen energy poses not just an exciting possibility but also major barriers. Its capacity to decarbonize a range of industries, alongside with its purpose as a medium for the storage of energy, putting its in an ideal position to be a vital component in the fight against climate change on a global basis. However, in order to fully realize it's potential, it is going to necessary to keep investing in technology, infrastructure, and public acceptability. This will be the case.

In the event that these challenges are conquered through the use of innovative solutions, hydrogen holds the chance to play a role in the energy landscape of tomorrow in a way that is genuinely revolutionary. As we move forward together, it will be vital to maintain a careful eye on developments in this rapidly emerging field in order to guarantee that hydrogen will contribute to a more comprehensive transition to a clean and sustainable energy system. This will be so that we can ensure that hydrogen will contribute to the transition. In spite of the fact that hydrogen energy has a great deal of potential, there are a number of significant challenges that impede its widespread usage. One of the most pressing issues that has to be solved right away is the current cost of manufacturing hydrogen, particularly green hydrogen. This is one among the most significant challenges. Despite the enormous developments that have been made in electrolysis technologies, these technologies continue to be expensive and have not yet reached a stage when they are economically comparable with conventional fossil fuels.

Another key difficulty that must be overcome is the lack of adequate infrastructure. Currently, the majority of the energy infrastructure is built on fossil fuels, and the establishment of a hydrogen distribution network calls for a substantial amount of financial investment. This included the installation of pipelines, refueling stations, and storage facilities, all of which demand considerable financial outlays and assistance from regulatory agencies. Moreover, all of these things require cooperation from the authorities.

In addition, there are worries surrounding the safety of hydrogen due to the fact that it is very flammable and requires stringent safety precautions to be followed. As a result of the public's perception of these concerns, there may be additional impediments to adoption. This is the reason why broad education and outreach activities are required.

The fact that a lot of countries are making substantial expenditures in research and development is a clear indication that the future of hydrogen energy is bright. There is a growing number of governments that are beginning to recognize the potential of hydrogen as a component of their energy strategy. As a direct consequence of this, aggressive goals for the implementation of hydrogen are being created in regions like Europe, Japan, and Australia. Initiatives like the Hydrogen Roadmap are now being developed with the intention of accelerating technological advancements and expanding production capacity. This is the current state of affairs.

Cooperation between governments, private corporations, and research institutes is also vital for the goal of expanding the use of hydrogen that is now being developed. There is a possibility that novel finance structures and public-private collaboration could be of considerable aid in overcoming initial investment barriers and fostering a shared interest in the creation of a hydrogen economy.

It is important to make considerable efforts to overcome the obstacles that hydrogen energy provides in terms of both economics and infrastructure, as stated by Bykara (2018). Although hydrogen energy does provide a feasible path for achieving sustainable energy goals, it is necessary to make these efforts that are necessary. As technology continues to advance and investments continue to rise, there is a considerable likelihood that hydrogen will play a crucial role in the decarbonization of the energy environment. This possibility is significant because these two factors are expected to continue. For the purpose of completely capitalizing on the advantages that hydrogen energy has to offer, it will be important to adopt a collaborative strategy that takes into account public awareness, technological innovation, and regulations that are supportive. In the event that we continue to demonstrate our dedication to hydrogen and make strategic preparations for its incorporation into our future energy system, there is a powerful likelihood that hydrogen will become an indispensable component.

## TECHNOLOGIES CURRENTLY IN USE FOR THE GENERATION OF ENERGY

1. Sources of Energy
    - Renewable: According to Needa (2024), solar power has witnessed a stratospheric growth in popularity, particularly photovoltaic (PV) cells and solar thermal technologies being at the vanguard of this revolution. Solar power is a renewable energy source. As a result of the rapid reduction in the price of photovoltaic panels, solar energy has emerged as a practical alternative that can be utilized in a variety of situations, including commercial and residential

facilities. Bifacial solar panels and floating solar farms are two examples of developments that are helping to further increase the efficiency of solar energy as well as its potential for land utilization.

- Wind power: According to Prashant (2023), wind power has a different leading contender in the realm of renewable energy systems. Wind power is proving to be an increasingly popular option. As a result of advancements in turbine technology, both onshore and offshore wind farms have undergone tremendous expansion. These advancements have made it feasible to produce turbines that are both higher and more efficient than these turbines that were before available. It is possible to ascribe, at least in part, the growing contribution of wind energy to the overall energy production of the world to the fact that it can be harnessed in a broad variety of geographical regions.

  Despite the fact that it is one of the oldest forms of renewable energy, hydropower continues to be a significant contributor to the electric grid. This is the case even though it has been around for generations. On the other side, the ecological impact of huge dams has led to a movement towards smaller, run-of-river systems which are suited to producing energy yet leaving a lesser footprint on the surrounding environment. This tendency has been brought about as a result of the environmental impact of major dams.

- Geothermal energy and biomass energy are both well-known for their capacity to deliver a continuous and consistent source of energy with their respective capabilities. There is another name for geothermal energy, which is geothermal energy. Geothermal energy and biomass energy are both intriguing prospects for integrated energy systems. This is due to the fact that both of these forms of energy make use of waste materials, and geothermal energy is able to capture the heat that is trapped inside the ground.

2. *The Energy of Nuclear:*
   - The use of nuclear power is a substantial contributor to the overall energy mix, particularly in countries that are moving towards the accomplishment of carbon neutrality targets. It continues to be a significant contributor to the overall energy mix while also providing a low-carbon alternatives to fossil fuels with which to compete. Small modular reactors (SMRs) and sophisticated nuclear systems are examples of recently developed technologies that carry the potential of enhancing safety, reducing waste, and lowering initial capital costs. Despite this, there are still significant obstacles to overcome in terms of the public's view and concerns around the disposal of garbage and safety.
3. Fossil fuels that are equipped with carbon capture:

Despite the fact that the long-term goal is to transition away from fossil fuels, technologies that focus on carbon capture and storage (also known as CCS) emerged as prospective transitional solutions. This is the case despite the fact that such technologies have emerged as prospective answers. The goal of these technologies is to reduce the pollutants of greenhouse gases typically get generated by power plants that rely on coal and natural gas. In addition to solving climate-related challenges, these technologies are also intended to reduce emissions. The significance of these plants in the energy landscape could potentially be expanded as a result of this phenomenon.

In spite of the significant advancements that have been made in the production of electric energy, here continuing to be a great deal of obstacles to encounter:

- There are a variety of challenges that arise as a result of the intermittent nature of renewable sources, especially solar and wind power. The implementation of energy storage solutions like batteries and pumped hydro is needed in order to guarantee the reliability and stability of the grid.
- Infrastructure: A significant number of the energy infrastructures presently established have no capacity to deal with the rise of renewable energy sources. For grids to be able to accept decentralized generation, it is essential required to improve them in such a way that they become increasingly intelligent and more resilient.
- Policies and Regulations: It is essential to have regulations and incentives that can be considered both effective and efficient in order to stimulate innovation and investment in clean energy technologies. This is accomplished through the implementation of policies and regulations. On the other side, regulatory frameworks might not be able to keep up with the advancements in technology, which might be a barrier to growth.
- Acceptance by the people: It is still difficult to win over the support of the general people whenever it comes to new energy projects. Engagement by communities and providing solutions to the problems that they face are both essential elements that must be present for an implementation to be effective.

It is anticipated that the future of electric energy generation will be diverse and dynamic, and it is likely that the development of technology and changes in policy around the world will have a considerable impact upon the future of electric energy generation. By serving as a clean fuel for transportation and industry, the hydrogen economy has the possibility to revolutionize energy storage and serve as a fuel for the hydrogen economy. Renewable energy sources can be used to produce green hydrogen, which has the potential to completely revolutionize energy storage. Green hydrogen is produced through electrolysis.

- Smart Grids and the IoT-Internet of Things will be discussed. As stated by Kazi K. S. (2024), the implementation of intelligent technology has the ability to optimize the distribution of energy, greatly enhance efficiency, and give customers with improved control over the amount of energy they consume.
- Decentralized Generation: The growth of microgrids and other forms of localized energy systems has made it possible for groups of people to become more self-sufficient, hence lessening their dependence on centralized power plants. This has resulted in a decrease in the amount of energy that is required to generate electricity.

It is possible to draw the conclusion that the production of electric energy happens to be at a crucial juncture that is characterized by rapid development and transition. The continued absorption of renewable sources, the emergence of storage and grid technologies, and a coordinated global commitment to sustainability all point to a promising future for the generation of electric energy. This is evidenced by the fact that the future appears to be bright. While we are navigating these shifts, it is absolutely vital to face the difficulties head-on in order to ensure that everyone is given equal access to clean energy. This is the only way to ensure that everyone has the same opportunities.

4. Hydrogen as a Potential Source of Energy:
    - The concept of hydrogen as a potential solution to the shifting energy landscape has garnered a lot of interest over the course of the past several years. When the world is struggling to cope

with the simultaneous issues of climate change and the need for sustainable energy sources, hydrogen is emerging as a clean, adaptable fuel that offers a pathway towards decarbonization. This comes at a time when the world is struggling to deal with both of these concerns simultaneously.

- For a substantial amount of time, hydrogen has been praised for its potential to serve as a source of clean energy. Nevertheless, the significance of this issue is progressively coming into the foreground as worries over climate change continue to grow and the necessity of transitioning to sustainable energy sources becoming more critical. This study explores the practicality of hydrogen as a potential source of energy, as well as its advantages, disadvantages, and potential for the future. Additionally, the study looks into the pros and cons of hydrogen.
- There are many different uses that hydrogen can be used for, including fuel cells for automobiles, stationary power production, and as a feedstock in chemical processes. Hydrogen can be used in all of these applications. Hydrogen is the element that may be found in the highest quantity throughout the entirety of the cosmos. Because it is produced by a reaction between oxygen and hydrogen, which results in the generation of electricity, hydrogen is a very clean fuel when compared to fossil fuels. This is because the process results in the production of power. Water vapour is the only product that may be derived from this reaction.

Advantages of adaptation of Hydrogen energy:

- Because of its clean-burning properties, hydrogen is an environmentally benign fuel source because it does not produce any emissions of greenhouse gases at the moment of consumption. In the event that a product is generated from renewable sources, it is possible to keep carbon emissions to a minimum for the entirety of its overall existence. One of the most alluring characteristics of hydrogen is that it is capable of producing energy without releasing any carbon emissions. This is one about the most advantageous aspects of hydrogen. The one form of waste which is produced when hydrogen is used is water vapor instead of any other trash. Due to the fact that hydrogen possesses these features, it is a renewable replacement to fossil fuels. The worldwide attempts that have been made to reach carbon neutrality aspirations are in good alignment with this attribute, which contributes to the overall alignment.
- Energy Density: Hydrogen has a high energy density by weight, which makes it an enticing choice for applications that demand considerable energy outputs. Hydrogen is a good candidate for these kinds of applications. Applications as aviation and heavy-duty vehicles are two examples of these kinds of applications.
- Hydrogen possesses the ability to be utilized in a wide range of industries, which brings about its versatility. Additionally, it has the potential to be utilized in terms of heating, industrial processes, and energy storage solutions, that could render it possible to incorporate renewable energy sources. This is additionally to its potential use in transportation, which is one of its prospective applications. Hydrogen, the element which occurs in the highest abundance across the universe, describes the universe. There is a possibility that it is present in water, natural gas, and a wide range of biological substances. Electricity can be produced from a wide range of sources, including renewable energy through electrolysis, biomass, and even fossil fuels through reforming processes. It is feasible to create it from a wide number of sources. Because of the variety of the

technologies that are used to produce hydrogen, it is easy to seamlessly incorporate these methods into energy infrastructures that are already in existence.

- When contrasted with electricity, hydrogen comes a more adaptable choice for power systems that must maintain a balance between supply and demand. This is due to the fact that hydrogen can be stored and transferred with a relatively minimal degree of trouble. Hydrogen has the potential to function as an effective medium for energy storage, which is particularly relevant in light of the fact that renewable energy sources like solar and wind are inherently intermittent. Electrolysis is a method that can be used to convert excess power into hydrogen whenever it is produced. After that, this hydrogen can be stored, and at a later time, when there is a greater demand for it, it can be turned back into electricity or heat. This capability, that helps to stabilize the flow of energy, can be utilized to generate a more stable energy grid, which in turn helps to provide more steady energy.
- Fuel Cell Technology: Hydrogen fuel cells offer an efficient method of converting energy, delivering electricity with a high efficiency and requiring only a short length of time so that they can be refilled. In light of the fact that they may be utilized in a wide number of applications, ranging from stationary power generation to transportation, they offer a versatile platform for the distribution of energy.

A couple of the challenges are as follows:

The production of hydrogen involves the use of processes that utilize fossil fuels, like steam methane reforming, resulting in enormous volumes of carbon emissions. At the present time, a considerable portion of hydrogen is produced using these processes. It is required to transition to ecologically friendly means of creating hydrogen, like as electrolysis involving the utilization of renewable energy, in order to completely utilise the potential of hydrogen. This is necessary in order to fully utilize the potential of hydrogen. The combustion of hydrogen is a process that is friendly to the environment; however, not all hydrogen can be generated in a manner that is environmentally friendly. Steam reforming methane, which is the most common method employed today, can result in significant emissions of carbon dioxide. This is because steam reforming is the most common approach. In order to lessen the quantity of carbon dioxide emissions that hydrogen generates, it is vital to make the transition to manufacturing processes that are less harmful to the environment. One example of such a method is electrolysis, which is powered by renewable energy.

The existing infrastructure for hydrogen is badly insufficient, and a large amount of investment is required in order to establish distribution networks, storage facilities, and filling stations. All of these things are necessary in order to meet the demand for hydrogen. The general implementation of the technology is significantly hampered as a result of this significant impediment. In the present moment, there is a deficiency in the infrastructure that is necessary for the production, storage, transportation, and distribution of hydrogen. This deficiency is one of the most significant impediments that is preventing the widespread use of hydrogen. For policymakers and industrial stakeholders within the industry, the creation of the infrastructure that is required to maintain a hydrogen economy requires a major investment of both time and money. This provides a problem for the implementation of the infrastructure.

There is a possibility that the processes that are involved in the creation, storage, and use of hydrogen could result in the loss of energy. This is a potential issue from an efficiency standpoint. In the case of electrolysis, for example, the process is not entirely efficient, and the transformations that are associated with hydrogen add even more inefficiencies. There is a possibility that efficiency losses will occur during the process of manufacturing, storing, and converting hydrogen back into energy or heat sources once it has been converted. When considering the role that hydrogen will play in a future energy system that is sustainable, it is essential to compare the overall energy efficiency of hydrogen to what associated with other renewable options, like direct electrification. This comparison is the most important thing to do.

• Opinions of the General Public and the Existence of Risk: There are worries regarding the safety of hydrogen due to the fact that it is highly flammable. The acceptance of hydrogen and its incorporation into energy systems may become more challenging as a result of these considerations. Educating the general population and implementing safety measures that have been demonstrated to be effective are both necessary in order to ease these worries. In addition to being exceedingly flammable, hydrogen can be challenging to work with because of its great flammability. Because it is of the greatest significance to ensure safety during the entire process of storage, transportation, and utilization, it is necessary to set rigors limits through the utilization of technology.

It would appear that hydrogen has a promising future as a source of energy, provided that we are able to achieve success in overcoming the challenges it now finds itself confronting. There are currently a variety of international programs that are being carried out with the objectives of expanding hydrogen technology, boosting production from renewable sources, and enhancing the infrastructure that is necessary. In addition, the government's policies and investments are increasingly boosting clean hydrogen in their energy policies and strategies. This is a positive development.

Through international collaboration among states, the construction of a hydrogen economy might potentially facilitate the encouragement of research and development, the dissemination of best practices, and the codification of standardized norms. This would be accomplished on a worldwide scale. As the costs of renewable energy continue to decrease, there is a possibility that green hydrogen will become more economically viable in the future.

Because it has the potential to assist in addressing the important shift away from fossil fuels, the use of hydrogen as a potential cornerstone of a sustainable energy future is particularly notable. This is something that should be taken into consideration. It is likely that in the near future, technological advancements and backing from the government will make it feasible to realize its vast potential, despite the fact that it is still presented with significant hurdles, particularly in terms of manufacturing processes and infrastructure. As the global community has been moving towards decarbonization, hydrogen has the potential to play a big role in the process of constructing a cleaner and more resilient energy environment. This is because hydrogen has the ability to contribute significantly to the process. The potential returns make it an operation that's worthwhile trying, despite the fact that it will be a difficult path to travel.

## GENERATION OF HYDROGEN ENERGY

When it comes to finding solutions which are both environmentally friendly and sustainable when speaking of energy, the production of hydrogen energy has become an increasingly relevant topic of discussion. At a time when the world is trying to meet the urgent demand to combat climate change and reduce its dependence on fossil fuels, hydrogen stands out as a potentially valuable alternative that may be utilized. The purpose of this study is to investigate the various methods that can be utilized to produce hydrogen, as well as its benefits and drawbacks, and the possible part that it might have in the energy landscape of tomorrow.

Techniques for the Generation of Hydrogen:

- The production of hydrogen can be accomplished through a variety of approaches, each of which includes its own environmental footprint, feasibility, and level of technological proficiency. The following are the primary methods:
- Steam methane reforming (SMR) is the most widely used technology at the moment. It is characterized by the reaction of natural gas with steam, which results in the production of hydrogen and carbon dioxide respectively. Although this approach is economical, it does not adhere to the principle of carbon neutrality, which raises concerns regarding emissions of greenhouse gases.
- The process of electrolysis involves the utilization of electricity in order to separate water into hydrogen and oxygen. An option that is regarded to be both environmentally friendly and sustainable is "green hydrogen," which can be produced through electrolysis when it is fueled by renewable energy sources as solar or wind. With that being said, the expense of electrolysis at the moment continues to be a barrier to its general use.
- The third step is called gasification, and it involves applying heat in an atmosphere with a low oxygen content in order to transform organic molecules into hydrogen and carbon monoxide. While it is possible to make use of waste materials, the technology is still in the process of being developed and frequently encounters economic hurdles.
- The Conversion of Biomass Hydrogen can also be created from biomass through a variety of biochemical processes. Despite the fact that this approach has the potential to be sustainable, it is contingent on the availability of feedstock and may be in competition with the production of food.

There are a number of compelling advantages associated with hydrogen energy:

- In fuel cells, hydrogen is a clean energy source because it generates electricity with water as the only by-product. This makes hydrogen an environmentally benign energy carrier. Fuel cells are used to generate electricity.
- Energy Storage: Hydrogen has the ability to store excess renewable energy, which makes it possible to have a more consistent supply of energy from sources that are intermittent, as solar and wind.
- Hydrogen's versatility allows it to be utilized in a wide variety of applications, like transportation, heating, and industrial operations, which contributes to the decarbonization of a number of different industries.

Hydrogen energy generation has significant hurdles, despite the fact that it has a lot of potential:

- Cost: The manufacturing and delivery of hydrogen can be more expensive than other forms of hydrogen, particularly green hydrogen." When it comes to making hydrogen competitive with traditional fuels, investments in infrastructure are very necessary.
- Storage and Transportation: Because hydrogen has a low energy density, it must be kept at high pressures or in liquid form. This adds complexity and costs to the logistics of hydrogen's transportation.

- According to the general public, hydrogen energy is associated with a number of dangers, including the potential for explosion and flammability. For the purpose of achieving acceptance, it is vital to educate the general population and address concerns around safety.
- Policy and Regulation: Incentives and policies enacted by the government have the potential to play a significant part in facilitating the transition to a hydrogen economy and furthering development of hydrogen technology.

In order to make the transition must an alternative energy future, the creation of hydrogen energy becomes a vital piece of the puzzle that needs to be put together. It offers a choice that is not only environmentally friendly but also adaptable, and it can be utilized when combined with renewable energy sources whenever possible. Despite the fact that there are challenges to be conquered, particularly those that are connected to the expense and the infrastructure, ongoing technology advancements and investments have the ability to paves ways for more widespread application. In order to realize the full potential of hydrogen, it is necessary for policymakers, industry players, and academics to work together. Because of this, hydrogen will be able to play a significant part in the process of achieving carbon neutrality on a global basis. There is the potential for hydrogen energy to become a key component of a future that is more ecologically friendly and sustainable if a well-defined plan and a clear commitment are put into place.

## Methods and Procedures for the Generation of Hydrogen Energy

Hydrogen energy has emerged as a key force in the area as a result of the exploration of energy solutions that are both sustainable and kind to the environment. Because of its adaptability and the potential applications it has across a wide range of industries, including transportation, industrial, and home energy, hydrogen stands out as a particularly noteworthy substance. While the world is striving to meet the urgent need to combat climate change and reduce emissions of greenhouse gases, this is particularly essential at a time when the world is struggling to fulfil this need. This article provides an overview of the fundamental methods that are used to generate hydrogen. Particular attention is paid to the stages that are involved in each process, as well as the sustainability, efficiency, and present practicality of these methods' implementation.

The generation of hydrogen can be accomplished by a number of different techniques, each of which varies in terms of its efficiency and its capacity to be economically viable:

*Steam Methane Reforming (SMR)*

The Steps in the Chemical Process:

- The natural gas, which is mostly methane, is combined with steam and then heated to temperatures ranging from 700 to 1000 degrees Celsius.
- In addition to a trace amount of carbon dioxide, the reaction results in the production of hydrogen and carbon monoxide.
- In a shift reactor, the carbon monoxide undergoes additional reactions with steam, which results in the production of more hydrogen and carbon dioxide.
- In order to make use of the hydrogen, it is separated and purified.

Although small-scale modular reactors (SMR) are the most widely utilized technique in the world today, they are not sustainable in the long term since they are dependent on fossil fuels and produce considerable amounts of carbon dioxide emissions.

*The Electrolysis Process Steps are as follows*:

- An electric current is used to separate water into its component parts, hydrogen and oxygen.
- In order to improve its environmental friendliness, this process can make use of a variety of electrical sources, including renewables (like wind, solar, and hydro).
- The process of splitting is carried out by electrolyzes, which can be of a variety of sorts, including solid oxide, alkaline, and PEM.

Electrolysis is a viable and sustainable technique of production since it can produce "green hydrogen" with virtually no carbon emissions when it is fueled by renewable energy. This makes it a promising technology.

*Gasification: Instructions for the Process:*

- In an atmosphere with a low oxygen content, high temperatures are applied to carbonaceous materials or biomass, as coal.
- The feedstock is transformed into syngas, which is a mixture of hydrogen, carbon monoxide, and carbon dioxide, by this technological process.
- Additional processing can be performed on the syngas in order to isolate and purify hydrogen.
- Sustainability: Gasification has the ability to utilize waste materials and biomass, which makes it a possible contender for the production of sustainable hydrogen. However, emissions are dependent on the source material and the effectiveness of the process.

*Water Splitting Through Photoelectrochemical Processes o Steps in the Process:*

- Materials made of semiconductors are able to absorb sunlight and take advantage of the energy to split water molecules.
- The method in question brings about the direct transformation of solar energy into chemical energy in the form of hydrogen.
- Regarding sustainability, it is a relatively new technology that provides a novel technique to directly capture sunlight for the creation of hydrogen. However, it is still in the developmental stage with regard to scalability and efficiency.

*The procedures involved in the production of biological hydrogen:-*

- There are specific microorganisms, like bacteria and algae, that are capable of producing hydrogen through the process of biological reactions (for example, fermentation and photolysis).
- Sustainability: This technique has the potential to make use of organic waste, which provides a contribution to both the management of waste and the creation of hydrogen; nevertheless, its commercial feasibility is still limited.

The generation of hydrogen energy is a multifaceted subject that can be approached in a variety of different methods, each of which possesses varying degrees of efficiency and sustainability. It is essential to invest investments for research, the development of infrastructure, and policy assistance so as to unlock the full potential of hydrogen as a key player in a future energy system that is sustainable.

The reason for this is because there is an increasing need for renewable energy solutions all around the world. It is reasonable to suppose that hydrogen, with the assistance of technological advancements like electrolysis and biological techniques, having the potential to make a big contribution to reducing our reliance on fossil fuels and minimizing the consequences of climate change. This is because hydrogen has the ability to be produced through the process of electrolysis.

## ADOPTION OF HYDROGEN ENERGY: BACKGROUND

As the globe struggles to come to terms with the pressing need to lessen the impacts of climate change and make significant progress towards the transition to sustainable energy sources, hydrogen energy has emerged as a credible contender in the hunt for cleaner alternatives. This comes at a time when the world is struggling to cope with the urgent need to decrease the consequences of climate change. In order to research the history of the utilization of hydrogen energy, the objective of this study is to conduct an analysis of its historical context, technological advancements, and the challenges that are still to come.

In the early 19th century, chemist William Nicholson was the first person to show the electrolysis of water, which separated hydrogen and oxygen. This was the beginning of the concept of hydrogen as a source of energy, which has been around for quite some time. Over the course of the 20th century, hydrogen gained appeal as a potential fuel, particularly during the space race, when it was a crucial component in the propulsion of rockets. This increase in popularity occurred during the time period. In spite of the fact that hydrogen energy holds a great deal of potential, it has been mostly disregarded due to the fact that fossil fuels, which have been around for a longer period of time, have garnered more attention.

The resurgence of interest in hydrogen, which did not start to emerge until the early 2000s, was mostly driven by concerns about energy security, rising oil prices, and the urgent need to limit greenhouse gas emissions. These factors were the driving forces behind the rebirth of interest in hydrogen. For the purpose of laying the groundwork for the contemporary application of hydrogen technology, governments, research institutes, and private firms from all over the world have started making significant investments in the technology.

When it comes to the search for environmentally friendly energy options, hydrogen energy is typically recognized as a major competitor. As a result of the fact that it offers a means to significantly cut emissions of greenhouse gases and dependency on fossil fuels, it is an appealing alternative for prospective solutions. In the midst of governments wrestling with the compelling need to address climate change and assure energy independence, the adaptation of hydrogen as a viable energy carrier is gathering pace all throughout the world. In order to analyze the historical context, current trends, impediments, and potential future consequences of hydrogen energy adaptation, the goal of this study is to investigate these topics.

Since the beginning of the $19^{th}$ Century, when scientists recognized hydrogen for the first time as a viable source of energy, the concept of hydrogen fuel has been around. On the other hand, significant advancements did not take place until the latter half of the 20th century, particularly in the research and development of fuel cell technology. Hydrogen is converted into energy by a process known as fuel cell technology, which only results in the production of water as a byproduct of this conversion. Early adopters of hydrogen fuel cells in space missions, like the National Aeronautics and Space Administration (NASA), revealed the promise that these cells may bring to the field of space exploration.

As a direct consequence of the energy crisis that took place in the 1970s, there has been a rise in interest in alternative energy sources, like hydrogen. In contrast, hydrogen energy did not begin to be seriously considered as a basic component of a sustainable energy future until the 2000s, when technical improvements as well as worries about climate change began to arise. This was the time when hydrogen energy was first brought to the forefront of the conversation.

The generation, storage, and consumption of hydrogen have all seen substantial advancements in technology over the course of the past twenty years. These advancements have been made possible by the development of new technologies. Initially, the majority of hydrogen was created from natural gas through a process known as steam methane reforming. This was the primary method of hydrogen production. In spite of the fact that this approach was successful, it was responsible for a considerable amount of carbon emissions into the atmosphere. On the other hand, the advent of green hydrogen, which is produced by electrolyzing water using renewable energy sources, has the potential to radically alter the sector. Green hydrogen is made by electrolyzing that water. The decreasing costs of renewable energy sources have made environmentally friendly hydrogen production more viable than it has ever been before. This has resulted in the hydrogen being a more appealing alternative than it has ever been before.

It has been said by Murray (2009) that the development of hydrogen fuel cells has not only made it possible to have an alternative to battery electric vehicles, but it has also made it possible to have new prospects in the transportation sector. Using hydrogen fuel cells, it is possible to transfer the chemical energy that is carried by hydrogen into electricity. The only consequence of this conversion is the production of water vapor. By addressing some of the limitations of battery technology, as longer recharging periods and limitations in range, this technology not only promises to reduce emissions but also addresses some of the limitations that are associated with battery technology.

The most important factor that has led to the current shift towards hydrogen energy is the implementation of policy initiatives with the objective of reducing carbon emissions in economies. A number of nations are making substantial investments in hydrogen infrastructure because they recognize the flexibility of hydrogen across a variety of industries, including transportation, industrial operations, and energy storage. These governments are making these investments because they recognize the many applications of hydrogen. One particularly interesting alternative that is in line with the global goal of lowering carbon emissions is the creation of green hydrogen, which can be accomplished by electrolysis with the utilization of renewable energy sources. This is an alternative that stands out as particularly desirable.

Some countries, as Germany, Japan, and Australia, are at the forefront of the push to adopt hydrogen. These countries are adopting national policies with the intention of establishing themselves as hydrogen economies. The development of hydrogen technology has resulted in a reduction in prices and an improvement in efficiency as a consequence of expenditures made by both the public sector and private businesses in research and development. There has been an additional contribution made by this to the practicability of hydrogen fuel.

The adoption of hydrogen energy faces a number of challenges, despite the fact that it contains a great deal of potential. One of the most significant challenges that still needs to be conquered is the demand for significant expenditures in infrastructure. For the purpose of producing, transporting, and storing hydrogen, it is essential to create a dedicated network of pipelines, fuelling stations, and storage facilities. On the other hand, this network is just beginning to emerge in many different locations. In addition, the production of hydrogen needs a considerable amount of energy, which raises questions about the overall efficiency of hydrogen in compared to other alternatives that are derived from sustainable sources of energy.

In addition, the high cost of manufacturing green hydrogen at the present moment is another challenge that must be overcome. It is still challenging to attain cost parity with conventional fuels, despite the fact that there has been a general trend towards lower prices. The successful completion of this challenging undertaking is dependent not only on the development of new technologies but also on the existence of favorable legislative frameworks. The provision of incentives by governments for the study and development of hydrogen technologies as well as their commercialization is necessary for the development of a robust market. Commercialization of these technologies is also required.

It is supported by a strong historical background as well as major technological advancements that have the potential to revolutionize the energy landscape. The deployment of hydrogen energy is supported by both of these factors. Despite this, it is extremely necessary for stakeholders, policymakers, and industry leaders to collaborate in order for hydrogen to play a substantial part in a future energy system that is sustainable. This is the only way that hydrogen can be used. If hydrogen is able to meet the requirements for energy infrastructure and overcome the challenges connected with production, it has the potential to transition from a viable alternative into a dominant player in the global energy portfolio. This transformation would be possible if hydrogen is able to overcome the obstacles involved with manufacturing.

In conclusion, with the potential benefits of hydrogen energy, which include fewer emissions, greater energy security, and economic growth, it is an endeavor that is worth pursuing, despite the fact that the path towards widespread adoption of hydrogen energy is fraught with challenges. A archetypal example of human brilliance and invention, hydrogen energy is becoming increasingly crucial as the world moves towards a future that is both cleaner and more sustainable. Hydrogen energy is a classic example of human ingenuity and innovation.

In view of the fact that the climate situation is getting worse and there is an immediate need to transition away from fossil fuels, the quest for renewable energy sources has never been more crucial than it is right now. Because of its capacity to reduce emissions of greenhouse gases and its ability to provide a variety of techniques for energy storage, hydrogen energy has emerged as a potentially beneficial alternative. This is owing to the fact that it has the ability to minimize emissions. In order to shed light on the feasibility of hydrogen energy in both our current and future energy landscapes, the study on hydrogen energy adoption explores the opportunities and obstacles that are linked with the promotion of this clean energy source. The goal of this study is to shed light on the viability of hydrogen energy.

## PROBLEM STATEMENT

In spite of the fact that it is generally known that hydrogen energy technologies have a great deal of benefits, the primary concern that is investigated in this research is the delayed adoption of these technologies. There are a number of various factors that have contributed to this problem. These factors include the high production costs, the lack of infrastructure, the technological barriers, the regulatory gaps, and the socio-economic repercussions of transitioning to systems that are based on hydrogen. It is necessary to have a more in-depth awareness of the challenges that are associated with the implementation of hydrogen energy to address this complicated problem. In addition to this, it necessitates an exhaustive analysis of all significant factors that have an impact on stakeholders, as decision-makers in government, various industry entities, and consumers. The ranking will be made for further study using WASPAS method.

## GOALS

The overarching goal of the study that is being carried out to investigate the adoption of hydrogen energy is to identify and evaluate the key drivers and barriers that have consequences for the integration of hydrogen energy inside the existing energy framework. This is the objective of the study that is being carried out. In order to provide policymakers and other stakeholders with information about the necessary steps that require be taken in order to accelerate the adoption of hydrogen in an approach that not only demonstrates that it is financially viable, but also that it is socially equitable and environmentally sustainable, the purpose of the study is to provide this information.

## OBJECTIVES

To accomplish the stated purpose, the study outlines a number of distinct and concentrated goals, including the following:

- Evaluate the current level of hydrogen production, storage, and distribution technologies, finding innovation gaps and areas for improvement that could assist wider use. This is the first step in determining whether or not the hydrogen industry is ready for widespread adoption.
- Conduct an analysis of economic factors: In order to gain an understanding of how economic viability can be reached, it is necessary to investigate the economic consequences of increasing the production of hydrogen. These concerns include cost structures, investment requirements, and the possibility of market disruptions.
- Conduct an evaluation of policy frameworks: In this step, you will examine the policies and regulations that are currently in place across different regions in relation to hydrogen energy. You will look for supportive measures and regulatory barriers that could either encourage or discourage adoption.
- Determine the Perspectives of Stakeholders: Collect information from primary stakeholders, as government agencies, industry leaders, researchers, and consumers, regarding their perspectives on hydrogen energy, the risks they perceive, and their willingness to invest.
- Formulate Recommendations: Formulate recommendations that may be put into action in order to encourage the adoption of hydrogen energy. These recommendations should emphasize the necessity of coordinated efforts among stakeholders in order to establish conditions that are conducive to innovation, investment, and broad public acceptance.

To enable to acquire a more comprehensive understanding of the complex dynamics that are connected to the incorporation of hydrogen into the energy mix, it is necessary to do research on the utilization of hydrogen energy. Through the comprehensive specification of the issue statement, aims, and specific objectives, the study builds a solid foundation for the development of knowledge that may lead to the making of informed decisions. At the end of the day, the utilization of hydrogen energy has the potential to be an essential component in the accomplishment of a sustainable energy future. This research will give stakeholders with critical insights that will be of great value to them as they work to handle the obstacles that lie ahead and maximize the potential of hydrogen as a source of clean energy. These insights will be provided by the findings of this research.

## RESEARCH METHOD

With regard to the transition to a low-carbon economy, the idea of hydrogen energy as a crucial component surrounding the next energy landscape has gained a lot of interest. This is especially true in the context of the transition. Because governments and businesses all over the globe are looking for sustainable alternatives to fossil fuels, hydrogen is becoming more recognized not just as a clean energy carrier but also as a versatile feedstock for a number of industrial processes. This is because hydrogen is a versatile feedstock that can be used in a wide range of industrial processes. To provide an explanation of the current research orientations in the field of hydrogen energy adoption, the objective of this study is to highlight critical challenges and recommend prospective topics for further investigation. This will be accomplished by providing an explanation of existing research orientations.

These are the current research directions for the adoption of hydrogen energy:

- Production Technologies: The primary focus of research is on the development of technologies for the production of hydrogen, namely green hydrogen, which is produced through electrolysis when driven by renewable energy sources. In order to make the production of hydrogen more efficient and cost-effective, there is a growing interest in innovations as better catalysts, membrane technologies, and bio photolysis.
- Storage and Distribution: Because of its low density and strong reactivity, hydrogen requires innovative storage and distribution methods which are not currently available. Research is now being conducted to investigate a variety of materials for solid-state hydrogen storage, cryogenic storage systems, and pipelines for gaseous hydrogen. The purpose of these research is to overcome the safety and economic problems that are connected with transportation.
- Applications at the end of the chain: transportation, heating, and industrial operations are the most important factors in pushing the adoption of hydrogen. Technologies including as fuel cells, hydrogen combustion engines, and direct hydrogen applications are now being investigated for use in a variety of industries, including heavy industry and public transportation, among others. As a particular area of study, the potential of hydrogen to achieve net-zero emissions in industries that are difficult to decarbonize is being investigated.
- Policy and Economic Frameworks: The creation of policy frameworks that are supportive of hydrogen proliferation and economic incentives is of the utmost importance. Subsidies, tax incentives, and regulatory measures are being evaluated by researchers to determine how beneficial they are in attracting investment in hydrogen technology and infrastructure. Additionally, research on the socio-economic effects of hydrogen adoption in various regions is gaining pace and is becoming increasingly popular.
- The Perception and Acceptance of the Public It is essential to have a solid understanding of the attitudes that society has regarding hydrogen energy in order to ensure its acceptance and adoption. The research conducted in this field makes use of questionnaires, focus groups, and interviews with stakeholders in order to determine how the general public feels about issues of economic viability, environmental impact, and safety. It is essential to have this knowledge in order to develop policies that are informed and to encourage community engagement.
- Life Cycle Assessment (LCA): In the face of competing energy technologies, LCA studies are required in order to analyze the environmental implications of hydrogen in comparison to fossil fuels across its entire life cycle. It is helpful for policymakers to have an understanding of the

trade-offs involved with the various techniques of producing hydrogen so that they may make informed judgments regarding energy mix policies.

In spite of the several prospective pathways for the adoption of hydrogen, certain problems continue to exist:

- Cost Competitiveness: Despite the fact that there has been progress made, the production of hydrogen continues to be expensive in comparison to the production of traditional energy sources. Continued research into manufacturing processes and scaling technologies that are efficient and cost-effective is absolutely necessary.
- Limitations in Infrastructure: The current energy infrastructure is primarily designed for fossil fuels, which creates obstacles in the process of shifting to hydrogen. In order to overcome the logistical and financial obstacles that must be overcome, comprehensive hydrogen network development must be undertaken.
- Safety issues: The flammability of hydrogen causes safety issues, which demand intensive research into safety measures, standards, and public education campaigns. The goal of this research is to assuage worries and ensure the safe adoption of hydrogen technology.

Research Opportunities in the Field of Hydrogen Energy in the Future:

- Interdisciplinary Collaboration Due to the complex nature of hydrogen energy systems, it is necessary to collaborate across a variety of disciplines, including as chemistry, engineering, economics, and the social sciences. It would be beneficial for future study to incorporate the knowledge gained from these domains in order to address systemic difficulties.
- Integration with Renewable Energy: As part of the process of developing a sustainable energy ecosystem, it will be essential to investigate the potential synergies that exist between the production of hydrogen and the integration of renewable energy. The research that is being done on energy storage technologies and smart grids that integrate hydrogen is very promising.
- Perspectives from Around the World: When conducting research on hydrogen, adopting a global perspective can provide valuable insights into the various ways in which countries are handling hydrogen infrastructure and technologies. It is possible that comparative research will reveal the most effective approaches and encourage international collaborations.
- Studies of Long-term Impact While studies of immediate adoption are extremely important, it is also crucial for sustainable development to investigate the long-term socio-economic and environmental impacts of hydrogen energy systems.

Research on the utilization of hydrogen energy is growing at a rapid speed, encouraged by the urgent need for alternatives to energy sources that are less harmful to the environment. Nevertheless, in spite of the fact that the path that lies ahead is laden with challenges, the fact which the research that is now being conducted is multidisciplinary gives reason to be optimistic about the possibility of developing fresh solutions. If researchers focus their efforts on production, infrastructure, policy, and social involvement, they have the potential to make a large contribution to the effective incorporation of hydrogen energy into the global energy scene. This is the case if they can successfully incorporate hydrogen energy into the energy landscape. As we move towards a future that becomes more sustainable, it is feasible that hydrogen may play a significant role in maintaining energy security and combating climate change. This would constitute a significant contribution.

Because of the research directions that are currently accessible and the challenges that are associated with the implementation of hydrogen energy, we have developed a number of questions, and our analysis of the implementation was predicated on the answers to these questions. In order to carry out a study, a comprehensive literature review is carried out on the challenges that are related with the implementation of hydrogen energy as a source of energy, and a collection of twenty-one questionnaires is prepared. In order to carry out the survey, it is distributed to a large number of individuals, which includes academics, research scholars, and energy professionals. Following the distribution of approximately twenty-one questions that were divided into yes and no categories (yes = 1 and no = 0), which received a response rate of sixty percent, statistical analysis was carried out, and the WASPAS (Weight aggregated sum and product assessment method) was utilized to rank the challenges that are associated with hydrogen energy. The survey was taken from researchers and academics who had conducted it. We had inserted the values of the number of Yes and No in the answer of the questioner, and the survey had 143 research scholars participate. The results of the survey relied on the responses of these researchers.

*Table 1. Question*

| **Question Required** | **Yes=1** | **No=0** |
|---|---|---|
| Advance technology and more research is required for **hydrogen production?** | | |
| Production of hydrogen by electrolysis method is very costly? | | |
| For production, storage, distribution, and dispensing, good infrastructure and facilities are required? | | |
| For storing heavy tank requirements, which are technically difficult! | | |
| For transportation of hydrogen is requires heavy quality pipes and transportation systems as highly flammable? | | |
| An investigation of new fuel cell is essential? | | |
| Large-scale energy demands require significant investment and scaling, so subsidizing is required? | | |
| Tax incentives for Generation is required? | | |
| Hydrogen energy generation is safe? | | |
| Environmental impact study is required? | | |
| Hydrogen production cost is high ? | | |
| Cost effective generation methods are possible? | | |
| Government support required? | | |
| Government requires to make some policies? | | |
| Integrated Infrastructure development is essential for production, distribution networks, refuelling stations, and storage solutions, which requires high investment? | | |
| Public education is required to alleviate fears and ensure the safe adoption of hydrogen technologies? | | |
| Fuel cell vehicles are in their initial stage and very costly? | | |
| Special Safety standards to be developed for use of hydrogen fuels? | | |
| Design of hydrogen combustion engines required for vehicles required with any specifications? | | |
| Does Hydrogen energy generation required more water? | | |
| In Future, Hydrogen Vehicles emission problem? | | |

## CHROBANCH'S ALPHA

In the field of model evaluation, a novel metric that has recently come into existence is referred to as Chrobanch's Alpha. This metric offers researchers and practitioners a fresh perspective on performance measurement, which is beneficial to both groups. Individuals who are interested in predictive analytics, machine learning, and statistical modelling will find this metric to be particularly beneficial. This is due to the fact that it was intended to handle the problems that are inherent in evaluating the dependability and robustness of models.

It is the most significant feature of Chrobanch's Alpha that it is able to evaluate not merely the degree of accuracy additionally the degree of agreement which exists among a number of various models or raters. This is one of the most distinguished aspects of Chrobanch's Alpha. When it comes to applications that demand multi-rater assessment scenarios or ensemble approaches in machine learning, both of which are examples of situations in which agreement is essential, this is of the utmost importance. The measure provides a comprehensive evaluation of how effectively a model functions, taking into account the fact that different evaluators or models have a tendency to have varying levels of variability. This comprehensive evaluation is supplied by the measure.

In regard to methodology, Chrobanch's Alpha is founded on statistical foundations that have been established in the past; yet, it also features creative adjustments that increase its application and interpretability. Due to the fact that the formula incorporates characteristics of both reliability and validity, it is a tool that may be utilized for two different purposes. This information can be used to inform both the actual deployment of models and the theoretical knowledge of how well they perform. It can provide information that can be utilized to inform both. When it comes to situations in which conventional measurements as accuracy, precision, or memory might not be able to provide a thorough view of the situation, this is extremely important.

In light of the development of statistical software programs that offer support for the calculation, the feedback that was obtained from users indicates that Chrobanch's Alpha is rather easy to compute. This is especially true in light of the fact that the calculation is possible. Due to the fact that it is not difficult to obtain, its use is more democratic. This is because it enables professionals in a wide variety of disciplines, as education and healthcare, to adopt it without the requirement for considerable statistical training.

When it comes to performance evaluation, the technique encourages a shift towards a more holistic perspective, which is an additional advantage of the strategy. This method, which takes into account both unanimous agreement and reliability, presents a challenge to the conventionally narrow focus that is placed on the predictive power of a single model. On the other hand, it is feasible that this change will result in more robust decision-making, particularly in complex systems where it is required to take into consideration a number of factors and the interactions between them.

Chrobanch's Alpha, on the other hand, has a number of limitations that need to be taken into consideration by users, just like any other statistical indicator. Some people who disagree with it have stated that it might be susceptible to overfitting, particularly in circumstances in which the number of raters is restricted or when the datasets are substantially smaller. Furthermore, the interpretation of its scores may require a specific level of expertise in order to correctly appreciate the repercussions for the performance of the model. This is because the model's performance is affected by the scores.

The conclusion is that Chrobanch's Alpha represents a significant advancement in the toolkit that statisticians and data scientists have at their disposal. In addition to this, it fosters collaboration between different raters or models, which ultimately results in a more comprehensive comprehension of the evaluation of models. Taking into consideration a number of different variables of performance allows for this to be accomplished. It has the potential to redefine norms in the assessment of model reliability, which could ultimately lead to outputs in predictive analytics that are more trustworthy and informed. This measure is being used by an increasing number of academics and practitioners. With everything taken into consideration, individuals who are interested in enhancing the techniques they use to evaluate models have to give serious consideration to utilizing this metric.

## WASPAS METHOD: A MULTI-CRITERIA DECISION-MAKING APPROACH

An approach known as WASPAS, which is an acronym that stands for Weighted Aggregated Sum Product Assessment, has developed as a powerful instrument within the realm of Multi-Criteria Decision-Making (MCDM). It is possible for decision-makers to properly balance complex and frequently competing criteria with the help of WASPAS, which provides a comprehensive framework offering this capability. In order to achieve this goal, it is necessary to incorporate elements from both the weighted sum model (WSM) and the weighted product model (WPM).

As a result of its ability to include a wide range of considerations into a unified decision-making process, WASPAS is appealing to those who are confronted with complex difficulties in fields as environmental research, manufacturing, logistics, and finance. The initial step in the process is the collection of the input data, which includes both quantitative and qualitative categories of criteria. Following that, a weight is assigned to each criterion that indicates the relevance of that criterion in connection to the decision that is currently being made.

In order to construct the WASPAS algorithm, there are two basic processes that are involved:

- Aggregation: At this point in the process, the approach computes the weighted sum and weighted product for a group of different options under consideration. The utilization of this dual technique makes it possible to conduct an exhaustive evaluation of every possibility with regard to all criteria. The results of the aggregate are then combined in order to arrive at a final score for each alternative individually.
- After the options have been aggregated, the next step is to rank them according to the scores that they have been computed. On the basis of the criteria that have been set, the alternatives that have been ranked highest are determined to be the most desirable options.

Benefits of the WASPAS approach include:

- Adaptability: The versatility of the WASPAS method is likely one of the most notable advantages of using this approach. It is able to easily adapt to a wide variety of decision-making scenarios, which makes it useful for a wide variety of applications, ranging from environmental impact assessments to supply chain management.
- By combining both additive and multiplicative interactions among criteria, WASPAS reduces the risk of oversimplifying difficult options, so presenting a more nuanced perspective of the alternatives. This is accomplished through the process of balanced decision-making.
- User-Friendly: The method's organized approach makes it accessible even to individuals who may not have a strong experience in quantitative analysis. This is something that makes the method

user-friendly. The stages are easy to understand, which enables users to concentrate on the process of making decisions rather than becoming lost in computations that are difficult to understand.

- Comprehensive Criteria Handling: WASPAS is able to successfully accommodate various forms of criteria, which enables decision-makers to incorporate both qualitative and quantitative data without compromising the integrity of the decision-making model.

Concerns Regarding the WASPAS Method:

- Weight Sensitivity: WASPAS, like many other MCDM approaches, is sensitive to the weights that are allocated to various variables. There is a possibility that the ultimate judgment will be skewed if the weights do not appropriately reflect the significant value of the criterion. In light of this, it is necessary to give weight allocations considerable attention and demonstrate reason.
- The complexity of high-dimensional data: When dealing with situations that involve a large number of different options and criteria, the procedure might become difficult to use. The dual aggregation strategy is advantageous; however, it may result in greater complexity and computing demands, which may discourage its application in high-dimensional datasets for which it is not suitable.
- Assumption of Independence: The WASPAS technique operates under the assumption that particular criteria are independent, which is frequently not the case in situations that occur in the real world. In the event that interdependencies are significant but not taken into consideration, this could result in findings that are not accurate.

The WASPAS technique stands out in the realm of mission-critical decision making (MCDM) because of its ability to provide a decision-making strategy which is both comprehensive and well-balanced. It provides a versatile framework that may be applied to a wide variety of domains and applications due to the fact that it combines both weighted sum models and product models. On the other hand, users need to be aware of its limitations, particularly with regards to the presumption of independence across criteria and the tendency to be sensitive to weight. The WASPAS technique offers the potential to become an indispensable resource for decision-makers whose are attempting to navigate the complexities of situations that involve several sides. This potential is realized when the technique is utilized with due deliberation on the part of the decision-makers.

## RESULTS AND DISCUSSION

In the fight against climate change, it has become an imperative requirement to switch to energy sources that damage the environment less than traditional ones. One of the most important aspects of this change is the utilization of hydrogen energy, which is widely recognized for its ability to effectively reduce carbon emissions. An investigation into the controversy surrounding the use of hydrogen energy is carried out in this study with the use of Chrobanch's Alpha and the WASPAS (Weighted Aggregated Sum Product Assessment) methodology. Evaluation of the practicability of hydrogen energy technologies and the inclusion of these technologies into pre-existing energy systems can be accomplished through the use of these decision-making frameworks, which provide a logical approach to the evaluation process.

The utilization of Chrobanch's Alpha in the context of presentations relevant to hydrogen energy makes it feasible to do an in-depth examination of various hydrogen technologies, including electrolyzes, fuel cells, and storage systems. It is possible for decision-makers to determine which technologies represent the greatest potential for investment and development by employing a method that involves assigning weights to each criterion based on the preferences of stakeholders. For the sake of our planned work, we had conducted a survey with 143 researchers and academics. Figure 1 depicts the results of the results of the sample survey.

*Figure 1. Survey form sample*

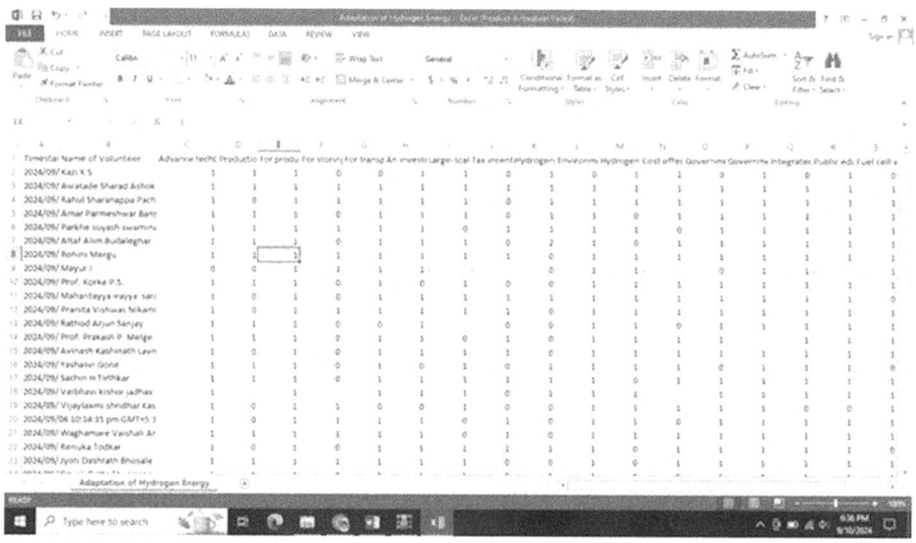

One of the most significant advantages of the WASPAS method is that it is quite versatile, and it also provides a comprehensive analysis of a number of different alternatives. The incorporation of qualitative and quantitative data, that in turn enables it feasible to develop a comprehensive perspective on the utilization of hydrogen energy, is made possible as a result of this, which makes it easier for stakeholders to engage.

Decision-makers are provided with a robust analytical framework as a result of the incorporation of Chrobanch's Alpha and WASPAS into discussions regarding the deployment of hydrogen energy. When these tactics are utilized, it is possible for stakeholders to acquire a more comprehensive comprehension of the complexities that are associated with hydrogen energy technology. The utilization of this all-encompassing strategy promotes the promotion of informed conversations, which, in the long run, may lead to more effective policymaking and investments that are directed towards a specific aim.

Nevertheless, there are challenges that are associated with the employing of these technological advancements. When it comes to the accuracy of the analysis, the availability of data and the reliability of the criteria that were used for input are both potential factors that could have an effect. There is also the possibility that the outcomes will be distorted due to the bias of the stakeholders in the processes of allocating weights. To ensure that all perspectives that are relevant to the assessment process are heard, it is vital for decision-makers to approach the process with transparency and inclusivity. This will maximize the likelihood that all perspectives will be heard.

In addition, despite the fact that these methods provide valuable insights, they should be supported with prototype projects that are being carried out in the real world and continuous feedback loops and adjustments. The theoretical evaluations are brought into line with the actual outcomes through the use of this iterative process, which ultimately results in decisions that are more well-informed.

The factor analysis is carried out after the replies to the problems connected with the adoption of hydrogen energy have been gathered under five dimensions in accordance with the goal of our study that is currently being carried out. Only 21 of the items that were selected had an alpha value that was greater than 0.5, according to Chrobanch methodology. Every element that has a value that is less than 0.5 is eliminated from the system. Presented in Table.2 are the findings that were obtained from the factor analysis. In the subsequent step, the WASPAS Method is applied in order to list the challenges that have been encountered.

*Table 2. Factor analysis of challenges*

| Dimensions | Questionnaire | Factor |
|---|---|---|
| **Technological Readiness** | Advance technology and more research is required for **hydrogen production?** | 0.532 |
| | Production of hydrogen by electrolysis method is very costly? | 0.527 |
| | For production, storage, distribution, and dispensing, good infrastructure and facilities are required? | 0.514 |
| | For storing heavy tank requirements, which are technically difficult? | 0.509 |
| | For transportation of hydrogen is requires heavy quality pipes and transportation systems as highly flammable? | 0.538 |
| | Integrated Infrastructure development is essential for production, distribution networks, refuelling stations, and storage solutions, which requires high investment? | 0.542 |
| **Economic Factors** | An investigation of new fuel cell is essential? | 0.503 |
| | Large scale energy demands require significant investment and scaling, so subsidizing is required? | 0.519 |
| | Hydrogen production cost is high? | 0.625 |
| | Cost effective generation methods are possible? | 0.611 |
| | Fuel cell vehicles are in their initial stage and very costly | 0.567 |
| **Policy Frameworks** | Tax incentives for Generation is required? | 0.549 |
| | Hydrogen energy generation is safe? | 0.533 |
| | Environmental impact study is required? | 0.510 |
| | Government support required? | 0.713 |
| | Government requires to make some policies? | 0.584 |
| | Public education is required to alleviate fears and ensure the safe adoption of hydrogen technologies? | 0.557 |

continued on following page

*Table 2. Continued*

| Dimensions | Questionnaire | Factor |
|---|---|---|
| Formulate Recommendations | Special Safety standards to be developed for use of hydrogen fuels? | 0.562 |
| | Design of hydrogen combustion engines required for vehicles required with any specifications? | 0.527 |
| | Does Hydrogen energy generation required more water? | 0.634 |
| | In Future, Hydrogen Vehicles emission problem? | 0.572 |

The choice matrix for the selected surveys and the selected factor is displayed using Table 3 and Figure 2, respectively. The Normalized decision matrix of the same kind is displayed in Figure 3 and Table 4, respectively, and was generated by WASPAS. The order of the factors that were chosen is presented in Table 5.

*Table 3.Decision matrix*

| Factor | Sections | | |
|---|---|---|---|
| | Commercial | Industrial | Residential |
| Technological Readiness | 0.88 | 0.97 | 0.82 |
| Economic Factors | 0.81 | 0.786 | 0.62 |
| Policy Frameworks | 0.35 | 0.97 | 0.8 |
| Formulate Recommendations | 0.55 | 0.5 | 0.5 |

*Table 4. Normalized decision matrix*

| Factor | Sections | | |
|---|---|---|---|
| | Commercial | Industrial | Residential |
| Technological Readiness | 0.8823 | 1 | 0.84331 |
| Economic Factors | 1 | 0.96923 | 0.820044 |
| Policy Frameworks | 0.40047 | 1 | 0.88971 |
| Formulate Recommendations | 0.48201 | 0.60929 | 1 |

*Table 5. Final rank*

| Factor | Sections | | | Rank |
|---|---|---|---|---|
| | Commercial | Industrial | Residential | |
| Technological Readiness | 0.90103 | 0.90871 | 0.904835 | 3 |
| Economic Factors | 0.92026 | 0.92719 | 0.923786 | 2 |
| Policy Frameworks | 0.75503 | 0.71147 | 0.73325 | 4 |
| Formulate Recommendations | 0.93541 | 0.943848 | 0.93955 | 1 |

*Figure 2. Decision matrix*

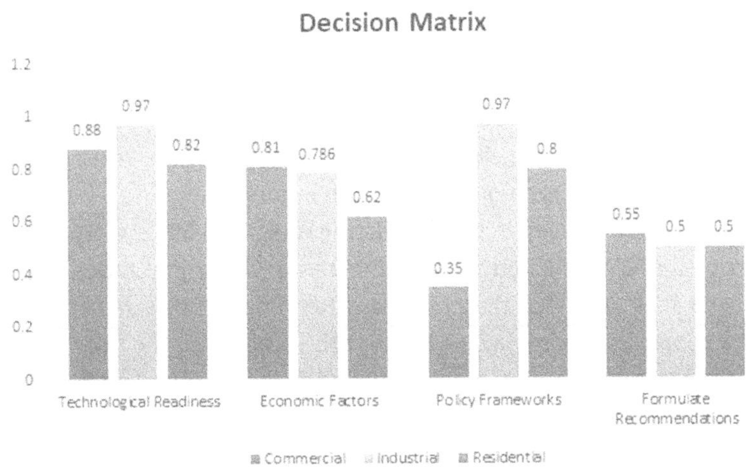


*Figure 3. Normalized decision matrix*

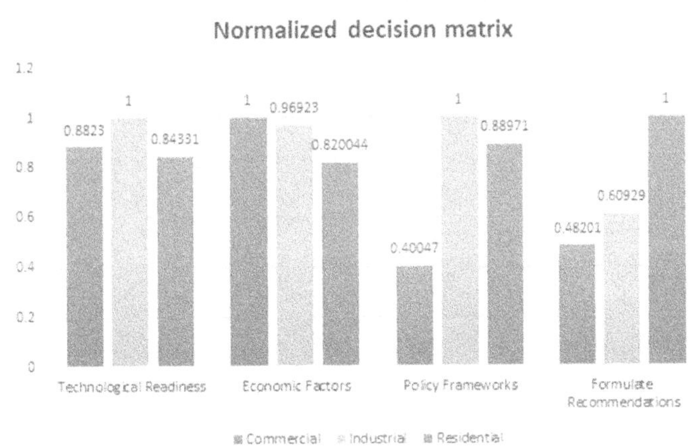


Following the application of the WASPAS method to the evaluation of the challenges, it is evident that the Formulate Recommendations problem is the one that is ranked first. The next task is the Economic Factors challenge, which is ranked second, followed by the Technological Readiness challenge, that is ranked third, and finally the Policy Frameworks challenge, which is ranked fourth. Therefore, in order for hydrogen energy to fulfil the conditions of general adoption, the necessary framework needs to be constructed in order to alleviate the environmental challenge as soon as possible.

## CONCLUSION

The discourse surrounding the use of hydrogen energy has been framed by the usage of Chrobanch's Alpha and the WASPAS approach. A major step toward a future with more sustainable energy is this conversation. If stakeholders conduct a thorough examination of the many technologies that use hydrogen and its implications, they will be able to make informed decisions that support the efficient integration of hydrogen into the worldwide mix of energy. It is undeniable that these strategies have the power to inspire meaningful conversation and promote cooperation across several stakeholders, even in the face of challenges. In the end, using hydrogen as the foundation of sustainable energy initiatives will require a well-balanced approach that blends practical implementation with rigorous analysis. This is essential to the adoption's success. It is clear that the Formulate Recommendations problem is the top-ranked task after the WASPAS approach was applied to evaluate the problems. The following assignment is the second-ranked Economic Factors challenge, that is followed by the third-ranked Technological Readiness challenge and the fourth-ranked Policy Frameworks challenge. Therefore, building the required structure is imperative in order to alleviate the environmental difficulty. This is required for hydrogen energy to meet the requirements for widespread use.

**CONFLICT:** No

**FUNDING:** Nil

## REFERENCES


Baykara, S. Z. (2018). Hydrogen: A brief overview on its sources, production and environmental impact. *International Journal of Hydrogen Energy*, *43*(23), 10605–10614. DOI: 10.1016/j.ijhydene.2018.02.022

Chakraborty, S., & Zavadskas, E. K. (2014). Applications of WASPAS method in manufacturing decision making. *Informatica (Vilnius)*, *25*(1), 1–20. DOI: 10.15388/Informatica.2014.01

Endo, N., Goshome, K., Tetsuhiko, M., Segawa, Y., Shimoda, E., & Nozu, T. (2021). Thermal management and power saving operations for improved energy efficiency within a renewable hydrogen energy system utilizing metal hydride hydrogen storage. *International Journal of Hydrogen Energy*, *46*(1), 262–271. DOI: 10.1016/j.ijhydene.2020.10.002

Ewan, B. C. R., & Allen, R. W. K. (2005). A figure of merit assessment of the routes to hydrogen. *International Journal of Hydrogen Energy*, *30*(8), 809–819. DOI: 10.1016/j.ijhydene.2005.02.003

Farrell, A. E., Keith, D. W., & Corbett, J. J. (2003). A strategy for introducing hydrogen into transportation. *Energy Policy*, *31*(13), 1357–1367. DOI: 10.1016/S0301-4215(02)00195-7

Kasat, K., Shaikh, N., Rayabharapu, V. K., & Nayak, M. (2023). Implementation and Recognition of Waste Management System with Mobility Solution in Smart Cities using Internet of Things, *2023 Second International Conference on Augmented Intelligence and Sustainable Systems (ICAISS)*, Trichy, India, 2023, pp. 1661-1665, DOI: 10.1109/ICAISS58487.2023.10250690

Kazi, K. (2024a). AI-Driven IoT (AIIoT) in Healthcare Monitoring. In Nguyen, T., & Vo, N. (Eds.), *Using Traditional Design Methods to Enhance AI-Driven Decision Making* (pp. 77–101). IGI Global., available at https://www.igi-global.com/chapter/ai-driven-iot-aiiot-in-healthcare-monitoring/336693, DOI: 10.4018/979-8-3693-0639-0.ch003

Kazi, K. (2024a). Machine Learning (ML)-Based Braille Lippi Characters and Numbers Detection and Announcement System for Blind Children in Learning. In Sart, G. (Ed.), *Social Reflections of Human-Computer Interaction in Education, Management, and Economics*. IGI Global., DOI: 10.4018/979-8-3693-3033-3.ch002

Kazi, K. (2024b). Modelling and Simulation of Electric Vehicle for Performance Analysis: BEV and HEV Electrical Vehicle Implementation Using Simulink for E-Mobility Ecosystems. *In L. D., N. Nagpal, N. Kassarwani, V. Varthanan G., & P. Siano (Eds.), E-Mobility in Electrical Energy Systems for Sustainability (pp. 295-320). IGI Global.* Available at: https://www.igi-global.com/gateway/chapter/full-text-pdf/341172DOI: 10.4018/979-8-3693-2611-4.ch014

Kazi, K. S. (2024). Artificial Intelligence (AI)-Driven IoT (AIIoT)-Based Agriculture Automation. In Satapathy, S., & Muduli, K. (Eds.), *Advanced Computational Methods for Agri-Business Sustainability* (pp. 72–94). IGI Global., DOI: 10.4018/979-8-3693-3583-3.ch005

Kazi, K. S. (2024). Machine Learning-Based Pomegranate Disease Detection and Treatment. In Zia Ul Haq, M., & Ali, I. (Eds.), *Revolutionizing Pest Management for Sustainable Agriculture* (pp. 469–498). IGI Global., DOI: 10.4018/979-8-3693-3061-6.ch019

Kazi, K. S. (2024a). Computer-Aided Diagnosis in Ophthalmology: A Technical Review of Deep Learning Applications. In Garcia, M., & de Almeida, R. (Eds.), *Transformative Approaches to Patient Literacy and Healthcare Innovation* (pp. 112–135). IGI Global., Available at https://www.igi-global.com/chapter/computer-aided-diagnosis-in-ophthalmology/342823, DOI: 10.4018/979-8-3693-3661-8.ch006

Kazi, K. S. (2024b). IoT Driven by Machine Learning (MLIoT) for the Retail Apparel Sector. In Tarnanidis, T., Papachristou, E., Karypidis, M., & Ismyrlis, V. (Eds.), *Driving Green Marketing in Fashion and Retail* (pp. 63–81). IGI Global., DOI: 10.4018/979-8-3693-3049-4.ch004

Kazi, S. S. L. (2023). PV Power Control for DC Microgrid Energy Storage Utilisation. *Journal of Digital Integrated Circuits in Electrical Devices*, *8*(3), 1–8.

Kazi, S. S. L. (2023). IoT in Electrical Vehicle: A Study. *Journal of Control and Instrumentation Engineering*, *9*(3), 15–21.

Kutubuddin, K. (2024c). Vehicle Health Monitoring System (VHMS) by Employing IoT and Sensors, *Grenze International Journal of Engineering and Technology,* Vol 10, Issue 2, pp- 5367-5374. Available at: https://thegrenze.com/index.php?display=page&view=journalabstract&absid=3371&id=8

Kutubuddin, K. (2024e). A Novel Approach on ML based Palmistry, *Grenze International Journal of Engineering and Technology,* Vol 10, Issue 2, pp- 5186-5193. Available at: https://thegrenze.com/index.php?display=page&view=journalabstract&absid=3344&id=8

Kutubuddin, K. (2024e). IoT based Boiler Health Monitoring for Sugar Industries, *Grenze. IACSIT International Journal of Engineering and Technology*, *10*(2), 5178–5185. https://thegrenze.com/index.php?display=page&view=journalabstract&absid=3343&id=8

Liyakat, K. K. S. (2023).Detecting Malicious Nodes in IoT Networks Using Machine Learning and Artificial Neural Networks, *2023 International Conference on Emerging Smart Computing and Informatics (ESCI)*, Pune, India, 2023, pp. 1-5, DOI: 10.1109/ESCI56872.2023.10099544

Liyakat, K. K. S. (2023). Machine Learning Approach Using Artificial Neural Networks to Detect Malicious Nodes in IoT Networks. In Shukla, P. K., Mittal, H., & Engelbrecht, A. (Eds.), *Computer Vision and Robotics. CVR 2023. Algorithms for Intelligent Systems*. Springer., DOI: 10.1007/978-981-99-4577-1_3

Liyakat, K. K. S. (2024). Machine Learning Approach Using Artificial Neural Networks to Detect Malicious Nodes in IoT Networks. In Udgata, S. K., Sethi, S., & Gao, X. Z. (Eds.), *Intelligent Systems. ICMIB 2023. Lecture Notes in Networks and Systems* (Vol. 728). Springer., available at https://link.springer.com/chapter/10.1007/978-981-99-3932-9_12, DOI: 10.1007/978-981-99-3932-9_12

Liyakat, K. K. S. (2024). Explainable AI in healthcare. *Explainable Artificial Intelligence in Healthcare Systems*, *2024*, 271–284.

Liyakat Kazi, K. S. (2024). ChatGPT: An Automated Teacher's Guide to Learning. In Bansal, R., Chakir, A., Hafaz Ngah, A., Rabby, F., & Jain, A. (Eds.), *AI Algorithms and ChatGPT for Student Engagement in Online Learning* (pp. 1–20). IGI Global., DOI: 10.4018/979-8-3693-4268-8.ch001

Murray, L. J., Dincă, M., & Long, J. R. (2009). Hydrogen storage in metal–organic frameworks. *Chemical Society Reviews*, *38*(5), 1294–1314. DOI: 10.1039/b802256a PMID: 19384439

Nagrale, M., Pol, R. S., Birajadar, G. B., & Mulani, A. O. (2024). Internet of Robotic Things in Cardiac Surgery: An Innovative Approach. *African Journal of Biological Sciences*, *6*(6), 709–725. DOI: 10.33472/AFJBS.6.6.2024.709-725

Neeraja, P., Kumar, R. G., Kumar, M. S., Liyakat, K. K. S., & Vani, M. S. (2024). *DL-Based Somnolence Detection for Improved Driver Safety and Alertness Monitoring. 2024 IEEE International Conference on Computing, Power and Communication Technologies (IC2PCT)*. Greater Noida., Available at https://ieeexplore.ieee.org/document/10486714, DOI: 10.1109/IC2PCT60090.2024.10486714

Nerkar, P. M., & Dhaware, B. U. (2023). Predictive Data Analytics Framework Based on Heart Healthcare System (HHS) Using Machine Learning, *Journal of Advanced Zoology*, 2023, Volume 44, Special Issue -2. *Page*, *3673*, 3686.

Pradeepa, M., . . .. (2022). Student Health Detection using a Machine Learning Approach and IoT, *2022 IEEE 2nd Mysore sub section International Conference (MysuruCon)*, 2022.

Prasad, K. R., Karanam, S. R., Ganesh, D., Liyakat, K. K. S., Talasila, V., & Purushotham, P. (2024, May). AI in public-private partnership for IT infrastructure development. *The Journal of High Technology Management Research*, *35*(1), 100496. DOI: 10.1016/j.hitech.2024.100496

Sayyad Liyakat, K. K., & Magadum, P. K. (2024). Machine Learning for Predicting Wind Turbine Output Power in Wind Energy Conversion Systems. Grenze International Journal of Engineering & Technology (GIJET), 10.

Shaikh, N. N., Chavan, M. D., Shirshikar, V. G., & Kutubuddin, K. (2023). PV Penetrations in Conventional Power System and Generation of Harmonic and Power Quality Issues: A Review. *International Journal of Power Electronics Controllers and Converters*, *9*(2), 12–19.

Singh, R., Altaee, A., & Gautam, S. (2020). Nanomaterials in the advancement of hydrogen energy storage. *Heliyon*, *6*(7), e04487. DOI: 10.1016/j.heliyon.2020.e04487 PMID: 32743097

Veena, C., Sridevi, M., Liyakat, K. K. S., Saha, B., Reddy, S. R., & Shirisha, N. (2023). *HEECCNB: An Efficient IoT-Cloud Architecture for Secure Patient Data Transmission and Accurate Disease Prediction in Healthcare Systems, 2023 Seventh International Conference on Image Information Processing*. ICIIP., Available at https://ieeexplore.ieee.org/document/10537627, DOI: 10.1109/ICIIP61524.2023.10537627

Yang, J., Sudik, A., Wolverton, C., & Siege, D. J. (2010). High capacity hydrogen storage materials: Attributes for automotive applications and techniques for materials discovery. *Chemical Society Reviews*, *39*(2), 656–675. DOI: 10.1039/B802882F PMID: 20111786

# Chapter 14
# Initial Tests on Various Formulations and the Recommended Dose of the Moroccan Trichoderma asperellum-Based Product for Stimulating and Protecting Tom

**Abdellatif Ouazzani Chahdi**
https://orcid.org/0009-0004-9880-8311
*Ibn Tofail University, Morocco*

**Amina Ouazzani Touhami**
https://orcid.org/0000-0001-9970-7724
*Ibn Tofail University, Morocco*

**Karima Selmaoui**
*Ibn Tofail University, Morocco*

**Rachid Benkirane**
*Ibn Tofail University, Morocco*

**Allal Douira**
*Ibn Tofail University, Morocco*

## ABSTRACT

*Treatment of tomato plants with T. asperellum-based products formulated with the T asperellum's conidia suspension, without talc (P1) and with added talc (P2) reduced verticillium wilt. The Dwarfing index (SI) of plants treated with P1 and P2 were -4.25% and -6.95% respectively, compared to 6.178% for plants inoculated with V dahliae alone. Treatment with P2 promoted the development of the agronomic parameters followed by P1 in comparison with the water control and the inoculated control. The average*

DOI: 10.4018/979-8-3693-8980-5.ch014

*weight of tomatoes harvested from each of the plants inoculated with V. dahliae and treated with P1 and P2 was respectively 194.17 g and 235.43 g / 148.40 g and 90.03 g. On the other hand, treatment of the plants with 40 ml of the conidial suspension volume, prepared from a fresh culture or product P2, increased all the agronomic parameters of the plants compared to the other tested volumes and the water control. The aerial part length is 111.17 cm and 115 cm/ 97.16 cm. The fresh roots weighed 118 g and 139.23 g / 42.07 g. The tomatoes weighed 122.07 g and 148.77 g/12.56 g.*

## 1 INTRODUCTION

The synthesis of a bioproduct for agricultural purposes must be characterized by a formulation that gives it a strong capacity to control pathogens and stimulate the agronomic parameters of cultivated plants. A number of studies have shown the importance of formulation in the efficacy of products based on microorganisms such as Trichoderma spp. as it's known to persist all climatic stages due to their ability to tolerate different ecological conditions (Benitez et al., 2004).

The capacity of Trichoderma spp. as a biological control agent against phytopathogenic fungi in several crops is well known, particularly against Fusarium spp. and Rhizoctonia spp. (Poddar et al., 2004 ; Rojo et al., 2007). In this case, different formulations have been used to control soil pathogens either by fungal spores (Harman et al., 1980) or by powdery preparations of fungal mycelium (Latunde-Dad 1993). The commercial Trichoderma-based preparations marketed were used for crop protection against plant pathogens or to promote different crops' plant growth and productivity (Rojo et al., 2007 ; Hanhong et al., 2011 ; Zancan et al., 2012).

The living fungal conidia (active substance) is incorporated in various formulations as pure conidia suspensions in liquid culture filtrates, and can be integrated with various inert components and kept for months while still maintaining its efficiency (Harman and Kubicek 1998 ; Harman et al., 2010). It can be used in several different ways: a foliar spray, applied pre-planting to seed or propagation material, a post-pruning treatment, incorporated in the soil during seeding or transplant, added during irrigation or applied as a root drench or dip (Woo et al., 2014).

In India, talc-based formulations of *T. viride* have been developed at the Tamil Nadu University of Agriculture, Coimbatore, for the treatment of legume and rice seeds (Jeyarajan et al., 1994). And several private industries produce large quantities of talc-formulated products to supply farmers (Sanjeev et al., 2014). The annual requirement of Trichoderma to cover 50% of the area in India has been estimated at 5000 tonnes (Jeyarajan et al., 2006).

Caron et al (2002) have shown that the *T. harzianum* strain MAUL-20 can have certain beneficial effects on the growth of cucumber and tomato plants when added to growing substrates.

The objective of this work is to designate the adequate formulation of a laboratory made bioproduct based on the conidia suspension of the Moroccan *Trichoderma asperellum* strain by adding an inert component and find the conidia suspension volume that can provide the greatest growth stimulating effect and protective power against *Verticillium dahliae* in tomato plants.

## 2 MATERIALS AND METHODS

### 2.1 Fungal Material and Bioproducts Formulation

The tested pathogen Verticillium dahliae (OMV5) isolate was isolated from the roots of an olive plant originating from the Meknes region. The isolate was grown on a potato saccharose agar PSA medium (Potato Saccharose Agar: 200 g potato, 15 g Agar-agar, 20 g sucrose and 1000 ml distilled water) at 28°C in the dark for 7 days.

The *Trichoderma asperellum* KU987252 strain isolated from compost brought back from the Missour region (East Morocco) was maintained on a PSA medium at 25°C in the dark

The *T. asperellum* conidial suspension was prepared by scraping the surface of seven-day-old cultures with a sterile metal spatula, and the resulting conidial suspension was filtered through muslin cloth. The concentration was estimated using a Malassez slide and adjusted by adding distilled water to $10^7$ conidia.ml$^{-1}$.

The *T. asperellum* based product P2 was formulated by adding to 1 L of the suspension adjusted to a concentration of $2\times10^9$ conidia.mL$^{-1}$ the following components: Sucrose, $CaCl_2$, Tween 20, Gelatin, Glycerin, Paraffin and Talc according to the protocol described by Ouazzani Chahdi et al., (2019).

The *T. asperellum* based product P1 was prepared in the same way as product P2 but without the Talc powder.

### 2.2 In Vivo Tests of *Trichoderma Asperellum* Based Products P1 and P2 Against *Verticillium Dahliae* on Tomato Plants

Tomato (Lycopersicon esculentum) seedlings of the THURAYA variety at the 4-leaf stage and a *V. dahliae* conidia suspension at a concentration $10^6$ conidies.mL$^{-1}$ are prepared according to the method described by Ouazzani Chahdi et al., (2014). Tomato seedlings at the four-leaf stage are grown in polyethylene pots 20 cm in diameter and filled with 2.8 kg of substrate (1.4 kg of peat + 1.4 kg of Mamora soil).

The roots seedlings removed from the peat were inoculated by dipping in a 360 mL conidia suspension of *Verticillium dahliae* for half hour.

The treatment of the two Trichoderma based products P1 and P2 at a concentration of $10^7$ conidia. mL$^{-1}$ with 40 mL of the slurries was carried out by irrigation of plants.

Different treatments were conducted, each with three replicates:

a- The first treatment was carried out by inoculating the seedlings with a conidial suspension of pathogen, or by treating them with a slurry of the *Trichoderma asperellum* based product P1 or P2. These plants constituted the positive control.
b- The second treatment was with water alone. These plants were the negative control.
c- In the third treatment, the roots with the peat surrounding them were treated, using a slurry of one of the *Trichoderma asperellum* based products: P1 or P2. 48 hours later, the roots were removed from the peat and inoculated with the conidial suspension of *Verticillium dahliae.*

The inoculated and treated plants were then placed under black plastic covers for 48 hours, sprayed with sterile water to maintain a high relative humidity. The plants were treated every 4 weeks for 3 months.

Dwarfing index (DI), is the reduction in size of the inoculated seedlings, inoculated and treated with the product compared to controls (Douira and Lahlou 1989).

$$DI = \frac{M - X}{M} \times 100$$

DI: Dwarfing index;
X: Length of the inoculated plants epicotyl;
M: Average length of control plants epicotyl.

Leaf alteration degree is noted according to the scale of Douira and Lahlou (1989) below: 0: leaves with healthy appearance; 1: wilting or chlorosis of the cotyledons; 2: loss of thecotyledons; 3: wilting or chlorosis of the true leaves; 4: necrosis of the true leaves; 5: loss of the true leaves. The total score of all the leaves constituted the foliar index. An average was calculated for each lot of plants.

$$LAI = \frac{[\sum (i \times Xi)]}{6 \times NtF}$$

LAI: Leaf alteration index; i: Note of the leaves appearance 0 – 5 ; Xi: Number of leaves presenting the note i ; NTL: Total leaves number.

The tomato plants' agronomic parameters were estimated after 1, 2 and 3 months of the inoculation with the pathogen. The length of the aerial part was measured using a 30 cm ruler. The numbers of leaves, flowers and fruits were counted and the tomatoes harvested from each plant were weighed by a precision balance. At the end of the different treatments, the tomato plants were removed from their culture substrate and the fresh weight of the aerial part and underground parts were also weighed.

## 2.3 Effect of the Conidia Suspension's Volume on Plant Agronomic Parameters and Maintaining the *T. Asperellum* Population in Rhizosphere Soil

60 tomato seedlings of the THURAYA variety at the four-leaf stage were grown in 60 polyethylene pots 20 cm in diameter and filled with 2.8 kg of substrate (1.4 kg of peat + 1.4 kg of Mamora soil). And to check the development of the *Trichoderma asperellum* population in the soil, 2 g of 2 cm-long fragments of wheat straw were added to each pot.

These seedlings were treated with the conidial suspension of *T. asperellum* prepared from the *T. asperellum* based product P2 preserved for one month or from young cultures of the strain. The prepared suspension is adjusted to a concentration of $10^7$ conidia.mL$^{-1}$.

Seedlings were treated at the start of the culture and after four weeks with different doses of the prepared suspension: 0 mL (control), 2.5 mL, 5 mL, 10 mL, 15 mL, 20 mL, 25 mL, 30 mL, 35 mL and 40 mL. Each treatment was repeated 3 times. The cultures were kept in the greenhouse for 8 weeks.

Agronomic parameters were estimated on tomato plants after 4 and 8 weeks of the inoculation with the pathogen. The length of the root and aerial part, the numbers of leaves, flowers and fruits, and the weight of tomatoes were measured as described above. The diameter of the stem was measured by a caliper. At the end of the different treatments, tomato plants were removed from their culture substrate. The fresh weight of the aerial and root parts were weighed by using a precision balance and dry weight was also weighed after drying at 70°C for 24 h.

At the end of the trial, the straw fragments added to the growing medium were recovered to check that the *Trichoderma asperellum* population was being maintained in the soil. They were washed in a stream of water, superficially disinfected with 5% sodium hypochlorite for 2 min, rinsed and cut into 5 mm long fragments. For each pot, 100 of these fragments were cultured on PSA medium. After one week of incubation at 28°C in the darkness, the number of the colonized fragments was counted and the colonization percent was calculated. Three repetitions were provided for each treatment.

*T. asperellum* was isolated from the roots of tomato plants. The roots were cut into small pieces of 1 cm in Length, superficially disinfected with alcohol 95°, rinsed with sterile distilled water several times and dried on sterile filter paper. The fragments were deposited on Water agar and incubated in the darkness at 22°C. Five replicates were made for each treatment.

After 5 days, the isolation percentage (IP %) was obtained by applying the following formula:

$$IP\% \frac{Nx}{NT} \times 100$$

Nx: Number of segments containing the fungal species x.
NT: Total number of segments.

Soil plates technique (Warcup, 1950) was used to isolate *T. asperellum* from the soil. The culture soil for each treatment was dried at 30°C and grounded in a sterilized mortar. A quantity of 5 to 15 mg was placed on the PSA medium, humidified by one or two drops of sterile distilled water and spread out on the medium. Agar plates were incubated for 5 days in the darkness at 22°C. Five replicates were made for each treatment. After 5 days of incubation, the number of colony forming units (CFU) per gram of soil was estimated for each treatment.

### 2.4 Statistical Analysis

Data were analyzed by one-way analysis of variance (ANOVA) and LSD test at 5% level.

## 3. RESULTS

### 3.1 Effect of the Two *T. Asperellum* Products P1 and P2 on the Development of Tomato Plants and Their Protection Against *Verticillium Dahliae*

Both products P1 and P2 stimulated growth and protected seedlings against the pathogen *V. dahliae*. However, P2 was found to be the best.

Treatment of the plants with both products reduced the development of the disease caused by *V. dahliae* during 3 months of greenhouse cultivation. The results show that the Dwarfing index (DI) of plants treated with P1 and P2 were -4.25% and -6.95% respectively, compared to 6.178% for plants inoculated with *V. dahliae* alone. The leaf alteration index (LAI) of these plants were also reduced by 0.06 and 0.14 respectively compared with 0.41 for plants inoculated with *V. dahliae* alone, showing a reduction of more than 65% (Table 1, Figure 1).

Treating the plants with P1 and P2 also favored the development of several agronomic parameters. The length of the plants treated only with P2's aerial part was statistically different from the control reaching 103.33 cm compared to 86.33 cm. However, the other growth parameters showed no significant difference from the control. The fresh weight of plants treated with P2 and plants treated with P2 and then inoculated with *V. dahliae* was statistically significant compared to the control, 116.27 g and 123.93 g / 99.9 g respectively. The average weight of tomatoes harvested from plants inoculated with *V. dahliae* and treated with P1 and P2 was 194.17 g and 235.43 g respectively, compared to 148.40 g for the control and 90.03 for plants inoculated only with the pathogen (Table 1, Figure 1).

### 3.2 Effect of Conidia Suspension Volume on Plant Agronomic Parameters and Maintaining *T. Asperellum* in Rhizospheric Soil

The results show that treating tomato plants with 40 ml of inoculum at a concentration of $10^7$ conidia. $mL^{-1}$ maintains the *Trichoderma asperellum* population in the soil and therefore promotes the growth of tomato plants.

Treatment of the plants with the product or with the conidial suspension of T.asperellum stimulates their growth. The length of the aerial part of the plants, after four weeks of cultivation in the greenhouse, was 87.67 cm and 81.33 cm respectively, compared to 64.83 cm for the water-treated control (Table 2).

After eight weeks in the greenhouse, all the agronomic parameters increased with the conidia suspension volume applied to the soil. The average number of fruits per 40 mL of *T. asperellum* conidial suspension and product P2 was 5.67 and 6 respectively, compared with 1.67 for the control (Table 3).

It appears that treatment with *T. asperellum* stimulates the development of plant biomass produced in the fruit, aerial and root parts. At 40 mL of conidial suspension and product, the average weight of tomatoes harvested per plant was 122.07 g and 148.77 g respectively, compared with 12.56 g of the control. Similarly, the fresh weight of the aerial part was 140.17 g and 143.27 g /70.07 g and The fresh weight of the root part was 118 g and 139.23 g /42.07 g (Table 4, Figure 2).

In addition, *Trichoderma asperellum* was able to be maintained in the tomato plants' rhizosphere. The conidia were able to colonize the soil, straw fragments and roots to a high level, especially during the product treatment. The number of colonies in the soil and the colonization percentages of the straw and roots after eight weeks of cultivation were respectively in 40 mL of conidial suspension and the product treatments $10.73\times10^3$ CFU/g, 96.08 - 92.59% and $14.27\times10^3$ CFU/g, 98.04% - 100% (Table 5, Figure 3).

## 4. DISCUSSION AND CONCLUSION

The in vivo tests showed that coating the conidia with talc allowed the P2 product, especially when the conidial suspension was applied in increasing volumes, to give optimal growth stimulation, fruit production and plant protection against *V. dahliae*. In addition, talc maintained and developed the *T. asperellum* population in the soil and rhizosphere of the tomato plants, protecting the root system throughout the cultivation period.

These results agree with those of Naseby et al, (2000) who stated that several studies have shown that the addition of specific isolates of *Trichoderma* to the rhizosphere can promote plant growth.

The effect of *Trichoderma asperellum* products is not only limited to protecting plants against fungal diseases, but they also stimulate plant development and yield. These results coincide with those of Singh and Singh (2004) ; Rojo et al, (2007) who highlighted the reduction in disease incidence and increase in yield of different crops after treatment with *Trichoderma harzianum*.

Besnard and Davet (1993) found that certain strains of *Trichoderma* have stimulating properties. This stimulation results in better germination, increased development of the aerial and underground parts and greater flowering in vitro and in the greenhouse. They also showed that Trichoderma strains AGI and BR105 significantly improved the emergence of tomato and cucumber plants, with a relative gain of up to 30%.

The stimulating effect of *Trichoderma* is not limited to the vegetative apparatus but can also increase plant yield (Lynch et al., 1991).

The results show that the stimulatory effect of *T. asperellum* is remarkable for all doses compared to the control, with a clear advantage for the highest dose, 40 ml at a concentration of $10^7$ conidia.ml$^{-1}$. Similarly, Caron et al (2002) found that the highest concentration ($10^7$ CFU.g$^{-1}$) of *T. harzianum* MAUL-20 spores in the substrate improved tomato plant germination. They also found that the height and fresh weight of tomato and cucumber plants' aerial parts significantly increased in the presence of concentrations of $10^5$ CFU.g$^{-1}$ or more. Additionally, Chet and Backer (1979) demonstrated that the minimum *Trichoderma* concentration required for soils should be around $10^6$ CFU.g$^{-1}$ soil.

Ramakrishnan et al (1994) showed that the talc-based formulation of Trichoderma significantly reduced root rot in beans. The talc-based formulation of T. viride was effective in reducing the incidence of root rot caused by *Macrophomina phaseolina* (Ramakrishnan et al., 1994).

In bio-efficacy tests, even after storage for 1 year, talc formulations of T. harzianum with the addition of glycerol at 3 or 6% in the production medium could protect the tomato plants from Fusarium wilt incidence by 44 - 50% (Sriram et al., 2011).

In conclusion, the study showed that the addition of talc to the product based on T. asperellum conidia increased its effectiveness at the root level of tomato plants, stimulating their development and protecting them against V. dahliae, especially when it was applied in increasing volumes. More advanced tests need to be carried out in open-ground cultivation under greenhouse or in the field. This will open up the prospect of its wider application in agriculture as an alternative to chemical plant protection products.

## REFERENCES


Benítez, T., Rincón, A. M., Limón, M. C., & Codón, A. C. (2004). Biocontrol mechanisms of *Trichoderma* strains. *International Microbiology*, *7*(4), 249–260. PMID: 15666245

Besnard, O., & Davet, P. (1993). Mise en évidence de souches de *Trichoderma* spp à la fois antagonistes de *Pythium ultimum* et stimulatrices de la croissance des plantes. Agronomie. *Agronomie*, *13*(5), 413–421. DOI: 10.1051/agro:19930506

Caron, J., Laverdière, L., Thibodeau, P. O., & Bélanger, R. R. (2002). Utilisation d'une souche indigène de *Trichoderma harzianum* contre cinq agents pathogènes chez le concombre et la tomate de serre au Québec. *Phytoprotection*, *83*(2), 73–87. DOI: 10.7202/706230ar

Chet, I., & Baker, R. (1979). *Biological control of soilborne plant pathogens by Trichoderma harzianum. Schippers*. Royaume-Uni.

Douira, A., & Lahlou, H. (1989). Variabilité de la spécificité parasitaire chez *Verticillium albo-atrum* Reinke et Berthold frome à microsclérotes. *Cryptogamie. Mycologie*, *10*(1), 19–32. DOI: 10.5962/p.354573

Douira, A., & Lahlou, H. (1989). Variabilité de la spécificité parasitaire chez Verticillium albo-atrum Reinke et Berthold frome à microsclérotes. *Cryptogamie. Mycologie*, *10*(1), 19–32. DOI: 10.5962/p.354573

Hanhong, B. (2011). Trichoderma species as abiotic and biotic Stress quenchers in plants. *Research Journal of Biotechnology*, *6*(3), 73–79.

Harman, G. E., Chet, I., & Baker, R. (1980). *Trichoderma hamatum* effects on seed and seedling disease induced in Radish and pea by *Pythium* spp. or *Rhizoctonia solani.Phytopathology*, *70*(12), 1167–1172. DOI: 10.1094/Phyto-70-1167

Harman, G. E., & Kubicek, C. P. (Eds.). **1998.** Trichoderma and Gliocladium. Volume 2. Enzymes, biological control and commercial applications. Published by CRC Press. ISBN 9780748408054, 393 Pages DOI: 10.1201/9781482267945

Harman, G. E., Obregón, M. A., Samuels, G., & Lorito, M. (2010). Changing models of biocontrol in the developing and developed world. *Plant Disease*, *94*(8), 928–939. DOI: 10.1094/PDIS-94-8-0928 PMID: 30743493

Jeyarajan, R. (2006). Prospects of indigenous mass production and formulation of Trichoderma.

Jeyarajan, R., Ramakrishnan, G., Dinakaran, D., & Sridar, R. (1994). Development of products of *Trichoderma viride* and *Bacillus subtilis* for biocontrol of root rot diseases. In Dwivedi, B. K. (Ed.), *Biotechnology in India* (pp. 25–36). Bioved Research Society.

Kumar, S., Thakur, M., & Rani, A. (2014). *Trichoderma*: Mass production, formulation, quality control, delivery and its scope in commercialization in India for the management of plant diseases. *African Journal of Agricultural Research*, *9*(53), 3838–3852.

Latunde-Dada, A. O. (1993). Biological control of southern blight disease of tomato caused by *Sclerotium rolfsii* with simplified mycelial formulation of *Trichoderma koningii.Plant Pathology*, *42*(4), 522–529. DOI: 10.1111/j.1365-3059.1993.tb01532.x

Lynch, J. M., Wilson, K. L., Ousley, M. A., & Whipps, J. M. (1991). Response of lettuce to *Trichoderma* treatment. *Letters in Applied Microbiology*, *12*(2), 59–61. DOI: 10.1111/j.1472-765X.1991.tb00503.x

Naseby, D. C., Pascual, J. A., & Lynch, J. M. (2000). Effect of biocontrol strains of *Trichoderma* on plant growth, *Pythium ultimum* populations, soil microbial communities and soil enzyme activities. *Journal of Applied Microbiology*, *88*(1), 161–169. DOI: 10.1046/j.1365-2672.2000.00939.x PMID: 10735255

Ouazzani Chahdi, A., Chliyeh, M., Mouria, B., Dahmani, J., Ouazzani Touhami, A., Benkirane, R., Achbani, E. H., & Douira, A. (2014). *In vitro* and *in vivo* effect of salinity on the antagonist potential of *Trichoderma harzianum* and sensitivity of tomato to Verticillium wilt. *International Journal of Recent Scientific Research*, *5*(4), 780–791.

Ouazzani Chahdi, A., Ouazzani Touhami, A., Khirallah, W., Benkirane, R., & Douira, A. **2019.** Production, formulation et recyclage d'un produit biofongicide et biostimulant à base de *Trichoderma asperellum*. MA 41534 Date de publication: 31/05/2019.url:https://patentregister.ompic.ma/SearchPatent/searchByDepot?typeNum=AP&numDepot=41534&count=0&lang=FR

Poddar, R. K., Singh, D. V., & Dubey, S. C. (2004). Integrated application of *Trichoderma harzianum* mutants and carbendazim to manage chickpea wilt (*Fusarium oxysporum* f.sp. *ciceri*). *Indian Journal of Agricultural Sciences*, *74*, 346–348.

Ramakrishnan, G., Jeyarajan, R., & Dinakaran, D. (1994). Talc-Based Formulation of *Trichoderma viride* for Biocontrol of *Macrophomina phaseolina*. *Journal of Biological Control*, *8*(i), 41–44.

Rojo, F. G., Reynoso, M. M., Ferez, M., Chulze, S. N., & Torres, A. M. (2007). Biological control by Trichoderma species of Fusarium solani causing peanut brown root rot under field conditions. *Crop Protection (Guildford, Surrey)*, *26*(4), 549–555. DOI: 10.1016/j.cropro.2006.05.006

Singh, A., & Singh, H. B. (2004). Control of collar rot of mint (*Mint* spp.) caused by *Sclerotium rolfsii* using biological means. *Current Science*, *87*, 362–366.

Sriram, S., Roopa, K. P., & Savitha, M. J. (2011). Extended shelf-life of liquid fermentation derived talc formulations of Trichoderma harzianum with the addition of glycerol in the production medium. *Crop Protection (Guildford, Surrey)*, *30*(10), 1334–1339. DOI: 10.1016/j.cropro.2011.06.003

Warcup, J. H. (1950). The soil-plate method for isolation of fungi from soil. *Nature*, *166*(4211), 117–118. DOI: 10.1038/166117b0 PMID: 15439172

Woo, S. L., Ruocco, M., Vinale, F., Nigro, M., Marra, R., Lombardi, N., & Lorito, M. (2014). Trichoderma-based products and their widespread use in agriculture. *The Open Mycology Journal*, *8*(1), 71–126.

Zancan, W. L. A., Machado, J. D., de Sousa, B. F. M., & de Matos, C. D. M. (2012). Mycelial growth, production and germination of sclerotia of sclerotinia sclerotiorum in the presence of fungicides and Trichoderma harzianum. *Bioscience Journal*, *28*(5), 782–789.

## APPENDIX

*Table 1. The effect of the two Trichoderma asperellum products P1 and P2 against Verticillium dahliae on tomato plants during 3 months of greenhouse cultivation.*

| Time | Treatments | AP Length (cm) | Number of TL | Number of CL | Number of flowers | Fresh AP weight (g) | Fresh root weight (g) | Number of fruits | Weight of fruits (g) | DI (%) | LAI |
|---|---|---|---|---|---|---|---|---|---|---|---|
| After 1 month | P2+V | 59.00a | 11.00a | 0.33d | 12.33ab | 63.67ab | 5.83abc | 0.00a | 0.00a | -4.734c | 0.051b |
| | P2 | 63.33a | 13.33a | 0.67c | 8.00cd | 77.03a | 6.43a | 0.00a | 0.00a | - | - |
| | P1+V | 53.22a | 12.33a | 1b | 13.67a | 46.07c | 4.33bc | 0.00a | 0.00a | 5.52b | 0.043b |
| | P1 | 63a | 11.88a | 1b | 10.33bc | 60.30b | 6.30a | 0.00a | 0.00a | - | - |
| | Control | 56.33a | 10.67a | 2.00a | 6.00de | 40.37cd | 3.63c | 0.00a | 0.00a | - | - |
| | V. dahliae | 41.67b | 11.00a | 1.00b | 4.00e | 31.57d | 3.33c | 0.00a | 0.00a | 26.036a | 0.207a |
| After 2 months | P2+V | 82.33a | 15.67cd | 1.67a | 15.67b | 109.87ab | 52.00b | 4.67a | 119.40ab | -4.661b | 0.059b |
| | P2 | 82.67a | 22.67ab | 1.00c | 20.33a | 118.97ab | 59.33ab | 5.00a | 121.00a | - | - |
| | P1+V | 83.66a | 26.67a | 1c | 17.33ab | 104.73b | 35.97c | 5a | 89.7c | -6.34c | 0.058b |
| | P1 | 83.33a | 20.67bc | 1.33b | 17.67ab | 125.63a | 64.1a | 4a | 124.57a | - | - |
| | Control | 78.67a | 14.67d | 0.00e | 15.33b | 81.00c | 33.00c | 4.00a | 109.07b | - | - |
| | V. dahliae | 77.33a | 15.67cd | 0.33d | 14.00b | 50.00d | 20.50d | 3.00a | 57.60d | 1.695a | 0.283a |
| After 3 months | P2+V | 92.33ab | 15.67ab | 0.00a | 12.67ab | 123.93a | 72.23ab | 5.67ab | 235.43b | -6.950c | 0.147b |
| | P2 | 103.33a | 17.33a | 0.00a | 9.67bc | 116.27ab | 81.10a | 4.67b | 339.53a | - | - |
| | P1+V | 90ab | 20.33a | 0.00a | 14.67a | 102.96bc | 60.57b | 7.67a | 194.17c | -4.25b | 0.06c |
| | P1 | 92ab | 18a | 0.00a | 15.67a | 103.4bc | 60.43b | 7.67a | 249.66b | - | - |
| | Control | 86.33b | 10.33bc | 0.00a | 7.00c | 99.90c | 74.23a | 3.67b | 148.40d | - | - |
| | V. dahliae | 81.00b | 9.67c | 0.00a | 7.33c | 67.90d | 43.20c | 3.33b | 90.03e | 6.178a | 0.41a |

**AP:** Aerial Part ; **TL:** True leaves ; **CL:** Cotyledonary leaves ; **DI:** Dwarfing index ; **LAI**: Leaf alteration index

Values to which the same letter was assigned within a column do not differ significantly (P < 0.05) according to the LSD test.

*Table 2. The effect of fertigation with increasing volumes of conidial suspension prepared from young cultures of Trichoderma asperellum and product P2 on tomato plants' agronomic parameters after four weeks of greenhouse cultivation.*

| Conidia suspension Volume (mL) | Trichoderma asperellumyoung cultures | | | | | | Trichoderma asperellumbased Product P2 | | | | | |
|---|---|---|---|---|---|---|---|---|---|---|---|---|
| | AP Length (cm) | Stem diameter (cm) | Number of CL | Number of TL | Number of flowers | Number of fruits | Length of AP (cm) | Stem diameter (cm) | Number of CL | Number of TL | Number of flowers | Number of fruits |
| 0 | 64.83b | 0.69b | 2a | 10.00a | 5.33bc | 0a | 64.83b | 0.69b | 2.00a | 10.00c | 5.33d | 0a |
| 2.5 | 72.00a | 0.68b | 2a | 10.33a | 4.00c | 0a | 63.00b | 0.73b | 1.67b | 11.33bc | 8.00cd | 0a |
| 5 | 75.33a | 0.70ab | 2a | 10.67a | 5.67bc | 0a | 69.67b | 0.73b | 1.67b | 11.33bc | 9.00bc | 0a |
| 10 | 76.83a | 0.68b | 2a | 11.67a | 6.00bc | 0a | 70.33b | 0.78ab | 2.00a | 11.33bc | 9.00bc | 0a |
| 15 | 77.67a | 0.70ab | 2a | 11.67a | 6.67bc | 0a | 70.33b | 0.77ab | 2.00a | 12.00bc | 9.67bc | 0a |

continued on following page

*Table 2. Continued*

| Conidia suspension Volume (mL) | Trichoderma asperellumyoung cultures | | | | | | Trichoderma asperellumbased Product P2 | | | | | |
|---|---|---|---|---|---|---|---|---|---|---|---|---|
| | AP Length (cm) | Stem diameter (cm) | Number of CL | Number of TL | Number of flowers | Number of fruits | Length of AP (cm) | Stem diameter (cm) | Number of CL | Number of TL | Number of flowers | Number of fruits |
| 20 | 77.50a | 0.72ab | 2a | 11.33a | 6.00bc | 0a | 75.00ab | 0.87a | 0.67c | 13.00abc | 11.67a | 0a |
| 25 | 81.33a | 0.73ab | 2a | 12.00a | 7.00b | 0a | 75.67ab | 0.87a | 1.67b | 13.67ab | 11.67a | 0a |
| 30 | 81.67a | 0.73ab | 2a | 11.67a | 7.00b | 0a | 86.33a | 0.85a | 1.67b | 14.67ab | 10.67abc | 0a |
| 35 | 81.33a | 0.73ab | 2a | 12.00a | 7.00b | 0a | 86.00a | 0.87a | 1.67b | 16.00a | 11.33a | 0a |
| 40 | 81.83a | 0.80a | 2a | 13.33a | 9.33a | 0a | 87.67a | 0.88a | 2.00a | 16.00a | 11.67a | 0a |

**AP:** Aerial Part ; **TL:** True leaves ; **CL:** Cotyledonary leaves

Values to which the same letter was assigned within a column do not differ significantly ($P < 0.05$) according to the LSD test.

*Table 3. The effect of fertigation with increasing volumes of the conidial suspension prepared from young cultures of Trichoderma asperellum and product P2 on tomato plants' agronomic parameters after eight weeks of greenhouse cultivation.*

| Conidia suspension Volume (mL) | Trichoderma asperellumyoung cultures | | | | | | Trichoderma asperellumbased Product P2 | | | | | |
|---|---|---|---|---|---|---|---|---|---|---|---|---|
| | AP Length (cm) | Stem diameter (cm) | Number of CL | Number of TL | Number of flowers | Number of fruits | Number of fruits | Length of AP (cm) | Stem diameter (cm) | Number of CL | Number of TL | Number of flowers | Number of fruits | Number of fruits |
| 0 | 97.16a | 36.66a | 0.77b | 1c | 13.33b | 8.33bcd | 1.67e | 97.16b | 36.66a | 0.77cd | 1c | 13.33b | 8.33e | 1.67d |
| 2.5 | 103.00a | 39.50a | 0.80ab | 1.33b | 13.33b | 4.33d | 2.00de | 94.83b | 38.33a | 0.75d | 0.00e | 13.00b | 10.00de | 2.67cd |
| 5 | 105.33a | 42.33a | 0.80ab | 0.33d | 13.67b | 4.33d | 1.33e | 100.33ab | 37.17a | 0.83cd | 0.67d | 14.33b | 10.67de | 2.67cd |
| 10 | 102.50a | 40.50a | 0.78b | 1.33b | 14.00b | 5.33cd | 2.67cde | 100.33ab | 37.00a | 0.83cd | 0.00e | 14.00b | 14.00bcd | 3.33cd |
| 15 | 106.00a | 43.67a | 0.80ab | 2.00a | 15.33b | 11.33ab | 3.33bcde | 99.83b | 35.83a | 0.82cd | 0.00e | 15.33b | 15.00bc | 3.00cd |
| 20 | 105.33a | 44.00a | 0.78b | 0.00e | 16.00ab | 12.33a | 4.33bc | 105.17ab | 37.67a | 0.88bcd | 0.00e | 19.00a | 21.00a | 4.33abc |
| 25 | 106.00a | 41.83a | 0.82ab | 1.33b | 16.67ab | 9.33abc | 4.67abc | 106.50ab | 36.33a | 0.88bcd | 1.33b | 19.33a | 17.33ab | 4.00bc |
| 30 | 108.00a | 41.33a | 0.83ab | 1.00c | 16.33ab | 5.33cd | 4.67abc | 106.67ab | 37.50a | 0.90bc | 1.67a | 18.67ab | 13.67bcd | 5.00ab |
| 35 | 109.67a | 42.50a | 0.83ab | 0.33d | 16.67ab | 5.00d | 5.00ab | 111.00ab | 40.83a | 0.97ab | 1.33b | 20.33a | 13.33cd | 5.33a |
| 40 | 111.17a | 43.17a | 0.92a | 1.33b | 19.67a | 7.00cd | 5.67a | 115.00a | 41.83a | 1.00a | 1.33b | 20.33a | 12.33cde | 6.00a |

**AP:** Aerial Part ; **TL:** True leaves ; **CL:** Cotyledonary leaves

Values to which the same letter was assigned within a column do not differ significantly ($P < 0.05$) according to the LSD test.

*Table 4. The effect of fertigation with increasing volumes of conidial suspension prepared from young cultures of Trichoderma asperellum and product P2 on the tomato plants' biomass after eight weeks of greenhouse cultivation.*

| Conidia suspension Volume (mL) | Trichoderma asperellumyoung cultures | | | | | Trichoderma asperellumbased Product P2 | | | | |
|---|---|---|---|---|---|---|---|---|---|---|
| | **Weight of fruits (g)** | **Weight of fresh AP (g)** | **Weight of fresh RP (g)** | **Weight of dry AP (g)** | **Weight of dry RP (g)** | **Weight of fruits (g)** | **Weight of fresh AP (g)** | **Weight of fresh RP (g)** | **Weight of dry AP (g)** | **Weight of dry RP (g)** |
| 0 | 12.56g | 70.07d | 42.07c | 10.33c | 5.48c | 12.56e | 70.07cd | 42.07e | 10.33f | 5.48d |
| 2.5 | 19.63fg | 69.43d | 41.67c | 13.20bc | 6.60c | 16.07e | 68.77d | 44.60e | 14.13ef | 7.73cd |
| 5 | 29.10efg | 72.83d | 43.50c | 13.37bc | 6.90c | 36.37d | 85.07c | 53.77e | 17.37cdef | 8.20cd |
| 10 | 25.50fg | 88.17c | 49.37c | 16.87bc | 7.37bc | 27.97de | 78.63cd | 49.33e | 15.30def | 9.37cd |
| 15 | 31.00def | 82.03cd | 48.57c | 15.80bc | 8.37bc | 38.57d | 83.50cd | 52.03e | 15.93def | 8.80cd |
| 20 | 48.00cd | 86.67cd | 42.33c | 16.30bc | 8.60bc | 91.03c | 103.23b | 74.10d | 23.27cd | 12.70bc |
| 25 | 46.20cde | 105.90b | 48.07c | 20.47b | 11.40bc | 89.77c | 105.50b | 91.77c | 21.97cde | 15.40b |
| 30 | 64.20bc | 103.63b | 49.23c | 19.83b | 11.20bc | 123.73b | 104.13b | 113.10b | 26.00bc | 17.50ab |
| 35 | 68.10b | 108.67b | 71.67b | 20.00b | 13.23ab | 130.73b | 118.43b | 115.20b | 32.33ab | 18.50ab |
| 40 | 122.07a | 140.17a | 118.00a | 28.37a | 18.10a | 148.77a | 143.27a | 139.23a | 35.73a | 23.30a |

**AP:** Aerial Part ; **RP:** root part

Values to which the same letter was assigned within a column do not differ significantly ($P < 0.05$) according to the LSD test.

*Table 5. The ability of Trichoderma asperellum to survive in the tomato rhizosphere after eight weeks in the greenhouse.*

| Conidia suspension Volume (mL) | Trichoderma asperellumyoung cultures | | | Trichoderma asperellumbased Product P2 | | |
|---|---|---|---|---|---|---|
| | **Root colonization (%)** | **Straw colonization (%)** | **Soil Colonisation (×103CFU/g)** | **Root colonization (%)** | **Straw colonization (%)** | **Soil colonization (×103CFU/g)** |
| 0 | 0.00d | 0.00h | 0.00d | 0.00f | 0.00e | 0.00d |
| 2.5 | 3.70d | 9.80f | 0.60d | 7.41f | 5.88de | 1.27d |
| 5 | 22.22c | 19.61f | 1.27d | 37.04e | 11.76d | 2.20d |
| 10 | 48.15b | 33.33e | 1.27d | 51.85d | 35.29c | 2.20d |
| 15 | 66.67b | 74.51c | 2d | 55.56d | 66.67b | 3.40cd |
| 20 | 51.85b | 47.06d | 2.33cd | 77.78c | 76.47b | 6.20cd |
| 25 | 85.19a | 78.43bc | 4.47bc | 81.48bc | 74.51b | 6.53bc |
| 30 | 81.48a | 84.31abc | 6.40b | 92.59ab | 96.08a | 9.67bc |
| 35 | 85.19a | 88.24ab | 8.20ab | 96.30a | 100.00a | 11.87ab |
| 40 | 92.59a | 96.08a | 10.73a | 100a | 98.04a | 14.27a |

Values to which the same letter was assigned within a column do not differ significantly ($P < 0.05$) according to the LSD test.

*Figure 1. Protective and growth-enhancing effects of Trichoderma asperellum based products (P1 and P2) on tomato plants. A: Control; B: Verticillium dahliae; C: product P1; D: product P2; E: product P1 + Verticillium dahliae; F: product P2 + Verticillium dahliae.*

A B C D E F

*Figure 2. The effect of conidial suspension volumes of Trichoderma asperellum prepared from young cultures and the product P2 on stimulating growth of tomato root part.* **A**: *0 mL ;* **B**: *2.5 mL ;* **C**: *5 mL ;* **D**: *10 mL ;* **E**: *15 mL ;* **F**: *20 mL ;* **G**: *25 mL ;* **H**: *30 mL ;* **I**: *35 mL ;* **J**: *40 mL ;* **P2**: *Product 2;* **YC**: *Young cultures.*

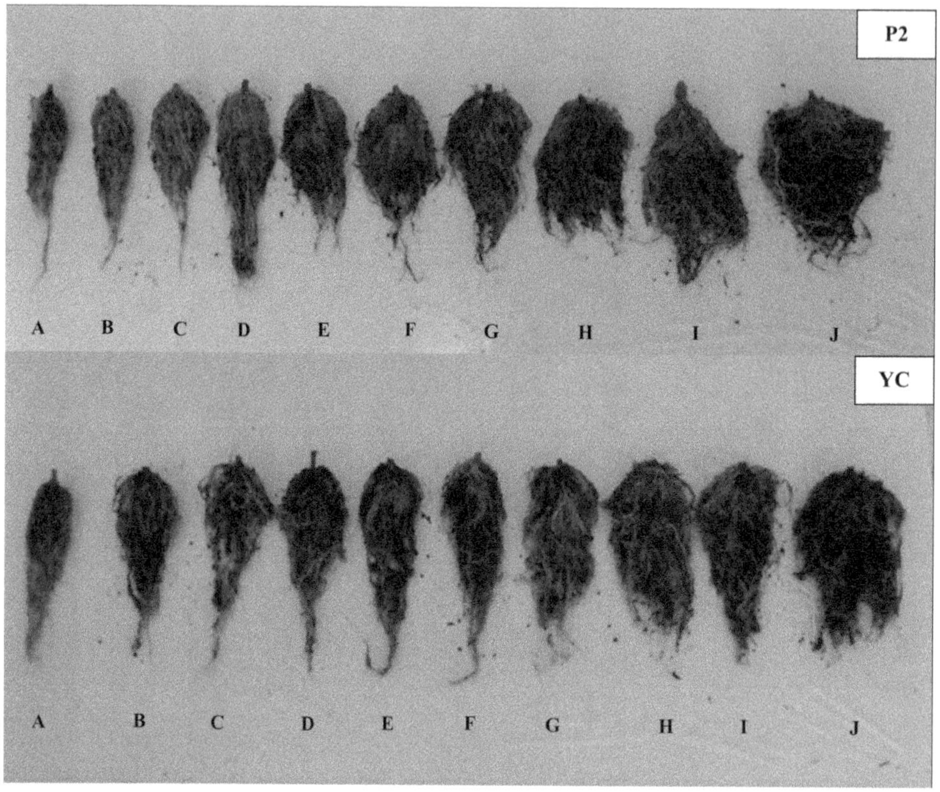

*Figure 3. Reisolation of Trichoderma asperellum from the rhizosphere of tomato plants treated with P2. A: Roots; B: Soil.*

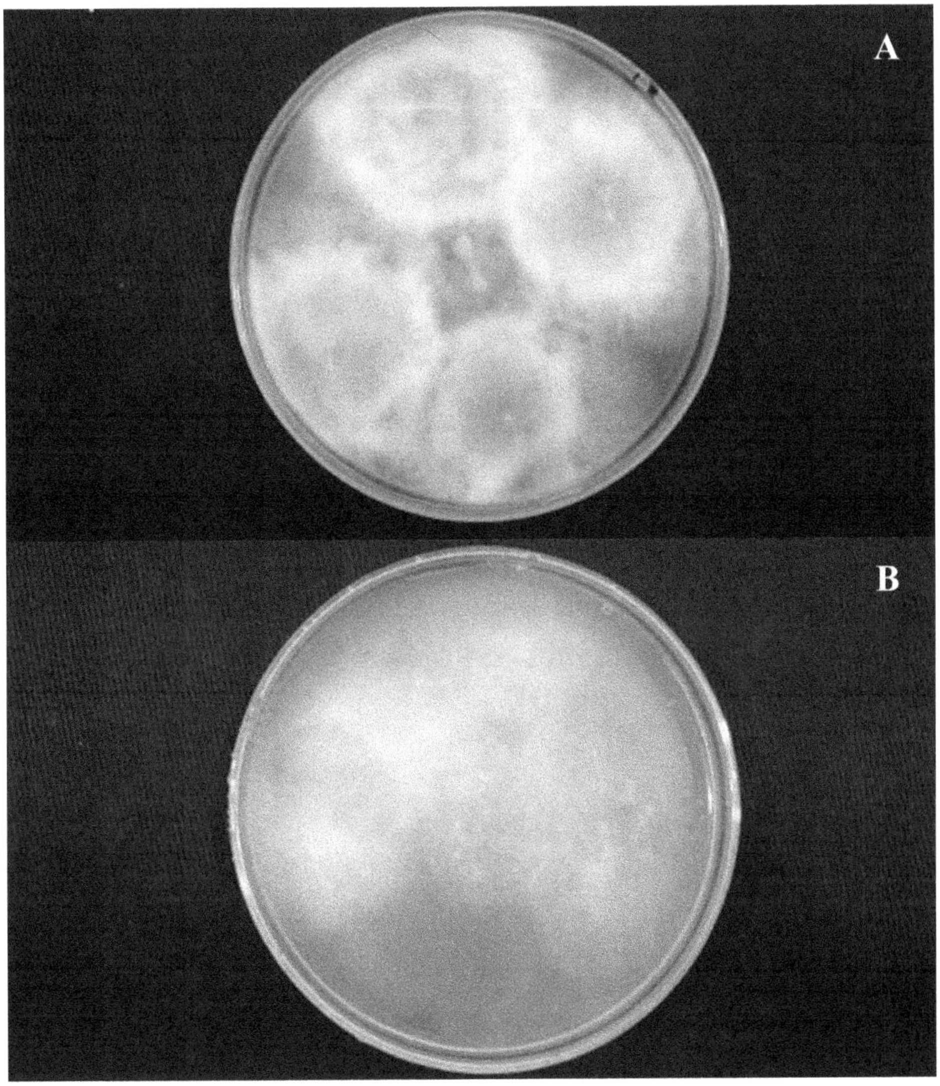

# Chapter 15
# Integration of Renewable Energy in Smart Cities Towards a Clean and Sustainable Future

**Salima Boudraham**
*Mohammed V University in Rabat, Morocco*

**Ilham Ait Oujallal**
*Mohammed V University in Rabat, Morocco*

**Anass Ariss**
*Mohammed V University in Rabat, Morocco*

**Souad Nasrdine**
https://orcid.org/0009-0004-7021-9005
*Mohammed V University in Rabat, Morocco*

**JAMAL Mabrouki**
https://orcid.org/0000-0002-3841-7755
*Mohammed V University in Rabat, Morocco*

**Younes Abrouki**
*Mohammed V University in Rabat, Morocco*

## ABSTRACT

*In the face of rapid urbanization and growing environmental challenges,integrating renewable energy into smart cities has become a global priority. This chapter examines the challenges, opportunities and future prospects of this major energy transition. By harnessing energy sources such as solar,wind, biomass and hydroelectricity, smart cities are diversifying their energy supply, reducing their dependence on fossil fuels.This diversification not only responds to environmental issues,but also stimulates economic development.However, this transition is not without its challenges.The variability of renewable energy sources requires innovative energy storage and management solutions to ensure a stable and reliable power supply.Despite these challenges,the integration of renewable energy in smart cities has considerable potential to promote sustainable urban development by reducing carbon emissions, improving energy resilience and fostering more inclusive and eco-responsible communities.*

## 1. INTRODUCTION

The gradual increase in demand for energy is a consequence of technological advances that improve comfort and access to technology, as well as population growth. This has led to a requirement for an expansion of the power distribution capacity. Current electricity systems rely primarily on fossil fuels

DOI: 10.4018/979-8-3693-8980-5.ch015

(such as coal, gas, and oil) or nuclear energy, which pose significant and irreversible risks to the environment by emitting greenhouse gases (Mahmood et al., 2021). At present, urban areas account for approximately 80% of the world's energy consumption and are responsible for more than 60% of global greenhouse gas emissions (United Nations, 2024). In order to respond effectively to the urban energy demand and achieve the sustainable development goals, there is a growing development of smart and sustainable cities ("Waste Heat Recovery from Diesel Engines Based on Organic Rankine Cycle," 2024). Over the past two decades, the concept of the smart city has evolved considerably, becoming a significant and increasingly relevant solution for achieving urban sustainability ("Recommender Systems for Smart Cities," 2024). A number of cities, municipalities, and local communities around the globe are developing strategies to transform their region's energy supply into a sustainable energy system for the future (Thellufsen et al., 2020). Energy is a significant factor in the advancement of intelligent concepts for the development of urban areas and industrial sectors ("Transition vers un système énergétique durable pour les villes et les industries intelligentes," 2024). The energy systems of smart cities must be capable of meeting the growing demand for energy for heat and power generation. In order to meet the aforementioned demand, it is necessary for cities to implement energy conservation practices and utilize clean energy generation technologies, such as renewable energy sources (including solar, wind, biomass and hydroelectric power) in order to achieve a more environmentally-friendly approach. The integration of renewable systems into smart cities has a direct impact on the well-being and satisfaction of residents, as it contributes to the creation of more beneficial urban environments. It is therefore imperative that these technologies are adopted in order to guarantee the sustainable production of energy without the emission of greenhouse gases, thus promoting a healthier and more sustainable living environment ("A Review of Artificial Intelligence-Based Optimization Techniques for the Sizing of Integrated Renewable Energy Systems in Smart Cities," 2024).

## 2. ROLE OF SMART ENERGY SYSTEM BASED ON RENEWABLE ENERGIES

The energy system plays a pivotal role in the heart of smart cities, serving as a foundational element for the transition to a sustainable urban lifestyle. It is of the utmost importance to guarantee a dependable, effective, and environmentally conscious energy supply in order to facilitate the advancement of smart cities ("Energy Management and Planning in Smart Cities," 2024). The implementation of smart grids plays a pivotal role in the modernization of modern public life, encompassing three key areas. Firstly, they enhance the efficacy of the electrical system by integrating automation, monitoring, remote control, micro grid and self-repair capabilities. Furthermore, they furnish users with precise data regarding their energy consumption, expenditures, and the array of energy alternatives available. Finally, they facilitate the integration and use of distributed and renewable energy resources within existing systems ("Hybrid-Cloud-Based Data Processing for Power System Monitoring in Smart Grids," 2024 ; "Méthodes de contrôle des micro-réseaux pour parvenir à une gestion énergétique durable," 2024). The primary objective of smart energy is to fulfill current and future energy demand through the use of renewable energy sources that are sustainable and environmentally friendly, while simultaneously ensuring the sustainability of non-renewable sources and minimizing the environmental impact of energy production and consumption (Thellufsen et al., 2020). Although renewable energies represent one of the most promising clean resources for meeting the growing demand for energy, numerous scientific studies and research initiatives are currently focused on the integration of these renewable resources in smart cities and their implementation within smart grids. The direct integration of energy at the point of consumption has been shown to have several significant benefits for the environmental and socio-economic aspects of urban sustainability (Xia et al., 2021), This integration can reduce carbon emissions and investment costs, promote energy independence and increased network security and also facilitate the promotion of social capital ("Socio-Technical Evolution of Decentralized Energy Systems," 2024). The primary objectives of planning and constructing smart cities based on renewable energies include facilitating the transition to low-carbon energy resources, implementing comprehensive land use and management policies, and enhancing effective supply and demand management and response (Sharifi & Yamagata, 2016). Consequently, renewable energy and energy-efficient technologies play a pivotal role in the global energy transition, with the objective of establishing clean, energy-efficient, and climate-friendly smart cities.

## 3. INTEGRATION OF RENEWABLE ENERGIES IN SMART ENERGY SYSTEMS FOR SMART CITIES

### 3.1. Renewable Energy Sources for Smart Cities

In light of the depletion of fossil fuels and their detrimental impact on the environment, it becomes imperative to transition towards more environmentally sustainable energy sources. In accordance with the regulations that have been in place since 2020, all new construction projects are required to generate a surplus of energy, exceeding their own consumption. In order to enhance the administration of this consumption and optimize the energy efficiency of urban areas and buildings, smart energy relies on the comprehensive integration of renewable energies. Fig 1 depicts the principal categories of these energy sources (Boudraham et al., 2024). In recent years, the proportion of electricity generated from renewable sources has increased significantly, with the majority of this growth being driven by the development of new renewable energy technologies. This is illustrated in Fig 2.

*Figure 1. Classification renewable energy sources*

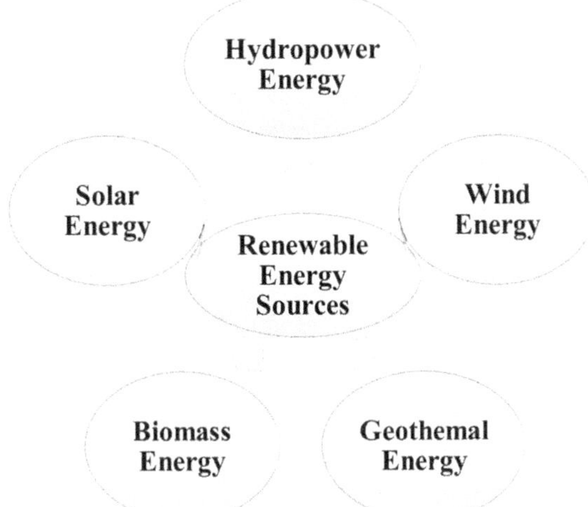

*Figure 2. Share of each energy source for global electricity production ("EI Stat Review," n.d.).*

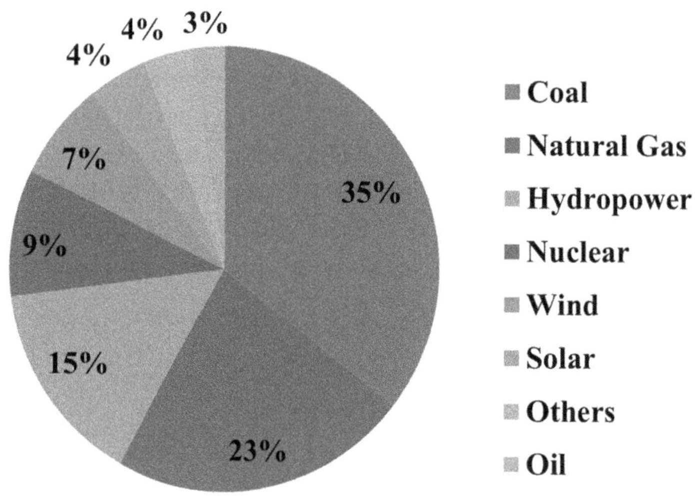


## Solar Energy

Solar energy is employed to generate electricity and heat through the photovoltaic effect, which creates voltage and electric current in a photovoltaic (PV) cell when exposed to sunlight (Gagliano et al., 2019). The integration of photovoltaic solar panels enables these systems to meet the energy demands of buildings, providing electricity for lighting, heating, cooling, and charging electric vehicles (Marinić-Kragić et al., 2018). Empirical evidence indicates that rooftop solar photovoltaic panels could potentially meet a significant portion of the morning peak demand of urban areas with a high solar energy potential, ranging from 47.7% to 94.1% ("Rooftop Solar Photovoltaic Potential in Cities," 2024). Large-scale concentrated solar energy is also an effective means of producing heat and thermal energy. However, the availability of solar resources is dependent on location and annual direct solar irradiation (Hamdan et al., 2020). Another innovative technology is seasonal storage of solar thermal energy, which collects and stores thermal energy for solar heating in summer and then releases it in winter (Hoang et al., 2021). Nevertheless, there are structural and design elements of urban environments that continue to influence the exploitation of solar energy and the integration of this source in smart cities.

## Wind Energy

Wind energy is a renewable resource generated by the movement of wind. It is utilized for electricity generation through urban wind turbines installed on tall buildings and skyscrapers, as well as through large wind turbines found in onshore and offshore wind farms. There are two main types of wind turbines: Horizontal-axis wind turbines and vertical-axis wind turbines (Rose & Benifa, 2021). In urban areas, the latter are distinguished by their superior performance compared to others. They offer a number of advantages, including reduced dependence on wind direction, lower manufacturing costs, the management of turbulence, and the ease of equipment maintenance (Hoang et al., 2021). In addition to its status as a renewable energy source that has seen a surge in recent years, urban wind energy offers a number of advantages, including natural ventilation and the dispersion of air pollutants. This helps to mitigate the urban heat island effect ("Energy Management and Planning in Smart Cities," 2024). However, it remains inadequately integrated into the urban environment. The primary impediments to the wider implementation of urban wind energy include the size of the facilities, the inability to capture turbulent flows and ("How Smart Cities Can Use Renewable Energy Sources," 2024).

### Biomass Energy

Biomass is a form of renewable energy that encompasses forest and agricultural residues, municipal waste, and biogas. This resource can be directly utilized for combustion to produce heat, or alternatively, it can be converted into gaseous or liquid biofuel to generate electricity at competitive costs (Assmann, 2012). It provides a stable and reliable source of energy while reducing emissions of greenhouse gases and other pollutants. The density of biomass available varies according to local climatic conditions and plant types (Jandačka et al., 2017). However, the availability of raw materials is subject to fluctuations depending on the waste management system in place. Furthermore, the operation of bioenergy facilities, such as incinerators, can impact air quality and public health ("How Smart Cities Can Use Renewable Energy Sources," 2024 ; "Converting Municipal Waste to Energy through the Biomass Chain," 2024). It is therefore of the utmost importance to implement effective coordination between waste management and energy production in order to minimize the negative impacts.

### Geothermal Energy

Geothermal energy plays a pivotal role in the electricity and heat network. However, it is more efficient in its use as a heat source ("A European Perspective of the Development of Deep Geothermal in Urban Areas," 2024). For intelligent energy systems, shallow geothermal energy represents a fundamental source of energy. Intelligent heat and power grids are interconnected through underground thermal storage systems. This guarantees a cost-effective and dependable supply of heating and cooling for urban and rural areas alike ("Factsheet-Smart-Cities," 2024). These systems employ either heat pumps to heat the air extracted from the ground, which is then used to produce electricity and heat, thereby offering installation flexibility (Hoang et al., 2021). Conversely, the technology of thermal energy storage in boreholes or aquifers allows for the recovery and storage of energy when it is available, and its subsequent use when required ("Seasonal Thermal Energy Storage," 2024).

## 3.2. Design of an Intelligent Energy System in the Context of Smart Cities

Modeling urban energy systems is a complex undertaking that requires careful planning and consideration of numerous factors. It is essential to consider a number of key elements in order to achieve effective modeling. Fig 3 presents a general diagram of the inputs and expected results for an energy model. There are several planning methodologies, including analytical, iterative, and hybrid methods ("A Review on Configurations, Control and Sizing Methodologies of Hybrid Energy Systems," 2024), while input quality is crucial to system performance. Each input, although having varied impacts, is essential for the final result Among the key criteria for a smart energy system is the integration of new technologies, which enables the use of the latest innovations to improve existing techniques, planning, and efficient management of electricity networks. This, in turn, reduces environmental impacts by minimizing greenhouse gas emissions, as well as reliability and stability, ensuring a consistent and resilient energy supply (Hoang et al., 2021).

*Figure 3. General design model for smart energy system [8].*

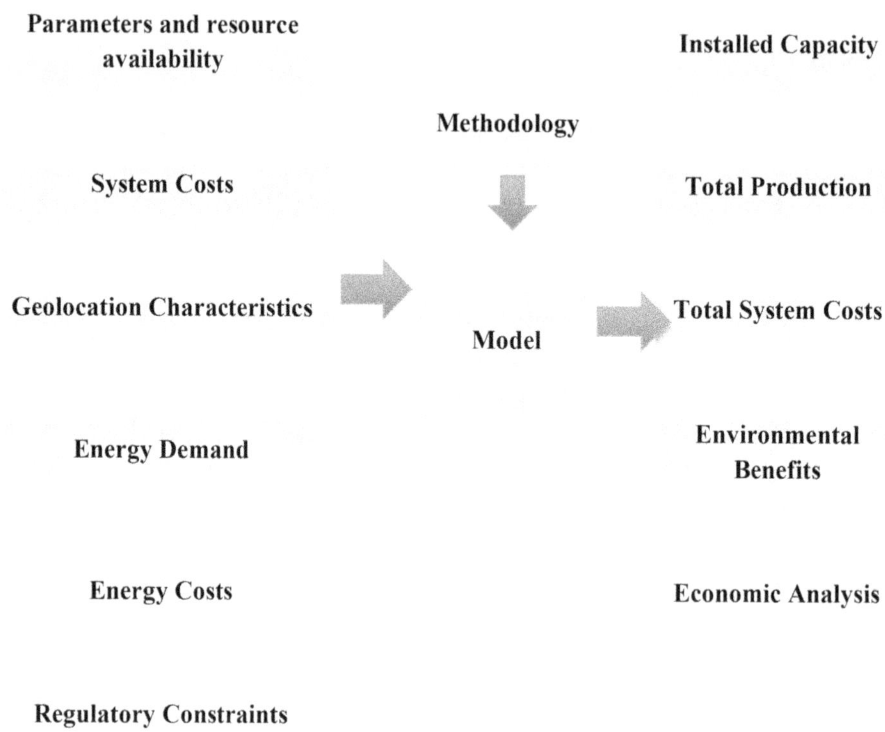


## 4. CHALLENGES OF SMART ENERGY SYSTEMS IN SMART CITIES

Although smart energy systems offer significant environmental, economic, and social benefits, their integration into smart cities is impeded by a number of significant obstacles. The continued growth of domestic energy demand represents a significant challenge, requiring effective management to optimize resource use and reduce consumption ("Reducing Energy Demand: A Review of Issues, Challenges and Approaches," 2024). The maintenance of system stability in power grids is contingent upon the implementation of effective voltage regulation, particularly in the context of an increasing reliance on renewable sources ("Internet of Things (IoT): Opportunities, Issues and Challenges Towards a Smart and Sustainable Future," 2024). Furthermore, the resilience of energy systems to disruptions and their ability to recover quickly are essential to ensure a reliable energy supply. The search for efficient energy storage solutions to manage fluctuations in renewable generation is also a significant challenge. Similarly, the harmonious integration of renewable technologies to diversify and stabilize the urban energy mix is a crucial endeavor (Sharifi & Yamagata, 2016). The interaction with other urban infrastructures offers potential benefits, but it is necessary to coordinate closely in order to achieve the most optimal synergies. Finally, it is imperative to conduct continuous assessments of the environmental and economic impacts of smart energy systems throughout their life cycles in order to ensure their sustainability and efficiency (Hoang et al., 2021). These challenges underscore the necessity for strategic planning and continuous innovation to overcome obstacles and effectively integrate smart energy systems into smart cities in order to sustainably and effectively address future energy needs.

## 5. CONCLUSION AND PERSPECTIVES FUTURES

The deployment of smart energy systems represents a pivotal step in the advancement of sustainable energy solutions, which are of paramount importance in achieving the global objectives of reducing greenhouse gas emissions and preserving the climate. Indeed, the necessity for smart cities to integrate renewable energies in order to reduce carbon emissions and toxic pollutants is becoming increasingly apparent, thus promoting more sustainable and clean production processes. The integration of renewable sources, such as solar and wind, not only optimizes energy efficiency and urban resilience but also maximizes economic and environmental benefits, thereby supporting the development of a robust bioeconomy. Nevertheless, this transition necessitates the implementation of sophisticated optimization solutions that are capable of meeting rigorous criteria, such as low emissions, efficient cost management, and operational reliability. This is despite the complex challenges of managing and interconnecting energy systems in expanding urban environments.

## REFERENCES


Adil, A. M., & Ko, Y. (2016). Socio-technical evolution of Decentralized Energy Systems: A critical review and implications for urban planning and policy. *Renewable & Sustainable Energy Reviews*, *57*, 1025–1037.

Assmann, D. (2012). *Renewable Energy: A Global Review of Technologies, Policies and Markets*. Routledge. DOI: 10.4324/9781849772341

Boudraham, S., Mabrouki, J., & Abrouki, Y. (2024). Application of Renewable Energy Systems for Seawater Desalination: The Future of Desalination. In Mabrouki, J., & Mourade, A. (Eds.), *Technical and Technological Solutions Towards a Sustainable Society and Circular Economy* (pp. 347–355). Springer Nature Switzerland., DOI: 10.1007/978-3-031-56292-1_28

Calvillo, C. F., Sánchez-Miralles, A., & Villar, J. (2016). Energy management and planning in smart cities. *Renewable & Sustainable Energy Reviews*, *55*, 273–287.

Castellanos, S., Sunter, D. A., & Kammen, D. M. (2017). Rooftop solar photovoltaic potential in cities: How scalable are assessment approaches? *Environmental Research Letters*, *12*(12), 125005.

Dumas, P. (2016). A European perspective of the development of deep geothermal in urban areas: Smart thermal grids, geothermal integration into smart cities. *Geomechanics and Tunnelling*, *9*(5), 447–450.

Gagliano, A., Tina, G. M., Aneli, S., & Nižetić, S. (2019, May). Comparative assessments of the performances of PV/T and conventional solar plants. *Journal of Cleaner Production*, *219*, 304–315. DOI: 10.1016/j.jclepro.2019.02.038

Hamdan, M., Sebastia-Saez, D., Hamdan, M., & Arellano-Garcia, H. (2020). CFD analysis of the use of desert sand as thermal energy storage medium in a solar powered fluidised bed harvesting unit. [). Elsevier.]. *Computer-Aided Chemical Engineering*, *48*, 349–354.

Hoang, A. T. (2018). Waste heat recovery from diesel engines based on Organic Rankine Cycle. *Applied Energy*, *231*, 138–166.

Hoang, A. T., Pham, V. V., & Nguyen, X. P. (2021, July). Integrating renewable sources into energy system for smart city as a sagacious strategy towards clean and sustainable process. *Journal of Cleaner Production*, *305*, 127161. DOI: 10.1016/j.jclepro.2021.127161

Jandačka, J., Mičieta, J., Holubčík, M., & Nosek, R. (2017, March). Experimental Determination of Bed Temperatures during Wood Pellet Combustion. *Energy & Fuels*, *31*(3), 2919–2926. DOI: 10.1021/acs.energyfuels.6b03071

Kanase-Patil, A. B., Kaldate, A. P., Lokhande, S. D., Panchal, H., Suresh, M., & Priya, V. (2020). A review of artificial intelligence-based optimization techniques for the sizing of integrated renewable energy systems in smart cities. *Environmental Technology Reviews*, *9*(1), 111–136.

Mahmood, D., Javaid, N., Ahmed, G., Khan, S., & Monteiro, V. (2021, June). A review on optimization strategies integrating renewable energy sources focusing uncertainty factor – Paving path to eco-friendly smart cities. *Sustainable Computing : Informatics and Systems*, *30*, 100559. DOI: 10.1016/j.suscom.2021.100559

Marinić-Kragić, I., Nižetić, S., Grubišić-Čabo, F., & Papadopoulos, A. M. (2018, February). Analysis of flow separation effect in the case of the free-standing photovoltaic panel exposed to various operating conditions. *Journal of Cleaner Production*, *174*, 53–64. DOI: 10.1016/j.jclepro.2017.10.310

Nations, U. (2024). Generating power », United Nations. Consulté le: 30 juin. [En ligne]. Disponible sur: https://www.un.org/en/climatechange/climate-solutions/cities-pollution

Nižetić, S., Šolić, P., Gonzalez-De, D. L. D. I., & Patrono, L. (2020). Internet of Things (IoT): Opportunities, issues and challenges towards a smart and sustainable future. *Journal of Cleaner Production*, *274*, 122877.

Quijano-Sánchez, L., Cantador, I., Cortés-Cediel, M. E., & Gil, O. (2020). Recommender systems for smart cities. *Information Systems*, *92*, 101545.

Rose, J. B. R., & Benifa, J. V. B. (2021). Energy Management in Smart Cities by Novel Wind Turbine Configurations. In *Handbook of Green Engineering Technologies for Sustainable Smart Cities*. CRC Press. DOI: 10.1201/9781003093787-7

Sharifi, A., & Yamagata, Y. (2016, July). Principles and criteria for assessing urban energy resilience: A literature review. *Renewable & Sustainable Energy Reviews*, *60*, 1654–1677. DOI: 10.1016/j.rser.2016.03.028

Talaat, M., Alsayyari, A. S., Alblawi, A., & Hatata, A. Y. (2020). Hybrid-cloud-based data processing for power system monitoring in smart grids. *Sustainable Cities and Society*, *55*, 102049.

Thellufsen, J. Z., Lund, H., Sorknæs, P., Østergaard, P. A., Chang, M., Drysdale, D., Nielsen, S., Djørup, S. R., & Sperling, K. (2020, September). Smart energy cities in a 100% renewable energy context. *Renewable & Sustainable Energy Reviews*, *129*, 109922. DOI: 10.1016/j.rser.2020.109922

Upadhyay, S., & Sharma, M. P. (2014). A review on configurations, control and sizing methodologies of hybrid energy systems. *Renewable & Sustainable Energy Reviews*, *38*, 47–63.

Vrabie, C. (2021). Converting municipal waste to energy through the biomass chain, a key technology for environmental issues in (Smart) cities. *Sustainability*, *13*(9), 4633.

Xia, X., & Wu, X., BalaMurugan, S., & Karuppiah, M. (2021). Effect of environmental and social responsibility in energy-efficient management models for smart cities infrastructure. *Sustainable Energy Technologies and Assessments*, *47*, 101525.

Chapter 16

# Navigating the Digital Frontier Key Challenges and Success Factors for Entrepreneurs in the Age of Technology

**El hassania Rahou**
https://orcid.org/0009-0000-5640-1813
*Hassan First University of Settat, Morocco*

**Abdelkrim El Assfouri**
https://orcid.org/0009-0001-8071-6599
*Ibn Tofaïl University, Morocco*

**Mouna Hajjaj**
*Hassan First University of Settat, Morocco*

**Yassin Allammari**
https://orcid.org/0009-0004-9461-435X
*Abdelmalek Essaadi University, Morocco*

**Chama Jaride**
https://orcid.org/0009-0004-5459-0162
*Abdelmalek Essaadi University, Morocco*

## ABSTRACT

*Technology has significantly accelerated digital entrepreneurship, enabling rapid business establishment. As digital firms increase, understanding the obstacles their founders face and the factors contributing to their success becomes crucial. This study identifies key barriers and highlights the role of human and social capital in overcoming them. Major challenges include difficulty in finding skilled workers, rising competition, limited financial resources, data security concerns, and the need for consumer engagement beyond digital platforms. These issues can hinder business growth and sustainability. By leveraging networks, skills, and relationships, entrepreneurs can better navigate these obstacles. Understanding these barriers and utilizing their resources effectively allows digital entrepreneurs to develop strategies*

DOI: 10.4018/979-8-3693-8980-5.ch016

*for success. This study offers valuable insights to help them identify challenges and find practical solutions for their ventures.*

## 1. INTRODUCTION

Entrepreneurship has long been recognized as a transformative force in driving poverty alleviation and fostering socioeconomic development, as highlighted by Bruton et al. (2013). By creating jobs, generating wealth, and empowering communities, entrepreneurial endeavors contribute to economic progress and social upliftment. In recent years, the convergence of technology and global digital connectivity has given rise to a new paradigm: digital entrepreneurship. This emerging field harnesses the power of technology to redefine business models, offering unparalleled opportunities for innovation, scalability, and economic growth.

Unlike traditional entrepreneurship, digital entrepreneurship transcends geographic and physical constraints, enabling entrepreneurs to operate in a global marketplace with relative ease. The advent of cutting-edge tools, platforms, and digital ecosystems provides entrepreneurs with the flexibility to establish and manage ventures online, reducing the barriers to entry. These technological advancements facilitate the rapid deployment of innovative ideas, fostering an environment where creativity and experimentation thrive (Hanane et al., 2018).

The unique characteristics of digital entrepreneurship are particularly advantageous for young entrepreneurs. Often adept at navigating the digital landscape, young innovators bring adaptability, tech-savviness, and fresh perspectives to their ventures. Their ability to embrace emerging technologies, engage with diverse audiences, and iterate quickly on their business models positions them well to succeed in the competitive and fast-paced digital economy. Furthermore, digital platforms and social media networks offer young entrepreneurs unprecedented opportunities to amplify their reach, engage with customers, and build brand loyalty (BenSaga & Elmqaddem, 2022).

The significance of digital entrepreneurship extends beyond individual success. It plays a pivotal role in reshaping industries, driving technological adoption, and contributing to sustainable economic development. By democratizing access to resources and markets, digital entrepreneurship has the potential to level the playing field, empowering underrepresented groups and fostering inclusivity in the entrepreneurial ecosystem.

Thus, digital entrepreneurship represents a powerful evolution of traditional business practices, leveraging technology to unlock new possibilities for innovation, growth, and social impact. For young entrepreneurs, the digital landscape offers a dynamic and fertile ground to turn visionary ideas into reality, contributing not only to their personal success but also to broader economic and social advancement (Phillipe Mignaval, 2008).

Governments worldwide, including Morocco, have increasingly recognized digital entrepreneurship as a cornerstone for stimulating job creation, fostering economic diversification, and driving technological innovation. With the global economy rapidly transitioning into a digital-first era, policymakers are prioritizing the development of ecosystems that empower entrepreneurs to thrive in this evolving landscape. Morocco, in particular, has demonstrated a strong commitment to supporting digital entrepreneurship through targeted initiatives and strategic investments.

One notable initiative is the "Innov Fund Invest," a program designed to catalyze the growth of innovative tech start-ups by providing substantial financial support and resources. With funding opportunities reaching up to 700 million MAD, this initiative underscores Morocco's ambition to position itself as a regional hub for digital innovation. These resources not only alleviate the financial constraints faced by budding entrepreneurs but also enable them to scale their operations, invest in research and development, and bring cutting-edge solutions to market (Jafari et al., 2019; Rahou et al., 2024).

Beyond funding, the Moroccan government has implemented complementary policies to enhance the digital entrepreneurial ecosystem. These include creating business incubators and accelerators, fostering public-private partnerships, and improving access to training programs that equip entrepreneurs with the technical and managerial skills needed to navigate the complexities of the digital economy. Moreover, the government has prioritized digital infrastructure development, such as expanding high-speed internet access and supporting e-government initiatives, which collectively lay the groundwork for a thriving digital economy.

Such programs highlight a broader trend among governments seeking to harness the transformative potential of digital entrepreneurship. By fostering innovation, enhancing competitiveness, and creating high-value jobs, these initiatives contribute to long-term economic growth and resilience. In Morocco's case, efforts like "Innov Fund Invest" signal a forward-thinking approach to integrating digital entrepreneurship into its broader development agenda, reflecting an understanding of the sector's potential to catalyze progress across industries.

However, the path to success in digital entrepreneurship is not without its formidable challenges. Young entrepreneurs entering this dynamic field face a complex array of obstacles that test their resilience, adaptability, and ingenuity. Intense competition is a defining feature of the digital marketplace, where countless players vie for customer attention in an environment characterized by rapid technological advancements and shifting consumer preferences. In such a crowded landscape, standing out requires more than just a good product or service—it demands exceptional branding, strategic differentiation, and a keen understanding of market trends (Boussetta, 2013).

Access to financial resources remains another significant hurdle for many aspiring digital entrepreneurs. Traditional funding avenues, such as bank loans, often require extensive documentation and collateral, which young entrepreneurs may lack. Venture capital and angel investments, while promising, are highly competitive and frequently skewed towards more established enterprises or those operating in specific high-growth sectors. This financial gap can hinder entrepreneurs from scaling their businesses, investing in necessary technology, or even maintaining day-to-day operations.

Beyond competition and funding, digital entrepreneurs must grapple with pressing concerns about data security and confidentiality. The digital nature of their businesses exposes them to cyber threats, data breaches, and stringent regulatory requirements surrounding data protection. Establishing robust cybersecurity measures and earning consumer trust are critical yet resource-intensive tasks that can strain a fledgling business's capacities.

The digital marketplace also demands relentless innovation. In an ecosystem where consumer preferences evolve rapidly, products and services that once captivated audiences can quickly become obsolete. Entrepreneurs must continuously refine their offerings, integrate emerging technologies, and adapt to new business models to stay relevant. This necessity for ongoing innovation requires not only financial investment but also a culture of creativity and a willingness to embrace change.

Moreover, customer engagement and retention pose significant challenges in this ever-evolving environment. Digital consumers have high expectations for personalized experiences, seamless interactions, and immediate responses. Entrepreneurs must develop sophisticated strategies to connect with their audiences, leveraging social media, data analytics, and targeted marketing campaigns to build loyalty and drive repeat business.

To navigate these multifaceted challenges successfully, entrepreneurs must cultivate a comprehensive understanding of the critical success factors unique to digital enterprises. This includes mastering digital tools and platforms, fostering strategic partnerships, and maintaining a proactive approach to market research. Furthermore, building a robust support network of mentors, advisors, and peers can provide valuable insights and guidance.

In essence, while digital entrepreneurship offers immense potential for innovation and economic growth, it is a field that demands exceptional preparation, continuous learning, and a relentless drive to overcome obstacles. Entrepreneurs who rise to these challenges are not only positioned to succeed but also to shape the future of the digital economy.

Responding to calls from scholars to explore the nuances of digital entrepreneurship (Nambisan, 2016), this study aims to identify the opportunities and challenges faced by young entrepreneurs in this domain. By examining the processes involved in launching, managing, and scaling digital ventures, we seek to uncover the strategies and resources that drive success. The study focuses particularly on the Moroccan context, where digital start-ups are increasingly prominent, and provides an in-depth analysis of the factors influencing their performance and growth.

This article is structured as follows: First, we explore the unique advantages of digital entrepreneurship for young entrepreneurs. Second, we delve into the success factors critical to thriving in this dynamic field. Finally, based on findings from an exploratory study, we examine the challenges faced by young digital entrepreneurs and propose actionable solutions to help them succeed. Through this comprehensive approach, the article seeks to offer valuable insights and practical guidance for aspiring digital entrepreneurs navigating the complexities of the modern technological frontier.

## 2. OPPORTUNITIES OF DIGITAL ENTREPRENEURSHIP FOR YOUNG ENTREPRENEURS

Over several decades, entrepreneurial research has expanded significantly, evolving into a diverse and multifaceted field with often conflicting perspectives (Tounes, 2003). These varying perspectives highlight the dynamic nature of the field and its adaptation to societal and economic changes. The methodologies employed in entrepreneurial research reflect this diversity, illustrating how conceptions evolve and how the challenges of entrepreneurship shift over time, placing researchers within different paradigms (Rahou & Taqi, 2021b).

The second half of the 20th century marked a pivotal shift as academics began to explore the interplay between technology and entrepreneurship. This was enabled by the advent of digital technologies, which transformed economies previously centered on agriculture and industrialization into digital economies. Investments in websites, mobile applications, and online platforms have since become the cornerstone of modern entrepreneurial ventures (Cohen et al., 2000).

These technological innovations have fundamentally reshaped the corporate environment, creating a wealth of opportunities for entrepreneurs across diverse sectors (Hull et al., 2007). Digital entrepreneurship, defined as the use of the internet and ICTs to establish new businesses or enhance existing organizational practices, has become a central pillar of this transformation (Rashidi et al., 2013).

The digital revolution has not only altered the way we live and work but has also democratized access to entrepreneurial opportunities. Digital entrepreneurship enables young individuals to start businesses with minimal capital investment. Thanks to digitization, activities such as creating an online sales platform or developing a mobile application require relatively low initial expenditures (Hull et al., 2007; Sussan & Acs, 2017). This accessibility allows entrepreneurs to quickly test various business models, pivot based on outcomes, and scale their ventures efficiently.

While traditional businesses may offer greater financial stability and credibility with clients, digital entrepreneurship stands out for its flexibility and global reach. Kollmann (2006) examined how electronic economic activities within digital data networks can generate value independently of physical value chains. By reducing transaction costs, enhancing communication channels, and improving marketing strategies, digital entrepreneurs can achieve unprecedented scalability. Amazon, initially a seller of physical goods, serves as a prime example of how a business can transition into a digital entrepreneurship model (Bensaid & Azdimousa, 2019).

This paradigm shift has opened doors for young entrepreneurs to enter high-growth industries such as e-commerce, mobile app development, online service platforms, and digital marketing. These industries present numerous opportunities for growth and development, with studies demonstrating the positive impact of social media on company performance, customer relationship management, and innovation (Pérez-González et al., 2017; Scuotto et al., 2017, as cited in Agrebi, 2021).

Moreover, digital entrepreneurship empowers young individuals to achieve professional independence. By establishing and running their own businesses, they can work remotely, become their own bosses, and reach global audiences through online platforms. The ability to access diverse markets reduces dependency on any single market, thus mitigating risks and diversifying revenue streams.

In conclusion, digital entrepreneurship represents a compelling opportunity for young entrepreneurs to establish businesses in high-growth industries with minimal costs, exceptional flexibility, and global access. However, this field is not without its challenges, requiring a robust marketing strategy and a clear differentiation to stand out in competitive markets. Aspiring digital entrepreneurs must carefully assess the advantages and drawbacks of digital entrepreneurship to position themselves for sustained success.

## 3. THE KEY SUCCESS FACTORS FOR YOUNG ENTREPRENEURS IN DIGITAL ENTREPRENEURSHIP

In Morocco, digital entrepreneurship is flourishing due to a combination of factors, including strong anticipated demand, attractive profit margins, and the relatively nascent stages of technological life cycles (Nigam, Benetti, Johan, 2020, cited by Sanaa & Nayme, 2022). This ecosystem operates within digitally enabled entrepreneurial environments that are still evolving and governed by institutional frameworks. The success and growth of digital enterprises in Morocco depend on a multitude of interrelated factors.

## The Role of Entrepreneurs and Their Characteristics

Research has consistently shown that the success or failure of a business is heavily influenced by the founder's ability to mobilize resources, establish professional networks, and navigate digital challenges (Ye, 2017; Jafari et al., 2019). Central to this is the entrepreneur's human capital, social capital, and individual attributes.

In the knowledge-driven economy, human capital has emerged as one of the most critical assets for organizations (Envick, 2004). Human capital encompasses both formal education and informal learning experiences, such as on-the-job training and work experience. Metrics like management experience, industry expertise, and hands-on knowledge are often used to evaluate human capital and its contribution to organizational objectives (Arenius & Clercq, 2005). In the context of digital businesses, adaptability, creativity, and responsiveness to market demands are indispensable traits (Envick, 2004; Luthans et al., 2004).

## Technical Skills and Their Importance

Technical expertise is particularly crucial in digital entrepreneurship. Founders need to not only understand but also effectively leverage emerging technologies to deliver relevant products and services. Proficiency in areas like user interface design, data security solutions, and software development can significantly enhance a start-up's competitive edge. Studies have shown that in countries like Australia, a majority of digital start-up founders possess backgrounds in computer science, engineering, mathematics, or information technology (Zaheer et al., 2018). Similarly, in Cameroon, Ngoasong (2017) found that a passion for ICT or a formal education in computer science greatly facilitates entrepreneurs in exploiting digital opportunities.

## Management Skills

In addition to technical expertise, effective management skills are essential for digital entrepreneurs. Leading a team, engaging with stakeholders, managing finances, and understanding regulatory requirements are fundamental aspects of running a digital business. Kollmann (2006) highlights that experience in internet-based industries, operational management, and the information economy is invaluable for those aspiring to succeed in digital entrepreneurship.

## Social Capital as a Complement to Human Capital

While human capital is vital, its value is often amplified when paired with robust social capital. Social capital refers to the networks, relationships, and social structures that entrepreneurs can leverage to achieve their goals. Burt (1995) and Schuller (2007) emphasize that social capital is crucial for attracting talent, finding co-founders, and adapting to technological advancements. Zaheer et al. (2018) further argue that strong social networks are instrumental in securing resources and navigating the complexities of digital entrepreneurship.

## Risk-Taking and Entrepreneurial Orientation

Digital entrepreneurship is inherently dynamic, requiring entrepreneurs to take calculated risks. Whether entering new markets, investing in advanced technologies, or expanding operations, risk-taking is a defining characteristic of successful digital entrepreneurs. However, this must be balanced with thorough risk-benefit analysis. Research on entrepreneurial orientation highlights the significance of traits such as creativity, proactivity, and risk tolerance in enhancing an organization's financial performance (Zhao et al., 2011; Satar & John, 2016).

## Monitoring Market Trends and Competitors

To thrive in a competitive digital landscape, entrepreneurs must remain vigilant about market trends and competitor activities. This involves analyzing data, studying emerging patterns, and adapting strategies to stay ahead. By continuously refining their approaches, digital entrepreneurs can better position themselves for long-term success.

# 4. METHODOLOGY: INVESTIGATING CHALLENGES AND SUCCESS FACTORS IN DIGITAL FIRMS

The study investigates the challenges entrepreneurs face in operationalizing and scaling their digital firms, along with the elements that contribute to their success. Employing a multi-case exploratory research design, this investigation aims to capture the nuanced issues and characteristics of digital entrepreneurs. This approach is well-suited to addressing the study's central question: "How do digital entrepreneurs succeed?" Moreover, the qualitative methodology is justified as an initial inquiry into a field that currently lacks comprehensive theoretical frameworks (Beliaeva et al., 2019).

## Data Collection and Sample Selection

Semi-structured interviews formed the core of our data collection strategy, involving three digital entrepreneurs from distinct sectors: manufacturing, e-commerce, and banking. These sectors were chosen based on the literature on digital entrepreneurship, which emphasizes "a new type of technology use and the transformation of existing businesses" (Union, 2016). Additionally, we conducted interviews with the founder of a Moroccan start-up investment firm and seasoned industry experts. To further enrich our understanding, we analyzed web publications, periodicals, and videos that reflected the perspectives and experiences of these stakeholders.

This multi-pronged approach enabled a comprehensive exploration of the challenges faced by digital entrepreneurs and the determinants of business success and growth. By triangulating data from diverse sources, we ensured a robust understanding of the complexities inherent in digital entrepreneurship.

## Analytical Framework

The analysis focused on identifying recurring themes across interviews and secondary data sources. Key areas of investigation included the profiles of digital entrepreneurs, the nature of the challenges they encounter, and the strategies they employ to overcome these hurdles. The analytical process involved coding and categorizing data to uncover patterns related to human capital, social capital, technological competencies, and market strategies.

## Insights from Interviews

Interviews with the selected entrepreneurs revealed a common set of challenges, including limited access to skilled talent, regulatory hurdles, and the need to continuously adapt to rapidly evolving technologies. The entrepreneurs also highlighted the significance of building professional networks, leveraging digital tools for marketing and operations, and maintaining a customer-centric approach.

Experts in the investment and digital sectors provided additional insights, emphasizing the critical role of mentorship, funding access, and government support in fostering a thriving digital entrepreneurial ecosystem. Their perspectives underscored the importance of aligning business strategies with emerging market trends and customer needs.

## Contribution to Literature

This study contributes to the growing body of literature on digital entrepreneurship by offering empirical insights into the Moroccan context. The findings provide a foundation for developing theoretical frameworks that address the unique challenges and success factors associated with digital firms. Moreover, the multi-case exploratory approach serves as a model for future research seeking to deepen our understanding of this dynamic field.

Through an in-depth exploration of the challenges and success factors in digital entrepreneurship, this study highlights the critical importance of adaptability, innovation, and strategic resource utilization. By leveraging human and social capital, embracing technological advancements, and fostering collaborative networks, digital entrepreneurs can overcome barriers and achieve sustainable growth in the competitive digital economy.

# 5. RESEARCH RESULTS AND DISCUSSION

Digital entrepreneurship is rapidly becoming a cornerstone of the modern economy, offering boundless opportunities for young entrepreneurs eager to harness the transformative power of technology. The proliferation of digital tools, platforms, and resources has democratized access to entrepreneurial ventures, enabling individuals with innovative ideas to establish businesses with relatively low barriers to entry. From e-commerce and app development to digital marketing and content creation, the digital landscape presents a fertile ground for those willing to innovate and adapt.

Young digital entrepreneurs often face a dynamic and unpredictable environment where technological advancements outpace the ability of businesses to adapt, requiring constant vigilance and upskilling. Financial constraints, limited industry experience, and intense competition from both local and global players compound the difficulties. Moreover, the need to establish credibility and build trust in an often skeptical market adds another layer of complexity.

This essay delves into the most pressing challenges young digital entrepreneurs encounter, such as managing financial risks, attracting and retaining skilled talent, and leveraging social and human capital effectively. It also explores the importance of technological adaptability and strategic innovation in overcoming these obstacles. By providing actionable insights and practical strategies, this analysis aims to equip aspiring entrepreneurs with the tools they need to not only survive but thrive in the comp.

In the realm of e-commerce, payment methods play a pivotal role in shaping consumer behavior and determining the viability of digital enterprises. In Morocco, the preference for cash-on-delivery (COD) as the dominant offline payment method reflects a broader consumer trend driven by factors such as limited trust in online transactions, lack of familiarity with digital payment systems, and cultural preferences for tangible exchanges. While this approach offers convenience to customers and alleviates concerns about fraudulent transactions, it also imposes significant financial and operational challenges on businesses.

One of the most critical risks associated with COD is the potential for financial losses due to failed deliveries, rejected orders, or logistical inefficiencies. For digital entrepreneurs operating on tight margins, such losses can jeopardize the sustainability and scalability of their ventures. As highlighted by Interviewee 2, retaining customers and persuading them to transition to online payments remain daunting tasks, yet these steps are crucial for the survival and growth of digital firms. The reliance on COD, while culturally entrenched, often hampers the ability of e-commerce businesses to optimize their cash flow and streamline their operations.

Despite these challenges, offline payment methods remain indispensable in Morocco's e-commerce landscape. For many consumers, particularly those without access to banking services or credit cards, COD provides a practical and inclusive solution. Recognizing this, e-commerce enterprises strive to balance the convenience of offline payment options with the need to promote secure and efficient digital payment systems.

To mitigate the risks associated with COD and encourage a gradual shift toward online transactions, digital entrepreneurs can employ several strategies:

**Consumer Education and Awareness**: Campaigns highlighting the safety, speed, and ease of online payments can help dispel misconceptions and build trust among hesitant customers.
**Incentives for Online Payments**: Offering discounts, cashback rewards, or loyalty points for choosing digital payment methods can serve as powerful motivators for behavioral change.
**Hybrid Payment Models**: Allowing partial upfront payments through online gateways while maintaining COD for the balance can help bridge the trust gap.
**Collaboration with Fintech Providers**: Partnering with financial technology companies to develop localized, user-friendly payment solutions tailored to Moroccan consumers can accelerate the adoption of online payments.

While offline payment methods remain significant in Morocco, the long-term success of digital entrepreneurs hinges on their ability to innovate and adapt. By addressing consumer concerns and introducing gradual changes to the payment ecosystem, they can navigate the challenges of COD while laying the foundation for a more sustainable and digitally integrated future.

Hiring suitable workers is one of the most significant challenges faced by digital entrepreneurs, especially in a rapidly evolving technological landscape. Our analysis revealed a unanimous consensus among respondents: attracting and retaining qualified individuals with the necessary skills remains a daunting task. The inventive and creative nature of digital firms inherently demands a workforce that is not only proficient in current technologies but also capable of innovating and adapting to continuous advancements (Korgaonkar & O'Leary, 2006)

The dynamic environment of digital enterprises is characterized by frequent software updates, shifts in operating system application portals, and the emergence of new tools and platforms. These changes require employees who can quickly assimilate knowledge and apply it effectively in real-time scenarios. Firms that fail to secure such talent risk falling behind competitors and may face operational inefficiencies or even obsolescence. As Interviewee 3 highlighted, "It's hard to locate skilled workers who can adapt to changing technology. There are competent individuals, but recruiting and convincing them to join your organization is challenging."

Several factors contribute to this difficulty. First, the global demand for tech-savvy professionals often outpaces supply, creating intense competition among employers. Larger corporations, with their attractive salary packages and comprehensive benefits, often draw top talent away from smaller digital startups. Moreover, startups frequently lack the brand recognition or resources to compete effectively in this talent market, further complicating their recruitment efforts.

Additionally, many young entrepreneurs find themselves at a disadvantage due to their limited networks and lack of experience in talent acquisition. For those attempting to build a team, the absence of a well-established hiring strategy can result in mismatches between the skills required and the capabilities of recruits

To address these challenges, digital entrepreneurs must adopt innovative approaches to attract and retain talent:

**Leverage Remote Work Opportunities**: Expanding the candidate pool by embracing remote work can help entrepreneurs access skilled workers from a global talent base, bypassing local limitations.
**Develop Competitive Compensation Packages**: While financial constraints may limit salary offerings, startups can offer equity, flexible working conditions, or opportunities for professional growth as alternative incentives.
**Invest in Upskilling and Training:** Providing in-house training programs can bridge skill gaps and cultivate a workforce capable of adapting to technological advancements.
**Strengthen Employer Branding:** Building a compelling narrative around the company's vision, culture, and growth potential can attract like-minded individuals who are passionate about contributing to a meaningful mission.
**Foster Partnerships with Educational Institutions:** Collaborating with universities and coding boot camps can create a pipeline of fresh talent trained in the specific skills required by the firm.

Ultimately, digital entrepreneurs must recognize that their team is their greatest asset. By adopting a proactive, flexible, and strategic approach to talent acquisition, they can build a workforce capable of driving innovation, ensuring adaptability, and achieving long-term success in the competitive digital landscape.

Young entrepreneurs often face significant hurdles when embarking on digital ventures, particularly due to a lack of technical skills such as programming, software development, and data analytics. These competencies are essential for navigating the complex technological landscape that underpins digital businesses, ranging from e-commerce platforms and mobile applications to artificial intelligence and blockchain technologies (Hajjaj & Rahou, 2023). Without these foundational abilities, young entrepreneurs may struggle to turn their innovative ideas into functional, scalable, and market-ready solutions.

To overcome this skills gap, one practical approach is to hire experienced developers, engineers, or technical consultants who can provide the expertise needed to design, build, and maintain the technological infrastructure of the enterprise. By delegating these critical tasks to specialists, entrepreneurs can focus on other aspects of their business, such as strategy, marketing, and customer engagement. However, hiring skilled professionals often comes with significant costs, which can be a barrier for startups operating with limited resources.

Another increasingly accessible solution is leveraging online training courses and digital learning platforms. Resources such as Coursera, Udemy, Codecademy, and Khan Academy offer a wide array of affordable courses tailored to varying skill levels, from beginner to advanced. These platforms cover topics like web development, app design, data science, and cybersecurity, enabling entrepreneurs to acquire essential technical skills at their own pace. Additionally, many of these courses are project-based, allowing learners to build practical applications that can directly benefit their businesses.

Entrepreneurs can also participate in coding boot camps or specialized workshops, which offer intensive, hands-on training in a condensed time frame. Such programs often emphasize practical application and problem-solving, providing entrepreneurs with the confidence to manage technical aspects of their ventures. Networking opportunities within these programs can also be invaluable for connecting with like-minded individuals and potential collaborators.

Furthermore, technological proficiency does not require complete mastery in every domain. For young entrepreneurs, understanding the fundamentals of programming and technology management can enhance their ability to communicate effectively with technical teams, make informed decisions, and identify the right tools or frameworks for their business needs.

This essay underscores the pivotal role of human capital in the success of digital organizations, highlighting its value as a cornerstone for innovation, adaptability, and sustained growth. In the digital economy, while technical expertise can provide a competitive edge, it is not always a prerequisite for entrepreneurial success. Digital entrepreneurs can achieve remarkable results even without a formal education in information and communication technologies (ICT) or a deep programming background. Instead, a strong vision, problem-solving skills, and a willingness to learn can often bridge the gap in technical knowledge.

However, a certain level of technical proficiency is undeniably beneficial, particularly for businesses operating in advanced fields such as the Internet of Things (IoT), artificial intelligence (AI), and machine learning. These industries demand a deeper understanding of complex algorithms, data management, and system integration, making technical expertise indispensable. As Interviewee 2 aptly observed, "Yes, it is possible to initiate a project without an ICT background, with the exception of companies that deal with advanced technologies like artificial intelligence." This distinction underscores the diversity within the digital sector, where different industries have varying demands for technical knowledge.

For entrepreneurs without technical skills, success often hinges on their ability to build and lead multidisciplinary teams. By surrounding themselves with skilled professionals, such as developers, engineers, and data scientists, they can compensate for their knowledge gaps while focusing on strategy, vision, and business development. Collaboration and delegation are critical skills that enable non-technical founders to thrive in the digital landscape.

Moreover, accessible learning resources, such as online courses, webinars, and coding boot camps, have democratized access to technical education. Entrepreneurs with limited technical backgrounds can acquire foundational knowledge to better understand their industry and communicate effectively with their teams. This understanding can also help them evaluate technological solutions, manage projects, and ensure that their vision aligns with the technical implementation of their products or services.

It is also worth noting that many successful digital ventures have emerged from a combination of technical and non-technical expertise. Entrepreneurs who excel in areas like marketing, customer experience, or finance often partner with technically skilled individuals to create balanced leadership teams. This synergy between different skill sets fosters innovation and ensures that all aspects of the business are addressed effectively.

In conclusion, while technical expertise can be advantageous, it is not an insurmountable barrier for aspiring digital entrepreneurs. A strong network, a commitment to learning, and strategic collaboration can empower individuals without ICT backgrounds to launch and grow successful digital enterprises. By recognizing the value of human capital and leveraging diverse talents, entrepreneurs can navigate the challenges of the digital frontier and create impactful, scalable solutions.

Past industry experience is a critical factor for success in the digital enterprise landscape. It equips entrepreneurs with the nuanced understanding needed to navigate the complexities of their chosen industry, master relevant legal frameworks, and establish a robust network of stakeholders and loyal customers. Without such experience, entrepreneurs may face significant hurdles in recognizing opportunities, mitigating risks, and achieving sustainable growth.

Industry expertise provides insight into market trends, consumer behavior, and competitive dynamics, enabling entrepreneurs to design products and services that cater effectively to the needs of their target audience. As one respondent, Interviewee 1, observed, "2–3 years of expertise in a specialized area is recommended when beginning a digital firm with a product. This expertise helps you grasp the characters, difficulties, and stakes of the industry and build a client base from the platform's introduction." Such familiarity allows entrepreneurs to anticipate challenges, refine their offerings, and create a strong value proposition that sets them apart from competitors.

In addition to industry knowledge, social capital plays an equally vital role in the success of digital entrepreneurs. Social capital refers to the networks, relationships, and connections that provide access to resources, mentorship, and opportunities. Research by Jafari et al. (2019) and Majda & Hassan (2013) highlights the transformative impact of social capital, noting that relationships with influential individuals can significantly enhance an entrepreneur's chances of success. For example, informal connections with bankers or investors can facilitate access to external funding, while relationships with skilled professionals can ease the process of hiring top talent.

The value of social capital becomes particularly evident in the face of challenges such as recruiting suitable workers. Many respondents emphasized the difficulty of finding individuals with the right skills and mindset to thrive in a fast-paced digital environment. By leveraging their networks, entrepreneurs can tap into a pool of talent and gain recommendations that streamline the hiring process.

Financial resources, while not the sole determinant of success, are undeniably crucial for scaling a digital business. Interviewee 3 highlighted a practical progression many startups follow: "Our firm started with a simple website and self-funding. After product validation and revenue, we sought investor investment. Our earnings might support us, but we need more to grow our platform and produce a mobile app. Funding will speed the process." This trajectory underscores the importance of achieving product-market fit and generating initial traction before seeking external investment. Once these milestones are achieved, additional funding can accelerate growth, enable technological advancements, and expand market reach.

In conclusion, a combination of past industry experience, strong social capital, and strategic financial management lays the foundation for success in digital entrepreneurship. Industry expertise fosters a deep understanding of market needs and operational challenges, while social capital opens doors to valuable resources and collaborations. Financial support, though secondary to a sound business model, provides the means to scale and compete effectively. Together, these elements empower digital entrepreneurs to build resilient businesses capable of thriving in an increasingly competitive landscape

Young digital entrepreneurs have a variety of strategies to build their human and social capital, which are crucial for success in the competitive digital landscape. By leveraging these opportunities, they can enhance their knowledge, skills, and network, ensuring they are well-equipped to navigate the challenges of their industries.

Networking is a foundational strategy for building social capital. Engaging with other entrepreneurs, attending industry events, and participating in startup meetups provide invaluable opportunities to establish connections with like-minded individuals, potential partners, and investors. These interactions often lead to collaborations, mentorship opportunities, and access to insider knowledge about industry trends and best practices.

Training programs—both online and in-person—are also essential for human capital development. These programs allow young entrepreneurs to acquire technical and business skills relevant to their ventures. From coding boot camps and digital marketing courses to workshops on financial management and leadership, training programs offer a structured way to build the expertise needed to drive a business forward.

Online discussion groups and forums, such as those found on LinkedIn, Reddit, or industry-specific platforms, provide a space for entrepreneurs to share ideas, ask questions, and gain insights from experienced professionals. These groups can also serve as a sounding board for new business concepts, offering feedback that helps refine strategies and identify potential pitfalls.

Mentorship is another powerful tool for personal and professional growth. By engaging with experienced mentors, young entrepreneurs gain access to valuable advice, encouragement, and guidance. Mentors can help entrepreneurs navigate complex challenges, provide introductions to influential contacts, and share lessons from their own experiences, helping avoid common pitfalls.

Social media serves a dual purpose: building an online profile and expanding one's network. Platforms like Twitter, Instagram, and LinkedIn allow entrepreneurs to showcase their expertise, promote their ventures, and connect with potential clients, partners, and investors. Consistent and strategic use of social media can significantly boost visibility, enhance credibility, and attract opportunities.

By participating in these activities, young entrepreneurs not only develop essential skills but also build a network of critical contacts. These connections can open doors to collaborations, partnerships, and funding opportunities, which are vital for scaling their businesses. Additionally, these engagements foster a better understanding of their industry, enabling entrepreneurs to stay updated on market trends and technological advancements.

In a competitive digital market, human and social capital are not just assets—they are necessities. Entrepreneurs who actively invest in these areas gain a significant advantage, positioning themselves as knowledgeable, resourceful, and connected individuals. Through consistent effort and strategic engagement, young digital entrepreneurs can create a strong foundation for sustained success and innovation in their fields

Young digital entrepreneurs must deliver distinctive goods or services to carve out a space in an increasingly competitive market. Differentiation is key in a landscape where consumers have a wealth of choices, and only those businesses that provide unique value propositions are able to stand out. To achieve this, entrepreneurs must undertake comprehensive market research to gain a deep understanding of customer needs, preferences, and pain points. This process enables them to identify underserved niches and tailor their offerings to fill these gaps effectively.

A strong value proposition is the cornerstone of attracting and retaining clients. Entrepreneurs must clearly communicate what sets their products or services apart, whether it's superior quality, innovative features, exceptional customer experience, or a more affordable price point. Crafting and emphasizing this distinctiveness not only attracts customers but also builds brand loyalty in a competitive market.

Despite their innovative ideas, young entrepreneurs often face significant challenges in turning their vision into reality due to the high costs associated with technology, marketing, and product development. From purchasing cutting-edge software and hardware to running advertising campaigns and refining product designs, the financial demands of starting and growing a digital enterprise can be overwhelming. These challenges are often exacerbated by a lack of experience, which can make traditional funding avenues such as bank loans harder to secure.

To overcome these hurdles, digital entrepreneurs can explore alternative funding sources. One promising option is seeking support from angel investors, who provide not only financial backing but also mentorship and industry connections. Similarly, applying for grants or joining incubation programs can offer crucial financial assistance along with access to valuable resources, networking opportunities, and expert guidance. These programs often serve as launchpads for startups, accelerating their growth and increasing their chances of long-term success.

For those who prefer grassroots funding, crowdsourcing platforms such as Kickstarter or Indiegogo can be an effective way to raise capital. These platforms enable entrepreneurs to showcase their ideas to a global audience and secure funding directly from consumers who believe in their vision. Crowdsourcing also serves as a validation tool, demonstrating market demand and building an initial customer base before the product is even launched.

While bank loans remain a traditional funding option, they often require a strong credit history or collateral, which can be a barrier for young entrepreneurs. However, those who meet the criteria can benefit from the stability and structured repayment plans that bank financing offers.

In conclusion, young digital entrepreneurs must be strategic in differentiating their offerings and securing funding to navigate the challenges of a competitive market. By focusing on delivering unique value, understanding their customers through rigorous market research, and leveraging a variety of funding sources, they can position themselves for success. With persistence, creativity, and resourcefulness, these entrepreneurs can turn their innovative ideas into thriving businesses that capture market share and build lasting impact.

Digital entrepreneurship offers young entrepreneurs a wealth of opportunities, including the potential to innovate, reach global markets, and create meaningful impact through technology-driven solutions. However, the path to success is fraught with distinct challenges that require strategic planning, adaptability, and resilience. Navigating these obstacles demands a clear focus on several critical areas, ensuring that entrepreneurs can harness the benefits of the digital economy while mitigating its inherent risks.

One key to success lies in the ability to define and refine a unique value proposition. In a competitive and saturated market, standing out requires more than just a good product or service—it requires offering something distinctive that resonates deeply with target customers. This could be an innovative solution to a persistent problem, exceptional customer service, or a business model that prioritizes sustainability and social impact. Entrepreneurs who understand their niche and communicate their value effectively are better positioned to attract and retain a loyal customer base.

Securing funding is another significant hurdle, as the costs of technology, marketing, and scaling operations can be prohibitive. Young entrepreneurs must explore diverse funding avenues, such as angel investors, venture capital, grants, or crowdsourcing, while demonstrating the viability and scalability of their business models. Building financial acumen and crafting compelling business plans are essential for convincing potential backers to invest in their vision.

Cultivating human and social capital is equally critical. Human capital refers to the skills, expertise, and innovation-driven mindset of the entrepreneurial team. Entrepreneurs must prioritize hiring talented individuals who can adapt to the dynamic demands of the digital landscape. At the same time, building social capital—through networks, mentorships, and partnerships—can open doors to opportunities, resources, and collaborations that might otherwise be inaccessible. By fostering relationships with industry leaders, peers, and stakeholders, young entrepreneurs can gain valuable insights and support to propel their ventures forward.

Staying current with industry trends and technological advancements is vital in a field defined by constant change. Entrepreneurs who proactively embrace innovation, invest in continuous learning, and adapt to evolving customer needs are better equipped to maintain a competitive edge. Keeping pace with trends such as artificial intelligence, blockchain, or immersive technologies can create opportunities to innovate and redefine industry standards.

To expand their reach and influence, young digital entrepreneurs should consider forming strategic alliances and collaborations. Partnerships with complementary businesses, tech providers, or industry influencers can enhance market exposure, streamline operations, and amplify the impact of their efforts. Collaborations can also foster knowledge-sharing and resource optimization, enabling entrepreneurs to achieve goals more efficiently.

In this dynamic and ever-changing landscape, success requires a blend of creativity, dedication, and innovation. Creativity fuels the development of groundbreaking ideas and solutions, dedication ensures consistent effort and perseverance through challenges, and innovation drives the continual improvement of products, services, and processes. By embodying these qualities, young entrepreneurs can not only overcome the obstacles they face but also thrive as leaders in the digital economy.

Ultimately, while the digital business landscape offers immense potential, it demands strategic thinking, adaptability, and a commitment to lifelong learning. Entrepreneurs who focus on their unique strengths, build robust networks, and embrace change with enthusiasm will be well-positioned to flourish and make a lasting impact in their industries.

## 6. CONCLUSION

In conclusion, digital entrepreneurship is a complex and highly competitive field that presents young entrepreneurs with both significant opportunities and formidable challenges. This study emphasizes that a comprehensive understanding of the market and customer needs, a well-defined and unique value proposition, effective access to funding, the development of human and social capital, and staying abreast of technological advancements are critical to success in this arena. Additionally, young entrepreneurs must demonstrate agility in responding to rapid market changes, actively engage with industry peers, and foster a culture of creativity and innovation. These attributes, when combined, provide the foundation for thriving in a constantly evolving digital landscape.

The aim of this research is not only to support individual entrepreneurial success but also to contribute to the broader field of entrepreneurship, particularly digital entrepreneurship, by shedding light on the factors that drive success in this fast-paced domain. Digital entrepreneurship is characterized by intense competition, rapid technological innovation, and evolving market dynamics. To navigate these challenges, continuous research and the development of new insights are essential. These efforts will empower entrepreneurs to adapt more effectively to market demands, anticipate challenges, and harness emerging opportunities.

This study highlights several critical determinants of success for young digital entrepreneurs. By exploring key areas such as market understanding, funding strategies, the cultivation of human and social capital, and the integration of cutting-edge technological trends, we aim to provide actionable insights for entrepreneurs seeking to establish and scale their digital ventures. The findings underline the importance of lifelong learning, resilience, and strategic decision-making in achieving long-term success in the digital economy.

Moreover, this research aspires to deepen our understanding of the nuances of digital entrepreneurship, offering valuable guidance for both practitioners and scholars. For entrepreneurs, the study provides a roadmap for navigating the complexities of the digital marketplace. For researchers, it lays the groundwork for further exploration of emerging trends and challenges in this dynamic field.

In conclusion, this research serves as a stepping stone toward enhancing the field of digital entrepreneurship. By equipping new entrepreneurs with the tools and knowledge they need to succeed, it aims to foster innovation, adaptability, and growth in a constantly changing digital market. The insights presented here underscore the importance of combining strategic vision with practical execution, empowering entrepreneurs to not only survive but thrive in the digital age. As the digital landscape continues to evolve, ongoing research and collaboration will remain crucial in shaping the future of entrepreneurship and driving meaningful progress in the global economy.

## Implications of This Research

The insights from this research have several practical and academic implications. For practitioners, it offers a roadmap to navigate the intricacies of the digital marketplace, emphasizing the importance of strategic planning, skill acquisition, and adaptive learning. Policymakers can also use these findings to craft supportive policies and initiatives aimed at nurturing entrepreneurial ecosystems, particularly in developing regions. For educators, this research underscores the need for curriculum development that bridges the gap between theoretical knowledge and practical entrepreneurial skills, especially in technology-driven industries.

## Prospects of This Research

The prospects of this research lie in its potential to inspire further investigation into emerging trends and challenges within digital entrepreneurship. With the ongoing advancement of technologies like artificial intelligence, blockchain, and the Internet of Things (IoT), the entrepreneurial landscape is bound to evolve, necessitating continuous adaptation. Future research can build upon these findings to explore specific technological applications, cross-industry collaborations, and the role of global markets in shaping digital entrepreneurship.

## Limitations of This Research

While this study offers valuable insights, it also has limitations. The findings are based on a specific set of interviews and literature, which may not capture the full spectrum of experiences across different industries and regions. Additionally, the dynamic nature of the digital economy means that the relevance of certain strategies may shift over time. Future research could address these limitations by incorporating a larger sample size, longitudinal studies, and comparative analyses across diverse geographic and economic contexts.

In conclusion, this research serves as a stepping stone toward enhancing the field of digital entrepreneurship. By equipping new entrepreneurs with the tools and knowledge they need to succeed, it aims to foster innovation, adaptability, and growth in a constantly changing digital market. The insights presented here underscore the importance of combining strategic vision with practical execution, empowering entrepreneurs to not only survive but thrive in the digital age. As the digital landscape continues to evolve, ongoing research and collaboration will remain crucial in shaping the future of entrepreneurship and driving meaningful progress in the global economy.

## REFERENCES

Agrebi, M. (2021). Entrepreneuriat sur les médias sociaux : Cas des entrepreneurs « pure players » sur le réseau social Facebook. ICSB 2021 WORLD CONGRES, July.

Arenius, P., & De Clercq, D. (2005). A network-based approach on opportunity recognition. *Small Business Economics*, *24*(3), 249–265. DOI: 10.1007/s11187-005-1988-6

BenSaga, A., & Elmqaddem, N. (2022). Usage de la réalité virtuelle et de la réalité augmentée dans l'enseignement supérieur : De l'expérimentation à l ' intégration effective dans les cursus de formation. *Journal of Information Science*, *21*(2), 194–202.

BenSaga, A., & Elmqaddem, N. (2022). Usage de la réalité virtuelle et de la réalité augmentée dans l'enseignement supérieur : De l'expérimentation à l ' intégration effective dans les cursus de formation. *Journal of Information Science*, *21*(2), 194–202.

Bensaid, W., & Azdimousa, H. (2019). L ' ENTREPRENEURIAT DIGITAL : REVUE DE LITTÉRATURE ET NOUVELLE DÉFINITION. *Revue Internationale Du Marketing et Management Stratégique*, *1*(13), 1–15.

Bhupatiraju, S., Nomaler, Ö., Triulzi, G., & Verspagen, B. (2012). Knowledge flows - Analyzing the core literature of innovation, entrepreneurship and science and technology studies. *Research Policy*, *41*(7), 1205–1218. DOI: 10.1016/j.respol.2012.03.011

Boussetta, M. (2013). Entrepreneuriat des Jeunes et Développement de l ' Esprit d ' Entreprise au Maroc : l ' Expérience de Moukawalati. In Rapport de Recherche du FR-CIEA N ° 54 / 13.

Burt, R. S. (1995). Le capital social, les trous structuraux et l'entrepreneur. *Revue Française de Sociologie*, *36*(4), 599. DOI: 10.2307/3322448

Cohen, S., de Long, B., & Zysman, J. (2000). Tools for thought: what is new and important about the e-conomy? Berkeley Roundtable on the International Economy. University of California at Berkeley., January.

Envick, B. R. (2004). Beyond Human and Social Capital: the Importance of Positive Psychological Capital for Entrepreneurial Success. *Allied Academies International Conference*, 0, 1–20. DOI: 10.1016/j.trb.2017.09.007

Fahmi, S., & Farah, N. (2022). Les déterminants de la survie et de la croissance des startups digitales au M. REVUE DE L'ENTREPRENEURIAT ET DE L'INNOVATION, 4(13).

Hajjaj, M., & Rahou, E. H. (2023). Indigenous perspective to Moroccan leadership values in the 2022 FIFA World Cup context: A grounded-theory approach. *International Journal of Work Innovation*, *4*(4), 356–373. DOI: 10.1504/IJWI.2023.134986

Hanane, B., Wazani Youssef, E., & Malika, S. (2018). Entrepreneuriat Digital : Revue De Litterature Digital. *Public & Nonprofit Management Review PNMReview*, *3*(1), 155–165.

Hanane, B., Wazani Youssef, E., & Malika, S. (2018). Entrepreneuriat Digital : Revue De Litterature Digital. *Public & Nonprofit Management Review PNMReview*, *3*(1), 155–165.

Jafari, A., Meskine, L. E. L., & Chakir, A. (2019). L'entrepreneuriat digital au Maroc : Les entrepreneurs face aux défis actuels. *Revue de l'Entrepreneuriat et de l'Innovation, II*(8), 1–10.

Kollmann, T. (2006). How technology and entrepreneurship are shaping the development of industries and companies. In Technology and Entrepreneurship are Shaping the Development of Industries and Companies (pp. 1–384). DOI: 10.4337/9781781951828

Korgaonkar, P., & O'Leary, B. (2006). Management, market, and financial factors separating winners and losers in e-business. *Journal of Computer-Mediated Communication, 11*(4), 1128–1149. DOI: 10.1111/j.1083-6101.2006.00311.x

Luthans, F., Luthans, K. W., & Luthans, B. C. (2004). Positive psychological capital: Beyond human and social capital. *Business Horizons, 47*(1), 45–50. DOI: 10.1016/j.bushor.2003.11.007

Majda, E. A., & Hassan, B. (2013). Pratiques d 'accompagnement, succès entrepreneurial et accès au financement externe : Cas des entrepreneurs naissants marocains.

Ngoasong, M. Z. (2017). Digital entrepreneurship in a resource-scarce context. *Journal of Small Business and Enterprise Development, 25*(3), 483–500. DOI: 10.1108/JSBED-01-2017-0014

Philippe, M. (2008). *Propositions pour favoriser le développement des petites et moyennes entreprises africaines*. Haut Conseil de la Coopération Internationale.

Rahou, E. H., Hajjaj, M., Taqi, A., Jaride, C., & Allammari, Y. (2024). Examining the Effect of Moroccan Informal Microenterprises' Attributes on the Decision to Transition into A Formal Status. *Journal of Developmental Entrepreneurship, 29*(1), 2450004. Advance online publication. DOI: 10.1142/S1084946724500043

Rahou, E. H., Hajjaj, M., Taqi, A., Jaride, C., & Allammari, Y. (2024a). Examining the Effect of Moroccan Informal Microenterprises' Attributes on the Decision to Transition into A Formal Status. *Journal of Developmental Entrepreneurship, 29*(1), 2450004. Advance online publication. DOI: 10.1142/S1084946724500043

Rahou, E. H., & Taqi, A. (2021a). Informal micro-enterprises: What impact does the business environment have on the decision of formalization? E3S Web of Conferences, 234, 1–6. DOI: 10.1051/e3sconf/202123400052

Rahou, E. H., & Taqi, A. (2021b). The Conversion From Informality to A Formal Entity: Links To The Nature Of The Owner-Manager. *Journal of Developmental Entrepreneurship, 26*(4), 1–20. DOI: 10.1142/S1084946721500278

Schuller, T. (2007). Reflections on the use of social capital. In Review of Social Economy (Vol. 65, Issue 3). DOI: 10.1080/00346760601132162

Sussan, F., & Acs, Z. J. (2017). The digital entrepreneurial ecosystem. *Small Business Economics, 49*(1), 55–73. DOI: 10.1007/s11187-017-9867-5

Tounès, A. (2003). L'intention entrepreneuriale. Une étude comparative entre des étudiants d'écoles de management et gestion suivant des programmes ou des formations en entrepreneuriat et des étudiants en DESS CAAE. Unpublished doctoral dissertation, Université de Rouen.

Union, E. (2016). *Accelerating the digital transformation and enterprises of European industry*. Issue March.

Ye, Q. (2017). BOOTSTRAPPING AND NEW-BORN STARTUPS PERFORMANCE : THE ROLE OF FOUNDING TEAM HUMAN CAPITAL. *Global Journal of Entrepreneurship*, *1*(2), 53–71. DOI: 10.47177/GJE.01.02.2017.053

Zaheer, H., Breyer, Y., Dumay, J., & Enjeti, M. (2018). Straight from the horse? s mouth: Founders? perspectives on achieving? traction? in digital start-ups. *Computers in Human Behavior*. Advance online publication. DOI: 10.1016/j.chb.2018.03.002

Zhao, Y., Li, Y., Hoon, S. L., & Chen, L. B. (2011). Entrepreneurial Orientation, Organizational Learning, and Performance: Evidence From China. *Entrepreneurship Theory and Practice*, *757*(2), 293–317. DOI: 10.1111/j.1540-6520.2009.00359.x

Chapter 17

# Opportunities and Challenges for a Green Hydrogen-Based Economy

**Oumaima El Ghammad**
https://orcid.org/0009-0004-6076-8801
*Mohammed V University in Rabat, Morocco*

**Imane Ennejjai**
*Mohammed V University in Rabat, Morocco*

**Driss Azdem**
*Mohammed V University in Rabat, Morocco*

**Anass Ariss**
*Mohammed V University in Rabat, Morocco*

**Jamal Mabrouki**
https://orcid.org/0000-0002-3841-7755
*Mohammed V University in Rabat, Morocco*

## ABSTRACT

*Hydrogen has a strong potential to accelerate the decarbonization of industries, surpassing other renewable alternatives. However, its overall development remains limited, with analyzes often focused on specific aspects (production, storage, transport). This study provides an overview of the challenges and opportunities of the hydrogen economy. Innovations such as wastewater recycling and electrolysis via microbes provide promising solutions. Technological advances are crucial to secure transport and storage, and an extensive network of refueling stations is necessary for its adoption in transport. Finally, government policies and incentives are essential for the transition to green hydrogen.*

DOI: 10.4018/979-8-3693-8980-5.ch017

## 1. INTRODUCTION

A substantial portion of global energy needs is still fulfilled by fossil fuels, leading to annual CO2 emissions of 37 gigatonnes (Gt)(Gür, 2024) (Azdem et al., 2025.). If greenhouse gas (GHG) emissions continue at this rate, the Earth's ecosystem may become inhospitable for humans by the end of the 21st century(Freedman et al., 2024) (Ghizlane et al., 2022). Among the various clean energy alternatives, hydrogen stands out for several reasons: (a) it serves as a fuel, an energy storage medium, and an energy carrier; (b) it has the potential to decarbonize heavy industries; and (c) it supports decentralized energy production(Laimon & Yusaf, 2024) (Mgalaa et al., 2022). With the significant decrease in costs associated with renewable energy sources (RES), there is a renewed surge of interest in green hydrogen as an energy carrier(Marouani et al., 2023) (Azdem, Fatni, et al., 2025).

The production of green hydrogen is still in its early stages(Squadrito et al., 2023) (Al-Jadabi et al., 2023), but it is essential for achieving net zero emissions (NZE) targets by 2050(Anika et al., 2022) (Azdem, Mabrouki, Moufti, Hajjaji, et al., 2024). Currently, approximately 95% of industrial hydrogen is produced through steam methane reforming (SMR)(Zolghadri et al., 2024) (Benchrifa, Elouardi, et al., 2023), which is the predominant method for hydrogen production. Although SMR has significant environmental drawbacks, it is preferred due to its relatively low production costs, which range from 2 to 4 USD/kg(Khalil, 2025) (Azdem, et al 2024). In contrast, hydrogen generated from renewable energy sources is priced between 6 and 10 USD/kg. New technologies, such as methane pyrolysis, have the potential to produce hydrogen with zero CO2 emissions or even achieve a net reduction in emissions(Pangestu et al., 2024) (Rachiq et al., 2021). While these innovations are still in the developmental phase, large-scale applications could lead to sustainable approaches for green hydrogen production. Additionally, the extraction of hydrogen from geological reservoirs is a viable option(Massarweh et al., 2023) (El Mouedden et al., 2025), although its sustainability over the long term is still in question. Furthermore, the considerable amount of fresh water needed for electrolysis in hydrogen production poses additional challenges(Kim et al., 2025) (El Alouani et al., 2024).

Looking back, approximately 43 liters of water are required to produce one kilogram of hydrogen through solar energy. In contrast, the extraction of oil and gas uses about 133 liters of water for each kilogram of oil produced (Okere et al., 2024) (Fahdi, Mabrouki, et al., 2024).While the water requirement for hydrogen production is relatively low, many countries with significant renewable energy (RES) potential are often grappling with acute drinking water shortages(Benchenina et al., 2025) (Mabrouki, Benbouzid, et al., 2022). However, utilizing domestic or industrial wastewater, which is abundant in minerals and microorganisms, could offer a sustainable alternative for hydrogen production(Swaminathan et al., 2024) (Azdem, et al., 2025). This method not only promotes a circular economy but also aligns with (UN-SDGs). In this light, the hydrogen economy could emerge as a vital energy component across various industrial sectors, thereby aiding in the reduction of global carbon emissions(Evro et al., 2024) (Mabrouki, Azrour, et al., 2020).

Numerous recent studies have explored the potential effects of transitioning from fossil fuels to green hydrogen. However, most of these studies have concentrated on specific aspects of the hydrogen value chain, including production, storage, applications, and economic factors(Azadnia et al., 2023) (Fatni et al., 2025). There is a lack of research that offers a holistic view of the advancements, opportunities, and challenges associated with the complete hydrogen value chain.

This chapter seeks to bridge this gap by providing an in-depth analysis of the potential of the hydrogen economy on a global scale. It will highlight recent advancements in production, storage, and utilization, while also identifying key milestones required for the global adoption of hydrogen as a means of energy transmission.

## 2. STATUS AND CHALLENGES OF HYDROGEN PRODUCTION

Hydrogen production, the starting point of its value chain, is predominantly achieved through two main methods: (i) using fossil energy sources, resulting in carbon-emitting hydrogen (blue/grey), and (ii) using renewable resources to generate carbon-neutral hydrogen (green)(Monteiro, et al 2023) (Loukili et al., 2021). Blue/grey hydrogen is frequently used across various industries due to its lower production costs, often paired with Carbon Capture and Storage (CCS) technology(Gunathilake et al., 2024) (Azdem, et al., 2024). In contrast, green hydrogen is produced from biomass and water using processes such as thermal decomposition, light-driven decomposition, photoelectrocatalytic reactions, and electrochemical splitting(Boddula et al., 2024) (Benchrifa & Mabrouki, 2022).

Current production processes demonstrate relatively reduced efficiency of conversion, leading to low readiness level of technology (TRLs) when compared to large-scale commercial production. Given the seasonal variability of renewable energy sources, hydrogen storage emerges as a cost-effective, time-efficient, and environmentally friendly solution compared to other energy storage methods. However, storing hydrogen in gaseous or liquid forms demands specialized tanks, and maintaining necessary storage conditions (70 MPa and 20K) consumes considerable energy(Dauletbay, 2024) (Fahdi, et al., 2024).

The growing need to protect the environment is driving the energy transition, encouraging the shift from fossil fuel-powered vehicles to electric vehicles (EVs) for land transportation. Despite this shift, EVs are mainly used for urban travel due to their limited battery capacities. On the other hand, recent advancements in hydrogen-powered vehicles, which utilize hydrogen fuel cells, are promising. These vehicles provide longer driving distances, faster refueling, and produce zero exhaust emissions.

Nevertheless, consumer adoption of hydrogen fuel cells remains sluggish, mainly because of worries about safety and the elevated overall ownership costs, adoption has been slow.

In other transportation sectors like aviation and maritime, hydrogen integration has progressed at a slower pace because of stringent safety regulations, long design phases, and substantial development costs. Additionally, major industries face the challenge of making further capital investments in green hydrogen to meet carbon-neutral (NZE) goals, potentially raising costs for consumers(Nnabuife et al., 2023) (Mabrouki, Benchrifa, et al., 2022).

The assessment reveals several significant challenges to building a green hydrogen economy.

- Hydrogen Production Performance: Green hydrogen production suffers from low efficiency, characterized by insufficient technology readiness levels (TRLs) and high costs associated with its extraction from renewable energy sources (RES).
- Hydrogen Infrastructure Network: Insufficient infrastructure for transporting hydrogen over long distances poses a significant challenge. Additionally, affordable materials and optimal conditions for safe and cost-effective storage are lacking.

- Investment Costs and Capital Requirements: The technological advances required to design and develop systems for all modes of transportation require significant investments, as well as supportive government policies.
- Innovation and Sustainability: Scientific progress is essential to encourage the use of alternative raw materials, such as wastewater, industrial by-products, biomass, and enzymes, to create a hydrogen-focused circular economy.

## 3. FUTURE PROSPECTS

Green hydrogen production techniques from renewable sources include alkaline electrolysis, photovoltaic (PV), as well as thermochemical, photochemical, dark fermentation (DF), solid-oxygen electrolysis (SOE), photofermentation, and photoelectrolytic electrochemistry (PEC) methods. These methods are currently considered the most promising.

The main obstacle of these techniques is their low energy efficiency. For example, the energy efficiency of photovoltaic (PV), water electrolysis and solid-oxygen electrolysis (SOE) is 45%, 61% and 80%, respectively(Sun et al., 2022) (Mabrouki, Moufti, et al., 2020). This low performance results in significant thermal losses and higher production costs. Although alkaline electrolysis has reduced energy efficiency, it remains a reliable, scalable and cost-effective method, and has a Technology Readiness Level (TRL) of 9. Similarly, photovoltaic technology has reached a comparable maturity, with a TRL of 9.

Their combined potential, including using excess photovoltaic electricity to power electrolysis, offers promising prospects for improving the overall efficiency and sustainability of hydrogen production(N-nabuife et al., 2024) (Bencheikh et al., 2020). By optimizing the electrolysis cell design and operational conditions, thermal losses and electrolyte convection can be minimized. Thermal insulation of the cell can reduce these heat losses, while electrolyte convection can be mitigated by adjusting the flow rate and installing baffles to limit turbulence in the solution(Xing et al., 2022) (Benchrifa, Mabrouki, et al., 2023).

Heat recovery from electrolysis could improve energy efficiency by preheating the water and electrolyte. A hybrid system, combining electrolysis and reforming of natural gas or biogas, could also reduce energy consumption. The efficiency of photovoltaic (PV) cells could be optimized by using monocrystalline or polycrystalline cells combined with a continuous flow electrolyzer. The addition of advanced catalysts, high-performance membranes and optimization of the electrolyte pH would further increase this efficiency, with supports such as alumina, silica or carbon nanotubes (CNT)(Tao et al., 2024) (Bencheikh et al., 2021).

Hydrogen obtained via SOE technology can be fully renewable and carbon-free when produced from electricity from renewable energy sources (RES). A pilot plant, equipped with Topsoe's SOEC electrolyser in Denmark, illustrates the increased potential of this technology by combining it with the use of industrial surplus heat(Axelsson et al., 2024).

SOE technology is divided into two categories: H+ SOE, which is still in the laboratory stage, and O2− SOE, which is in the demonstration phase. Both types have significant advantages, such as improved kinetics, reduced energy consumption, and lower investment costs. Although SOE shows higher efficiency than water electrolysis, it faces electrode delamination problems, which can be mitigated by incorporating metal oxides into the electrolyte. Currently, SOE technology has a Technology Readiness Level (TRL) of 5, and its efficiency can be enhanced by preventing electrode passivation, which can be achieved through adequate surface treatments and optimized gas flow management(Li et al., 2024).

Thermochemical processes, which use substances such as metal oxides, metal chlorides and sulfuric acid, offer an advantage in terms of scalability with respect to specific installations or conditions

Geological requirements for hydrogen production. The three-stage thermochemical dissociation cycle exhibits efficiencies of 42%, 52% and 60% at thermal radiation temperatures of 827 °C, 900 °C and 1000 °C, respectively. However, no industrial scale demonstration has been carried out so far, due to the corrosive conditions associated with these processes(Gonuguntla et al., 2024).

Thermochemical water splitting benefits from advances in thermal energy storage and concentrated solar power, which enhances its market potential. Conversely, photocatalytic splitting methods are confined to laboratory environments due to their high initial investment cost and scalability challenges, which hinder their commercial deployment for hydrogen production. Furthermore, the selection of stable photocatalysts and cost-effective, non-toxic sacrificial agents remains insufficiently advanced, with a Technology Readiness Level (TRL) below 4 (Meißnitzer, 2023).

Optimization of photoreactor engineering can be achieved by employing stable and affordable catalysts to improve their surface properties and increase quantum efficiency under natural solar irradiation. To ensure efficient charge separation and transfer, it is essential to tune the energy band structure of semiconductor surfaces in photoelectrodes, which promotes higher photovoltage and better energy conversion.

On the other hand, dark fermentation (DF) stands out as an efficient method for biohydrogen production, using microorganisms capable of generating hydrogen from organic substrates without the need for light (see Figure 1). Its main advantage lies in its flexibility and sustainability, allowing the transformation of various organic wastes into hydrogen. However, this approach faces scaling barriers due to process limitations and economic viability issues. Despite a Technology Readiness Level (TRL) of 7, DF faces challenges related to its limited energy yields and high chemical oxygen demand. Combining DF with other hydrogen production techniques could contribute to improving the overall biohydrogen yield.

*Figure 1. Assessment of technology readiness levels (TRLs) for green hydrogen production through various methods: PEC, photofermentation, SOE, dark fermentation, photochemical, thermochemical, alkaline electrolysis and photovoltaic.*

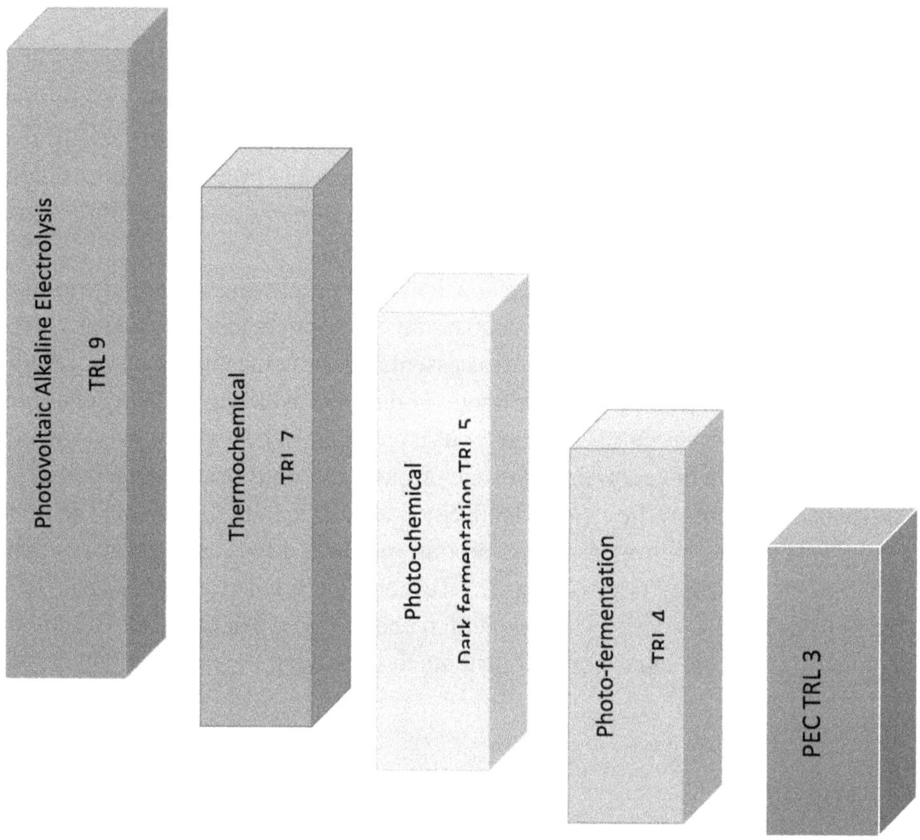


## 3.1. Storage and Transportation

Hydrogen is capable of being stored in substantial quantities, from megawatt-hours (MWh) to terawatt-hours (TWh), using tanks, salt formations, or pressure vessels. However, cyclic conversion efficiency poses a considerable challenge, varying between 20% and 36%, which is significantly lower than the efficiency of hydroelectric storage, which ranges from 70% to 85%. In comparison, compressed air energy storage operates at 65% to 80% efficiency, while batteries achieve efficiencies of 85% to 95%. Consequently, the cost of hydrogen for consumers hinges on a trade-off between storage duration and the losses incurred during the conversion process(Glenk & Reichelstein, 2022).

Hydrogen is available for storage in three forms: gas, liquid, or solid, with systems designed for both mobile and stationary applications. Typically, hydrogen gas is compressed to about 70 MPa in stainless steel tanks, which are heavy for mobile use. To reduce this weight, reinforced carbon composites, although more expensive, may be a viable alternative due to their ability to withstand high pressures and shocks(Khan et al., 2024).

Liquefied hydrogen presents an appealing alternative, with a volumetric density nearly 1.8 times greater than that of pressurized hydrogen gas. However, its storage necessitates cryogenic temperatures of 20 K, requiring specialized containers with reinforced insulation to prevent leaks. Additionally, this method incurs an energy loss of approximately 30% to maintain such low temperatures, which diminishes its overall appeal.

In this context, liquid organic hydrogen carriers (LOHCs) are emerging as a promising option, providing a safe and cost-effective method for storing hydrogen at room temperature for extended periods while enabling transport similar to that of conventional fuels. Storage using LOHCs depends on reversible chemical reactions of hydrogenation and dehydrogenation, necessitating the use of a catalyst. However, these solutions are still in their early stages, and further research is required to evaluate their actual market potential.

Hydrogen may be stored in solid form through two processes: physisorption and chemisorption. In physisorption, molecular hydrogen adheres to various carbon-based substances, but this method has a very limited capacity for storage under normal conditions. Conversely, chemisorption involves a reaction between atomic hydrogen and certain metals to create hydrides, which can release hydrogen as needed via hydrogenation and dehydrogenation processes. Metal hydrides exhibit a significantly higher density for hydrogen storage, achieving approximately 6.5 atoms per $cm^3$ for magnesium hydride, while other forms contain roughly 4.5 atoms and 0.99 atoms, respectively, in the same amount of space.

Hydrides can be categorized into two main groups: those operating at lower temperatures (20–100 °C), which include certain elements and compounds, and those functioning at elevated temperatures (300–400 °C), primarily composed of magnesium-based alloys. The high-temperature variants provide significantly greater hydrogen storage capacity, reaching up to 8%, while the lower-temperature options only achieve capacities of 1–3%(Rasul et al., 2022). The primary challenges associated with these systems relate to thermodynamic and kinetic limitations. Recently, an innovative method has been developed that integrates a hydride system with a latent heat storage device, aiming to enhance power density and thermal management progressively. This approach utilizes the benefits of high storage capacity while maintaining the heat of fusion at a nearly constant temperature, making the integration of hydrides and materials that undergo phase changes particularly promising. Furthermore, it is crucial to conduct in-depth research on the properties of paraffin-like substances employed for phase transitions(Zhan et al., 2023).

Metal oxides represent a promising solution for hydrogen storage, alternative to metal hydrides, and contribute to the idea of a circular economy. This approach relies on reversible chemical interactions, allowing hydrogen to be absorbed and released. More specifically, iron-containing metal oxides can react with hydrogen to produce metal hydrides, which are able to efficiently store hydrogen within their structure.

This mechanism operates at relatively low temperatures and pressures, making it an efficient storage method. Once hydrogen is stored, it can be released by adjusting these parameters, providing a clean energy source suitable for various applications. Therefore, the ability of metal oxide materials to transition from a storage state to a hydrogen release state is in line with the sustainability principles of the circular economy, reducing the need for single-use storage options.

## 3.2. Areas of Industrial Application

Globally, the transport sector is responsible for about a quarter of global greenhouse gas (GHG) emissions(Soni et al., 2025). Among the different categories in this sector, land transport plays a significant role.

### 3.2.1. Land Transport Systems

Land transportation has traditionally relied on internal combustion engines (ICEs) as a foundational technology. In contrast, hydrogen-powered engines have the potential to achieve peak efficiencies of around 50%, while providing higher power density than engines running on conventional fuels. However, further research into hydrogen storage and injection systems, as well as the costs associated with these technologies, is essential(Sadeq et al., 2024).

According to a report from the International Energy Agency (IEA), sales of hydrogen-powered vehicles (FCEVs) are projected to reach 150 million units by 2050, representing 25% of the road transportation market [40]. In light of this trend, several major original equipment manufacturers (OEMs) are adjusting their future plans to prioritize both battery electric vehicles (BEVs) and FCEVs. The primary difference between these two vehicle types lies in the amount of energy they can store for each kilogram of weight.

Hydrogen fuel cell vehicles (FCEVs) stand out for their increased range and faster charging times compared to battery electric vehicles (BEVs), with the most efficient models offering a range of 400–550 km per charge. Due to their high energy density, hydrogen fuel cell vehicles (FCEVs) are also being considered for heavy-duty applications, including forklifts and construction vehicles. Furthermore, the use of green hydrogen could reduce greenhouse gas emissions by up to 50% compared to current levels of internal combustion engines. However, the adoption of hydrogen fuel cell vehicles (FCEVs) is hampered by three main challenges: large upfront investments, insufficient infrastructure for the hydrogen supply chain, and an unstable policy framework(Hassan et al., 2023).

By 2022 (Figure 2), approximately 800 hydrogen refueling stations (HRS) were publicly available worldwide. The high cost of their installation is a major obstacle to their rapid development. Hydrogen is being considered not only for passenger vehicles in developed countries such as the United States, Japan, China, and Germany, but also for applications in light vans, heavy trucks, buses, as well as in the marine and rail sectors(Brenna et al., 2020).

The transition to hydrogen energy in various modes of transportation is not economically viable in the short term, but is expected to become more cost-effective as its adoption increases, leading to market expansion.

*Figure 2. Description of the existing global distribution of hydrogen refueling stations (HRS).*

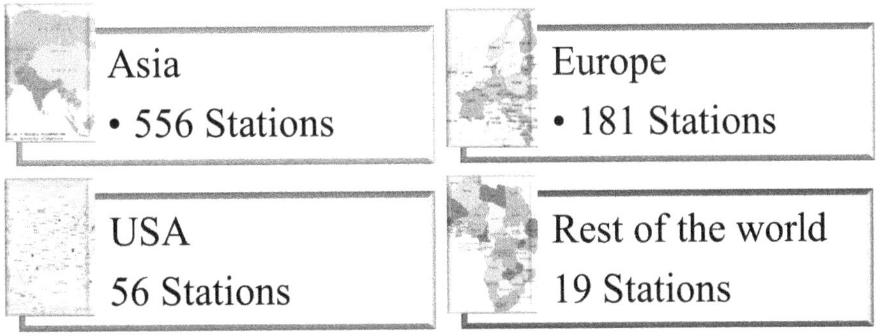


### 3.2.2. Aviation and Alternative Energies

The fossil fuel-dependent aviation sector is responsible for approximately 918 million tonnes of CO2 emissions per year, representing between 10 and 15% of total transport emissions. Decarbonizing the aviation sector is challenging due to stringent safety standards and weight and space constraints, which limit the integration of sustainable alternative fuels. With a net calorific value of 120 MJ/kg and a conversion efficiency of between 60 and 70%, hydrogen could power electric aircraft with an energy density of around 20 kWh/kg. This feature makes it a more cost-effective option than batteries for long-haul routes. However, it is important to highlight that to generate equivalent thrust, the amount of hydrogen required is four times that of kerosene, which is a major challenge(Al-Mohannadi et al., 2024).

In contrast to fossil fuels, which can be housed directly in aircraft wings, hydrogen storage requires specialized tanks with insulation to maintain a temperature of 20 K. Several fuselage configurations have been suggested to facilitate the integration of hydrogen into aircraft. However, implementation of these designs requires detailed research and testing(Adler & Martins, 2023).

Hydrogen is being explored as an energy option for aviation; however, its application is restricted by stringent safety regulations. This shift will necessitate modifications in aircraft design, refueling processes, and airport facilities, along with research, investment, and suitable regulatory frameworks. In addition to propulsion, other operational systems, such as the Auxiliary Power Unit (APU), cabin pressure management, air conditioning, instrumentation, wing de-icing, and landing gear operation, also demand energy. Studies have shown that employing fuel cells for the APU could lead to a reduction in fuel consumption of up to 75% under stationary conditions.

### 3.2.3. Navigability and Sustainable Development in Maritime Transport

Shipping is essential to global trade, but growing demand requires larger ships, increasing greenhouse gas (GHG) emissions, which could reach 17% of global emissions by 2050. The (IMO) is committed to reducing ship emissions by 50% by 2050 compared to 2008 levels(Lagdami & Baig, 2024). To achieve this goal, it is focusing on improving the energy efficiency of new ship engines. In addition, hydrogen has been identified as a promising alternative fuel that can optimize operational efficiency while reducing emissions.

Hydrogen on board ships represents a promising solution, especially when renewable energies, such as solar and ocean thermal energy, are used to produce green hydrogen from salt water or wastewater.

Hydrogen may additionally be employed in liquefied natural gas (LNG) tankers, allowing for a complementary thermodynamic process. This approach improves the energy efficiency of ships while mitigating negative environmental impacts, making shipping more environmentally friendly. Through these innovations, the maritime sector can contribute to broader greenhouse gas emission reduction goals.

Hydrogen storage on board ships is becoming increasingly important in the quest for sustainable energy solutions for the maritime sector. Unlike land-based storage systems, where volume and weight constraints can be critical, ships offer greater flexibility in this regard. The amount of hydrogen to be stored depends mainly on the size of the ship, allowing storage systems to be adapted to the specific needs of each vessel.

To optimize the use of hydrogen on board, several approaches are being considered. The integration of hydrogen fuel cells is an effective solution to convert hydrogen into electricity, thus powering the engines and equipment on board. At the same time, the exploitation of renewable energies, such as solar and wind energy, makes it possible to produce green hydrogen from salt water or waste water, thus creating a sustainable energy source.

Recovering waste heat from engines also offers an opportunity to improve energy efficiency. This heat, which would otherwise be lost, can be reused to generate hydrogen, thus contributing to a more virtuous energy cycle. However, the transition to hydrogen in the maritime sector faces challenges, particularly due to the low rate of production of new ships each year and the costs associated with these technologies. As a result, the widespread adoption of hydrogen in maritime transport is expected to be gradual, with significant progress expected by 2030(Alavi-Borazjani et al., 2025). This transition period will be crucial to develop the necessary infrastructure and improve the economic viability of hydrogen-based solutions.

### 3.2.4. Residential Applications

However, further research is needed to analyse the risks and safety associated with higher hydrogen concentrations. Furthermore, hydrogen could play a key role in distributed energy systems, enabling households to become energy self-sufficient, especially in remote areas with renewable energy resources. Residential and commercial buildings require a significant amount of energy to maintain indoor comfort. Currently, the natural gas pipeline network extends about 2.91 million kilometers, while the hydrogen pipeline network covers only 16,000 kilometers. To efficiently transport hydrogen to consumers, one approach is to blend it with natural gas and leverage existing infrastructure. This method could help reduce greenhouse gas emissions in the residential sector. Indeed, it is possible to incorporate up to 10% hydrogen into natural gas without requiring changes to existing infrastructure (see Fig. 3).

However, further research is essential to understand the implications of higher hydrogen concentrations. These studies should focus on safety, potential combustion risks and effects on existing equipment. Careful management of these aspects will ensure safe integration of hydrogen into the grid. In addition, hydrogen plays a key role in the development of distributed energy systems, where each household can become an energy producer. This is particularly advantageous in remote areas where access to conventional energy sources is limited. Thanks to hydrogen, these households can use available renewable energies, such as wind or solar, to produce their own energy, thus increasing their energy autonomy. This approach not only helps reduce the carbon footprint, but also strengthens the energy resilience of communities.

*Figure 3. A sustainable hydrogen solutions model for home energy and mobility requirements*

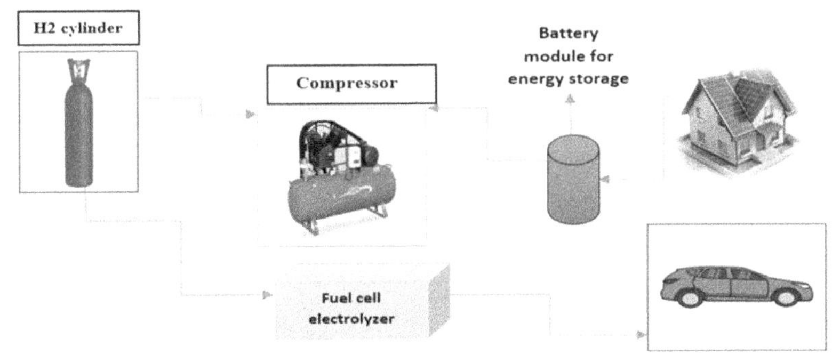


## 4. GLOBAL STRATEGIES AND INSIGHTS ON CIRCULAR ECONOMY PRACTICES

A linear economic model, defined by the extraction, utilization, and disposal of resources, contributes to the depletion of resources and the degradation of ecosystems. Conversely, a circular economy seeks to repurpose materials to reduce waste and emissions. At present, merely 8.6% of the global economy operates within a circular framework. In Europe, approximately 10 billion cubic meters of hydrogen are wasted annually in industrial wastewater. Generating biohydrogen from waste through microbial processes is an essential approach to fostering a sustainable circular economy. This strategy could support the achievement of the UN Sustainable Development Goals (SDGs), although obstacles persist related to waste management, storage, transportation, and safety concerns(Aravindaraj & Chinna, 2022).

In the long term, hydrogen production by electrolysis is expected to surpass other techniques, due to its increasing accessibility and the decreasing costs of renewable energy sources. Green hydrogen will become a key element for energy storage, heat generation, electricity, as well as for transportation systems in a hydrogen-based economy.

## 5. CONCLUSION

The future of hydrogen as a power source in addressing global warming and achieving net-zero emissions by 2050 looks promising. A holistic approach to the hydrogen value chain is essential, as current production methods need to become more environmentally sustainable. While steam methane reforming remains cost-effective, advancements in catalyst materials and renewable energy technologies can enhance green hydrogen production. Seasonal variations in renewable energy limit energy supply, highlighting the need for improved hydrogen storage solutions. High costs of hydrogen refueling stations necessitate supportive government policies to encourage adoption. Additionally, developing infrastructure for transporting green hydrogen is crucial. Scientific progress is required to utilize alternative water sources for hydrogen production, while critical industries begin their transition to hydrogen, especially in aviation and shipping. To maximize hydrogen's potential as a clean energy source, combining technological advancements with favorable policies is vital. Government subsidies have already helped reduce costs, and stable energy policies are needed to support innovation and investment in the hydrogen sector.

## REFERENCES

Adler, E. J., & Martins, J. R. (2023). Hydrogen-powered aircraft: Fundamental concepts, key technologies, and environmental impacts. *Progress in Aerospace Sciences*, *141*, 100922. DOI: 10.1016/j.paerosci.2023.100922

Al-Jadabi, N., Laaouan, M., El Hajjaji, S., Mabrouki, J., Benbouzid, M., & Dhiba, D. (2023). The dual performance of Moringa oleifera seeds as eco-friendly natural coagulant and as an antimicrobial for wastewater treatment: A review. *Sustainability (Basel)*, *15*(5), 4280. DOI: 10.3390/su15054280

Al-Mohannadi, A. A., Ertogral, K., & Erkoc, M. (2024). Alternative Fuels in Sustainable Logistics—Applications, Challenges, and Solutions. *Sustainability (Basel)*, *16*(19), 8484. DOI: 10.3390/su16198484

Alavi-Borazjani, S. A., Adeel, S., & Chkoniya, V. (2025). Hydrogen as a Sustainable Fuel: Transforming Maritime Logistics. *Energies*, *18*(5), 1231. DOI: 10.3390/en18051231

Anika, O. C., Nnabuife, S. G., Bello, A., Okoroafor, E. R., Kuang, B., & Villa, R. (2022). Prospects of low and zero-carbon renewable fuels in 1.5-degree net zero emission actualisation by 2050: A critical review. *Carbon Capture Science & Technology*, *5*, 100072. DOI: 10.1016/j.ccst.2022.100072

Aravindaraj, K., & Chinna, P. R. (2022). A systematic literature review of integration of industry 4.0 and warehouse management to achieve Sustainable Development Goals (SDGs). *Cleaner Logistics and Supply Chain*, *5*, 100072. DOI: 10.1016/j.clscn.2022.100072

Axelsson, L., Edvall, M., Harvey, S., Roshan Kumar, T., Jannasch, A.-K., Westin, J., Starfelt, F., Pettersson, L., Pira Edman, A., & Hellesöy, F. (2024). *Future hydrogen supply in Stenungsund: Pre-study of a SOEC pilot plant and analysis of large-scale integration of SOEC and ammonia cracking plants.*

Azadnia, A. H., McDaid, C., Andwari, A. M., & Hosseini, S. E. (2023). Green hydrogen supply chain risk analysis: A european hard-to-abate sectors perspective. *Renewable & Sustainable Energy Reviews*, *182*, 113371. DOI: 10.1016/j.rser.2023.113371

Azdem, D., Fatni, A., Mabrouki, J., & El Hajjaji, S. (2025). The Desalination Strategy and Economic Integration of Morocco: Water Scarcity, Prospects, and Challenges. *Circular Economy Applications in Energy Policy*, 257–268. Azdem, D., Mabrouki, J., & El Hajjaji, S. (1 C.E., January 1). *The Use of Artificial Intelligence in a Desalination Plant Powered by Renewable Energy* [Chapter]. Https://Services.Igi-Global.Com/Resolvedoi/Resolve.Aspx?Doi=10.4018/979-8-3373-1220-0.Ch004Global, I. G. I. https://www.igi-global.com/chapter/the-use-of-artificial-intelligence-in-a-desalination-plant-powered-by-renewable-energy/373103

Azdem, D., Mabrouki, J., & El Hajjaji, S. (2024). Methods and Technologies for the Treatment and Disposal of Brine in Seawater Desalination Plants. In Mabrouki, J., & Azrour, M. (Eds.), (pp. 79–87). Advances in Environmental Engineering and Green Technologies. IGI Global., DOI: 10.4018/979-8-3693-3807-0.ch007

Azdem, D., Mabrouki, J., Fahdi, G., & Hajjaji, S. E. (2025). ENVIRONMENTAL SUSTAINABILITY OF DESALINATION: A CRITICAL REVIEW OF BRINE TREATMENT CHALLENGES AND MITIGATION STRATEGIES. *Water Conservation and Management, 9*(1), 55–60. *Scopus.* Advance online publication. DOI: 10.26480/wcm.01.2025.55.60

Azdem, D., Mabrouki, J., Moufti, A., El hajjaji, S., & Fatni, A. (2024). Assessment of heavy metal contamination in seawater in Agadir coastline, Morocco. *Desalination and Water Treatment, 317*, 100129. DOI: 10.1016/j.dwt.2024.100129

Azdem, D., Mabrouki, J., Moufti, A., Hajjaji, S. E., & Fatni, A. (2024). Present Situation and Environmental Impact of Seawater Desalination Plants in Morocco. In Mabrouki, J., & Mourade, A. (Eds.), *Technical and Technological Solutions Towards a Sustainable Society and Circular Economy* (pp. 581–592). Springer Nature Switzerland., DOI: 10.1007/978-3-031-56292-1_46

Bencheikh, I., Azoulay, K., Mabrouki, J., El Hajjaji, S., Dahchour, A., Moufti, A., & Dhiba, D. (2020). The adsorptive removal of MB using chemically treated artichoke leaves: Parametric, kinetic, isotherm and thermodynamic study. *Scientific African, 9*, e00509. DOI: 10.1016/j.sciaf.2020.e00509

Bencheikh, I., Azoulay, K., Mabrouki, J., El Hajjaji, S., Moufti, A., & Labjar, N. (2021). The use and the performance of chemically treated artichoke leaves for textile industrial effluents treatment. *Chemical Data Collections, 31*, 100597. DOI: 10.1016/j.cdc.2020.100597

Benchenina, Y., Zemmit, A., Bouzaki, M. M., Loukriz, A., Elsayed, S. K., Alzaed, A., Ali, G., & Ghoneim, S. S. (2025). Advancing green hydrogen production in Algeria with opportunities and challenges for future directions. *Scientific Reports, 15*(1), 5559. DOI: 10.1038/s41598-025-90336-1 PMID: 39953199

Benchrifa, M., Elouardi, M., Fattah, G., Mabrouki, J., & Tadili, R. (2023). Identification, simulation and modeling of the main power losses of a photovoltaic installation and use of the internet of things to minimize system losses. In *Advanced technology for smart environment and energy* (pp. 49–60). Springer. DOI: 10.1007/978-3-031-25662-2_4

Benchrifa, M., & Mabrouki, J. (2022). Simulation, sizing, economic evaluation and environmental impact assessment of a photovoltaic power plant for the electrification of an establishment. *Advances in Building Energy Research, 16*(6), 736–753. DOI: 10.1080/17512549.2022.2096693

Benchrifa, M., Mabrouki, J., Elouardi, M., Azrour, M., & Tadili, R. (2023). Detailed study of dimensioning and simulating a grid-connected PV power station and analysis of its environmental and economic effect, case study. *Modeling Earth Systems and Environment, 9*(1), 53–61. DOI: 10.1007/s40808-022-01457-9

Boddula, R., Lee, Y.-Y., Masimukku, S., Chang-Chien, G.-P., Pothu, R., Srivastava, R. K., Sarangi, P. K., Selvaraj, M., Basumatary, S., & Al-Qahtani, N. (2024). Sustainable hydrogen production: Solar-powered biomass conversion explored through (Photo) electrochemical advancements. *Process Safety and Environmental Protection, 186*, 1149–1168. DOI: 10.1016/j.psep.2024.04.068

Brenna, M., Bucci, V., Falvo, M. C., Foiadelli, F., Ruvio, A., Sulligoi, G., & Vicenzutti, A. (2020). A review on energy efficiency in three transportation sectors: Railways, electrical vehicles and marine. *Energies, 13*(9), 2378. DOI: 10.3390/en13092378

Dauletbay, A. (2024). Transportation of Hydrogen: Hydrogen Usage. In *Hydrogen Technologies-Advances, Insights, and Applications*. IntechOpen.

El Alouani, M., Aouan, B., Rachdi, Y., Alehyen, S., El Herradi, E. H., Saufi, H., Mabrouki, J., & Barka, N. (2024). Porous geopolymers as innovative adsorbents for the removal of organic and inorganic hazardous substances: A mini-review. *International Journal of Environmental Analytical Chemistry*, *104*(16), 4784–4796. DOI: 10.1080/03067319.2022.2115891

El Mouedden, M., Azdem, D., Mabrouki, J., & El Hajjaji, S. (2025). MarocFarm: A digital revolution for agriculture in Morocco. In *Circular Economy Applications in Energy Policy* (pp. 141–155). Scopus. DOI: 10.4018/979-8-3693-6680-6.ch010

Evro, S., Oni, B. A., & Tomomewo, O. S. (2024). Carbon neutrality and hydrogen energy systems. *International Journal of Hydrogen Energy*, *78*, 1449–1467. DOI: 10.1016/j.ijhydene.2024.06.407

Fahdi, G., Azdem, D., Lamchaimech, A., Benrhanem, M., Mabrouki, J., & El Hajjaji, S. (2024). ANALYSIS OF THE IMPACT OF CLIMATE CHANGE ON WATER RESOURCES: CASE OF THE TENSIFT BASIN (MOROCCO). *Water Conservation and Management, 8*(4), 461–465. *Scopus.* Advance online publication. DOI: 10.26480/wcm.04.2024.461.465

Fahdi, G., Mabrouki, J., Lamchaimech, A., Azdem, D., Benrhanem, M., & Hajjaji, S. E. (2024). IMPACTS OF CLIMATE CHANGE ON GROUNDWATER QUALITY AND RECHARGE IN THE TENSIFT BASIN, MOROCCO. *Water Conservation and Management, 8*(4), 466–474. *Scopus.* Advance online publication. DOI: 10.26480/wcm.04.2024.466.474

Fatni, A., Addich, M., Essekri, A., & Baraka, E. L. (2025). Characterization and mitigation strategies for inorganic scaling in reverse osmosis system treating brackish groundwater. *Sustainable Chemistry One World*, *5*, 100035. DOI: 10.1016/j.scowo.2024.100035

Freedman, B., Dorsey, W., Frazier, A., Kambhampati, M., Galiotos, J., & Mukherjee, S. (2024). ~ Global Climate and Greenhouse Gases. *Environmental Sciences (Ruse)*.

Ghizlane, F., Mabrouki, J., Ghrissi, F., & Azrour, M. (2022). Proposal for a high-resolution particulate matter (PM10 and PM2. 5) capture system, comparable with hybrid system-based internet of things: Case of quarries in the western rif, Morocco. *Pollution*, *8*(1), 169–180.

Glenk, G., & Reichelstein, S. (2022). Reversible power-to-gas systems for energy conversion and storage. *Nature Communications*, *13*(1), 2010. DOI: 10.1038/s41467-022-29520-0 PMID: 35440135

Gonuguntla, S., Jaksani, B., Jamma, A., Vennapoosa, C. S., Chatterjee, D., & Pal, U. (2024). Design principle of anti-corrosive photocatalyst for large-scale hydrogen production. *Wiley Interdisciplinary Reviews. Energy and Environment*, *13*(4), e530. DOI: 10.1002/wene.530

Gunathilake, C., Soliman, I., Panthi, D., Tandler, P., Fatani, O., Ghulamullah, N. A., Marasinghe, D., Farhath, M., Madhujith, T., Conrad, K., Du, Y., & Jaroniec, M. (2024). A comprehensive review on hydrogen production, storage, and applications. *Chemical Society Reviews*, *53*(22), 10900–10969. DOI: 10.1039/D3CS00731F PMID: 39421882

Gür, T. M. (2024). Giga-ton and tera-watt scale challenges at the energy-climate crossroads: A global perspective. *Energy*, *290*, 129971. DOI: 10.1016/j.energy.2023.129971

Hassan, Q., Azzawi, I. D., Sameen, A. Z., & Salman, H. M. (2023). Hydrogen fuel cell vehicles: Opportunities and challenges. *Sustainability (Basel)*, *15*(15), 11501. DOI: 10.3390/su151511501

Khalil, Y. F. (2025). *A Comparative Science-Based Viability Assessment Among Current and Emerging Hydrogen Production Technologies.*

Khan, F., Hossain, N., Mim, J. J., Rahman, S. M., Iqbal, M. J., Billah, M., & Chowdhury, M. A. (2024). Advances of composite materials in automobile applications–A review. *Journal of Engineering Research*. Advance online publication. DOI: 10.1016/j.jer.2024.02.017

Kim, J., Seo, J. H., Lee, J. K., Oh, M. H., & Jang, H. W. (2025). Challenges and strategies in catalysts design towards efficient and durable alkaline seawater electrolysis for green hydrogen production. *Energy Materials*, *5*(7), N/A, N/A. DOI: 10.20517/energymater.2024.220

Lagdami, K., & Baig, M. Z. (2024). The IMO and the future of shipping. In *The Elgar Companion to the Law and Practice of the International Maritime Organization* (pp. 244–263). Edward Elgar Publishing. DOI: 10.4337/9781802206883.00023

Laimon, M., & Yusaf, T. (2024). Towards energy freedom: Exploring sustainable solutions for energy independence and self-sufficiency using integrated renewable energy-driven hydrogen system. *Renewable Energy*, *222*, 119948. DOI: 10.1016/j.renene.2024.119948

Li, Y., Lin, R., O'Shea, R., Thaore, V., Wall, D., & Murphy, J. D. (2024). A perspective on three sustainable hydrogen production technologies with a focus on technology readiness level, cost of production and life cycle environmental impacts. *Heliyon*, *10*(5), e26637. DOI: 10.1016/j.heliyon.2024.e26637 PMID: 38444498

Loukili, H., Mabrouki, J., Anouzla, A., Kouzi, Y., Younssi, S. A., Digua, K., & Abrouki, Y. (2021). Pretreated Moroccan natural clays: Application to the wastewater treatment of textile industry. *Desalination and Water Treatment*, *240*, 124–136. DOI: 10.5004/dwt.2021.27644

Mabrouki, J., Azrour, M., Farhaoui, Y., & El Hajjaji, S. (2020). Intelligent system for monitoring and detecting water quality. *Big Data and Networks Technologies*, *3*, 172–182. DOI: 10.1007/978-3-030-23672-4_13

Mabrouki, J., Benbouzid, M., Dhiba, D., & El Hajjaji, S. (2022). Simulation of wastewater treatment processes with Bioreactor Membrane Reactor (MBR) treatment versus conventional the adsorbent layer-based filtration system (LAFS). *International Journal of Environmental Analytical Chemistry*, *102*(19), 7458–7468. DOI: 10.1080/03067319.2020.1828394

Mabrouki, J., Benchrifa, M., Ennouhi, M., Azoulay, K., Bencheikh, I., Rachiq, T., El-Moustaqim, K., Al-Jadabi, N., Azrour, M., & Fatima-zahra, A. (2022). Geographic information system for the study of water resources in Chaâba El Hamra, Mohammedia (Morocco). *The International Conference on Artificial Intelligence and Smart Environment*, 469–474.

Mabrouki, J., Moufti, A., Bencheikh, I., Azoulay, K., El Hamdouni, Y., & El Hajjaji, S. (2020). Optimization of the coagulant flocculation process for treatment of leachate of the controlled discharge of the city Mohammedia (Morocco). *Advanced Intelligent Systems for Sustainable Development (AI2SD'2019) Volume 7-Advanced Intelligent Systems for Sustainable Development Applied in Energy and Electrical Engineering*, 200–212.

Marouani, I., Guesmi, T., Alshammari, B. M., Alqunun, K., Alzamil, A., Alturki, M., & Hadj Abdallah, H. (2023). Integration of renewable-energy-based green hydrogen into the energy future. *Processes (Basel, Switzerland)*, *11*(9), 2685. DOI: 10.3390/pr11092685

Massarweh, O., Al-khuzaei, M., Al-Shafi, M., Bicer, Y., & Abushaikha, A. S. (2023). Blue hydrogen production from natural gas reservoirs: A review of application and feasibility. *Journal of CO2 Utilization, 70*, 102438.

Meißnitzer, M. (2023). *Feasibility of semi-artificial photosynthesis.*

Mgalaa, S., Mabrouki, J., Elouardi, M., El Azzouzi, L., Moufti, A., El Hajjaji, S. E. L., Azzouzi, M., & El Belghiti, M. A. (2022). Study and evaluation of the degradation of procion blue dye by the ozonation method: Parametric and isothermal study. *Nanotechnology for Environmental Engineering*, *7*(3), 691–697. DOI: 10.1007/s41204-022-00262-0

Monteiro, E., & Brito, P. S. (2023). Hydrogen supply chain: Current status and prospects. *Energy Storage*, *5*(7), e466. DOI: 10.1002/est2.466

Nnabuife, S. G., Hamzat, A. K., Whidborne, J., Kuang, B., & Jenkins, K. W. (2024). Integration of renewable energy sources in tandem with electrolysis: A technology review for green hydrogen production. *International Journal of Hydrogen Energy*.

Nnabuife, S. G., Oko, E., Kuang, B., Bello, A., Onwualu, A. P., Oyagha, S., & Whidborne, J. (2023). The prospects of hydrogen in achieving net zero emissions by 2050: A critical review. *Sustainable Chemistry for Climate Action*, *2*, 100024. DOI: 10.1016/j.scca.2023.100024

Okere, C. J., Sheng, J. J., & Ikpeka, P. M. (2024). Maximizing hydrogen yield: Pioneering gas injection for enhanced light oil reservoir utilization. *SPE Western Regional Meeting*, D021S011R006.

Pangestu, M. R. G., Malaibari, Z., Muhammad, A., Al-Rowaili, F. N., & Zahid, U. (2024). Comprehensive review on methane pyrolysis for sustainable hydrogen production. *Energy & Fuels*, *38*(15), 13514–13538. DOI: 10.1021/acs.energyfuels.4c01551

Rachiq, T., Abrouki, Y., Mabrouki, J., Samghouli, N., Fersi, C., Rahal, S., & El Hajjaji, S. (2021). Evaluation of the efficiency of different materials to remove specific pollutants from landfill leachate. *Desalination and Water Treatment*, *238*, 240–250. DOI: 10.5004/dwt.2021.27779

Rasul, M. G., Hazrat, M. A., Sattar, M. A., Jahirul, M. I., & Shearer, M. J. (2022). The future of hydrogen: Challenges on production, storage and applications. *Energy Conversion and Management*, *272*, 116326. DOI: 10.1016/j.enconman.2022.116326

Sadeq, A. M., Homod, R. Z., Hussein, A. K., Togun, H., Mahmoodi, A., Isleem, H. F., Patil, A. R., & Moghaddam, A. H. (2024). Hydrogen energy systems: Technologies, trends, and future prospects. *The Science of the Total Environment*, *939*, 173622. DOI: 10.1016/j.scitotenv.2024.173622 PMID: 38821273

Soni, R., Dvivedi, A., & Kumar, P. (2025). Carbon neutrality in transportation: In the context of renewable sources. *International Journal of Sustainable Transportation*, *19*(1), 1–15. DOI: 10.1080/15568318.2024.2447999

Squadrito, G., Maggio, G., & Nicita, A. (2023). The green hydrogen revolution. *Renewable Energy*, *216*, 119041. DOI: 10.1016/j.renene.2023.119041

Sun, Y., Qin, Z., Tang, Y., Liao, C., Liu, Y., & Ma, X. (2022). Techno-environmental-economic assessment on municipal solid waste to methanol coupling with/without solid oxygen electrolysis cell unit. *Process Safety and Environmental Protection*, *161*, 611–628. DOI: 10.1016/j.psep.2022.03.060

Swaminathan, P., Ghosh, A., Sunantha, G., Sivagami, K., Mohanakrishna, G., Aishwarya, S., Shah, S., Sethumadhavan, A., Ranjan, P., & Prajapat, R. (2024). A Comprehensive Review of Microbial Electrolysis Cells Integrated with Wastewater Treatment for Hydrogen Generation. *Process Safety and Environmental Protection*, *190*, 458–474. DOI: 10.1016/j.psep.2024.08.032

Tao, Z., Zhao, Y., Wang, Y., & Zhang, G. (2024). Recent advances in carbon nanotube technology: Bridging the gap from fundamental science to wide applications. *C*, *10*(3), 69.

Xing, S., Zhao, C., Zou, J., Zaman, S., Yu, Y., Gong, H., Wang, Y., Chen, M., Wang, M., Lin, M., & Wang, H. (2022). Recent advances in heat and water management of forced-convection open-cathode proton exchange membrane fuel cells. *Renewable & Sustainable Energy Reviews*, *165*, 112558. DOI: 10.1016/j.rser.2022.112558

Zhan, H., Mahyuddin, N., Sulaiman, R., & Khayatian, F. (2023). Phase change material (PCM) integrations into buildings in hot climates with simulation access for energy performance and thermal comfort: A review. *Construction & Building Materials*, *397*, 132312. DOI: 10.1016/j.conbuildmat.2023.132312

Zolghadri, S., Kiani, M. R., & Rahimpour, M. R. (2024). Enhanced hydrogen production in steam methane reforming: Comparative analysis of industrial catalysts and process optimization. *Journal of the Energy Institute*, *113*, 101541. DOI: 10.1016/j.joei.2024.101541

# Chapter 18
# Opportunities and Challenges of Integrating Methanization and Biogas Technologies Into the Green Hydrogen Economy

**Soukaina Agoud**
https://orcid.org/0009-0009-8848-4719
*Mohammed V University in Rabat, Morocco*

**Driss Azdem**
*Mohammed V University in Rabat, Morocco*

**Hajar Khabouche**
https://orcid.org/0009-0003-4466-5784
*Mohammed V University in Rabat, Morocco*

**Khadija Elmoustaqim**
https://orcid.org/0009-0007-9193-6638
*Ibn Tofail University, Morocco*

**Jamal Mabrouki**
https://orcid.org/0000-0002-3841-7755
*Mohammed V University in Rabat, Morocco*

**Miloudia Slaoui**
*Mohammed V University in Rabat, Morocco*

## ABSTRACT

*The transition to a green hydrogen economy offers a path toward sustainable energy but faces challenges in production efficiency and cost. This chapter explores integrating methanization and biogas technologies within the green hydrogen landscape, focusing on synergies and obstacles. Methanization, which converts organic waste into methane-rich biogas, could supplement hydrogen production via methods like steam methane reforming and methane pyrolysis. The chapter examines pathways for using biogas-derived methane in hydrogen production, highlighting opportunities to enhance yield and reduce reliance on fossil fuels. It also discusses economic implications, policy frameworks, and offers strategic recommendations for advancing the green hydrogen economy and supporting the global energy transition.*

## 1. INTRODUCTION

The green hydrogen economy is increasingly recognized as a key solution to reducing carbon emissions and fostering global energy sustainability. Green hydrogen, produced using renewable energy sources, is essential for decarbonizing sectors such as transportation, industry, and power generation

DOI: 10.4018/979-8-3693-8980-5.ch018

(Emetere et al., 2024). However, its large-scale implementation faces significant challenges, particularly in terms of production costs, efficiency, and scalability. Emerging research suggests that integrating methanization and biogas technologies into hydrogen production could enhance the feasibility of green hydrogen, providing a sustainable, low-carbon solution to these challenges (Basso et al., 2023). Methanization, the anaerobic digestion of organic waste to produce methane-rich biogas, offers a circular economy approach to hydrogen production. This biogas can be upgraded into biomethane and converted into hydrogen through various methods, such as steam methane reforming (SMR) or methane pyrolysis (Ongis et al., 2023). These processes offer an opportunity to utilize organic waste to supplement hydrogen production, reducing dependency on fossil fuels and lowering carbon emissions. While the integration of methanization and biogas technologies into the hydrogen economy presents promising synergies, it also poses several technical, economic, and policy-related challenges that must be addressed to unlock its full potential (Giocoli et al., 2023). Green hydrogen, produced from water electrolysis powered by renewable energy sources, is considered one of the cleanest forms of hydrogen. It has the potential to decarbonize industries that are difficult to electrify, such as steel production and aviation (Nemmour et al., 2023). However, as of 2024, the cost of green hydrogen production remains significantly higher than hydrogen produced using fossil fuels, primarily due to the high costs associated with renewable energy and electrolysis technology. For instance, the cost of producing green hydrogen can be 2-3 times that of grey hydrogen, which is derived from natural gas (Elhaus et al., 2024). As a result, there is a pressing need to explore alternative methods, such as biogas integration, to make green hydrogen more cost-competitive and scalable. Methanization involves the biological decomposition of organic matter, such as agricultural waste, food waste, and sewage sludge, into biogas (Bertasini et al., 2023). This biogas, composed primarily of methane ($CH_4$) and carbon dioxide ($CO_2$), can be upgraded to biomethane, which is chemically identical to natural gas. Biomethane can then be used in SMR or methane pyrolysis to produce hydrogen (Murphy et al., 2024). SMR is a widely used process that reacts methane with steam to produce hydrogen and $CO_2$. Methane pyrolysis, on the other hand, generates hydrogen and solid carbon, providing a carbon-neutral pathway for hydrogen production. These methods offer a sustainable way to generate hydrogen, especially when the methane is derived from biogenic sources. Despite the technical feasibility of producing hydrogen from biogas, several challenges persist (Lanni et al., 2023). The efficiency of methane conversion technologies, the need for carbon capture and storage (CCS) to mitigate $CO_2$ emissions in SMR, and the economic viability of biogas-to-hydrogen systems remain significant obstacles. Moreover, the development of robust policy frameworks is crucial to incentivizing investments in biogas technologies and integrating them into the hydrogen economy (Ahmad et al., 2024). Techno-economic assessments highlight that while the initial costs of biogas-to-hydrogen technologies are high, they present long-term cost savings through the reduction of waste disposal fees, production of biofertilizers, and the creation of additional revenue streams. Furthermore, supportive policies, such as renewable energy credits and subsidies for biogas infrastructure, can make these technologies more attractive to investors. As governments worldwide, including the European Union and the United States, continue to push for renewable energy adoption, the integration of methanization and biogas technologies is expected to gain traction in the coming years.

## 2. MATERIALS AND METHODS

This section outlines the methodologies and materials used to investigate the integration of methanization and biogas technologies into the green hydrogen economy. The study focused on evaluating technical, economic, and policy aspects through a combination of experimental analysis, literature review, and techno-economic modeling. The approach included assessing biogas production through methanization, hydrogen production via steam methane reforming (SMR) and methane pyrolysis, and analyzing the associated costs, environmental impacts, and regulatory frameworks

### 2.1 Methanisation Process

The first step involved selecting suitable organic waste materials for the methanization process. Various feedstocks, including agricultural residues, food waste, and municipal solid waste, were chosen based on their methane production potential, carbon to nitrogen (C/N) ratio, and availability (Agoud et al., 2024). The key materials used included agricultural waste, food waste and sewage sludge from wastewater treatment plants (Azoulay et al., 2024). The feedstocks were pretreated to optimize their methane yield, following the established procedures for mechanical shredding and anaerobic digestion. The pretreatment phase aimed to enhance the hydrolysis rate of complex organic matter, facilitating faster microbial digestion (Jamal, 2022). The feedstocks were introduced into a lab scale anaerobic digester operating under mesophilic conditions (35-40°C). The digester's parameters, such as hydraulic retention time (HRT) and organic loading rate (OLR), were adjusted based on feedstock characteristics (El-Moustaqim et al., 2024). Gas production was continuously monitored using gas flow meters, while the methane content was measured using gas chromatography. The biogas produced was composed mainly of methane ($CH_4$) and carbon dioxide ($CO_2$) (Mabrouki et al., 2021).

*Figure 1. Biogas to hydrogen production via steam methane reforming (SMR) process.*

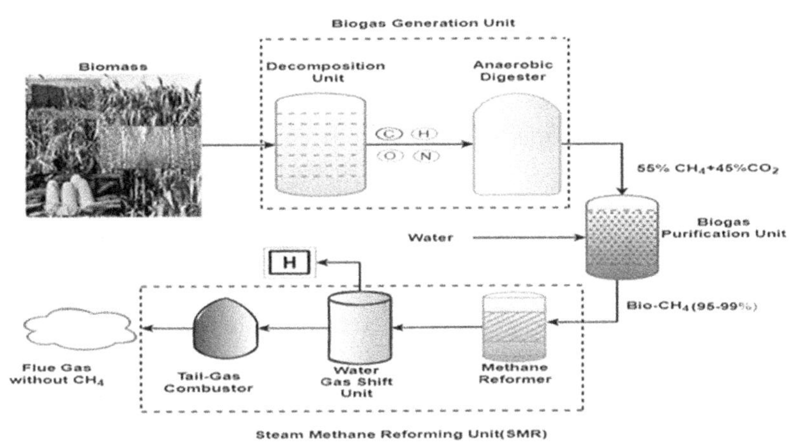


The figure 1 illustrates the process of converting biomass into hydrogen through anaerobic digestion and steam methane reforming (SMR). Biomass undergoes decomposition, generating biogas composed of 55% methane and 45% carbon dioxide. This biogas is purified to nearly pure methane, which then

enters the SMR system. In this system, methane is reformed with steam to produce hydrogen and carbon dioxide. The process is enhanced by a water-gas shift reaction to maximize hydrogen yield. Any unconverted gases are combusted, minimizing emissions. This approach represents a sustainable pathway for hydrogen production from renewable biomass.

## 2.2 Hydrogen Production Pathways

Two primary pathways for converting biogas into hydrogen were examined: steam methane reforming and methane pyrolysis. In the SMR process, the biomethane produced during methanization was reformed with steam at high temperatures (700-1000°C) to produce hydrogen and carbon dioxide (Maniscalco et al., 2024). The process occurred in a tubular reactor system, equipped with a nickel-based catalyst to enhance the reforming reaction. A mass balance was performed to quantify the hydrogen yield and $CO_2$ emissions. Additionally, carbon capture and storage (CCS) technologies were considered to reduce the overall carbon footprint of this pathway (Onwuemezie et al., 2024). Methane pyrolysis was used as an alternative pathway to SMR, where methane was decomposed at elevated temperatures (1000°C) in the absence of oxygen to produce hydrogen and solid carbon (Mokheimer et al., 2024). This process was performed in a pyrolysis reactor, and the solid carbon byproduct was collected and weighed. Hydrogen output was measured and compared to the SMR pathway to determine the efficiency and environmental benefits (Henriksen et al., 2024).

## 2.3 Techno Economic Modeling

To evaluate the economic viability of integrating methanization and biogas technologies into the green hydrogen economy, a techno-economic model was developed. The model incorporated capital and operational expenditures (CAPEX and OPEX), methane conversion efficiencies, hydrogen production costs, and potential revenue from byproducts such as biofertilizers and carbon (Kumar & Kumar, 2024; Gholizadeh et al., 2024). Sensitivity analyses were conducted to assess the impact of various factors, including feedstock cost, hydrogen market prices, and government subsidies, on the overall profitability of the system (Ma et al., 2024). Key Parameters Considered in the Model such as feedstock acquisition costs, capital costs for methanization and hydrogen production technologies, operating and maintenance costs, the hydrogen production rates finally the revenue from byproducts (biofertilizers, solid carbon) (Mohammadpour et al., 2024).

## 2.4 Environmental Impact Assessment

An environmental impact assessment was conducted using a life cycle analysis (LCA) framework to quantify the emissions, energy consumption, and resource utilization associated with biogas-to-hydrogen conversion pathways (Daş et al., 2024). The assessment focused on Global warming potential (GWP) in terms of $CO_2$ equivalent emissions and energy return on investment (EROI) for hydrogen production also the water usage during feedstock processing and hydrogen generation (Zheng et al., 2024). The LCA followed ISO 14040/44 standards and included a cradle-to-gate analysis, covering feedstock collection, methane production, and hydrogen generation (Liu et al., 2024).

## 2.5 Policy and Regulatory Analysis

A comprehensive analysis of current policies and regulatory frameworks was conducted to understand the legal and financial incentives for biogas and hydrogen production. The study focused on analyzing policies from the European Union (RED II) and North America (U.S. Department of Energy's Hydrogen and Fuel Cell Technologies Office) that promote renewable hydrogen and biomethane integration (Sesini et al., 2024). The study also reviewed emerging carbon pricing schemes, renewable energy credits, and subsidies for renewable energy projects (Emetere et al., 2024). These policy tools were evaluated for their effectiveness in driving investment in methanization and biogas-to-hydrogen systems.

# 3. DISCUSSION OF RESULTS

The results from the integration of methanization and biogas technologies into the green hydrogen economy offer valuable insights into the potential benefits and challenges associated with this approach. This section discusses the findings from the methanization process, hydrogen production via steam methane reforming and methane pyrolysis, techno-economic modeling, environmental impact assessment, and policy analysis.

## 3.1 Methanisation Process

The anaerobic digestion of organic waste materials in the methanization process yielded promising results, with the biogas containing a high methane concentration (60-65%). This result aligns with previous research indicating that well-optimized methanization can generate methane yields that are sufficient for conversion into hydrogen (Chelvam et al., 2024). Feedstock characteristics, particularly the carbon to nitrogen (C/N) ratio, played a significant role in determining biogas yield, with agricultural residues and food waste exhibiting superior performance compared to sewage sludge. One of the key findings is that the efficiency of the methanization process is influenced by both feedstock type and operating conditions (Awad et al., 2024). Pretreatment of feedstocks, such as shredding and ensuring proper mixing in the digester, improved the rate of methane production. However, the study highlighted the importance of further optimization, particularly regarding HRT and organic loading rate OLR, to maximize methane yields in large-scale operations (Lanni et al., 2023).

## 3.2 Hydrogen Production Pathways

The two hydrogen production pathways SMR and methane pyrolysis produced varying results in terms of efficiency, carbon emissions, and overall feasibility. SMR produced high hydrogen yields, but the process was accompanied by significant $CO_2$ emissions (Mio et al., 2024). While carbon capture and storage (CCS) technologies were considered to mitigate these emissions, their integration increased operational costs and energy demand. Despite these challenges, SMR remains a well-established technology, and when combined with biomethane from methanization, it offers a more sustainable alternative to traditional grey hydrogen production from fossil fuels (Mokheimer et al., 2024). The study demonstrated that the carbon footprint of biogas based SMR could be reduced by approximately 50% compared to conventional natural gas based SMR, assuming successful CCS implementation (Joo et al., 2024).

Methane pyrolysis, in contrast, emerged as a promising carbon-neutral option for hydrogen production. The process decomposes methane into hydrogen and solid carbon, eliminating the need for $CO_2$ capture. The solid carbon byproduct can be repurposed for industrial applications or sequestered, further reducing the environmental impact (Ratnakar & Balakotaiah, 2024). However, the pyrolysis process requires high temperatures and advanced reactor designs, which increases capital and operational expenditures. In this study, the hydrogen yield from methane pyrolysis was comparable to that of SMR, but with significantly lower greenhouse gas emissions. The main challenge remains the high energy input required for pyrolysis, which affects the overall energy return on investment.

## 3.3 Techno Economic Modeling

The techno-economic model provided a comprehensive evaluation of the costs associated with integrating methanization and biogas technologies into the green hydrogen economy. The analysis revealed several critical insights While both hydrogen production pathways demonstrated technical feasibility, their economic viability depended heavily on feedstock availability, energy prices, and market conditions for hydrogen and byproducts (Nemitallah et al., 2024). The cost of producing hydrogen from biogas was estimated to be higher than grey hydrogen but significantly lower than pure green hydrogen derived from water electrolysis. For SMR, carbon capture and storage contributed to higher operational costs, while methane pyrolysis faced high initial capital expenditures (Witte et al., 2024). However, with the right policy incentives, such as carbon credits and renewable energy subsidies, these technologies could become more competitive (Sharma et al., 2024). Methanization generated not only biogas but also digestate, a nutrient rich byproduct that can be used as a biofertilizer. This offers an additional revenue stream, improving the overall economics of the process (Hussam et al., 2024). Similarly, the solid carbon produced during methane pyrolysis presents opportunities for sale in various industries, such as construction and manufacturing, thereby enhancing the financial viability of the hydrogen production process (Bube et al., 2024).

*Table 1. Key techno-economic parameters for hydrogen production technologies (IEA, n.d.; Jamieson, 2022; "Direct Conversion of Methane to Hydrogen by Pyrolysis," 2022).*

| Technology | CAPEX (USD/kW H2) | OPEX (% of CAPEX) | Efficiency (LHV) | Full-Load Hours | CO2 Capture Rate | Fuel Price (USD/GJ) | CO2 Price (USD/tCO2) | Lifetime (years) |
|---|---|---|---|---|---|---|---|---|
| SMR (without CCS) | 910 (2019, 2050) | 4.7% | 76% | 8,322 | N/A | 1.4-6.3 (2019) | 0-15 (2019) | 30 |
| SMR (with CCS) | 1,583 (2019) | 3.0% | 69% | 8,322 | 95% | 1.4-6.3 (2019) | 0-15 (2019) | 30 |
| Coal (without CCS) | 2,672 (2019, 2050) | 5.0% | 60% | 8,322 | N/A | 1.6-3.8 (2019) | 0-15 (2019) | 30 |
| Coal (with CCS) | 2,783 (2019, 2050) | 5.0% | 58% | 8,322 | 90% | 1.6-3.8 (2019) | 0-15 (2019) | 30 |
| Electrolysis (Low-Carbon) | 872 (2019), 269 (2050) | 2.2% (2019), 1.5% (2050) | 64% (2019), 74% (2050) | 3,000-4,000 (2019), 2,000-3,000 (2050) | N/A | 36-116 (2019) | 0-15 (2019) | 30 |

The table 01 highlights key distinctions between hydrogen production technologies in terms of costs, efficiency, and environmental impact. SMR with CCS offers a balance between capital expenditure (CAPEX) and efficiency, with a high CAPEX of 1,583 USD/kW H2 and a CO2 capture rate of 95%. Coal gasification with CCS is less efficient (58%) and more expensive (2,783 USD/kW H2). Electrolysis using renewable energy is seen as a future key player, with significantly lower CAPEX projected for 2050 (269 USD/kWe) and higher efficiency (74%), making it more sustainable (Patel et al., 2024).

### 3.4 Environmental Impact Assessment

The environmental impact assessment, conducted through an LCA, demonstrated that integrating biogas into hydrogen production significantly reduced the carbon footprint compared to traditional hydrogen production methods (Bakkaloglu & Hawkes, 2024). The global warming potential of biogas-based SMR with CCS was estimated to be approximately 2.8 kg $CO_2$ eq per kg of hydrogen, compared to 10-12 kg $CO_2$ eq for grey hydrogen (Patel et al., 2024; Ortiz et al., 2024). Methane pyrolysis, with its solid carbon byproduct, showed an even lower GWP of approximately 0.5 kg $CO_2$ eq per kg of hydrogen, making it one of the most environmentally friendly hydrogen production methods evaluated in this study (Molina et al., 2024). Additionally, the study highlighted the reduced reliance on fossil fuels when biogas is used as a feedstock. By utilizing organic waste to produce biomethane, the overall resource efficiency of the hydrogen production process was significantly improved (Rabbani et al., 2024). However, the water usage in biogas upgrading and SMR processes remained a concern, emphasizing the need for further research into water-efficient technologies.

### 3.5 Policy and Regulatory Analysis

The policy analysis revealed that supportive regulations are crucial for the large-scale adoption of biogas-to-hydrogen technologies. In regions where renewable energy credits, carbon pricing, and subsidies for biogas infrastructure were available, the integration of methanization into hydrogen production was more economically viable. For example, in the European Union, the Renewable Energy Directive RED II provides incentives for biomethane production, making it more attractive for hydrogen producers (Hydrogen and Fuel Cell Technologies Office, n.d.). Similarly, the U.S. Department of Energy's Hydrogen and Fuel Cell Technologies Office promotes biogas to hydrogen projects through grants and subsidies (Jamieson, 2022). However, the study identified gaps in the regulatory framework, particularly regarding the certification of hydrogen derived from biogas. Establishing clear guidelines for green hydrogen that includes biogas as a feedstock would provide greater market clarity and encourage further investment.

## 4. CONCLUSION

This study provides a thorough comparative analysis of two hydrogen production methods, SMR with CCS and methane pyrolysis, both pivotal for advancing a low-carbon hydrogen economy. SMR, widely adopted due to its technical maturity and integration into existing industrial processes, presents a solution for scaling hydrogen production. However, its reliance on fossil fuels and the need for carbon capture pose significant challenges for long-term environmental sustainability. Methane pyrolysis, by contrast, offers an innovative approach with the potential to achieve near-zero emissions. Yet, this technology is still in its early stages, facing technological hurdles, such as reactor efficiency, carbon material handling, and economic viability, particularly for large-scale implementation. The results emphasize that while SMR with CCS can offer an immediate response to hydrogen demand under stricter carbon regulations, methane pyrolysis represents a longer-term solution for cleaner hydrogen production. Furthermore, this analysis underscores the necessity of integrating economic incentives, supportive policies, and technological advancements to address both methods' existing limitations. In a future hydrogen economy, hybrid or complementary solutions may emerge, combining the best attributes of both SMR with CCS and methane pyrolysis. This underscores the importance of continued research, infrastructure development, and policy frameworks aimed at enhancing methane pyrolysis' feasibility while optimizing SMR's efficiency with reduced emissions. Overall, the study reveals that achieving sustainable hydrogen production will require a balanced, multi-dimensional strategy, uniting technical innovation with economic and environmental considerations to support the global energy transition.

## REFERENCES


Agoud, S., Benchrifa, M., Mabrouki, J., Moufti, A., & Slaoui, M. (2024). Study of Energy Production by the Anaerobic Digestion System of Waste Management. In Mabrouki, J., & Mourade, A. (Eds.), *Technical and Technological Solutions Towards a Sustainable Society and Circular Economy* (pp. 67–76). Springer Nature Switzerland., DOI: 10.1007/978-3-031-56292-1_5

Ahmad, A., Khan, S., Chhabra, T., Tariq, S., Javed, M. S., & Li, H.. (2024). Synergic impact of renewable resources and advanced technologies for green hydrogen production: Trends and perspectives. *International Journal of Hydrogen Energy*, *67*, 788–806. DOI: 10.1016/j.ijhydene.2023.06.337

Awad, M., Said, A., Saad, M. H., Farouk, A., Mahmoud, M. M., Alshammari, M. S., Alghaythi, M. L., Abdel Aleem, S. H. E., Abdelaziz, A. Y., & Omar, A. I. (2024). A review of water electrolysis for green hydrogen generation considering PV/wind/hybrid/hydropower/geothermal/tidal and wave/biogas energy systems, economic analysis, and its application. *Alexandria Engineering Journal*, *87*, 213–239. DOI: 10.1016/j.aej.2023.12.032

Azoulay, K., Bencheikh, I., Mohammed, B., Rachiq, T., & Mabrouki, J. (2024). Methanization—Transforming Waste to Energy for a Sustainable Future: Mini Review. In Mabrouki, J., & Mourade, A. (Eds.), *Advanced Technology for Smart Environment and Energy* (pp. 321–331). Springer Nature Switzerland., DOI: 10.1007/978-3-031-50871-4_22

Bakkaloglu, S., & Hawkes, A. (2024). A comparative study of biogas and biomethane with natural gas and hydrogen alternatives. *Energy & Environmental Science*, *17*(4), 1482–1496. DOI: 10.1039/D3EE02516K

Basso, G. L., Pastore, L. M., Mojtahed, A., & de Santoli, L. (2023). From landfill to hydrogen: Techno-economic analysis of hybridized hydrogen production systems integrating biogas reforming and Power-to-Gas technologies. *International Journal of Hydrogen Energy*, *48*(96), 37607–37624. DOI: 10.1016/j.ijhydene.2023.07.130

Bertasini, D., Battista, F., Rizzioli, F., Frison, N., & Bolzonella, D. (2023). Decarbonization of the European natural gas grid using hydrogen and methane biologically produced from organic waste: A critical overview. *Renewable Energy*, *206*, 386–396. DOI: 10.1016/j.renene.2023.02.029

Bube, S., Sens, L., Drawer, C., & Kaltschmitt, M. (2024). Power and biogas to methanol–A techno-economic analysis of carbon-maximized green methanol production via two reforming approaches. *Energy Conversion and Management*, *304*, 118220. DOI: 10.1016/j.enconman.2024.118220

Chelvam, K., Hanafiah, M. M., Woon, K. S., & Al Ali, K. (2024). A review on the environmental performance of various hydrogen production technologies: An approach towards hydrogen economy. *Energy Reports*, *11*, 369–383. DOI: 10.1016/j.egyr.2023.11.060

Daş, İ. T. Ö., Özmıhçı, S., & Büyükkamacı, N. (2024). Environmental impact analysis of different wastes to biohydrogen, biogas and biohytane processes. *International Journal of Hydrogen Energy*, *56*, 1446–1463. DOI: 10.1016/j.ijhydene.2023.12.184

El-Moustaqim, K., Mabrouki, J., & Hmouni, D. (2024). *From Microalgal Biomass to Biohydrogen: A Novel Approach to Environmentally Sustainable Technology*. Advancements in Climate and Smart Environment Technology. DOI: 10.4018/979-8-3693-3807-0.ch003

Elhaus, N., Volkmann, M., Kolb, S., Schindhelm, L., Herkendell, K., & Karl, J. (2024). Techno-economic evaluation of anaerobic digestion and biological methanation in Power-to-Methane-Systems. *Energy Conversion and Management*, *315*, 118787. DOI: 10.1016/j.enconman.2024.118787

Emetere, M. E., Oniha, M. I., Akinyosoye, D. A., Elughi, G. N., & Afolalu, S. A. (2024). Progress and challenges of green hydrogen gas production: Leveraging on the successes of biogas. *International Journal of Hydrogen Energy*, *79*, 1071–1085. DOI: 10.1016/j.ijhydene.2024.07.115

Gholizadeh, T., Ghiasirad, H., Skorek-Osikowska, A., & Arabkoohsar, A. (2024). Techno-economic optimization and working fluid selection of a biogas-based dual-loop bi-evaporator ejector cooling cycle involving power-to-hydrogen and water facilities. *International Journal of Hydrogen Energy*, *76*, 247–262. DOI: 10.1016/j.ijhydene.2024.04.104

Giocoli, A., Motola, V., Scarlat, N., Pierro, N., & Dipinto, S. (2023). Techno-economic viability of renewable electricity surplus to green hydrogen and biomethane, for a future sustainable energy system: Hints from Southern Italy. *Renewable and Sustainable Energy Transition*, *3*, 100051. DOI: 10.1016/j.rset.2023.100051

Henriksen, M. S., Matthews, H. S., White, J., Walsh, L., Grol, E., Jamieson, M., & Skone, T. J. (2024). Tradeoffs in life cycle water use and greenhouse gas emissions of hydrogen production pathways. *International Journal of Hydrogen Energy*, *49*, 1221–1234. DOI: 10.1016/j.ijhydene.2023.08.079

Hussam, W. K., Abdul-Niby, M., & Sheard, G. J. (2024). Techno-economic analysis and optimization of hydrogen production from renewable hybrid energy systems: Shagaya renewable power plant-Kuwait. *International Journal of Hydrogen Energy*, *58*, 56–68. DOI: 10.1016/j.ijhydene.2024.01.153

Jamal, M. (2022). Mathematical modelling of biogas production in a controlled landfill: Characterization, valorization study and energy potential. *Sustainability (Basel)*, *14*(23), 15490. DOI: 10.3390/su142315490

Jamieson, M. (2022). *Steam Methane Reforming (SMR) operations with carbon capture (CCS). National Energy Technology Laboratory (NETL)*. Pittsburgh, PA, Morgantown, WV, and Albany., DOI: 10.2172/1922946

Joo, C., Lee, J., Kim, Y., Cho, H., Gu, B., & Kim, J. (2024). A novel on-site SMR process integrated with a hollow fiber membrane module for efficient blue hydrogen production: Modeling, validation, and techno-economic analysis. *Applied Energy*, *354*, 122227. DOI: 10.1016/j.apenergy.2023.122227

Kumar, R., & Kumar, A. (2024). Recent advances of biogas reforming for hydrogen production: Methods, purification, utility and techno-economics analysis. *International Journal of Hydrogen Energy*, *76*, 108–140. DOI: 10.1016/j.ijhydene.2024.02.143

Lanni, D., Minutillo, M., Cigolotti, V., & Perna, A. (2023). Biomethane production through the power to gas concept: A strategy for increasing the renewable sources exploitation and promoting the green energy transition. *Energy Conversion and Management*, *293*, 117538. DOI: 10.1016/j.enconman.2023.117538

Liu, B., Han, B., Liang, X., & Liu, Y. (2024). Hydrogen production from municipal solid waste: Potential prediction and environmental impact analysis. *International Journal of Hydrogen Energy*, *52*, 1445–1456. DOI: 10.1016/j.ijhydene.2023.11.027

Ma, C., Yu, H., Monticone, G., Ma, S., & Wang, L. (2024). Techno-economic evaluation of biogas-fed SOFC systems with novel biogas purification and carbon capture technologies. *Renewable Energy*, *235*, 121302. DOI: 10.1016/j.renene.2024.121302

Mabrouki, J., Azrour, M., Fattah, G., Dhiba, D., & El Hajjaji, S. (2021). Intelligent monitoring system for biogas detection based on the Internet of Things: Mohammedia, Morocco city landfill case. *Big Data Mining and Analytics*, *4*(1), 10–17. DOI: 10.26599/BDMA.2020.9020017

Maniscalco, M. P., Longo, S., Cellura, M., Miccichè, G., & Ferraro, M. (2024). Critical Review of Life Cycle Assessment of Hydrogen Production Pathways. *Environments (Basel, Switzerland)*, *11*(6), 108. DOI: 10.3390/environments11060108

Mio, A., Barbera, E., Pavan, A. M., Bertucco, A., & Fermeglia, M. (2024). Sustainability analysis of hydrogen production processes. *International Journal of Hydrogen Energy*, *54*, 540–553. DOI: 10.1016/j.ijhydene.2023.06.122

Mohammadpour, M., Ashjaee, M., Mohammadpour, A., & Houshfar, E. (2024). Hybrid CFD/techno-economic assessments of carbon capturing combustion system integrated with PEM electrolyzer for efficient hydrogen production. *Journal of Cleaner Production*, *469*, 143240. DOI: 10.1016/j.jclepro.2024.143240

Mokheimer, E. M. A., Shakeel, M. R., Harale, A., Paglieri, S., & Mansour, R. B. (2024). Fuel reforming processes for hydrogen production. *Fuel*, *359*, 130427. DOI: 10.1016/j.fuel.2023.130427

Molina, S., Gomez-Soriano, J., Lopez-Juarez, M., & Olcina, M. (2024). Evaluation of the environmental impact of HCNG light-duty vehicles in the 2020–2050 transition towards the hydrogen economy. *Energy Conversion and Management*, *301*, 117968. DOI: 10.1016/j.enconman.2023.117968

Murphy, J. D., Rusmanis, D., Gray, N., & O'Shea, R. (2024). Circular economy approaches to integration of anaerobic digestion with Power to X technologies. *IEA Bioenergy Task*, *37*, 1.

Nemitallah, M. A., Alnazha, A. A., Ahmed, U., El-Adawy, M., & Habib, M. A. (2024). Review on techno-economics of hydrogen production using current and emerging processes: Status and perspectives. *Results in Engineering*, *21*, 101890. DOI: 10.1016/j.rineng.2024.101890

Nemmour, A., Inayat, A., Janajreh, I., & Ghenai, C. (2023). Green hydrogen-based E-fuels (E-methane, E-methanol, E-ammonia) to support clean energy transition: A literature review. *International Journal of Hydrogen Energy*, *48*(75), 29011–29033. DOI: 10.1016/j.ijhydene.2023.03.240

Ongis, M., Di Marcoberardino, G., Manzolini, G., Gallucci, F., & Binotti, M. (2023). Membrane reactors for green hydrogen production from biogas and biomethane: A techno-economic assessment. *International Journal of Hydrogen Energy*, *48*(51), 19580–19595. DOI: 10.1016/j.ijhydene.2023.01.310

Onwuemezie, L., Darabkhani, H. G., & Montazeri-Gh, M. (2024). Pathways for low carbon hydrogen production from integrated hydrocarbon reforming and water electrolysis for oil and gas exporting countries. *Sustainable Energy Technologies and Assessments*, *61*, 103598. DOI: 10.1016/j.seta.2023.103598

Ortiz, P. S., Lin, J.-W., Mariano, A. P., Maciel Filho, R., & Jocher, A. (2024). Environmental Assessment of Liquid Hydrogen Production Routes. *Chemical Engineering Transactions*, *109*, 583–588.

Patel, A., Gami, B., Patel, B., Parmar, V., & Patel, P. (2024). Cost benefit and environmental impact assessment of compressed biogas (CBG) production from industrial, agricultural, and community organic waste from India. *Biomass Conversion and Biorefinery*, *14*(3), 4123–4137. DOI: 10.1007/s13399-022-02780-y

Patel, G. H., Havukainen, J., Horttanainen, M., Soukka, R., & Tuomaala, M. (2024). Climate change performance of hydrogen production based on life cycle assessment. *Green Chemistry*, *26*(2), 992–1006. DOI: 10.1039/D3GC02410E

Rabbani, M., Hosseini, A., Karim, M. A., Fahimi, A., Karimi, K., & Vahidi, E. (2024). Environmental impact assessment of a novel third-generation biorefinery approach for astaxanthin and biofuel production. *The Science of the Total Environment*, *912*, 168733. DOI: 10.1016/j.scitotenv.2023.168733 PMID: 38013097

Ratnakar, R. R., & Balakotaiah, V. (2024). Sensitivity analysis of hydrogen production by methane reforming using electrified wire reactors. *International Journal of Hydrogen Energy*, *49*, 916–926. DOI: 10.1016/j.ijhydene.2023.07.352

Sesini, M., Cretì, A., & Massol, O. (2024). Unlocking European biogas and biomethane: Policy insights from comparative analysis. *Renewable & Sustainable Energy Reviews*, *199*, 114521. DOI: 10.1016/j.rser.2024.114521

Sharma, P., Parakh, S. K., Tsui, T. H., Bano, A., Singh, S. P., Singh, V. P., Lam, S. S., Nadda, A. K., & Tong, Y. W. (2024). Synergetic anaerobic digestion of food waste for enhanced production of biogas and value-added products: Strategies, challenges, and techno-economic analysis. *Critical Reviews in Biotechnology*, *44*(6), 1040–1060. DOI: 10.1080/07388551.2023.2241112 PMID: 37643972

Witte, J., Madi, H., Elber, U., Jansohn, P., & Schildhauer, T. J. (2024). Grid-neutral hydrogen mobility: Dynamic modelling and techno-economic assessment of a renewable-powered hydrogen plant. *International Journal of Hydrogen Energy*, *78*, 52–67. DOI: 10.1016/j.ijhydene.2024.05.331

Zheng, X., Wang, J., Huang, J., Xu, X., Tang, J., Hou, P., Han, W., & Li, H. (2024). Environmental impact assessment of a combined bioprocess for hydrogen production from food waste. *Waste Management (New York, N.Y.)*, *173*, 152–159. DOI: 10.1016/j.wasman.2023.11.009 PMID: 37989014

Chapter 19
# Overview of Water Electrolysis Technologies for Green Hydrogen

**P. Selvakumar**
https://orcid.org/0000-0002-3650-4548
*Department of Science and Humanities, Nehru Institute of Technology, Coimbatore, India*

**Sonali R. Dhokpande**
https://orcid.org/0000-0003-0476-7033
*Datta Meghe College of Engineering, Airoli, India*

**S. Seenivasan**
https://orcid.org/0000-0002-2155-9781
*Rathinam Technical Campus, India*

**Santosh Kumar Nathsharma**
https://orcid.org/0000-0003-3935-2938
*Stewart Science College, India*

**M. Jegadeeswaran**
https://orcid.org/0009-0009-2099-742X
*Rathinam Technical Campus, India*

**Mohit Sharma**
https://orcid.org/0009-0007-2280-8077
*Maharshi Dayanand University, Rohtak, India*

**T. C. Manjunath**
https://orcid.org/0000-0003-2545-9160
*Rajarajeswari College of Engineering, India*

## ABSTRACT

*These systems are well-established and benefit from relatively low costs and durability, making them suitable for large-scale hydrogen production. However, they may have limitations in terms of response time and efficiency at lower power densities.In contrast, PEM electrolyzers utilize a solid polymer electrolyte membrane to separate the anode and cathode, operating at higher efficiencies and with faster response times. They are more compact and can operate under higher pressure, which can be advantageous for integrating with renewable energy sources and optimizing hydrogen storage. On the one hand, the variability of renewable energy generation, such as intermittent wind or solar power, necessitates advances in electrolyzer technology and energy storage solutions to ensure a stable and reliable hydrogen supply. Additionally, hydrogen can serve as an energy carrier and storage medium, enabling the efficient storage of excess renewable energy and its subsequent use when energy demand exceeds generation capacity.*

DOI: 10.4018/979-8-3693-8980-5.ch019

## INTRODUCTION TO WATER ELECTROLYSIS FOR GREEN HYDROGEN PRODUCTION

These systems are well-established and benefit from relatively low costs and durability, making them suitable for large-scale hydrogen production. However, they may have limitations in terms of response time and efficiency at lower power densities.In contrast, PEM electrolyzers utilize a solid polymer electrolyte membrane to separate the anode and cathode, operating at higher efficiencies and with faster response times. They are more compact and can operate under higher pressure, which can be advantageous for integrating with renewable energy sources and optimizing hydrogen storage. On the one hand, the variability of renewable energy generation, such as intermittent wind or solar power, necessitates advances in electrolyzer technology and energy storage solutions to ensure a stable and reliable hydrogen supply. Additionally, hydrogen can serve as an energy carrier and storage medium, enabling the efficient storage of excess renewable energy and its subsequent use when energy demand exceeds generation capacity.Advancements in electrolyzer technology, coupled with supportive policies and investments, are crucial for scaling up water electrolysis and green hydrogen production. Furthermore, efforts to reduce the energy consumption of the electrolysis process and integrate it with renewable energy systems are key to achieving economic competitiveness.In conclusion, water electrolysis represents a cornerstone technology in the pursuit of green hydrogen production, offering a pathway to cleaner and more sustainable energy systems. Its ability to produce hydrogen from water using renewable electricity aligns with global climate goals and supports the development of a hydrogen economy. As technological advancements and supportive policies drive progress in this field, water electrolysis has the potential to play a transformative role in achieving a low-carbon future and addressing the pressing challenges of climate change and energy security.

## PRINCIPLES OF WATER ELECTROLYSIS: FUNDAMENTALS AND REACTIONS

Water electrolysis is a fundamental electrochemical process that serves as a cornerstone for sustainable hydrogen production. The principles of water electrolysis are grounded in basic electrochemical reactions and Faraday's laws of electrolysis, which describe how electric current drives chemical change (Ramakrishnan T., et al.,2022). The electrolytic cell consists of two electrodes, typically made from conductive materials such as platinum, iridium, or other metals resistant to corrosion, and an electrolyte that facilitates the flow of ions between the electrodes.The efficiency of water electrolysis is influenced by several factors, including the overpotentials required to drive the reactions and the energy input. Overpotentials are extra voltage requirements beyond the theoretical decomposition voltage to overcome various resistances and kinetic barriers. Achieving high efficiency in electrolysis involves minimizing these overpotentials through optimized electrode materials and designs.Applications and Implications:Water electrolysis is integral to producing green hydrogen, particularly when powered by renewable energy sources like wind, solar, or hydroelectric power (Abdelshafy AM, et al., 2018). The process occurs at two electrodes within an electrolytic cell and is influenced by various factors, including the type of electrolyzer and the efficiency of the reactions. As technology advances, water electrolysis holds significant promise for sustainable energy solutions and the development of a green hydrogen economy (Abohamzeh E, et al., 2021).

Alkaline Electrolysis: History, Status, and Future DirectionsAlkaline electrolysis, a process for generating hydrogen from water, has a rich history and continues to be a significant technology in the realm of sustainable energy. The method dates back to the early 19th century and has evolved considerably over time (Agaton CB, et al.,2022). Its historical development, current status, and future directions reflect both the advances in electrochemical science and the growing demand for clean energy solutions.Historical Background.The principles of alkaline electrolysis were first demonstrated by the British scientist Sir Humphry Davy in the early 1800s. Davy's work laid the foundation for the electrolysis of water using an electric current to split water into hydrogen and oxygen gases. By the mid-19th century, the technology had become more refined with the introduction of better electrode materials and electrolytes (Sugumaran K. M, D., et al., 2024). Modern alkaline electrolyzers have seen numerous advancements, including improved catalysts, better electrode designs, and enhanced cell configurations.

These improvements have led to increased efficiency, reduced energy consumption, and longer operational lifespans.Current alkaline electrolyzers typically operate at temperatures ranging from 60°C to 90°C and pressures up to 30 bar. Future Directions:As the world moves towards a more sustainable energy future, the role of alkaline electrolysis is expected to grow. Several key areas are being explored to advance the technology and enhance its viability (Babatunde OM, et al.,2022) Integration with Renewable Energy The integration of alkaline electrolysis with renewable energy sources is a key area of focus. Developing systems that can efficiently handle variable power inputs from sources like wind and solar is crucial for maximizing the production of green hydrogen and ensuring the reliability of supply. Operational Efficiency Enhancing operational efficiency involves reducing the energy consumption of the electrolysis process and minimizing overpotentials (Balta MT, et al.,2016). Research into optimizing operating conditions, improving electrolyte management, and reducing energy losses are critical for achieving higher efficiency and lower operating costs.Scalability and Cost Reduction Efforts are being made to scale up alkaline electrolysis systems to meet growing hydrogen demands and reduce costs. Innovations in manufacturing processes, economies of scale, and modular designs are being explored to make large-scale hydrogen production more economically viable.Environmental and Sustainability Considerations.

## Proton Exchange Membrane (PEM) Electrolysis: Advantages and Challenges

However, it also faces specific challenges that impact its broader adoption and implementation. Advantages of PEM Electrolysis High Efficiency and Compact Design PEM electrolyzers are known for their high energy efficiency and compact design. The solid polymer membrane enables high current densities, which translates to a higher rate of hydrogen production relative to the size of the electrolyzer (Vijayakumar G, et al., 2024). This compactness is advantageous for integration into various applications, from small-scale on-site hydrogen generation to large-scale industrial operations.Rapid Response to Power Fluctuations One of the significant advantages of PEM electrolyzers is their ability to respond quickly to changes in power input. This feature is particularly beneficial when integrating with intermittent renewable energy sources such as solar or wind power (Bernal-Agustín et al.,2010). Operation Under Higher Pressures.PEM electrolyzers can operate effectively at higher pressures, which can be advantageous for hydrogen storage and transportation. Producing hydrogen at higher pressures directly from the electrolyzer can reduce the need for additional compression stages, thereby lowering overall costs and energy requirements.Lower Operating Temperature PEM electrolyzers typically operate at lower temperatures compared to alkaline electrolyzers. This characteristic simplifies system design and integration, as well as potentially extending the lifespan of components by reducing thermal stresses(Cai D, et al.,2020).

The lower operating temperature also contributes to faster start-up times and operational stability. Challenges of PEM Electrolysis Cost of Materials and Components A significant challenge for PEM electrolysis is the high cost of materials, particularly the catalysts and membrane components. Energy Consumption and Efficiency (Carmo M, et al.,2013)While PEM electrolyzers are efficient, they still face challenges related to energy consumption. The process requires significant electrical energy to drive the electrochemical reactions, and any inefficiencies in the system can lead to increased operational costs. Research is ongoing to improve the energy efficiency of PEM electrolyzers and reduce the overall energy consumption of the process.Integration with Renewable Energy Although PEM electrolyzers are capable of responding to power fluctuations, integrating them effectively with variable renewable energy sources remains a challenge (Paramasivam Selvakumar, et al.,2024). Ensuring consistent and reliable hydrogen production while managing the variability of renewable energy input requires advanced control systems and energy management strategies.Future Directions Development of Alternative Catalysts Research is focused on finding alternative, less expensive catalysts to replace platinum and iridium. Progress in this area could significantly lower the cost of PEM electrolyzers and make the technology more accessible. Improving Membrane Technology Advances in membrane materials and manufacturing techniques aim to enhance membrane durability and reduce degradation. New materials with improved chemical and mechanical properties are being explored to extend the operational life of PEM electrolyzers.

**Solid Oxide Electrolysis (SOE): High-Temperature Electrolysis for Efficiency Solid Oxide** Electrolysis (SOE) is an advanced high-temperature electrolysis technology that offers significant advantages in hydrogen production efficiency. SOE operates at elevated temperatures, typically between 700°C and 1,000°C, leveraging the unique properties of solid oxide electrolytes to enhance the electrolysis process (Carton JG, et al.,2010). This synergy allows for the utilization of waste heat to drive the electrolysis process, further enhancing overall system efficiency and reducing operational costs. The ability to capture and use waste heat makes SOE an attractive option for industrial applications with significant thermal outputs.Scalability and Versatility SOE systems can be scaled up for large-scale hydrogen production or tailored for smaller, decentralized applications. The technology is also versatile, capable of producing both hydrogen and valuable chemical feedstocks such as carbon monoxide, which can be used in various industrial processes, including synthetic fuel production and chemical manufacturing.Potential for Low-Cost Materials The solid oxide electrolytes and electrodes used in SOE systems are often made from relatively inexpensive materials compared to those used in other electrolysis technologies. While high-temperature operation presents material challenges, advances in material science are improving the durability and cost-effectiveness of these components.Challenges Facing SOE Technology Material Durability and Longevity Operating at high temperatures poses significant challenges for the durability of materials used in SOE cells. The solid oxide electrolytes and electrodes must withstand thermal stresses, chemical reactions, and potential degradation over time. Research is ongoing to develop more robust materials and coatings that can enhance the longevity and performance of SOE systems (Dabar OA, et al.,2022). Effective integration strategies are essential for optimizing the performance and cost-effectiveness of SOE systems (David M, et al.,2019). Future Directions Material Innovations Developing advanced materials with enhanced thermal and chemical stability is a key focus for improving SOE technology. Research into new electrolyte compositions, electrode materials, and protective coatings aims to increase the durability and reduce the costs associated with high-temperature operation.System Integration and

Optimization Future efforts will focus on optimizing system design and integration strategies to maximize efficiency and reduce costs. This includes improving thermal management, developing efficient heat exchange systems, and integrating SOE technology with renewable energy and industrial heat sources. Economic Feasibility Studies Comprehensive economic analyses and feasibility studies will be essential for determining the cost-effectiveness of SOE technology in various applications. These studies will help identify the most promising markets and applications for SOE systems, as well as guide investments and policy decisions.Scaling Up and Deployment:Scaling up SOE technology for commercial deployment will require addressing technical and economic challenges. Demonstration projects and pilot plants will play a crucial role in validating the technology at larger scales and developing strategies for widespread adoption.In conclusion, Solid Oxide Electrolysis represents a promising high-temperature technology for efficient hydrogen production. Its advantages include high efficiency, potential for integration with industrial heat sources, and versatility in producing various chemical products. Despite facing challenges related to material durability, system design, and economic viability, ongoing research and development efforts are poised to address these issues and advance SOE technology towards broader application and commercialization.

## Anion Exchange Membrane (AEM) Electrolysis: Emerging Technology for Green Hydrogen

Anion Exchange Membrane (AEM) electrolysis is an emerging technology poised to play a significant role in the production of green hydrogen (Selvakumar, P.,2023)..Fundamentals of AEM Electrolysis In AEM electrolysis, the electrochemical cell consists of an anion exchange membrane situated between two electrodes—an anode and a cathode(Eldesoukey A, et al.,2019). Instead, AEM electrolyzers often utilize more cost-effective materials, such as nickel-based catalysts, which can lower overall system costs.Higher Efficiency than Alkaline Electrolysis AEM electrolysis can achieve higher efficiencies compared to traditional alkaline electrolysis (El-Emam RS, et al.,2019). Challenges Facing AEM Electrolysis Membrane Durability and Cost The anion exchange membranes used in AEM electrolysis must withstand harsh chemical conditions and high voltages. Ensuring the durability and longevity of these membranes is a significant challenge. Reducing the costs of membranes, catalysts, and other components is essential for making AEM technology competitive with other electrolysis methods and ensuring its widespread adoption.Future Directions Material Innovations Advances in membrane and catalyst materials are crucial for improving the performance and reducing the costs of AEM electrolysis systems (Faid AY, et al.,2018). This includes efforts to lower the cost of materials, streamline manufacturing processes, and develop scalable solutions that can drive down production costs.In conclusion, Anion Exchange Membrane (AEM) electrolysis represents a promising emerging technology for the production of green hydrogen. Its advantages include reduced reliance on precious metals, high efficiency, operational flexibility, and compatibility with various electrolytes. Despite challenges related to membrane durability, catalyst development, and economic viability, ongoing research and development efforts are set to address these issues and advance AEM electrolysis towards broader adoption. As the world shifts towards sustainable energy solutions, AEM electrolysis has the potential to play a significant role in the development of a green hydrogen economy.

## Comparative Analysis of Electrolysis Technologies: Efficiency, Cost, and Scalability

Alkaline Electrolysis Alkaline electrolysis has been a well-established technology for decades. It operates efficiently under moderate temperatures (60-90°C) and pressures (Fan L, et al.,2021). The theoretical energy efficiency of alkaline electrolyzers is around 70-80%, but practical efficiencies are often slightly lower due to losses associated with overpotentials and internal resistances. Operating at lower temperatures (around 50-80°C), PEM electrolyzers achieve higher energy efficiencies due to their ability to handle high current densities and rapidly respond to fluctuating power inputs. The practical efficiency of PEM systems can exceed 80%, with some advanced systems approaching 90% (Ganeshan IS, et al.,2016). This efficiency is attributed to the solid polymer electrolyte's excellent ionic conductivity and the system's compact design, which minimizes energy losses.AEM Electrolysis AEM electrolysis represents a newer technology with promising efficiency potential. Operating at ambient to moderate temperatures, AEM systems can achieve efficiencies comparable to or slightly lower than PEM electrolyzers. The use of an anion exchange membrane allows for effective ionic transport, leading to high efficiency in the electrolysis process. However, the practical efficiency of AEM systems is still being optimized, with current values ranging from 60-80%. The technology's efficiency is expected to improve with advances in membrane and catalyst materials.Cost Alkaline Electrolysis Alkaline electrolysis is relatively cost-effective, primarily due to the use of inexpensive materials such as nickel for electrodes and alkaline solutions for the electrolyte (Genç MS, et al.,2012). The capital costs are lower compared to PEM systems, making alkaline electrolysis an attractive option for large-scale hydrogen production. However, operational costs can increase with the need for frequent maintenance and the management of corrosive alkaline solutions.PEM Electrolysis PEM electrolysis is generally more expensive than alkaline electrolysis, primarily due to the high cost of materials such as platinum and iridium used in the catalysts and the solid polymer membrane. AEM Electrolysis AEM electrolysis is emerging as a scalable technology with the potential to offer modular and flexible solutions. The ability to produce hydrogen efficiently at moderate temperatures and the potential for reduced costs make AEM systems suitable for various scales of operation. Ongoing research and development are focused on improving the scalability of AEM technology to match or exceed the capabilities of alkaline and PEM systems.Conclusion In summary, each electrolysis technology—alkaline, PEM, and AEM—offers unique advantages and challenges related to efficiency, cost, and scalability. Alkaline electrolysis remains a cost-effective and scalable option with proven efficiency, suitable for large-scale hydrogen production. PEM electrolysis provides high efficiency and compact design but comes with higher costs and material challenges. AEM electrolysis, as an emerging technology, shows promise in reducing costs and offering scalability, with ongoing advancements expected to enhance its efficiency and economic feasibility. The choice of technology depends on specific application requirements, including efficiency needs, cost constraints, and scalability goals, with each technology contributing to the development of a sustainable hydrogen economy.

## Electrode Materials and Design for Enhanced Electrolysis Performance

Electrode materials and design play a crucial role in enhancing the performance of electrolysis systems, influencing efficiency, durability, and overall system cost. In electrolysis, electrodes are the sites where electrochemical reactions occur, and their properties directly impact the effectiveness of hydrogen and oxygen production. For alkaline, Proton Exchange Membrane (PEM), and Anion Exchange Membrane (AEM) electrolysis, the selection of materials and design approaches are tailored to optimize performance under different operating conditions.Alkaline Electrolysis Electrode Materials Alkaline electrolyzers traditionally use nickel-based materials for electrodes due to their good electrochemical properties and cost-effectiveness. Nickel is often alloyed with other metals, such as cobalt or iron, to enhance its performance and longevity (Hassan IA, et al.,2021). The anode typically uses nickel or nickel oxide, while the cathode is usually nickel or a nickel-based alloy. These materials offer a balance of conductivity, stability, and resistance to corrosion in the alkaline electrolyte environment. Recent advancements also explore the use of more robust materials and coatings to improve electrode life and efficiency.Design Considerations In alkaline electrolysis, electrode design focuses on maximizing surface area to improve reaction kinetics and efficiency. Electrode structures may include porous layers or incorporate advanced coatings to improve reaction kinetics and durability. Optimizing the electrode geometry and incorporating efficient gas and ion transport pathways are essential to achieve high performance and long-term stability.

General Trends and Future Directions Catalyst Development Across all electrolysis technologies, there is a strong emphasis on developing new catalysts that can enhance performance while reducing costs. Research into non-precious metal catalysts, alloyed materials, and composite structures aims to make electrolysis more economically viable and sustainable. Innovations in catalyst design focus on improving activity, stability, and resistance to degradation.Durability and Stability Enhancing the durability and stability of electrode materials is critical for long-term performance (Hassan AA, et al.,2022). This involves developing materials and coatings that can withstand harsh operating conditions, such as high temperatures, corrosive environments, and high current densities. Advances in material science are expected to lead to more robust and reliable electrode materials.Advanced Fabrication Techniques The fabrication of electrodes using advanced techniques, such as nanostructuring, deposition methods, and 3D printing, is an area of active research. These techniques can improve the uniformity, surface area, and performance of electrodes, leading to more efficient and cost-effective electrolysis systems.Integrated System Design Future developments in electrode design will increasingly focus on integrating electrodes with other system components to enhance overall performance. This includes optimizing the interaction between electrodes, membranes, and flow fields, as well as developing integrated systems that can efficiently manage reactants, products, and energy inputs.In summary, electrode materials and design are pivotal in optimizing electrolysis performance. For alkaline, PEM, and AEM electrolysis technologies, advancements in materials and design approaches aim to improve efficiency, reduce costs, and enhance durability. Ongoing research and development efforts are focused on developing new catalysts, enhancing material stability, and utilizing advanced fabrication techniques to push the boundaries of electrolysis technology and contribute to a more sustainable hydrogen economy.

## System Integration and Balance of Plant (BoP) for Electrolysis Systems

System integration and the Balance of Plant (BoP) are critical aspects of electrolysis systems that significantly impact their efficiency, reliability, and overall cost-effectiveness. Effective integration of electrolysis units with supporting components and processes ensures seamless operation and maximizes performance. The BoP encompasses all the ancillary systems required to support the primary electrolysis process, including utilities, control systems, and safety measures (He W, et al.,2020). Here's an in-depth examination of these aspects System Integration Integration with Power Sources Electrolysis systems require a stable and efficient power supply to operate. Integration with power sources, whether they are renewable (solar, wind, hydropower) or conventional (grid electricity, combined heat and power systems), is crucial. Renewable energy sources, which are often intermittent, necessitate advanced energy management systems to ensure consistent electrolysis operation. This can involve using energy storage systems, such as batteries or pumped hydro storage, to buffer fluctuations and provide a stable power supply. Effective power integration also includes the design of power electronics and inverters to match the voltage and current requirements of the electrolyzer.Heat Management Electrolysis processes, particularly those operating at high temperatures (such as Solid Oxide Electrolysis), generate substantial heat. Efficient heat management is essential to maintain operational stability and efficiency. This involves integrating heat recovery systems that capture and utilize excess thermal energy for other processes, such as pre-heating water or generating additional power (Ishaq H, et al.,2020). For low-temperature systems, maintaining thermal stability and managing the heat dissipation effectively can prevent operational issues and improve overall efficiency.Water Supply and Management The availability and quality of water are critical for electrolysis. The integration of water treatment systems, such as filtration, deionization, or desalination, ensures that the water supplied to the electrolyzer meets the required purity standards. Efficient water management also involves handling and recycling water to minimize waste and operational costs. This includes integrating systems for managing feed water and handling by-products, such as oxygen or hydrogen.Hydrogen and Oxygen Handling Proper integration of gas handling systems is vital for the safe and efficient operation of electrolysis systems (Joshi AS, et al.,2010). This includes compression, storage, and distribution of hydrogen and oxygen. Advanced compression technologies and storage solutions must be integrated to handle the high pressures and volumes of gases produced. Safety systems, such as gas detection and leakage prevention, are also crucial to mitigate risks associated with hydrogen, which is highly flammable.Control and Monitoring Systems Advanced control and monitoring systems are essential for optimizing electrolysis performance and ensuring safe operation. Integration of real-time monitoring systems allows for the continuous assessment of key parameters such as voltage, current, temperature, and pressure. Control systems automate adjustments to operational parameters based on real-time data, enhancing efficiency and responding to fluctuations in power supply or system conditions.

Balance of Plant (BoP) Utility Systems The BoP includes various utility systems that support the core electrolysis process. These systems include electrical supply systems, cooling systems, and compressed air systems. Electrical systems must be designed to handle the power demands of the electrolyzer, including transformers, switchgear, and protective devices. Cooling systems, whether liquid or air-based, manage the temperature of the electrolyzer and associated equipment to ensure optimal performance. Compressed air systems may be used for purging or cleaning purposes.Safety and Maintenance Safety systems are a fundamental part of the BoP, ensuring the safe operation of electrolysis plants (Joshi AS, et al.,2011). This includes explosion-proof equipment, gas detection systems, emergency shutdown mechanisms, and fire suppression systems. Regular maintenance routines, including inspections and preventive measures, are integral to ensuring the longevity and reliability of both the electrolyzer and the ancillary systems. Maintenance strategies must be integrated into the operational workflow to minimize downtime and ensure efficient operation.Infrastructure and Layout The physical layout and infrastructure of an electrolysis facility play a significant role in system integration. This includes designing the plant layout to optimize space utilization, ensure ease of access for maintenance, and facilitate efficient flow of materials and gases. Infrastructure considerations also include the integration of pipelines, storage tanks, and transfer systems for handling reactants and products.Environmental Considerations Integrating environmental management systems ensures that the electrolysis process adheres to regulatory requirements and minimizes environmental impact (Khalilnejad A, et al.,2014). This includes systems for managing emissions, waste disposal, and energy consumption. Environmental considerations also involve assessing the lifecycle impacts of the electrolysis system and implementing measures to reduce the ecological footprint.Economic and Operational Efficiency System integration and BoP design must also focus on economic and operational efficiency. This includes optimizing the balance between capital expenditure and operational costs, implementing energy-efficient technologies, and maximizing the return on investment. Cost-effective integration solutions, such as modular designs or scalable systems, can enhance the economic viability of electrolysis projects.Future Directions Advanced Integration Technologies Future advancements in system integration may involve the use of digital technologies, such as Internet of Things (IoT) and artificial intelligence (AI), to enhance real-time monitoring, predictive maintenance, and automated control. These technologies can improve the efficiency and reliability of electrolysis systems by providing deeper insights into system performance and enabling more precise control.Innovative Heat Management Solutions As electrolysis technologies evolve, innovative heat management solutions will be critical. This may include the development of advanced heat exchangers, integration with district heating systems, or novel thermal energy storage solutions to improve overall energy efficiency and reduce operational costs.Enhanced Safety Protocols Continuous improvements in safety protocols and technologies will be essential to manage the risks associated with electrolysis systems. This includes advancements in gas detection technologies, emergency response systems, and safety training for operators.Sustainable Water Management Future developments in water management for electrolysis will focus on enhancing water recycling and treatment technologies to reduce water consumption and environmental impact. Innovations in desalination and water purification will be key to ensuring a sustainable water supply for electrolysis processes.In conclusion, effective system integration and the Balance of Plant (BoP) are vital for optimizing the performance, reliability, and cost-effectiveness of electrolysis systems. Integrating power sources, managing heat, handling water and gases, and implementing advanced control and safety systems are crucial for the successful operation of electrolysis technologies. Ongoing advancements in these areas will drive improvements in efficiency, reduce costs, and support the broader adoption of electrolysis as a key technology for sustainable hydrogen production.

## Water Quality and Pretreatment for Electrolysis: Importance and Strategies

Water quality and pretreatment are essential components in the operation and efficiency of electrolysis systems, impacting both the performance and longevity of electrolyzers. The purity of the water used in electrolysis directly affects the efficiency of the hydrogen production process, the integrity of the electrolyzer components, and the overall operational costs. Ensuring high-quality water and implementing effective pretreatment strategies are crucial for optimizing electrolysis performance and achieving sustainable hydrogen production.Importance of Water Quality in Electrolysis Electrolyzer Performance High-quality water is critical for maintaining the efficiency of the electrolysis process. Impurities such as minerals, organic compounds, and particulates can interfere with the electrochemical reactions occurring at the electrodes. These contaminants can lead to increased electrical resistance, reduced reaction efficiency, and lower hydrogen production rates (Khan AA, et al.,2022). The need for additional cleaning and descaling processes, as well as the potential for reduced hydrogen output, can increase the overall cost of hydrogen production. Ensuring high-quality water can help mitigate these costs and improve the economic viability of electrolysis systems (Lagorse J, et al.,2008).Pretreatment Strategies for Electrolysis Filtration The first step in water pretreatment is filtration to remove particulate matter such as sediments, rust, and organic debris. Filtration systems, including coarse filters, cartridge filters, and bag filters, are used to protect downstream components from physical damage and clogging. Proper filtration ensures that large particles do not enter the electrolyzer, maintaining its performance and reducing the frequency of maintenance. Softening typically involves ion exchange or chemical treatment methods to replace hardness ions with sodium ions. This process helps prevent scaling on electrodes and other components, improving system efficiency and longevity (Marefati M, et al.,2019).pH Control Maintaining the correct pH level of the water is essential for optimal electrolysis performance.

Deviations from the recommended pH range can affect the electrochemical reactions and the stability of the electrolyzer components. pH adjustment systems, including acid or base dosing, are used to control and maintain the desired pH level of the water.Deaeration Dissolved gases, particularly oxygen and carbon dioxide, can interfere with the electrolysis process and contribute to corrosion. Deaeration systems, such as vacuum deaerators or degassing membranes, are used to remove dissolved gases from the water before it enters the electrolyzer. Proper waste management practices, including recycling and disposal, are essential to minimize environmental impact and ensure compliance with regulatory requirements. Smart Monitoring and Control Implementing smart monitoring and control systems for water quality management can improve the precision and efficiency of pretreatment processes. Real-time sensors, data analytics, and automated control systems can optimize pretreatment performance and ensure consistent water quality.Sustainability and Resource Efficiency Future developments in water pretreatment will focus on enhancing sustainability and resource efficiency (Martinez-Burgos WJ, et al.,2021). This includes reducing water and energy consumption, minimizing waste, and exploring methods for water recycling and reuse.In conclusion, water quality and pretreatment are fundamental to the successful operation of electrolysis systems. Ensuring high-quality water through effective pretreatment strategies is crucial for maintaining the performance, durability, and cost-effectiveness of electrolyzers.

## Energy Efficiency and Optimization in Water Electrolysis

This focus on efficiency is essential as the global push for cleaner energy sources continues.Fundamentals of Energy Efficiency in Electrolysis Electrolysis Thermodynamics The efficiency of water electrolysis is fundamentally constrained by thermodynamic principles. The theoretical minimum energy required to split water into hydrogen and oxygen is approximately 1.23 volts at standard conditions. Optimization Strategies for Electrolysis Systems Advanced Catalysts and Electrode Materials Enhancing the performance of catalysts and electrode materials is essential for improving energy efficiency. In alkaline electrolysis, nickel-based catalysts are commonly used, but ongoing research aims to develop more efficient and durable alternatives. For PEM electrolysis, platinum and iridium are used as catalysts, but efforts are focused on reducing their usage through the development of platinum-group metal-free catalysts. In AEM electrolysis, non-precious metal catalysts are being explored to achieve similar or better performance at lower costs (Nadaleti WC,et al.,2019). The electrolyte's ionic conductivity affects the overall energy efficiency of the electrolysis process. In alkaline electrolysis, optimizing the concentration and composition of the alkaline solution can improve conductivity and reduce energy losses (Naseri A, et al.,2017). For PEM and AEM electrolysis, ensuring the proper hydration of the membrane and minimizing resistance through the use of high-quality electrolytes are critical for enhancing efficiency. System Integration and Energy Management Effective system integration and energy management are key to optimizing the overall efficiency of electrolysis systems. Integrating electrolyzers with renewable energy sources, such as wind or solar power, requires efficient energy management strategies to handle fluctuations in power supply. Using energy storage systems, such as batteries or thermal storage, can help balance supply and demand and ensure consistent operation. Additionally, employing advanced control systems to optimize power input and operational parameters can further enhance efficiency. Heat Recovery and Utilization Electrolysis processes generate heat, which, if not managed effectively, represents a loss of energy. Implementing heat recovery systems to capture and utilize waste heat can improve overall energy efficiency (Olateju B, et al.,2014). For high-temperature electrolysis processes, such as Solid Oxide Electrolysis (SOE), recovering and reusing thermal energy can significantly enhance system performance (Okundamiya MS 2020). Heat exchangers and thermal storage systems can help integrate recovered heat into other processes or use it to preheat water, reducing the energy required for electrolysis. Integration with Renewable Energy The intermittent nature of renewable energy sources poses challenges for energy efficiency in electrolysis systems.

## Scalability and Deployment of Electrolysis Technologies for Green Hydrogen

Alkaline electrolysis is well-established and scalable due to its mature technology and relatively lower costs. PEM and AEM electrolysis, while offering higher efficiency and performance in certain applications, face challenges related to material costs, durability, and system integration. SOE technology, with its high efficiency, is still in the early stages of commercialization, with scaling efforts focusing on improving system robustness and reducing costs (Papadopoulos V, et al.,2018). Each technology must demonstrate scalability through successful pilot projects and continuous development to transition effectively to commercial scales.Capital and Operating Costs (Olateju B, et al.,2016). The capital expenditure required for deploying electrolysis systems at scale can be substantial, including costs associated with purchasing, installing, and maintaining electrolyzers..Deployment Strategies and Consideration pilot Projects and Demonstrations Conducting pilot projects and demonstration plants is a key strategy for validating the scalability of electrolysis technologies. These projects provide valuable insights into the performance, reliability, and operational challenges of electrolyzers at larger scales (Puig-samper G, et al.,2022). Emerging technologies, such as next-generation SOE systems or hybrid electrolysis technologies, may offer new opportunities for scalability and deployment.Integration with Renewable Energy: (Qureshy AMMI, et al.,2020) The integration of electrolysis systems with renewable energy sources is a key focus for future development. Advances in energy storage, grid management, and hybrid systems can help address the challenges of intermittent renewable energy and support the scalability of electrolysis. Research into integrated systems that combine electrolysis with renewable energy generation and storage solutions will be crucial for optimizing performance and achieving long-term sustainability.

## Cost Reduction Strategies for Electrolysis Technologies

Effective cost reduction strategies focus on decreasing capital and operational expenditures, improving efficiencies, and leveraging economies of scale. Here, we explore comprehensive approaches to reducing costs across various aspects of electrolysis technologies.Technological Advancements Material Innovations One of the most direct ways to reduce costs in electrolysis is through the development of more affordable and efficient materials. For instance, catalysts used in electrolysis, such as platinum and iridium in Proton Exchange Membrane (PEM) electrolysis, are expensive and scarce. Research into alternative catalysts, such as non-precious metal-based or low-platinum alloys, can significantly lower material costs. Similarly, advancements in membrane technology—such as the development of cost-effective and durable ion exchange membranes—can reduce expenses associated with maintenance and replacements.Efficiency Improvements Enhancing the energy efficiency of electrolysis systems can lower the operational costs associated with energy consumption. Innovations in electrode design, electrolyte composition, and operational parameters can lead to reductions in the energy required for hydrogen production (Ratlamwala TAH, et al.,2011). Modular designs also facilitate easier maintenance and upgrades, further reducing operational costs.

Operational Efficiency Maintenance and Durability Reducing maintenance requirements and extending the lifespan of electrolysis components can significantly lower operational costs. Innovations in materials and design that enhance the durability of electrodes, membranes, and other critical components can lead to fewer replacements and reduced maintenance intervals. This can be achieved through better material science, protective coatings, and improved design practices (Razi F, et al.,2020). Manufacturing Efficiency Enhancing manufacturing processes to improve efficiency and reduce waste can lower production costs. Adopting lean manufacturing techniques, optimizing production workflows, and implementing quality control measures can contribute to cost savings. Automation and advanced manufacturing technologies can also enhance efficiency and reduce labor costs (Saadi A, et al.,2016). Public Awareness and Adoption Raising public awareness about the benefits of green hydrogen and its potential applications can drive demand and support market development. Educational initiatives, industry demonstrations, and outreach efforts can help build confidence in electrolysis technologies and encourage adoption across various sectors.Conclusion In summary, reducing costs for electrolysis technologies requires a multi-faceted appoach that encompasses technological advancements, economies of scale, operational efficiency, and strategic policy support. Innovations in materials, manufacturing processes, and energy management can significantly lower both capital and operational expenses. Collaborative research, optimized supply chains, and supportive policies further contribute to cost reduction and market growth (Temiz M, et al.,2021). By addressing these areas, the electrolysis industry can achieve greater cost competitiveness, facilitate widespread adoption, and support the transition to a sustainable hydrogen economy.

## Future Directions and Emerging Trends in Water Electrolysis

The future directions and emerging trends in water electrolysis are shaping the next phase of hydrogen production, with a focus on improving efficiency, reducing costs, and integrating advanced technologies to support a sustainable energy transition (Temiz M, et al.,2022). As the demand for green hydrogen increases, driven by global efforts to decarbonize various sectors, innovations in electrolysis are expected to play a crucial role in meeting this demand. The following sections outline key future directions and trends in water electrolysis technology. AEM electrolysis offers potential benefits such as reduced use of precious metals and lower operational costs (Ursúa A, et al.,2016). Innovations in catalyst design, including nanostructured materials and composite catalysts, are expected to enhance the efficiency of electrolysis systems.Membrane Technologies Advances in membrane technologies are essential for improving the performance and longevity of electrolysis systems. Research is focused on developing membranes with higher ionic conductivity, better chemical stability, and lower resistance (Valdés R, et al.,2012). These technologies enable predictive maintenance, optimize operational parameters, and enhance overall system efficiency by adjusting to varying conditions and demands.

Automation and Process Optimization Automation of electrolysis systems, including robotic handling and automated process control, is improving operational efficiency and reducing labor costs..Circular Economy and Recycling The adoption of circular economy principles in electrolysis technology development is gaining traction (van der Roest E, et al.,2020). Research is focused on improving recycling processes for critical materials, such as catalysts and membranes, and minimizing waste generation. Incentives, subsidies, and regulatory frameworks that support research, pilot projects, and commercial-scale deployments are essential for driving innovation and market expansion.Global and Regional Deployment International Collaboration International collaboration is key to advancing electrolysis technologies and addressing global hydrogen needs. Collaborative research initiatives, joint ventures, and international partnerships are fostering knowledge exchange, accelerating technology development, and facilitating the deployment of electrolysis systems across different regions.Regional Adaptation Tailoring electrolysis technologies and deployment strategies to regional contexts is important for maximizing their impact. Regional factors such as available renewable energy resources, infrastructure, and market demand influence the design and implementation of electrolysis systems. Adapting technologies and strategies to local conditions ensures effective and efficient hydrogen production.Conclusion The future of water electrolysis is characterized by ongoing advancements in technology, materials science, and system integration.

## Conclusion: Water Electrolysis Technologies for a Sustainable Green Hydrogen Future

In conclusion, water electrolysis technologies stand at the forefront of the transition to a sustainable green hydrogen future, offering a promising pathway for producing hydrogen with minimal environmental impact. PEM electrolysis, known for its high efficiency and compact design, is advancing with innovations in catalyst reduction and durability. AEM electrolysis represents a promising new frontier, with ongoing research focused on reducing material costs and improving performance. SOE technology, with its high-temperature operation, offers the potential for superior efficiency but is still under development to enhance its commercial viability (Wu B, et al.,2021) Current Challenges and Solutions Despite technological advancements, several challenges must be addressed to fully realize the potential of water electrolysis. High capital costs and the expense of critical materials, such as precious metals used in PEM and AEM electrolysis, remain significant barriers.

Global and Regional Impact The global and regional deployment of electrolysis technologies will be instrumental in meeting diverse hydrogen needs and achieving climate objectives. International collaboration and regional adaptation will play a key role in scaling up technologies and addressing specific local conditions and market demands. By leveraging regional renewable resources and tailoring deployment strategies to local contexts, electrolysis systems can be effectively integrated into various energy and industrial applications Conclusion In summary, water electrolysis technologies offer a critical solution for producing green hydrogen in a sustainable manner, playing a pivotal role in the global energy transition. Advances in technology, materials science, and system integration are driving improvements in efficiency, cost-effectiveness, and scalability. Addressing current challenges through innovation, economic strategies, and supportive policies will be essential for realizing the full potential of electrolysis. As these technologies continue to evolve, they hold the promise of transforming the hydrogen economy, supporting decarbonization efforts, and contributing to a sustainable and resilient energy future. The journey towards a green hydrogen economy is complex, but with continued commitment to research, development, and collaboration, water electrolysis technologies are well-positioned to lead the way in achieving a cleaner, more sustainable energy landscape.

## CONCLUSION

In conclusion, water electrolysis technologies represent a transformative approach to producing green hydrogen, offering a sustainable pathway for hydrogen generation that aligns with global efforts to mitigate climate change and transition to renewable energy systems. Alkaline electrolysis, the most established technology, has a proven track record of reliability and cost-effectiveness. It operates efficiently with low-cost materials and is well-suited for large-scale applications. However, its lower current density and slower response to dynamic power inputs are areas of ongoing improvement. Advances in catalyst and membrane technologies continue to enhance the performance and reduce the costs of alkaline electrolysis, making it a key player in the green hydrogen landscape.PEM electrolysis offers high efficiency and compact design, making it particularly suitable for applications where space and quick start-up times are critical. The technology's reliance on precious metal catalysts and its sensitivity to impurities in water present challenges that are being addressed through ongoing research into alternative materials and improved system designs. PEM electrolysis is poised for growth as advancements in material science and manufacturing techniques help lower costs and enhance durability.AEM electrolysis is an emerging technology that promises significant advantages in terms of reducing the reliance on precious metals and improving overall cost-effectiveness. The development of robust, high-performance AEM systems is an area of active research, with promising innovations aimed at enhancing membrane durability and catalyst performance. As these technologies mature, AEM electrolysis could become a competitive option for a broader range of applications, contributing to the diversification of green hydrogen production methods.SOE technology, with its high-temperature operation, offers exceptional efficiency by utilizing thermal energy to assist in water splitting. While SOE systems are still in the early stages of commercialization, they hold great potential for integration with industrial processes and high-temperature heat sources. Ongoing efforts to improve the materials used in SOE systems and reduce their costs are critical for advancing this technology from pilot projects to widespread deployment.The integration of these electrolysis technologies with renewable energy sources is crucial for achieving sustainable hydrogen production. As the world shifts towards increasing reliance on solar, wind, and other renewable energy sources, electrolysis systems must be optimized to operate efficiently under variable energy conditions. Innovations in energy management, storage solutions, and hybrid systems are essential for maximizing the benefits of renewable energy integration and ensuring the reliability of hydrogen production.

Economic considerations, including capital and operational costs, are central to the widespread adoption of electrolysis technologies. Reducing costs through technological advancements, economies of scale, and efficient manufacturing processes is vital for making green hydrogen competitively priced. Supportive policies and incentives play a significant role in driving investment, fostering market development, and accelerating the commercialization of electrolysis technologies.Moreover, the future of water electrolysis will be shaped by advancements in digitalization and automation, which promise to enhance system efficiency, optimize operations, and reduce labor costs. With ongoing advancements and a growing focus on efficiency, cost reduction, and system integration, electrolysis holds the promise of transforming the energy landscape and contributing to global climate goals. The successful deployment of these technologies will depend on continued innovation, supportive policies, and collaborative efforts to overcome challenges and maximize their impact. As we advance towards a cleaner, more sustainable future, water electrolysis technologies will be instrumental in shaping a resilient and low-carbon energy system.

## REFERENCES


Abdelshafy, A. M., Hassan, H., & Jurasz, J. (2018). Optimal design of a grid-connected desalination plant powered by renewable energy resources using a hybrid PSO–GWO approach. *Energy Conversion and Management*, *173*, 331–347. DOI: 10.1016/j.enconman.2018.07.083

Abohamzeh, E., Salehi, F., Sheikholeslami, M., Abbassi, R., & Khan, F. (2021). Review of hydrogen safety during storage, transmission, and applications processes. *Journal of Loss Prevention in the Process Industries*, *72*, 104569. DOI: 10.1016/j.jlp.2021.104569

Agaton, C. B., Batac, K. I. T., & Reyes, E. M.Jr. (2022). Prospects and challenges for green hydrogen production and utilization in the Philippines. *International Journal of Hydrogen Energy*, *47*(41), 17859–17870. DOI: 10.1016/j.ijhydene.2022.04.101

Babatunde, O. M., Munda, J. L., & Hamam, Y. (2022). Hybridized off-grid fuel cell/wind/solar PV / battery for energy generation in a small household: A multi-criteria perspective. *International Journal of Hydrogen Energy*, *47*(10), 6437–6452. DOI: 10.1016/j.ijhydene.2021.12.018

Balta, M. T., Kizilkan, O., & Yilmaz, F. (2016). Energy and exergy analyses of integrated hydrogen production system using high temperature steam electrolysis. *International Journal of Hydrogen Energy*, *41*(19), 8032–8041. DOI: 10.1016/j.ijhydene.2015.12.211

Bernal-Agustín, J. L., & Dufo-López, R. (2010). Techno-economical optimization of the production of hydrogen from PV-wind systems connected to the electrical grid. *Renewable Energy*, *35*(4), 747–758. DOI: 10.1016/j.renene.2009.10.004

Bhandari, R., Trudewind, C. A., & Zapp, P. (2014). Life cycle assessment of hydrogen production via electrolysis - a review. *Journal of Cleaner Production*, *85*, 151–163. DOI: 10.1016/j.jclepro.2013.07.048

Cai, D., Bamisile, O., & Adebayo, V.. (2020). *Integration of wind turbine with heliostat based CSP/ CPVT system for hydrogen production and polygeneration: a thermodynamic comparison*. Int J Hydrog Energy., DOI: 10.1016/j.ijhydene.2020.11.106

Carmo, M., Fritz, D. L., Mergel, J., & Stolten, D. (2013). A comprehensive review on PEM water electrolysis. *International Journal of Hydrogen Energy*, *38*(12), 4901–4934. DOI: 10.1016/j.ijhydene.2013.01.151

Carton, J. G., & Olabi, A. G. (2010). Wind/hydrogen hybrid systems: Opportunity for Ireland's wind resource to provide consistent sustainable energy supply. *Energy*, *35*(12), 4536–4544. DOI: 10.1016/j.energy.2010.09.010

Dabar, O. A., Awaleh, M. O., Waberi, M. M., & Adan, A.-B. I. (2022). Wind resource assessment and techno-economic analysis of wind energy and green hydrogen production in the Republic of Djibouti. *Energy Reports*, *8*, 8996–9016. DOI: 10.1016/j.egyr.2022.07.013

David, M., Ocampo-Martínez, C., & Sánchez-Peña, R. (2019). Advances in alkaline water electrolyzers: A review. *Journal of Energy Storage*, *23*, 392–403. DOI: 10.1016/j.est.2019.03.001

El-Emam, R. S., & Özcan, H. (2019). Comprehensive review on the techno-economics of sustainable large-scale clean hydrogen production. *Journal of Cleaner Production*, *220*, 593–609. DOI: 10.1016/j.jclepro.2019.01.309

Eldesoukey, A., & Hassan, H. (2019). 3D model of thermoelectric generator (TEG) case study: Effect of flow regime on the TEG performance. *Energy Conversion and Management*, *180*, 231–239. DOI: 10.1016/j.enconman.2018.10.104

Faid, A. Y., Barnett, A. O., Seland, F., & Sunde, S. (2018). Highly active nickel-based catalyst for hydrogen evolution in anion exchange membrane electrolysis. *Catalysts*, *8*(12), 614. DOI: 10.3390/catal8120614

Fan, L., Tu, Z., & Chan, S. H. (2021). Recent development of hydrogen and fuel cell technologies: A review. *Energy Reports*, *7*, 8421–8446. DOI: 10.1016/j.egyr.2021.08.003

Ganeshan, I. S., Manikandan, V. V. S., Ram Sundhar, V., Sajiv, R., Shanthi, C., Kottayil, S. K., & Ramachandran, T. (2016). Regulated hydrogen production using solar powered electrolyser. *International Journal of Hydrogen Energy*, *41*(24), 10322–10326. DOI: 10.1016/j.ijhydene.2015.05.048

Genç, M. S., Çelik, M., & Karasu, I. (2012). A review on wind energy and wind-hydrogen production in Turkey: A case study of hydrogen production via electrolysis system supplied by wind energy conversion system in Central Anatolian Turkey. *Renewable & Sustainable Energy Reviews*, *16*(9), 6631–6646. DOI: 10.1016/j.rser.2012.08.011

Hassan, A. A., Elwardany, A. E., Ookawara, S., & Hassan, H. (2022). Energy, exergy, economic and environmental (4E) assessment of hybrid solar system powering adsorption-parallel/series ORC multigeneration system. *Process Safety and Environmental Protection*, *164*, 761–780. DOI: 10.1016/j.psep.2022.06.024

Hassan, I. A., Ramadan, H. S., Saleh, M. A., & Hissel, D. (2021). Hydrogen storage technologies for stationary and mobile applications: Review, analysis and perspectives. *Renewable & Sustainable Energy Reviews*, *149*, 111311. DOI: 10.1016/j.rser.2021.111311

He, W., Namar, M. M., Li, Z., Maleki, A., Tlili, I., & Safdari Shadloo, M. (2020). Thermodynamic analysis of a solar-driven high-temperature steam electrolyzer for clean hydrogen production. *Applied Thermal Engineering*, *172*, 115152. DOI: 10.1016/j.applthermaleng.2020.115152

IRENA2022 Workbook. IRENA_Electrolysers_Patents_Insights. https://public.tableau.com/views/IRENA_Electrolysers_Patents_Insights/Electrolysers_patent_insight?:showVizHome=no&embed=yes. Accessed 29 Jul 2022

Ishaq, H., & Dincer, I. (2020). Evaluation of a wind energy based system for co-generation of hydrogen and methanol production. *International Journal of Hydrogen Energy*, *45*(32), 15869–15877. DOI: 10.1016/j.ijhydene.2020.01.037

Joshi, A. S., Dincer, I., & Reddy, B. V. (2010). Exergetic assessment of solar hydrogen production methods. *International Journal of Hydrogen Energy*, *35*(10), 4901–4908. DOI: 10.1016/j.ijhydene.2009.09.067

Joshi, A. S., Dincer, I., & Reddy, B. V. (2011). Solar hydrogen production: A comparative performance assessment. *International Journal of Hydrogen Energy*, *36*(17), 11246–11257. DOI: 10.1016/j.ijhydene.2010.11.122

Kanipriya, M., Sugumaran, D., Pandiyarajan, V., Yogananda, B. S., Selvakumar, P., & Mohan, S. (2024, January). Analysis of Electromyography Signals for Control of Mechanical Prosthesis using Machine Learning Techniques. In 2024 2nd International Conference on Intelligent Data Communication Technologies and Internet of Things (IDCIoT) (pp. 1059-1067). IEEE.

Khalilnejad, A., & Riahy, G. H. (2014). A hybrid wind-PV system performance investigation for the purpose of maximum hydrogen production and storage using advanced alkaline electrolyzer. *Energy Conversion and Management*, *80*, 398–406. DOI: 10.1016/j.enconman.2014.01.040

Khan, A. A., Khan, S. U., Ali, M. A. S., Safi, A., Gao, Y., & Luo, J. (2022). Identifying impact of international trade and renewable energy consumption on environmental quality improvement and their role in global warming. *Environmental Science and Pollution Research International*, *29*(23), 33935–33944. DOI: 10.1007/s11356-022-18574-8 PMID: 35032271

Lagorse, J., Simões, M. G., Miraoui, A., & Costerg, P. (2008). Energy cost analysis of a solar-hydrogen hybrid energy system for stand-alone applications. *International Journal of Hydrogen Energy*, *33*(12), 2871–2879. DOI: 10.1016/j.ijhydene.2008.03.054

Lahoussine Ouali, H. A., Moussaoui, M. A., & Mezrhab, A. (2020). Hydrogen production from two commercial dish/Stirling systems compared to the photovoltaic system-case study: Eastern Morocco. [English Transl Geliotekhnika]. *Applied Solar Energy*, *56*(6), 466–476. DOI: 10.3103/S0003701X20060080

Marefati, M., Mehrpooya, M., & Mousavi, S. A. (2019). Introducing an integrated SOFC, linear Fresnel solar field, Stirling engine and steam turbine combined cooling, heating and power process. *International Journal of Hydrogen Energy*, *44*(57), 30256–30279. DOI: 10.1016/j.ijhydene.2019.09.074

Martinez-Burgos, W. J., de Souza, C. E., & Pedroni Medeiros, A. B.. (2021). Hydrogen: Current advances and patented technologies of its renewable production. *Journal of Cleaner Production*, *286*, 124970. DOI: 10.1016/j.jclepro.2020.124970

Nadaleti WC, Borges dos Santos G, Lourenço VA (2019) The potential and economic viability of hydrogen production from the use of hydroelectric and wind farms surplus energy in Brazil: a national and pioneering analysis. Int J Hydrog Energy 5:0–11. DOI: 10.1016/j.ijhydene.2019.08.199

Naseri, A., Bidi, M., Ahmadi, M. H., & Saidur, R. (2017). Exergy analysis of a hydrogen and water production process by a solar-driven transcritical CO2 power cycle with Stirling engine. *Journal of Cleaner Production*, *158*, 165–181. DOI: 10.1016/j.jclepro.2017.05.005

Okundamiya, M. S. (2021). Size optimization of a hybrid photovoltaic/fuel cell grid connected power system including hydrogen storage. international journal of hydrogen energy, 46(59), 30539-30546.

Olateju, B., Kumar, A., & Secanell, M. (2016). A techno-economic assessment of large scale wind-hydrogen production with energy storage in Western Canada. *International Journal of Hydrogen Energy*, *41*(21), 8755–8776. DOI: 10.1016/j.ijhydene.2016.03.177

Olateju, B., Monds, J., & Kumar, A. (2014). Large scale hydrogen production from wind energy for the upgrading of bitumen from oil sands. *Applied Energy*, *118*, 48–56. DOI: 10.1016/j.apenergy.2013.12.013

Papadopoulos, V., Desmet, J., Knockaert, J., & Develder, C. (2018). Improving the utilization factor of a PEM electrolyzer powered by a 15 MW PV park by combining wind power and battery storage – feasibility study. *International Journal of Hydrogen Energy*, *43*(34), 16468–16478. DOI: 10.1016/j.ijhydene.2018.07.069

Qureshy, A. M. M. I., & Dincer, I. (2020). A new integrated renewable energy system for clean electricity and hydrogen fuel production. *International Journal of Hydrogen Energy*, *45*(41), 20944–20955. DOI: 10.1016/j.ijhydene.2020.04.218

Ramakrishnan, T., Mohan Gift, M. D., Chitradevi, S., Jegan, R., Hency Jose, P. S., Nagaraja, H. N., & Hailegiorgis, S. M. (2022). Study of numerous resins used in polymer matrix composite materials. *Advances in Materials Science and Engineering*, *2022*(1), 1088926.

Ratlamwala, T. A. H., Gadalla, M. A., & Dincer, I. (2011). Performance assessment of an integrated PV/T and triple effect cooling system for hydrogen and cooling production. *International Journal of Hydrogen Energy*, *36*(17), 11282–11291. DOI: 10.1016/j.ijhydene.2010.11.121

Saadi, A., Becherif, M., & Ramadan, H. S. (2016). Hydrogen production horizon using solar energy in Biskra, Algeria. *International Journal of Hydrogen Energy*, *41*(47), 21899–21912. DOI: 10.1016/j.ijhydene.2016.08.224

Selvakumar, P. (2023). Biomedical applications of polymers in biosensors, cancer vaccines and drug delivery systems. *Biopolimery i Kletka*, *39*(4), 245–256. DOI: 10.7124/bc.000AA0

Selvakumar, P., Mariappan, P., Shunmugam, S., Paul, R., & Jabez, S. (2024). Removal of chemicals from effluent using photobioreactor technology to improve environmental and health impacts. *Zeitschrift für Physikalische Chemie*, *2024*(7), 1295–1302. Advance online publication. DOI: 10.1515/zpch-2023-0450

Temiz, M., & Dincer, I. (2021). Concentrated solar driven thermochemical hydrogen production plant with thermal energy storage and geothermal systems. *Energy*, *219*, 119554.

Temiz, M., & Dincer, I. (2022). A unique bifacial PV and hydrogen-based cleaner energy system with heat recovery for data centers. *Applied Thermal Engineering*, *206*, 118102. DOI: 10.1016/j.applthermaleng.2022.118102

Ursúa, A., Barrios, E. L., Pascual, J., San Martín, I., & Sanchis, P. (2016). Integration of commercial alkaline water electrolysers with renewable energies: Limitations and improvements. *International Journal of Hydrogen Energy*, *41*(30), 12852–12861. DOI: 10.1016/j.ijhydene.2016.06.071

Valdés, R., Rodríguez, L. R., & Lucio, J. H. (2012). Procedure for optimal design of hydrogen production plants with reserve storage and a stand-alone photovoltaic power system. *International Journal of Hydrogen Energy*, *37*(5), 4018–4025. DOI: 10.1016/j.ijhydene.2011.11.133

van der Roest, E., Snip, L., Fens, T., & van Wijk, A. (2020). Introducing power-to-H3: Combining renewable electricity with heat, water and hydrogen production and storage in a neighbourhood. *Applied Energy*, *257*, 114024. DOI: 10.1016/j.apenergy.2019.114024

Wang, Z., Zhang, X., & Rezazadeh, A. (2021). Hydrogen fuel and electricity generation from a new hybrid energy system based on wind and solar energies and alkaline fuel cell. *Energy Reports*, *7*, 2594–2604. DOI: 10.1016/j.egyr.2021.04.060

Wu, B., Zhai, B., & Mu, H.. (2021). (2021) Evaluating an economic application of renewable generated hydrogen: A way forward for green economic performance and policy measures. *Environmental Science and Pollution Research International*, *2910*(29), 15144–15158. DOI: 10.1007/s11356-021-16770-6 PMID: 34628612

Xiao, P., Hu, W., Xu, X., Liu, W., Huang, Q., & Chen, Z. (2020). Optimal operation of a wind-electrolytic hydrogen storage system in the electricity/hydrogen markets. *International Journal of Hydrogen Energy*, *45*(46), 24412–24423. DOI: 10.1016/j.ijhydene.2020.06.302

Zhang, F., Zhao, P., Niu, M., & Maddy, J. (2016). The survey of key technologies in hydrogen energy storage. *International Journal of Hydrogen Energy*, *41*(33), 14535–14552. DOI: 10.1016/j.ijhydene.2016.05.293

# Chapter 20
# Physico-Chemical Quality Control of Network Water From Different Drinking Water Supply Points in the Province of Salé

**Rachiq Toufik**
https://orcid.org/0009-0001-8930-3328
*Mohammed V University in Rabat, Morocco*

**karima Azoulay**
*Mohammed V University in Rabat, Morocco*

**Ayoub Ghailane**
*Mohammed V University in Rabat, Morocco*

**Driss Azdem**
*Mohammed V University in Rabat, Morocco*

**Jamal Mabrouki**
https://orcid.org/0000-0002-3841-7755
*Mohammed V University in Rabat, Morocco*

## ABSTRACT

*Drinking water is one of the most important elements on which our everyday life depends, because its importance lies in its multiple uses in different domains. The town of SALE is supplied by the REDAL water towers, which are supplied by the ONEP, which processes the waters of the Bourgreg complex coming from the Sidi Mohamed Ben Abdellah Dam. This work consists in carrying out quality checks of drinking water of the province of SALE, We have carried out sampling, and then physico-chemical analysis in the laboratory, of the distribution points of the castles. According to the standard of the quality of the water intended for the food said that for a water to be potable it must be potable on the bacteriological and physico-chemical side in this report the emphasis is placed solely on the physico-chemical analyses. From this study allowed us to deduce that the physico-chemical quality of the waters*

DOI: 10.4018/979-8-3693-8980-5.ch020

*of the province of SALE follows the Moroccan standard of potability.*

## 1. INTRODUCTION

Water is one of the most important substances on earth. Every living being needs water for its survival. Without water, plants, animals, everything would perish. So, water is a natural resource at the basis of life and an essential commodity for the majority of human economic activities (Rachiq et al., 2021) (Mabrouki,et al 2021). However, it is subject to various pollutions due to the negative effects of human activities which can alter its quality and make it undrinkable and therefore a potential source of diseases. According to a report by the World Health Organization, five million infants and children die each year from diarrheal diseases due to contaminated food or drinking water, requiring appropriate and adequate treatment water before consumption (Rachiq, et al., 2022) (Mabrouki, et al 2021).

In Morocco, groundwater constitutes a strategic resource. It represents approximately 20% of the country's water resources potential. Of the 130 aquifers, 32 are deep aquifers and 98 are superficial. At the current state of knowledge, the exploitable potential of groundwater resources is approximately 3.9 billion $m^3$, with a minimum of 22 million $m^3$/year recorded in the Sakia El Hamra and Oued Eddahab basins and a maximum of 1.11 billion $m^3$/year in the Sebou basin (Rachiq, Samghouli, et al., 2022).

The potential water resources relative to the population in 2020 is estimated at 620 $m^3$/inhabitant/year. This ratio, commonly accepted to compare countries with each other, classifies Morocco among the countries with water stress, being less than 1,000 $m^3$/inhabitant/year(Mabrouki, Moufti, et al., 2020) (Fahdi, Azdem, et al., 2024). The needs of social and economic development require resorting to water development to meet the needs of populations. These needs are themselves continually growing, often competing, which makes the water management process very complex and difficult to implement. Indeed, drinking water must meet a certain number of characteristics making it suitable for human consumption (Bencheikh et al., 2021) (Azdem, Mabrouki, Moufti, et al., 2024). The standards from this point of view are extremely different depending on the uses and the situation(Bencheikh et al., 2020) (Benchrifa, Mabrouki, et al., 2023).

## 2. MATERIALS AND METHODS

### 2.1. The Study Area and Sampling Sites

The city of Salé, located on the Atlantic coast at the mouth of the Bouregreg River, is part of the Rabat-Salé-Kénitra region and covers approximately 95.48 $km^2$. It is bordered by the Atlantic Ocean to the west, Khemisset to the east, Kenitra to the north, and Rabat to the south(Mabrouki, Benchrifa, et al., 2022) (Fatni et al., 2025). The area experiences a Mediterranean climate with oceanic influence, marked by mild winters (≈8°C) and moderate summers (≈30°C), with annual rainfall ranging from 500–600 mm. The rainy season extends from October to March (Azdem et al., 2025) (Benchrifa & Mabrouki, 2022).

*Figure 1. Geographical location of the Bouregreg River and the Sidi Mohammed Ben Abdellah Dam supplying the cities of Rabat and Salé with drinking water.*

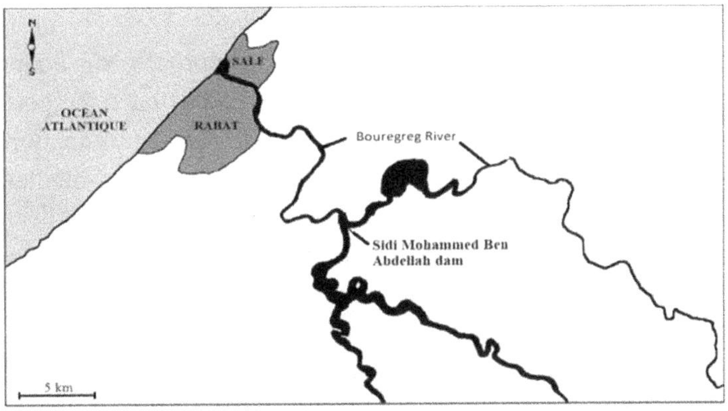


The city's drinking water is supplied from the Bouregreg complex, sourced from the Sidi Mohammed Ben Abdellah dam (**Figure.1**), with a storage capacity of 630 million $m^3$ (Ezzaouini et al., 2020). Treated water is transported via pipelines delivering 280,000 $m^3$/day, then distributed by REDAL. All samples were collected under strict aseptic conditions following a standardized protocol(Cao et al., 2024). Polyethylene bottles were first rinsed with distilled water, then conditioned with the sample water itself. Prior to collection, the tap was allowed to run for 2 to 3 minutes to flush out stagnant water(Cutter et al., 2010). The bottles were then filled completely to avoid air entrapment and sealed hermetically. Samples were immediately stored under appropriate conditions until analysis (Fahdi, et al., 2024).

## 2.2. Physicochemical Parameters

The analysis of physicochemical indicators is imperative in order to comprehend the quality of the water and its environmental effects(Loukili et al., 2021) (El Mouedden et al., 2025). In the present study, the focus was on the examination of four significant indicators: pH, electrical conductivity, turbidity, and the concentration of chloride **(Figure.2).** Each parameter was assessed by using standard procedures under controlled laboratory conditions in order to guarantee accuracy and reliability (Abood et al., 2014) (Azdem, et al 2024). The pH value of the water sample was measured using a pH meter that had been previously calibrated. It is imperative to note that, given its role as a highly sensitive indicator of temperature, with its effect on dissociation equilibrium in water being a crucial factor, the temperature was meticulously recorded in every measurement (Boguniewicz-Zablocka et al., 2019) (Azdem et al., 2025). The samples were transferred into clean beakers, after which the electrode was immersed within the solution. Following the attainment of a stable reading, the pH was recorded. Subsequently, the probe was subjected to a process of rinsing with distilled water, which was repeated between each test (Mabrouki, Azrour, et al., 2020). This procedure was implemented with the objective of preventing contamination. The electrical conductivity (EC), which is a measure of the ionic concentration present in water, was measured using a conductivity meter (Mabrouki, Benbouzid, et al., 2022). Following the calibration process, the probe was inserted into the water samples. The results were then recorded once the necessary stabilisation had been achieved. Prior to and subsequent to each measurement, the probe was meticulously rinsed with distilled water to guarantee the precision of the measurements. The measurement of turbidity, defined as the presence of suspended particles, was conducted utilising a turbidimeter, a device specifically designed for the quantification of turbidity. Following the calibration process, water samples were inserted into tubes for measurement. Turbidity measurements were then read in Nephelometric Turbidity Units (NTU). A meticulous approach was adopted to guarantee that the tubes were free from air bubbles and residue, as these elements have the potential to compromise the precision of the measurements (El Alouani et al., 2024). The determination of chloride content was accomplished through the utilisation of the mercuric-nitrate titration method. A 100 mL sample was meticulously transferred into an Erlenmeyer flask. Subsequently, a pH indicator was incorporated, along with nitric acid (N/3), until the solution exhibited a dark yellow hue, indicative of a pH value of 3.6. The titration process was executed by employing N/10 mercuric-based nitrate until the endpoint exhibited a persistent purple colouration. As the endpoint approached the culmination of the titration process, a deceleration of the process was initiated, accompanied by vigorous agitation of the solution to ensure precision. It is important to note that each test was meticulously replicated for each designated sampling point. Moreover, the instruments were meticulously cleaned and calibrated when necessary. The employment of this rigorous methodology permitted meaningful comparisons to be made across all of the collected data.

*Figure 2. Instruments used for physicochemical analysis of water samples: (a) pH meter during calibration, (b) conductivity meter in operation, (c) chloride titration setup using mercuric nitrate, (d) portable turbidimeter with accessories*

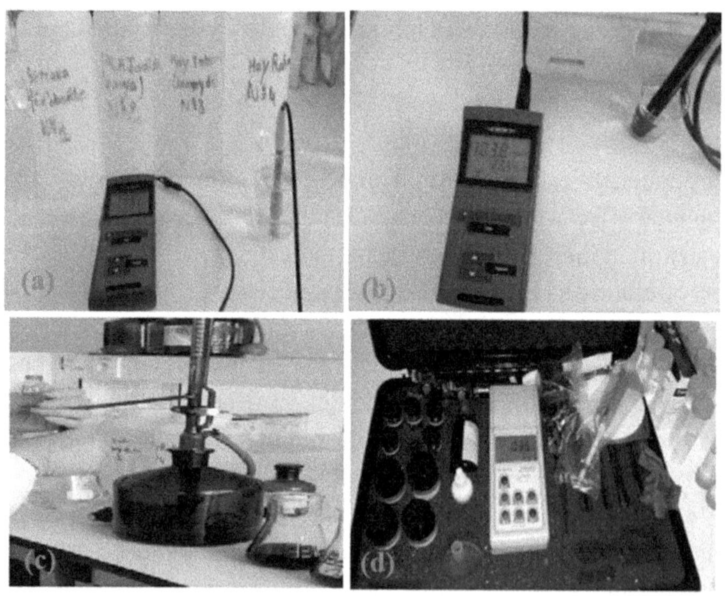

## 2.3. Spectrometric Analyses

In this study, molecular absorption spectrometry and flame emission spectrometry were employed to determine key parameters, namely nitrates ($NO_3^-$), nitrites ($NO_2^-$), ammonium ($NH_4^+$), sulfates ($SO_4^{2-}$), sodium ($Na^+$), and potassium ($K^+$).

**UV-Visible spectrophotometry** was used to quantify nitrates, nitrites, ammonium, and sulfate ions. These ions are essential indicators of nutrient pollution, eutrophication potential, and the biochemical health of aquatic systems (Rachiq et al., 2021) (Benchrifa, Elouardi, et al., 2023). Nitrate levels were determined based on molecular absorption at 415 nm, while nitrites were assessed using the sulfanilamide colorimetric method at 540 nm. The presence of ammonium was measured by a colorimetric reaction yielding a green complex, with readings performed at 655 nm. Sulfate concentrations were obtained through complexation with barium chloride, forming a turbid solution measured at 650 nm.

Additionally, **flame emission spectrometry (FES)** was applied for the quantification of sodium and potassium ions, both of which are major constituents of natural waters and sensitive markers of salinity and anthropogenic inputs (Al-Jadabi et al., 2023) (Mgalaa et al., 2022). This technique relies on the emission of light by excited atoms in a flame, with the intensity of the emitted radiation directly proportional to the concentration of the element in the sample. Sodium and potassium were detected using specific emission lines and calibration curves derived from standard solutions.

*Figure 3. Laboratory equipment used for spectrometric analysis: (a) UV-Visible spectrophotometer and drying oven, (b) Flame photometer setup for sodium and potassium determination*

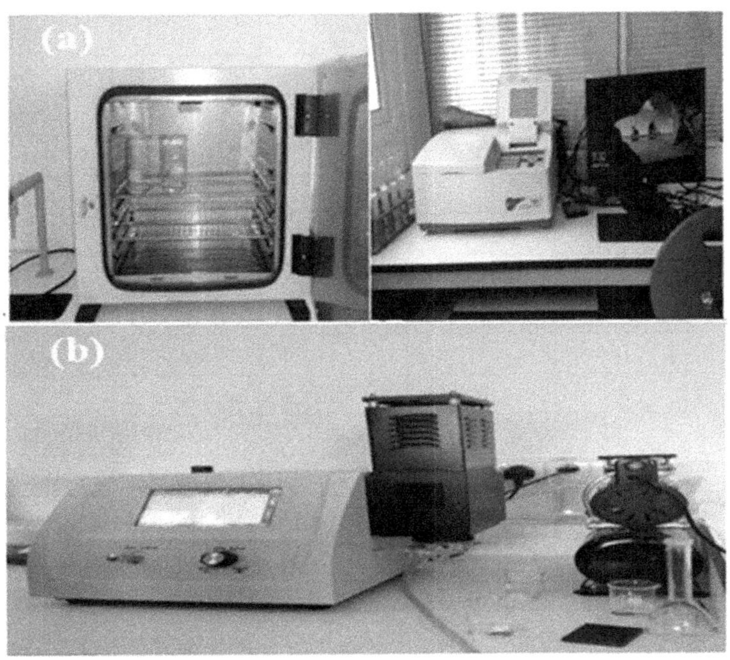

## 3. RESULTS AND DISCUSSION

As illustrated in **Table 1**, a comprehensive overview of the most significant physicochemical properties of water samples collected from four distinct locations is provided. The parameters that were measured comprised major ions ($Ca^{2+}$, $Mg^{2+}$, $Na^{+}$, $K^{+}$, $Cl^{-}$, $SO^{42-}$ and $HCO^{3-}$) as well as broad indicators of water quality (pH, conductivity, temperature, turbidity) and nutrient-related compounds (nitrates, nitrites and ammonium). A comprehensive evaluation of all parameters was conducted in accordance with the established national and international drinking standards for water (Tsaridou & Karabelas, 2021) (Ghizlane et al., 2022). The results indicate that the majority of parameters are within acceptable limits, suggesting that the water is of generally good quality. However, slight variations in sulfate and sodium levels were observed in the samples.

*Table 1. Physicochemical characteristics of water samples compared to standards*

| Physicochemical Parameters | Sample 1 | Sample 2 | Sample 3 | Sample 4 | Standard |
|---|---|---|---|---|---|
| Calcium (mg/l) | 47.3 | 46.73 | 46.9 | 47.29 | 200 |
| Chloride (mg/l) | 142 | 138.45 | 142 | 138.45 | 250 |
| Hardness TH (mg/l $CaCO_3$) | 20.2 | 19.7 | 19.2 | 19.2 | 60 |
| TA | 0 | 0 | 0 | 0 | - |
| TAC (mg/l) | 12.25 | 12.25 | 12.25 | 12.25 | $\geq 2.5$ |
| Temperature (°C) | 21 | 21 | 23 | 23 | $\leq 25$ |
| pH | 7.56 | 7.60 | 7.61 | 7.66 | $6.5 \leq pH \leq 9$ |
| Conductivity (µS/cm at 25°C) | 763 | 763 | 767 | 764 | 2700 |
| Dry residue at 110°C (mg/l) | 534 | 534 | 536.9 | 534.8 | - |
| Dry residue at 525°C (mg/l) | 373 | 373 | 375.83 | 347.36 | - |
| Turbidity (NTU) | 0.36 | 0.13 | 0.15 | 0.14 | 5 |
| Potassium (mg/l) | 8.5 | 8.4 | 8.4 | 8.4 | 12 |
| Sodium (mg/l) | 74.4 | 74.8 | 75.8 | 76.6 | 200 |
| Sulfates (mg/l) | 53.2 | 51.4 | 52.4 | 57 | 250 |
| Magnesium (mg/l) | 20.42 | 4.02 | 19.55 | 4 | 150 |
| Nitrates (mg/l) | 0.25 | 0.11 | 0.42 | 0.886 | 50 |
| Nitrites (mg/l) | 0 | 0 | 0 | 0 | 0.1 |
| Ammonium (mg/l) | - | 0 | 0 | 0 | 0.01 |
| Bicarbonates (mg/l) | 155.55 | 155.5 | 155.5 | 155.5 | 500 |

The physico-chemical analysis of tap water samples collected from different supply points in the province of Salé demonstrates overall compliance with national and international drinking water standards. The pH (**Figure 4 (a)**), of the water was found to be slightly basic, ranging between 7.56 and 7.66, suggesting that the water is of surface origin and chemically stable. The Total Alkalinity Concentration (TAC) was constant across all samples (12.25 mg/L), consistent with stable bicarbonate ($HCO_3^-$) presence. The bicarbonate buffering capacity is essential for pH stabilization and protects against sudden acidification.

*Figure 4. Comparative analysis of key physicochemical parameters in water samples (S1–S4) relative to international drinking water standards*

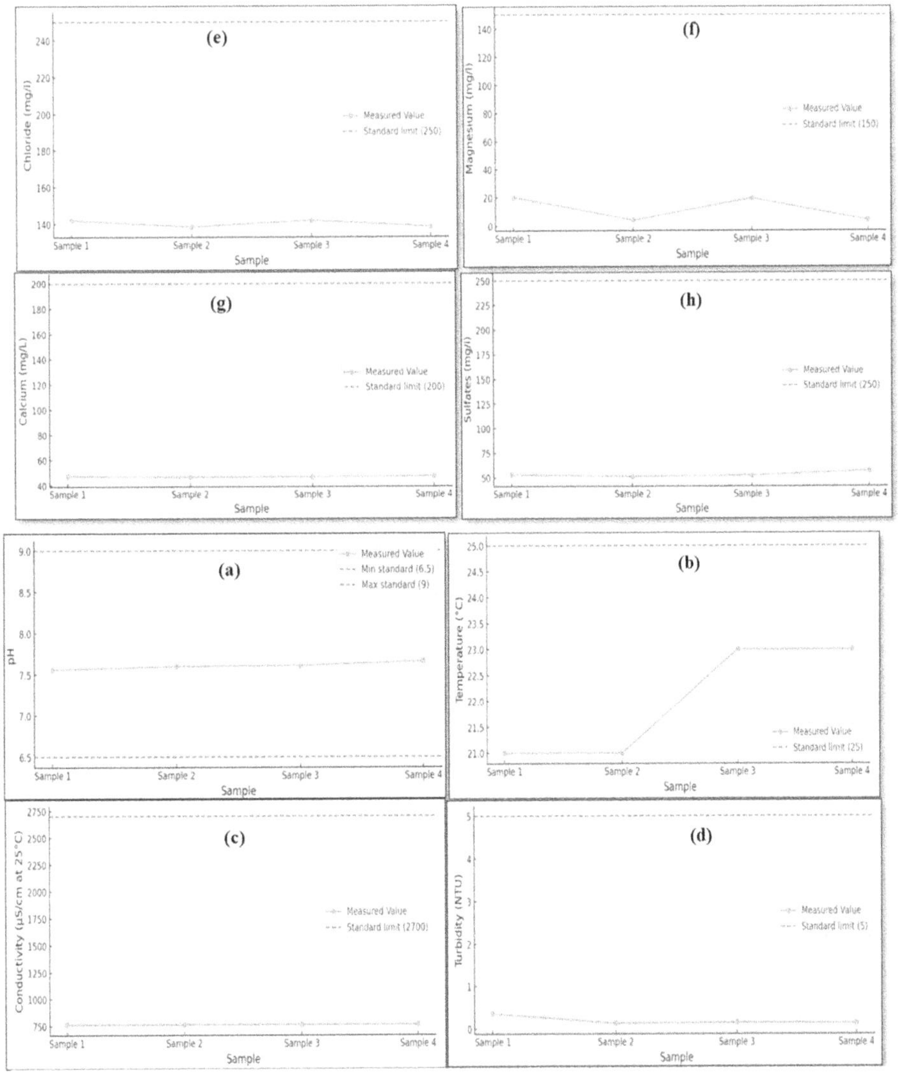


The pH ((**Figure 4 (a)**) of the water was found to be slightly basic, ranging between 7.56 and 7.66, suggesting that the water is of surface origin and chemically stable. The Total Alkalinity Concentration (TAC) was constant across all samples (12.25 mg/L), consistent with stable bicarbonate ($HCO_3^-$) presence. The bicarbonate buffering capacity is essential for pH stabilization and protects against sudden acidification.

Temperature values ranged between 21°C and 23°C, remaining well below the maximum allowable limit of 25°C, thus ensuring comfort in domestic use and microbial stability ((**Figure 4 (b)**). Electrical conductivity varied slightly between 763 and 767 µS/cm, indicating moderate mineralization and confirming that the water remains within the potability threshold of 2700 µS/cm (**Figure 4 (C)**). Although elevated conductivity may suggest a high ionic load, the values observed reflect water that is still acceptable for consumption. Turbidity levels were low across all samples (0.13 to 0.36 NTU), confirming good physical clarity and absence of significant particulate matter, in accordance with Moroccan standards which set the upper limit at 5 NTU (**Figure 4 (d)**), Chloride concentrations were relatively high (138.45–142 mg/L), yet remained within the recommended maximum of 250 mg/L (**Figure 4 (e)**), These levels could be attributed to the geological nature of the catchment area or the proximity to coastal influences but do not indicate anthropogenic contamination.

Magnesium concentrations ranged from 4 to 20.42 mg/L and calcium levels from 46.73 to 47.3 mg/L **(Figures 4 (f) and (g)),** both remaining well below their respective threshold values (150 mg/L and 200 mg/L). These ions contribute to total water hardness, which was found to vary between 19.2 and 20.2 mg/L $CaCO_3$, classifying the water as soft. Such levels are generally favorable, minimizing scaling in domestic appliances while providing essential minerals. The presence of $Ca^{2+}$ ions can be attributed to the dissolution of carbonate ($CaCO_3$) and/or gypsum ($CaSO_4{\cdot}2H_2O$) formations, through the following equilibria:

- $CaCO_3 \rightleftharpoons Ca^{2+} + CO_3^{2-}$
- $CaSO_4{\cdot}2H_2O \rightleftharpoons Ca^{2+} + SO_4^{2-} + 2H_2O$

Sulfate concentrations ranged between 51.4 and 57 mg/L, significantly below the limit of 250 mg/L (**Figure 4 (h)**). This suggests a limited influence of sulfate-bearing rocks or industrial discharge in the study area. Similarly, nitrate concentrations were very low (0.11–0.88 mg/L), far below the guideline value of 50 mg/L, indicating the absence of agricultural runoff or wastewater infiltration.

The results show that the water quality of the Sale network is generally acceptable from a physico-chemical point of view. The parameters measured do not just conform to regulatory standards, but also show an excellent mineral balance and a low human-caused impact. This demonstrates the efficiency of the water distribution and treatment infrastructures in the region.

## 4. CONCLUSION

As the primary element of life, has to be protected with extreme attention to detail, not only to ensure adequate supply for the needs of humans, but also to ensure its quality. Although it is a vital element, it also can become an agent of disease when not properly controlled. This dual nature is the reason of our study. Our research was focused on analyzing the physicochemical qualities of tap water that is distributed across four major supply zones within Sale. The city has four supply zones Sale (Bettana, Sala Jadida, Hay Rahma, and Hay Inbiaat), all of which are nourished via the Bouregreg water complex. The analysis was based on a thorough analysis of the physico-chemical parameters employing standard methods for the determination of major minerals and trace mineral ions. The results showed that the most prevalent ions in all the samples were chlorides, magnesium and calcium. These are elements typically found in natural water. However, ammonium and nitrate levels were extremely low, which suggests that there is no contamination by domestic or agricultural waste sources. The consistency of the measurements across the various sampling sites also supports the idea that there is a common source for water. In the end, based on the examination of all parameters with Moroccan standard for drinking water and international guidelines, the water that is distributed within the Province of Sale is classified as potable and chemically safe for consumption by humans. This study demonstrates the need for regular monitoring and provides a reliable baseline for future assessments of the quality of water within the region.

## REFERENCES


Abood, A. R., Bao, J., Du, J., Zheng, D., & Luo, Y. (2014). Non-biodegradable landfill leachate treatment by combined process of agitation, coagulation, SBR and filtration. *Waste Management (New York, N.Y.)*, *34*(2), 439–447. DOI: 10.1016/j.wasman.2013.10.025 PMID: 24287299

Al-Jadabi, N., Laaouan, M., El Hajjaji, S., Mabrouki, J., Benbouzid, M., & Dhiba, D. (2023). The dual performance of Moringa oleifera seeds as eco-friendly natural coagulant and as an antimicrobial for wastewater treatment: A review. *Sustainability (Basel)*, *15*(5), 4280. DOI: 10.3390/su15054280

Azdem, D., Mabrouki, J., & El Hajjaji, S. (2024). Methods and Technologies for the Treatment and Disposal of Brine in Seawater Desalination Plants. In Mabrouki, J., & Azrour, M. (Eds.), (pp. 79–87). Advances in Environmental Engineering and Green Technologies. IGI Global., DOI: 10.4018/979-8-3693-3807-0.ch007

Azdem, D., Mabrouki, J., & El Hajjaji, S. (1 C.E., January 1). *The Use of Artificial Intelligence in a Desalination Plant Powered by Renewable Energy* [Chapter]. Https://Services.Igi-Global.Com/Resolvedoi/Resolve.Aspx?Doi=10.4018/979-8-3373-1220-0.Ch004Global, I. G. I. https://www.igi-global.com/chapter/the-use-of-artificial-intelligence-in-a-desalination-plant-powered-by-renewable-energy/373103

Azdem, D., Mabrouki, J., Fahdi, G., & Hajjaji, S. E. (2025). ENVIRONMENTAL SUSTAINABILITY OF DESALINATION: A CRITICAL REVIEW OF BRINE TREATMENT CHALLENGES AND MITIGATION STRATEGIES. *Water Conservation and Management*, *9*(1), 55–60. *Scopus.* Advance online publication. DOI: 10.26480/wcm.01.2025.55.60

Azdem, D., Mabrouki, J., Moufti, A., El hajjaji, S., & Fatni, A. (2024). Assessment of heavy metal contamination in seawater in Agadir coastline, Morocco. *Desalination and Water Treatment*, *317*, 100129. DOI: 10.1016/j.dwt.2024.100129

Bencheikh, I., Azoulay, K., Mabrouki, J., El Hajjaji, S., Dahchour, A., Moufti, A., & Dhiba, D. (2020). The adsorptive removal of MB using chemically treated artichoke leaves: Parametric, kinetic, isotherm and thermodynamic study. *Scientific African*, *9*, e00509. DOI: 10.1016/j.sciaf.2020.e00509

Bencheikh, I., Azoulay, K., Mabrouki, J., El Hajjaji, S., Moufti, A., & Labjar, N. (2021). The use and the performance of chemically treated artichoke leaves for textile industrial effluents treatment. *Chemical Data Collections*, *31*, 100597. DOI: 10.1016/j.cdc.2020.100597

Benchrifa, M., Elouardi, M., Fattah, G., Mabrouki, J., & Tadili, R. (2023). Identification, simulation and modeling of the main power losses of a photovoltaic installation and use of the internet of things to minimize system losses. In *Advanced technology for smart environment and energy* (pp. 49–60). Springer. DOI: 10.1007/978-3-031-25662-2_4

Benchrifa, M., & Mabrouki, J. (2022). Simulation, sizing, economic evaluation and environmental impact assessment of a photovoltaic power plant for the electrification of an establishment. *Advances in Building Energy Research*, *16*(6), 736–753. DOI: 10.1080/17512549.2022.2096693

Benchrifa, M., Mabrouki, J., Elouardi, M., Azrour, M., & Tadili, R. (2023). Detailed study of dimensioning and simulating a grid-connected PV power station and analysis of its environmental and economic effect, case study. *Modeling Earth Systems and Environment*, *9*(1), 53–61. DOI: 10.1007/s40808-022-01457-9

Boguniewicz-Zablocka, J., Klosok-Bazan, I., & Naddeo, V. (2019). Water quality and resource management in the dairy industry. *Environmental Science and Pollution Research International*, *26*(2), 1208–1216. DOI: 10.1007/s11356-017-0608-8 PMID: 29098587

Cao, X., Xiong, H., Fan, Y., & Xiong, L. (2024). Comparing the effects of two culture methods to determine the total heterotrophic bacterial colony count in hospital purified water. *Journal of Epidemiology and Global Health*, *14*(1), 184–192. DOI: 10.1007/s44197-023-00186-1 PMID: 38358615

Cutter, G., Andersson, P., Codispoti, L., Croot, P., Francois, R., Lohan, M. C., Obata, H., & Rutgers vd Loeff, M. (2010). *Sampling and sample-handling protocols for GEOTRACES cruises.*

El Alouani, M., Aouan, B., Rachdi, Y., Alehyen, S., El Herradi, E. H., Saufi, H., Mabrouki, J., & Barka, N. (2024). Porous geopolymers as innovative adsorbents for the removal of organic and inorganic hazardous substances: A mini-review. *International Journal of Environmental Analytical Chemistry*, *104*(16), 4784–4796. DOI: 10.1080/03067319.2022.2115891

El Mouedden, M., Azdem, D., Mabrouki, J., & El Hajjaji, S. (2025). MarocFarm: A digital revolution for agriculture in Morocco. In *Circular Economy Applications in Energy Policy* (pp. 141–155). Scopus. DOI: 10.4018/979-8-3693-6680-6.ch010

Ezzaouini, M. A., Mahé, G., Kacimi, I., & Zerouali, A. (2020). Comparison of the MUSLE model and two years of solid transport measurement, in the Bouregreg Basin, and impact on the sedimentation in the Sidi Mohamed Ben Abdellah Reservoir, Morocco. *Water (Basel)*, *12*(7), 1882. DOI: 10.3390/w12071882

Fahdi, G., Azdem, D., Lamchaimech, A., Benrhanem, M., Mabrouki, J., & El Hajjaji, S. (2024). ANALYSIS OF THE IMPACT OF CLIMATE CHANGE ON WATER RESOURCES: CASE OF THE TENSIFT BASIN (MOROCCO). *Water Conservation and Management, 8*(4), 461–465. *Scopus.* Advance online publication. DOI: 10.26480/wcm.04.2024.461.465

Fahdi, G., Mabrouki, J., Lamchaimech, A., Azdem, D., Benrhanem, M., & Hajjaji, S. E. (2024). IMPACTS OF CLIMATE CHANGE ON GROUNDWATER QUALITY AND RECHARGE IN THE TENSIFT BASIN, MOROCCO. *Water Conservation and Management, 8*(4), 466–474. *Scopus.* Advance online publication. DOI: 10.26480/wcm.04.2024.466.474

Fatni, A., Addich, M., Essekri, A., & Baraka, E. L. (2025). Characterization and mitigation strategies for inorganic scaling in reverse osmosis system treating brackish groundwater. *Sustainable Chemistry One World*, *5*, 100035. DOI: 10.1016/j.scowo.2024.100035

Ghizlane, F., Mabrouki, J., Ghrissi, F., & Azrour, M. (2022). Proposal for a high-resolution particulate matter (PM10 and PM2. 5) capture system, comparable with hybrid system-based internet of things: Case of quarries in the western rif, Morocco. *Pollution*, *8*(1), 169–180.

Loukili, H., Mabrouki, J., Anouzla, A., Kouzi, Y., Younssi, S. A., Digua, K., & Abrouki, Y. (2021). Pretreated Moroccan natural clays: Application to the wastewater treatment of textile industry. *Desalination and Water Treatment*, *240*, 124–136. DOI: 10.5004/dwt.2021.27644

Mabrouki, J., Azrour, M., Farhaoui, Y., & El Hajjaji, S. (2020). Intelligent system for monitoring and detecting water quality. *Big Data and Networks Technologies*, *3*, 172–182. DOI: 10.1007/978-3-030-23672-4_13

Mabrouki, J., Azrour, M., Fattah, G., Dhiba, D., & El Hajjaji, S. (2021). Intelligent monitoring system for biogas detection based on the Internet of Things: Mohammedia, Morocco city landfill case. *Big Data Mining and Analytics*, *4*(1), 10–17. DOI: 10.26599/BDMA.2020.9020017

Mabrouki, J., Azrour, M., & Hajjaji, S. E. (2021). Use of internet of things for monitoring and evaluating water's quality: A comparative study. *International Journal of Cloud Computing*, *10*(5–6), 633–644. DOI: 10.1504/IJCC.2021.120399

Mabrouki, J., Benbouzid, M., Dhiba, D., & El Hajjaji, S. (2022). Simulation of wastewater treatment processes with Bioreactor Membrane Reactor (MBR) treatment versus conventional the adsorbent layer-based filtration system (LAFS). *International Journal of Environmental Analytical Chemistry*, *102*(19), 7458–7468. DOI: 10.1080/03067319.2020.1828394

Mabrouki, J., Benchrifa, M., Ennouhi, M., Azoulay, K., Bencheikh, I., Rachiq, T., El-Moustaqim, K., Al-Jadabi, N., Azrour, M., & Fatima-zahra, A. (2022). Geographic information system for the study of water resources in Chaâba El Hamra, Mohammedia (Morocco). *The International Conference on Artificial Intelligence and Smart Environment*, 469–474.

Mabrouki, J., Moufti, A., Bencheikh, I., Azoulay, K., El Hamdouni, Y., & El Hajjaji, S. (2020). Optimization of the coagulant flocculation process for treatment of leachate of the controlled discharge of the city Mohammedia (Morocco). *Advanced Intelligent Systems for Sustainable Development (AI2SD'2019) Volume 7-Advanced Intelligent Systems for Sustainable Development Applied in Energy and Electrical Engineering*, 200–212.

Mgalaa, S., Mabrouki, J., Elouardi, M., El Azzouzi, L., Moufti, A., El Hajjaji, S. E. L., Azzouzi, M., & El Belghiti, M. A. (2022). Study and evaluation of the degradation of procion blue dye by the ozonation method: Parametric and isothermal study. *Nanotechnology for Environmental Engineering*, *7*(3), 691–697. DOI: 10.1007/s41204-022-00262-0

Rachiq, T., Abrouki, Y., Mabrouki, J., Samghouli, N., Fersi, C., Rahal, S., & El Hajjaji, S. (2021). Evaluation of the efficiency of different materials to remove specific pollutants from landfill leachate. *Desalination and Water Treatment*, *238*, 240–250. DOI: 10.5004/dwt.2021.27779

Rachiq, T., Mabrouki, J., El Hajjaji, S., & Rahal, S. (2022). Simulation of the treatment performance of a purification plant for a dairy effluent. In *IoT and smart devices for sustainable environment* (pp. 19–27). Springer. DOI: 10.1007/978-3-030-90083-0_2

Rachiq, T., Samghouli, N., Mabrouki, J., El-Moustaqim, K., Rahmani, M., Al-Jadabi, N., Hajjaji, S. E., & Rahal, S. (2022). Study and modeling of dairy effluents treatment using design of experiments methodology. *Desalination and Water Treatment, 261*, 131–140. *Scopus.* Advance online publication. DOI: 10.5004/dwt.2022.28524

Tsaridou, C., & Karabelas, A. J. (2021). Drinking water standards and their implementation—A critical assessment. *Water (Basel)*, *13*(20), 2918. DOI: 10.3390/w13202918

Chapter 21

# Preliminary In Vitro Tests to Demonstrate Various Formulations' Effect on the Stability of the Moroccan Trichoderma asperellum-Based Product's Biology

**Abdellatif Ouazzani Chahdi**
https://orcid.org/0009-0004-9880-8311
*Ibn Tofail University, Morocco*

**Fadoua Berber**
*Ibn Tofail University, Morocco*

**Amina Ouazzani Touhami**
https://orcid.org/0000-0001-9970-7724
*Ibn Tofail University, Morocco*

**karima Selmaoui**
*Ibn Tofail University, Morocco*

**Rachid Benkirane**
*Ibn Tofail University, Morocco*

**Allal Douira**
*Ibn Tofail University, Morocco*

## ABSTRACT

*The selection of the T asperellum-based product required several in vitro tests in order to preserve its efficacy during the formulation process, while maintaining the viability of the conidia and safeguarding their antagonistic properties. The addition of talc to all of the compounds added to the formulation of T asperellum's conidia, Moroccan strain isolated from compost, preserved their viability, stability and antagonistic power during the entire duration under storage conditions. P1 and P2's conidia germination percentages while stored for 38 weeks at room temperature were 0% - 12% and at 4°C were 31% - 46.33% respectively. The antagonistic effect of P1 and P2 against V dahliae by direct contact is estimated by the T asperellum colonization percentages, the mycelial growth of V dahliae's inhibition percentages, the mycelial growth of V dahliae's inhibition percentages and the mycelial inhibition percentages by antibiosis are respectively 0 - 79.58%, 0 - 39.24% and 3.25 - 39.84% at room temperature and 80.22 - 81.49%, 39.83 - 44.99% and 42.27 - 43.09% at 4°C.*

DOI: 10.4018/979-8-3693-8980-5.ch021

## 1. INTRODUCTION

Biological control is an important promising avenue to control plant pathogenic diseases and reduce the damage caused by the intensive use of chemical pesticides.

In this context, several researchers have declared that species of the *Trichoderma* genus and in particular *Trichoderma viride* and *Trichoderma harzianum* are the best antagonists against several soil and seed phypathogens (Poddar *et al.,* 2004).

The biological control agent formulated product's shelf life plays an important role in the success of commercialization (Sanjeev et al., 2014). And to maintain the duration of viability of biological control agents, several works were carried out using different adjuvants at different formulations. Among these works, talc, peat, lignite and kaolin preparations of *Trichoderma* have a shelf life of 3-4 months (Sanjeev et al., 2014).

The PDBC (Project Directorate of Biological Control) in Bangalore is working on increasing the shelf life of talc-based *Trichoderma* formulations by using various ingredients (chitin and glycerol) in the production medium and heat shock at the end of the fermentation phase which allows shelf life to be extended up to one year (Sriram et al., 2010, 2011).

The objective of this study is to demonstrate, *in vitro,* the effects of various formulations on the viability and antagonistic power of *Trichoderma asperellum* based product.

## 2. MATERIALS AND METHODS

### 2.1 Effect of the Protectant on the Viability and Antagonistic Power of *Trichoderma Asperellum* During Storage

Two organic products based on a Moroccan strain of *T. asperellum*'s conidia registered under the code "Banklt1902509 SMis 1 KU987252 RAB 95369", one of which is formulated using talc (protectant) registered under patent MA 41534B2 (Ouazzani Chahdi et al., 2019). named P2 and the other without the protectant named P1. With the exception of the protectant, the other additives are similar in both products.

### 2.2 *In vitro* Tests of the Two Biological Products P1 and P2 Against *Verticillium Dahliae* Testing the Viability of Formulated P1 and P2 Conidia

Conidial suspensions from the two products formulated based on *Trichoderma asperellum*: P1 and P2 were prepared and adjusted to a concentration of $10^3$ conidia/ml using the Malassez slide.

In boxes containing 20 ml of agar water (15 g of Agar-agar, 1000 ml distilled water), 0.1 ml of each of the two suspensions was spread, with three repetitions for each product. After 24 hours of incubation in the dark at 28°C, the conidia's germination percentage was determined by counting the spores that had germinated in function of the total number of spores.

## 2.3 Direct Confrontation Test

In a Petri dish containing PSA medium, a 5 mm diameter disk made up of the inoculum of a *Verticillium dahliae* strain known for its aggressiveness on tomatoes (Douira and Lahlou, 1989), and a sterilized round of filter paper 5 mm in diameter, soaked with the *Trichoderma* product alone (P1 or P2) or combined with phytosanitary products were placed 40 mm from each other, symmetrically in relation to the center of the box. For the control, a mycelial disk of the pathogen was placed in the center of the box.

After 8 days of incubation in the dark at 28°C, the colonization percentage of colonization C (%) of the space was evaluated according to the Comporota method (1985):

$$C\,(\%) = (DT/DE) \times 100$$

C (%): Percentage of *Trichoderma* colonization;
DT: The distance traversed by the antagonist;
DE: The distance separating the two explants.

The inhibition percentage IC (%) of the pathogen's mycelial growth by the antagonists was evaluated according to the Sy method (1976):

$$IC(\%) = \frac{D_T - D_{PA}}{D_T} \times 100$$

$D_T$: diametrical growth of the control;
$C_{PA}$: diametrical mycelial growth of the pathogen in the presence of the antagonist.

## 2.4 Volatile Substance Test

It was highlighted by the Dennis and Webster technique (1971a). In two Petri dish bases containing the PSA medium, a 5 mm diameter mycelial disk of *Verticillium dahliae* is added to the center of a base and a round of filter paper of 5 mm in diameter soaked in the *Trichoderma asperellum* product (Banklt1902509 SMis 1 KU987252 RAB 95369) alone or combined is placed in the center of the other petri dish's base. The two juxtaposed bases (the antagonist at the bottom and the pathogen at the top) are closed with cellophane. The base of the petri dish containing the medium alone is placed below the base of the petri dish containing the pathogen, which constitutes the control (Berber et al., 2009). After 8 days of incubation at 28°C and in the dark, the inhibition of mycelial growth is estimated as above.

## 2.5 Comparison Between Solid and Viscous Formulation

To highlight the most appropriate formulation to preserve the viability and antagonistic power of *Trichoderma asperellum* during storage, three products: The conidial suspension obtained from a young culture, Product P1 and Product P2 underwent freeze-drying just after their preparation and they were then compared with a P2 Product kept for a month.

Three in vitro tests (germination test, antagonism test by direct contact and antagonism test by antibiosis using volatile substances) were carried out and three repetitions were carried out for each treatment. The pathogen chosen was (OMV5) from *Verticillium dahliae.*

## 2.6 Statistical Analysis

Data processing focused on variance analysis. A comparison of the means was carried out by the PPDS test (least significant difference) if a significant difference was recorded at a probability threshold of less than 5%.

# 3. RESULTS

## 3.1 Effect of Protectant (talc) on the Viability and Antagonistic Power of *Trichoderma Asperellum* During Storage In Vitro Tests of the Two Biological Products P1 and P2 Against *Verticillium Dahliae*

### Germination Test

The addition of the protectant makes it possible to maintain *T. asperellum* conidia's viability for more than 38 weeks regardless of the storage temperature. The germination percentages of P1 and P2 conidia after 20 weeks of storage at room temperature were respectively 27.33% and 52.33%, while after 28 weeks they were respectively 0% and 12% (Table 1, Figure 1). Storing the two products at 4°C makes it possible to keep the conidia's viability almost identical with a clear advantage in P2's case. The germination percentages after 38 weeks are 31% and 46.33% respectively (Table 1, Figure 1).

### Antagonism Test

The protectant's addition is essential to maintaining the product's antagonistic potential for a long time in ambient conservation conditions.

The direct comparison's results show that the two products P1 and P2 stored at 4°C. show no significant difference even after 38 weeks of storage. The colonization percentages are respectively 80.22% and 81.49% and the inhibition percentages are respectively 39.83% and 44.99% (Table 2, Figure 1).

Whereas at room temperature, the colonization and inhibition percentages are nil for the P1 product after 30 weeks of storage. On the other hand, the P2 product can withstand storage conditions of more than 38 weeks. The colonization and inhibition percentages are 79.58% and 39.24% respectively (Table 2, Figure 1).

The same results are observed in *Verticillium dahliae*'s inhibition by antibiosis using volatile substances and in the pathogen's sporulation reduction. The inhibition of the pathogen, by volatile substances, allowed the products P1 and P2 to be stored for 38 weeks with inhibition percentages reaching respectively 3.25% and 39.84% at room temperature and 42.27% and 43.09% at 4°C (Table 3, Figure 1).

With the exception of the P1 product stored at room temperature, the other products showed a pathogen conidia production inhibition percentage of more than 99.9% after 38 weeks of storage (Table 4).

### Comparison Between Solid and Viscous Formulation

The viscous formulation shows better protection of viability and antagonistic power depending on the duration and conditions of storage compared to the solid formulation.

The results show a huge difference between the germination power and the antagonistic power of the lyophilized products and the P2 product (Tables 5 and 6, Figure 2).

The germination percentages of *T. asperellum* for freeze-dried products are between 0 and 0.67%, while for product P2, whose formulation is viscous, stored for one month, the germination percentage is 90.33%.(Table 5).

The results in Table 6 further show that the conidial suspension and the lyophilized P1 product do not exhibit any pathogen colonization or inhibition power either by direct contact or by antibiosis. For the lyophilized P2 product, the colonization and inhibition percentages, in the case of direct contact, are respectively 20.83% and 9.64% while in the case of antibiosis the inhibition percentage is 7.61%. As for the P2 product, the colonization and inhibition percentages, in the case of direct contact, are respectively 87.24% and 69.62% while in the case of antibiosis the inhibition percentage is 68.8% (Table 6).

## 4. DISCUSSION AND CONCLUSION

In order to preserve a biological control agent for as long as possible, several works have been carried out. Nashwa *et al.*, (2008) demonstrated that the powder formulation of *Trichoderma spp.* allowed it to retain a viability over 40% after 120 days of conservation at room temperature.

The addition of a protectant and in particular talc to a conidia formulation, made it possible to increase the conidia's viability -during storage. Bhat *et al.*, (2009) produced, by fermentation, a liquid product of *T. harzianum* to which the addition of talc in $2\times10^6$ CFU made it possible to preserve for more than 150 days of storage. Similarly, Sriram *et al.* (2011) showed that the addition of chitin and glycerol to talc-based formulations increased the viability of *Trichoderma viride* conidia for up to one year. Sanjeev *et al.*, (2014) showed that talc-based formulations of *Trichoderma* have a shelf life of 3 to 4 months.

Bhat *et al.*, (2009) stated that the shelf life of the talc formulation was up to 180 days under ambient conditions.

A talc-based preparation maintains the viability of *T. virens* conidia up to 82% at 5°C after 6 months of storage in the refrigerator, while at room temperature the same percentage of viability was observed for a period of 3 months (Chaube *et al.*, 2002).

Bhat et al., (2009) found a number of $10^5$ to $10^8$ CFU/g of *T. harzianum* conidia preserved on talc with the addition of carboxymethylcellulose after six months of storage.

The study of the stability of *Trichoderma harzianum* and *Trichoderma virens* during storage by Gade *et al.* (2009) showed that the addition of talc to the biopreparation leads to $10^7$ CFU/g after six months of storage conservation.

The addition of talc not only affects the viability of the conidia but also the antagonistic power of the fungus. In this context, Ramakrishnan *et al.,* (1994) showed that seed treatment with *T. viride* formulated with talc effectively reduced the incidence of bean root rot even 120 days after storage.

And as part of determining the most suitable formulation to maintain the potential for viability and antagonism during storage whatever the conditions; Butler (1980); Tommerup and Kidby (1979); Dahmen *et al.* (1983) showed that the classic techniques most often used in the laboratory to preserve the conidiospores of filamentous collection fungi are: freezing at -18°C, conservation in liquid nitrogen at -196° C and freeze-drying. However, all of these methods apply to very small quantities of conidiospores and require a significant investment in equipment or operation (Roussos, 1985).

Roussos (1985) showed that the preservation method by drying under vacuum at a temperature of +40°C using a rotavapor makes it possible to maintain 40% of viable conidiospores.

Grzegorczyk *et al.,*(2018) found that the stability of *Trichoderma* strains freeze-dried and stored for 3 months at room temperature varies between 0.08 and 95%.

Panahian *et al.,* (2012) noted a reduction in the survival rate of freeze-dried *T. atroviride* and *T. koningii* strains after three months of storage, reaching 9% compared to the control.

From these results, it can be concluded that the liquid formulation has a very significant potential in increasing the biocontrol agent 's stability, viability and effectiveness while stored for a long time either at room temperature or at low temperature.

## REFERENCES


Berber, F., Ouazzani Touhami, A., Badoc, A., & Douira, A. **2009.**In vitro and in vivo antagonism of two Trichoderma against four species of Bipolaris pathogenic on Sorghum. Bull. Soc. Pharm. Bordeaux, 2009, 148, 93-114 pp.

Bhat, K. A., Anwar, A., Lone, G. M., Hussain, K., & Nazir, G. (2009). Shelf life of liquid fermented product of Trichoderma harzianum in talc. *Journal of Mycology and Plant Pathology*, *39*(2), 263.

Butler, E. E. (1980). A method for long-time culture storage of *Rhizoctonia solani.Phytopathology*, *22*(8), 820–821. DOI: 10.1094/Phyto-70-820

Chaube, H. S., Mishra, D. S., Varshney, S., & Singh, U. S. (2002). *Biological control of plant pathogens by fungal antagonists: Historical background, present status and future prospect, Annu.* Rev. Plant Pathol.

Comporota, P. (1985). *In vitro* Antagonism of *Trichoderma spp.* against *Rhizoctonia solani* Kühn. *Agronomie*, *5*(7), 613–620. DOI: 10.1051/agro:19850707

Dahmen, H., Staub, Th., & Schwinn, F. J. (1983). Technique for long-term preservation of phytopathogenic fungi in liquid nitrogen. *Phytopathology*, *73*(2), 241–246. DOI: 10.1094/Phyto-73-241

Dennis, C., & Webster, J. (1971a). Antagonistic properties of species-group of *Trichoderma*. I. Production of non-volatile antibiotics. *Transactions of the British Mycological Society*, *57*, 25–39. DOI: 10.1016/S0007-1536(71)80077-3

Douira, A., & Lahlou, H. (1989). Variability of parasite specificity in *Verticillium albo-atrum* Reinke and Berthold frome with microsclerotia. *Cryptogamie. Mycologie*, *10*, 19–32. DOI: 10.5962/p.354573

Gade, R. M., Wardhe, S. R., & Armarker, S. V. (2009). Shelf life study of *Trichoderma spp.* in different carrier materials. *Journal of Maharashtra Agricultural Universities*, *34*(2), 181–182.

Grzegorczyk, M., Kancelista, A., Łaba, W., Piegza, M., & Witkowska, D. (2018). The effect of lyophilization and storage time on the survival rate and hydrolytic activity of *Trichoderma* strains. *Folia Microbiologica*, *63*(4), 433–441. DOI: 10.1007/s12223-017-0581-0 PMID: 29305688

Kumar, S., Thakur, M., & Rani, A. (2014). *Trichoderma*: Mass production, formulation, quality control, delivery and its scope in commercialization in India for the management of plant diseases. *African Journal of Agricultural Research*, *9*(53), 3838–3852.

Nashwa, M. A. S., Abo-Elyousr, K. A. M., & Hassan, M. A. E. (2008). Evaluation of *Trichoderma* Species as Biocontrol Agents for Damping-Off and Wilt Diseases of Phaseolus vulgaris L. and Efficacy of Suggested Formula. *Egypt. J. Phytopathol.*, *36*(1-2), 81–93.

Ouazzani Chahdi, A., Ouazzani Touhami, A., Khirallah, W., Benkirane, R., & Douira, A. **2019.** Production, formulation et recyclage d'un produit biofongicide et biostimulant à base de *Trichoderma asperellum*. MA 41534 Date de publication: 31/05/2019.url:https://patentregister.ompic.ma/SearchPatent/searchByDepot?typeNum=AP&numDepot=41534&count=0&lang=FR

Panahian, R., Rahnama, K., & Jafari, M. (2012). Mass production of *Trichoderma ssp.* and application. *International Research Journal of Applied and Basic Sciences*, *3*(2), 292–298.

Poddar, R. K., Singh, D. V., & Dubey, S. C. (2004). Integrated application of *Trichoderma harzianum* mutants and carbendazim to manage chickpea wilt (*Fusarium oxysporum f.sp. ciceri*). *Indian Journal of Agricultural Sciences*, *74*, 346–348.

Ramakrishnan, G., Jeyarajan, R., & Dinakaran, D. (1994). Talc-Based Formulation of *Trichoderma viride* for Biocontrol of *Macrophomina phaseolina*. *Journal of Biological Control*, *8*(i), 41–44.

Roussos, S. (1985). Growth of *Trichoderma harzianum* by fermentation in a solid medium: physiology, Sporulation and production of cellulases, doctoral thesis; University of Provence, Aix-Marseille.

Sriram, S., Palanna, K. B., & Ramanujam, B. (2010). Effect of chitin on the shelf life of *Trichoderma harzianum* in talc formulation. *Indian Journal of Agricultural Sciences*, *80*, 930–932.

Sriram, S., Roopa, K. P., & Savitha, M. J. (2011). Extended shelf life of liquid fermentation derived talc formulations of *Trichoderma harzianum* with the addition of glycerol. *Crop Protection (Guildford, Surrey)*, *30*(10), 1334–1339. DOI: 10.1016/j.cropro.2011.06.003

Tommerup, I. C., & Kidby, D. K. (1979). Preservation of spores of vesicular-arbuscular endophytes by L-drying. *Applied and Environmental Microbiology*, *37*(5), 831–835.

## APPENDIX

*Table 1. Evolution of Trichoderma asperellum's germinative power formulated in the two products P1 and P2 as a function of storage time and temperature*

| The duration of the conversation | Germination percentage (%) | | | |
|---|---|---|---|---|
| | P1 | | P2 | |
| | Ambient temperature | 4°C | Ambient temperature | 4°C |
| 4 weeks | **89.33a** | **90.67a** | **87.67a** | **90.33a** |
| 8 weeks | 87.33a | 89.33a | 88.33a | 88.67a |
| 12 weeks | 61b | 83.33 ab | 68.33 b | 83.67a |
| 16 weeks | 49.33 b | 73 b | 61.67 bc | 77.67 ab |
| 20 weeks | 27.33c | 66.33c | 52.33c | 72.67b |
| 30 weeks | 0 d | 50.33 d | 32d | 63.67b |
| 38 weeks | 0 d | 31e | 12e | 46.33c |

Two results read on the same column present no significant difference at the 5% threshold if they were assigned the same letter.

*Table 2. Antagonistic effect of Trichoderma asperellum formulated in the two products P1 and P2 against Verticillium dahliae by evaluation of the colonization percentage (C) and the inhibition percentage (I), as a function of storage time and temperature.*

| The duration of the conversation | P1 | | | | P2 | | | |
|---|---|---|---|---|---|---|---|---|
| | Ambient temperature | | 4°C | | Ambient temperature | | 4°C | |
| | VS (%) | I (%) | VS (%) | I (%) | VS (%) | I (%) | VS (%) | I (%) |
| 4 weeks | **87.41a** | **69.62 a** | **87.5a** | **69.62 a** | **86.90 a** | **68.80 a** | **87.24 a** | **69.62 a** |
| 8 weeks | 84.92a | 65.15ab | 83.61a | 68.18a | 84.95 a | 57.31ab | 85.99 a | 60.59a |
| 12 weeks | 85.88a | 49.3bc | 87.14a | 56.22ab | 85.15a | 50.74bc | 85.40 a | 58.13ab |
| 16 weeks | 82.07a | 45.81c | 83.07a | 47.45bc | 83.42a | 46.63bc | 83.78a | 48.28bc |
| 20 weeks | 83.21a | 43.63c | 83.66a | 48.47bc | 82.87 a | 41.71c | 84.31a | 49.10bc |
| 30 weeks | 0 b | 0 d | 82.77 a | 41.46 c | 82.17a | 40.89 c | 83.88 a | 45.81bc |
| 38 weeks | 0 b | 0 d | 80.22a | 39.83c | 79.58 a | 39.24c | 81.49 a | 44.99c |

Two results read on the same column present no significant difference at the 5% threshold if they were assigned the same letter.

*Table 3. Effect of volatile substances produced by Trichoderma asperellum formulated in the two products P1 and P2 on the reduction of mycelial growth of Verticillium dahliae by estimation of the percentage of inhibition (%).*

| The duration of the conversation | P1 | | P2 | |
|---|---|---|---|---|
| | Ambient temperature | 4°C | Ambient temperature | 4°C |
| 4 weeks | 68.8a | 68.8a | 68.80 a | 68.80 a |
| 8 weeks | 66.67a | 68.18a | 65.91a | 66.67a |
| 12 weeks | 58.52 ab | 62.36a | 57.76a | 63.13a |
| 16 weeks | 41.7 bc | 43.34b | 40.07 b | 43.35 b |
| 20 weeks | 38.8c | 42.83 b | 43.64b | 47.67b |
| 30 weeks | 4.87d | 41.46b | 43.09b | 43.90b |
| 38 weeks | 3.25d | 42.27b | 39.84b | 43.09b |

Two results read on the same column present no significant difference at the 5% threshold if they were assigned the same letter.

*Table 4. Verticillium dahliae sporulation inhibition by Trichoderma asperellum formulated in the two products P1 and P2*

| The duration of the conversation | Percentage inhibition of sporulation (%) | | | | | | | |
|---|---|---|---|---|---|---|---|---|
| | P1 (without protectant) | | | | P2 (with protectant) | | | |
| | Direct contact | | antibiosis | | Direct contact | | antibiosis | |
| | AT | 4°C | AT | 4°C | AT | 4°C | AT | 4°C |
| 4 weeks | **99.96a** | **99.95a** | **99.94a** | **99.94a** | **99.96a** | **99.96a** | **99.95a** | **99.96a** |
| 8 weeks | 99.9a | 99.92a | 99.91 a | 99.93 a | 99.91 a | 99.92a | 99.92a | 99.95a |
| 12 weeks | 99.96a | 99.93 a | 99.93 a | 99.93 a | 99.96a | 99.95a | 99.96a | 99.96a |
| 16 weeks | 99.93 a | 99.91 a | 99.94a | 99.95a | 99.95a | 99.96a | 99.95a | 99.95a |
| 20 weeks | 99.95a | 99.95a | 99.94a | 99.95a | 99.94a | 99.96a | 99.94a | 99.95a |
| 30 weeks | 0 b | 99.94a | 0 b | 99.92a | 99.91 a | 99.92a | 99.92a | 99.92a |
| 38 weeks | 0 b | 99.92a | 0 b | 99.9a | 99.92a | 99.92a | 99.91 a | 99.90 a |

**AT:**Ambient temperature

Two results read on the same column show no significant difference at the 5% threshold if they were assigned the same letter.

*Table 5. Effect of solid and viscous formulations on the viability of Trichoderma asperellum conidia by determining the germination percentage.*

| Products | Germination percentage (%) | | |
|---|---|---|---|
| | After 24 hours | After 48h | After 72 hours |
| Freeze-dried SC | 0 b | 0 b | 0 b |
| Freeze-dried P1 | 0 b | 0 b | 0 b |
| Freeze-dried P2 | 0 b | 0.67b | 0.67b |
| P2 | 90.33a | 90.33a | 90.33a |

**SC:** Conidial suspension; P1: Product without prot

*Figure 1. In vitro, the antagonistic effect of the two Trichoderma asperellum products (P1 and P2) against Verticillium dahliae after 10 days of incubation. A: Direct confrontation of P1; B: Direct confrontation of P2; C: Antagonism by antibiosis of P1; D: Antagonism by antibiosis of P1; E: Verticillium dahliae control; An: Antagonist (Trichoderma asperellum); Pa: Pathogen (Verticillium dahliae).*

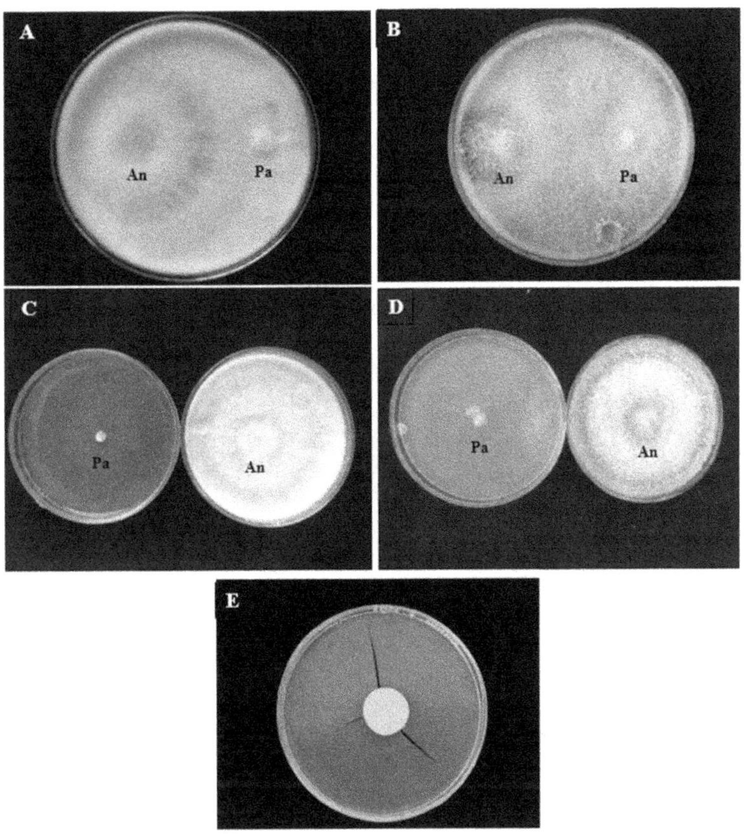

*Figure 2. In vitro, the antagonistic effect of Trichoderma asperellum against Verticillium dahliae after 10 days of incubation by antibiosis using volatile substances. A: Freeze-dried P2 product; B: Freeze-dried product P1; C: Freeze-dried conidial suspension; D: Product P2 stored for one month (control); An: Antagonist (Trichoderma asperellum); Pa: Pathogen (Verticillium dahliae).*

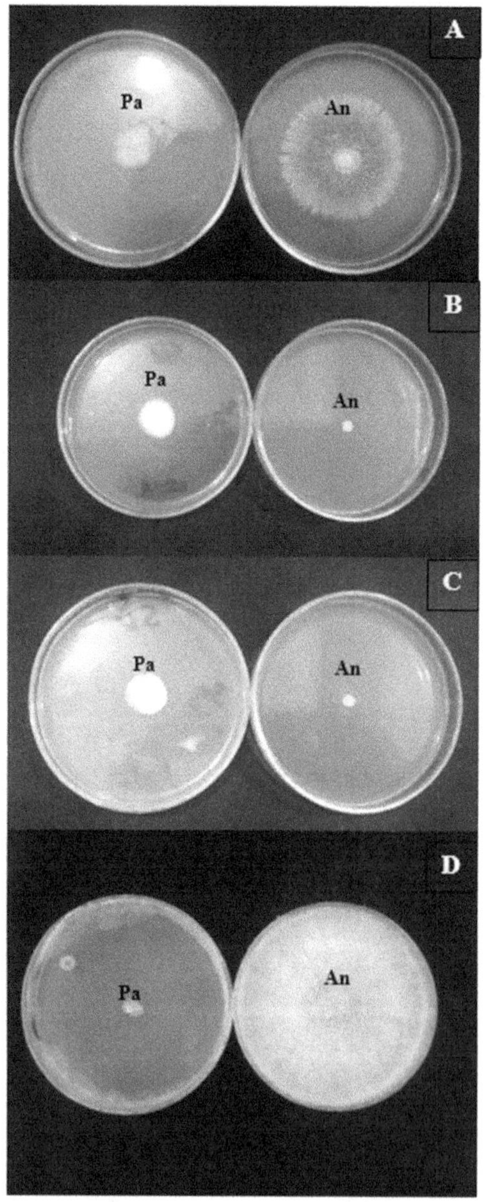

# Chapter 22
# Prospects and Challenges for Green Hydrogen as a Renewable Energy Source

**Shivesh Kumar Uttam**
https://orcid.org/0009-0005-6466-9517
*Panjab University, Chandigarh, India*

**Alka Bali**
https://orcid.org/0000-0002-8034-3518
*Panjab University, Chandigarh, India*

**Pallavi Negi**
*Panjab University, Chandigarh, India*

**Sahil Kumar**
https://orcid.org/0009-0005-1046-8779
*Panjab University, Chandigarh, India*

## ABSTRACT

*In recent years, several forms of sustainable or renewable energy like hydroelectric power, solar energy, nuclear fission, and wind energy have been developed. Hydrogen energy is a new form of energy considered to be more useful and sustainable because of its versatility, storage capabilities, potential for energy security, and its role in achieving global climate goals. Currently, more attention is being paid to production of green hydrogen, which can be regarded as the substitute for other widespread methods of hydrogen production. This has created the opportunity for the utilization of renewable energy sources into the energy basket. This chapter captures a detailed discussion on the production of green hydrogen, its potential uses, advantages and disadvantages, inherent challenges, thus comprehensively exploring its importance in the global energy transition process and the potential that the green hydrogen can bring with a view to catalysing positive socio-economic and environmental returns.*

DOI: 10.4018/979-8-3693-8980-5.ch022

## INTRODUCTION

Growing concern about climate change and the environmental impact of human activities have driven a rising demand for sustainable energy sources (Hassan et al., 2024). According to the report of NASA, average surface temperature of earth in 2024 was the warmest on record since recordkeeping began in 1880 (Fig. 1) (*NASA*, n.d.). According to multi-model projections, if the increase in global temperature continuous at the present rate, it is projected that global warming will warm to the level of 1.5 degree Celsius by the year 2030 to 2052 (*IPCC*, 2021). State of Global Air report 2024, pollution accountant for 8.1 million deaths globally in 2021, becoming the second leading cause for death (*UNICEF,2024*). It is predicted by WHO that climate change will lead to a rise of approximately to 250,000 excess deaths per year between 2030 to 2050 due to heat stress, pollution and global warming (*WHO*, n.d.).

*Figure 1. Maximum temperature rise in degree celsius over the years 2000 to 2024 (Retrieved November 17, 2024, from https://climate.nasa.gov/vital-signs/global-temperature/?intent=121)*

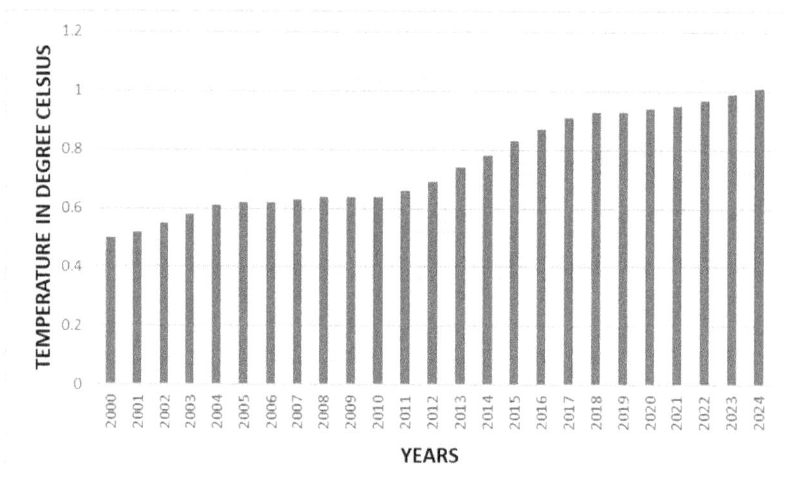


One of the main reasons of climate change is extensive use of fossil fuels and emission of carbon dioxide. In the last three centuries, mankind has experienced an ever-increasing need for fossil fuels for the purposes, including industrialization and urbanization (Narayanan & Sahu, 2014). Further, with growing population, energy demand is also increasing according to the International Energy Agency (IEA) the global energy consumption raised form 13,647 Mtoe on 2015 to 14,477 Mtoe on 2021(Fig. 2) (*IEA*, 2022). So there is an urgent need of decarbonization and to search for new source of energy which will be renewable, sustainable and least polluting (Ghaffour et al., 2015)

*Figure 2. Global energy consumption in mtoe (Million Tonnes of Oil Equivalent) in the Years 2015 to 2021*

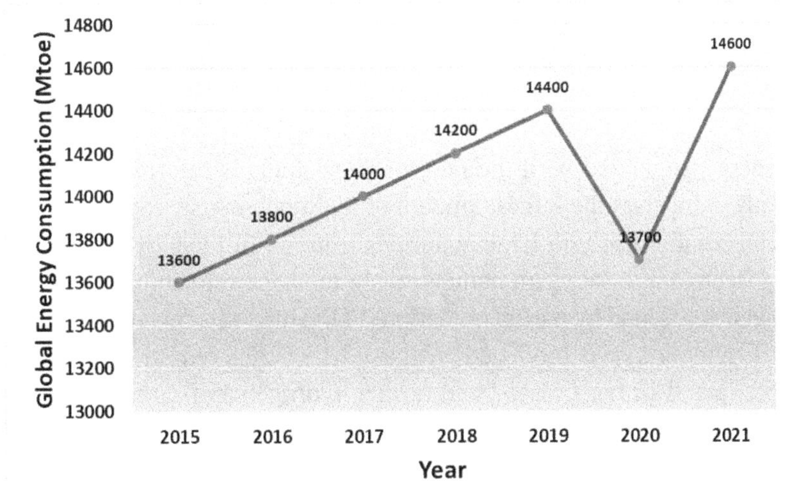


## GREEN HYDROGEN

To reduce the emission of greenhouse gases emission and to lessen the involvement of fossil fuels in the energy market, most countries in the world are focusing on the development of renewable energy sources to reduce their dependence on external supplies and to drive the energy transition (Maka & Mehmood, 2024). One of these alternatives is "green hydrogen", which may actually be a key element in this transition to a greener future.

The term "green hydrogen" has been applied to hydrogen that meets certain sustainability criteria. As there is no international green hydrogen standard, there is no widely accepted definition of green hydrogen (Velazquez Abad & Dodds, 2020). But some researches define green hydrogen as "the hydrogen obtained from renewable sources" (W. Su et al., 2021, 2022). In others words, green hydrogen is a type of hydrogen produced using a renewable energy source, such as solar energy or wind energy, for the production of electric current that is used in the process of conversion" (Atilhan et al., 2021; Vasseghian et al., 2022a). Green hydrogen is a hydrogen gas that is produced through electrolysis using solar, wind or hydroelectric power. Green hydrogen differs from the traditional methods which involve creating hydrogen from natural gas, thus green hydrogen is one of the cleanest and most sustainable form of hydrogen because it emits zero greenhouse gases in the production process and during its use (Gracia et al., 2018; Hassan, Abdulrahman, et al., 2023).

**Hydrogen and types of hydrogen** Hydrogen is the lightest of the known elements and the most abundant gas in the universe, accounting for around 90% of the visible universe. Even though hydrogen is the most abundant element, it is not easy to get hydrogen in its elemental form (Abdin et al., 2020). Hydrogen can be produced by variety of methods. Steam methane reforming is the most widely used approach for producing hydrogen. It can also be produced from renewable energy sources like wind, solar, and water, through electrolysis (Vidas & Castro, 2021). Once produced, hydrogen can be used in a variety of applications, including fuel for transportation, heating, and electricity generation (Ferraren-

De Cagalitan & Abundo, 2021). Hydrogen energy is differentiated into different colour shades based on the technique used to produce hydrogen, the energy source used and its effects on the environment, which will be discussed below.

## Grey Hydrogen

The hydrogen energy generated with the help of fossil fuels is referred as "Grey Hydrogen" and currently, this actually comprises the largest amount of hydrogen in use. (Shafiqah et al., 2022). 'Grey' hydrogen is produced commercially in large quantities from fossil fuels by a hydrocarbon reformation technique involving steam and/or oxygen; namely steam methane reforming (SMR) or autothermal reforming (ATR) of natural gas, and by partial oxidation (POX) of coal or heavy oil. In each case a mixture of hydrogen and carbon monoxide is produced (syngas), which then requires the carbon monoxide to be removed *via* the water gas shift reaction to yield further hydrogen and carbon dioxide (Newborough & Cooley, 2020; Park et al., 2022). Grey hydrogen is mostly used in the petrochemicals industry and for ammonia production (Yukesh Kannah et al., 2021). In general, around 6% of the worldwide extracted natural gas and 2% of coal are used for the production of grey hydrogen per year (Ajanovic et al., 2022). Grey hydrogen is obtained by producing greenhouse gases higher than 36.4 gr $CO_2$ emissions per MJ, renewable or non-renewable (Qureshi et al., 2022a). However, the main problem is that, the $CO_2$ produced during production is not captured and is released into the atmosphere (Fig. 3) (Nikolaidis & Poullikkas, 2017). Also, Carlo Zorzoli, the head of business development at Enel Green Power, has indeed labelled grey hydrogen as a "climate killer" primarily because the production is not environmentally friendly.

*Figure 3. Difference between grey, blue and green hydrogen.*

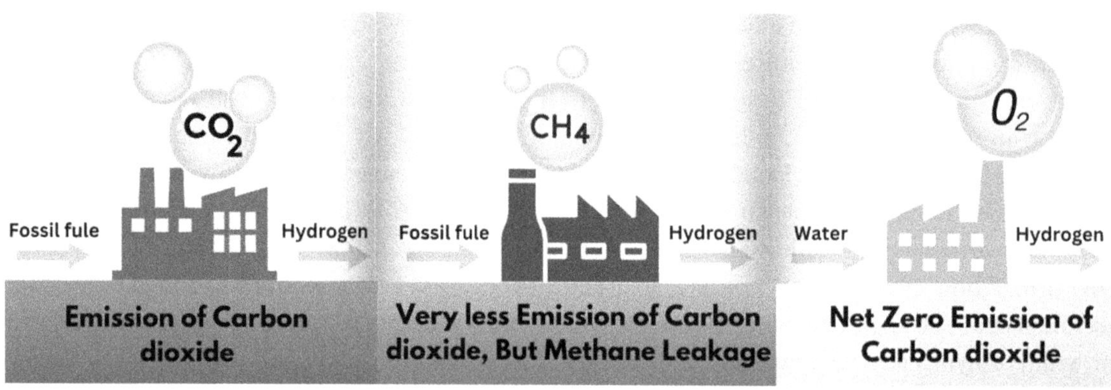


## Blue Hydrogen

Blue hydrogen, similar to grey hydrogen, is produced through steam reforming or gasification. However, the most common blue hydrogen production technique is the steam reformation of methane (natural gas) with Carbon Capture and Sequestration (M. Yu et al., 2021a). In contrast to the grey hydrogen generation, in this case, the $CO_2$ emissions generated during the production process are captured and stored instead of being put into the atmosphere. Thus, it is more eco-friendly than grey hydrogen (M. Yu et al., 2021a). This process of Carbon capturing is known as Carbon Capture Usage and Storage (CCUS; Carbon Capture and Sequestration (CCS) in the United States). Some methods of producing blue hydrogen can be carried out by the carbon absorption method, but it is not inherently free of carbon (Antzaras & Lemonidou, 2022). It has the advantage of being less costly than green hydrogen. Blue hydrogen presents a practical solution, for industries to leverage while still utilizing fossil fuel resources with a reduction, in $CO_2$ emissions (Rooijers & van Cappellen, 2018). One of the main drawbacks of blue hydrogen is methane leaking during the production process, and methane is a potent greenhouse gas, with a 20-year global warming potential (GWP) more than 80 times that of carbon dioxide (Schlissel & Juhn, n.d.). Also, at present, the initial carbon dioxide capture facilities linked to hydrogen generation are only able to capture around 50–60% of the carbon dioxide emissions generated. This limitation arises from their focus, on capturing carbon dioxide from syngas in steam methane reformer processes and not, from the combustion by-products themselves (Bauer et al., 2022).

## Green Hydrogen

One option of obtaining 'zero-carbon' hydrogen is carrying out the hydrogen production from renewables and in the future, renewables can help reduce conflicts linked to oil and gas (Scholten et al., 2020). Green hydrogen can be created through electrolysis using renewable energy sources, like solar power, wind power, hydroelectricity or from biofuels such as biogas in an environmentally friendly manner. When looking at the impact with Life Cycle Assessment, wind and hydroelectricity for electrolysis stands out as a choice, among hydrogen production methods (Bhandari et al., 2014). Today, various methods have been introduced to produce green hydrogen. One of the best and growing methods used for hydrogen production that has received particular attention in recent years is hydrogen generation by electrolysis (d'Amore-Domenech et al., 2020). Water electrolysis is much more economical than green hydrogen generation through synthetic natural gas (Fasihi & Breyer, 2020), but the energy and the production in 2030, is still high cost-competitive with their current equivalent methods (*S&P Global*, n.d.; Hou et al., 2019). Technically usable today and soon emerging as a global economic competitor, hydrogen produced from renewable energy can considerably improve the output of renewable energy (Zhao et al., 2021). Without causing any harm, we can use the green hydrogen and release oxygen into the environment. Although the technique of using energy to split water is expensive, it is far more environmentally benign than the process of creating grey hydrogen (Qureshi et al., 2022b), thus, making green hydrogen the cleanest option (Hassan, Abdulateef, et al., 2023).

## Pink Hydrogen

Pink hydrogen is a type of hydrogen which is produced through thermochemical electrolysis of water, a process that harnesses the heat generated by nuclear reactors to break down water and obtain hydrogen (Fernández-Arias et al., 2024). The main difference between pink hydrogen and green hydrogen constitutes the source of the electrical power supply necessary for electrolysis. Green hydrogen makes use of renewable resources to draw electrical energy (solar, wind or hydroelectric) while pink hydrogen sources electrical energy from nuclear power plants (Arcos & Santos, 2023; Vasseghian et al., 2022b; Y. Zhou et al., 2022). The generation process in both technologies is the same, water electrolysis, and the greenhouse gas emissions generated *via* water electrolysis are zero. There are several advantages of pink hydrogen as outlined below.

- Efficiency: Nuclear energy promulgation is associated with the production of large volumes of heat as a residual of the fission of the atomic nucleus (Dehner et al., 2023a; Horvath & Rachlew, 2016). This heat can be used for nuclear electrolysis which significantly improves system efficiency by harnessing thermal energy that would otherwise be lost.
- Net zero emissions: The production of hydrogen by nuclear electrolysis does not generate direct greenhouse gas emissions, which contributes to the sustainability and reduction of the carbon footprint of nuclear power plants (Dehner et al., 2023b).
- Stable supply: Nuclear plants are known to produce continuous and stable electricity. With this, through electrolysis, the production of hydrogen shall be steady as it is needed in industrial processes and energy application (Lechtenböhmer & Samadi, 2013).

However, the development of pink hydrogen projects may face challenges such as public perception of nuclear energy and high initial capital costs. The clean hydrogen economy holds promise for a net-zero environment, but it is still in its early stages and requires significant changes to become a viable energy source (*The Future of Pink Hydrogen as a Fossil-Fuel Alternative*, n.d.).

## White Hydrogen

Similar to oil and gas, white hydrogen is naturally occurring. Generated by continuous geochemical reactions in hard rock, white hydrogen's characteristics differ from hydrocarbon molecules in that they are small and light (Hood, 2024). The processes by which natural hydrogen or white hydrogen is generated are poorly understood. Such hypotheses include: hydrogen degassing from the earth's core; water reacting with ultrabasic rocks (i.e., serpentinization) or with reducing agents in the mantle; natural radiolysis (i.e., water dissociation by uranium or plutonium); and decomposition of organic matter (Arcos & Santos, 2023).

## Black Hydrogen/ Brown Hydrogen

Black or Brown hydrogen refers to hydrogen produced by coal gasification. The black and brown colours sometimes indicate the coal type: bituminous (black) and lignite (brown). This process generates significant $CO_2$ emissions (De Blasio, 2024). It is regarded as the least environmentally friendly hydrogen production method, creating as much $CO_2$ as burning the source fuel would have in the first place. Around 20 kg of $CO_2$ is released for every kg of brown/black hydrogen produced (Singla et al., 2024). This is a very widely used method of hydrogen production, as coal is the fossil energy source with the largest worldwide reserves. Notably, China produces large amounts of hydrogen through coal gasification because of the high natural gas prices and large coal reserves (Ji & Wang, 2021).

## Aqua Hydrogen

The University of Calgary and Proton Technologies, both based in Western Canada, have developed a new technology for extracting hydrogen from oil sands (natural bitumen) and conventional oil fields without carbon emissions. It involves injecting oxygen into heavy oil reservoirs or oil sands deep underground (Phys.org, 2019). This hydrogen can thus be a zero-carbon means of enhancing the recovery of energy from oil and gas resources in Canada.

The aqua hydrogen technology allows for oxygen injection into a sealed fuel deposit between grains of rock, using un-swept petroleum as the fuel and the vessel ground as the reactor. The process starts with separating atmospheric air into nitrogen and oxygen by using air separation units. Oxygen is then sent to an underground reservoir, the temperature of which is elevated, promotes water-gas shift, hot gasification, and aqua thermolysis reactions within the fuel reservoir thus producing syngas. Lastly, $H_2$ is recovered using membranes in the production wells (Arcos & Santos, 2023). These membranes are made out of a palladium alloy, in which carbon oxides stay underground, and $H_2$ reacts with palladium. The process is highly efficient since all the conversion of energy takes place underground. This technology could be a clean method of producing hydrogen with the aid of technologically unrecoverable oil reserves, bringing only hydrogen to the surface (National Observer, 2020).

There are several challenges in the production of Aqua Hydrogen. An important challenge lies in scaling up of the process as it is still in the pilot stage, so expanding production is difficult. Environmental risks are a cause of concern on account of underground $CO_2$ storage and potential leaks. Further, lack of awareness and preference for green hydrogen breeds significant public resistance as well (M. Yu et al., 2021).

**Turquoise hydrogen:** Turquoise hydrogen *via* methane-pyrolysis produce solid carbon in the form of filamentous carbon or carbon nanotubes (Amin et al., 2011), that can be used for the further production process or might be easier to store, hence, having a lower carbon footprint. The prices for solid carbon-based materials vary significantly. Coke costs between €150 and €400 per ton, carbon black ranges from €500 to €1,000 per ton, and activated carbon is priced from €1,500 to €1,800 per ton. Carbon filaments, on the other hand, can reach up to €1 million per ton. These differences reflect variations in the materials' production processes, quality, and their use in industries such as energy, manufacturing, and advanced technologies. (Leal Pérez et al., 2021). Pyrolysis has not yet been commercialized from the point of view of hydrogen production.

**Yellow hydrogen:** Yellow hydrogen is a relatively new phrase for hydrogen made through electrolysis using solar power. However, it is not explored to depth yet ((Newborough & Cooley, 2020)).

*Table 1. Types of hydrogen colours, advantages and disadvantages of different types of hydrogen*

| Type of Hydrogen Colors | Definition by technologies used | Advantages | Drawbacks |
|---|---|---|---|
| | It is naturally occurring hydrogen molecules formed through fracking and found in underground deposits. | Naturally occurring hydrogen. | – |
| | Hydrogen is produced by water electrolysis using renewable energy sources such as solar, wind, and hydro. | Increase recycling, promote bioenergy, Does not emit GHGs, Produce from clean electricity, Emits no carbon dioxide. | High production costs, Energy losses, Lack of dedicated infrastructure, Lack of value recognition. |
| | Hydrogen is produced by SMR from fossil fuels (other than coal). | Most common form of hydrogen production. | Emits carbon dioxide, emits other GHG, continued use of some fossil fuels. |
| | Hydrogen from natural gas is produced through the SMR process to capture and bury GHGs. | Could capture and remove GHGs using carbon capture and storage (CCS) technology, also known as low-carbon hydrogen | Produce carbon dioxide as a byproduct and methane leaking. |
| | Hydrogen is produced by coal gasification. | Produce liquefied hydrogen for low-emission usage. | This process emits large amounts of carbon dioxide, carbon monoxide, and other GHGs. (Opposite of green hydrogen). |
| | | | |
| | Hydrogen is produced using nuclear energy to power water electrolysis. | The steam generated from the nuclear reactors could be used for other hydrogen productions for more efficient electrolysis | Use very high temperatures from nuclear reactors, generate a small amount of GHGs. |
| | Hydrogen is produced from the energy grid via water electrolysis. | Hydrogen is produced through electrolysis using solar power. | Carbon emissions vary greatly depending on the sources used to power the grid. |
| | Hydrogen is derived from the pyrolysis of methane (thermal splitting) | Produce hydrogen and solid carbon. In the future, it may be valued as low-emission hydrogen, depending on several factors. | Generate a small number of GHGs |
| | Produced from waste plastics through gasification or pyrolysis with CCS.. | Low raw material costs. Requires less energy.. | At an early stage, more research is needed.. |

*Figure 4. Uses of green hydrogen*

Applications of Green Hydrogen:

- **Power Generation:** Green hydrogen can be utilized to generate electricity (Fig. 4) either through fuel cells or by burning it in combustion turbines. This offers a clean and efficient method for producing power, suitable for residential, commercial, and industrial applications (Hassan et al., 2022).
- **Transportation:** Green hydrogen is a promising fuel for various modes of transportation, including cars, buses, trucks, trains, and even ships (Fig. 5). Utilizing hydrogen fuel cells in vehicles offers a zero-emission alternative to fossil fuels, with water being the only byproduct (Agaton et al., 2022; Hassan & Jaszczur, 2021).
- **Industrial Processes:** Many industries, including steel, chemicals, and refining, depend heavily on hydrogen as a feedstock. Hydrogen is utilized in chemical production, such as ammonia and methanol, as well as in refining processes like hydrocracking. It also plays a critical role in the iron and steel industry for low-carbon steelmaking (Hassan & Jaszczur, 2021; Undertaking, 2019). Green hydrogen has the potential to replace conventionally produced grey or blue hydrogen, thereby, significantly reducing greenhouse gas emissions in these sectors.
- **Synthetic fuels**: Green hydrogen can be combined with captured carbon dioxide to produce synthetic fuels like methane, methanol, ammonia, and other hydrocarbons. These fuels can be utilized in transportation, power generation, or as feedstock for various industries (Hassan et al., 2022).

*Figure 5. Classification of different hydrogen production methods*

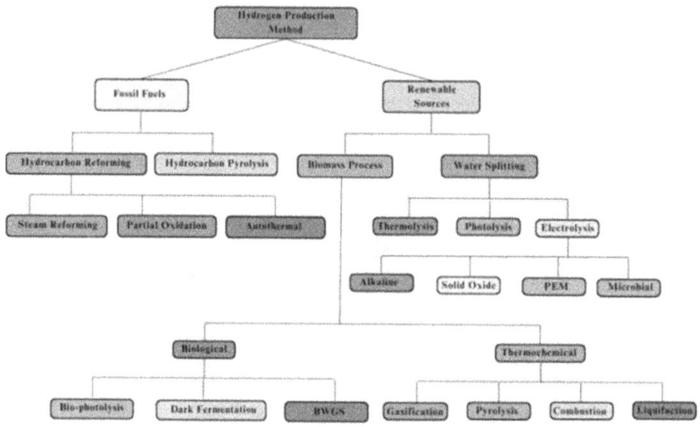


- **Energy Source:** Green hydrogen can be used to store surplus renewable energy, such as solar or wind power. When renewable energy generation exceeds demand, hydrogen can be produced through electrolysis and stored for later use. It can then be converted back into electricity using fuel cells or combustion turbines (Agaton et al., 2022; Widera, 2020). This process helps balance energy supply and demand, enhances grid stability, and enables the integration of more renewable energy into the power system, making it a crucial component of a sustainable energy infrastructure.
- **Green Hydrogen for Heating Solutions:** Green hydrogen can be utilized for space and water heating across residential, commercial, and industrial structures. It can be burned in boilers or utilized in fuel cells to generate both heat and electricity simultaneously, offering a clean and efficient heating solution. Hydrogen-based heating systems can significantly lower the carbon footprint of buildings and aid in the de-carbonization of the heating sector (Longoria et al., 2021; Novotny, 2023).
- **Agriculture:** Green hydrogen can be utilized in the production of ammonia, a crucial element in fertilizers (Fig. 4). By using green hydrogen for ammonia synthesis, the environmental footprint of the agricultural sector can be significantly reduced. Ammonia production traditionally relies on natural gas, resulting in high $CO_2$ emissions (Armijo & Philibert, 2020). Shifting to green hydrogen, which is produced from renewable energy, helps decrease these emissions, contributing to more sustainable agricultural practices. This transition also promotes the use of cleaner energy sources in fertilizer manufacturing, addressing both food security and environmental concerns.
- **Cleaner Cooking Fuel:** Green hydrogen can serve as a cleaner alternative for cooking, especially in rural regions that currently depend on biomass (such as wood and charcoal) and fossil fuels like liquefied natural gas. By adopting green hydrogen, local air pollution can be reduced, leading to fewer health issues related to smoke and harmful emissions. The use of traditional biomass for cooking contributes to indoor air pollution, a major health risk in many developing areas (Agaton et al., 2022; Schmidt Rivera et al., 2018). Transitioning to green hydrogen not only improves air

quality but also reduces deforestation and greenhouse gas emissions. This shift supports both environmental sustainability and improved public health.

**Methods for Production of Green Hydrogen:** There are several methods employed for the production of green hydrogen (Fig. 5), which will be discussed further.

A. Water Splitting

I. **Water Electrolysis:** Water electrolysis is a method of splitting water using electricity to produce hydrogen without emitting any pollutants into the environments air. The basic reaction of water electrolysis is as outlined below (Shiva Kumar & Lim, 2022):

$$1\ H_2O + \text{Electricity}\ (237.2\ \text{kJ mol}-1) + \text{Heat}\ (48.6\ \text{kJ mol}-1) \rightarrow H_2 + 1/2\ O_2$$

The above reaction requires 1.23 V theoretical thermodynamic cell voltage to split the water into hydrogen and oxygen at room temperature. However, experimentally, the required cell voltage for efficient water splitting is 1.48 V. The additional voltage is required to overcome the kinetics and ohmic resistance of the electrolyte and cell components of the electrolyser (Shiva Kumar et al., 2018). For the past two centuries, water electrolysis has been considered as a well-established technology for the production of 'green' hydrogen. However, globally only 4% of hydrogen is produced from water electrolysis because of its economy issues. Most of this hydrogen is produced as by-product from the Chlor-alkali industry (Shiva Kumar & Himabindu, 2019). Water electrolysis can be classified into four types based on their electrolyte, operating conditions, and ionic agents ($OH^-$, $H^+$, $O_2$). These four methods include (i) Alkaline Water Electrolysis (AWE), (ii) Solid Oxide Electrolysis (SOE), (iii) Microbial Electrolysis Cells (MEC) and (iv) Proton Exchange Membrane (PEM) Water Electrolysis. However, the operating principles are the same for all the cases (F. Yu et al., 2018).

1. **Alkaline water electrolysis:** Alkaline electrolysis has already been employed for many years in some industries, such as chlor-alkali production, and has proven to be very reliable (Zeng & Zhang, 2010). The phenomenon of alkaline water electrolysis was first introduced by Troostwijk and Diemann in 1789. Following various developments, the first large-scale industrial alkaline water electrolyzer plant, with a capacity of 10,000 N $m^3$ $H_2$ $h^{-1}$, began operations in 1939 (Allebrod et al., 2013).Alkaline electrolysis is the process of producing hydrogen gas from water, through an electrolyte solution containing hydroxide ions, $OH^-$. The process takes place in an electrolytic cell, which is a container consisting of two electrodes, *viz,* the anode and the cathode, that are submerged in an alkaline electrolyte solution, usually the solution of potassium hydroxide or sodium hydroxide dissolved in water (Fig. 6) (Sánchez et al., 2020; Santos et al., 2013).

*Figure 6. Alkaline electrolysis*

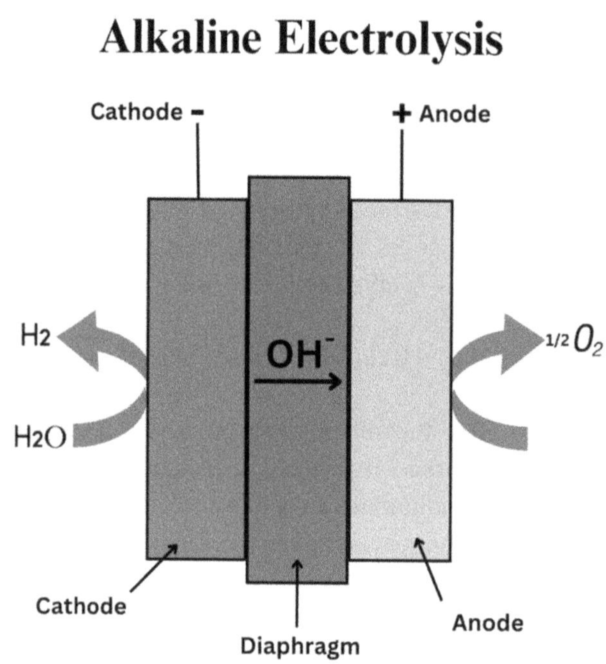


The overall reaction for alkaline electrolysis is as follows:

$2\ H_2O\ (l) \rightarrow 2\ H_2\ (g) + O_2\ (g)$

This reaction occurs in two half-reactions, one at the anode and one at the cathode.
At the anode (oxidation): $\mathbf{4\ OH^- \rightarrow 2\ H_2O\ (l) + O_2 + 4\ e^-}$
At the cathode (reduction): $\mathbf{4\ H_2O + 4\ e^- \rightarrow 2\ H_2\ (g) + 4\ OH^-\ (aq)}$
The key components and variables of alkaline electrolysis:

**Electrolyte:** In alkaline electrolysis, an electrolyte is prepared which consists of either KOH or NaOH in water, often at concentration levels of about 20 to 40 wt.%. The role of the electrolyte is to enhance the conductivity of the solution and to create conditions favourable to the flow of ions between the electrodes (Xiao et al., 2012).
**Electrodes:** Usually, anode and cathode are made of materials having corrosion resistance and important catalytic properties, such as nickel, stainless steel, or noble metals such as platinum, held inside the solution (Zhou et al., 2020).
**Electric Current:** The driving current allows for electrolysis because it provides the cell with the energy to fragment the chemical bonds existing within water molecules. The amount of hydrogen and oxygen produced is proportional to the electric charge that passes through the cell (Ganley, 2009). This proportionality can be expressed in terms of Faraday's laws of electrolysis:

$Q = n \times F$

Where, Q is the electric charge (in coulombs), n is the number of moles of the substance produced or consumed, and F is Faraday constant (approximately 96,485 C/mol) (Chi & Yu, 2018). In case of hydrogen production from water electrolysis: $\boldsymbol{Q} = 2 \times \boldsymbol{n}(\boldsymbol{H}2) \times \boldsymbol{F}$

The current density (A/m2) and cell voltage are important parameters that affect the efficiency and cost of the electrolysis process. The efficiency of alkaline electrolysis is influenced by several factors, including the electrode material, electrolyte concentration, temperature, and pressure. The energy efficiency of the process can be calculated using the following equation:

$$\textit{Energy efficiency } (\%) = (\textit{Theoretical cell voltage/Actual cell voltage}) \times 100.$$

Alkaline electrolysis typically has an energy efficiency of 60 to 80% (Dobó & Palotás, 2017).

**2. Solid Oxide Electrolysis (SOE):** The solid oxide electrolysis (SOE) first introduced by Donitz and Erdle in the1980s (Donitz & Erdle, 1985). The Solid Oxide Electrolysis (SOE) process, frequently referred to as high-temperature electrolysis, represents a technique for generating hydrogen gas ($H_2$) from water ($H_2O$) through the utilization of a solid oxide electrolysis cell (Carmo et al., 2013). Solid oxide electrolysis operates at high temperatures (500–850 °C) and high pressures, using steam as the water source (Liang et al., 2009). The SOE cell comprises a solid oxide electrolyte, usually made from a ceramic material like yttria-stabilized zirconia (YSZ). It features an anode and a cathode, typically constructed from a porous nickel-YSZ composite (Fig. 7) (Cinti et al., 2016).

*Figure 7. Solid oxide electrolysis*

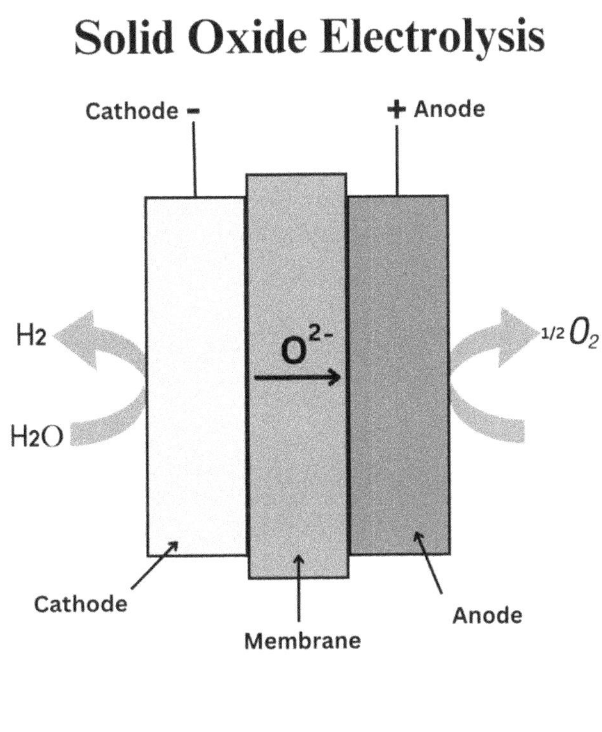

Anode reaction (Oxidation): At the anode, steam ($H_2O$) is introduced into the cell, where it reacts with the nickel anode material to generate hydrogen gas ($H_2$) and oxygen ions ($O^{2-}$).

$$H2O\ (g) + 2\ e\ \rightarrow H2\ (g) + 2\ O^2\ (aq)$$

The oxygen ions generated at the anode can migrate through the solid oxide electrolyte because of its operation at high temperatures (Fallah Vostakola et al., 2023).

Cathode reaction (Reduction): At the cathode, oxygen ions ($O^{2-}$) react with electrons ($e^-$) from an external circuit to produce oxygen gas ($O_2$):

$$2\ O^2\ (aq) + 4\ e\ \rightarrow O2\ (g)$$

Overall reaction: The overall process combines the anode and cathode reactions, showing the production of hydrogen gas and oxygen gas from the electrolysis of water:

$$H\ O\ (g) \rightarrow H\ \ (g) + O\ \ (g)$$

The hydrogen gas formed at the anode and the oxygen gas formed at the cathode can be separately collected for further use or storage. SOE typically operates at high temperatures, allowing for the movement of oxygen ions through the solid oxide electrolyte. This process requires a source of heat or electrical energy to maintain the high-temperature conditions (Biswas et al., 2023) (Wang et al., 2022a). The elevated temperatures also improve the efficiency of water electrolysis and hydrogen production. In addition, SOE can use excess renewable energy or waste heat from industrial processes, which means it has a promising technology for sustainable hydrogen generation.

**3. Proton Exchange Membrane (PEM) Electrolysis:** The initial concept of PEM water electrolysis was introduced by Grubb during the early 1950s, and it was subsequently advanced by General Electric Co. in 1966 to overcome the constraints associated with alkaline water electrolysis (S. Baykara & Atmaca, 2019; Grubb, n.d.; Hassan, 2020). The PEM electrolysis cell consists of two electrodes, an anode and a cathode, and a proton exchange membrane or PEM in between them. Between the electrode, the electrolyte with water or even an aqueous acidic solution is incorporated into the cell (Fig. 8) (Adrien et al., n.d.; Danilovic et al., 2016).

*Figure 8. Proton exchange membrane (PEM) electrolysis*

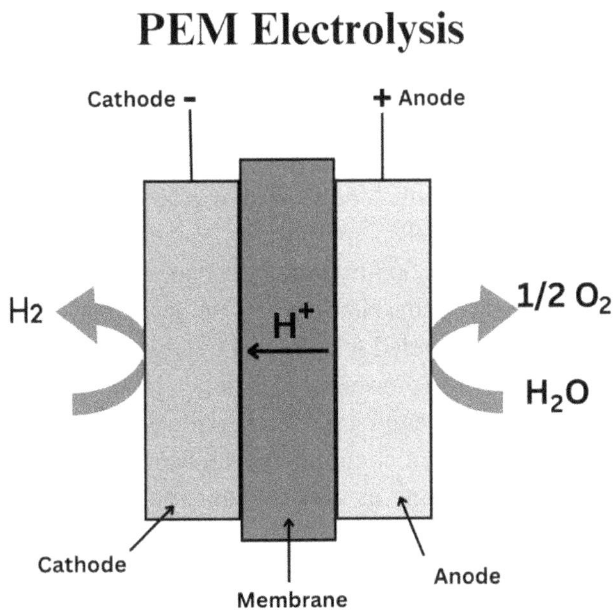


Anode reaction (Oxidation): At the anode, water molecules are oxidized, leading to the release of oxygen gas, protons, and electrons:

$$2\,\boldsymbol{H}\,\boldsymbol{O}\,(l) \rightarrow \boldsymbol{O}\ (g) + 4\,\boldsymbol{H}\ (aq) + 4\,\boldsymbol{e}$$

Proton transport through the membrane: The PEM selectively allows protons ($H^+$) to pass while blocking the flow of electrons ($e^-$). As a result, only protons can migrate through the membrane from the anode to the cathode compartment (Hu et al., 2014).

Cathode reaction (Reduction): At the cathode, the protons received from the anode combine with electrons supplied from an external circuit to produce hydrogen gas.

$$4\,\boldsymbol{H}\ (aq) + 4\,\boldsymbol{e}\ \rightarrow 2\,\boldsymbol{H}\ (g)$$

The overall reaction combines the anode and cathode processes, illustrating the conversion of water into hydrogen and oxygen gases.

$$2\,\boldsymbol{H}\,\boldsymbol{O}\,(l) \rightarrow 2\,\boldsymbol{H}\ (g) + \boldsymbol{O}\ (g)$$

The hydrogen gas produced at the cathode and the oxygen gas generated at the anode can be collected separately for further use. The PEM electrolysis process requires an external power source, typically electricity, to drive the electrolysis reaction. This power source provides the necessary energy to split water molecules into hydrogen and oxygen gases (Wang et al., 2022b). Importantly, this method of

PEM electrolysis is efficient and clean hydrogen production, especially if fuelled by renewable energy sources. The proton exchange membrane has the ability to selectively allow protons to be transported, which helps in the separation of hydrogen and oxygen gases. This makes PEM electrolysis a promising technology in the generation of hydrogen as a sustainable energy carrier (Hassan et al., 2024).

**4. Microbial Electrolysis:** Hydrogen is generated through the application of Microbial Electrolysis Cell technology on renewable biomass and wastewater from organic substrates. Microbial Electrolysis Cell technology is correlated with significant importance with Microbial Fuel Cells, though the operational mechanism is basically reverse to the one seen in Microbial Fuel Cells (Lozada-Hidalgo et al., 2018). The first Microbial Electrolysis Cell

(MEC) approach was introduced in 2005 by two independent research organizations: Penn State University and Wageningen University in the Netherlands (Kadier, Simayi, et al., 2016; Liu et al., 2005). In MECs, electrical energy can be converted to chemical energy in order to generate hydrogen from organic material using electricity. In the process of microbial electrolysis, a substrate first undergoes oxidation by microbes on the anode, generating carbon dioxide ($CO_2$), protons, and electrons (Fig. 9). The electrons then travel along an external circuit to the cathode; however, the protons are transferred to the cathode through a proton-conducting membrane, sometimes called an electrolyte. Hydrogen is formed by combining the protons.

*Figure 9. Microbial electrolysis*

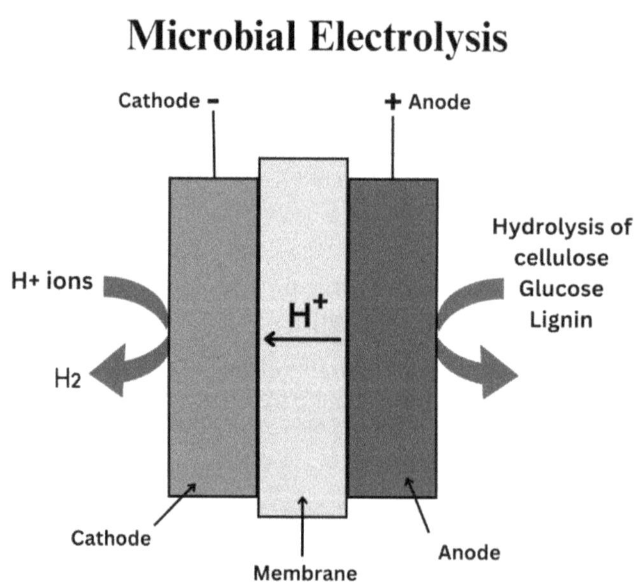


The MEC process produces a certain amount of electrochemical potential during the oxidation at the anode, but the potential is not usually sufficient to provide the required voltage for the hydrogen evolution reaction at the cathode. Hence, an additional voltage of between 0.2 V and 1.0 V is required (Rathinam et al., 2018). The MEC process thus requires less external voltage than conventional water electrolysis. Nevertheless, the MEC technology remains in a developmental phase and encounters numerous obstacles, such as the rates of hydrogen production, elevated internal resistance, the properties of electrode materials, and intricate designs that require resolution prior to commercialization (Kadier et al., 2016).

**II. Plasma Arc Decomposition**: Plasma is an ionized state of matter, which is composed of atomic particles and electrons in a high energy state. The existence of the charged particles allows plasma to serve as a medium for high-voltage electric current dissipation. Natural gas that is primarily composed of methane can be decomposed through thermal plasma. Thermal plasma is characterized by the co-existence of ionized atoms and electrons at the same thermodynamic temperature. Methane breaks up into hydrogen and carbon black, or soot, when it passes through a plasma arc. Soot can be recovered from carbon black as sediment at the bottom, and the hydrogen will be gaseous (Dincer, 2012). The decomposition of natural gas (methane) to hydrogen and carbon result in the production of hydrogen gas.

$$CH4 \rightarrow C + 2\,H2\,(g)\,;\,\Delta H\;=\;74.6\,MJ/kmol$$

This reaction is conducted under high-temperature thermal plasma in a reactor with three electrodes connected to a three-phase voltage, where plasma gas is injected at two of the electrodes and methane is introduced from above (Fulcheri et al., 2002). A 100% yield of hydrogen is achieved with no $CO_2$ release, and carbon black is formed at the bottom of the reactor. This plasma dissociation process operates as a high-temperature pyrolysis. The cost of hydrogen generated through plasma cracking is reduced by at least 5% compared to large-scale production via steam methane reforming with carbon dioxide sequestration (Gaudernack & Lynum, 1998).

**III. Water Thermolysis:** The single-step thermal dissociation of water is referred to as water thermolysis, which can be represented as:

$$H2O \rightarrow H2\,(g) + 0.5\,O2$$

This reaction requires a high-temperature heat source, typically above 2500 K, to achieve a reasonable degree of dissociation (U.S. Department of Energy, n.d.). The main drawback of the process is the requirement for an efficient process to enable the $H_2$ and $O_2$ gases to separate. This should not result in the possibility of a flammable mixture. In this, semipermeable membranes made from materials such as $ZrO_2$ and other high-temperature material may be used at temperatures up to 2500 K. Separation may also be carried out at lower temperature after the product gas mixture has cooled (Baykara et al., 2004). The conclusions are that residence time in the reactor should be 1 ms at 2500 K to attain 90% of equilibrium. Re-combination of hydrogen and oxygen can be avoided by rapidly cooling the product gas through sharp temperature drops of 1500-2000 K in a few milliseconds. Effective hydrogen separation may then be carried out by palladium membranes (Dincer, 2012).

**IV. Photo-Electrolysis/Photolysis:** Photo-electrolysis is a technique that uses sunlight to create hydrogen from water in a photo-electrochemical cell (Pino et al., 2002). Photo-electrolysis is the most economic and efficient means of producing hydrogen from renewable sources. The process involves systems that are almost the same as those used in photovoltaic technology, which also uses semiconductor materials. In photo-electrolysis, a photoelectrode—a photo-electrochemical device is used to capture light to perform water electrolysis. The photoelectrode is made of a semiconducting material that absorbs solar energy and generates the voltage required to split water molecules into hydrogen and oxygen (Kalamaras & Efstathiou, 2013). If a photoelectrode is immersed in an aqueous electrolyte and exposed to solar radiation, it will generate enough electrical energy to produce $H_2$. The reaction can be presented as:

$$H2O + hv \rightarrow H2\,(g) + 1/2\ O2\,(g)$$

The reaction is affected by the semiconductor material types and the intensity of sunlight, producing a current density ranging from 10 to 30 mA/cm. Electrolysis requires a voltage of around 1.35 V when these current densities are present (Armaroli & Balzani, 2011).

**B. Biomass technologies:** Biomass is a renewable energy source obtained from plant and animal materials. This includes residues from energy crops, waste from forests, industrial by-products, and both animal and human waste, among other sources (Hoang et al., 2021; Zamri et al., 2021). Biomass is rapidly becoming one of the cleanest sources of alternative energy because of renewability, low carbon dioxide, and low sulfur contents (van Rossum et al., 2009). Biomass conversion can broadly be divided into two distinct pathways: thermochemical and biological. Thermochemical methods convert biomass into hydrogen and hydrogen-rich gases via methods such as pyrolysis, co-pyrolysis, gasification, combustion, and liquefaction (Bridgwater, 1999; Osman et al., 2021). On the biological side, hydrogen synthesis pathways include bio-photolysis, direct and indirect; photo-fermentation; dark fermentation; and biological shift reactions.

Biomass technologies may be divided further into thermochemical and biological technologies.

Thermochemical technologies

1. **Pyrolysis or co-pyrolysis:** The hydrogen generation process is suitable if pyrolysis takes place since pyrolysis has some desirable properties like easy storage and transportation, with all types of equipment, for instance, turbines, incinerators, containers, or appliances in use. This process allows biological materials, heated by degassing from 500 to 900 °C at 0.1 to 0.5 MPa oxygen-free environments (Ni et al., 2006). During the transformation process, temperatures can vary from 300 to 500 °C and then from 700 to 800 °C (G. Su et al., 2022). The products of pyrolysis undergo secondary reactions that, depending on the reactor conditions, can yield hydrogen gas and other by-products (López et al., 2011). In addition to hydrogen, the gas stream also contains measurable levels of methane ($CH_4$) and lighter hydrocarbons, which can be further steam-reformed for extra hydrogen. Moreover, the CO in the gas can be further converted into hydrogen and carbon dioxide through a subsequent water-gas shift (WGS) reaction (Iribarren et al., 2014a; Parthasarathy & Narayanan, 2014). Only very recently, the co-pyrolysis of a mixture of coal with organic waste has been explored in industrialized countries as a way of addressing waste management issues (Bičáková & Straka, 2012).
2. **Biomass Gasification:** Gasification is an efficient technology in which biomass is converted into power, heat, and chemical compounds (Campoy et al., 2008). It is carried out through a two-stage process, first, pyrolysis in which the biomass undergoes thermal decomposition by heat and gasifying agent into volatile hydrocarbons and char and second, gasification that converts these products to syngas. This can be achieved at a temperature between 500 to 1400 °C by a gasifying agent, either air, oxygen, or steam (Zhang et al., 2010). The syngas composition varies based on the type of biomass, as well as the type of gasification. In most cases, air is used for gasification; this dilutes syngas by the nitrogen and lowers its energy content to 3.7–6.4 MJ/$m^3$ at STP. However, in the case of oxygen and steam, syngas contains higher energy content (9.2 to 16.5 MJ/$m^3$ at STP) (Iribarren et al., 2014b). Gasification, on the other hand, is a two-stage process where it transforms biomass energy into combustible gas rather than one step like combustion. Syngas obtained contains carbon monoxide, hydrogen, carbon dioxide, and other trace components; thus, it can power engines, tur-

bines, or as a feedstock for liquid fuels like ammonia and methanol (Huber et al., 2006). Gasification includes not only biochemical processes, which are anaerobic digestions, but also includes thermochemical methods that use elevated temperatures above 800 °C (Eloffy et al., 2022; Mckendry, n.d.) Supercritical gasification can rapidly convert various agricultural and other wastes to high-pressure gas, the major products of which consist of hydrogen and carbon dioxide (Liao et al., 2013). This method uses supercritical water as a solvent to increase the solubility of the components of biomass, usually requiring 600–800 °C without a chemical catalyst or less than 600 °C with a catalyst (Kumar et al., 2023).

## Biological Technologies

**1. Bio-photolysis;** The bio-photolysis process generates a theoretical ratio of 2:1 $H_2$ to $O_2$ without the need for fixing $CO_2$ or other metabolic actions (Veeravalli et al., 2019). This clean energy technique involves cyanobacteria and green microalgae in the decomposition process of water into $H_2$ and $O_2$ through photosynthesis using sunlight (Kosourov et al., 2017). Hydrogenase enzymes in these microorganisms cause the process through photosynthesis without any dependency on extrinsic energy sources, while sunlight and $CO_2$ would be the energy-generating resources (Ghirardi et al., 2000).

The main advantage of bio-photolysis is that it does not require nutrient substrates because water acts as the main electron donor. It may be either direct or indirect. In direct biophotolysis, the absorption of light energy through a photosynthetic device is used to couple the splitting of water so that it forms a low-potential reductant known as ferredoxin, which catalyzes hydrogen formation. In indirect bio-photolysis, both PSII and PSI split water, but reduction of ferredoxin is used in the fixation of $CO_2$ (Veeravalli et al., 2019).

**2. Dark Fermentation:** Dark fermentation is a biological process for hydrogen production using anaerobic microorganisms, such as microalgae or specific bacteria, in dark, at temperatures between 25 and 80 °C or higher, depending on the strain. The process involves several biochemical phases, starting with the enzymatic hydrolysis of high molecular weight organics into water-soluble compounds, followed by the conversion of simple organics into volatile fatty acids (VFA), hydrogen, and carbon dioxide (Argun et al., 2009; Kapdan & Kargi, 2006).

Hydrogen is mainly produced from the anaerobic metabolism of pyruvates formed during carbohydrate breakdown (Hallenbeck & Benemann, 2002). Key reactions include:

$$C6H12O6 \rightarrow 2\ H2 + CH3CH2CH2COOH + 2\ CO2$$

However, the presence of minor gases reduces hydrogen yield compared to photolysis (Levin et al., 2004). Dark fermentation techniques are advantageous as they can continuously produce hydrogen without relying on solar energy, making them potentially more profitable than photo-fermentation methods (Vijayaraghavan et al., 2006).

**3. Biological Water Gas Shift Reaction (BWGS):** It can be defined as the fermentation of synthesis gas to give hydrogen. It employs photosynthetic bacteria, preferably carboxydotrophs belonging to the genus Rhodospirillaceae, which can utilise carbon monoxide as its carbon source (Veeravalli et al., 2019). These microorganisms ferment glucose in the dark to evolve hydrogen and carbon dioxide while oxidizing carbon monoxide and reducing water through the enzyme-catalysed reaction shown in (Tanksale et al., 2010):

Some commonly used microorganisms are *Rhodospirillum rubrum* and *Carboxydothermus hydrogenoformans* (Hosseini et al., 2015; Logan et al., 2006). The BWGS process employs lower temperatures/pressures compared to thermochemical water gas shift reactions, typically operating in the range of 30–70 °C depending on the strain and the ambient pressure. However, several parameters need optimization including controlling carbon monoxide levels from the feedstock, improving microbial concentration in the bioreactor, and addressing gas-to-liquid mass transfer challenges (Xia et al., 2020).

**4. Bio-catalysed Electrolysis**: Bio-catalysed electrolysis is a biological hydrogen production technology that converts many dissolved organic compounds in wastewater into hydrogen efficiently. The process involves electrochemically active microorganisms that create electrical current from the oxidation of organic materials. The biological anode connected through a power supply to the proton-reducing cathode produces hydrogen.

There are two significant types of biocatalyzed electrolysis systems: Microbial Fuel Cells (MFCs) and Microbial Electrolysis Cells (MECs). In the MFC, cathodes are exposed to air for electricity generation and oxygen reduction reaction (Logan et al., 2006). In the MEC, a minor external energy input is used to accelerate hydrogen generation at the cathode (Rozendal et al., 2006).

Example reaction of the biocatalytic electrolysis of acetate:

Hydrogen synthesis from acetate, under normal conditions, requires an input of 104.6 kJ/mol, or an applied voltage of 0.14 V (Thauer et al., 1977).

Different technologies have differences in their efficiencies. Steam reforming is the most efficient among all of these, with a 70–85% efficiency level (David et al., 2019; Holladay et al., 2009). In partial oxidation and autothermal reforming, 60–75% is realized. Plasma reforming is found to have the largest efficiency range at 9–85% (Paulmier & Fulcheri, 2005). Aqueous phase reforming is expected to have around 35–55% efficiency and biomass gasification 35–50% efficiency. Dark fermentation has an efficiency of 60–80% while that of MEC is about 78%. The process of photo fermentation is poor at 0.1%. The efficiencies of Alkaline and PEM electrolysers are about 50–60% and 55–70% respectively (Turner et al., 2008). The expected mid-term efficiency of the SOECs is of the order of 40–60% and 12.4% by photoelectrochemical water splitting (Turner et al., 2008). These diverse technologies therefore present a spectrum of efficiencies critical for their integration into the hydrogen energy landscape.

## Challenges and Barriers to Green Hydrogen

**1. Economic viability:** The main issues with green hydrogen (GH) are its high cost (Bade et al., 2024; Longden et al., 2022). The GH market is still in the nascent stage globally, and GH is not yet economically competitive compared to hydrogen produced from other sources. For GH to become economically viable, it needs to become cost-competitive with grey hydrogen (M. Yu et al., 2021b). IRENA further observes that the overall cost is highly dependent on the electrolyser price, size, and availability of renewable electricity. The direct utilization of GH or its conversion into ammonia, synthetic fuels, or methane is cheaper and easier to transport and store, and industries are seeking alternatives for direct use or conversion. This would also save costs if the natural gas infrastructure is utilized (Scita et al., 2020). In addition, GH suffers from efficiency losses, which increase its cost and lowers its competitiveness with other sources of energy (Longden et al., 2022).

According to IRENA (Longden et al., 2022), all parts of the hydrogen supply chain are reported to impact the final price. For example, technology, distance between production and storage, and the requirements of the end product are key cost factors. Electrolysers are widely used in many chemical processes, but GH production involves a very high investment cost that is quite comparable to fossil fuel-based technologies, so minimization of cost within the sector remains the focus. Another critical aspect is regional access to cheap renewable electricity. The scale of production of GH needs to be escalated to make it more economical as well. This will again require a lot of support from the public and many incentives from the government also.

**2. Infrastructure Development:** The infrastructure for hydrogen production, storage, and distribution is less developed compared to the conventional fuels. It takes a lot of investment to build pipelines, refuelling stations, and storage facilities to enable a transition towards green hydrogen (*Energy - European Commission*, n.d.). Currently, the hydrogen infrastructure is still very limited and is basically designed for conventional hydrogen sources rather than for green sources. According to the Hydrogen Council, a solid hydrogen infrastructure will be necessary to scale up the sector. Without a network of infrastructure that would be comprehensive and accessible, potential users in transport and industries may not invest in hydrogen technologies (*Homepage | Hydrogen Council*, n.d.).

**3. Lack of Value Chain Infrastructure**: The limited number of GH projects and the dominance of fossil fuels in the hydrogen value chain underscore the need for new infrastructure. Building a robust global GH value chain requires significant investment in production, transport, and storage infrastructure, as well as cross-border cooperation. This infrastructure development is essential to create a global market for GH, as large-scale transportation of GH is challenging and costly due to its low density (Ratnakar et al., 2021a; Yáñez et al., 2018a).

**4. Infrastructure Requirements**: The infrastructure for GH production, storage, and transport is underdeveloped, which hinders the establishment of a global GH market. Expanding pipelines, storage facilities, and refuelling stations will require extensive public-private partnerships, large capital investments, and coordinated efforts across countries. Due to GH's low density, it requires specialized infrastructure for transport and storage, adding to the complexity and cost of GH adoption (Ratnakar et al., 2021b; Yáñez et al., 2018b)

**5. Regulatory and Policy Gaps**: The absence of global standards for GH production, safety, and environmental impact hinders its expansion. Regulatory frameworks vary widely across regions, creating an inconsistent environment for GH development. A globally unified framework, with standardized product definitions, operational rules, and safety requirements, is necessary to facilitate broader adoption of GH. Currently, countries are forced to create their own localized norms, adding to the complexity of GH expansion [Irena].

**6. Infrastructure development:** Green hydrogen is produced through the electrolysis of water in which electricity produced from renewable energy splits the water into hydrogen and oxygen. It has proven technology but has various limitations. Among these limitations are improvement in electrolysis efficiency, enhancement in system durability, and enhancement of the energy density of storage solutions of hydrogen (Bade et al., 2024). Scale also emerges as another serious challenge in large-scale applications, wherein a lot of investments are needed for infrastructure building of renewable energy to accommodate the massive applications of green hydrogen. This requirement may not be feasible in the short term, hindering the widespread adoption of green hydrogen as a sustainable energy solution. Addressing these technological limitations is crucial for realizing the full potential of green hydrogen in the energy transition (Hassan et al., 2024).

Green Hydrogen Initiatives Around the World:

**1. India:** India has announced long term low emission development strategy, that focuses on climate justice, sustainable lifestyles, and equity, at the ongoing UN climate summit in Egypt, joining a select group of fewer than 60 nations to do so (*India's Initiatives on Green Hydrogen Could Help Global Decarbonisation*, n.d.). India has introduced several policies and guidelines like Green Hydrogen Policy, Harnessing Green Hydrogen by Niti Ayog, National Green Hydrogen Mission (NGHM) (*Green Hydrogen Adoption in India – Opportunities, Challenges and the Way Ahead, ET EnergyWorld*, n.d.). The nation's green hydrogen mission (NGHM), announced in 2021, will see a rapid expansion in production, making India a green hydrogen hub, the long-term strategy said. The South Asian nation plans a massive expansion of green hydrogen production to curb dependence on crude oil imports and wean its rapidly expanding economy from planet-warming fossil fuels. The central government has set a target of an annual production capacity of 25 million tonnes by 2047 (*National Green Hydrogen Mission Portal of India*, n.d.).

Also under the newly launched Green Energy Corridor Project, any new renewable energy plants for green hydrogen production till July 2025 would be provided free power for 25 years. The seven states participating in the project are Gujarat, Himachal Pradesh, Karnataka, Kerala, Rajasthan, Tamil Nadu, and Uttar Pradesh that would initially receive funding to build 20 GW capacity renewable energy power plants in the Indian state of Avaada Group will invest more than $5 billion in building an integrated green hydrogen and ammonia factory with captive RE capacity. Avaada aims to achieve 11 GW by 2025 and 30 GW by 2030.(*WordPress › Error*, n.d.).

**2. Australia:** Australia is positioning itself as a key player in the global green hydrogen market, leveraging its abundant renewable energy resources, especially solar and wind. The Australian government has developed a strategic roadmap focused on producing and exporting green hydrogen, aiming to harness these vast resources to drive economic growth and energy innovation (Renew Economy, 2024). The National Hydrogen Strategy, launched in November 2019, outlines the vision and priorities for the development of a clean, innovative, competitive, and safe hydrogen industry in Australia. The strategy focuses on producing clean hydrogen for domestic use and export, creating jobs, and reducing emissions (ARENA, 2018). Key projects include the Asian Renewable Energy Hub, which plans to produce green hydrogen using wind and solar power (International Energy Agency, 2024)

, and the Woodside Energy H2TAS project in Tasmania, aims to become a global pro ducer of green hydrogen by 2030, leveraging the island abundant hydropower and wind resources (State Growth, n.d.). Additionally, the Australian Renewable Energy Agency (ARENA) supports research and development for green hydrogen technologies, assessing export market opportunities and prioritizing investment areas.

**3. The European Union:** The European Union (EU) has developed an ambitious Hydrogen Strategy to facilitate the transition to a climate neutral economy by 2050. The strategy aims to scale up the production and use of green hydrogen across the EU, focusing on its potential to decarbonize various sectors, such as industry, transport, power generation, and buildings. The strategy sets clear targets and outlines a range of measures to support the development and deployment of green hydrogen technologies (*EUR-Lex - 52020DC0301 - EN - EUR-Lex*, n.d.). The EU aims to install 6 GW of green hydrogen electrolysers by 2024 and at least 40 GW by 2030. The strategyalsosetsatargettoproduceupto10million-metric tons of renewable hydrogen by 2030. These targets are designed to help the EU meet its climate and energy goals, including the European Green Deal, which aims to achieve climate neutrality by 2050.

**4. United States:** The U.S. Department of Energy (DOE) has initiated the Hydrogen Program, focusing on the research, development, and deployment of hydrogen technologies, particularly green hydrogen production via water electrolysis. The program's goals include lowering production costs, enhancing efficiency, and improving the overall performance of hydrogen systems (*Natural Resources Canada*, n.d.). $H_2$@Scale is a U.S. Department of Energy (DOE) initiative that brings together stakeholders to advance affordable hydrogen production, transport, storage, and utilization to increase revenue opportunities in multiple energy sectors. This initiative explores the potential for wide scale hydrogen production and utilization in the United States (HYDROGEN AND FUEL CELL TECHNOLOGIES OFFICE Emerging Hydrogen Markets, n.d.).

**5. Canada:** Canada has launched the Hydrogen Strategy which was released in December 2020, and positions Canada as a world-leading producer, user and exporter of low-carbon hydrogen and associated technologies. Natural Resources Canada engaged with stakeholder groups, provincial and territorial governments, and Indigenous partners to develop a strategy that will help set us on the path to meet our climate change goals of becoming net-zero by 2050 (*The Hydrogen Strategy*, n.d.).

**6. Japan:** Japan vision for a hydrogen-based society is outlined in its Basic Hydrogen Strategy, which was released in December 2017. The strategy provides a comprehensive roadmap for the development, adoption, and integration of hydrogen technologies to create a sustainable and low-carbon energy system. Japan strategic approach is driven by its commitment to address climate change, reduce energy imports, and foster economic growth through the hydrogen industry (METI, n.d.). In June 2023, the Japanese government revised its Basic Hydrogen Strategy to support such corporate initiatives. This strategy identifies nine key technologies, including fuel cells and water electrolysis devices and has decided to invest over JPY 15 trillion ($98.8 billion) over the next 15 years. It also aims to increase hydrogen usage to 12 million tons annually by 2040 (World Economic Forum, 2024).

Other nations are also investing into green hydrogen technology like Germany, China, Britain, South Korea etc.

## CONCLUSION

Green hydrogen can play a transformative role in the reduction of overall greenhouse gas emissions in this world and help set up a resilient low-carbon economy. This chapter discussed the methods of green hydrogen production from electrolysis of water to highly advanced biomass and photolysis technologies that hold promise for efficiency and sustainability. From powering heavy industry to fueling residential heating, an alternative zero-emission resource to fossil fuels, the use of green hydrogen contributes to energy independence and protection of the environment.

However, a few barriers stand in its way to mainstream use as an energy solution. Production costs are high, infrastructural facilities for transportation and storage are limited, and standardized global regulations are non-existent. Initiatives at the global level and national policies, including those of the EU, India, Australia, and the United States, will address these concerns in due course of time, making green hydrogen both economically viable and accessible to everyone. Finally, scaling up green hydrogen production will require innovative public-private partnerships and investment in dedicated infrastructure. It will all depend on persistent efforts from governments, industries, and communities toward this global pivot toward green hydrogen. Continuously improving technology and pro-green policy frameworks will do the trick in ensuring that sustainable energy systems will be placed on the right footing when it comes to long-term climate goals and paving the future toward a cleaner and better-resilient world for everyone.

## REFERENCES


Abdin, Z., Zafaranloo, A., Rafiee, A., Mérida, W., Lipiński, W., & Khalilpour, K. R. (2020). Hydrogen as an energy vector. *Renewable & Sustainable Energy Reviews*, *120*, 109620. DOI: 10.1016/j.rser.2019.109620

Agaton, C. B., Batac, K. I. T., & Reyes, E. M. (2022). Prospects and challenges for green hydrogen production and utilization in the Philippines. In *International Journal of Hydrogen Energy* (Vol. 47, Issue 41, pp. 17859–17870). Elsevier Ltd. DOI: 10.1016/j.ijhydene.2022.04.101

Ajanovic, A., Sayer, M., & Haas, R. (2022). 10. *International Journal of Hydrogen Energy*, *47*(57), 24136–24154. DOI: 10.1016/j.ijhydene.2022.02.094

Allebrod, F., Chatzichristodoulou, C., & Mogensen, M. B. (2013). Alkaline electrolysis cell at high temperature and pressure of 250 °c and 42 bar. *Journal of Power Sources*, *229*, 22–31. DOI: 10.1016/j.jpowsour.2012.11.105

Amin, A. M., Croiset, E., & Epling, W. (2011). Review of methane catalytic cracking for hydrogen production. In *International Journal of Hydrogen Energy* (Vol. 36, Issue 4, pp. 2904–2935). DOI: 10.1016/j.ijhydene.2010.11.035

Amore-Domenech, R., Santiago, Ó., & Leo, T. J. (2020). Multicriteria analysis of seawater electrolysis technologies for green hydrogen production at sea. *Renewable & Sustainable Energy Reviews*, *133*, 110166. DOI: 10.1016/j.rser.2020.110166

Antzaras, A. N., & Lemonidou, A. A. (2022). Recent advances on materials and processes for intensified production of blue hydrogen. *Renewable & Sustainable Energy Reviews*, *155*, 111917. DOI: 10.1016/j.rser.2021.111917

Arcos, J. M. M., & Santos, D. M. F. (2023). The Hydrogen Color Spectrum: Techno-Economic Analysis of the Available Technologies for Hydrogen Production. *Gases*, *3*(1), 25–46. DOI: 10.3390/gases3010002

Argun, H., Kargi, F., & Kapdan, I. K. (2009). Effects of the substrate and cell concentration on bio-hydrogen production from ground wheat by combined dark and photo-fermentation. *International Journal of Hydrogen Energy*, *34*(15), 6181–6188. DOI: 10.1016/j.ijhydene.2009.05.130

Armaroli, N., & Balzani, V. (2011). The hydrogen issue. In *ChemSusChem* (Vol. 4, Issue 1, pp. 21–36). Wiley-VCH Verlag. DOI: 10.1002/cssc.201000182

Armijo, J., & Philibert, C. (2020). Flexible production of green hydrogen and ammonia from variable solar and wind energy: Case study of Chile and Argentina. *International Journal of Hydrogen Energy*, *45*(3), 1541–1558. DOI: 10.1016/j.ijhydene.2019.11.028

Atilhan, S., Park, S., El-Halwagi, M. M., Atilhan, M., Moore, M., & Nielsen, R. B. (2021). Green hydrogen as an alternative fuel for the shipping industry. In *Current Opinion in Chemical Engineering* (Vol. 31). Elsevier Ltd., DOI: 10.1016/j.coche.2020.100668

Australian Renewable Energy Agency (ARENA). (2018). *Opportunities for Australia from hydrogen exports*. Retrieved November 16, 2024, from https://www.arena.gov.au/assets/2018/08/opportunities-for-australia-from-hydrogen-exports.pdf

Bade, S. O., Tomomewo, O. S., Meenakshisundaram, A., Ferron, P., & Oni, B. A. (2024). Economic, social, and regulatory challenges of green hydrogen production and utilization in the US: A review. In *International Journal of Hydrogen Energy* (Vol. 49, pp. 314–335). Elsevier Ltd., DOI: 10.1016/j.ijhydene.2023.08.157

Bauer, C., Treyer, K., Antonini, C., Bergerson, J., Gazzani, M., Gencer, E., Gibbins, J., Mazzotti, M., McCoy, S. T., McKenna, R., Pietzcker, R., Ravikumar, A. P., Romano, M. C., Ueckerdt, F., Vente, J., & van der Spek, M. (2022). On the climate impacts of blue hydrogen production. *Sustainable Energy & Fuels*, *6*(1), 66–75. DOI: 10.1039/D1SE01508G

Baykara, S., & Atmaca, M. (2019). The relationship between self mutilative behavior, suicide attempt history and impulsivity and some clinical variables in male opiate-dependent patients. *Anadolu Psikiyatri Dergisi*, *20*(1), 29–37. DOI: 10.5455/apd.299899

Baykara, S. Z. (2004). Experimental solar water thermolysis. *International Journal of Hydrogen Energy*, *29*(14), 1459–1469. DOI: 10.1016/j.ijhydene.2004.02.011

Bhandari, R., Trudewind, C. A., & Zapp, P. (2014). Life cycle assessment of hydrogen production via electrolysis - A review. In *Journal of Cleaner Production* (Vol. 85, pp. 151–163). Elsevier Ltd., DOI: 10.1016/j.jclepro.2013.07.048

Bičáková, O., & Straka, P. (2012). Production of hydrogen from renewable resources and its effectiveness. In *International Journal of Hydrogen Energy* (Vol. 37, Issue 16, pp. 11563–11578). DOI: 10.1016/j.ijhydene.2012.05.047

Bridgwater, A. V. (1999). Principles and practice of biomass fast pyrolysis processes for liquids. In *Journal of Analytical and Applied Pyrolysis* (Vol. 51). DOI: 10.1016/S0165-2370(99)00005-4

Campoy, M., Gómez-Barea, A., Villanueva, A. L., & Ollero, P. (2008). Air-steam gasification of biomass in a fluidized bed under simulated autothermal and adiabatic conditions. *Industrial & Engineering Chemistry Research*, *47*(16), 5957–5965. DOI: 10.1021/ie800220t

Carmo, M., Fritz, D. L., Mergel, J., & Stolten, D. (2013). A comprehensive review on PEM water electrolysis. In *International Journal of Hydrogen Energy* (Vol. 38, Issue 12, pp. 4901–4934). DOI: 10.1016/j.ijhydene.2013.01.151

Chi, J., & Yu, H. (2018). Water electrolysis based on renewable energy for hydrogen production. In *Cuihua Xuebao/Chinese Journal of Catalysis* (Vol. 39, Issue 3, pp. 390–394). Science Press. DOI: 10.1016/S1872-2067(17)62949-8

Cinti, G., Discepoli, G., Bidini, G., Lanzini, A., & Santarelli, M. (2016). Co-electrolysis of water and CO2 in a solid oxide electrolyzer (SOE) stack. *International Journal of Energy Research*, *40*(2), 207–215. DOI: 10.1002/er.3450

*Climate change*. (n.d.). Retrieved October 29, 2024, from https://www.who.int/news-room/fact-sheets/detail/climate-change-and-health

Danilovic, N., Ayers, K. E., Capuano, C., Renner, J. N., Wiles, L., & Pertoso, M. (2016). (Plenary) Challenges in Going from Laboratory to Megawatt Scale PEM Electrolysis. *ECS Transactions*, *75*(14), 395–402. DOI: 10.1149/07514.0395ecst

David, M., Ocampo-Martínez, C., & Sánchez-Peña, R. (2019). Advances in alkaline water electrolyzers: A review. In *Journal of Energy Storage* (Vol. 23, pp. 392–403). Elsevier Ltd., DOI: 10.1016/j.est.2019.03.001

De Blasio, N. (2024, July 8). *The colors of hydrogen*. Environment and Natural Resources Program, Belfer Center for Science and International Affairs. https://www.belfercenter.org/research-analysis/colors-hydrogen

Dehner, G., McBeth, M. K., Moss, R., & van Woerden, I. (2023). A Zero-Carbon Nuclear Energy Future? Lessons Learned from Perceptions of Climate Change and Nuclear Waste. *Energies*, *16*(4), 2025. Advance online publication. DOI: 10.3390/en16042025

*Department of State Growth*. (n.d.). Retrieved October 30, 2024, from https://www.stategrowth.tas.gov.au/

Dincer, I. (2012). Green methods for hydrogen production. *International Journal of Hydrogen Energy*, *37*(2), 1954–1971. DOI: 10.1016/j.ijhydene.2011.03.173

Dobó, Z., & Palotás, Á. B. (2017). Impact of the current fluctuation on the efficiency of Alkaline Water Electrolysis. *International Journal of Hydrogen Energy*, *42*(9), 5649–5656. DOI: 10.1016/j.ijhydene.2016.11.142

Donitz, W., & Erdle, E. (1985). HIGH-TEMPERATURE ELECTROLYSIS OF WATER VAPOR-STATUS OF DEVELOPMENT AND PERSPECTIVES FOR APPLICATION. *International Journal of Hydrogen Energy*, *10*(5), 291–295. DOI: 10.1016/0360-3199(85)90181-8

Eloffy, M. G., Elgarahy, A. M., Saber, A. N., Hammad, A., El-Sherif, D. M., Shehata, M., Mohsen, A., & Elwakeel, K. Z. (2022). Biomass-to-sustainable biohydrogen: Insights into the production routes, and technical challenges. In *Chemical Engineering Journal Advances* (Vol. 12). Elsevier B.V., DOI: 10.1016/j.ceja.2022.100410

*Energy - European Commission*. (n.d.). Retrieved October 30, 2024, from https://energy.ec.europa.eu/index_en

*Essential Intelligence | S&P Global*. (n.d.). Retrieved October 30, 2024, from https://www.spglobal.com/en

*EUR-Lex - 52020DC0301 - EN - EUR-Lex*. (n.d.). Retrieved October 30, 2024, from https://eur-lex.europa.eu/legal-content/EN/TXT/?uri=CELEX%3A52020DC0301

Fallah Vostakola, M., Ozcan, H., El-Emam, R. S., & Amini Horri, B. (2023). Recent Advances in High-Temperature Steam Electrolysis with Solid Oxide Electrolysers for Green Hydrogen Production. In *Energies* (Vol. 16, Issue 8). MDPI. DOI: 10.3390/en16083327

Fasihi, M., & Breyer, C. (2020). Baseload electricity and hydrogen supply based on hybrid PV-wind power plants. *Journal of Cleaner Production*, *243*, 118466. Advance online publication. DOI: 10.1016/j.jclepro.2019.118466

Fernández-Arias, P., Antón-Sancho, Á., Lampropoulos, G., & Vergara, D. (2024). Emerging Trends and Challenges in Pink Hydrogen Research. *Energies*, *17*(10), 2291. Advance online publication. DOI: 10.3390/en17102291

Ferraren-De Cagalitan, D. D. T., & Abundo, M. L. S. (2021). A review of biohydrogen production technology for application towards hydrogen fuel cells. *Renewable & Sustainable Energy Reviews*, *151*, 111413. DOI: 10.1016/j.rser.2021.111413

Fulcheri, L., Probst, N., Flamant, G., Fabry, F., Grivei, E., & Bourrat, X. (2002). Plasma processing: a step towards the production of new grades of carbon black. In *Carbon* (Vol. 40). DOI: 10.1016/S0008-6223(01)00169-5

Ganley, J. C. (2009). High temperature and pressure alkaline electrolysis. *International Journal of Hydrogen Energy*, *34*(9), 3604–3611. DOI: 10.1016/j.ijhydene.2009.02.083

Gaudernack, B., & Lynum, S. (1998). Hydrogen from natural gas without release of CO2 to the atmosphere. *International Journal of Hydrogen Energy*, *23*(12), 1087–1093. DOI: 10.1016/S0360-3199(98)00004-4

Ghaffour, N., Bundschuh, J., Mahmoudi, H., & Goosen, M. F. A. (2015). Renewable energy-driven desalination technologies: A comprehensive review on challenges and potential applications of integrated systems. *Desalination*, *356*, 94–114. DOI: 10.1016/j.desal.2014.10.024

Ghirardi, M. L., Zhang, L., Lee, J. W., Flynn, T., Seibert, M., Greenbaum, E., & Melis, A. (2000). Microalgae: A green source of renewable H2. *Trends in Biotechnology*, *18*(12), 506–511. DOI: 10.1016/S0167-7799(00)01511-0 PMID: 11102662

Gracia, L., Casero, P., Bourasseau, C., & Chabert, A. (2018). Use of hydrogen in off-grid locations, a techno-economic assessment. *Energies*, *11*(11), 3141. Advance online publication. DOI: 10.3390/en11113141

*Green Hydrogen adoption in India – Opportunities, challenges and the way ahead, ET EnergyWorld.* (n.d.). Retrieved October 30, 2024, from https://energy.economictimes.indiatimes.com/news/renewable/green-hydrogen-adoption-in-india-opportunities-challenges-and-the-way-ahead/100177937

Grubb, W. T. (n.d.). *Batteries with Solid Ion Exchange Electrolytes I.* Secondary Cells Employing Metal Electrodes. DOI: 10.1149/1.2427329

Hallenbeck, P. C., & Benemann, J. R. (2002). Biological hydrogen production; fundamentals and limiting processes.[]. *International Journal of Hydrogen Energy*, *27*(11-12), 1185–1193. www.elsevier.com/locate/ijhydene. DOI: 10.1016/S0360-3199(02)00131-3

Hassan, Q. (2020). Optimisation of solar-hydrogen power system for household applications. *International Journal of Hydrogen Energy*, *45*(58), 33111–33127. DOI: 10.1016/j.ijhydene.2020.09.103

Hassan, Q., Abdulateef, A. M., Hafedh, S. A., Al-samari, A., Abdulateef, J., Sameen, A. Z., Salman, H. M., Al-Jiboory, A. K., Wieteska, S., & Jaszczur, M. (2023). Renewable energy-to-green hydrogen: A review of main resources routes, processes and evaluation. In *International Journal of Hydrogen Energy* (Vol. 48, Issue 46, pp. 17383–17408). Elsevier Ltd. DOI: 10.1016/j.ijhydene.2023.01.175

Hassan, Q., Abdulrahman, I. S., Salman, H. M., Olapade, O. T., & Jaszczur, M. (2023). Techno-Economic Assessment of Green Hydrogen Production by an Off-Grid Photovoltaic Energy System. *Energies*, *16*(2), 744. Advance online publication. DOI: 10.3390/en16020744

Hassan, Q., Algburi, S., Sameen, A. Z., Salman, H. M., & Jaszczur, M. (2024). 1. *International Journal of Hydrogen Energy*, *50*, 310–333. DOI: 10.1016/j.ijhydene.2023.08.321

Hassan, Q., & Jaszczur, M. (2021). Self-consumption and self-sufficiency improvement for photovoltaic system integrated with ultra-supercapacitor. *Energies*, *14*(23), 7888. Advance online publication. DOI: 10.3390/en14237888

Hassan, Q., Jaszczur, M., Teneta, J., Abbas, M. K., Hasan, A., & Al-Jiboory, A. K. (2022). Experimental investigation for the estimation of the intensity of solar irradiance on oblique surfaces by means of various models. *Energy Harvesting and Systems*, *9*(2), 227–237. DOI: 10.1515/ehs-2021-0087

Hassan, Q., Pawela, B., Hasan, A., & Jaszczur, M. (2022). Optimization of Large-Scale Battery Storage Capacity in Conjunction with Photovoltaic Systems for Maximum Self-Sustainability. *Energies*, *15*(10), 3845. Advance online publication. DOI: 10.3390/en15103845

Hoang, A. T., Ong, H. C., Fattah, I. M. R., Chong, C. T., Cheng, C. K., Sakthivel, R., & Ok, Y. S. (2021). Progress on the lignocellulosic biomass pyrolysis for biofuel production toward environmental sustainability. In *Fuel Processing Technology* (Vol. 223). Elsevier B.V., DOI: 10.1016/j.fuproc.2021.106997

Holladay, J. D., Hu, J., King, D. L., & Wang, Y. (2009). An overview of hydrogen production technologies. In *Catalysis Today* (Vol. 139, Issue 4, pp. 244–260). DOI: 10.1016/j.cattod.2008.08.039

*Homepage | Hydrogen Council*. (n.d.). Retrieved October 30, 2024, from https://hydrogencouncil.com/en/

Hood, R. (2024, August 29). *White hydrogen: 5 critical questions answered*. World Economic Forum. https://www.weforum.org/agenda/2024/08/white-hydrogen-5-critical-questions-answered/

Horvath, A., & Rachlew, E. (2016). Nuclear power in the 21st century: Challenges and possibilities. *Ambio*, *45*(S1), 38–49. DOI: 10.1007/s13280-015-0732-y PMID: 26667059

Hosseini, S. S., Aghbashlo, M., Tabatabaei, M., Younesi, H., & Najafpour, G. (2015). Exergy analysis of biohydrogen production from various carbon sources via anaerobic photosynthetic bacteria (Rhodospirillum rubrum). *Energy*, *93*, 730–739. DOI: 10.1016/j.energy.2015.09.060

Hou, X., Wang, Y., Yang, Y., Hu, R., Yang, G., Feng, L., & Suo, G. (2019). Microstructure evolution and controlled hydrolytic hydrogen generation strategy of Mg-rich Mg-Ni-La ternary alloys. *Energy*, *188*, 116081. DOI: 10.1016/j.energy.2019.116081

Hu, S., Lozada-Hidalgo, M., Wang, F. C., Mishchenko, A., Schedin, F., Nair, R. R., Hill, E. W., Boukhvalov, D. W., Katsnelson, M. I., Dryfe, R. A. W., Grigorieva, I. V., Wu, H. A., & Geim, A. K. (2014). Proton transport through one-atom-thick crystals. *Nature*, *516*(7530), 227–230. DOI: 10.1038/nature14015 PMID: 25470058

Huber, G. W., Iborra, S., & Corma, A. (2006). Synthesis of transportation fuels from biomass: Chemistry, catalysts, and engineering. In *Chemical Reviews* (Vol. 106, Issue 9, pp. 4044–4098). DOI: 10.1021/cr068360d

*HYDROGEN AND FUEL CELL TECHNOLOGIES OFFICE Emerging Hydrogen Markets*. (n.d.). https://www.energy.gov/eere/fuelcells/h2-scale

*Hydrogen is developing fast in Japan, closer to societal use | World Economic Forum*. (n.d.). Retrieved October 30, 2024, from https://www.weforum.org/agenda/2024/04/hydrogen-japan/

*Hydrogen Production: Thermochemical Water Splitting | Department of Energy*. (n.d.). Retrieved October 30, 2024, from https://www.energy.gov/eere/fuelcells/hydrogen-production-thermochemical-water-splitting

*India's initiatives on green hydrogen could help global decarbonisation*. (n.d.). Retrieved October 30, 2024, from https://india.mongabay.com/2022/11/indias-initiatives-on-green-hydrogen-could-help-global-decarbonisation/

Intergovernmental Panel on Climate Change (IPCC). (2021, August 9). *Climate change widespread, rapid, and intensifying – IPCC*. Retrieved October 29, 2024, from https://www.ipcc.ch/2021/08/09/ar6-wg1-20210809-pr/

International Energy Agency. (2024, April 18). Clean energy is boosting economic growth. *International Energy Agency*. https://www.iea.org/commentaries/clean-energy-is-boosting-economic-growth

Iribarren, D., Susmozas, A., Petrakopoulou, F., & Dufour, J. (2014). Environmental and exergetic evaluation of hydrogen production via lignocellulosic biomass gasification. *Journal of Cleaner Production*, *69*, 165–175. DOI: 10.1016/j.jclepro.2014.01.068

Ji, M., & Wang, J. (2021). Review and comparison of various hydrogen production methods based on costs and life cycle impact assessment indicators. In *International Journal of Hydrogen Energy* (Vol. 46, Issue 78, pp. 38612–38635). Elsevier Ltd. DOI: 10.1016/j.ijhydene.2021.09.142

Kadier, A., Kalil, M. S., Abdeshahian, P., Chandrasekhar, K., Mohamed, A., Azman, N. F., Logroño, W., Simayi, Y., & Hamid, A. A. (2016). Recent advances and emerging challenges in microbial electrolysis cells (MECs) for microbial production of hydrogen and value-added chemicals. In *Renewable and Sustainable Energy Reviews* (Vol. 61, pp. 501–525). Elsevier Ltd., DOI: 10.1016/j.rser.2016.04.017

Kadier, A., Simayi, Y., Abdeshahian, P., Azman, N. F., Chandrasekhar, K., & Kalil, M. S. (2016). A comprehensive review of microbial electrolysis cells (MEC) reactor designs and configurations for sustainable hydrogen gas production. In *Alexandria Engineering Journal* (Vol. 55, Issue 1, pp. 427–443). Elsevier B.V. DOI: 10.1016/j.aej.2015.10.008

Kalamaras, C. M., & Efstathiou, A. M. (2013). Hydrogen Production Technologies: Current State and Future Developments. *Conference Papers in Energy*, *2013*, 1–9. DOI: 10.1155/2013/690627

Kapdan, I. K., & Kargi, F. (2006). Bio-hydrogen production from waste materials. In *Enzyme and Microbial Technology* (Vol. 38, Issue 5, pp. 569–582). DOI: 10.1016/j.enzmictec.2005.09.015

Kosourov, S., Murukesan, G., Seibert, M., & Allahverdiyeva, Y. (2017). Evaluation of light energy to H2 energy conversion efficiency in thin films of cyanobacteria and green alga under photoautotrophic conditions. *Algal Research*, *28*, 253–263. DOI: 10.1016/j.algal.2017.09.027

Kumar, G., Eswari, A. P., Kavitha, S., Kumar, M. D., Kannah, R. Y., How, L. C., Muthukaruppan, G., & Banu, J. R. (2023). Thermochemical conversion routes of hydrogen production from organic biomass: processes, challenges and limitations. In *Biomass Conversion and Biorefinery* (Vol. 13, Issue 10, pp. 8509–8534). Springer Science and Business Media Deutschland GmbH. DOI: 10.1007/s13399-020-01127-9

Leal Pérez, B. J., Medrano Jiménez, J. A., Bhardwaj, R., Goetheer, E., van Sint Annaland, M., & Gallucci, F. (2021). Methane pyrolysis in a molten gallium bubble column reactor for sustainable hydrogen production: Proof of concept & techno-economic assessment. *International Journal of Hydrogen Energy*, *46*(7), 4917–4935. DOI: 10.1016/j.ijhydene.2020.11.079

Lechtenböhmer, S., & Samadi, S. (2013). Blown by the wind. Replacing nuclear power in German electricity generation. *Environmental Science & Policy*, *25*, 234–241. DOI: 10.1016/j.envsci.2012.09.003

Levin, D. B., Pitt, L., & Love, M. (2004). Biohydrogen production: Prospects and limitations to practical application. *International Journal of Hydrogen Energy*, *29*(2), 173–185. DOI: 10.1016/S0360-3199(03)00094-6

Liang, M., Yu, B., Wen, M., Chen, J., Xu, J., & Zhai, Y. (2009). Preparation of LSM-YSZ composite powder for anode of solid oxide electrolysis cell and its activation mechanism. *Journal of Power Sources*, *190*(2), 341–345. DOI: 10.1016/j.jpowsour.2008.12.132

Liao, B., Guo, L., Lu, Y., & Zhang, X. (2013). Solar receiver/reactor for hydrogen production with biomass gasification in supercritical water. *International Journal of Hydrogen Energy*, *38*(29), 13038–13044. DOI: 10.1016/j.ijhydene.2013.03.113

Liu, H., Grot, S., & Logan, B. E. (2005). Electrochemically assisted microbial production of hydrogen from acetate. *Environmental Science & Technology*, *39*(11), 4317–4320. DOI: 10.1021/es050244p PMID: 15984815

Logan, B. E., Hamelers, B., Rozendal, R., Schröder, U., Keller, J., Freguia, S., Aelterman, P., Verstraete, W., & Rabaey, K. (2006). Microbial fuel cells: Methodology and technology. In *Environmental Science and Technology* (Vol. 40, Issue 17, pp. 5181–5192). DOI: 10.1021/es0605016

Longden, T., Beck, F. J., Jotzo, F., Andrews, R., & Prasad, M. (2022). 'Clean' hydrogen? – Comparing the emissions and costs of fossil fuel versus renewable electricity based hydrogen. *Applied Energy*, *306*, 118145. Advance online publication. DOI: 10.1016/j.apenergy.2021.118145

Longoria, G., Lynch, M., & Curtis, J. (2021). Green hydrogen for heating and its impact on the power system. *International Journal of Hydrogen Energy*, *46*(53), 26725–26740. DOI: 10.1016/j.ijhydene.2021.05.171

López, A., De Marco, I., Caballero, B. M., Laresgoiti, M. F., & Adrados, A. (2011). Dechlorination of fuels in pyrolysis of PVC containing plastic wastes. *Fuel Processing Technology*, *92*(2), 253–260. DOI: 10.1016/j.fuproc.2010.05.008

Lozada-Hidalgo, M., Zhang, S., Hu, S., Kravets, V. G., Rodriguez, F. J., Berdyugin, A., Grigorenko, A., & Geim, A. K. (2018). Giant photoeffect in proton transport through graphene membranes. *Nature Nanotechnology*, *13*(4), 300–303. DOI: 10.1038/s41565-017-0051-5 PMID: 29358638

Maka, A. O. M., & Mehmood, M. (2024). 2. In *Clean Energy* (Vol. 8, Issue 2, pp. 1–7). Oxford University Press. DOI: 10.1093/ce/zkae012

*Massive Asian Renewable Energy Hub grows to 26GW of wind and solar | RenewEconomy*. (n.d.). Retrieved October 30, 2024, from https://reneweconomy.com.au/massive-asian-renewable-energy-hub-grows-to-26gw-of-wind-and-solar-49343/

Mckendry, P. (n.d.). *Energy production from biomass (part 3): gasification technologies*.

Ministry of Economy, Trade, and Industry (METI). (n.d.). *Title of the page*. Retrieved November 16, 2024, from https://www.meti.go.jp/

Narayanan, K., & Sahu, S. K. (2014). 13. *IIM Kozhikode Society &. Management Review*, *3*(2), 123–133. DOI: 10.1177/2277975214544014

*National Aeronautics and Space Administration (NASA). (n.d.). Global temperature. NASA Climate Change: Vital Signs of the Planet. Retrieved October 29, 2024, from*https://climate.nasa.gov/vital-signs/global-temperature/?intent=121

*National Green Hydrogen Mission Portal of India*. (n.d.). Retrieved October 30, 2024, from https://nghm.mnre.gov.in/

National Observer. (2020, July 14). *Saskatchewan pilots hydrogen fuel for the future*. National Observer. Retrieved from https://www.nationalobserver.com/2020/07/14/news/saskatchewan-pilots-hydrogen-fuel-future

*Natural Resources Canada*. (n.d.). Retrieved October 30, 2024, from https://natural-resources.canada.ca/home

Newborough, M., & Cooley, G. (2020). Developments in the global hydrogen market: The spectrum of hydrogen colours. *Fuel Cells Bulletin*, *2020*(11), 16–22. DOI: 10.1016/S1464-2859(20)30546-0

Ni, M., Leung, D. Y. C., Leung, M. K. H., & Sumathy, K. (2006). An overview of hydrogen production from biomass. *Fuel Processing Technology*, *87*(5), 461–472. DOI: 10.1016/j.fuproc.2005.11.003

Nikolaidis, P., & Poullikkas, A. (2017). A comparative overview of hydrogen production processes. *Renewable & Sustainable Energy Reviews*, *67*, 597–611. DOI: 10.1016/j.rser.2016.09.044

Novotny, V. (2023). Blue hydrogen can be a source of green energy in the period of decarbonization. *International Journal of Hydrogen Energy*, *48*(20), 7202–7218. DOI: 10.1016/j.ijhydene.2022.11.095

Osman, A. I., Mehta, N., Elgarahy, A. M., Al-Hinai, A., Al-Muhtaseb, A. H., & Rooney, D. W. (2021). Conversion of biomass to biofuels and life cycle assessment: a review. In *Environmental Chemistry Letters* (Vol. 19, Issue 6, pp. 4075–4118). Springer Science and Business Media Deutschland GmbH. DOI: 10.1007/s10311-021-01273-0

Park, C., Koo, M., Woo, J. R., Hong, B., & Shin, J. (2022). Economic valuation of green hydrogen charging compared to gray hydrogen charging: The case of South Korea. *International Journal of Hydrogen Energy*, *47*(32), 14393–14403. DOI: 10.1016/j.ijhydene.2022.02.214

Parthasarathy, P., & Narayanan, K. S. (2014). Hydrogen production from steam gasification of biomass: Influence of process parameters on hydrogen yield - A review. In *Renewable Energy* (Vol. 66, pp. 570–579). Elsevier Ltd., DOI: 10.1016/j.renene.2013.12.025

Paulmier, T., & Fulcheri, L. (2005). Use of non-thermal plasma for hydrocarbon reforming. *Chemical Engineering Journal*, *106*(1), 59–71. DOI: 10.1016/j.cej.2004.09.005

Phys.org. (2019, August 19). *Scientists extract hydrogen gas from oil and bitumen, giving potential pollution-free energy*. Phys.org. Retrieved from https://phys.org/news/2019-08-scientists-hydrogen-gas-oil-bitumen.html

Pino, L., Recupero, V., Beninati, S., Kumar Shukla, A., Subraya Hegde, M., & Bera, P. (2002). Catalytic partial-oxidation of methane on a ceria-supported platinum catalyst for application in fuel cell electric vehicles. In *Applied Catalysis A* (Vol. 225). General. DOI: 10.1016/S0926-860X(01)00734-7

Qureshi, F., Yusuf, M., Kamyab, H., Zaidi, S., Junaid Khalil, M., Arham Khan, M., Azad Alam, M., Masood, F., Bazli, L., Chelliapan, S., & Abdullah, B. (2022). Current trends in hydrogen production, storage and applications in India: A review. *Sustainable Energy Technologies and Assessments*, *53*, 102677. DOI: 10.1016/j.seta.2022.102677

Rathinam, N. K., Sani, R. K., & Salem, D. (2018). Rewiring extremophilic electrocatalytic processes for production of biofuels and value-added compounds from lignocellulosic biomass. In *Extremophilic Microbial Processing of Lignocellulosic Feedstocks to Biofuels, Value-Added Products, and Usable Power* (pp. 229–245). Springer International Publishing. DOI: 10.1007/978-3-319-74459-9_12

Ratnakar, R. R., Gupta, N., Zhang, K., van Doorne, C., Fesmire, J., Dindoruk, B., & Balakotaiah, V. (2021). Hydrogen supply chain and challenges in large-scale LH2 storage and transportation. In *International Journal of Hydrogen Energy* (Vol. 46, Issue 47, pp. 24149–24168). Elsevier Ltd. DOI: 10.1016/j.ijhydene.2021.05.025

Rooijers, F., & van Cappellen, L. (2018). *Feasibility study into blue hydrogen*. CE Delft. Retrieved from https://cedelft.eu/publications/feasibility-study-into-blue-hydrogen/

Rozendal, R. A., Hamelers, H. V. M., Euverink, G. J. W., Metz, S. J., & Buisman, C. J. N. (2006). Principle and perspectives of hydrogen production through biocatalyzed electrolysis. *International Journal of Hydrogen Energy*, *31*(12), 1632–1640. DOI: 10.1016/j.ijhydene.2005.12.006

Sánchez, M., Amores, E., Abad, D., Rodríguez, L., & Clemente-Jul, C. (2020). Aspen Plus model of an alkaline electrolysis system for hydrogen production. *International Journal of Hydrogen Energy*, *45*(7), 3916–3929. DOI: 10.1016/j.ijhydene.2019.12.027

Santos, D. M. F., Sequeira, C. A. C., & Figueiredo, J. L. (2013). HYDROGEN PRODUCTION BY ALKALINE WATER ELECTROLYSIS. *Quimica Nova*, *36*(8), 1176–1193. DOI: 10.1590/S0100-40422013000800017

Schlissel, D., & Juhn, A. (n.d.). *September 2023*.

Schmidt Rivera, X. C., Topriska, E., Kolokotroni, M., & Azapagic, A. (2018). Environmental sustainability of renewable hydrogen in comparison with conventional cooking fuels. *Journal of Cleaner Production*, *196*, 863–879. DOI: 10.1016/j.jclepro.2018.06.033

Scholten, D., Bazilian, M., Overland, I., & Westphal, K. (2020). The geopolitics of renewables: New board, new game. *Energy Policy*, *138*, 111059. Advance online publication. DOI: 10.1016/j.enpol.2019.111059

Scita, R., Raimondi, P. P., & Noussan, M. (2020). *Working Paper*.

Shafiqah, M. N. N., Siang, T. J., Kumar, P. S., Ahmad, Z., Jalil, A. A., Bahari, M. B., Van Le, Q., Xiao, L., Mofijur, M., Xia, C., Ahmed, S. F., & Vo, D. V. N. (2022). Advanced catalysts and effect of operating parameters in ethanol dry reforming for hydrogen generation. A review. In *Environmental Chemistry Letters* (Vol. 20, Issue 3, pp. 1695–1718). Springer Science and Business Media Deutschland GmbH. DOI: 10.1007/s10311-022-01394-0

Shiva Kumar, S., & Himabindu, V. (2019). Hydrogen production by PEM water electrolysis – A review. In *Materials Science for Energy Technologies* (Vol. 2, Issue 3, pp. 442–454). KeAi Communications Co. DOI: 10.1016/j.mset.2019.03.002

Shiva Kumar, S., & Lim, H. (2022). An overview of water electrolysis technologies for green hydrogen production. In *Energy Reports* (Vol. 8, pp. 13793–13813). Elsevier Ltd., DOI: 10.1016/j.egyr.2022.10.127

Shiva Kumar, S., Ramakrishna, S. U. B., Rama Devi, B., & Himabindu, V. (2018). Phosphorus-doped carbon nanoparticles supported palladium electrocatalyst for the hydrogen evolution reaction (HER) in PEM water electrolysis. *Ionics*, *24*(10), 3113–3121. DOI: 10.1007/s11581-018-2471-0

Singla, M. K., Gupta, J., Beryozkina, S., Safaraliev, M., & Singh, M. (2024). The colorful economics of hydrogen: Assessing the costs and viability of different hydrogen production methods - A review. *International Journal of Hydrogen Energy*, *61*, 664–677. DOI: 10.1016/j.ijhydene.2024.02.255

Su, G., Ong, H. C., Fattah, I. M. R., Ok, Y. S., Jang, J. H., & Wang, C. T. (2022). State-of-the-art of the pyrolysis and co-pyrolysis of food waste: Progress and challenges. *The Science of the Total Environment*, *809*, 151170. DOI: 10.1016/j.scitotenv.2021.151170 PMID: 34699825

Su, W., Liu, P., Cai, C., Ma, H., Jiang, B., Xing, Y., Liang, Y., Cai, L., Xia, C., Van Le, Q., Sonne, C., & Lam, S. S. (2021). Hydrogen production and heavy metal immobilization using hyperaccumulators in supercritical water gasification. *Journal of Hazardous Materials*, *402*, 123541. Advance online publication. DOI: 10.1016/j.jhazmat.2020.123541 PMID: 32745873

Su, W., Zhao, M., Xing, Y., Ma, H., Liu, P., Li, X., Zhang, H., Wu, Y., & Xia, C. (2022). Supercritical water gasification of hyperaccumulators for hydrogen production and heavy metal immobilization with alkali metal catalysts. *Environmental Research*, *214*, 114093. Advance online publication. DOI: 10.1016/j.envres.2022.114093 PMID: 35998690

Tanksale, A., Beltramini, J. N., & Lu, G. Q. M. (2010). A review of catalytic hydrogen production processes from biomass. In *Renewable and Sustainable Energy Reviews* (Vol. 14, Issue 1, pp. 166–182). DOI: 10.1016/j.rser.2009.08.010

Thauer, R. K., Jungermann, K., & Decker, K. (1977). *Energy Conservation in Chemotrophic Anaerobic Bacteria* (Vol. 41, Issue 1). https://journals.asm.org/journal/br

*The Colors of Hydrogen | The Belfer Center for Science and International Affairs*. (n.d.). Retrieved October 30, 2024, from https://www.belfercenter.org/research-analysis/colors-hydrogen

*The Future of Pink Hydrogen as a Fossil-Fuel Alternative*. (n.d.). Retrieved October 30, 2024, from https://www.azocleantech.com/article.aspx?ArticleID=1718

*The Hydrogen Strategy*. (n.d.). Retrieved October 30, 2024, from https://natural-resources.canada.ca/climate-change/canadas-green-future/the-hydrogen-strategy/23080

Turner, J., Sverdrup, G., Mann, M. K., Maness, P. C., Kroposki, B., Ghirardi, M., Evans, R. J., & Blake, D. (2008). Renewable hydrogen production. *International Journal of Energy Research*, *32*(5), 379–407. DOI: 10.1002/er.1372

UNICEF. (2024). *State of global air report 2024*. Retrieved November 16, 2024, from https://knowledge.unicef.org/CEED/resource/state-global-air-report-2024

van Rossum, G., Potic, B., Kersten, S. R. A., & van Swaaij, W. P. M. (2009). Catalytic gasification of dry and wet biomass. *Catalysis Today*, *145*(1–2), 10–18. DOI: 10.1016/j.cattod.2008.04.048

Vasseghian, Y., Le, V. T., Joo, S. W., Dragoi, E. N., Kamyab, H., Chelliapan, S., & Klemeš, J. J. (2022). Spotlighting graphene-based catalysts for the mitigation of environmentally hazardous pollutants to cleaner production: A review. In *Journal of Cleaner Production* (Vol. 365). Elsevier Ltd., DOI: 10.1016/j.jclepro.2022.132702

Veeravalli, S. S., Shanmugam, S. R., Ray, S., Lalman, J. A., & Biswas, N. (2019). Biohydrogen production from renewable resources. In *Advanced Bioprocessing for Alternative Fuels, Biobased Chemicals, and Bioproducts: Technologies and Approaches for Scale-Up and Commercialization* (pp. 289–312). Elsevier. DOI: 10.1016/B978-0-12-817941-3.00015-2

Velazquez Abad, A., & Dodds, P. E. (2020). Green hydrogen characterisation initiatives: Definitions, standards, guarantees of origin, and challenges. *Energy Policy*, *138*, 111300. Advance online publication. DOI: 10.1016/j.enpol.2020.111300

Vidas, L., & Castro, R. (2021). Recent developments on hydrogen production technologies: State-of-the-art review with a focus on green-electrolysis. In *Applied Sciences (Switzerland)* (Vol. 11, Issue 23). MDPI. DOI: 10.3390/app112311363

Vijayaraghavan, K., Ahmad, D., & Ibrahim, M. K. (2006). Biohydrogen generation from jackfruit peel using anaerobic contact filter. *International Journal of Hydrogen Energy*, *31*(5), 569–579. DOI: 10.1016/j.ijhydene.2005.06.006

Wang, T., Cao, X., & Jiao, L. (2022). PEM water electrolysis for hydrogen production: fundamentals, advances, and prospects. In *Carbon Neutrality* (Vol. 1, Issue 1). Springer. DOI: 10.1007/s43979-022-00022-8

Widera, B. (2020). Renewable hydrogen implementations for combined energy storage, transportation and stationary applications. *Thermal Science and Engineering Progress*, *16*, 100460. Advance online publication. DOI: 10.1016/j.tsep.2019.100460

*WordPress › Error*. (n.d.). Retrieved October 30, 2024, from https://www.pv-magazine.com/2022/08/26/avaada-to-invest-5-billion-in-green-hydrogen-ammonia-plant-in-india/

*World Energy Outlook 2022 – Analysis - IEA*. (n.d.). Retrieved October 30, 2024, from https://www.iea.org/reports/world-energy-outlook-2022

Xia, S., Fang, L., Meng, Y., Zhang, X., Zhang, L., Yang, C., & Ni, Z. (2020). Water-gas shift reaction catalyzed by layered double hydroxides supported Au-Ni/Cu/Pt bimetallic alloys. *Applied Catalysis B: Environmental*, *272*, 118949. Advance online publication. DOI: 10.1016/j.apcatb.2020.118949

Xiao, L., Zhang, S., Pan, J., Yang, C., He, M., Zhuang, L., & Lu, J. (2012). First implementation of alkaline polymer electrolyte water electrolysis working only with pure water. *Energy & Environmental Science*, *5*(7), 7869–7871. DOI: 10.1039/c2ee22146b

Yáñez, M., Ortiz, A., Brunaud, B., Grossmann, I. E., & Ortiz, I. (2018). Contribution of upcycling surplus hydrogen to design a sustainable supply chain: The case study of Northern Spain. *Applied Energy*, *231*, 777–787. DOI: 10.1016/j.apenergy.2018.09.047

Yu, F., Yu, L., Mishra, I. K., Yu, Y., Ren, Z. F., & Zhou, H. Q. (2018). Recent developments in earth-abundant and non-noble electrocatalysts for water electrolysis. In *Materials Today Physics* (Vol. 7, pp. 121–138). Elsevier Ltd., DOI: 10.1016/j.mtphys.2018.11.007

Yu, M., Wang, K., & Vredenburg, H. (2021). Insights into low-carbon hydrogen production methods: Green, blue and aqua hydrogen. *International Journal of Hydrogen Energy*, *46*(41), 21261–21273. DOI: 10.1016/j.ijhydene.2021.04.016

Yukesh Kannah, R., & Kavitha, S. Preethi, Parthiba Karthikeyan, O., Kumar, G., Dai-Viet, N. V., & Rajesh Banu, J. (2021). Techno-economic assessment of various hydrogen production methods – A review. In *Bioresource Technology* (Vol. 319). Elsevier Ltd. DOI: 10.1016/j.biortech.2020.124175

Zamri, M. F. M. A., Hasmady, S., Akhiar, A., Ideris, F., Shamsuddin, A. H., Mofijur, M., Fattah, I. M. R., & Mahlia, T. M. I. (2021). A comprehensive review on anaerobic digestion of organic fraction of municipal solid waste. In *Renewable and Sustainable Energy Reviews* (Vol. 137). Elsevier Ltd., DOI: 10.1016/j.rser.2020.110637

Zeng, K., & Zhang, D. (2010). Recent progress in alkaline water electrolysis for hydrogen production and applications. In *Progress in Energy and Combustion Science* (Vol. 36, Issue 3, pp. 307–326). DOI: 10.1016/j.pecs.2009.11.002

Zhang, L., Xu, C., & Champagne, P. (2010). Overview of recent advances in thermo-chemical conversion of biomass. *Energy Conversion and Management*, *51*(5), 969–982. DOI: 10.1016/j.enconman.2009.11.038

Zhao, H., Lu, D., Wang, J., Tu, W., Wu, D., Koh, S. W., Gao, P., Xu, Z. J., Deng, S., Zhou, Y., You, B., & Li, H. (2021). Raw biomass electroreforming coupled to green hydrogen generation. *Nature Communications*, *12*(1), 2008. Advance online publication. DOI: 10.1038/s41467-021-22250-9 PMID: 33790295

Zhou, D., Li, P., Xu, W., Jawaid, S., Mohammed-Ibrahim, J., Liu, W., Kuang, Y., & Sun, X. (2020). Recent Advances in Non-Precious Metal-Based Electrodes for Alkaline Water Electrolysis. In *ChemNanoMat* (Vol. 6, Issue 3, pp. 336–355). Wiley-VCH Verlag. DOI: 10.1002/cnma.202000010

Zhou, Y., Li, R., Lv, Z., Liu, J., Zhou, H., & Xu, C. (2022). 4. In *Chinese Journal of Chemical Engineering* (Vol. 43, pp. 2–13). Materials China., DOI: 10.1016/j.cjche.2022.02.001

# Chapter 23
# Rediscovery and Typification of Amanita From the Cork Groves of Morocco

**Abdelkrim El Assfouri**
https://orcid.org/0009-0001-8071-6599
*Ibn Tofaïl University, Morocco*

**Amina Ouazzani Touhami**
https://orcid.org/0000-0001-9970-7724
*Ibn Tofaïl University, Morocco*

**Allal Douira**
*Ibn Tofaïl University, Morocco*

## ABSTRACT

*Objective: The purpose of this work is to study, to rediscover, and to classify the Amanita mushrooms found in the cork oak forests of Morocco. Methods and Results: Surveys were conducted in various regions of Morocco (Mamora, Northwest, Rif) allowing for the collection of infor-mation on the Amanita species encountered. In the laboratory, macroscopic and microscopic identification criteria for these species were established. In total, twenty-nine species (29) werwe recorded. Among these, five were not found (Amanita codinae, Amanita eliae, Amanita excelsa, Amanita gemmata, and Amanita porphyria). The remaining twenty-four species (24) are con-sistent with the vegetation formations of Morocco. The systematic position, ecology, edibility, and toxicity risks of all the studied species were discussed. Conclusion and Application of the Research: The results of this study contribute to determining the diversity of the Amanita genus in Morocco highlighting that the Moroccan forest ecosystem still holds many mysteries to uncover.*

DOI: 10.4018/979-8-3693-8980-5.ch023

## INTRODUCTION

Amanita are fungi belonging to the Amanitaceae family (Basidiomycotina): they generally have a ring at the top of the foot, a volva at the base of the foot, free white lamellae, and scales on the cap (Jérôme, 2002).

These elegant, slender, fleshy, terrestrial mushrooms are highly putrescent (their tissues are rich in proteolytic ferments of the rennet group). The cap, easily separable from the stem, is sometimes white, but more often colored (bright red, orange-red, reddish, greenish, pale yellow) (Malençon & Bertault, 1970).

The lamellae, which are generally white, are of two kinds. For example, some comes right up to the foot without fusing to it; white others or lamellae, starts from the periphery and stop about halfway upthe cap; they end obliquely, as if cut by scissors (Deysson et *al.*, 1980).

The development of Amanita begins underground where they appear as small white eggs. Then, the membrane enveloping the fungus, called the volva, partially tears, sometimesremaining at the base of the foot, and can be very friable (Isabelle, 1994).

Amanita are generally mycorrhizal but often not very specific, where all have a mycorrhizal requirement with *Quercus suber* (Abourouh, 2000).

According to previous works (Maire & Werner 1937, Malençon & Bertault 1970, Bertault 1980.) and Nounsi A., 2014, El kholfy and *al.*, 2011); Larouz and *al.*, 2012) in Morocco, the genus *Amanita* is represented by 32 species. Eleven of these species have been encountered in the Mamora, 10 in the Middle Atlas, 6 in the Rif and 5 in the Central Plateau (El Assfouri, 2006). Amanites is a very important genus containing some good edibles but above all the most toxic fungi: *Amanita phalloïdes*, *A. verna* and *A. virosa* (Pirot P., 2010).

## MATERIALS AND METHODS

To help determine the fungal diversity of Macromycetes in the different regions of Morocco, surveys were carried out in certain areas (Mamora, North-West). Investigations were carried out on various substrates, including *Quercus suber* and various *Eucalyptus* species. In the field, information on the season of basidiocarp appearance and the climatic conditions that characterize the periods of this appearance were noted. Macroscopic identification features are observed with the naked eye and a hand-held magnifying glass. The dimensions of the carpophores and their anatomical parts were measured. Basidiocarp dimensions in cm are determined according to their diameters projected on a horizontal plane. This method consists in expressing the dimensions of basidiomas by the diameter of the imperfect circle drawn

by the outermost part of the cap's periphery (Josserand, 1983). In addition, the appearance of the basidiomata, the color of the cap and hymenophore, the type of hymenophore, the thickness and consistency of the flesh, the ornamentation of the specimens, their odour and their smell were determined. The various criteria were studied on fresh specimens. The information gathered was recorded in situ on the harvesting notebook. In the laboratory, observations of the various parts of the carpophores were made in depth using a binocular magnifying glass. The spore was obtained by depositing mature carpophores on sheets of paper of different colors and on glass slides. The natural color of the spores was determined by comparing the color of the spore with the color of white chalk. According to Gilbert (1927), white chalk is the reference color for white spores. Spore amyloidity is tested by adding a drop of iodized water to a pinch of spore. Macrochemical tests, which may give colored reactions, are carried out on the flesh. These included iodine water, NaOH (10%), FeSO4 (10%) and HNO3 (6%). Microscopic criteria for species identification were determined from microscopic sections of anatomical structures.

Sections of the coating, flesh and hymenophore were taken and observed under the microscope. Observation of spores and determination of their dimensions were carried out on the basis of spores. Using the tip of a razor blade or lanceolate needle, we scrape a pinch of spores from the spore. This is then placed in a drop of water or mounting liquid. Spores dimensions were established for 20 spores and given in micrometers (µm). Microscopic observations weremade first in distilled water, then in cotton blue, iodized water, aqueous ammonia solution (20%) and ammoniacal Congo red (20%). The dimensions of the microscopic structures were determined using an ocular micrometer. Magnifications used were 100, 400 and 600. Spore shape was determined using the sporic quotient of Bas (1969).

Photographs were taken for each species. The biotope, basidiomes, macroscopic and microscopic criteria of the fungus were photographed in situ and in the laboratory. The systematic study of the species collected was based on the habitat, ecology and macroscopic characteristics of the specimens of each species. Determination criteria are carefully noted. In the laboratory, the study of the samples is completed by microscopic data and macro- and microchemical reactions. The species encountered were determined using the determination keys (Romagnési, 1995; Courtecuisse and Duhem, 2000, Poumarat, 2001 and Kirk M. P. and *al.* 2008). The distribution of the species in Morocco and worldwide, the substrates on which they grow, the importance of each species for ecosystems and their gastronomic and pharmaco-medical importance were discussed. In this work, the species are studied separately.

## RESULTS

*Amanita pantherina* (DC) Krombholz (1828). Figure 1.

Under *Quercus suber* (Malençon and Bertault 1970). In the Mamora,under *Quercus suber* (Sidi Taibi) and under *Eucalyptus* in the Kasbah of Mehdia and in Sidi Taybi (Outcoumit, 2011). Under *Quercus suber* in the Mamora (El Assfouri, 2006 a). It is very common in the Mkhenza subéraie (Abourouh, 2000), and also occurs in the Lalla Mimouna site (Bakkali Yakhlef et *al.*, 2009 a).

The cap (6 to 14 cm in diameter) is hemispherical, convex, then plano-convex to planar, thin, with a short furrowed margin and a separable lining, sown with pure white (milk-drop) flakes of mealy consistency on a brown, ochre-brown, tawny-brown background, lightening slightly from the margin. The stalk (8-13 × 1-2 cm) is floury, then hollow, with a bulb clearly marked by a volval bead with a narrow rebore, topped with tiered bracelets resembling supernumeraryrings, white, appressed ring, sometimes deciduous, white. Blades are free, tight, white, with flaky edges. 6 mm thick flesh.

white, not reddening or smearing when cut (this is the main difference between *Amanita pantherina* and *Amanita rubescens*, with which it can be confused), with a faint, unpleasant odor. $NH_3$, KOH and $H_2SO_4$ have no effect on fresh samples, but it browns slightly under the action of NaOH. The spores are white (10-12.5 × 7.7-8 µm), but the most numerous (10 ×7.5 µm) are hyaline, smooth, broadly elliptical, non-amyloid, with a clear apicle. Microscopically, the slides show an inverted weft and a hymenium with tetrasporic, cylindrical basidia (37 × 10 µm) on average. The basidia show a number of bright, rounded to subcircular globules and dense granulation under ammonia.

*Amanita muscaria* (L.) Lam (1783), = *Amanita muscaria* var. *muscaria* (L.) Lam (1783). Figure 2.

Species growing under Eucalyptus, in Sidi Taibi (Malençon and Bertault 1970) cited by El Assfouri 2006a); (Abourouh, 2000); under *Eucalyptus* in Mamora (Outcoumit, 2011).

The cap (10 cm in diameter) is convex, then spreading to depressed with age, with a fluted margin in adults, separable lining, sown with white, detersile warty flakes on a red background, bright red that can fade to orange-yellow. The stipe (9 × 1-2.7 cm) is white, more or less bulbous, stuffed then hollow, with a membranous, fleeting ring and a crenellated margin of white deciduous flakes. The volva is made up of layers of white flakes arranged in annular beads. Blades (absence of lamellae) are free, white or tinted with light citrine cream, thicker than the flesh. The flesh is white, yellow under the cuticle of the cap, with no particular odor or flavor. Microscopically, the slides show an inverted weave and basidia (37.5 × 10µm) that are cylindrical, sometimes slightly obese or, on the contrary, narrowed and tapered. Spores white, spores (9-12 × 6-7.5 µm) elliptical, non-amyloid, with lateral apicles. *Amanita crocea* (Quélet) Singer (1951). Figure 3.

Found at the foot of *Quercus faginea* in the Jaaba forest and at the Lalla Mimouna site (Malençon et Bertault 1970). This species is cited by Larouz (2007) and Bakkali Yakhlef et *al*., 2009a) and described by El Assfouri (2006a).

Terricolous species, growing under *Quercus suber* in Sidi Taibi and Sidi Allal El Bahraoui. Was harvested on 7-05 and 12-12-2004.

The cap (over 6 cm in diameter) is convex to spreading, mamelonized, saffron-yellow to orange-yellow, visciduose (confluent flakes), whitish especially in the center, smooth elsewhere, with a separable cuticle and a fluted, straight margin. The stipe (10-13 × 1-2 cm) is cylindro-clavate, solid, but appears hollow, white, mottled with russet-orange yellow. It has a fleeting ring on the Thick volva (2.5 cm high and 2.5 cm wide), creamy-orange inside. Blades (0.4 cm thick), lamellae and lamellae are loose, loosely packed, white to pale yellowish, with flaky edges. The flesh (0.2 cm thick) is white and unchanging. NH3 and NaOH have no effect on fresh sporophore. White spore, spores (7.33-8.32 × 6.66 µm) are subglobose, smooth. Basidia tetrasporic (9.99-16.65 × 13.32 µm), sterigmata (1.66-3.33 µm).

*Amanita boudieri* Barla (1887), = *Tricholoma saponaceum* var. *lavedanum* Rolland (1891). Figure 4.

Common from mid-winter to early spring in sandy open forests of *Qercus suber* and *Pinus pinea* in the coastal Meseta (Rabat, Kénitra, Larache, Tangier). Also found under *Quercus rotundifolia*, but rarely, in the Middle Atlas (Azrou, alt. 1500m) (Malençon et Bertaut, 1970; El Assfouri, 2006a).

This mushroom grows alone in the clearings under *Quercus suber* in the Mamora forest (Sidi Taibi). Harvesting takes place in March 2003.

The cap measures up to 15 cm in diameter, convex, then spread to subdepressed in the center, with a smooth, appendiculate margin, separable lining, covered with white pyramidal warts on a white background, dirtier white in the center. The stipe (6 × 3 cm) is white, fibrous, cylindrical, dilated at the base, which extends into a napiform bulb 4 to 6 cm long, solid, entirely pruinose white beneath a drooping ring, attached beneath the blades, membranous, fragile. The volva is white and crumbly, forming soft annular beads at the base of the foot, usually fragmented and fleeting, often disappearing in older specimens. Blades, lamellae and lamellae are ventral, moderately tight, broad, rounded at the front, subliberal, lamellae cut squarely at the back and often fused to the blade, white with cream highlights and a concolored edge. Flesh is white, firm in the cap, a little fibrous in the foot, unchangeable. White spore, spores (10-13.75 × 4-6.25 µm) are amyloid, oblong to cylindrical with rounded apex and large lateral basilar apical.

*Amanita strobiliformis* (Paulet ex Vittard.) Bertillon (1866). Figure 5.

Terricolous, solitary species under *Quercus suber* in Mamora. Harvested on 4-04-2004.

The cap (8 to 14 cm in diameter) is fleshy, white, more café au lait in the center, hemispherical to convex, with a slightly appendiculate margin, separable lining, scattered with patches of veil remnants and labile polyhedral warty flakes. The stipe (9 × 1.5-2.5 cm) is whitish with a swollen base and a more or less radicating bulb with staggered beads and a fleeting ring. Blades, lamellae and lamellae are white, moderately tight, adnate, with flaky edges. Flesh is creamy- white, slightly thick. NH3 and NaOH have no effect. on fresh specimens. Spores white, spores (9.99-13.32 × 6.66-9 µm) broadly elliptic to elliptic, amyloid.

*Amanita phalloïdes* (Vail. Ex Fr.) Link. (1833). Figure 6.

Widespread but not very common in all oak forests (*Quercus suber*) of the coastal Meseta and Central Plateau; also in the mountains (Rif, Tazekka, Middle Atlas) under *Quercus rotundifolia* and *Q. pyrenaica*; in autumn and winter (Malençon and Bertault, 1970 cited by Ait Aguil (2005) and El Assfouri (2006a); under *Q. suber* in the Mamora (Outcoumit, 2011). A humicolous species that grows solitarily in autumn (January), under cork oaks.

The cap (8 cm in diameter) is typically yellowish-green to olive-brown, conical then convex, with a separable cuticle, shiny when dry, finely striped with radiating fibrils, and a thin, regular, unstriped margin. The flesh is white, unchanging, tinged with greenish-yellow under the cuticle, fragile, with a faded rose odor, then a wormy, unpleasant odor as it dries or in old age, with a lilac-purple reaction to sulfuric acid on the blades. Blades, lamellae and lamellae are white with greenish highlights, free-standing. The stipe (10 × 1- 2cm) is whitish with greenish highlights, often mottled with greenish-gray, bulbous, full, with a white, persistent, hanging membranous ring. Well-formed, lobed white volva. White spore species. Spores (8- 10 × 6.66-7.5 µm) are subglobose to broadly elliptic, amyloïdes, smooth and hyaline.

*Amanita verna* Sensu Rea (1922), = *Amanita virosa* (Fr.) Bertill (1866). Figure 7.

Quite common in early spring, under the oaks of the Mamora. Harvested 09-03-2003. The cap (6cm in diameter) is convex then plano-convex, spreading at the end, white except for the center, which has ochre or carmine tints, smooth and glabrous, with a regular, smooth margin, its cuticle is separable and not radially stretchy. Flesh is white, soft at the base. Blades are medium-tight, white with bluntly truncated, fairly tight, attenuated-adnexed lamellae. The stipe

(11.5 × 0.8-1, cm) is separable, full then hollow, cylindrical, bulbous, white, smooth, with a membranous, persistent ring, white with a hanging blade, striated on the upper surface by the imprint of the blades. The volva is white, membranous, rounded and irregularly torn. The flesh is white, with an unpleasant odor at the end, and no reaction to potash. Spores (8.5-11.5 × 6.- 9 µm) are smooth, amyloid, broadly elliptical to ovoid. *Amanita gilbertii* Beauseign (1926).

In sandy clearings in Quercus woods and Pinus plantations, in the Tangier region and around Larache, in spring (Malençon and Bertault 1970), in *Quercus suber* clearings in the Mamora (El Assfouri, 2006a).

This mushroom grows in sandy clearings of *Quercus suber*. Harvested in spring (March). The cap (10 cm in diameter) is convex with a rounded margin, its cuticle separable, silky, whitish, becoming ochre-stained in the center in old age. The stipe (10 × 2-3 cm) is robust, deeply embedded in the sand, cylindrical above a typically ovoid bulb, full then stuffed, with a finely fluffy white surface. The bulb has a drooping, membranous, wide, persistent ring attached to the top of the stipe. The volva is membranous, thin and white. Blades, lamellae and lamellae are white with beige-pink highlights, broad, thin, free, attenuated at both ends. Flesh is white, unchanging and fungus-like. The spore is white. Spores (10-17.5 × 5-7.5 m) are amyloid, smooth, long-elliptical to cylindrical.

*Amanita curtipes* E.-J. Gilbert (1941). Figure 8.

In the open forests and sandy wastelands of the coastal Meseta; a fairly exceptional harvest in the Rif (Bab-Tariouentz, alt. 1300 m) (Malençon and Bertault 1970) cited by Ait Aguil (2005); under *Quercus suber* in the Mamora (El Assfouri, 2006a); under *Q. suber* and among rockroses in the Mamora and at Khmiss-es-Sahel (Outcoumit, 2011; Outcoumit et *al.*, 2013). This small, russet-like amanita mushroom grows solitarily under *Quercus suber* (Sidi Taibei) and Pines (Sidi Allal El Bahraoui). Harvested on 12-12, 8-04-2004 and 16-02-2006.

The cap (4 to 6.5 cm in diameter) is convex then flat, with a curved margin, fluted at the end, and initially covered with fine flakes or patches of the general veil, reddish on a whitish background that soon turns pinkish-brown. The stipe (4-5 × 0.4-1.4 cm) is covered in white, then light pinkish-ochre, powdery to fibrillose, full. The stipe has a torn ring, leaving only a scar. The volva (3 cm high) is sheathed, white then carmine-pink and finally russet-brown. Blades (0.7 cm thick), lemellae and lamellae are spaced, subliberal, cream-colored and russet when rubbed. The flesh (0.4 cm thick) is whitish, turning pink and then hazelnut when cut. NH3 and NaOH remain negative on fresh carpophore. White spore, spores (10-15 × 5-8 µm) are hyaline, amyloid, oblong to cylindrical. Cystidia (9.99 × 6.66 µm), basidia (13.22 × 7 µm), sterigmata (2.22 µm).

*Amanita rubescens* Pers. (1797). Figure 9.

Solitary, but sometimes scattered under *Quercus suber* in Mkhenza (Malençon et Bertaut, 1970). In sidi Allal El Bahraoui and Sidi Taibi (El Assfouri, 2006a). Common under *Quercus suber* in the Mamora (Outcoumit 2011; Outcoumit et *al.*, 2014).

Humicole species growing solitary, but sometimes in clumps (2) under *Quercus suber* (Sidi Allal El Bahraoui and Sidi Taibi). Harvested on 27-12-2003; 15-03-2004 and 5-01-2005. The cap (7 to 14.5 cm in diameter) is hemispherical then plano-convex, with an eccentric margin. The lining is separable, sown with more or less confluent flakes, whitish to pale grayish, more rarely dirty yellowish with carmine highlights, on a pinkish-brown, reddish- brown, vinous background, sometimes very pale. The flesh (1 cm thick) is white, becoming wine-red with exposure to air, especially in the bulb, thick and firm. Blades, lamellae and lamellae are tightly packed, white with wine-red spots, and free. The foot (10-14 × 2-3.7 cm) is white, but tends to become veiny purple-brown, especially at the base. It is swollen into a progressive bulb, bearing a volva, minus a white, membranous ring, hanging and ample. NH3, NaOH have no effect on the flesh of the foot, but brownish on the cap lining. White spore species. Spores (7.32-9.99 × 4.99-6.6 µm) are subelliptic, rarely ovoid, amyloid.

*Amanita caesarea* (Scop. Ex Fr.) Quélet.

Abundant in autumn under *Quercus suber* around Larache; common under oaks or pines mixed with *Quercus coccifera* in the Tangier region. Quite rare, but present under *Quercus suber* in the rif at Chaouene and Dardara and under *Quercus pyrenaica* at Bab-Tariouentz. Also found under *Quercus suber* in the Tazekka massif (Forêt de Bab-Azhar) (Malençon et Bertaut, 1970, cited by Ait Aguil, 2005).

*Amanita citrina* (Schaeff.) Pers. 1797. In the subéraie of the Rabat region (Abourouh, 2000).

*Amanita citrina* f. *citrina* (Pers.) Pers. Figure 10.

In January, under *Quercus suber* and among Cysts in the Khmis-es-Sahel forest, Mamora (Malençon and Bertault, 1970; Outcoumit, 2011).

*Amanita citrina* var. *alba* (Pers.) Quél. & Bataille.

In January under *Quercus suber* and in the Mamora, on January 5, 2010, under the same plant species (Malençon and Bertault, 1970; Outcoumit, 2011).

*Amanita lactea* G. Malençon, H. Romagnési and D.A. Reid).

Isolated species in open subéraie (Mamora forest near Rabat) (Malençon and Bertault, 1970);under *Quercus suber* in the Mamora (Outcoumit, 2011).

*Amanita spissa* (Fr.) Quélet.

Under *Eucalyptus* and *Quercus* trees in Morocco: Tangier, Larache, in winter and spring (Malençonand Bertaut, 1970). Under *Quercus suber* in the Mamora (Outcoumit, 2011).

*Amanita vaginata* var. *grisea* (DC.) Quél. et Bataille 1902. Figure 11.

At Khmis-es- Sahel (Larache region.), among Cistes (Malençon and Bertaut, 1970; Outcoumit,2011).

*Amanita lividopallescens* (Secr. ex Boud.) Kühner et Romagn. 1931.

under *Abies pinsapo* mixed with *Quercus faginea*, (LNS) (Malençon and Bertault, 1961 and1970); cited by (Ait Aguil, 2005; El Assfouri, 2006a).

*Amanita inaurata* Secr. ex Gillet. 1874 (syn *Amanita ceciliae* (Berk. Et Broome) Bas 1983. Under *Quercus rotundifolia*, (LNS) (Malençon and Bertault, 1961 and 1970) cited by Ait Aguil (2005).

*Amanita ovoidea* (Bull.) Link. Figure 12.

Under *Pinus halepensis* (Bni-Dercul) (Malençon et Bertault, 1960, 1970) cited by Ait Aguil(2005).

*Amanita umbrinolutea* (Secr. ex Gillet) Bertill (1910) (= *Amanita battarrae* (Boud.) Bon 1985.

Under *Quercus rotundifolia* (Akarrate, opposite Chaouene) (Malençon and Bertault 1960 and1970 cited by Ait Aguil (2005).

*Amanita vaginata* (Bull.) Lam. 1783. Figure 21.

In the Dardara site (Malençon and Bertaut, 1970); cited by Ait Aguil (2005); Bakkali Yakhlefet *al.*, 2009a).

*Amanita codinae* (Maire) R. Bertault 1955.

Isolated or in circles in wastelands and open forests (*Quercus*, *Eucalyptus*) along the coast or in low mountains, in heavy, slightly damp soil: Témara, Rabat, Tangier, Khénifra, Auguelmane Azigza (alt. 1800 m). Quite rare everywhere, though relatively frequent and well established at Cherf-el-Akab near Tangier (Malençon et Bertault, 1970).

*Amanita eliae* Quélet.

Rare, but constant near Tangier, at Jbel Kbir, under *Pinus pinea*, in spring (Malençon et Bertaut, 1970).

*Amanita excelsa* Fr.

Uncommon, under Eucalyptus and *Quercus* near Tangier (Jbel Kbir), under *Quercus pyrenaica* in the Rif (Bab-Tariouentz): Autumn and spring (Malençon and Bertault, 1970).

*Amanita gemmata* (Fr.) Gillet

Under *Acacia*, *Eucalyptus* and *Pinus*, frequent from Tangier to Larache; rarer in the Rif (Ktama) and under *Cedrus* and *Pinus* in the Middle Atlas (Azrou) and Tazekka under *Cedrus*. Autumn-winter and sometimes spring (Maire, 1937; Malençon et Bertaut, 1970).

*Amanita porphyria* Alb. & Schwein.1805 A single collection, under *Pinus pinea*, at Jbel Kbir near Tangier in May 1957 (Malençon et Bertaut, 1970).

## DISCUSSION AND CONCLUSION

In Morocco, the genus *Amanita* is represented by thirty-two species (32) (Malençon & Bertault, 1970). In this study, twenty-nine of these species (29) were recorded, 5 of which were not found (*Amanita codinae, Amanita eliae, Amanita excelsa, Amanita gemmata,* and *Amanita porphyria*).

Amanita are widespread in forests, often forming mycorrhizae with certain trees. According to Courtecuisse & Duhem (2000), they are generally ground-dwelling and ecto- mycorrhizic.

Oak groves, cedar groves and pine and *Eucalyptus* plantations are the most common plant formations in Morocco, where ectomycorrhizal fungi species that are faithful to these ecosystems, particularly Amanita, are found.

The amanita is of great importance in protecting plants against pathogens and improving the efficiency of water use by mycorrhizal plants. Understanding these species and their behavior in their environment as well as developing strategies to manage their use, is essential for bothmankind and ecosystems.

The number of mycorrhizal fungi varieties listed (26) seems rather limited. In fact, mycological research in Morocco is still incomplete and many new species, previously unknown in this region, have been discovered in recent years (El Assfouri and *al.*, 2024).

In general, Amanita have a bad reputation. Although some are delicious and safe to eat. However, there are some more dangerous, even deadly species. In fact, 90% of mushroom poisoning results from the consumption of *Amanita phalloides* (Courtecuisse & Duhem,2000). This species is public enemy number one, whose poisoning accidents can be severe and, in some cases, even fatal. According to Aoued (2016), it contains two types of toxins, phallotoxins and amatoxins, whose structures are well established. The two toxins act synergistically, causing widespread digestive, renal and nervous system damage.

The dreaded phalloid syndrome, of the delayed type, manifests itself through digestive signs, with diarrhea, vomiting and cold sweats, six to twenty-four hours after ingestion. This syndrome also leads to liver cell damage, resulting in respiratory disturbance and acute gastroenteritis (Courtecuisse & Duhem, 2000).

There's no way to recognize *Amanita phalloïdes* other than its botanical characteristics: white blades, the presence of a volva and a ring (carpophore in good condition), the green, brown, olive, yellowish or whitish color of the cap, no odor or flavor, stipe streaked with fine cracks. Nevertheless, when a mushroom is harvested, it can be collected without care, and the volva is often left in the ground, or even the ring may well have been torn off by some accident of vegetation or the voracity of a slug.

In fact, confusion usually comes from Lepiotes and especially *Agaricus* (= Psalliotes), both of which never have a volva. A closerlook reveals a number of easily identifiable characteristics. *Amanita verna* closely resembles *Amanita phalloides*, but the latter always has a network of radiating fibrils on the cap and appressed flakes on the stipe, between the ring and the volva, which is lacking in *Amanita verna*, which is also lethal (Romagnesi, 1995).

*Amanita rubescens* is a sought-after edible, even sold in markets. Yet, it contains a hemolytic substance destroyed by cooking and should not be eaten raw; indeed, some people dislike itor digest it badly (Romagnesi, 1995). Any confusion must be avoided with the dangerous *Amanita pantherina*, which you must learn to recognize. This mushroom can be fatal causing intoxication. neuropsychological effects, at least if absorbed in sufficient quantity. But once the threshold isexceeded, it can cause serious accidents. Symptoms appear within 30 minutes.

What's more, this fungus doesn't show the slightest trace of reddening, and its cap scales resemble milk drops, remaining white well into old age. *Amanita rubescens, on the* other hand, tends to take on a winey purple-brown color, especially at the base of the stipe and in the larval punctures, with plate-like, browning, fibrillated scales on the cap.

Concerning *Amanita muscaria*, it's certainly the most magnificent, because of the bright red color of the cap. It's a poisonous species, producing atropine-like intoxication (mainly nervous disorders), but it's not very dangerous, and is even eaten in some regions (Italy). The same is not true for flies: simply leave a specimen of this *Amanita* on a table and the next day, inanimate flies will be falling all around it (Romagnesi, 1995).

*A. muscaria* can be confused with the excellent edible *Amanita caesarea*, absent from the Mamora, but present in other parts of Morocco. It is characterized by a cap covered with white scales (which may disappear in the rain), white stipe, bulbous swelling, but for *Amanita caesarea,* which has a naked cap from the start, a yellow stipe, and an un-swollen bulb.

*Amanita curtipes is a* small mushroom in the group, with a russula-like appearance, but unlike Russulas, it has a volva at the base of the stipe. During our surveys, only two specimens were collected, one in 2004 and the other in 2006, which means that this species is very rare in the Mamora forest.

*Figure 1. Carpophores of Amanita pantherina during hatching*

*Figure 2. Aspect of Amanita muscaria cap, basidiospores (× 400)*

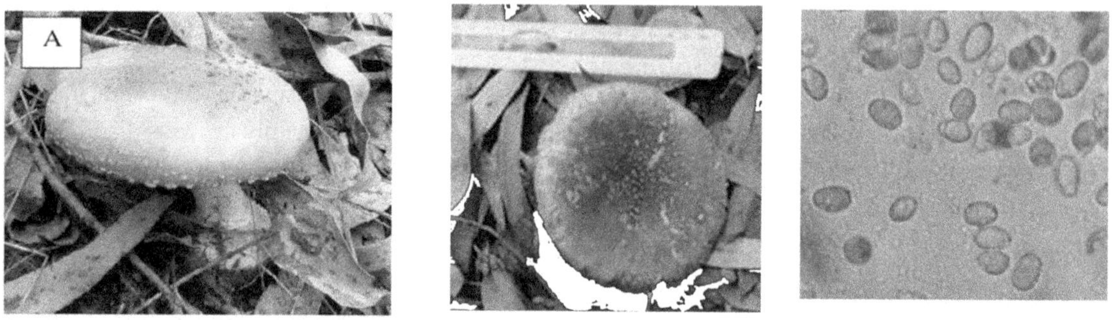


*Figure 3. Appearance of Amanita crocea cap and stipe, spores (× 1000)*

*Figure 5. Sporophore and basidiospores (× 600) of Amanita strobiliformis*

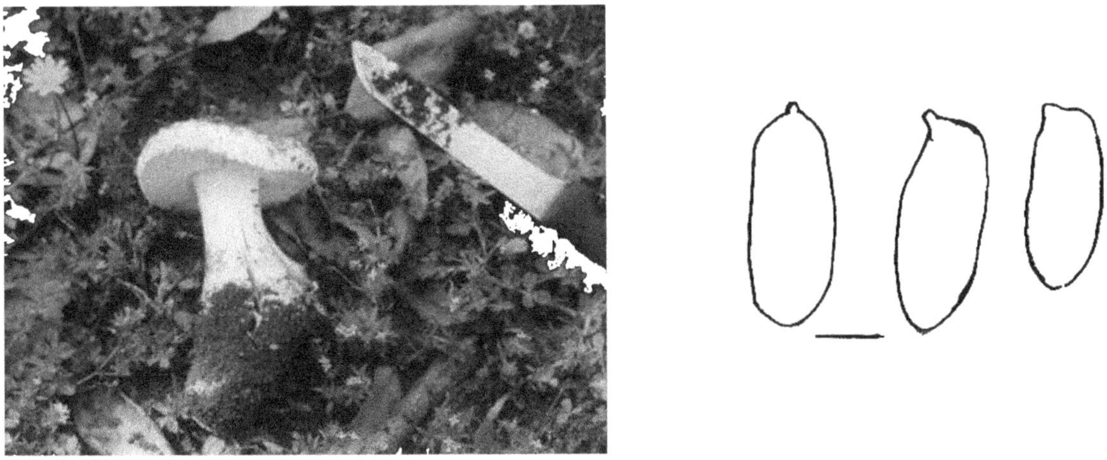

*Figure 6. Aspect of cap and stipe with ring of Amanita phalloïdes*

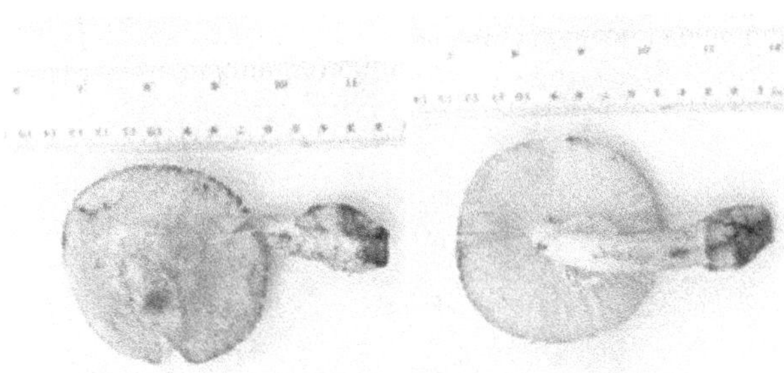

*Figure 7. Aspect of the volva, spores (× 1000) and cystid (× 400) of Amanita verna.*

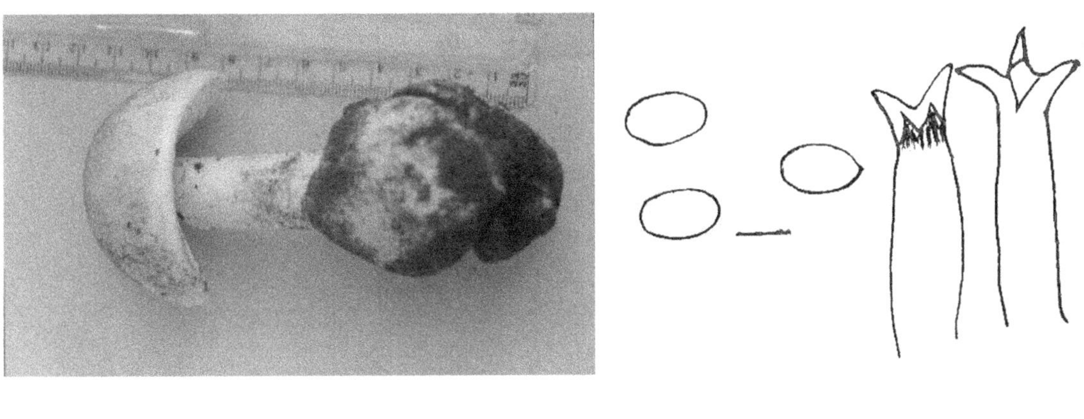

*Figure 8. Young and adult carpophore, spores (× 400) and (× 1000) of Amanita curtipes.*

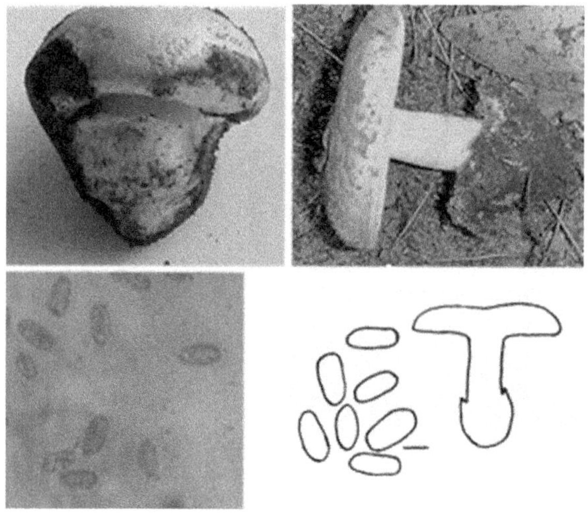

*Figure 9. Amanita rubescens*

*Figure 10. Amanita citrina f. citrina*

*Figure 12. Amanita ovoidea*

## REFERENCES


Abourouh, M. (2000). Mycorrhizae and mycorrhization of the main forest species in Morocco. PhD thesis, Université Mohammed V, Fac. des Sciences, Rabat, Morocco, 175 p.

Ait Aguil, F. (2005). Contribution à l'étude de la biodiversité de la flore fongique des forêts marocaines: cas des Basidiomycètes du Rif Oriental et du Rif Occidental et des Gastéromycètes de la Mamora. Doctoral thesis, Univérsité Ibn Tofail, Faculté des Sciences, Kénitra, Morocco, 127p.

Bakkali Yakhlef, S., Kerdouh, B., Mousain, D., Ducousso, M., Duponnois, R., & Abourouh, M. (2009a). Phylogenetic diversity of Moroccan cork oak woodlands fungi. *Biotechnol. Agron. Soc.*, *13*(4), 521–528.

Bas, C. (1969). Morphology and subdivision of Amanita and a monograph of its section Lepidella. *Persoonia*, *5*(4), 285–588.

Bertault, R. (1980). Amanites du Maroc (Troisième contribution). *Bulletin de la Société Mycologique, France*, *96*(3), 271–287.

Channieras, I. (1994). Amanita and Lepiota from the Limousin region. Thesis n°59, for the Diplôme d'Etat de Docteur en pharmacie. University of Limoges, Faculty of Pharmacy. 124 p

Courtecuisse, R., & Duhem, B. (2000). *Guide des champignons de France et d'Europe. Delachaux et Niestlé S. A. Lausane*. Switzerlande.

Deysson, G., & Delcourt, A. (1980). *Cryptogamie: Mycologie générale et appliquée*. Edition CDU and SEDES.

El-Assfouri, A. (2006). Biodiversité fongique du Maroc: Inventaire et étude des Basidiomycètes de la forêt de la Mamora. Doctoral thesis, Université Ibn Tofail, Faculté des Sciences de Kénitra, Morocco, 301p.

El-Assfouri, A., Ouazzani-Touhami, A., & Douira, A. (2024). A new mycorrhizal of association *Quercus suber* (Morocco): *Lactarius glaucescens* Crossl., 1900. Global Summit Symposium on Energy and Environment. AI2SD Symposia Series N° 5. will be published automatically in indexed Proceedings within Studies in computational intelligence Book Serie (as *regular papers)*.

El Kholfy, S., Aït Aguil, F., Ouazzani Touhami, A., Benkirane, A., & Douira, A. (2011). Bibliographic inventory of Moroccan Rif's fungi: Catalog of rifain fungal flora. *Journal of Animal & Plant Sciences*, *12*(1), 1493–1526.

El-Kholfy, S., Larouz, B., Touhami, A. O., Benkirane, R., & Douira, A. (2012). Bibliographic inventory of middle atlas fungi: catalogue of Middle Atlas fungal flora.

Gilbert, E. J. (1927). - The mycologist's books. Tome I: la spore des champignons supérieurs, couleur, forme, ornementation, terminologie, valeur taxonomique. Eds. Librairie E. Le François, Paris, 219 p.

Jérôme G. (2002). Appui scientifique et technique sur les risques liés à la présence d'Amanite panthère dans un lot de champignons secs (2002-SA-0215). Maisons-Alfort, November 25, 2002.

Josserand, M. (1983). *La description des champignons supérieurs (Basidiomycètes)* (2nd ed.). Lechevalier.

Kirk, M. P., Cannon, P. F., Minter, D. W., & Stalpers, J. A. (2008). *Dictionary of the Fungi* (Kirk, P. M., Ed.; 10th ed.). International Mycological Institute.

Larouz, B. (2007). Biodiversité fongique du Maroc: Etude des champignons supérieurs du Moyen Atlas. Doctoral thesis, Université Ibn Tofaïl, Faculté des Sciences, Kénitra, Morocco, 322p.

Malençon, G., & Bertault, R. (1975). *Flore des champignons supérieurs du Maroc* (Vol. II). Faculté des Sciences de Rabat.

Nounsi, A., Outcoumit, A., Selmaoui, K., Ouazzani Touhami, A., Benkirane, R., & Douira, A. (2014). Inventory of ectomycorrhizal fungi ofMorocco. *Journal of Applied Biosciences*, *79*, 6826–6854. DOI: 10.4314/jab.v79i1.10

Outcoumit, A. (2011). Contribution à l'étude de la diversité fongique du Maroc et mise en évidence de quelques espèces funongicoles et de l'importance des espèces sécotïdes dans la systématique des Basidiomycètes. PhD thesis, Université Ibn Tofaïl, Faculté des Sciences, Kénitra, Morocco, 497p.

Outcoumit, A., Kholfy, S. E., Touhami, A. O., & Douira, A. (2014). Bibliographic inventory of tangier fungi: catalogue of the basidiomycetes fungal flora.

Outcoumit, A., Ouazzani Touhami, A., & Douira, A. (2013). First contribution to the study of Basidiomycetes from the Lalla Mimouna region (Northwest Morocco). *Journal of Applied Biosciences*, *66*, 5024–5039. DOI: 10.4314/jab.v66i0.94999

Pirot, P. (2010). *Are so-called edible mushrooms still edible? Mycolux 2010, 2: 13-26. Poumarat S., 2001. Clé des Gastéromycètes épigés d'Europe: Lycoperdales, Nidulariales, Phallales, Sclérodematales, Tulostomatales. Monographies Mycologiques de la F.A.M.M., no. 2.* Editions Fédération des Associations Mycologiques Méditerranéennes.

Romagnesi, H. (1995). *Atlas des champignons d'Europe*. Bordas.

# Chapter 24
# Study of Performance and Improvement of the Quality of Treatment of the Purification Station of Wastewater

**Rachiq Toufiq**
https://orcid.org/0009-0001-8930-3328
*Mohammed V University in Rabat, Morocco*

**Imane Bencheikh**
*Mohammed V University in Rabat, Morocco*

**Aziza Lamchaimech**
https://orcid.org/0009-0002-8393-7167
*Mohammed V University in Rabat, Morocco*

**Ayoub Ghailane**
*Mohammed V University in Rabat, Morocco*

**Jamal Mabrouki**
https://orcid.org/0000-0002-3841-7755
*Mohammed V University in Rabat, Morocco*

## ABSTRACT

*Water constitutes an essential element of the nation's common heritage. Its protection, enhancement, and sustainable development of its usable resources, while respecting natural balances, are of general interest. Water, vital for life and indispensable to all economic activities, is used in domestic, industrial, and agricultural sectors. Unfortunately, it is often contaminated by these various uses. Discharging these polluted waters into the natural environment can cause serious short-term and long-term problems, both for public health and hygiene, as well as for the ecosystem. Therefore, it is essential to treat these wastewater to restore their quality for reuse. This wastewater comes from food processing and conservation industries as well as domestic activities, amplified by the intense tourism activity in the region. Treating wastewater in the city of Agadir appears to be a useful solution to prevent degradation*

DOI: 10.4018/979-8-3693-8980-5.ch024

*of receiving environments and simultaneously allows the production of purified water for irrigation.*

## 1. INTRODUCTION

Water is now a key factor conditioning any industrial growth or even any possible human development. However, water resources are experiencing overexploitation, increasingly accentuated, exacerbated by the water shortage declared on a global scale as a direct consequence of global climate change (Rachiq et al. 2021; Mabrouki et al. 2021).

In order to overcome this shortage, the reuse of treated wastewater can be presented as a sustainable alternative for specific purposes, in particular for watering green spaces and golf courses, large consumers of available water resources... In addition, wastewater treatment plants help protect receiving environments. In fact, the discharge of wastewater, loaded with pollutants, into the receiving environment without any prior treatment continues to threaten human health and the environment (Rachiq, Mabrouki, et al. 2022; Rachiq, Samghouli, et al. 2022). In this context, the M'ZAR Wastewater Treatment Plant (WWTP) uses an existing infiltration-percolation system. It is designed for the treatment of domestic water from the city of Agadir, before its discharge into the natural environment. Part of this treated water is stored in a tank for reuse in watering green spaces and golf courses. The other part will be discharged into the sea, while ensuring tertiary treatment by ultraviolet which attacks the pathogenic germs existing in the waters. In this context, the present project aims to make an overview of the treatment works of the M'ZAR STEP, then a study of purification performance and the determination of the reduction rates of organic and mineral matter via the performance of laboratory analyses to determine the physicochemical parameters of the STEP, as well as a project which contains 3 components object of realization which aims at the collection and transfer of industrial wastewater by the construction of two pumping stations MZ1 and MZ2 and discharge and gravity pipes its object of routing to the M'ZAR STEP for their treatment in the decontamination sector at the existing station of infiltration percolation system then the discharge to the sea, with the construction of a new system called reuse component which aims at the treatment of domestic wastewater by the activated sludge process and the use of this water treated for purposes watering and industrial.

## 2. MATERIALS AND METHODS

### 2.1 M'ZAR Wastewater Treatment Plant

The M'ZAR wastewater treatment plant is located 8.5 km south of the city of Agadir on the left bank of Oued Souss at a latitude between 30 and 31°N, on the coastal dune of M'ZAR at a distance of 1.40 km from the Atlantic coast (Figure 1). The M'ZAR wastewater treatment plant is located in the hunting reserve located in the commune of Lqliaa bordering the commune of Ait Melloul and 3 km from the Oued Souss (the outlet of all the watersheds in the area up to Biougra).

The bridge crossing the Oued closest to the site is that of Aït Melloul, which will place the station approximately 21 km from the city centre of Agadir. The western limit of the site is approximately 1500m from the sea and the north of it is an equivalent distance from the bed of the Souss (Mabrouki, Azrour, et Hajjaji 2021).

The existing M'ZAR treatment plant is composed of a primary treatment plant in the form of decanters with a capacity of 75,000 m3/day, a secondary treatment plant using the infiltration-percolation process on sand with a capacity of 30,000 m3/day and a tertiary treatment plant for disinfection of purified water by ultraviolet rays which has a capacity of 30,000 m3/day, and an outfall at sea at M'Zar with a length of 700 m.

*Figure 1. Location of the STEP (location of the SP on the map)*

## 2.2 Experimental Protocol for Wastewater Control

### a. Physical Chemical Parameters

These parameters are determined from one-off samples for raw water (settler outlet) and purified water from the M'ZAR station (filter outlet) twice a week, during the two months: April, May 2024.

The air temperature is measured using a thermometer that measures the wet bulb temperature. The pH provides information on the activity of the hydronium ions $H3O+$ and $H+$. It is measured using a pH meter in order to prevent corrosion in the pipes of the treatment facilities (Lhanafi et al. 2014).

It measures the ion concentration of a solution. It increases with temperature and the concentration of dissolved salts (Benjankar et Kafle 2021). The conductivity of water is expressed in microsiemens per centimeter (µS/cm), at 20°C. Figure 31 shows the conductivity meter used to measure the conductivity of samples taken from wastewater. Dissolved oxygen is a very important parameter in the biological process. It provides information on the presence of bacteria and, consequently, the degradation of organic matter, and is measured using an oximeter (Boguniewicz-Zablocka, Klosok-Bazan, et Naddeo 2019).

Refers to solid and volatile matter visible to the naked eye that is found in drinking water, groundwater and wastewater. To determine the concentration of suspended solids, the following is applied: MES apparatus. The gravimetric method describes a set of quantitative chemical analysis methods carried out by determining the mass of a product by weighing.

A portion of the sample is filtered through a borosilicate glass fiber filter, without binder, previously weighed. When filtration is complete, the residue is dried at 105 °C and weighed again. The weight of the suspended solids is obtained by the difference between the weights before and after drying.

Calculation and expression of results(Belghyti et al. 2009) To determine the MES, the result is expressed as follows:

$$Mg/L = (Pf - Pi) * 1000 * 1000 / Sample\ volume\ (ml) \quad (1)$$

- Pf: Final weight (Weight of the filter + solids + aluminum capsule) at 105°C (g)
- Pi: Initial weight (Weight of the empty filter + aluminum capsule) at 105°C (g)

This analysis is widely used to estimate the quality of organic matter in wastewater. It is a measure of the oxygen equivalent of organic and inorganic matter present in the effluents that can be oxidized by a powerful chemical oxidant, in this case potassium dichromate (Belghyti et al. 2009).

The COD makes it possible to assess the concentration of organic or mineral matter, dissolved or suspended in water, through the amount of oxygen required for their total chemical oxidation. Thus, by measuring the COD, it will be possible to assess the pollutant load of wastewater in organic matter before and after physical, chemical or biological treatment in order to control the operation of a WWTP and the activity of microorganisms.

The oxidizable material contained in the wastewater sample, of V = 3ml, is oxidized by a digestive solution contained in a tube already prepared, with a significant amount of potassium dichromate ($K_2Cr_2O_7$), with silver sulfate ($Ag_2SO_4$) as a catalyst, and mercuric sulfate ($HgSO_4$) which allows the chloride ions to be complexed. The calibration range is between 0 and 150 mg of O2/L for the low level method and 0 to 1500 mg of O2/L for the high level method.

After the mineralization step, at a temperature of T = 148°C for 2 hours, the concentration of green $Cr^{3+}$ ions is then measured by spectrophotometry at a wavelength of 600 nm (range: 0-1500 mg/L) and a wavelength of 420 nm (range: 0-150 mg/L).

The results are obtained directly by the spectrophotometer. These results are expressed in mg O2/L and are multiplied by a dilution factor, if applicable.

***COD in mg O2 / L = COD read * dilution factor***

## b. Biological Parameters

BOD5 corresponds to the quantity of oxygen O2 required to degrade and mineralize, in 5 days, biodegradable organic matter present in a liter of water, aerobically and through oxidation phenomena (Hassen et al. 2024). For successfully analyze BOD5, it must comply with operating conditions. BOD5 is measured:

- After 5 days of incubation;
- At a temperature of 20°C, because it is favorable to microbial activity that consumes O2;
- Away from light and air, to avoid any parasitic photosynthesis of algae.

And in order to measure the entire demand, oxygen O2 must not become a limiting factor in microbial activity. The glassware used must be strictly clean, free of toxic or biodegradable materials adsorbed, and must be kept away from contamination.

A well-defined sample volume is introduced into each bottle, according to the COD concentration. A magnetic bar and two sodium hydroxide (NaOH) tablets are added.

Before placing the brown bottles in the BOD5 chamber, the OXITOP must be set to 0; for 5 days, the sample is stirred continuously. The OXITOP automatically records a value every 24 hours; The values are stored after 5 days (S. Hazourli et al. 2007; Fateh Tarmoul 2007 ; A. Abood et al.2014; Bencheikh, I et al.2021; Benchrifa, M et al. Bencheikh, I et al.2023; Bencheikh, I et al.2020)

Thus, the result is expressed by the following relationship:

***BOD*5 *(mg of O*2*/L) = digit* dilution factor * D***

Where digit- value displayed by the OXITOP and D- Dilution of the sample before treatment.

The calculation of the reduction rate of a parameter, expressed as a percentage, is based on the following formula:

Reduction = ((E – S) / E) x 100

E: Concentration at the station inlet S: Concentration at the station outlet.

## 3. RESULTANTS AND DISCUSSIONS

The performance study, carried out as part of this project, focuses on the evaluation of said parameter at the inlet and outlet of the WWTP.

### 3.1 Temperature

Temperature is an important ecological factor of the environment. It allows to correct the analysis parameters whose values are linked to temperature (conductivity in particular).

In addition, it is important to know the water temperature precisely. Indeed, it plays a role in the solubility of salts and especially gases, in determining the pH and for knowing the origin of the water.

*Figure 2. Variation in temperature of raw and treated wastewater across the plant*

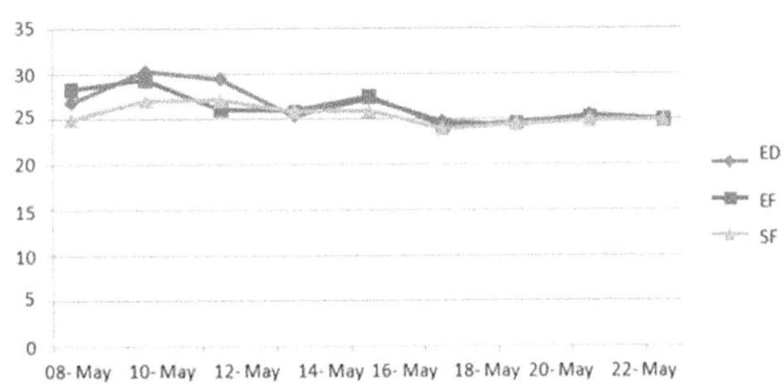

The values of the temperature of the wastewater between the filter vary between 24 and 30 °C, with an average of 27 °C. For the water leaving the filter (purified) the temperature varies between 24 and 26 °C, with an average of 25, °C. It is important to note that the temperature values recorded are lower than 30 °C considered as the limit value for direct discharge into the receiving environment. Similarly, these values are lower than 35 °C, considered as the limit value for the standard in Morocco for wastewater intended for irrigation (Mabrouki, J., Moufti, A et al.2020; Mabrouki, J., Benchrifa M et al.2022; Benchrifa, M et al.2022; Loukili, H et al.2021; Mabrouki, J., Azrour et al.2020; Mabrouki, J., Benbouzid et al 2022).

## 3.2 Hydrogen Potential

The pH (Hydrogen potential) measures the concentration of H+ ions in water. This parameter characterizes a large number of physicochemical equilibria, and depends on multiple factors including the origin of the water.

*Figure 3. Variation in pH of raw and treated wastewater across the plant*

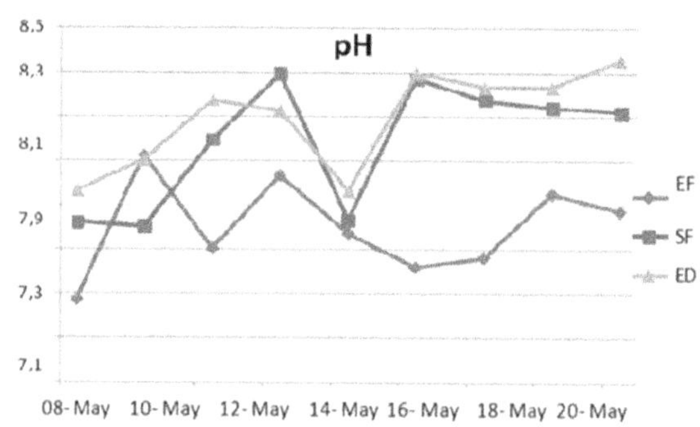


In fact, the alkaline environment promotes the precipitation of salts such as calcium carbonate with a decrease in the alkalimetric and hydrotimetric title. In addition, the pH is an important element for the interpretation of corrosion in the pipes of purification installations and also to define the aggressive or encrusting nature of water (Mabrouki, J., Benbouzid et al 2022; El Alouani, M., Aouan, B et al. 2024; Benchrifa, M., Elouardi, M et al.2023).

## 3.3. Electrical Conductivity

Electrical conductivity is one of the simplest and most important parameters for wastewater quality control. It provides a good indication of the concentration of mineral salts in water (relationship between the dissolved salt content of water and its ability to conduct an electric current). Conductivity varies with water temperature, it is more important when the temperature increases. Electrical conductivity is expressed in micro siemens /cm (µs/cm) (Benchrifa, M., Elouardi, M et al.2023; Al-Jadabi, N., Laaouan et al.2023; Mgalaa, S et al. 2022).

*Figure 4. Variation of electrical conductivity of raw and treated wastewater across the plant*

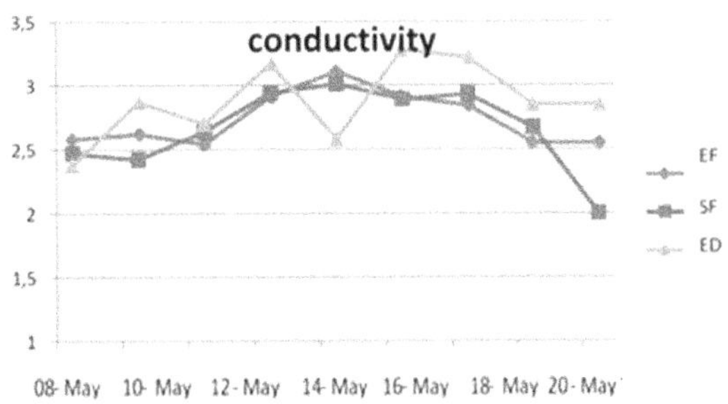


## 3.4. Suspended Solids

MES include organic fractions (MVS) and mineral fractions (MMS), not dissolved in pollution. They constitute an important parameter that clearly indicates the degree of pollution of an urban or even industrial effluent.

*Figure 5. Variation in the SS content of raw and treated wastewater across the plant*

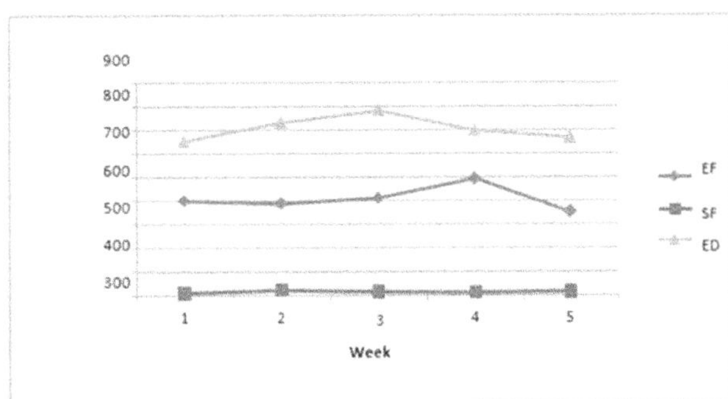


SS vary between 650 and 780 mg/L for raw water with an average of 715 mg/L. For treated water, SS vary between 9 and 23 mg/L, with an average of 8 mg/L. The reduction of SS is an average of 99.23%. Thus, the SS content of the treated water leaving the plant remains well below the limit value for direct discharges (50 mg/L).

## 3.3.5 Chemical Oxygen Demand

COD is the amount of oxygen consumed by the materials existing in the water and oxidizable under the defined operating conditions. In fact, the measurement corresponds to an estimate of the oxidizable materials, whatever their organic or mineral origin (ferrous iron, nitrite, ammonium, sulfide). It is expressed in mg of O2 per liter of solution.

*Figure 6. Variation of COD of raw and treated wastewater across the plant*

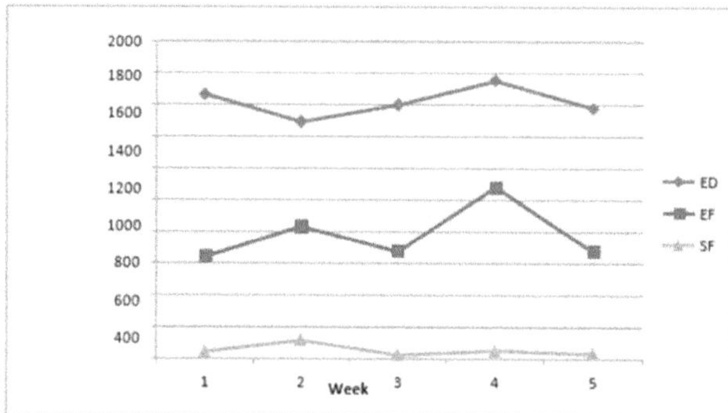


The COD values of the raw water entering the plant are very high and vary between 1488 and 1751 mg/L. The average is around 1619 mg/L. This high COD value reflects the significant quantity of oxidizable materials contained in this raw water. The purified water has COD values ranging from 25 to 150 mg/L. The average COD of this water is 87 mg/L. These COD concentrations are well below the limit for direct discharges into the natural environment, which is equal to 500 mg/L. The COD reduction is an average of 98.15%.

### 3.3.6 Biochemical Oxygen Demand

The biochemical oxygen demand (BOD5) expresses the potential oxygen consumption of the main pollutants (MES, aerobic microorganisms) of a receiving environment.

*Figure 7. Variation of BOD5 of raw and treated wastewater across the plant*

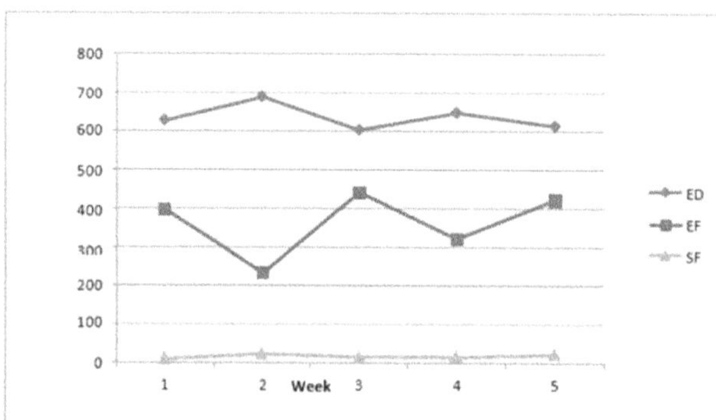


The BOD5 values of raw water range from 614 to 690 mg/L. The average BOD5 of raw water is around 652 mg/L. The purified water has very low BOD5 values ranging from 10 to 19 mg/L with an average of 14.5 mg/L. The average BOD5 reduction remains high and is around 99.08%. This sharp decrease in BOD5 values is explained by the activity of purifying micro-organisms that allow the degradation and transformation of organic matter, thus ensuring the elimination of organic pollution.

The purification yields of COD and BOD5 are very satisfactory with rates of 98.15% for COD and 99.08% for BOD5. These values are greater than 90%. According to Rejesk, these values characterize an effective wastewater treatment. The COD / BOD5 ratio determines the possibility of degradation that can be expected by a biological oxidation treatment. The average ratio of raw water is around 1.92. A value that remains below 3, which means that the effluent has good biodegradability.

## 4. CONCLUSION

The city of Agadir is characterized by a semi-arid climate, water resources are very limited and the reuse of purified water from the M'ZAR WWTP will make it possible to fill the water deficit that the city is experiencing, protect the environment and water resources and contribute to the local development of the area by improving the yields of agricultural production. The extension projects of the M'ZAR WWTP currently being carried out by RAMSA are part of the management of the purified water potential of the M'ZAR WWTP for the establishment of a system for the management and distribution of this water for the benefit of the industrial sector, watering of green spaces, and for local farmers, the latter are convinced of the positive impact of this effluent on the increase in agricultural production, unfortunately, this water exploited upstream of the WWTP without treatment presents fatal health risks. This work is a contribution to the valorization of wastewater by characterizing and evaluating the impact of these waters on watering in the Greater Agadir region

The results of analyses carried out on urban wastewater in Agadir indicate that pollution corresponds to domestic water that only requires biological treatment. As a result, the infiltration-percolation process remains suitable following the favorable climatic conditions of the region but with the new extension component of this WWTP. The activated sludge process remains a technical and strategic choice to meet the increased needs to produce purified water with better performance for watering the city's green spaces.

The residual concentrations of nitrogen and phosphorus nutrients present an interesting contribution of fertilization in nitrogen and phosphorus nutrients for reuse in watering green spaces. The concentrations of heavy metals in the urban waters of the city of Agadir are low and do not represent any threat to their reuse in irrigation and the majority of these metals are retained by the decantation of sludge at the bottom of the WWTP basins.

## REFERENCES


Abood, A., Bao, J., Du, J., Zheng, D., & Luo, Y. (2014). Non-biodegradable landfill leachate treatment by combined process of agitation, coagulation, SBR and filtration. *Waste Management (New York, N.Y.)*, *34*(2), 439–447. DOI: 10.1016/j.wasman.2013.10.025 PMID: 24287299

Al-Jadabi, N., Laaouan, M., El Hajjaji, S., Mabrouki, J., Benbouzid, M., & Dhiba, D. (2023). The dual performance of Moringa oleifera seeds as eco-friendly natural coagulant and as an antimicrobial for wastewater treatment: A review. *Sustainability (Basel)*, *15*(5), 4280. DOI: 10.3390/su15054280

Belghyti, D.. (2009). Caractérisation physico-chimique des eaux usées d'abattoir en vue de la mise en oeuvre d'un traitement adéquat: Cas de Kénitra au Maroc. *Afr. Sci. Rev. Int. Sci. Technol.*, *5*, 199–216.

Bencheikh, I., Azoulay, K., Mabrouki, J., El Hajjaji, S., Dahchour, A., Moufti, A., & Dhiba, D. (2020). The adsorptive removal of MB using chemically treated artichoke leaves: Parametric, kinetic, isotherm and thermodynamic study. *Scientific African*, *9*, e00509. DOI: 10.1016/j.sciaf.2020.e00509

Bencheikh, I., Azoulay, K., Mabrouki, J., El Hajjaji, S., Moufti, A., & Labjar, N. (2021). The use and the performance of chemically treated artichoke leaves for textile industrial effluents treatment. *Chemical Data Collections*, *31*, 100597. DOI: 10.1016/j.cdc.2020.100597

Benchrifa, M., Elouardi, M., Fattah, G., Mabrouki, J., & Tadili, R. (2023). Identification, simulation and modeling of the main power losses of a photovoltaic installation and use of the internet of things to minimize system losses. In *Advanced technology for smart environment and energy* (pp. 49–60). Springer International Publishing. DOI: 10.1007/978-3-031-25662-2_4

Benchrifa, M., & Mabrouki, J. (2022). Simulation, sizing, economic evaluation and environmental impact assessment of a photovoltaic power plant for the electrification of an establishment. *Advances in Building Energy Research*, *16*(6), 736–753. DOI: 10.1080/17512549.2022.2096693

Benchrifa, M., Mabrouki, J., Elouardi, M., Azrour, M., & Tadili, R. (2023). Detailed study of dimensioning and simulating a grid-connected PV power station and analysis of its environmental and economic effect, case study. *Modeling Earth Systems and Environment*, *9*(1), 53–61. DOI: 10.1007/s40808-022-01457-9

Benjankar, R., & Kafle, R. (2021). Salt Concentration Measurement Using Re-Usable Electric Conductivity–Based Sensors. *Water, Air, and Soil Pollution*, *232*(1), 13. DOI: 10.1007/s11270-020-04971-7

Boguniewicz-Zablocka, J., Klosok-Bazan, I., & Naddeo, V. (2019). Water quality and resource management in the dairy industry. *Environmental Science and Pollution Research International*, *26*(2), 1208–1216. DOI: 10.1007/s11356-017-0608-8 PMID: 29098587

El Alouani, M., Aouan, B., Rachdi, Y., Alehyen, S., El Herradi, E. H., Saufi, H., Mabrouki, J., & Barka, N. (2024). Porous geopolymers as innovative adsorbents for the removal of organic and inorganic hazardous substances: A mini-review. *International Journal of Environmental Analytical Chemistry*, *104*(16), 4784–4796. DOI: 10.1080/03067319.2022.2115891

Hassen, W., Hassen, B., El Ouaer, M., Kouki, S., & Hassen, A. (2024). Physicochemical and Biological Processing of Municipal Landfill Leachates. In Souabi, S., & Anouzla, A. (Eds.), *Landfill Leachate Treatment Techniques* (pp. 67–100). Springer Nature Switzerland., DOI: 10.1007/978-3-031-63157-3_4

Hazourli, S., Boudiba, L., & Ziati, M. "Caractérisation de la pollution des eaux résiduaires de la zone industrielle d'El-Hadjar, Annaba," LARHYSS J. ISSN 1112-3680, no. 6, 2007.

Lhanafi, S, R Aba-aaki, S Et-taleb, R Elhaouti, M Abbaz, et M Ez-zahery. 2014. « Caractérisation des effluents laitiers en vue de leur valorisation : Cas de lactosérum (Characterization of dairy effluents in view of their valorization: Case of whey) ».

Loukili, H., Mabrouki, J., Anouzla, A., Kouzi, Y., Younssi, S. A., Digua, K., & Abrouki, Y. (2021). Pre-treated Moroccan natural clays: Application to the wastewater treatment of textile industry. *Desalination and Water Treatment*, *240*, 124–136. DOI: 10.5004/dwt.2021.27644

Mabrouki, J., Azrour, M., & El Hajjaji, S. (2021). Use of internet of things for monitoring and evaluating water's quality: A comparative study. *International Journal of Cloud Computing*, *10*(5-6), 633–644. DOI: 10.1504/IJCC.2021.120399

Mabrouki, J., Azrour, M., Farhaoui, Y., & El Hajjaji, S. (2020). Intelligent system for monitoring and detecting water quality. In *Big Data and Networks Technologies 3* (pp. 172–182). Springer International Publishing. DOI: 10.1007/978-3-030-23672-4_13

Mabrouki, J., Azrour, M., Fattah, G., Dhiba, D., & El Hajjaji, S. (2021). Intelligent monitoring system for biogas detection based on the Internet of Things: Mohammedia, Morocco city landfill case. *Big Data Mining and Analytics*, *4*(1), 10–17. DOI: 10.26599/BDMA.2020.9020017

Mabrouki, J., Benbouzid, M., Dhiba, D., & El Hajjaji, S. (2022). Simulation of wastewater treatment processes with Bioreactor Membrane Reactor (MBR) treatment versus conventional the adsorbent layer-based filtration system (LAFS). *International Journal of Environmental Analytical Chemistry*, *102*(19), 7458–7468. DOI: 10.1080/03067319.2020.1828394

Mabrouki, J., Benchrifa, M., Ennouhi, M., Azoulay, K., Bencheikh, I., Rachiq, T., & El Hajjaji, S. (2022, November). Geographic information system for the study of water resources in Chaâba El Hamra, Mohammedia (Morocco). In *The international conference on artificial intelligence and smart environment* (pp. 469–474). Springer International Publishing.

Mabrouki, J., Moufti, A., Bencheikh, I., Azoulay, K., El Hamdouni, Y., & El Hajjaji, S. (2020). Optimization of the coagulant flocculation process for treatment of leachate of the controlled discharge of the city Mohammedia (Morocco). In Advanced Intelligent Systems for Sustainable Development (AI2SD'2019) Volume 7-Advanced Intelligent Systems for Sustainable Development Applied in Energy and Electrical Engineering (pp. 200-212). Springer International Publishing.

Mgalaa, S., Mabrouki, J., Elouardi, M., El Azzouzi, L., Moufti, A., El Hajjaji, S., EL Azzouzi, M., & El Belghiti, M. A. (2022). Study and evaluation of the degradation of procion blue dye by the ozonation method: Parametric and isothermal study. *Nanotechnology for Environmental Engineering*, *7*(3), 691–697. DOI: 10.1007/s41204-022-00262-0

Rachiq, T., Abrouki, Y., Mabrouki, J., Samghouli, N., Fersi, C., Rahal, S., & El Hajjaji, S. (2021). Evaluation of the efficiency of different materials to remove specific pollutants from landfill leachate. *Desalination and Water Treatment*, *238*(octobre), 240–250. DOI: 10.5004/dwt.2021.27779

Rachiq, T., Mabrouki, J., El Hajjaji, S., & Rahal, S. (2022). Simulation of the Treatment Performance of a Purification Plant for a Dairy Effluent. In Azrour, M., Irshad, A., & Chaganti, R. (Eds.), *IoT and Smart Devices for Sustainable Environment* (pp. 19–27). Springer International Publishing., DOI: 10.1007/978-3-030-90083-0_2

Rachiq, T., Samghouli, N., Mabrouki, J., El-Moustaqim, K., Rahmani, M., Al-Jadabi, N., El Hajjaji, S., & Rahal, S. (2022). Study and modeling of dairy effluents treatment using design of experiments methodology. *Desalination and Water Treatment*, *261*(juin), 131–140. DOI: 10.5004/dwt.2022.28524

Tarmoul, F. (2007). *Détermination de la pollution résiduelle d'une station d'épuration par lagunage naturel 'cas de la lagune de béni-messous*. Institut des Sciences de la Mer et de l'Amenagement du Littoral - DEUA.

# Chapter 25
# Sustainable and Renewable Strategies for Desalination

**Driss Azdem**
https://orcid.org/0009-0008-5691-5249
*Mohammed V University in Rabat, Morocco*

**Aziza Lamchaimech**
https://orcid.org/0009-0002-8393-7167
*Mohammed V University in Rabat, Morocco*

**Anass Ariss**
*Mohammed V University in Rabat, Morocco*

**Jamal Mabrouki**
https://orcid.org/0000-0002-3841-7755
*Mohammed V University in Rabat, Morocco*

## ABSTRACT

*Access to clean water and energy is essential for human well-being, but rapid population growth and industrialization are straining these resources. Desalination using renewable energy offers a promising solution. Sustainable practices like staged processes, waste heat recovery, and hybrid systems can ease the burden on water and energy supplies while reducing environmental impact. This paper examines both traditional and innovative desalination methods, focusing on integrating renewable energy and evaluating costs. It proposes combining renewable energy with energy and water recovery technologies to meet global needs and suggests a hybrid approach using grid power and renewables to lower costs in small-scale applications.*

## 1. INTRODUCTION

Freshwater and energy are two indispensable resources for human survival and socio-economic development (Azdem, Mabrouki, Moufti, Hajjaji, et al., 2024; Edegbene et al., 2024). However, the growing global population, coupled with industrialization and urbanization, has put immense pressure on these

DOI: 10.4018/979-8-3693-8980-5.ch025

vital resources(Azdem et al 2024; Swain, 2024). In regions such as North Africa, where water scarcity is already acute, the situation is further exacerbated by climate change, which intensifies the demand for both freshwater and energy (Azdem et al., 1apr. J.-C.). Morocco, for example, faces increasing challenges in maintaining adequate water supplies due to a significant decline in per capita water availability, which has dropped from 2,600 $m^3$ in 1960 to less than 620 $m^3$ in recent years (Fahdi, Azdem, et al., 2024) (Fatni et al., 2025), pushing the nation into a state of water stress. Projections suggest that this figure will fall below 500 $m^3$ per capita by 2030, further intensifying the need for alternative water sources. One of the most promising solutions to mitigate water shortages is desalination the process of converting saline or brackish water into potable water (Foghagen et al 2024). Desalination technologies, especially reverse osmosis (RO) and multi-stage flash distillation (MSF), are increasingly being adopted in Morocco and other parts of North Africa to address critical water shortages, particularly in coastal and arid regions (Azdem, 2024). The Agadir desalination plant, for example, is designed to provide up to 275,000 cubic meters of potable water per day(Chari et al., 2023), serving both urban and agricultural needs. However, desalination is an energy-intensive process, with traditional plants relying on fossil fuels, which not only increase greenhouse gas emissions but also raise concerns about long-term sustainability (Al-Obaidi et al., 2024; Zolghadr-Asli et al., 2023).

The environmental impact of conventional desalination methods is particularly relevant in the context of the global shift towards reducing carbon emissions and meeting sustainability goals (Panagopoulos, 2025) (Azdem, Fatni, et al., 2025). The energy demand for desalination processes is substantial, with estimates suggesting that the energy consumption of desalination plants accounts for up to 3-10 kWh per cubic meter of water produced, depending on the technology used. In Morocco, where reliance on imported fossil fuels for energy is still high, the environmental and economic implications of conventional desalination are significant. Consequently, integrating renewable energy sources such as solar and wind into desalination processes offers a more sustainable solution. Morocco (Azdem, 2024), in particular, is well-positioned to leverage renewable energy resources, given its abundant solar radiation and wind potential (Fahdi, et al., 2024). The country receives approximately 3,000 hours of sunlight annually(Benbba et al., 2024), making solar energy an attractive and viable option for powering desalination plants. Solar-powered desalination systems, such as those utilizing photovoltaic (PV) modules or solar thermal collectors, can significantly reduce the carbon footprint of desalination and lower operational costs over the long term (Maftouh et al., 2023). For instance, the Noor Solar Complex, one of the largest solar power plants in the world (Laaroussi et al., 2023), demonstrates Morocco's commitment to renewable energy and provides a blueprint for integrating solar power into desalination operations. In addition to solar energy, wind energy presents another renewable option, particularly along Morocco's Atlantic coast(Hajou et al., 2024; Mahdavi & Vera, 2023), where wind conditions are favorable for power generation. Wind-driven desalination plants, especially those utilizing reverse osmosis, have been shown to be highly efficient and can operate with lower energy inputs compared to fossil fuel-powered plants. A study conducted in coastal regions of Morocco indicated that wind energy could supply up to 35% of the energy required for desalination, with hybrid systems combining wind and solar power offering even greater efficiencies (Tashtoush et al., 2023). Hybrid systems, which blend renewable energy with conventional grid power, are becoming an increasingly practical solution to overcome the intermittency issues associated with renewable energy sources(Ahmed et al., 2024; Vaziri Rad et al., 2023). By combining grid electricity with solar or wind energy, hybrid systems ensure a continuous and reliable energy supply(Hassan et al., 2023), which is critical for the consistent operation of desalination plants.

The Agadir desalination plant, for example, is designed to operate using a combination of grid power and renewable energy, enabling it to meet water demand while minimizing energy costs and environmental impacts(Ait Lahoussine Ouali et al., 2023). Advances in desalination technologies themselves also contribute to more sustainable operations (Cai et al., 2023). Innovations such as energy recovery devices, which capture and reuse energy from high-pressure brine streams, have the potential to reduce energy consumption by up to 50% in reverse osmosis systems(Nurjanah et al, 2024). Furthermore, multi-stage desalination processes, waste heat recovery systems, and the integration of membrane distillation with renewable energy sources offer additional pathways to improve the efficiency of desalination technologies(Huang et al., 2024). These innovations not only reduce operational costs but also minimize the environmental footprint of desalination plants, making them more aligned with global sustainability goals.

In North Africa, particularly in Morocco, the integration of renewable energy into desalination processes is critical not only for addressing water scarcity but also for meeting the broader objectives of sustainable development (Arabi et al., 2024). Morocco's ambitious target of generating 52% of its electricity from renewable sources by 2030 is closely linked to its water security strategies (Fragkos, 2023), as the country seeks to reduce its dependence on fossil fuels and mitigate the environmental impact of desalination. In Morocco, where agriculture accounts for nearly 87% of water use (Cheikhaoui et al., 2024), treated wastewater reuse could provide a sustainable water source for irrigation, reducing the pressure on freshwater supplies. Combining desalination with water reuse systems can enhance overall water security by providing high-quality water for non-potable uses, such as agriculture and industrial processes, while desalinated water is reserved for drinking and domestic purposes. The integration of renewable energy with desalination offers a viable and sustainable approach to addressing the growing water and energy challenges in North Africa, particularly in Morocco. By adopting a holistic approach that includes energy recovery, process optimization, and hybrid systems, Morocco and other countries in the region can secure their water supplies while minimizing their environmental impact and contributing to global sustainability goals.

## 2. DESALINATION TECHNOLOGIES

Contemporary desalination techniques can be grouped into three primary categories(Harby et al., 2024). The specific features of different desalination processes are outlined in Table 1.

a. **Phase-change processes**: These techniques operate by raising the temperature of the feedwater (which can be seawater, brackish water, or other sources) until it reaches its boiling point under the system's pressure. The resulting steam is then transferred to a condenser, where it is cooled and converted back into freshwater. Examples of technologies utilizing this approach include solar distillation (SD), multi-effect distillation (MED), multi-stage flash distillation (MSF), mechanical vapor compression (MVC), and thermal vapor compression (TVC).
b. **Non-phase-change processes**: These methods involve the separation of dissolved salts from feedwater through mechanical, chemical, or electrical techniques. A membrane acts as a barrier, separating the saline feed from the resulting freshwater. Key technologies in this category include electrodialysis (ED) and reverse osmosis (RO).

c. **Hybrid processes**: These systems combine phase-change methods with membrane-based separation techniques, either within a single process or across multiple stages, to generate freshwater. Examples include membrane distillation (MD) and hybrid configurations that pair reverse osmosis with MSF or MED technologies.

*Table 1. Key characteristics of various desalination processes*

| Characteristic | Phase Change Processes (MED, MSF, MVC) | Non-Phase Change Processes (RO, ED) | Hybrid Processes (MD, MSF/RO, MED/RO) |
|---|---|---|---|
| Nature | Thermal processes involving evaporation and condensation | Membrane-based or electrochemical processes for salt separation | Combination of thermal and membrane-based processes |
| Feed Temperature | 60–120°C | <45°C | 40–80°C |
| Membrane Pore Size | Not applicable | 0.1–3.5 nm (RO), 0.2–0.6 mm (ED) | 0.2–3.5 nm |
| Driving Force | Thermal and concentration differences | Concentration and pressure gradient | Thermal and concentration differences |
| Energy Source | Primarily thermal, including steam or low-grade heat | Mechanical or electrical energy | Low-temperature heat sources or renewable energy |
| Product Water Quality | High-quality distillate, TDS <20 ppm | Potable water, TDS <500 ppm | High-purity distillate with TDS ranging from 20 to 500 ppm. |
| Cold Water Requirement | May be required | No cold water stream required | 20–25°C |

## 3. ENERGY CONSUMPTION AND CARBON EMISSIONS

Producing freshwater through desalination is inherently energy-intensive(Shokri & Sanavi Fard, 2023), as energy is required at various stages of the process, even when freshwater is readily available underground. Energy is necessary for tasks such as pumping water to the surface and treating it to meet potable standards. For instance, drawing groundwater requires between 0.14 to 0.24 kWh per cubic meter for a pumping height of 120 to 200 feet, while treating surface water typically consumes around 0.36 kWh per cubic meter(Gude, 2023). The minimum theoretical energy required for desalinating saline water, to remove salts and produce potable water, is estimated to be 0.706 kWh per cubic meter. However, in practice, the actual energy consumption of current desalination technologies is significantly higher. For example, approximately one ton of oil is needed to produce 20 tons of freshwater through desalination, a process that also releases carbon dioxide, thereby contributing to climate change (Table 2). Carbon dioxide emissions are calculated based on the assumption that producing one kilowatt-hour (kWh) of electricity generates 0.96 kg of CO2.

*Table 2. Energy consumption and carbon footprint of various desalination technologies*

| Process | Thermal Energy (kJ/kg) | Electrical Energy ($kWh/m^3$) | GHG Emissions (kg $CO2/m^3$ $H_2O$) | Max GHG Emissions (kg $CO2/m^3$) |
|---|---|---|---|---|
| Multi-Effect Solar Still (MESS) | 1500 | 0 | 0 | 0 |
| Multi-Stage Flash Distillation (MSF) | 250–300 | 3.5–5 | 15–25 | 24 |
| Multi-Effect Distillation (MED) | 150–220 | 1.5–2.5 | 8–20 | 19.2 |
| Mechanical Vapor Compression (MVC) | — | 11–12 | 11–12 | 11.5 |
| MED–TVC | 220–240 | 1.5–2 | 21.5–22 | 21 |
| Reverse Osmosis (RO) | — | 5–9 | 5–9 | 8.6 |
| Electrodialysis (ED) | — | 2.6–5.5 | 2.6–5.5 | 5.3 |

## 4. SUSTAINABLE MANAGEMENT OF WATER AND ENERGY

Many of the world's large desalination plants are still powered by fossil fuels(Almasoudi & Jamoussi, 2024), but this reliance is increasingly unsustainable. The depletion of these finite resources, coupled with the rising levels of greenhouse gas emissions, underscores the urgent need for a transition to more sustainable practices. To achieve this, renewable energy must play a central role in desalination processes(Alawad et al., 2023). Embracing hybrid systems can enhance efficiency, while the development of new technologies that are both energy-efficient and cost-effective is crucial. Additionally, encouraging water reuse practices is key to reducing pressure on natural water sources and promoting a more sustainable balance between water and energy consumption.

### a) The Importance of Renewable Energy Integration

The global consumption of primary energy is estimated at approximately $450\times10^{15}$ kJ. Currently, around 82% of this energy is sourced from non-renewable or exhaustible resources, while renewable energy accounts for about 18% (Amir et al., 2023). Projections for the role of renewable energy in global consumption extend up to 2045, as shown in Table 3. If current consumption trends persist, petroleum reserves could be exhausted within the next 45 years. To ensure a sustainable future, it is crucial to prioritize the development and expansion of renewable energy technologies.

*Table 3. Renewable energy contributions to global energy consumption projections*

| Year | Contribution of Renewable Energy |
|---|---|
| 2002 | 13% |
| 2011 | 18% |
| 2023 | 25% |
| 2041 | 50% |

**Figure 1** illustrates various potential methods by which commonly available renewable energy sources can be integrated with different desalination processes. Each approach utilizes specific technologies, with varying levels of efficiency and output. The following sections will explore these technologies in detail.

*Figure 1. Integration of renewable energy sources with desalination technologies*

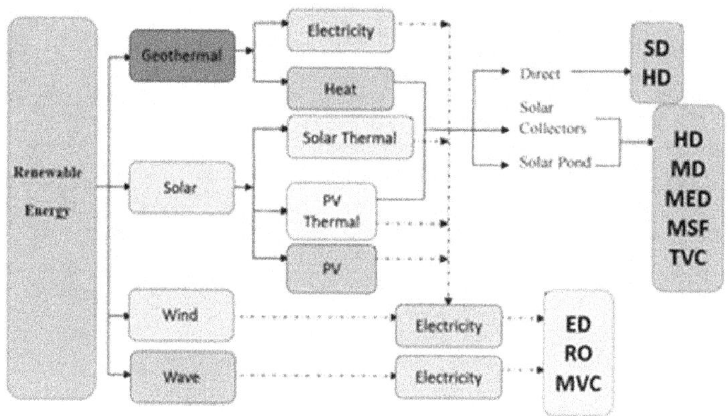


## Use of Solar Energy

The direct use of solar energy for desalination is well-demonstrated in solar stills. These devices capture solar energy to evaporate seawater, brackish water, or other suboptimal sources, offering fresh water for local communities and households. Solar stills are most effective in areas with limited access to advanced infrastructure, owing to their straightforward construction. The economic benefits of solar distillation are significant because it utilizes free solar energy(Kumar et al., 2023), thereby keeping operating costs exceptionally low. This method is particularly advantageous for small-scale desalination in regions receiving high levels of solar radiation. Its mechanical simplicity (few moving parts) makes it reliable and easy to maintain, requiring minimal technical skills. With proper operation, solar stills can effectively remove bacteria and both volatile and non-volatile contaminants, making them suitable for areas experiencing water shortages, particularly in drought-prone regions of Africa. Despite its promise, solar stills come with design challenges, such as managing the depth of the brine, ensuring vapor-tight seals, preventing distillate contamination, and optimizing thermal insulation. The standard efficiency for these stills hovers around 32%, with a productivity rate of about 3 to 3.5 liters per square meter. The substantial land area required for larger applications is another limiting factor that affects scalability.

## Solar Collectors

Solar collectors' harness solar energy and transform it into heat, which can be used in various desalination processes such as multi-stage flash (MSF), multi-effect distillation (MED), and membrane distillation (MD). Several studies have highlighted the effective integration of solar ponds into desalination systems. One noteworthy project was conducted by the University of Texas at El Paso (UTEP), in partnership with the U.S. Bureau of Reclamation, where a solar pond covering 3,200 square meters was utilized. This setup produced between 2.5 and 7.5 cubic meters of freshwater daily, with energy consumption ranging from 720 to 1,500 kJ/kg, depending on factors such as water salinity and cooling conditions. Potential combinations of solar collectors with desalination technologies are illustrated in Table 4.

*Table 4. Potential pairings of solar collectors and desalination techniques*

| Type of Solar Collector | Water Source | Desalination Technique |
|---|---|---|
| Direct Solar | Seawater, Brackish Water | Solar Evaporation Processes |
| Flat Plate Solar Collectors | Seawater | Low-Temperature Multi-Effect Distillation (LT-MED) |
| Evacuated Tube Collectors | Seawater | Multi-Stage Flash (MSF), Thermal Vapor Compression (TVC) |
| Parabolic Trough Collectors | Seawater | Multi-Effect Distillation (MED), Multi-Effect Boiling (MEB) |
| Photovoltaic Thermal Collectors | Seawater, Brackish Water | Low-Temperature Multi-Effect Distillation (LT-MED) |

## Wave Energy

Wave energy presents an excellent opportunity for desalination in coastal areas where seawater and wave power are abundant(Mahmoudi et al., 2023). With over 30 years of experimental history, wave energy conversion technologies have shown impressive potential. Unlike solar energy, which is limited by conversion efficiencies of around 87%, wave energy has the potential to reach near 100% efficiency in theory. Several technologies, such as buoy-based systems and oscillating water columns, have demonstrated full-scale operational success in powering reverse osmosis and vapor compression units. The DELBUOY system, for instance, used oscillating buoys to pump seawater through RO membranes, achieving an average freshwater production rate of 2.2 cubic meters per day. Although this system operated successfully for 18 months, its energy consumption was higher due to the absence of energy recovery mechanisms.

## Wind Energy

Wind energy is available across the globe, particularly in coastal areas, mountain regions, and islands, making it an ideal renewable source for desalination. The challenge with wind-powered desalination plants lies in managing the intermittent nature of wind, which can cause frequent start-ups and shutdowns. Technologies like reverse osmosis (RO) and mechanical vapor compression (MVC) are among the few desalination methods compatible with wind energy(Alawad et al., 2023). The MVC process, which can tolerate intermittent power supplies, has been successfully paired with wind energy. A plant located on Borkum Island used a 65 kW wind turbine to drive an MVC unit, producing between 0.35 and 2.1 cubic meters per hour of fresh water, with an energy consumption of 16.5 to 19.5 kWh per cubic meter. Another wind-powered MVC plant on the island of Rugen had a capacity of 370 cubic meters per day, supported by a wind energy production capacity of 310 kW (Fatni et al., 2025). Unlike conventional desalination plants, where energy efficiency is a primary concern, wind-powered desalination focuses on the fixed costs of the wind turbine, as there are no recurring fuel costs. This makes the capital investment in the wind turbine the major economic consideration, shifting the focus from operational efficiency to the overall economics of the system.

## Geothermal Energy

Geothermal energy is a proven technology primarily used for electricity production. However, its application in desalination is gaining traction due to its several advantages. One of the key benefits is that geothermal energy provides a steady and reliable heat source, essential for both thermal desalination processes and reverse osmosis. The technology for extracting geothermal heat from underground aquifers is mature, and geothermal source temperatures, typically between 68 to 92°C, are ideal for low-temperature MED desalination. Geothermal desalination systems are cost-effective, as they simultaneously generate electricity and fresh water, reducing reliance on imported fossil fuels(Azdem, al., 2025). These systems are also environmentally friendly, as they use renewable energy without emitting greenhouse gases or pollutants. The primary advantage of geothermal energy is that its output remains constant, without the intermittency issues seen in solar or wind power, making it an ideal fit for continuous desalination processes. Additionally, geothermal waters themselves can be used as feedwater in desalination, acting as both the energy source and the medium for desalination.

## b) Hybrid Process Integration

Desalination systems operating independently require considerable energy and financial investments. However, the concept of integrating electricity generation and desalination has gained significant acceptance in many regions, referred to as cogeneration. This method, particularly effective for large-scale projects, involves coupling power plants with desalination technologies such as Multi-Stage Flash (MSF) distillation (El Mouedden et al., 2025). For instance, the Gulf States initially adopted desalination through an MSF plant in Kuwait with a production capacity of approximately 4,600 $m^3$/day. When not connected to a power plant, the MSF system demands around 300 kJ/kg of energy, but this can be reduced to 165 kJ/kg when integrated with power generation.

## Coupling MSF with RO Process

MSF technology has long been considered a robust solution for desalination(Saleem et al., 2023), accounting for nearly 60% of global desalination capacity and up to 80% in the Middle East. However, Reverse Osmosis (RO) is gaining popularity due to its technological benefits. Integrating MSF and RO in hybrid systems can provide both technical and economic advantages. In a typical hybrid configuration, the cooling seawater from the MSF process is used to preheat the feedwater for the RO unit, resulting in a potential 42-46% increase in water recovery(Azdem et al., 2024). By sharing infrastructure like intake systems and reducing the replacement frequency of RO membranes, the overall operating costs decrease. Additionally, hybrid systems offer greater flexibility in power and water production, particularly under fluctuating operational demands. The coupling of these processes leads to reduced energy consumption and lower water production costs, especially when utilizing common facilities for power and desalination.

## Integrating Membrane Distillation (MD) with Other Desalination Processes

Membrane distillation (MD) is a promising technology that can be integrated with other desalination methods. Although its combination with MSF or Multi-Effect (ME) distillation is still under exploration, some studies have focused on coupling MD with RO. In such a system, the RO reject brine serves as the feed for the MD unit, reducing the volume of brine and minimizing environmental impacts. An economic analysis revealed that integrating MD with RO could reduce overall water production costs. For instance, an RO-only plant might cost \$1.30/m$^3$, while an integrated RO-MD system can double water output at a similar cost.

## Hybrid Membrane Process

In hybrid membrane processes, ultrafiltration (UF) is often used to remove particulates, bacteria, and viruses before the water is subjected to reverse osmosis (RO) or nanofiltration (NF)(Castro & Abejón, 2024). This approach enhances both water recovery and energy efficiency. Field trials in New South Wales demonstrated the successful production of clean drinking water from various feedwater sources, including brackish and high-turbidity water. The energy consumption in hybrid membrane systems varies depending on the feedwater salinity, ranging from 2.5 to 8.5 kWh per cubic meter. Table 5 highlights the key benefits of combining different desalination processes, showing how they increase water production capacity, reduce energy consumption, and save fuel annually.

*Table 5. Hybrid process benefits*

| Hybrid Process | Capacity (m$^3$/day) | Energy (kWh/m$^3$) | Fuel Savings (tons/year) |
|---|---|---|---|
| MSF-RO Hybrid | 15 | 9.2 | 5,2 |
| MD-RO Hybrid | 12 | 7.5 | 4,7 |
| MVC-MED Hybrid | 18,5 | 6.9 | 10,9 |

## c) Affordable and Energy-Efficient Desalination Technologies

Desalination technologies often require substantial energy input, accounting for up to 28% of the total cost of water production. The integration of desalination processes with low-grade waste heat sources offers a promising solution to improve both economic viability and efficiency. Heat sources with temperatures as low as 58°C can be effectively harnessed, making this approach particularly suitable for power plants or industrial settings where waste heat is abundant.

## Using Low-Grade Waste Heat Sources

In many desalination applications, utilizing low-grade waste heat from processes such as diesel cogeneration can significantly reduce operating costs. The Multi-Effect Distillation (MED) process can use the exhaust heat from diesel engines or power plants, increasing overall energy efficiency(He et al., 2022). In such systems, the recovered waste heat raises the thermal efficiency of power stations from around 42% to over 78%. Low-temperature MED (LT-MED) systems can function using waste heat streams as low as 58°C, making them ideal for integration with power generation facilities. A study demonstrated that by harnessing waste heat from a domestic air-conditioning system, it is possible to power a low-temperature desalination unit. The system, operating with thermal energy storage (TES) and solar power, produced an average of 4.7 kg of desalinated water per hour. The necessary solar panel area for this setup was 24 square meters. Additionally, using heat pumps to capture and utilize waste heat from industrial processes has proven effective for driving desalination with minimal heat loss to the environment.

## Low-Temperature Desalination Systems

Low-temperature desalination systems (LTDS) operate by taking advantage of modest temperature differentials between the seawater and the heat source. For example, a system implemented in the 1990s in Italy produced high-quality distilled water for industrial purposes at a cost of $0.64 per cubic meter. More recently, India's National Institute of Ocean Technology developed a Low-Temperature Thermal Desalination (LTTD) system(Jalihal, 2024), which exploits the temperature difference between surface seawater (typically 28-30°C) and deep seawater (7-15°C). This system requires only 7.9 kWh of energy to produce 1 cubic meter of fresh water, with a production cost of approximately $0.68 per cubic meter. Furthermore, innovative approaches utilizing low-cost building materials as passive heat collectors have also been explored. In one example, a 25 square meter roof area provided a consistent energy output of 9.5 kW, resulting in the production of 0.31 cubic meters of fresh water per day(Azdem, et al 2024).

## d) Reuse of Water Resources

In many regions worldwide, traditional water supplies are increasingly strained, and the necessity for water reuse has become more critical. Water reuse is especially relevant in areas facing water scarcity, such as North Africa. Population growth, urban expansion, and climate changes in these arid regions have heightened the demand for fresh water. As a result, water reuse presents an effective strategy for mitigating water shortages, particularly in Morocco, Tunisia, and Egypt. Recycled water can be employed in various non-potable applications, including agricultural irrigation, industrial processes, and landscaping, thus relieving pressure on natural freshwater sources.

The main sources for water reuse include municipal wastewater, industrial effluent, and agricultural runoff. These sources can be treated and reclaimed for non-drinking purposes, which helps preserve high-quality freshwater for essential uses. In Morocco, water reuse initiatives are still developing, but efforts are growing to reclaim treated wastewater for agriculture and industrial sectors, especially in regions where fresh water is scarce. In this context, North African countries have shown varying levels of commitment to water reuse, with some countries investing more heavily in infrastructure and technology than others. The table 6 outlines the levels of water reuse in several North African countries.

*Table 8. Water reuse practices in North African countries*

| Country | Total Water Supply (million $m^3$/year) | Percentage of Water Reuse (%) |
|---|---|---|
| Morocco | 5,6 | 4.2 |
| Algeria | 6,8 | 3.5 |
| Tunisia | 4,3 | 6.9 |
| Egypt | 58 | 8.1 |
| Libya | 1,7 | 2.3 |

## 5. ESTIMATING THE COSTS OF DESALINATION

Determining the capital and operational costs of desalination systems can be complex due to various factors such as location, energy prices, and material costs. These factors vary across different regions and affect the overall cost of desalination plants. Additionally, the salinity of the feed water and the size of the plant also play critical roles in determining the cost. The cost of desalinated water can be broken down into capital investment, operational costs, and energy consumption(Saboori, 2023). Different desalination processes have varying energy requirements, which significantly impacts the total cost. Renewable energy-powered desalination plants may offer long-term benefits but typically have high initial capital costs. The tables 7 provide estimated costs for different desalination processes.

*Table 9. Desalinated water costs using various processes*

| Desalination Process | Cost ($/m³) | Energy Source |
|---|---|---|
| Reverse Osmosis (RO) | 1.40 - 2.50 | Solar PV |
| Multi-Effect Distillation | 2.10 - 3.30 | Solar Thermal |
| Vapor Compression (MVC) | 1.80 - 3.20 | Wind Energy |
| MSF | 1.70 - 3.50 | Hybrid Wind/Solar |

## CONCLUSION

The demand for fresh water in arid and semi-arid regions like North Africa is growing rapidly, driven by population growth, urbanization, and industrialization. As traditional water resources become increasingly scarce, desalination emerges as a crucial solution. However, the energy-intensive nature of conventional desalination processes poses economic and environmental challenges. This has led to the exploration of renewable energy-powered desalination, which offers a sustainable and cost-effective alternative. In countries like Morocco, significant progress is being made to integrate renewable energy sources such as solar and wind power into desalination projects. Morocco's largest desalination plant in Agadir, which produces over 100,000 $m^3$ of fresh water per day, is a prime example of how renewable energy can support large-scale desalination efforts. The cost of desalinated water in the region has been reduced to as low as $0.80 per cubic meter, making it a viable option for water-stressed areas.

The use of hybrid systems that combine different desalination technologies and renewable energy sources, such as solar and wind, has proven to be particularly effective. These hybrid systems not only reduce energy consumption but also lower operational costs, making desalination more accessible in regions with limited resources. The integration of low-grade waste heat and low-temperature desalination processes also contributes to greater energy efficiency. In conclusion, North African countries, especially Morocco, are leading the way in adopting sustainable desalination technologies powered by renewable energy. These efforts are critical in addressing water scarcity and ensuring a stable supply of fresh water for future generations. As the cost of renewable energy continues to decline and desalination technologies advance, the combination of these two solutions will become even more essential for the region's water security.

## REFERENCES


Ahmed, M. M. R., Mirsaeidi, S., Koondhar, M. A., Karami, N., Tag-Eldin, E. M., Ghamry, N. A., ... & Sharaf, A. M. (2024). Mitigating uncertainty problems of renewable energy resources through effi

Ait Lahoussine Ouali, H., Soomro, M. I., Touili, S., Eltaweel, M., & Alami Merrouni, A. (2023). Performance investigation of seawater desalination system powered by central receiver concentrated solar thermal plant. *Applied Thermal Engineering*, *225*, 120165. DOI: 10.1016/j.applthermaleng.2023.120165

Al-Jadabi, N., Laaouan, M., El Hajjaji, S., Mabrouki, J., Benbouzid, M., & Dhiba, D. (2023). The dual performance of Moringa oleifera seeds as eco-friendly natural coagulant and as an antimicrobial for wastewater treatment: A review. *Sustainability (Basel)*, *15*(5), 4280. DOI: 10.3390/su15054280

Al-Obaidi, M., Alsarayreh, A. A., Rashid, F. L., Sowgath, M. T., Alsadaie, S., Ruiz-García, A., Khayet, M., Ghaffour, N., & Mujtaba, I. M. (2024). Hybrid membrane and thermal seawater desalination processes powered by fossil fuels : A comprehensive review, future challenges and prospects. *Desalination*, *583*, 117694. DOI: 10.1016/j.desal.2024.117694

Alawad, S. M., Mansour, R. B., Al-Sulaiman, F. A., & Rehman, S. (2023). Renewable energy systems for water desalination applications : A comprehensive review. *Energy Conversion and Management*, *286*, 117035. DOI: 10.1016/j.enconman.2023.117035

Almasoudi, S., & Jamoussi, B. (2024). Desalination technologies and their environmental impacts : A review. *Sustainable Chemistry One World*, *1*, 100002. DOI: 10.1016/j.scowo.2024.100002

Amir, M., Deshmukh, R. G., Khalid, H. M., Said, Z., Raza, A., Muyeen, S. M., Nizami, A.-S., Elavarasan, R. M., Saidur, R., & Sopian, K. (2023). Energy storage technologies : An integrated survey of developments, global economical/environmental effects, optimal scheduling model, and sustainable adaption policies. *Journal of Energy Storage*, *72*, 108694. DOI: 10.1016/j.est.2023.108694

Arabi, M., Mechkirrou, L., El Malki, M., Alaoui, K., Chaieb, A., Maaroufi, F., & Karmich, S. (2024). Overview of ecological dynamics in Morocco–Biodiversity, water scarcity, climate change, anthropogenic pressures, and energy resources–Navigating towards ecosolutions and sustainable development. In *E3S Web of Conferences* (Vol. 527, p. 01001). EDP Sciences.

Azdem, D., Mabrouki, J., & El Hajjaji, S. (2024). Methods and Technologies for the Treatment and Disposal of Brine in Seawater Desalination Plants. In Mabrouki, J., & Azrour, M. (Eds.), (pp. 79–87). Advances in Environmental Engineering and Green Technologies. IGI Global., DOI: 10.4018/979-8-3693-3807-0.ch007

Azdem, D., Mabrouki, J., Fahdi, G., & Hajjaji, S. E. (2025). ENVIRONMENTAL SUSTAINABILITY OF DESALINATION : A CRITICAL REVIEW OF BRINE TREATMENT CHALLENGES AND MITIGATION STRATEGIES. Water Conservation and Management, 9(1), 55-60. *Scopus*. Advance online publication. DOI: 10.26480/wcm.01.2025.55.60

Azdem, D., Mabrouki, J., & Hajjaji, S. E. (2024). Methods and Technologies for the Treatment and Disposal of Brine in Seawater Desalination Plants. In Advancements in Climate and Smart Environment Technology (p. 79-87). IGI Global. DOI: 10.4018/979-8-3693-3807-0.ch007

Azdem, D., Mabrouki, J., Moufti, A., El hajjaji, S., & Fatni, A. (2024). Assessment of heavy metal contamination in seawater in Agadir coastline, Morocco. *Desalination and Water Treatment*, *317*, 100129. DOI: 10.1016/j.dwt.2024.100129

Azdem, D., Mabrouki, J., Moufti, A., & Fatni, A. (2024). Assessment of heavy metal contamination in seawater in Agadir coastline, Morocco. *Desalination and Water Treatment*, *317*, 100129. DOI: 10.1016/j.dwt.2024.100129

Azdem, D., Mabrouki, J., Moufti, A., Hajjaji, S. E., & Fatni, A. (2024). Present Situation and Environmental Impact of Seawater Desalination Plants in Morocco. In *Technical and Technological Solutions Towards a Sustainable Society and Circular Economy* (pp. 581–592). Springer. DOI: 10.1007/978-3-031-56292-1_46

Azdem, D., Fatni, A., Mabrouki, J., & El Hajjaji, S. (2025). The Desalination Strategy and Economic Integration of Morocco: Water Scarcity, Prospects, and Challenges. Circular Economy Applications in Energy Policy, 257-268.

Benbba, R., Barhdadi, M., Ficarella, A., Manente, G., Romano, M. P., Hachemi, N. E., Barhdadi, A., Al-Salaymeh, A., & Outzourhit, A. (2024). Solar Energy Resource and Power Generation in Morocco : Current Situation, Potentials and Future Perspective (2024081025). *Preprints*. DOI: 10.20944/preprints202408.1025.v1

Benchrifa, M., Elouardi, M., Fattah, G., Mabrouki, J., & Tadili, R. (2023). Identification, simulation and modeling of the main power losses of a photovoltaic installation and use of the internet of things to minimize system losses. In *Advanced technology for smart environment and energy* (pp. 49–60). Springer International Publishing. DOI: 10.1007/978-3-031-25662-2_4

Cai, Y., Wu, J., Shi, S. Q., Li, J., & Kim, K.-H. (2023). Advances in desalination technology and its environmental and economic assessment. *Journal of Cleaner Production*, *397*, 136498. DOI: 10.1016/j.jclepro.2023.136498

Castro, K., & Abejón, R. (2024). Removal of Heavy Metals from Wastewaters and Other Aqueous Streams by Pressure-Driven Membrane Technologies : An Outlook on Reverse Osmosis, Nanofiltration, Ultrafiltration and Microfiltration Potential from a Bibliometric Analysis. *Membranes*, *14*(8), 8. Advance online publication. DOI: 10.3390/membranes14080180 PMID: 39195432

Chari, Z., Cherkaoui, E., Khamar, M., & Nounah, A. (2023, June). Evolution of Desalination in Morocco. In *The scientific conference on Geosciences and Environmental Management (GeoME)* (pp. 464–471). Springer Nature Switzerland.

Cheikhaoui, Y., Sadiki, M., Allouza, M., Chakiri, S., & Bouabdli, A. (2024). Estimation of irrigation water requirements in the Gharb-irrigated perimeter (north-western Morocco). *Water Science and Technology: Water Supply*, *24*(2), 436–452. DOI: 10.2166/ws.2024.012

Edegbene, A. O., Yilmaz, N., & Akamagwuna, F. (2024). Editorial : Biomonitoring freshwater ecosystems health in a changing world: the significance of socio-ecological approaches. *Frontiers in Ecology and Evolution*, *11*, 1342732. Advance online publication. DOI: 10.3389/fevo.2023.1342732

El Alouani, M., Aouan, B., Rachdi, Y., Alehyen, S., El Herradi, E. H., Saufi, H., Mabrouki, J., & Barka, N. (2024). Porous geopolymers as innovative adsorbents for the removal of organic and inorganic hazardous substances: A mini-review. *International Journal of Environmental Analytical Chemistry, 104*(16), 4784–4796. DOI: 10.1080/03067319.2022.2115891

El Mouedden, M., Azdem, D., Mabrouki, J., & El Hajjaji, S. (2025). MarocFarm : A digital revolution for agriculture in Morocco. In Circular Economy Applications in Energy Policy (p. 141-155). Scopus. DOI: 10.4018/979-8-3693-6680-6.ch010

Fahdi, G., Azdem, D., Lamchaimech, A., Benrhanem, M., Mabrouki, J., & El Hajjaji, S. (2024). ANALYSIS OF THE IMPACT OF CLIMATE CHANGE ON WATER RESOURCES : CASE OF THE TENSIFT BASIN (MOROCCO). Water Conservation and Management, 8(4), 461-465. *Scopus.* Advance online publication. DOI: 10.26480/wcm.04.2024.461.465

Fahdi, G., Mabrouki, J., Lamchaimech, A., Azdem, D., Benrhanem, M., & Hajjaji, S. E. (2024). IMPACTS OF CLIMATE CHANGE ON GROUNDWATER QUALITY AND RECHARGE IN THE TENSIFT BASIN, MOROCCO. Water Conservation and Management, 8(4), 466-474. *Scopus.* Advance online publication. DOI: 10.26480/wcm.04.2024.466.474

Fatni, A., Addich, M., Essekri, A., EL Baraka, N., Azdem, D., Ait Taleb, A., & EL Hammadi, A. (2025). Characterization and mitigation strategies for inorganic scaling in reverse osmosis system treating brackish groundwater. *Sustainable Chemistry One World, 5*, 100035. DOI: 10.1016/j.scowo.2024.100035

Foghagen, C., & Alriksson, S. (2024). Mitigation of water shortage impacts among tourism and farming companies on the islands of Öland and Gotland, Sweden. *Environment, Development and Sustainability, 26*(3), 7509–7527. DOI: 10.1007/s10668-023-03019-0

Fragkos, P. (2023). Assessing the energy system impacts of Morocco's nationally determined contribution and low-emission pathways. *Energy Strategy Reviews, 47*, 101081. DOI: 10.1016/j.esr.2023.101081

Ghizlane, F., Mabrouki, J., Ghrissi, F., & Azrour, M. (2022). Proposal for a high-resolution particulate matter (PM10 and PM2. 5) capture system, comparable with hybrid system-based internet of things: Case of quarries in the western rif, Morocco. *Pollution, 8*(1), 169–180.

Gude, V. G. (2023). Energy storage for sustainable desalination and renewable energy integration. In Gude, V. G. (Ed.), *Energy Storage for Multigeneration* (pp. 1–23). Academic Press., DOI: 10.1016/B978-0-12-821920-1.00008-X

Hajou, A., El Mghouchi, Y., & Chaoui, M. (2024). A new solar-wind complementarity index : An application to the climate of Morocco. *Renewable Energy, 227*, 120490. DOI: 10.1016/j.renene.2024.120490

Harby, K., Emad, M., Benghanem, M., Abolibda, T. Z., Almohammadi, K., Aljabri, A., Alsaiari, A., & Elgendi, M. (2024). Reverse osmosis hybridization with other desalination techniques : An overview and opportunities. *Desalination, 581*, 117600. DOI: 10.1016/j.desal.2024.117600

Hassan, Q., Algburi, S., Sameen, A. Z., Salman, H. M., & Jaszczur, M. (2023). A review of hybrid renewable energy systems : Solar and wind-powered solutions: Challenges, opportunities, and policy implications. *Results in Engineering, 20*, 101621. DOI: 10.1016/j.rineng.2023.101621

He, F., Lu, T., Pan, Z., Niu, Y., Zhou, G., & Lu, Y. (2022). Thermodynamic Analysis and Optimization for Exhaust Energy Splitting : An Innovative Method of Improving Exhaust Energy Grade for the Low-Speed Marine Diesel Engine. *Journal of Thermal Science and Engineering Applications*, *15*(031006), 031006. Advance online publication. DOI: 10.1115/1.4056301

Huang, J., Zheng, H., & Kong, H. (2024). Key pathways for efficient solar thermal desalination. *Energy Conversion and Management*, *299*, 117806. DOI: 10.1016/j.enconman.2023.117806

Jalihal, P. (2024). Energy and freshwater in the context of blue economy. Current Science (00113891), 126(2).

Kumar, Ch., Singh, S., Gupta, M. K., Nimdeo, Y. M., Raushan, R., Deorankar, A. V., Kumar, T. M. A., Rout, P. K., Chanotiya, C. S., Pakhale, V. D., & Nannaware, A. D. (2023). Solar energy : A promising renewable source for meeting energy demand in Indian agriculture applications. *Sustainable Energy Technologies and Assessments*, *55*, 102905. DOI: 10.1016/j.seta.2022.102905

Laaroussi, A., Laaroussi, O., & Bouayad, A. (2023). Environmental impact study of the NOOR 1 solar project on the Southern Region of Morocco. *Renewable Energy and Environmental Sustainability*, *8*, 9. DOI: 10.1051/rees/2023007

Mabrouki, J., Benbouzid, M., Dhiba, D., & El Hajjaji, S. (2022). Simulation of wastewater treatment processes with Bioreactor Membrane Reactor (MBR) treatment versus conventional the adsorbent layer-based filtration system (LAFS). *International Journal of Environmental Analytical Chemistry*, *102*(19), 7458–7468. DOI: 10.1080/03067319.2020.1828394

Maftouh, A., El Fatni, O., Bouzekri, S., Bahaj, T., Kacimi, I., El Hajjaji, S., & Malik, A. (2023). Solar Desalination : Current Applications and Future Potential in MENA Region – A Case Study. Journal of Sustainable Development of Energy. *Water and Environment Systems*, *11*(2), 1–26. DOI: 10.13044/j.sdewes.d10.04377

Mahdavi, M., & Vera, D. (2023). Importance of renewable energy sources and agricultural biomass in providing primary energy demand for Morocco. *International Journal of Hydrogen Energy*, *48*(88), 34575–34598. DOI: 10.1016/j.ijhydene.2023.05.246

Mahmoudi, A., Bostani, M., Rashidi, S., & Valipour, M. S. (2023). Challenges and opportunities of desalination with renewable energy resources in Middle East countries. *Renewable & Sustainable Energy Reviews*, *184*, 113543. DOI: 10.1016/j.rser.2023.113543

Mgalaa, S., Mabrouki, J., Elouardi, M., El Azzouzi, L., Moufti, A., El Hajjaji, S., EL Azzouzi, M., & El Belghiti, M. A. (2022). Study and evaluation of the degradation of procion blue dye by the ozonation method: Parametric and isothermal study. *Nanotechnology for Environmental Engineering*, *7*(3), 691–697. DOI: 10.1007/s41204-022-00262-0

Nurjanah, I., Chang, T.-T., You, S.-J., Huang, C.-Y., & Sean, W.-Y. (2024). Reverse osmosis integrated with renewable energy as sustainable technology : A review. *Desalination*, *581*, 117590. DOI: 10.1016/j.desal.2024.117590

Panagopoulos, A. (2025). Assessing the Energy Footprint of Desalination Technologies and Minimal/Zero Liquid Discharge (MLD/ZLD) Systems for Sustainable Water Protection via Renewable Energy Integration. *Energies*, *18*(4), 962. DOI: 10.3390/en18040962

Rachiq, T., Abrouki, Y., Mabrouki, J., Samghouli, N., Fersi, C., Rahal, S., & El Hajjaji, S. (2021). Evaluation of the efficiency of different materials to remove specific pollutants from landfill leachate. *Desalination and Water Treatment*, *238*, 240–250. DOI: 10.5004/dwt.2021.27779

Rad, V., Amin, M., Kasaeian, A., Niu, X., Zhang, K., & Mahian, O. (2023). Excess electricity problem in off-grid hybrid renewable energy systems : A comprehensive review from challenges to prevalent solutions. *Renewable Energy*, *212*, 538–560. DOI: 10.1016/j.renene.2023.05.073

Saboori, H. (2023). Hybrid renewable energy powered reverse osmosis desalination – Minimization and comprehensive analysis of levelized cost of water. *Sustainable Energy Technologies and Assessments*, *56*, 103065. DOI: 10.1016/j.seta.2023.103065

Saleem, H., Abounahia, N., Siddiqui, H. R., & Zaidi, S. J. (2023). Qatar desalination research : An overview. *Desalination*, *564*, 116802. DOI: 10.1016/j.desal.2023.116802

Shokri, A., & Fard, S. (2023). Water-energy nexus : Cutting edge water desalination technologies and hybridized renewable-assisted systems; challenges and future roadmaps. *Sustainable Energy Technologies and Assessments*, *57*, 103173. DOI: 10.1016/j.seta.2023.103173

Swain, C. K. (2024). Environmental pollution indices : A review on concentration of heavy metals in air, water, and soil near industrialization and urbanisation. *Discover Environment*, *2*(1), 5. DOI: 10.1007/s44274-024-00030-8

Tashtoush, B., Alyahya, W., Al Ghadi, M., Al-Omari, J., & Morosuk, T. (2023). Renewable energy integration in water desalination : State-of-the-art review and comparative analysis. *Applied Energy*, *352*, 121950. DOI: 10.1016/j.apenergy.2023.121950

Tanji, R., & Boutaibi, Y. (2023). Use of wind energy for desalination of seawater for irrigation. *World Water Policy*, *9*(4), 787–798.

Yatawatta, Y. J. M., & Sridarran, P. (2024). Investigating the barriers and strategies for establishing desalination plants to mitigate water scarcity in Sri Lankan dry zones. Sustainability Accounting. *Management and Policy Journal*, *15*(6), 1408–1441.

Zolghadr-Asli, B., McIntyre, N., Djordjevic, S., Farmani, R., & Pagliero, L. (2023). The sustainability of desalination as a remedy to the water crisis in the agriculture sector : An analysis from the climate-water-energy-food nexus perspective. *Agricultural Water Management*, *286*, 108407. DOI: 10.1016/j.agwat.2023.108407

# Chapter 26
# Technical Challenges and Innovations in Hydrogen Storage

**Oussama Aziz**
https://orcid.org/0000-0002-8057-0358
*Faculty of Sciences, Mohammed V University, Morocco*

**Abdelilah Benyoussef**
*Faculty of Sciences, Mohammed V University, Morocco*

**Hamid Ez-Zahraouy**
*Faculty of Sciences, Mohammed V University, Morocco*

## ABSTRACT

*The transition to a carbon-neutral economy necessitates a shift from fossil fuels to renewable energy sources, with hydrogen emerging as a promising solution due to its environmental benefits. Hydrogen-powered vehicles produce zero CO2 emissions, positioning hydrogen as a key player in decarbonizing the transportation sector and achieving net-zero goals. However, challenges in hydrogen generation, storage, and infrastructure development remain significant barriers to widespread adoption. This paper offers a comprehensive analysis of hydrogen storage technologies, including gaseous, liquid, and solid-state methods, while also focusing on porous media, which show great potential for enhanced storage capabilities. By summarizing recent advancements and examining various hydrogen storage systems, this research aims to support the growth of the hydrogen energy sector, ultimately contributing to sustainable development and addressing pressing energy challenges.*

## 1. HYDROGEN: A KEY SOLUTION TO GLOBAL ENERGY AND ENVIRONMENTAL CHALLENGES

The contemporary globe confronts an unparalleled energy dilemma due to the escalating worldwide energy demand (Swadi et al., 2024) and the pressing necessity to diminish carbon emissions (Ball & Wietschel, 2009). For decades, fossil fuels (Abas, Kalair, & Khan, 2015) have been the primary energy

DOI: 10.4018/979-8-3693-8980-5.ch026

source; nevertheless, their environmental repercussions—especially regarding greenhouse gas emissions (Gillingham & Stock, 2018)—have catalyzed the pursuit of cleaner, more sustainable alternatives. The shift to a carbon-neutral economy ("Aboveground Hydrogen Storage," 2024) has emerged as a worldwide imperative, as governments, corporations, and researchers endeavor to alleviate climate change while ensuring energy security (Iyke, 2024). In this scenario, hydrogen has emerged as a viable approach to address future energy requirements ("Emerging Technologies by Hydrogen,"2024).

Hydrogen is a multifaceted (Dillman & Heinonen, 2022) and eco-friendly energy carrier (Qureshi et al., 2023) with considerable potential to tackle energy and environmental issues. In contrast to fossil fuels, hydrogen combustion does not emit carbon dioxide (CO2) (Shadidi, Najafi, & Yusaf, 2021), rendering it a carbon-neutral alternative. Hydrogen, when utilized in fuel cells, produces energy with water as the sole byproduct ("Impact of Hydrogen on the Environment," 2024), rendering it an appealing choice for several uses, including transportation and stationary power generation (Hwang, Maharjan, & Cho, 2023). As the globe progresses towards a sustainable energy future ("Opportunities and Challenges for a Sustainable Energy Future," 2024), hydrogen is increasingly acknowledged as an essential element in the energy composition.

## 1.1 The Role of Hydrogen in the Transition to a Carbon-Neutral Economy

The increasing interest in hydrogen is primarily driven by its potential to decarbonize sectors (Maestre, Ortiz, & Ortiz, 2021) that are challenging to electrify, like heavy industrial and long-distance transportation. Hydrogen-powered vehicles ("Overview of Development and Challenges in Hydrogen Powered Vehicles," 2019) provide a substantial benefit compared to conventional internal combustion engines by generating no exhaust pollutants (Fayaz et al., 2012). Hydrogen is especially important for the transportation sector (Singh et al., 2015), as attaining net-zero emissions is a primary objective for the 21st century (Renné, 2022).

The environmental advantages of hydrogen are evident: substituting fossil fuels with hydrogen in diverse applications can diminish the carbon footprint of industries (Evro, Oni, & Tomomewo, 2024) and lead to significant emissions reductions (Ocko & Hamburg, 2022). This is particularly crucial in industries such as aviation (Yusaf et al., 2024), shipping (Atilhan et al., 2021), and heavy-duty trucks ("Hydrogen Fuel Cell Heavy-Duty Trucks," 2024), where the energy density needed is challenging to achieve with existing battery technology (Le et al., 2024). Hydrogen, being a high energy density per unit mass (McCay & Shafiee, 2020), is ideally suited for these uses and can facilitate the global transition from carbon-intensive energy sources (Agarwal, 2022).

## 1.2 The Challenges of Hydrogen Storage and Infrastructure

Notwithstanding its potential, hydrogen has substantial hurdles that must be resolved for it to emerge as a predominant energy carrier. A critical concern is the advancement of effective hydrogen storage devices (Zhang et al., 2015). Hydrogen possesses a minimal volumetric energy density in its natural gaseous form ("Hydrogen: The Future Energy Carrier," 2010), hence complicating its storage and transfer. Current storage techniques, including compression and liquefaction, exhibit constraints regarding cost, energy efficiency, and safety (Abe et al., 2019). This presents a considerable barrier to the broad adoption of hydrogen as a feasible fuel source.

In addition to storage challenges, the establishment of a comprehensive hydrogen infrastructure is crucial for the broad adoption of hydrogen (Kurtz, Sprik, & Bradley, 2019). The transportation of hydrogen to fueling stations and the development of refueling infrastructure for hydrogen-powered vehicles (Turoń, 2020) necessitate significant investment and ingenuity. The shift to hydrogen as a primary energy source will be slow and fragmented in the absence of an efficient and accessible infrastructure.

### 1.3 Hydrogen's Growing Global Importance

Hydrogen has drawn more and more interest from around the world as worries about energy security and climate change grow (Noussan et al., 2021). Numerous nations have formulated national hydrogen programs to incorporate hydrogen into their energy frameworks (Kovač, Paranos, & Marciuš, 2021). The European Union has established ambitious objectives for the production and utilization of green hydrogen (Pingkuo & Junqing, 2024), whereas nations such as Japan and South Korea have delineated strategies to shift their energy sectors towards hydrogen-based solutions (Jaunatre,2021).

The applicability of hydrogen transcends transportation, encompassing industries such as steel manufacture, chemical manufacturing, and power generation ("Hydrogen as an Energy Vector," 2019). The adaptability of hydrogen renders it an essential instrument in the endeavor to decarbonize many sectors of the economy.

In the transition to a carbon-neutral economy, hydrogen is poised to be essential. Its potential to serve as a clean, efficient, and diverse energy carrier establishes it as an essential element in efforts to reduce carbon emissions and achieve sustainability goals. However, the challenges associated with hydrogen storage, transportation, and infrastructure remain significant barriers to its widespread adoption. Confronting these technical obstacles requires continuous research, innovation, and investment in hydrogen technology. As the global energy landscape evolves, hydrogen is set to become a crucial component of the sustainable energy future.

## 2. HYDROGEN PRODUCTION AND ITS CHALLENGES

Hydrogen is frequently acclaimed as the fuel of the future due to its capacity to facilitate a sustainable, carbon-neutral energy system. Nonetheless, the effective, cheap, and environmentally benign production of hydrogen continues to pose a considerable problem (Zhang et al., 2021). This chapter examines the diverse techniques for hydrogen production, the obstacles related to the scaling of these technologies, The role of renewable energy in advancing green hydrogen.

### 2.1 Methods of Hydrogen Production

Numerous technologies exist for hydrogen production, each possessing distinct advantages, disadvantages, and environmental consequences ("Review on Hydrogen Production and Utilization," 2024). The primary processes utilized include steam methane reforming (SMR), electrolysis, and biomass gasification, among others.

### 2.1.1 Steam Methane Reforming (SMR)

Steam methane reforming (SMR) is the predominant technique for hydrogen production (Navas-Anguita et al., 2021), constituting over 48% of the worldwide hydrogen supply (Stenberg et al., 2020). In steam methane reforming (SMR), natural gas, primarily methane, reacts with steam under high temperatures and pressures, producing hydrogen and carbon dioxide as by-products. Notwithstanding the efficacy of SMR, its reliance on fossil fuels and significant CO2 emissions present environmental concerns ("Natural-Gas-Derived Hydrogen," 2017). Mitigation measures entail the incorporation of carbon capture and storage (CCS) technologies to reduce emissions (Wang et al., 2022).

### 2.1.2 Electrolysis

Electrolysis serves as a more environmentally friendly alternative to steam methane reforming (SMR) (Faheem et al., 2021), as it entails the dissociation of water into hydrogen and oxygen through the use of electricity. This process generates "green hydrogen" when fueled by renewable energy sources like sun or wind as seen in **figure 1**. Various forms of electrolysis exist, including:

- **Alkaline Electrolysis:** Utilizes an alkaline electrolyte and represents one of the most developed technologies (Zeng & Zhang, 2010).
- **Proton Exchange Membrane (PEM) water Electrolysis:** functions with enhanced efficiency and is more appropriate for the integration of renewable energy sources (Wang, Cao, & Jiao, 2022).
- **Solid Oxide Electrolysis:** A high-temperature method that has significant efficiency potential, although remains in the nascent phase pertaining to development Economic evaluation of hydrogen generation using high-temperature solid oxide electrolysis (Prosser et al., 2024).

Notwithstanding its ecological advantages, electrolysis presently incurs more expenses than steam methane reforming (SMR) owing to elevated power costs and the materials utilized in electrolyzers (Farhana, Mahamude, & Kadirgama, 2024).

*Figure 1. Green hydrogen production using electrolysis*

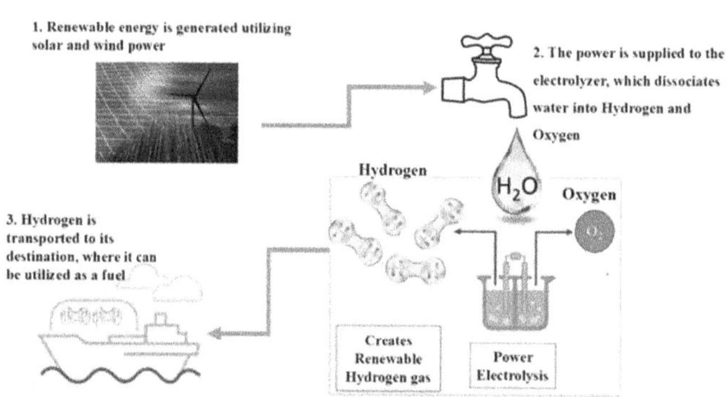

### 2.1.3 Biomass Gasification

Biomass gasification is the conversion of organic materials (including agricultural waste and wood chips) into hydrogen, carbon monoxide, and carbon dioxide by high-temperature processes ("Exploiting Agricultural Biomass," 2024). This technology is regarded as a renewable approach to hydrogen production, contingent upon the sustainable sourcing of biomass. Gasification presents potential, especially in areas rich in agricultural waste; yet, the technology requires additional advancement to enhance efficiency and reduce costs ("Enhanced Biomass Gasification," 2024).

### 2.1.4 Emerging Hydrogen Production Technologies

Recent improvements in hydrogen generation include enhancing efficiency and minimizing environmental impact. Among the nascent technologies are:

- **Photoelectrochemical Water Splitting**: Utilizes sunlight to directly dissociate water, eliminating the requirement for electricity ("Hydrogen from Sunlight and Water," 2021). This technology, while highly promising, is still in its nascent stage, exhibiting low efficiency and scalability challenges.
- **Thermochemical Water Splitting**: Utilizes heat, frequently derived from nuclear reactors or concentrated solar power to separate water molecules ("Concentrated Solar Energy with TES," 2021). This technology remains in development but may provide a feasible approach to large-scale hydrogen production in the future.

## 2.2 Obstacles in Expanding Hydrogen Production

Although hydrogen is set to be pivotal in the future energy framework, considerable obstacles persist in enhancing its production. The issues encompass elevated production costs, technological obstacles, and environmental factors (McDowall & Eames, 2006).

### 2.2.1 Cost and Efficiency

The primary obstacle to hydrogen manufacturing is its cost. Contemporary hydrogen generation systems, particularly green hydrogen derived from electrolysis, continue to be more costly than conventional fossil fuel-based energy sources (Marouani et al., 2023). The expense of energy, especially from renewable sources, significantly influences the total cost of hydrogen production.

Initiatives to decrease expenses encompass improvements in electrolyzer technology, economies of scale, and the reduction of renewable electricity costs. Advancements in catalysts, membranes, and system design are anticipated to lower the capital and operational expenses of electrolyzers (Yang, Zhang, Shao, & Zhang, 2023).

### 2.2.2 Carbon Emissions in Production

Although hydrogen is a clean fuel, its synthesis, especially by steam methane reforming (SMR), is linked to considerable carbon emissions. The integration of carbon capture and storage (CCS) technology in steam methane reforming (SMR) processes may mitigate certain emissions., yielding "blue hydrogen," which is more environmentally friendly while still dependent on fossil fuels. Nonetheless, CCS increases the expense and intricacy of hydrogen production (Cloete, Arnaiz del Pozo, & Jiménez Álvaro, 2022).

Green hydrogen, generated from renewable electricity through electrolysis, provides a genuinely zero-emission alternative ("Green Hydrogen Energy Systems," 2024). Scaling up green hydrogen production necessitates substantial quantities of renewable energy, presenting issues associated with land utilization, infrastructure, and energy storage ("Large-Scale Hydrogen Production and Storage Technologies," 2024).

### 2.2.3 Technological Barriers

Hydrogen generation technologies, especially innovative approaches such as photoelectrochemical and thermochemical water splitting, are in nascent stages of research (Verma, 2015). These technologies encounter obstacles regarding efficiency, durability, and scalability. Attaining the requisite advancements in these domains would necessitate continuous investment in research and development.

Moreover, electrolyzer technology, despite its maturity, continues to encounter hurdles concerning efficiency, durability, and integration with renewable energy sources (Ikuerowo et al., 2024). PEM electrolyzers, despite their high efficiency, depend on costly elements such as platinum and iridium ("PEM Electrolyzer Performance," 2024), which impede their cost-effectiveness at scale.

### 2.2.4 Infrastructure and Market Development

The hydrogen production industry is nascent., and the establishment of a comprehensive hydrogen infrastructure is essential for its extensive acceptance (Ball & Weeda, 2015). This include the development of hydrogen pipelines, refueling stations, and storage facilities, necessitating significant investment ("Industrial Status of Hydrogen Energy," 2024). Furthermore, establishing a hydrogen fuel market—especially in transportation, industry, and power production sectors—will be essential for realizing economies of scale (Maggio, Nicita, & Squadrito, 2019).

## 2.3 Renewable Energy Integration for Hydrogen Production

The utilization of renewable energy sources, including wind and sun., into hydrogen generation is essential for the advancement of green hydrogen ("Prospect of Green Hydrogen Generation," 2024). These energy sources supply the electricity necessary for water electrolysis, enabling hydrogen production devoid of carbon emissions.

### 2.3.1 Green Hydrogen from Renewable Energy Sources

Green hydrogen, produced by electrolysis from renewable electricity, is considered the most sustainable method of hydrogen production (Nnabuife et al., 2024). The decreasing expenses associated with renewable energy, particularly solar and wind, have rendered green hydrogen more economically viable (Taibi, Miranda, Carmo, & Blanco, 2020), despite ongoing obstacles.

The capacity of green hydrogen to significantly contribute to the decarbonization of energy-intensive sectors, including steel, cement, and chemicals, has attracted considerable interest. Additionally, green hydrogen functions as an energy storage system, facilitating the equilibrium of supply and demand in networks primarily reliant on intermittent renewable energy sources.

### 2.3.2 Hybrid Systems for Hydrogen Production

Hybrid systems integrating various renewable energy sources or energy storage techniques with hydrogen generation are being investigated to enhance efficiency and decrease expenses ("Hybrid Renewable Energy Systems: Review Analysis," 2024). For instance, integrating wind or solar farms with hydrogen manufacturing facilities can offer a method to store surplus energy generated during peak production intervals. Integrating battery storage with hydrogen production can provide a more reliable and continuous electrical source for electrolysis (Nnabuife et al., 2024).

## 2.4 The Future of Hydrogen Production

The future of hydrogen generation depends on addressing the difficulties of affordability, scalability, and environmental effect. Advancements in electrolyzer technology and the incorporation of renewable energy, and novel production techniques such as photoelectrochemical water splitting will be essential for establishing hydrogen as a predominant energy carrier. Additionally, legislative backing and governmental incentives will be essential to expedite the adoption of hydrogen technologies and foster the establishment of a global hydrogen economy ("Accelerating the Green Hydrogen Revolution," 2024) .

Hydrogen is anticipated to be crucial in the decarbonization of various industries, including transportation and industry. Realizing this promise necessitates ongoing innovation, investment, and collaboration across industries, governments, and research institutions.

# 3. HYDROGEN STORAGE TECHNOLOGIES

Hydrogen storage systems are essential for the progression of a sustainable hydrogen economy (Muthukumar et al., 2023). The principal challenge is to deliver efficient, secure, and cost-effective hydrogen storage systems applicable across various sectors, including transportation and energy storage. Hydrogen can be held in gaseous, liquid, or solid states, each possessing unique attributes and challenges (Tashie-Lewis & Nnabuife, 2021). This chapter examines the principal hydrogen storage methods, current developments, and their potential for extensive energy uses.

## 3.1 Storage of Gaseous Hydrogen

Gaseous hydrogen storage is a well-established technology owing to the simplicity of its storage design. Hydrogen gas is generally stored at high pressure in specialized tanks constructed from materials capable of enduring extreme pressures (Tarhan & Çil, 2021). The standard pressures for compressed hydrogen storage vary between 350 and 700 bar (Hua et al., 2011).

### 3.1.1 Advantages of Gaseous Hydrogen Storage

A principal advantage of storing hydrogen in its gaseous state is its facilitation of straightforward and immediate utilization in fuel cells without necessitating additional conversion. This is especially advantageous in automotive applications, where fuel cell electric cars (FCEVs) employ compressed hydrogen gas for electricity generation ("Fuel Cell Electric Vehicles and Hydrogen Infrastructure," 2012). Furthermore, the materials necessary for the production of high-pressure hydrogen tanks, including carbon fiber composites, are comparatively lightweight and durable ("Cost Analysis of Hydrogen Gas Storage Tanks," 2024).

### 3.1.2 Challenges in Gaseous Hydrogen Storage

Despite its benefits, gaseous hydrogen storage presents several challenges:

- **Energy costs:** Compressing hydrogen to elevated pressures requires substantial energy, which can represent 10-15% on the energy content of the stored hydrogen (Ball & Weeda, 2015 ; "Hydrogen Storage Technologies for Future Energy Systems," 2024).
- **Storage safety:** High-pressure tanks raise safety issues regarding potential leaks or ruptures. Although advancements in tank materials have alleviated some risks, they remain a significant concern for the widespread adoption of hydrogen (Muradov & Veziroğlu, 2008).
- **Low volumetric energy density:** Compressed gaseous hydrogen demonstrates a poor energy density relative to liquid fuels or solid-state hydrogen storage, limiting the practicality of storing substantial amounts in compact volumes ("Hydrogen Storage Methods: Review and Current Status," 2024).

### 3.1.3 Advances in Gaseous Hydrogen Storage

Recent studies have concentrated on enhancing the safety and efficiency of compressed hydrogen storage devices. The advancement of sophisticated materials, including fiber-reinforced polymers ("Glass Capillary Hybridized CFRP for Hydrogen Storage," 2024) (FRPs) and metal composites, has markedly improved the strength and longevity of hydrogen storage tanks (Okonkwo et al., 2023). Investigations into hydrogen adsorbents, comprising metal-organic frameworks (MOFs), have facilitated hydrogen storage at reduced pressures while maintaining capacity (Li & Xu, 2013 ; Suh et al., 2012).

## 3.2 Storage of Liquid Hydrogen

Liquid hydrogen storage entails chilling hydrogen gas to cryogenic temperatures (−253°C) ("Cryogenics and Liquid Hydrogen Storage," 2024) until it liquefies. This approach provides significantly greater volumetric energy density compared to gaseous storage, making it especially advantageous for large-scale applications and long-distance transportation.

### 3.2.1 Advantages of Liquid Hydrogen Storage

- **High volumetric density:** Liquid hydrogen possesses significantly greater energy density per unit volume compared to compressed hydrogen gas (Ustolin, Campari, & Taccani, 2022). This renders it a feasible choice for applications necessitating substantial quantities of hydrogen, such as aviation or shipping.
- **Transportation versatility:** Liquid hydrogen can be efficiently transported in insulated cryogenic tanks, rendering it appropriate for industries necessitating long-distance hydrogen conveyance (Stetson, McWhorter, & Ahn, 2016).

### 3.2.2 Challenges in Liquid Hydrogen Storage

The primary obstacles of liquid hydrogen storage arise from the severe conditions necessary for hydrogen liquefaction:

- **Cryogenic temperatures:** To sustain hydrogen in its liquid form, temperatures must be maintained below −253°C, necessitating intricate and expensive insulation systems ("Hydrogen-Based Energy Systems," 2024).
- **Boil-off losses:** Prolonged storage of liquid hydrogen can lead to evaporation caused by heat transfer, resulting in boil-off losses that reduce the overall efficiency of liquid hydrogen storage (Naquash, Agarwal, & Lee, 2024).
- **Significant energy consumption:** Liquefying hydrogen is an energy-intensive process, using up to 30% of the hydrogen's energy content (Ghafri et al., 2022). This diminishes the overall efficacy of the hydrogen storage system, especially for applications necessitating frequent storage and retrieval (Malki et al., 2024).

### 3.2.3 Advances in Liquid Hydrogen Storage

Research has concentrated on enhancing insulation materials and technologies to mitigate boil-off losses, hence addressing the constraints of liquid hydrogen storage. Double-walled cryogenic tanks including vacuum insulation represent one of the most promising methods for minimizing heat transfer and maintaining liquid hydrogen (Yin, Yang, & Ju, 2024). Moreover, There is growing interest in the amalgamation of liquid hydrogen storage with renewable energy systems, whereby excess energy from renewable sources can be employed for hydrogen liquefaction (Aziz, 2021 ; Wang, Kowal, Leuthold, & Sauer, 2012).

### 3.3 Solid-State Hydrogen Storage

Solid-state hydrogen storage offers a unique answer to the challenges posed by gaseous and liquid storage techniques. This technique entails the retention of hydrogen within solid substances by either chemical bonding or physical adsorption. Conventional solid-state storage materials include metal hydrides, complex hydrides, and porous materials like metal-organic frameworks (MOFs) and covalent organic frameworks (COFs) as seen in **figure 2** (Li & Xu, 2013 ; Suh et al., 2012 ; Han et al., 2008).

*Figure 2. Gravimetric vs volumetric density diagram for several hydrogen storage systems including the graphene-based ones.*

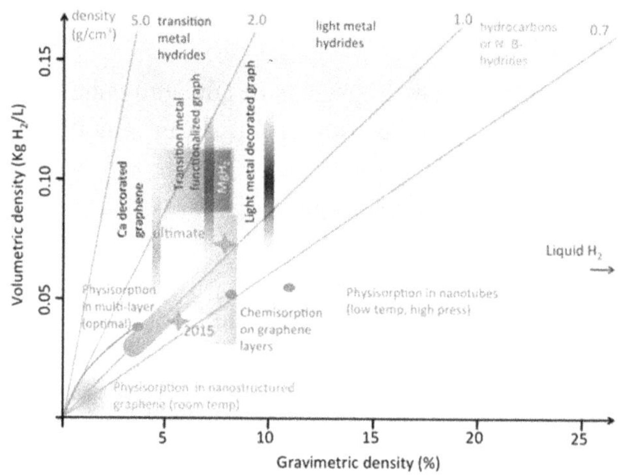


The orange line represents the optimal relationship for physisorption in nanotubes (the dots correspond to different sizes). The line tends to the value of liquid hydrogen for large sized nanotubes. In general, nanostructured physisorption based graphitic systems occupy the area below this line. The oblique shaded strip represents the optimal physisorption within graphene multilayers with spacing nearly double than in graphite (and density nearly one half). Different storage densities in this case correspond to different pressure and temperatures. The vertical dark red strips represent adsorption in decorated or functionalized graphene. These systems have been mostly studied at the level of a single layer, for this reason only the gravimetric density is well defined, while the volumetric density range has been roughly estimated considering variable inter-layer spacing 2-4 times that of graphite. The same criterion has been used to estimate the volumetric density for chemisorption in multilayers (blue rectangle); for this system the gravimetric density has a sharp right edge, corresponding to the maximum loading with 1:1 stoichiometry of C and H (~8%). The storage properties of systems based on material different from graphene (different metal hydrides (including $MgH_2$), hydrocarbons, N- and B- hydrides) are also reported as shaded areas in red, violet and green. The DOE targets (for 2015 and ultimate) are indicated with green stars. The constant density lines are in grey.

### 3.3.1 Metal Hydrides

Metal hydrides are intensively studied materials for solid-state hydrogen storage. In these systems, hydrogen is chemically absorbed by metals, culminating in the creation of a metal-hydrogen compound. Metal hydrides have numerous benefits:

- **High volumetric density:** Metal hydrides has the capability to store hydrogen at a density superior to that of both gaseous and liquid states, rendering them a compelling choice for stationary storage systems (Drawer, Lange, & Kaltschmitt, 2024; Klopčič et al., 2023).
- **Moderate working pressures and temperatures:** Metal hydrides generally store hydrogen at moderate pressures (below 10 bar) and temperatures (between 100°C and 300°C), which are less extreme than those necessary for gaseous or liquid storage ("Application of Hydrides in Hydrogen Storage and Compression," 2024 ; "Complex Metal Hydrides for Hydrogen, Thermal and Electrochemical Energy Storage," 2024).

Nonetheless, metal hydrides possess certain restrictions. The mass of the metal can diminish the total gravimetric energy density ("Thermodynamics of Metal Hydrides," 2024), and certain hydride materials exhibit sluggish absorption and desorption kinetics, which may impede their efficacy in dynamic applications such as fuel cell vehicles.

### 3.3.2 Complex Hydrides

Complex hydrides, such as alanates and borohydrides ("Complex Hydrides for Hydrogen Storage," 2024), have emerged as substitutes for metal hydrides in solid-state hydrogen storage. These materials provide elevated gravimetric hydrogen densities; Nonetheless, their practical implementation is constrained by elevated desorption temperatures and sluggish kinetics (Abd.Khalim Khafidz, Yaakob, Lim, & Timmiati, 2016). Recent studies have concentrated on enhancing the thermodynamic properties of these materials via catalysts and nanostructuring methods (Kim, 2018 ; Zhang et al., 2020 ; Lin et al., 2022).

### 3.3.3 Porous Materials

Porous substances, including Metal-Organic Frameworks (MOFs) and Covalent Organic Frameworks (COFs), have garnered considerable interest for hydrogen storage because to their large surface area and customizable pore structures. Hydrogen molecules are attracted to the surfaces of porous materials and are adsorbed without any chemical alteration. Physisorption is a low-energy phenomenon, rendering it especially appealing for reversible hydrogen storage. This mechanism, however, frequently necessitates freezing temperatures to optimize adsorption efficiency, particularly in materials such as MOFs and COFs.

Challenges:

porous materials frequently necessitate low temperatures to attain substantial hydrogen adsorption, hence constraining their applicability under ambient conditions (Thomas, 2007).

### 3.3.4 Advances in Solid-State Storage

Advancements in solid-state hydrogen storage have focused on improving the kinetics and thermodynamics of hydrogen absorption and desorption. Nanostructuring metal hydrides has been shown to reduce the diffusion channel for hydrogen atoms (Luo et al., 2020), hence enhancing both absorption and desorption rates. Doping complicated hydrides with catalysts like titanium has resulted in enhanced hydrogen release at reduced temperatures (Rusman & Dahari, 2016).

## 3.4 Comparative Analysis of Hydrogen Storage Technologies

The selection of hydrogen storage technique is contingent upon various aspects, including application, efficiency, cost, and safety considerations.

*Table 1. presents a comparative review of hydrogen storage systems in gaseous, liquid, and solid states.*

| Storage Type | Energy Density (volumetric) | Energy Density (gravimetric) | Cost | Operating Conditions | Safety |
|---|---|---|---|---|---|
| **Gaseous Storage** | Low | High | Moderate | High pressure (350-700 bar) | Moderate to High |
| **Liquid Storage** | High | High | High | Cryogenic temperatures (-253°C) | Low to Moderate |
| **Solid-State Storage** | Moderate to High | Moderate to Low | High | Moderate temperature and pressure | High |

## 4. HYDROGEN STORAGE IN POROUS MEDIA

Porous materials have emerged as feasible alternatives for hydrogen storage due to their substantial surface areas and distinctive chemical characteristics, facilitating the adsorption of hydrogen molecules. These materials possess considerable promise to mitigate the problems related to the reduced volumetric energy density of hydrogen gas. This chapter examines various sorts of porous media used for hydrogen storage, including covalent-organic frameworks (COFs), metal-organic frameworks (MOFs), zeolites, and activated carbons. This chapter analyzes current advancements in the field and investigates the mechanics of hydrogen adsorption and desorption.

## 4.1 Metal-Organic Frameworks (MOFs)

Metal-organic frameworks (MOFs) have attracted significant attention in hydrogen storage research due to their modular and customizable structures. Metal-Organic Frameworks (MOFs) are crystalline substances composed of metal ions or clusters linked by organic connectors (Langmi et al., 2014). Their extensive surface area and adjustable pore dimensions render them optimal candidates for hydrogen adsorption (Zhang et al., 2020).

### 4.1.1 Advantages of MOFs for Hydrogen Storage

- **Significant Surface Area:** Metal-organic frameworks (MOFs) exhibit some of the most extensive surface areas of any substance, hence augmenting their hydrogen storage capacity (Shet, Shanmuga Priya, Sudhakar, & Tahir, 2021).
- **Tunability:** The pore dimensions, functional characteristics, and metallic composition of MOFs can be adjusted to enhance hydrogen adsorption capabilities (Langmi et al., 2014).
- **Lightweight:** Metal-organic frameworks (MOFs) possess a lower weight in comparison to conventional storage materials, rendering them appropriate for mobile applications such fuel cell vehicles (FCVs) (Pramuanjaroenkij & Kakaç, 2023).

### 4.1.2 Limitations of MOFs for Hydrogen Storage

Notwithstanding their benefits, MOFs encounter numerous challenges:

- **Low Operating Temperatures**: Numerous MOFs necessitate freezing settings (about 77 K) to attain substantial hydrogen adsorption, constraining their practical applications (Thomas, 2007 ; Bobbitt, Chen, & Snurr, 2016).
- **Hydrogen Binding Energy**: The weak van der Waals interactions between hydrogen molecules and metal-organic framework (MOF) surfaces can lead to comparatively low hydrogen binding energy, hence influencing the storage capacity at ambient temperatures (Meduri & Nandanavanam, 2023).
- **Stability**: The stability of metal-organic frameworks (MOFs) under diverse environmental circumstances, such as humidity and temperature fluctuations, is a significant problem, as degradation of the framework can impair performance (Bao et al., 2023 ; "Current Research Trends and Perspectives on Materials-Based Hydrogen Storage Solutions," 2024).

## 4.2 Covalent-Organic Frameworks (COFs)

Covalent-organic frameworks (COFs) represent a distinct category of porous materials engineered for hydrogen storage. COFs are composed of light components, including carbon, hydrogen, nitrogen, and boron, interconnected by robust covalent bonds to create a porous crystalline framework (Tylianakis, Klontzas, & Froudakis, 2011). Their low density and high porosity make them attractive choices for hydrogen storage applications.

### 4.2.1 Mechanisms of Hydrogen Adsorption in COFs

Covalent organic frameworks (COFs) can adsorb hydrogen molecules by physisorption (Sinha & Pakhira, 2022), wherein hydrogen is tenuously attached to the framework's surface. The adsorption capacity of COFs can be enhanced by increasing the surface area and modifying the pore size, and incorporating functional groups into the framework.

### 4.2.2 Recent Advances in COF-based Hydrogen Storage

Recent research has concentrated on enhancing COF structures for hydrogen storage through:

- **Pore Size Regulation**: Accurate regulation of pore size has demonstrated an improvement in hydrogen adsorption at elevated pressures (Gogotsi et al., 2009).
- **Functionalization:** The incorporation of functional groups into the COF structure might augment the interaction between hydrogen molecules and the structure (Zou & Zhou, 2017), hence improving adsorption capacity. Although COFs exhibit significant potential for hydrogen storage, their efficacy under practical conditions continues to be a subject of ongoing investigation ("Current Research Trends and Perspectives on Solid-State Nanomaterials in Hydrogen Storage," 2021).

## 4.3 Zeolites

Zeolites are microporous aluminosilicate minerals utilized extensively in many industrial applications, such as catalysis and gas separation (Pérez-Botella, Palomino, Valencia, & Rey, 2019). Their crystalline architecture offers an extensive surface area and consistent pore dimensions, rendering them appropriate for hydrogen storage (Geng et al., 2015).

### 4.3.1 Hydrogen Storage in Zeolites

Zeolites adsorb hydrogen by physisorption ("Insights on the Molecular Mechanisms of Hydrogen Adsorption in Zeolites," 2014), akin to MOFs and COFs. Their natural prevalence, affordability, and stability under many situations render them a viable choice for hydrogen storage (Li, Li, & Yu, 2017). Nonetheless, their comparatively reduced surface area in relation to MOFs and COFs constrains their hydrogen storage capacity (Tylianakis, Klontzas, & Froudakis, 2011).

### 4.3.2 Enhancements in Zeolite Performance

Recent study has concentrated on augmenting the hydrogen storage capacity of zeolites through:

- Ion exchange: wherein the substitution of various cations within the zeolite framework can improve hydrogen adsorption (Prasanth et al., 2008).
- Nanocomposites: The integration of zeolites with materials like carbon nanostructures or metal hydrides can enhance their hydrogen storage efficacy.

## 4.4 Activated Carbons

Activated carbons, characterized by their extensive surface area and porous architecture, are extensively researched for hydrogen storage applications ("High Hydrogen Storage Capacity of Porous Carbons," 2008). These materials can be obtained from many carbonaceous sources and treated to form highly porous structures that can adsorb substantial quantities of hydrogen (Xia, Hu, & Jiang, 2014).

### 4.4.1 Properties of Activated Carbons

Activated carbons are characterized by their:

- **Significant Surface Area**: Activated carbon might possess a surface area of up to 3000 $m^2/g$, rendering it very efficient for gas adsorption (Rouquerol et al., 2013).
- **Affordability and Accessibility**: Activated carbon is cost-effective (Sevilla & Mokaya, 2014) and may be derived from plentiful materials such as coconut shells, coal, or wood (Zhao, Luo, Wang, & Fan, 2017).

### 4.4.2 Limitations of Activated Carbons

Activated carbons possess significant potential for hydrogen storage; yet, their efficacy is frequently constrained by:

- **Temperature Sensitivity:** Similar to other porous materials, activated carbons necessitate low temperatures to optimize hydrogen adsorption.
- **Pore Structure Concerns:** The hydrogen storage capacity of activated carbons is significantly influenced by their pore dimensions and distribution (Sethia & Sayari, 2016). Attaining a uniform and ideal pore structure during manufacture is difficult, frequently resulting in diminished storage efficiency.

## 4.5 Advances in Porous Media for Hydrogen Storage

The domain of hydrogen storage within porous media has experienced significant progress in recent years. Significant advancements encompass:

- **Hybrid Materials:** The amalgamation of several porous materials, including MOFs, COFs, and zeolites, can provide hybrid structures that exhibit improved hydrogen storage capacities.
- **Nanostructured Porous Materials:** Progress in nanotechnology has facilitated the creation of materials with meticulously regulated pore architectures at the nanoscale, enhancing hydrogen adsorption and desorption kinetics.
- **Computational Modeling:** Employing computational techniques to simulate hydrogen adsorption in porous materials has expedited the identification of novel materials with enhanced properties (Yang et al., 2024).

These improvements highlight the capacity of porous media to transform hydrogen storage methods, tackling the critical issues of volumetric energy density and operating efficiency.

## 4.6 Challenges and Future Directions

Regarding the advancements achieved in porous media for hydrogen storage, numerous problems persist:

- **Operating Conditions:** The majority of porous materials necessitate freezing temperatures to attain substantial hydrogen storage, hence constraining their practical applications (Yu et al., 2017).
- **Material Costs:** The expense associated The synthesis of particular porous materials, such as MOFs and COFs, may be prohibitively expensive for large-scale applications (Zhu & Liu, 2019).
- **Prolonged Stability:** Guaranteeing the enduring stability and resilience of porous materials under diverse environmental circumstances is a primary concern (Heinemann et al., 2021).

Future research is anticipated to concentrate on creating materials that function well at ambient temperatures (Aziz et al., 2024) and pressures, while also minimizing costs and enhancing the scalability of synthesis methods. Progress in material science, especially in nanotechnology and computer modeling, is expected to be pivotal in addressing these difficulties.

Porous media present promising opportunities for the enhancement of hydrogen storage methods. Metal-organic frameworks (MOFs), covalent organic frameworks (COFs), zeolites, and activated carbons possess distinct characteristics that render them viable options for hydrogen adsorption. As research advances in this domain, the improvement of these materials is anticipated to be important in facilitating hydrogen's emergence as a prevalent energy carrier, especially for applications necessitating compact and effective storage solutions.

## 5. INFRASTRUCTURE FOR HYDROGEN TRANSPORTATION AND DISTRIBUTION

### 5.1 Hydrogen Refueling Infrastructure

As hydrogen energy systems expand, the necessity for a comprehensive refueling infrastructure becomes imperative **figure 3** ("Hydrogen Energy Systems: Technologies, Trends, and Future Prospects," 2024). Hydrogen refueling stations, crucial for hydrogen-powered automobiles, are currently being established in multiple places ("Automotive Hydrogen Fuelling Stations," 2015), especially in nations such as Japan and Germany (Vijayakumar, Fulton, Shams, & Sperling, 2022). These stations must be interconnected to guarantee extensive accessibility and facilitate the use of hydrogen fuel. Both corporate and governmental initiatives are facilitating the spread of hydrogen refueling networks to address this difficulty.

*Figure 3. Hydrogen infrastructure (Pande, 2024).*

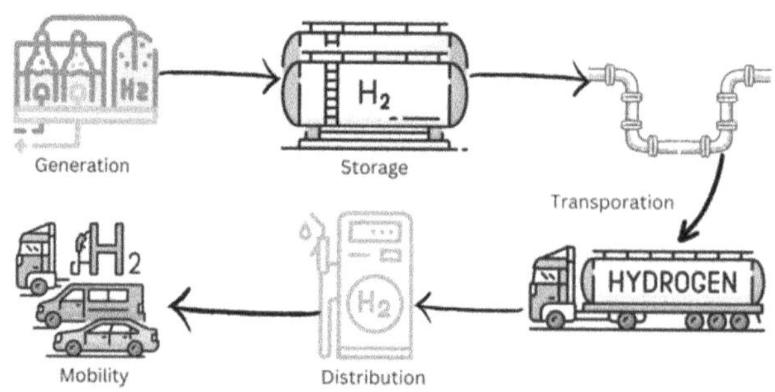

## 5.2 Transportation Methods: Pipelines, Tanker Trucks, and Ships

The transportation of hydrogen poses distinct obstacles in comparison to conventional fuels. Pipelines, while efficient, necessitate particular materials to accommodate hydrogen's diminutive molecule size (Raj et al., 2024), perhaps resulting in leakage. Tanker trucks and vessels encounter technical challenges, chiefly with the compression and liquefaction of hydrogen for effective transportation ("Transportation of Hydrogen: Hydrogen Usage," 2024). With the increasing need for hydrogen, various solutions are being devised to address these issues, such as enhanced storage containers and hybrid transportation systems ("Adaptive Multi-Temperature Control for Transport and Storage Containers," 2023).

## 5.3 Challenges in Hydrogen Distribution

Hydrogen Leakage, Compression Losses, and Safety Issues: The diminutive molecular size of hydrogen predisposes it to leakage, resulting in both losses and heightened safety issues. Compression losses exacerbate the transportation process, necessitating the compression of hydrogen to elevated pressures for effective storage and transmission (Tashie-Lewis & Nnabuife, 2021). Guaranteeing the safety and efficacy of hydrogen distribution represents a considerable obstacle in the advancement of the hydrogen economy.

Cost and Energy Considerations in Distribution: The energy necessary for the compression, storage, and transportation of hydrogen is considerable, frequently negating the environmental advantages of utilizing hydrogen as a sustainable energy source (Abdalla et al., 2018). Research is currently underway to create more energy-efficient transportation technologies, including cryogenic hydrogen storage and solid-state hydrogen carriers, which may substantially lower distribution expenses.

## 5.4 Developing a Global Hydrogen Infrastructure

To facilitate a global hydrogen economy, nations must collaborate to establish transnational hydrogen networks. Europe has been at the forefront of developing these networks, with the European Hydrogen Backbone initiative outlining ideas for a transcontinental hydrogen pipeline ("Strategies for Regional Deployment of Hydrogen Infrastructure," 2024). These networks will facilitate the importation and exportation of hydrogen, allowing regions rich in renewable energy to produce hydrogen and deliver it to energy-scarce areas. Policy incentives are crucial for facilitating the development of these infrastructures, so enabling hydrogen to emerge as an internationally traded energy commodity ("The Future European Hydrogen Market," 2024).

**Case Studies of Successful Hydrogen Infrastructure Models:** Countries like as Japan, Germany, and the Netherlands are now executing hydrogen infrastructure initiatives (Gielen, Taibi, & Miranda, 2019). These case studies offer significant perspectives on the obstacles and prospects associated with growth of hydrogen distribution networks. The insights gained from these efforts will be crucial for the global expansion of hydrogen infrastructure.

## 5.5 Policy Incentives for Hydrogen Infrastructure Development

Governments are progressively acknowledging the importance of hydrogen in achieving net-zero emissions objectives ("The Prospects of Hydrogen in Achieving Net Zero Emissions by 2050," 2023). Policy incentives, such as subsidies, tax incentives, and research money, are being implemented to advance hydrogen infrastructure development (Dolci et al., 2019). The viability of the hydrogen economy will significantly rely on continuous backing from policymakers to lower expenses, promote innovation, and guarantee the safety of hydrogen storage and transportation infrastructure.

## CONCLUSION

Hydrogen energy presents considerable potential in meeting the pressing global demand for greener, more sustainable energy alternatives. Its distinctive attributes—abundance, carbon neutrality, and high efficiency—establish it as a vital contributor to the shift towards a carbon-neutral economy. This book examines multiple facets of hydrogen technology, including production techniques, storage, and distribution issues, emphasizing hydrogen's significant potential to transform transportation, industrial processes, and energy storage sectors.

The chapters comprehensively analyze the issues associated with hydrogen generation, specifically focusing on efficiency, cost, and environmental sustainability. Conventional techniques like Steam Methane Reforming (SMR) prevail in contemporary hydrogen generation, however they generate significant CO2 emissions. The transition to sustainable manufacturing techniques for instance, electrolysis, energized by renewable energy, is essential for securing a green hydrogen future. Advancements in technologies such as photoelectrochemical and thermochemical water splitting are crucial advancements in expanding the frontiers of hydrogen production.

Hydrogen storage technologies have advanced considerably, with gaseous, liquid, and solid-state storage solutions developing to suit the distinct characteristics of hydrogen. Although gaseous hydrogen storage offers ease of use, it has issues related to compression and safety. Liquid hydrogen storage provides superior energy densities; but, it necessitates cryogenic settings, hence increasing complexity and expense. Solid-state storage, especially the application of metal hydrides with porous materials such as MOFs and COFs, demonstrates significant potential for delivering efficient and compact storage solutions. Nonetheless, the advancement of porous materials is still nascent, with low operational temperatures and scalability posing significant hurdles.

Moreover, the infrastructure essential for extensive hydrogen adoption—comprising transportation pipes, refueling stations, and safety regulations—demands substantial investment and development. The low density of hydrogen, coupled with the necessity for compression and liquefaction, renders its transportation more expensive and intricate than that of conventional fuels. International cooperation and policy frameworks are established to foster the development of a global hydrogen infrastructure, with leading nations spearheading successful hydrogen economies.

In conclusion, whereas hydrogen offers exceptional prospects for decarbonizing critical sectors and supplying sustainable energy alternatives, considerable obstacles persist in production efficiency, storage technology, and infrastructure development. Ongoing innovation, investment, and collaboration across businesses and governments will be crucial for unlocking hydrogen's complete potential. The future of hydrogen is unequivocally promising, providing a route to long-term sustainability and energy security.

## REFERENCES


Abas, N., Kalair, A., & Khan, N. (2015). Review of fossil fuels and future energy technologies. *Futures*, *69*, 31–49. DOI: 10.1016/j.futures.2015.03.003

Abdalla, A. M., Hossain, S., Nisfindy, O. B., Azad, A. T., Dawood, M., & Azad, A. K. (2018). Hydrogen production, storage, transportation and key challenges with applications: A review. *Energy Conversion and Management*, *165*, 602–627. DOI: 10.1016/j.enconman.2018.03.088

Abdin, Z., Zafaranloo, A., Rafiee, A., Mérida, W., Lipiński, W., & Khalilpour, K. R. (2020). Hydrogen as an energy vector. *Renewable & Sustainable Energy Reviews*, *120*, 109620.

Abd.Khalim Khafidz, N. Z., Yaakob, Z., Lim, K. L., & Timmiati, S. N. (2016). The kinetics of lightweight solid-state hydrogen storage materials: A review. *International Journal of Hydrogen Energy*, *41*(30), 13131–13151. DOI: 10.1016/j.ijhydene.2016.05.169

Abe, J. O., Popoola, A. P. I., Ajenifuja, E., & Popoola, O. M. (2019). Hydrogen energy, economy and storage: Review and recommendation. *International Journal of Hydrogen Energy*, *44*(29), 15072–15086. DOI: 10.1016/j.ijhydene.2019.04.068

Agarwal, R. (2022). Transition to a hydrogen-based economy: Possibilities and challenges. *Sustainability*, *14*(23), 15975. Advance online publication. DOI: 10.3390/su142315975

Ahmed, K. W., Jang, M. J., Park, M. G., Chen, Z., & Fowler, M. (2022). Effect of components and operating conditions on the performance of PEM electrolyzers: A review. *Electrochem*, *3*(4), 581–612.

Al-Bassam, A. M., Conner, J. A., & Manousiouthakis, V. I. (2018). Natural-gas-derived hydrogen in the presence of carbon fuel taxes and concentrated solar power. *ACS Sustainable Chemistry & Engineering*, *6*(3), 3029–3038.

Al Ghafri, S. Z., Munro, S., Cardella, U., Funke, T., Notardonato, W., Trusler, J. M., & May, E. F. (2022). Hydrogen liquefaction: A review of the fundamental physics, engineering practice and future opportunities. *Energy & Environmental Science*, *15*(7), 2690–2731.

Alazemi, J., & Andrews, J. (2015). Automotive hydrogen fuelling stations: An international review. *Renewable & Sustainable Energy Reviews*, *48*, 483–499.

Arsad, A. Z., Hannan, M. A., Al-Shetwi, A. Q., Mansur, M., Muttaqi, K. M., Dong, Z. Y., & Blaabjerg, F. (2022). Hydrogen energy storage integrated hybrid renewable energy systems: A review analysis for future research directions. *International Journal of Hydrogen Energy*, *47*(39), 17285–17312.

Ashfaq, M. M., Tüzemen, G. B., & Noor, A. (2024). Exploiting agricultural biomass via thermochemical processes for sustainable hydrogen and bioenergy: A critical review. *International Journal of Hydrogen Energy*, *84*, 1068–1084.

Atilhan, S., Park, S., El-Halwagi, M. M., Atilhan, M., Moore, M., & Nielsen, R. B. (2021). Green hydrogen as an alternative fuel for the shipping industry. *Current Opinion in Chemical Engineering*, *31*, 100668. DOI: 10.1016/j.coche.2020.100668

Aziz, M. (2021). Liquid hydrogen: A review on liquefaction, storage, transportation, and safety. *Energies*, *14*(18), 5917. DOI: 10.3390/en14185917

Aziz, O., Labrousse, J., Belasfar, K., Essajai, R., El Kenz, A., Benyoussef, A., & Ez-Zahraouy, H. (2024). Reversible and high-capacity hydrogen storage on two-dimensional monolayer C2N-h2D expected by first-principles calculations. *International Journal of Hydrogen Energy*, *50*, 586–596.

Ball, M., & Weeda, M. (2015). The hydrogen economy – Vision or reality? *International Journal of Hydrogen Energy*, *40*(25), 7903–7919. DOI: 10.1016/j.ijhydene.2015.04.032

Ball, M., & Wietschel, M. (2009). The future of hydrogen – Opportunities and challenges. *International Journal of Hydrogen Energy*, *34*(2), 615–627. DOI: 10.1016/j.ijhydene.2008.11.014

Bao, W., Yu, J., Chen, F., Du, H., Zhang, W., Yan, S., & Zhu, B. (2023). Controllability construction and structural regulation of metal-organic frameworks for hydrogen storage at ambient condition: A review. *International Journal of Hydrogen Energy*, *48*(92), 36010–36034.

Beck, R., & Prewitz, M. (2024). Experimental investigation of tensile properties of glass capillary hybridized carbon fiber reinforced plastic (GCRP) for structurally integrated hydrogen storage. *International Journal of Hydrogen Energy*, *62*, 321–330.

Bobbitt, N. S., Chen, J., & Snurr, R. Q. (2016). High-throughput screening of metal–organic frameworks for hydrogen storage at cryogenic temperature. *The Journal of Physical Chemistry. C, Nanomaterials and Interfaces*, *120*(48), 27328–27341. DOI: 10.1021/acs.jpcc.6b08729

Boretti, A. (2021). Concentrated solar energy with thermal energy storage for hydrogen production by three-step thermochemical water-splitting cycles. *Energy & Fuels*, *35*(13), 10832–10840.

Cheng, W. H., De La Calle, A., Atwater, H. A., Stechel, E. B., & Xiang, C. (2021). Hydrogen from sunlight and water: A side-by-side comparison between photoelectrochemical and solar thermochemical water-splitting. *ACS Energy Letters*, *6*(9), 3096–3113.

Chu, S., & Majumdar, A. (2012). Opportunities and challenges for a sustainable energy future. nature, 488(7411), 294-303.

Cloete, S., Arnaiz del Pozo, C., & Jiménez Álvaro, Á. (2022). System-friendly process design: Optimizing blue hydrogen production for future energy systems. *Energy*, *259*, 124954. DOI: 10.1016/j.energy.2022.124954

Dauletbay, A. (2024). *Transportation of Hydrogen*. Hydrogen Usage. In Hydrogen Technologies-Advances, Insights, and Applications. IntechOpen.

Deeg, K. S., Gutiérrez-Sevillano, J. J., Bueno-Pérez, R., Parra, J. B., Ania, C. O., Doblaré, M., & Calero, S. (2013). Insights on the molecular mechanisms of hydrogen adsorption in zeolites. *The Journal of Physical Chemistry. C, Nanomaterials and Interfaces*, *117*(27), 14374–14380.

Dillman, K. J., & Heinonen, J. (2022). A 'just' hydrogen economy: A normative energy justice assessment of the hydrogen economy. *Renewable & Sustainable Energy Reviews*, *167*, 112648. DOI: 10.1016/j.rser.2022.112648

Dolci, F., Thomas, D., Hilliard, S., Guerra, C. F., Hancke, R., Ito, H., & Lucchese, P. (2019). Incentives and legal barriers for power-to-hydrogen pathways: An international snapshot. *International Journal of Hydrogen Energy*, *44*(23), 11394–11401.

Dornheim, M. (2011). Thermodynamics of metal hydrides: tailoring reaction enthalpies of hydrogen storage materials. In *Thermodynamics-Interaction Studies-Solids*. Liquids and Gases. IntechOpen.

Drawer, C., Lange, J., & Kaltschmitt, M. (2024). Metal hydrides for hydrogen storage – Identification and evaluation of stationary and transportation applications. *Journal of Energy Storage*, *77*, 109988. DOI: 10.1016/j.est.2023.109988

Eberle, U., Müller, B., & Von Helmolt, R. (2012). Fuel cell electric vehicles and hydrogen infrastructure: Status 2012. *Energy & Environmental Science*, *5*(10), 8780–8798.

Evro, S., Oni, B. A., & Tomomewo, O. S. (2024). Carbon neutrality and hydrogen energy systems. *International Journal of Hydrogen Energy*, *78*, 1449–1467. DOI: 10.1016/j.ijhydene.2024.06.407

Faheem, H. H., Tanveer, H. U., Abbas, S. Z., & Maqbool, F. (2021). Comparative study of conventional steam-methane-reforming (SMR) and auto-thermal-reforming (ATR) with their hybrid sorption enhanced (SE-SMR & SE-ATR) and environmentally benign process models for hydrogen production. *Fuel*, *297*, 120769. DOI: 10.1016/j.fuel.2021.120769

Farhana, K., Mahamude, A. S. F., & Kadirgama, K. (2024). Comparing hydrogen fuel cost of production from various sources – A competitive analysis. *Energy Conversion and Management*, *302*, 118088. DOI: 10.1016/j.enconman.2024.118088

Fayaz, H., Saidur, R., Razali, N., Anuar, F. S., Saleman, A. R., & Islam, M. R. (2012). An overview of hydrogen as a vehicle fuel. *Renewable & Sustainable Energy Reviews*, *16*(8), 5511–5528. DOI: 10.1016/j.rser.2012.06.012

Geng, Z., Zhang, C., Wang, D., Zhou, X., & Cai, M. (2015). Pore size effects of nanoporous carbons with ultra-high surface area on high-pressure hydrogen storage. *Journal of Energy Chemistry*, *24*(1), 1–8. DOI: 10.1016/S2095-4956(15)60277-7

Gielen, D., Taibi, E., & Miranda, R. (2019, September). Hydrogen: A reviewable energy perspective. Report prepared for the 2nd Hydrogen Energy Ministerial Meeting in Tokyo, Japan. Retrieved September 29, 2024, from https://www.h2knowledgecentre.com/content/policypaper1305

Gillingham, K., & Stock, J. H. (2018). The cost of reducing greenhouse gas emissions. *The Journal of Economic Perspectives*, *32*(4), 53–72. DOI: 10.1257/jep.32.4.53

Gogotsi, Y., Portet, C., Osswald, S., Simmons, J. M., Yildirim, T., Laudisio, G., & Fischer, J. E. (2009). Importance of pore size in high-pressure hydrogen storage by porous carbons. *International Journal of Hydrogen Energy*, *34*(15), 6314–6319.

Gómez, J., & Castro, R. (2024). Green hydrogen energy systems: A review on their contribution to a renewable energy system. *Energies*, *17*(13), 3110.

Han, S. S., Furukawa, H., Yaghi, O. M., & Goddard, W. A. I. (2008). Covalent organic frameworks as exceptional hydrogen storage materials. *Journal of the American Chemical Society*, *130*(35), 11580–11581. DOI: 10.1021/ja803247y

Heinemann, N., Alcalde, J., Miocic, J. M., Hangx, S. J., Kallmeyer, J., Ostertag-Henning, C., & Rudloff, A. (2021). Enabling large-scale hydrogen storage in porous media–the scientific challenges. *Energy & Environmental Science*, *14*(2), 853–864.

Hua, T. Q., Ahluwalia, R. K., Peng, J. K., Kromer, M., Lasher, S., McKenney, K., & Sinha, J. (2011). Technical assessment of compressed hydrogen storage tank systems for automotive applications. *International Journal of Hydrogen Energy*, *36*(4), 3037–3049.

Hwang, J., Maharjan, K., & Cho, H. (2023). A review of hydrogen utilization in power generation and transportation sectors: Achievements and future challenges. *International Journal of Hydrogen Energy*, *48*(74), 28629–28648. DOI: 10.1016/j.ijhydene.2023.04.024

Ikuerowo, T., Bade, S. O., Akinmoladun, A., & Oni, B. A. (2024). The integration of wind and solar power to water electrolyzer for green hydrogen production. *International Journal of Hydrogen Energy*, *76*, 75–96. DOI: 10.1016/j.ijhydene.2024.02.139

Ishaq, H., Dincer, I., & Crawford, C. (2022). A review on hydrogen production and utilization: Challenges and opportunities. *International Journal of Hydrogen Energy*, *47*(62), 26238–26264.

Islam, A., Islam, T., Mahmud, H., Raihan, O., Islam, M. S., Marwani, H. M., & Awual, M. R. (2024). Accelerating the green hydrogen revolution: A comprehensive analysis of technological advancements and policy interventions. *International Journal of Hydrogen Energy*, *67*, 458–486.

Iyke, B. N. (2024). Climate change, energy security risk, and clean energy investment. *Energy Economics*, *129*, 107225. DOI: 10.1016/j.eneco.2023.107225

Jaunatre, M. (2021). *Renewable hydrogen: Renewable energy and renewable hydrogen APAC markets policies analysis*. Springer Nature.

Kim, K. C. (2018). A review on design strategies for metal hydrides with enhanced reaction thermodynamics for hydrogen storage applications. *International Journal of Energy Research*, *42*(4), 1455–1468. DOI: 10.1002/er.3919

Klopčič, N., Grimmer, I., Winkler, F., Sartory, M., & Trattner, A. (2023). A review on metal hydride materials for hydrogen storage. *Journal of Energy Storage*, *72*, 108456. DOI: 10.1016/j.est.2023.108456

Kovač, A., Paranos, M., & Marciuš, D. (2021). Hydrogen in energy transition: A review. *International Journal of Hydrogen Energy*, *46*(16), 10016–10035. DOI: 10.1016/j.ijhydene.2020.11.256

Kurtz, J., Sprik, S., & Bradley, T. H. (2019). Review of transportation hydrogen infrastructure performance and reliability. *International Journal of Hydrogen Energy*, *44*(23), 12010–12023. DOI: 10.1016/j.ijhydene.2019.03.027

Langmi, H. W., Ren, J., North, B., Mathe, M., & Bessarabov, D. (2014). Hydrogen storage in metal-organic frameworks: A review. *Electrochimica Acta*, *128*, 368–392. DOI: 10.1016/j.electacta.2013.10.190

Le, T. T., Sharma, P., Bora, B. J., Tran, V. D., Truong, T. H., Le, H. C., & Nguyen, P. Q. P. (2024). Fueling the future: A comprehensive review of hydrogen energy systems and their challenges. *International Journal of Hydrogen Energy*, *54*, 791–816.

Li, S.-L., & Xu, Q. (2013). Metal–organic frameworks as platforms for clean energy. *Energy & Environmental Science*, *6*(6), 1656–1683. DOI: 10.1039/C3EE40507A

Li, Y., Li, L., & Yu, J. (2017). Applications of zeolites in sustainable chemistry. *Chem*, *3*(6), 928–949. DOI: 10.1016/j.chempr.2017.10.009

Lin, H.-J., Lu, Y.-S., Zhang, L.-T., Liu, H.-Z., Edalati, K., & Révész, Á. (2022). Recent advances in metastable alloys for hydrogen storage: A review. *Rare Metals*, *41*(6), 1797–1817. DOI: 10.1007/s12598-021-01917-8

Luo, Y., Wang, Q., Li, J., Xu, F., Sun, L., Zou, Y., & Zhang, K. (2020). Enhanced hydrogen storage/sensing of metal hydrides by nanomodification. *Materials Today Nano*, *9*, 100071.

Maestre, V. M., Ortiz, A., & Ortiz, I. (2021). Challenges and prospects of renewable hydrogen-based strategies for full decarbonization of stationary power applications. *Renewable & Sustainable Energy Reviews*, *152*, 111628. DOI: 10.1016/j.rser.2021.111628

Maggio, G., Nicita, A., & Squadrito, G. (2019). How the hydrogen production from RES could change energy and fuel markets: A review of recent literature. *International Journal of Hydrogen Energy*, *44*(23), 11371–11384. DOI: 10.1016/j.ijhydene.2019.03.121

Malki, M. L., Chellal, H., Mao, S., Rasouli, V., & Mehana, M. (2024). A critical review of underground hydrogen storage: From fundamentals to applications, unveiling future frontiers in energy storage. *International Journal of Hydrogen Energy*, *79*, 1365–1394. DOI: 10.1016/j.ijhydene.2024.07.076

Marouani, I., Guesmi, T., Alshammari, B. M., Alqunun, K., Alzamil, A., Alturki, M., & Hadj Abdallah, H. (2023). Integration of renewable-energy-based green hydrogen into the energy future. *Processes (Basel, Switzerland)*, *11*(9), 2685.

McCay, M. H., & Shafiee, S. (2020). Hydrogen: An energy carrier. In Letcher, T. M. (Ed.), *Future energy* (3rd ed., pp. 475–493). Elsevier., DOI: 10.1016/B978-0-08-102886-5.00022-0

McDowall, W., & Eames, M. (2006). Forecasts, scenarios, visions, backcasts and roadmaps to the hydrogen economy: A review of the hydrogen futures literature. *Energy Policy*, *34*(11), 1236–1250. DOI: 10.1016/j.enpol.2005.12.006

Meduri, S., & Nandanavanam, J. (2023). Materials for hydrogen storage at room temperature – An overview. *Materials Today: Proceedings*, *72*, 1–8. DOI: 10.1016/j.matpr.2022.05.059

Molfetta, R. D. (2022). Strategies for regional deployment of hydrogen infrastructure (Master's thesis, Universitat Politècnica de Catalunya).

Møller, K. T., Sheppard, D., Ravnsbæk, D. B., Buckley, C. E., Akiba, E., Li, H. W., & Jensen, T. R. (2017). Complex metal hydrides for hydrogen, thermal and electrochemical energy storage. *Energies*, *10*(10), 1645.

Muradov, N. Z., & Veziroğlu, T. N. (2008). "Green" path from fossil-based to hydrogen economy: An overview of carbon-neutral technologies. *International Journal of Hydrogen Energy*, *33*(23), 6804–6839. DOI: 10.1016/j.ijhydene.2008.08.054

Muthukumar, P., Kumar, A., Afzal, M., Bhogilla, S., Sharma, P., Parida, A., & Jain, I. P. (2023). Review on large-scale hydrogen storage systems for better sustainability. *International Journal of Hydrogen Energy*, *48*(85), 33223–33259.

Naquash, A., Agarwal, N., & Lee, M. (2024). A review on liquid hydrogen storage: Current status, challenges and future directions. *Sustainability*, *16*(18), 8270. DOI: 10.3390/su16188270

Navas-Anguita, Z., García-Gusano, D., Dufour, J., & Iribarren, D. (2021). Revisiting the role of steam methane reforming with $CO_2$ capture and storage for long-term hydrogen production. *The Science of the Total Environment*, *771*, 145432. DOI: 10.1016/j.scitotenv.2021.145432

Nnabuife, S. G., Hamzat, A. K., Whidborne, J., Kuang, B., & Jenkins, K. W. (2024). Integration of renewable energy sources in tandem with electrolysis: A technology review for green hydrogen production. *International Journal of Hydrogen Energy*. Advance online publication. DOI: 10.1016/j.ijhydene.2024.06.342

Nnabuife, S. G., Oko, E., Kuang, B., Bello, A., Onwualu, A. P., Oyagha, S., & Whidborne, J. (2023). The prospects of hydrogen in achieving net zero emissions by 2050: A critical review. *Sustainable Chemistry for Climate Action*, *2*, 100024.

Noussan, M., Raimondi, P. P., Scita, R., & Hafner, M. (2021). The role of green and blue hydrogen in the energy transition—A technological and geopolitical perspective. *Sustainability*, *13*(1), 1. Advance online publication. DOI: 10.3390/su13010298

Ocko, I. B., & Hamburg, S. P. (2022). Climate consequences of hydrogen emissions. *Atmospheric Chemistry and Physics*, *22*(14), 9349–9368. DOI: 10.5194/acp-22-9349-2022

Okonkwo, P. C., Belgacem, I. B., Mansir, I. B., Aliyu, M., Emori, W., Uzoma, P. C., ... & Shakoor, R. A. (2023). A focused review of the hydrogen storage tank embrittlement mechanism process. international journal of hydrogen energy, 48(35), 12935-12948.

Olabi, A. G., Abdelghafar, A. A., Baroutaji, A., Sayed, E. T., Alami, A. H., Rezk, H., & Abdelkareem, M. A. (2021). Large-vscale hydrogen production and storage technologies: Current status and future directions. *International Journal of Hydrogen Energy*, *46*(45), 23498–23528.

Orimo, S. I., Nakamori, Y., Eliseo, J. R., Züttel, A., & Jensen, C. M. (2007). Complex hydrides for hydrogen storage. *Chemical Reviews*, *107*(10), 4111–4132.

Pande, N. (n.d.). Hydrogen infrastructure. Green H2 World. Retrieved September 29, 2024, from https://www.greenh2world.com/post/hydrogen-infrastructure

Pérez-Botella, E., Palomino, M., Valencia, S., & Rey, F. (2019). Zeolites and other adsorbents. In Kaneko, K., & Rodríguez-Reinoso, F. (Eds.), *Nanoporous materials for gas storage* (pp. 173–208). Springer., DOI: 10.1007/978-981-13-3504-4_7

Pingkuo, L., & Junqing, G. (2024). Comparative analysis on the development potential of green hydrogen industry in China, the United States and the European Union. *International Journal of Hydrogen Energy*, *84*, 700–717. DOI: 10.1016/j.ijhydene.2024.08.298

Pramuanjaroenkij, A., & Kakaç, S. (2023). The fuel cell electric vehicles: The highlight review. *International Journal of Hydrogen Energy*, *48*(25), 9401–9425. DOI: 10.1016/j.ijhydene.2022.11.103

Prasanth, K. P., Bhat, D. K., & Kamath, V. (2008). Adsorption of hydrogen in nickel and rhodium exchanged zeolite X. *International Journal of Hydrogen Energy*, *33*(2), 735–745. DOI: 10.1016/j.ijhydene.2007.10.047

Preuster, P., Alekseev, A., & Wasserscheid, P. (2017). Hydrogen storage technologies for future energy systems. *Annual Review of Chemical and Biomolecular Engineering*, *8*(1), 445–471.

Prosser, J. H.. (2024). Cost analysis of hydrogen production by high-temperature solid oxide electrolysis. *International Journal of Hydrogen Energy*, *49*, 207–227. DOI: 10.1016/j.ijhydene.2023.07.084

Qureshi, F., Yusuf, M., Ibrahim, H., Kamyab, H., Chelliapan, S., Pham, C. Q., & Vo, D. V. N. (2023). Contemporary avenues of the Hydrogen industry: Opportunities and challenges in the eco-friendly approach. *Environmental Research*, *229*, 115963.

Raj, A., Larsson, I. S., Ljung, A. L., Forslund, T., Andersson, R., Sundström, J., & Lundström, T. S. (2024). Evaluating hydrogen gas transport in pipelines: Current state of numerical and experimental methodologies. *International Journal of Hydrogen Energy*, *67*, 136–149.

Ren, J., Musyoka, N. M., Langmi, H. W., Mathe, M., & Liao, S. (2017). Current research trends and perspectives on materials-based hydrogen storage solutions: A critical review. *International Journal of Hydrogen Energy*, *42*(1), 289–311.

Renné, D. S. (2022). Progress, opportunities and challenges of achieving net-zero emissions and 100% renewables. *Solar Compass*, *1*, 100007. DOI: 10.1016/j.solcom.2022.100007

Rolo, I., Costa, V. A., & Brito, F. P. (2023). Hydrogen-based energy systems: Current technology development status, opportunities and challenges. *Energies*, *17*(1), 180.

Rouquerol, J., Rouquerol, F., Llewellyn, P., Maurin, G., & Sing, K. (2013). *Adsorption by powders and porous solids: Principles, methodology and applications*. Academic Press.

Rubinsin, N. J., Karim, N. A., Timmiati, S. N., Lim, K. L., Isahak, W. N. R. W., & Pudukudy, M. (2024). An overview of the enhanced biomass gasification for hydrogen production. *International Journal of Hydrogen Energy*, *49*, 1139–1164.

Rusman, N. A. A., & Dahari, M. (2016). A review on the current progress of metal hydrides material for solid-state hydrogen storage applications. *International Journal of Hydrogen Energy*, *41*(28), 12108–12126. DOI: 10.1016/j.ijhydene.2016.05.244

Sadeq, A. M., Homod, R. Z., Hussein, A. K., Togun, H., Mahmoodi, A., Isleem, H. F., & Moghaddam, A. H. (2024). Hydrogen energy systems: Technologies, trends, and future prospects. *The Science of the Total Environment*, •••, 173622.

Sarker, A. K., Azad, A. K., Rasul, M. G., & Doppalapudi, A. T. (2023). Prospect of green hydrogen generation from hybrid renewable energy sources: A review. *Energies*, *16*(3), 1556.

ScienceDirect. (2024a). Aboveground hydrogen storage – Assessment of the potential market relevance in a carbon-neutral European energy system. Retrieved September 27, 2024, from https://www.sciencedirect.com/science/article/pii/S0196890424002334

ScienceDirect. (2024b). Emerging technologies by hydrogen: A review. Retrieved September 27, 2024, from https://www.sciencedirect.com/science/article/pii/S0360319920317493

ScienceDirect. (2024c). Impact of hydrogen on the environment. Retrieved September 27, 2024, from https://www.sciencedirect.com/science/article/pii/S0360319911017319

ScienceDirect. (2024d). Hydrogen fuel cell heavy-duty trucks: Review of main research topics. Retrieved September 27, 2024, from https://www.sciencedirect.com/science/article/pii/S0360319922029068

Sethia, G., & Sayari, A. (2016). Activated carbon with optimum pore size distribution for hydrogen storage. *Carbon*, *99*, 289–294. DOI: 10.1016/j.carbon.2015.12.032

Sevilla, M., & Mokaya, R. (2014). Energy storage applications of activated carbons: Supercapacitors and hydrogen storage. *Energy & Environmental Science*, *7*(4), 1250–1280. DOI: 10.1039/C3EE43525C

Shadidi, B., Najafi, G., & Yusaf, T. (2021). A review of hydrogen as a fuel in internal combustion engines. *Energies*, *14*(19), 19. Advance online publication. DOI: 10.3390/en14196209

Shet, S. P., Shanmuga Priya, S., Sudhakar, K., & Tahir, M. (2021). A review on current trends in potential use of metal-organic framework for hydrogen storage. *International Journal of Hydrogen Energy*, *46*(21), 11782–11803. DOI: 10.1016/j.ijhydene.2021.01.020

Shin, H. K., & Ha, S. K. (2023). A review on the cost analysis of hydrogen gas storage tanks for fuel cell vehicles. *Energies*, *16*(13), 5233.

Singh, S., Jain, S., Tiwari, A. K., Nouni, M. R., Pandey, J. K., & Goel, S. (2015). Hydrogen: A sustainable fuel for future of the transport sector. *Renewable & Sustainable Energy Reviews*, *51*, 623–633.

Sinha, N., & Pakhira, S. (2022). $H_2$ physisorption on covalent organic framework linkers and metalated linkers: A strategy to enhance binding strength. *Molecular Systems Design & Engineering*, *7*(6), 577–591. DOI: 10.1039/D1ME00166C

Steinbach, S. A., & Bunk, N. (2024). The future European hydrogen market: Market design and policy recommendations to support market development and commodity trading. *International Journal of Hydrogen Energy*, *70*, 29–38.

Stenberg, V., Spallina, V., Mattisson, T., & Rydén, M. (2020). Techno-economic analysis of $H_2$ production processes using fluidized bed heat exchangers with steam reforming – Part 1: Oxygen carrier aided combustion. *International Journal of Hydrogen Energy*, *45*(11), 6059–6081. DOI: 10.1016/j.ijhydene.2019.10.202

Stetson, N. T., McWhorter, S., & Ahn, C. C. (2016). Introduction to hydrogen storage. In Gupta, R. B., Basile, A., & Veziroğlu, T. N. (Eds.), *Compendium of Hydrogen Energy* (Vol. 1, pp. 3–25). Woodhead Publishing., DOI: 10.1016/B978-1-78242-362-1.00001-8

Suh, M. P., Park, H. J., Prasad, T. K., & Lim, D.-W. (2012). Hydrogen storage in metal–organic frameworks. *Chemical Reviews*, *112*(2), 782–835. DOI: 10.1021/cr200274s

Swadi, M., Kadhim, D. J., Salem, M., Tuaimah, F. M., Majeed, A. S., & Alrubaie, A. J. (2024). Investigating and predicting the role of photovoltaic, wind, and hydrogen energies in sustainable global energy evolution. *Global Energy Interconnection*, *7*(4), 429–445. DOI: 10.1016/j.gloei.2024.08.009

Taibi, E., Miranda, R., Carmo, M., & Blanco, H. (2020). Green hydrogen cost reduction. Retrieved September 29, 2024, from https://www.h2knowledgecentre.com/content/researchpaper1613

Tarhan, C., & Çil, M. A. (2021). A study on hydrogen, the clean energy of the future: Hydrogen storage methods. *Journal of Energy Storage*, *40*, 102676. DOI: 10.1016/j.est.2021.102676

Tashie-Lewis, B. C., & Nnabuife, S. G. (2021). Hydrogen production, distribution, storage and power conversion in a hydrogen economy – A technology review. *Chemical Engineering Journal Advances*, *8*, 100172. DOI: 10.1016/j.ceja.2021.100172

Taylor & Francis. (2024). An overview of development and challenges in hydrogen powered vehicles. Retrieved September 27, 2024, from https://www.tandfonline.com/doi/full/10.1080/15435075.2019.1685999

Thomas, K. M. (2007). Hydrogen adsorption and storage on porous materials. *Catalysis Today*, *120*(3), 389–398. DOI: 10.1016/j.cattod.2006.09.015

Turoń, K. (2020). Hydrogen-powered vehicles in urban transport systems – Current state and development. *Transportation Research Procedia*, *45*, 835–841. DOI: 10.1016/j.trpro.2020.02.086

Tylianakis, E., Klontzas, E., & Froudakis, G. E. (2011). Multi-scale theoretical investigation of hydrogen storage in covalent organic frameworks. *Nanoscale*, *3*(3), 856–869. DOI: 10.1039/C0NR00505C

Usman, M. R. (2022). Hydrogen storage methods: Review and current status. *Renewable & Sustainable Energy Reviews*, *167*, 112743.

Ustolin, F., Campari, A., & Taccani, R. (2022). An extensive review of liquid hydrogen in transportation with focus on the maritime sector. *Journal of Marine Science and Engineering*, *10*(9), 1222. DOI: 10.3390/jmse10091222

Verma, S. (2015). A review on solar hydrogen production techniques, 4, February.

Vijayakumar, V., Fulton, L., Shams, M., & Sperling, D. (2022, August). Creating a global hydrogen economy: Review of international strategies, targets, and policies with a focus on Japan, Germany, South Korea, and California. https://doi.org/DOI: 10.7922/G2N014VF

Von Colbe, J. B., Ares, J. R., Barale, J., Baricco, M., Buckley, C., Capurso, G., ... & Dornheim, M. (2019). Application of hydrides in hydrogen storage and compression: Achievements, outlook and perspectives. international journal of hydrogen energy, 44(15), 7780-7808.

Wang, H., Gao, Q., & Hu, J. (2009). High hydrogen storage capacity of porous carbons prepared by using activated carbon. *Journal of the American Chemical Society*, *131*(20), 7016–7022.

Wang, T., Cao, X., & Jiao, L. (2022). PEM water electrolysis for hydrogen production: Fundamentals, advances, and prospects. *Carbon Neutrality*, *1*(1), 21. DOI: 10.1007/s43979-022-00022-8

Wang, Y., Kowal, J., Leuthold, M., & Sauer, D. U. (2012). Storage system of renewable energy generated hydrogen for chemical industry. *Energy Procedia*, *29*, 657–667. DOI: 10.1016/j.egypro.2012.09.076

Wang, Y., Pan, Z., Zhang, W., Borhani, T. N., Li, R., & Zhang, Z. (2022). Life cycle assessment of combustion-based electricity generation technologies integrated with carbon capture and storage: A review. *Environmental Research*, *207*, 112219. DOI: 10.1016/j.envres.2021.112219

Xia, K., Hu, J., & Jiang, J. (2014). Enhanced room-temperature hydrogen storage in super-activated carbons: The role of porosity development by activation. *Applied Surface Science*, *315*, 261–267. DOI: 10.1016/j.apsusc.2014.07.144

Yang, B., Zhang, R., Shao, Z., & Zhang, C. (2023). The economic analysis for hydrogen production cost towards electrolyzer technologies: Current and future competitiveness. *International Journal of Hydrogen Energy*, *48*(37), 13767–13779. DOI: 10.1016/j.ijhydene.2022.12.204

Yang, X., Li, Y., Liu, Y., Li, Q., Yang, T., & Jia, H. (2024). Crystal structure prediction and performance assessment of hydrogen storage materials: Insights from computational materials science. *Energies*, *17*(14), 14. Advance online publication. DOI: 10.3390/en17143591

Yin, L., Yang, H., & Ju, Y. (2024). Review on the key technologies and future development of insulation structure for liquid hydrogen storage tanks. *International Journal of Hydrogen Energy*, *57*, 1302–1315. DOI: 10.1016/j.ijhydene.2024.01.093

Yu, X., Tang, Z., Sun, D., Ouyang, L., & Zhu, M. (2017). Recent advances and remaining challenges of nanostructured materials for hydrogen storage applications. *Progress in Materials Science*, *88*, 1–48. DOI: 10.1016/j.pmatsci.2017.03.001

Yusaf, T., Mahamude, A. S. F., Kadirgama, K., Ramasamy, D., Farhana, K., Dhahad, H. A., & Talib, A. R. A. (2024). Sustainable hydrogen energy in aviation–A narrative review. international journal of hydrogen energy, 52, 1026-1045.

Zeng, K., & Zhang, D. (2010). Recent progress in alkaline water electrolysis for hydrogen production and applications. *Progress in Energy and Combustion Science*, *36*(3), 307–326. DOI: 10.1016/j.pecs.2009.11.002

Zhang, B., Zhang, S.-X., Yao, R., Wu, Y.-H., & Qiu, J.-S. (2021). Progress and prospects of hydrogen production: Opportunities and challenges. *Journal of Electronic Science and Technology*, *19*(2), 100080. DOI: 10.1016/j.jnlest.2021.100080

Zhang, X.. (2020). Optimization of the pore structures of MOFs for record high hydrogen volumetric working capacity. *Advanced Materials*, *32*(17), 1907995. DOI: 10.1002/adma.201907995

Zhang, X. L., Liu, Y. F., Zhang, X., Hu, J. J., Gao, M. X., & Pan, H. G. (2020). Empowering hydrogen storage performance of $MgH_2$ by nanoengineering and nanocatalysis. *Materials Today Nano*, *9*, 100064. DOI: 10.1016/j.mtnano.2019.100064

Zhang, Y., Jia, Z., Yuan, Z., Yang, T., Qi, Y., & Zhao, D. (2015). Development and application of hydrogen storage. *Journal of Iron and Steel Research International*, *22*(9), 757–770. DOI: 10.1016/S1006-706X(15)30069-8

Zhao, W., Luo, L., Wang, H., & Fan, M. (2017). Synthesis of bamboo-based activated carbons with super-high specific surface area for hydrogen storage. *BioResources*, *12*(1), 1246–1262. DOI: 10.15376/biores.12.1.1246-1262

Zheng, J., Wang, C. G., Zhou, H., Ye, E., Xu, J., Li, Z., & Loh, X. J. (2021). *Current research trends and perspectives on solid-state nanomaterials in hydrogen storage*. Research.

Zhou, X., Xu, X., & Huang, J. (2023). Adaptive multi-temperature control for transport and storage containers enabled by phase-change materials. *Nature Communications*, *14*(1), 5449.

Zhu, H., & Liu, D. (2019). The synthetic strategies of metal–organic framework membranes, films and 2D MOFs and their applications in devices. *Journal of Materials Chemistry. A, Materials for Energy and Sustainability*, *7*(37), 21004–21035. DOI: 10.1039/C9TA05383B

Zohuri, B. (2019). Cryogenics and liquid hydrogen storage. Hydrogen energy: challenges and solutions for a cleaner future, 121-139.

Zou, C., Li, J., Zhang, X., Jin, X., Xiong, B., Yu, H., & Pan, S. (2022). Industrial status, technological progress, challenges, and prospects of hydrogen energy. *Natural Gas Industry B*, *9*(5), 427–447.

Zou, L., & Zhou, H.-C. (2017). Hydrogen storage in metal-organic frameworks. In Chen, Y.-P., Bashir, S., & Liu, J. L. (Eds.), *Nanostructured materials for next-generation energy storage and conversion: Hydrogen production, storage, and utilization* (pp. 143–170). Springer., DOI: 10.1007/978-3-662-53514-1_5

Züttel, A., Remhof, A., Borgschulte, A., & Friedrichs, O. (2010). Hydrogen: the future energy carrier. Philosophical Transactions of the Royal Society A: Mathematical, Physical and Engineering Sciences, 368(1923), 3329-3342.

# Chapter 27
# Technical Challenges in Green Hydrogen Production

**Souad Nasrdine**
https://orcid.org/0009-0004-7021-9005
*Mohammed V University in Rabat, Morocco*

**Aziza Lamchaimech**
https://orcid.org/0009-0002-8393-7167
*Mohammed V University in Rabat, Morocco*

**Khadija Elmoustaqim**
https://orcid.org/0009-0007-9193-6638
*Ibn Tofail University, Morocco*

**Mohamed Benchrifa**
*Mohammed V University in Rabat, Morocco*

**Jamal Mabrouki**
https://orcid.org/0000-0002-3841-7755
*Mohammed V University in Rabat, Morocco*

**Miloudia Slaoui**
*Mohammed V University in Rabat, Morocco*

## ABSTRACT

*Green hydrogen, generated through the electrolysis of water from renewable energies, is seen as an essential energy component in the transition to a low-carbon economy. However, its large-scale manufacture faces a number of significant technical challenges. In this chapter, we examine these challenges, focusing on four key aspects. Firstly, it is necessary to improve the energy efficiency of current processes in order to minimize energy losses throughout the production chain. Secondly, the interruption of renewable energy sources poses a challenge, requiring energy storage and management solutions to ensure continuous hydrogen production. In addition, manufacturing costs remain high compared with grey hydrogen, making it essential to develop more competitive and cost-effective technologies. Finally, the lack of adequate infrastructure for transporting and storing hydrogen represents another brake on its large-scale use.*

## INTRODUCTION

Green hydrogen, also known as dihydrogen, represents a promising energy source for the transition to a greener future (Ma et al., 2023). Green hydrogen, produced by electrolysis of water using electricity generated solely from renewable sources such as wind and solar power, has the advantage of producing no greenhouse gas emissions during production and use (Oliveira, Beswick, & Yan, 2021). Compared

DOI: 10.4018/979-8-3693-8980-5.ch027

with grey hydrogen from fossil fuels, green hydrogen is a much more environmentally-friendly option (IFP Énergies nouvelles, n.d.). It is therefore essential to promote its development in order to decarbonize certain sectors such as industry, transport and power generation (Pollutec, n.d.). It is considered an essential element by European and national institutions to achieve carbon neutrality by 2050 (EcoAct, 2023). However, even if the potential is there, large-scale production of green hydrogen still faces a number of major technical challenges that need to be overcome in order to facilitate mass deployment and accelerate the transition (Nasrdine, Mabrouki, Ben-Lhachemi, & Slaoui, 2024). Improving the energy efficiency of electrolysis processes is essential to reduce energy losses and production costs. Another obstacle is the intermittency of renewable energies, requiring the creation of storage solutions to balance hydrogen production (Agaton, Batac, & Reyes Jr, 2022). It is also necessary to build an appropriate transmission and distribution infrastructure (Taibi, Miranda, Carmo, & Blanco, 2020). Finally, technological advances play a crucial role in reducing costs and making hydrogen competitive with fossil fuels (Hassan et al., 2024). Overcoming these technical challenges is crucial if green hydrogen is to play its full role in the ecological transition (Kumar & Lim, 2022). Investment and collaboration between the public and private sectors play an essential role in stimulating the development of this future industry (Gondal, Masood, & Khan, 2018).

## MATERIAL AND METHOD

## I. Description of production processes

### 1. Water Electrolysis Powered by Renewable Energies

One of the most popular methods for creating green hydrogen is water electrolysis (Rozain, 2013). It includes splitting water into dihydrogen (H2) and dioxygen (O2) by running an electric current through it. Since there is no carbon footprint associated with the hydrogen produced, it is referred to as green hydrogen when all electricity used is derived from renewable sources like solar or wind power. Three primary electrolyzer technologies exist (Le Bideau, 2021):

### 1.1 Electrolysers for Alkalinity

These electrolysers use a potassium hydroxyde solution as the electrolyte. They are known for being reliable and reasonably priced, although their effectiveness is not as high as that of other technologies (Chennouf, Settou, Negrou, Bouziane, & Dokkar, 2013).

*Figure 1. Principle of water electrolysis.*

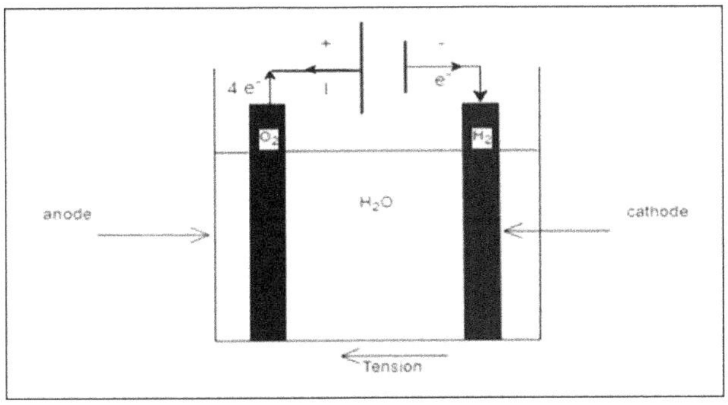


## 1.2 Proton Exchange Membrane (PEM) Electrolyzers

These devices have a higher current density and better efficiency since they use a solid polymer electrolyte. They are especially well-suited for applications that need quick starts (Audichon, 2014).

*Figure 2. Scheme of a PEM electrolyzer cell*

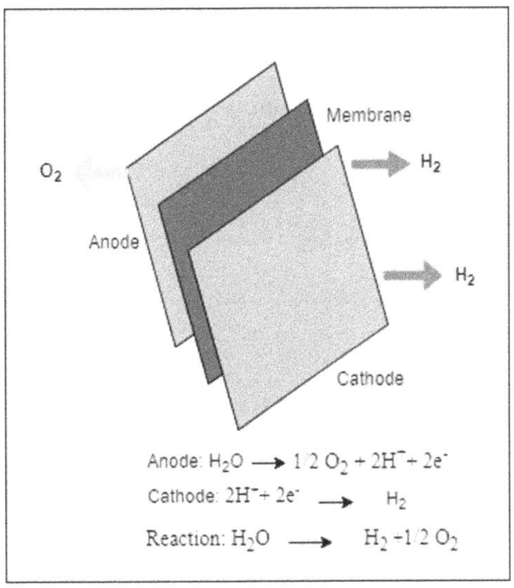


## 1.3 Solid Oxide Electrolyzers (SOEC)

These electrolyzers use a ceramic electrolyte and operate at high temperatures. They're still in the early stages of development, but their efficiency looks promising (Reytier et al., 2016).

## 2. Biomass Gasification

An further technique for generating green hydrogen is biomass gasification. During this process, biomass is heated in the presence of a small quantity of steam or oxygen, turning the organic matter into a combustible gas that is mostly hydrogen (H2) and carbon monoxide (CO). This gas is known as syngas (Ahmad, Zawawi, Kasim, Inayat, & Khasri, 2016).

### 2.1 Principle and Techniques

Biomass is heated to a high temperature in a reactor called a gasifier during the gasification process. Syngas is created during this process, and it can be refined to recover hydrogen (Puig-Arnavat, Bruno, & Coronas, 2010).

*Figure 3. Diagram of the gasification process*

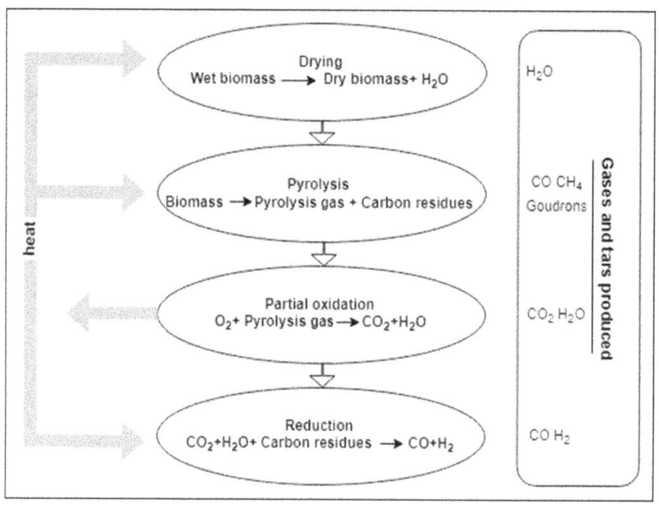


Since the CO2 released during syngas combustion has been absorbed by the plants during their growth, gasification effectively utilizes organic waste while also assisting in the reduction of CO2 emissions.

### 2.2 Biomass Supply Challenges

The sustainable supply of raw materials is one of the primary issues with biomass gasification. Seasons and geographical locations can have a significant impact on the availability and quality of biomass. Furthermore, it is crucial to make sure that neither the environment nor food security are harmed by the gathering and use of biomass (Molino, Chianese, & Musmarra, 2016).

## RESULTS AND DISCUSSION

### 1. Energy Efficiency

#### 1.1 Analysis of Process Efficiencies

One of the most important factors in assessing green hydrogen production technology is energy efficiency. The effectiveness of electrolyzers, which convert water into hydrogen and oxygen, varies greatly (Spitzer, Gicquel, Gurevsky, & Sanson, 2024):

**Alkaline electrolyzers:** The efficiency of these tried-and-true methods typically range from 60% to 70%. They are a popular choice due to their ruggedness and relative affordability, although internal resistance and heat losses reduce their efficiency (Guénan, 2024).
**Proton exchange membrane (PEM) electrolyzers:** These gadgets have greater efficiency, frequently ranging from 70% to 80%. They are perfect for applications demanding flexibility because of their quick start-up times and capacity to operate at greater pressures (Sebbahi, Nabil, Rachidi, El Ganaoui, & Benyoussef, 2022).
**Solid oxide electrolyzers (SOEC):** These electrolyzers, which are still in the development stage, can produce yields exceeding 80% by increasing process efficiency through the use of waste heat (Melgarejo, 2021).

#### 1.2 Comparison of Energy Losses in the Production Chain

One of the main obstacles in the process of hydrogen production is energy losses. Generally, the electrolysis process might result in losses of up to 30%. This comprises:

**Thermal losses:** A significant proportion of the energy used is lost in the form of heat (Demirci & Miele, 2010).
**Electrical losses:** The internal resistance of electrolyzers also contributes to energy losses.

These losses underline the importance of optimizing electrolysis systems and developing technologies to recover this lost energy (Phan, 2020).

### 2. Intermittency of Renewable Energies

#### 2.1 Impact on Hydrogen Production

The intermittent nature of renewable energy sources, such wind and solar power, presents a significant obstacle to the creation of green hydrogen. The creation of hydrogen is directly reliant on the supply of renewable electricity:
**Variability:** There may be times when electrolysis is neither technically nor commercially feasible due to fluctuations in the output of energy (Wiesenfeld, 2023).

## 2.2 Potential Solutions for Energy Storage and Management

To overcome these challenges, several solutions are being considered:

**Energy storage:** By storing extra electricity when it's available, the development of storage technologies (such batteries or pumped storage) can assist regulate the creation of hydrogen (Alexandre, 2022).

**Power-to-Gas:** With the use of this technology, excess electricity may be transformed into hydrogen, which can be stored for later use (Bouramdane, 2023).

# 3. Production Costs

## 3.1 Comparison Between Green and Grey Hydrogen

Production costs remain a major obstacle to the widespread adoption of green hydrogen. At present:

**Grey hydrogen**: is produced mostly from fossil fuels and is typically priced per kilogram around 1.50 to 2.50 USD (Phan, 2020).

**Green hydrogen:** Depending on the type of electrolyzer and the cost of renewable electricity, the price per kilogram can range from 4 to 6 USD. This distinction emphasizes how technical innovation is necessary to cut expenses (Nzihou et al., 2021).

## 3.2 Factors Influencing Costs (Energy, Technology)

**Cost of electricity:** One important factor is the cost of renewable electricity; if this keeps going down, green hydrogen might become more affordable (Al Bacha, 2020).

**Technological investment:** Cost-cutting measures will also heavily rely on economies of scale and improvements in electrolysis technologies (DJAFOUR, Aida, & Azoui, 2015).

# 4. Infrastructure Requirements

## 4.1 Assessment of Infrastructure Requirements for Transport and Storage

The development of suitable infrastructures is essential for the efficient distribution of green hydrogen:

**Transportation:** Liquid or pipeline-based hydrogen transportation is both possible. To manage this energy carrier, new infrastructure must be built or current infrastructure must be modified (Picard, 2022).

**Storage:** To ensure ongoing availability, creative long-term storage options, including subterranean storage, must be created (Barhoum, Zaiz, & DIB, 2024).

## 4.2 Logistical Challenges Associated With Setting Up These Infrastructures

Among the difficulties in logistics are:

**High initial costs**: Establishing infrastructure needs a significant financial commitment, which may deter market participants (Barhoum, Zaiz, & DIB, 2024).
**Safety:** Strict safety regulations must be followed when handling and transporting hydrogen to avoid mishaps (Abdallah, 2023).

## CONCLUSION

This chapter has highlighted the major technical challenges that need to be overcome to enable large-scale deployment of green hydrogen production (Boulanger, 2020). Although this technology has considerable potential to contribute to the energy transition (Zeino-Mahmalat & Verdo, 2024), several significant obstacles remain, including improving the energy efficiency of production processes, managing the intermittency of the renewable energies that power electrolyzers, reducing production costs to achieve competitiveness with grey hydrogen (Attou & Nadir, 2024), and building suitable infrastructures for transporting and storing green hydrogen (Nasrdine et al., 2024a). To meet these challenges, major technological innovations are essential (Nasrdine et al., 2024b), as research and development play a crucial role in improving the performance and lowering the costs of electrolyzers, optimizing hydrogen production and purification processes (Nasrdine, Benchrifa, Mabrouki, Qisse, & Slaoui, 2024), developing innovative seasonal storage solutions, and adapting existing infrastructures or designing new facilities dedicated to hydrogen transport and distribution (Nasrdine, Benchrifa, Mabrouki, & Slaoui, 2024). Despite these challenges, green hydrogen has immense potential to contribute to the decarbonization of the global economy (Nasrdine, Mabrouki, Ben-Lhachemi, & Slaoui, 2024), supported by many governments and institutions who see it as a key vector for achieving carbon neutrality (Nasrdine, Benchrifa, Ben-Lhachemi, Mabrouki, & Slaoui, 2024). To encourage its adoption, measures are needed, including massive investment in R&D and pilot projects to accelerate technological maturity (Nasrdine, Mabrouki, Ben-Lhachemi, & Slaoui, 2024), incentive-based public policies to stimulate the demand for and production of green hydrogen (Ben-Lhachemi et al., 2024a), international coordination to develop common standards and infrastructures, and the involvement of all players (industry (Ben-Lhachemi, Benchrifa, Nasrdine, Mabrouki, & Slaoui, 2024), researchers, public authorities) to meet these challenges (Ben-Lhachemi, Nasrdine, Mabrouki, & Slaoui, 2024). By overcoming these technical obstacles and accelerating innovation, green hydrogen has the potential to become an essential component of tomorrow's energy mix, helping to build a more sustainable, low-carbon future (Ben-Lhachemi et al., 2024b).

## REFERENCES


Abdallah, R. (2023). Optimisation prédictive des Modes de Fonctionnement des Systèmes d'Énergie Renouvelable: Application à une Station de Rechargement d'Hydrogène Vert.

Agaton, C. B., Batac, K. I. T., & Reyes, E. M.Jr. (2022). Prospects and challenges for green hydrogen production and utilization in the Philippines. *International Journal of Hydrogen Energy*, *47*(41), 17859–17870. DOI: 10.1016/j.ijhydene.2022.04.101

Ahmad, A. A., Zawawi, N. A., Kasim, F. H., Inayat, A., & Khasri, A. (2016). Assessing the gasification performance of biomass: A review on biomass gasification process conditions, optimization and economic evaluation. *Renewable & Sustainable Energy Reviews*, *53*, 1333–1347. DOI: 10.1016/j.rser.2015.09.030

Al Bacha, S. (2020). Générateur d'Hydrogène «vert» pour mobilité légère ou de courte distance (Doctoral dissertation, Université de Bordeaux; Université Libanaise).

Alexandre, P. (2022). Méthodologie de modélisation et de caractérisation en régime transitoire d'électrolyseurs pour la production d'hydrogène vert (Doctoral dissertation, Polytechnique Montréal).

Attou, S. E. Y., & Nadir, B. (2024). Émergence de l'hydrogène vert au Maroc: Perspectives juridiques. *Revue Internationale du Chercheur*, *5*(1).

Audichon, T. (2014). Synthèse et caractérisation de matériaux électrocatalytiques: activation anodique de l'eau dans un électrolyseur PEM (Doctoral dissertation, Poitiers).

Barhoum, H., Zaiz, I., & Encadré par DIB, D. (2024). Conception et Contrôle Optimale d'un Système de Production Electrique à Base d'Hydrogène Vert (Doctoral dissertation, UNIVERSITE DE ECHAHID CHEIKH LARBI TEBESSI).

Ben-Lhachemi, N., Benchrifa, M., Nasrdine, S., Mabrouki, J., & Slaoui, M. (2024). Intelligent temperature and humidity control system leveraging AI for precise climate management. In *Artificial intelligence systems in environmental engineering* (1st ed., p. 7). CRC Press. DOI: 10.1201/9781003436218-7

Ben-Lhachemi, N., Benchrifa, M., Nasrdine, S., Mabrouki, J., & Slaoui, M. (2024). Assessing the Environmental Effects Associated with Flood Protection Measures. In Mabrouki, J., & Azrour, M. (Eds.), *Integrated Solutions for Smart and Sustainable Environmental Conservation. Studies in Systems, Decision and Control* (Vol. 527). Springer., DOI: 10.1007/978-3-031-55787-3_9

Ben-Lhachemi, N., Benchrifa, M., Nasrdine, S., Mabrouki, J., Slaoui, M., & Azrour, M. a. (2024). Effect of IoT Integration in Agricultural Greenhouses. In Mabrouki, J., & Mourade, A. (Eds.), *Technical and Technological Solutions Towards a Sustainable Society and Circular Economy. World Sustainability Series*. Springer., DOI: 10.1007/978-3-031-56292-1_35

Ben-Lhachemi, N., Nasrdine, S., Mabrouki, J., & Slaoui, M. (2024). Integration of Bio-sensors-IoT System for the Evaluation and Control of Soil Quality. In Mabrouki, J., & Mourade, A. (Eds.), *Advanced Technology for Smart Environment and Energy. Environmental Science and Engineering*. Springer., DOI: 10.1007/978-3-031-50871-4_14

Boulanger, V. (2020). L'hydrogène vert, carburant de la transition? *Alternatives Économiques*, *405*(10), 62–65. DOI: 10.3917/ae.405.0062

Bouramdane, A. A. (2023). Production d'hydrogène vert au Maroc: Quelle technologie est la plus adaptée à différents niveaux de pénétration renouvelable. Énergie/Mines Carrières, 5.

Chennouf, N., Settou, N., Negrou, B., Bouziane, K., & Dokkar, B. (2013). Etude d'une Installation de production d'hydrogène solaire par électrolyse de l'eau dans la région d'Ouargla. Ann. Des. Sci. Et. Technol., 5.

Demirci, U. B., & Miele, P. (2010). Les méthodes de production d'hydrogène sont-elles vertes. *L'Actualité Chimique*, *341*, 5–12.

Djafour, A., Aida, M. S., & Azoui, B. (2015). Etude d'un système photovoltaïque de production d'énergie à base d'hydrogène vert.

EcoAct. (2023). Les défis de l'hydrogène pour la décarbonation. Retrieved [September 18, 2024], from https://eco-act.com/fr/blog/defis-hydrogene-decarbonation/

Gondal, I. A., Masood, S. A., & Khan, R. (2018). Green hydrogen production potential for developing a hydrogen economy in Pakistan. *International Journal of Hydrogen Energy*, *43*(12), 6011–6039. DOI: 10.1016/j.ijhydene.2018.01.113

Guénan, K. (2024). L'avion à hydrogène ZEROe: défis technologiques et impacts sur l'écosystème. In Annales des Mines-Réalités industrielles (No. 2, pp. 99-103). Cairn/Cairn.

Hassan, Q., Algburi, S., Sameen, A. Z., Salman, H. M., & Jaszczur, M. (2024). Green hydrogen: A pathway to a sustainable energy future. *International Journal of Hydrogen Energy*, *50*, 310–333. DOI: 10.1016/j.ijhydene.2023.08.321

IFP Énergies nouvelles. (n.d.). Tout savoir sur l'hydrogène. Retrieved [September 18, 2024], from https://www.ifpenergiesnouvelles.fr/enjeux-et-prospective/decryptages/energies-renouvelables/tout-savoir-lhydrogene

Kumar, S. S., & Lim, H. (2022). An overview of water electrolysis technologies for green hydrogen production. *Energy Reports*, *8*, 13793–13813. DOI: 10.1016/j.egyr.2022.10.127

Le Bideau, D. (2021). Étude de l'amélioration de la production d'hydrogène par le procédé d'électrolyse de l'eau alcaline: simulation avec mécanique des fluides numérique et optimisation génétique (Doctoral dissertation, Université de Bretagne Sud).

Ma, N., Zhao, W., Wang, W., Li, X., & Zhou, H. (2023). Large scale of green hydrogen storage: Opportunities and challenges. *International Journal of Hydrogen Energy*.

Melgarejo, S. B. (2021). Le rôle de l'hydrogène dans la transition énergétique (Doctoral dissertation, Pau).

Molino, A., Chianese, S., & Musmarra, D. (2016). Biomass gasification technology: The state of the art overview. *Journal of Energy Chemistry*, *25*(1), 10–25. DOI: 10.1016/j.jechem.2015.11.005

Nasrdine, S., Benchrifa, M., Ben-Lhachemi, N., Mabrouki, J., & Slaoui, M. (2024). Effect of various factors on the productivity of solar stills: Mini review. *Euro-Mediterranean Journal for Environmental Integration*, *9*(2), 977–987. DOI: 10.1007/s41207-024-00465-8

Nasrdine, S., Benchrifa, M., Ben-Lhachemi, N., Mabrouki, J., & Slaoui, M. (2024). Remote monitoring of an inclined solar distiller. In *Artificial intelligence systems in environmental engineering* (1st ed., p. 7). CRC Press. DOI: 10.1201/9781003436218-6

Nasrdine, S., Benchrifa, M., Ben-Lhachemi, N., Mabrouki, J., Slaoui, M., & Azrour, M. (2024). New Design of an Inclined Solar Distiller for Freshwater Production: Experimental Study. In Mabrouki, J., & Mourade, A. (Eds.), *Technical and Technological Solutions Towards a Sustainable Society and Circular Economy. World Sustainability Series*. Springer., DOI: 10.1007/978-3-031-56292-1_36

Nasrdine, S., Benchrifa, M., Mabrouki, J., Qisse, N., & Slaoui, M. (2024). Comparative study of the quality of water produced by a solar distiller and by reverse osmosis. *Desalination and Water Treatment*, *317*, 100264. DOI: 10.1016/j.dwt.2024.100264

Nasrdine, S., Benchrifa, M., Mabrouki, J., & Slaoui, M. (2024). The Application of Solar Distillation Technology as a Sustainable Solution for Seawater Desalination in the Arid Regions of the World. In Mabrouki, J., & Mourade, A. (Eds.), *Technical and Technological Solutions Towards a Sustainable Society and Circular Economy. World Sustainability Series*. Springer., DOI: 10.1007/978-3-031-56292-1_4

Nasrdine, S., Mabrouki, J., Ben-Lhachemi, N., & Slaoui, M. (2024). Design and Dimensioning of a Green Hydrogen Production System. In Mabrouki, J., & Mourade, A. (Eds.), *Advanced Technology for Smart Environment and Energy. Environmental Science and Engineering*. Springer., DOI: 10.1007/978-3-031-50871-4_3

Nasrdine, S., Mabrouki, J., Ben-Lhachemi, N., & Slaoui, M. (2024). Application of Machine Learning Algorithms for Air Quality Prediction. In Mabrouki, J., & Azrour, M. (Eds.), *Integrated Solutions for Smart and Sustainable Environmental Conservation. Studies in Systems, Decision and Control* (Vol. 527). Springer., DOI: 10.1007/978-3-031-55787-3_7

Nasrdine, S., Mabrouki, J., Ben-Lhachemi, N., & Slaoui, M. (2024). Design and Dimensioning of a Green Hydrogen Production System. In Mabrouki, J., & Mourade, A. (Eds.), *Advanced Technology for Smart Environment and Energy. Environmental Science and Engineering*. Springer., DOI: 10.1007/978-3-031-50871-4_3

Nzihou, A., Martínez, M. G., Minh, D. P., Millan, L. M. R., & Gourinat, Y. (2021). L’hydrogène vert en aéronautique-Production. Techniques de l'ingénieur, (trp4055), 20-p.

Oliveira, A. M., Beswick, R. R., & Yan, Y. (2021). A green hydrogen economy for a renewable energy society. *Current Opinion in Chemical Engineering*, *33*, 100701. DOI: 10.1016/j.coche.2021.100701

Phan, T. S. (2020). Élaboration, caractérisation et mise en œuvre d’un catalyseur dans le reformage du biogaz en vue de la production d’hydrogène vert (Doctoral dissertation, Ecole des Mines d'Albi-Carmaux).

Picard, V. (2022). Étude de catalyseurs hydrosolubles pour la génération d’hydrogène vert par méthodes photocatalytiques.

Pollutec. (n.d.). Guide de l'hydrogène vert: Une énergie renouvelable. Retrieved [September 18, 2024], from https://learnandconnect.pollutec.com/guide-hydrogene-vert/guide-hydrogene-vert-energie-renouvelable/

Puig-Arnavat, M., Bruno, J. C., & Coronas, A. (2010). Review and analysis of biomass gasification models. *Renewable & Sustainable Energy Reviews, 14*(9), 2841–2851. DOI: 10.1016/j.rser.2010.07.030

Reytier, M., Bernard, C., Chatroux, A., Di Iorio, S., Laurencin, J., Petitjean, M., ... & Mougin, J. (2016). Une production d'hydrogène à haut rendement. l'actualité chimique, (408-409), 40.

Rozain, C. (2013). Développement de nouveaux matériaux d'électrodes pour la production d'hydrogène par électrolyse de l'eau (Doctoral dissertation, Université Paris Sud-Paris XI).

Sebbahi, S., Nabil, N., Rachidi, S., El Ganaoui, M., & Benyoussef, A. (2022, November). Hydrogène vert au Maroc: état des lieux et perspectives. In Annales des Mines-Réalités industrielles (No. 4, pp. 148-153). Cairn/Softwin. DOI: 10.3917/rindu1.224.0148

Spitzer, V., Gicquel, C., Gurevsky, E., & Sanson, F. (2024, March). Dimensionnement de lots de production un jour à l'avance sous incertitudes éoliennes: application à la production d'hydrogène vert. In *25ème Congrès Annuel de la Société Française de Recherche Opérationnelle et d'Aide à la Décision (ROADEF 2024)*.

Taibi, E., Miranda, R., Carmo, M., & Blanco, H. (2020). Green hydrogen cost reduction.

Wiesenfeld, B. (2023). *L'hydrogène vert: Le défi de demain, pour une énergie inépuisable et décarbonée*. EDP Sciences.

Zeino-Mahmalat, T., & Verdo, Y. (2024). L'hydrogène vert pour l'OCP: une bénédiction stratégique. Le journal de l'école de Paris du management, (2), 37-43.

# Chapter 28
# Techno-Economic Analysis of Green Hydrogen Production for Industrial Use

**Shalini Jaiswal**
https://orcid.org/0000-0003-0137-7734
*Amity University, Greater Noida, India*

**Shaista Zarrin Khanam**
*University of Technology and Applied Science, Nizwa, Oman*

## ABSTRACT

*This chapter will explore the role of green hydrogen as a sustainable energy source for the industrial sector. The will provide an overview of the importance of decarbonizing industry and the challenges associated with conventional energy sources. This will also explain the fundamentals of green hydrogen production through electrolysis, using renewable energy sources like solar, wind, and hydropower. It will compare green hydrogen with grey, blue, and other forms of hydrogen, discussing the environmental impact and energy efficiency. This also focuses on specific industrial applications where green hydrogen has the most potential like Steel Production, Chemical Industry, etc. We also focus on challenges associated with green hydrogen, such as the cost of production, transportation, and storage. At the same we will examine the role of governments in promoting green hydrogen, exploring policies, subsidies, and international agreements. It will also discuss the alignment of green hydrogen initiatives with the UN's Sustainable Development Goals (SDGs).*

## 1. INTRODUCTION

The rise in global energy consumption has led to an increase in discourse around clean, inexpensive, and sustainable types of energy generation (Agyekum, 2020; Agyekum & Velkin,2020; Gyamfi et al., 2021). Exploiting these traditional fossil fuel resources will result in increased global warming and $CO_2$ emissions (Pour Azarm & Verma,2022). Power generation is critical to every country's industrial development. Renewable energy sources such as solar, wind, hydro, geothermal, ocean thermal energy conversion (OTEC), and biomass are among the most promising possibilities for fossil fuel replacement

DOI: 10.4018/979-8-3693-8980-5.ch028

(Ishaq & Dincer, 2020). For this reason, among others, researchers and experts are researching alternative kinds of energy generation that have a zero or lower negative impact on the environment (Adebayo et al., 2021; Tarhan & Çil, 2021; Yaqoob et al., 2021; Xhao et al., 2021). **Figure 1** illustrates the proportion of the global energy supply. Coal accounts for 35% of global energy supply, followed by gas (23%), hydropower (16%), and nuclear (10%). Renewable energy sources account for approximately 13%, with 7% coming from wind energy, 4% from solar, about 2% from hydrogen, and the remainder coming from biomass and biogas. Coal, oil, and gas combustion account for most worldwide fossil fuel emissions. In 2022, coal will account for 40% of total global fossil fuel emissions, more than any other fossil fuel. Gas production ranks third with a 21% contribution, followed by cement production at 4% and oil at 32% of fossil $CO_2$. Biogas can be purified and used as a fuel, or it can be injected into the national gas grid if it meets the purity standards after upgrading (Jaiswal, 2018; Dwivedi et al., 2022). Hydrogen's current contribution to the energy industry is small, and when hydrogen is produced from renewable energy sources, the net $CO_2$ emissions are far lower than for other fuels. **Figure 2** shows the annual $CO_2$ emissions from various types of fuels. **Figures 1** and **2** show that oil, coal, and natural gas are traditional energy sources that account for most of the global energy supply, power generation, and $CO_2$ emissions.

*Figure 1. Global share of energy sources 2021*

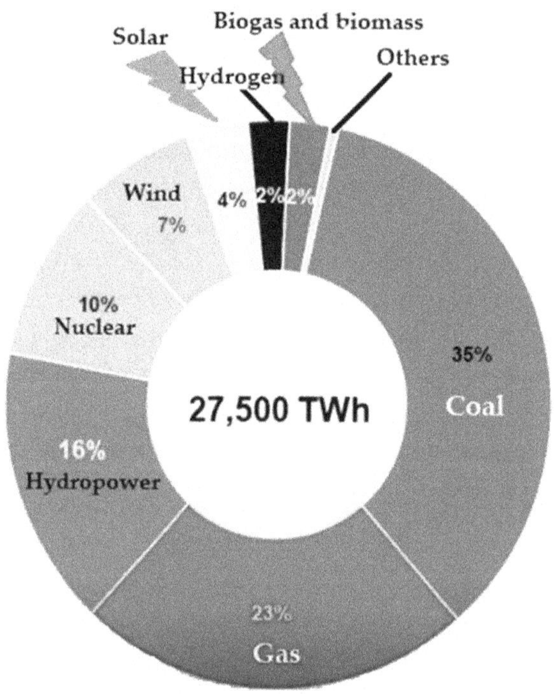

*Figure 2. Annual $CO_2$ emissions from different types of fuels*

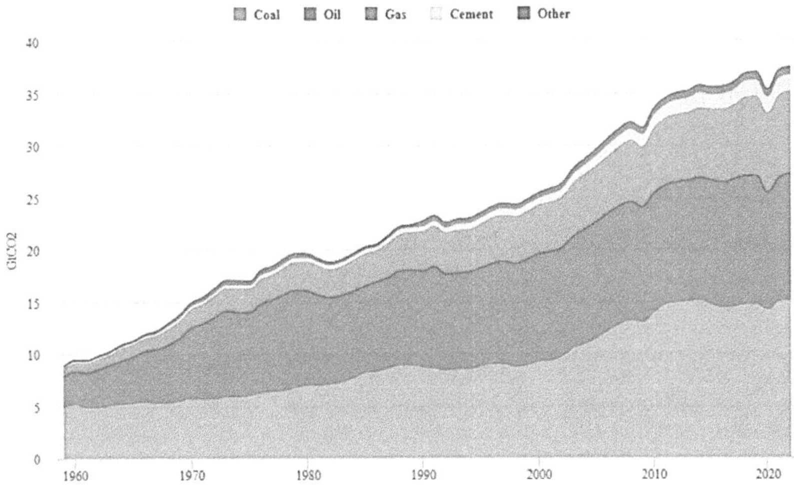


Carbon neutrality is a technique that creates a balance between emitting carbon and absorbing it from the environment in carbon sinks. Carbon sequestration is the process of extracting carbon dioxide from the atmosphere, for this we should focus on Innovative techniques to reduce carbon emissions (Singh et al., 2022). To achieve net-zero $CO_2$ emissions by 2050, it may also be necessary to decarbonize the major industries (Nieto, 2022). Because of its carbon-free solutions, hydrogen is gaining popularity as a distinct energy source and potential fuel on a global scale. Carbon neutrality by the mid-twentieth century is essential to limit global warming to 1.5°C, which the Intergovernmental Panel on Climate Change considers safe (Adedeji, 2014; (Thapa & Thapa, 2020). We should focus on Energy Generation through Integrated Sustainable Waste Management like in NAXOS 2018 (Jaiswal et al.,2018; Jaiswal et al., 2020: Jaiswal, 2024, pp .21-33).

## 2. TECHNOLOGY OF GREEN HYDROGEN PRODUCTION

The complexities of the production system, particularly the power source employed, determine whether hydrogen is green, blue, Gray, pink, or yellow is mentioned in **Figure 3**(IEA, (2019). Hydrogen production can be divided into three categories: green (based on renewable energy), blue (based on coal gasification and natural gas, combined with CCS hydrogen generation systems), and red (based solely on fossil fuels).

*Figure 3. Types of hydrogen*

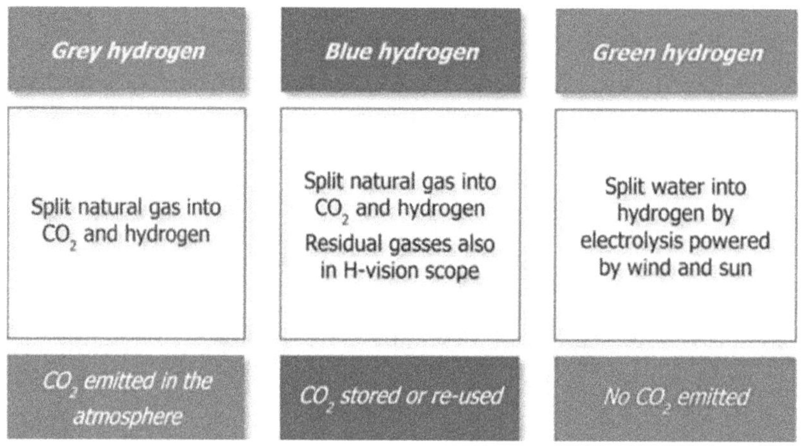


Clean hydrogen may be created at home using a range of energy sources, including nuclear (Nadaleti et al., 2022), natural gas (Bassam et al., 2018), coal gasification (George et al., 2016), and renewable energy sources such as solar (Hou et al., 2017), wind (Wang et al., 2021), biomass (Rathi et al., 2022), geothermal (Murray et al., 2020), hydro, and OTEC (Ahmadi et al., 2015). Recent literature (Boretti, 2020; Manna et al., 2021; Rabiee et al., 2021) has discussed the utilization of renewable energy resources for hydrogen generation. Innovation in Society can achieved through Eco-Friendly Technology (Jaiswal, S. & Mojahid,2020). Some of the key advantages of hydrogen include efficient energy conversion; production through water splitting with no carbon emissions; synthesis of various intermediate fuels, including synthetic fuels and ammonia, followed by various chemical reactions; storage availability; ability to use in existing infrastructure for long-distance transportation; and a high LHV in comparison to other fuels. Hydrogen is a modern, almost established energy source (Apostolou et al., 2019; Staffell et al.,2019; Falcone et al., 2021). **Table 1** compares the main technical properties of hydrogen with different fuels. (Kumar etv al., 2015; Dimitriou et al.,2019; Gheorghe et al., 2019Huang et al., 2020).

*Table 1. Technical properties of hydrogen vs other fuels .*

| **Parameter** | **Hydrogen** | **Diesel** | **Methane** |
|---|---|---|---|
| Density at STP ($kg/m^3$) | 0.089 | 830.0 | 0.720 |
| Volumetric energy at STP ($MJ/m^3$) | 1.07 × 10 | $3.5 \times 10^4$ | 3.3 × 10 |
| Net Lower Heating value (MJ/kg) | 119.9 | 42.5 | 45.8 |
| Boiling point (K) | 20.0 | 453–633 | 111.0 |
| Auto-ignition temperature (K) | 853 | ~523 | 813 |
| Minimum ignition energy in air at 1 bar & stoichiometry (mJ) | 0.020 | 0.240 | 0.290 |
| Stoichiometry air/fuel mass ratio | 34.4 | 14.5 | 17.2 |
| Quenching distance at NTP (mm) | 0.64 | - | 2.1 |
| Laminar flame speed in air at NTP (m/s) | 1.85 | 0.37–0.43 | 0.38 |

continued on following page

*Table 1. Continued*

| Parameter | Hydrogen | Diesel | Methane |
|---|---|---|---|
| Diffusion coefficient in air at STP ($m^2/s$) | $85 \times 10^{-7}$ | - | $19 \times 10^{-7}$ |
| Flammability limits in air (% vol) | 4–76 | 0.6–5.5 | 5.3–15 |
| Adiabatic flame temperature at NTP (K) | 2480 | ~2300 | 2214 |

Because the thermal Carnot cycle's thermal efficiency limits do not apply to hydrogen fuel cells, they provide a high-performance signal in terms of efficiency. Recent research (Ishaq & Dincer, 2020) has compared numerous strategies for creating hydrogen from renewable energy. To give an economic and environmental evaluation of hydrogen generation using various feedstocks, such as biomass, biogas, ethanol, and natural gas (Kaiwen et al., 2018). A solar-powered steam methane reforming method has been presented to implement a transition path to a solar-powered hydrogen economy and decarbonize fossil fuels. Two renewable methods for producing H2 are water and biomass (Ayodele & Munda,2019; Rambhujun et al., 2020; Kopteva et al., 2021; Pajak et al., 2021). Some important hydrogen generation techniques include landfill gas dry reformation, coal gasification, $H_2S$ methane reformation, naphtha reformation, natural gas pyrolysis, the steam-iron process, steam reforming of waste oil, the partial oxidation of heavy oil and coal, grid electrolysis of water, high-temperature water electrolysis, chloralkali electrolysis, solar and PV water electrolysis, photolysis of water, and biomass gasification. However, if we fully appreciate its potential as a full energy carrier, it can play an important role in a variety of other sectors. Most of the estimated 50 million metric tons produced globally each year are used as feedstock for ammonia synthesis, with oil refining accounting for 35% (Hanley et al., 2018). It could have significant applications in freight and passenger transport (i.e., internal combustion engines and fuel cell vehicles), storage (gaseous or liquid hydrogen), thermal (natural gas blending and solid oxide fuel cells) (Niaz et al., 2015), electricity generation, and power to gas generation (Bakenne et al., 2016), once its role as a flexible energy carrier is recognized, as shown in **Figure 4**.

*Figure 4. Effect of hydrogen production on other sectors*

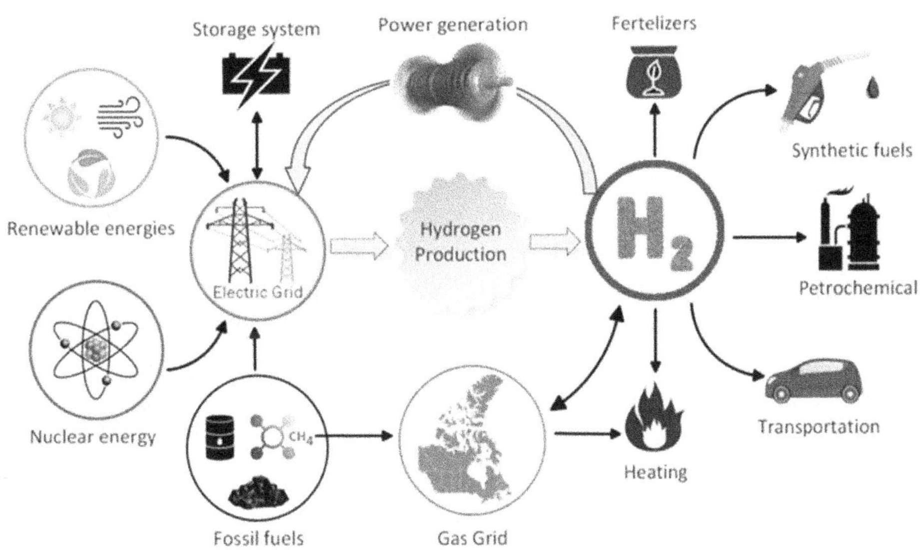

## 2.1 Hydrogen Production from Water

Water is the most abundant source for hydrogen production. Electrolysis, which uses an electrical current to separate water into oxygen and hydrogen, produces green hydrogen without emitting carbon dioxide. Renewable energy sources can be employed to generate the required electricity. Currently, electrolysis is the most well-established commercially available method of creating hydrogen from water. Water electrolysis is the process of dissolving water into its constituent elements, hydrogen, and oxygen, by employing an electric current. The electric potential attracts positive ions (H+) to the cathode and negative ions (OH-) to the anode. **Figure 5** depicts various water electrolysis methods, including alkaline water electrolysis (AEL), proton exchange membrane (PEM) water electrolysis, solid oxide water electrolysis (SOE), and alkaline anion exchange membrane (AEM) water electrolysis (Chi & Yu, 2018). They function on the same principle, albeit under different conditions and with different materials

*Figure 5. Water electrolysis processes to generate hydrogen*

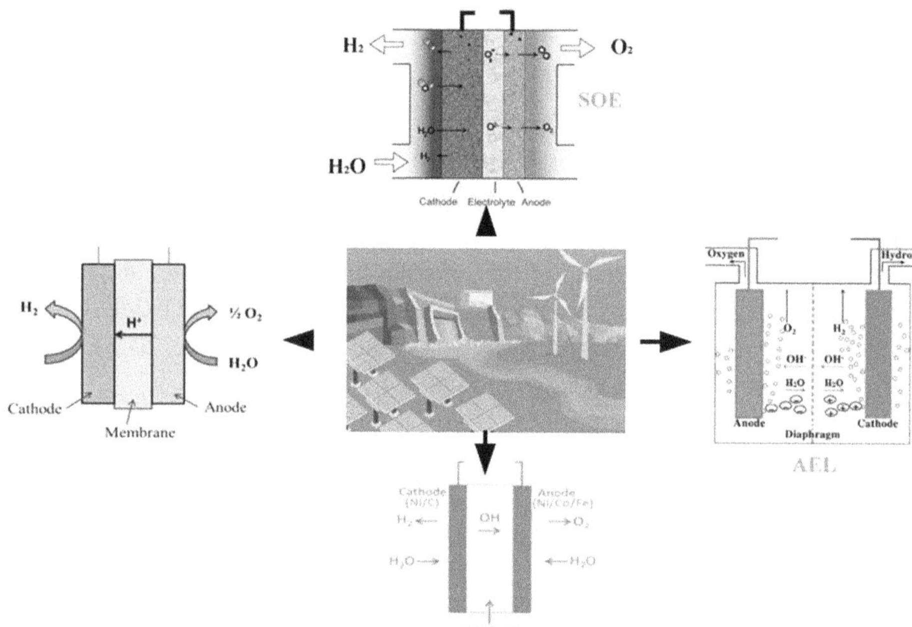


Using AEM water electrolysis, conventional noble metal electrocatalysts could be replaced with less expensive transition metals. Despite being a relatively new technology, AEM electrolysis has sparked interest due to its higher power efficiency, membrane stability, durability, simplicity of handling, and low-cost hydrogen generation (Vincent & Bessarabov, 2018). In addition to the rise in electrolysis voltage caused by the bubbles formed during the electrolysis process, excessive energy consumption is a barrier to producing hydrogen from water (Hu et al., 2019). To save energy, hydrocarbons can be used in water electrolysis. Future electrodes are predicted to be made of low-cost metals or nonmetal composite materials like Ni. Comparative analysis of different water electrolysis processes to generate hydrogen (Hydrogen-Production-By-Electrolysis- Ann-Cornell, 2017;Hydrogen-Production-Through-Electrolysis, 2017; Article landing, 2020) is tabulated in **Table 2**.

*Table 2. Comparative analysis of different types of water electrolysis process to generate hydrogen.*

| **Technology** | **AEL** | **PEM** | **SOE** | **AEM** |
|---|---|---|---|---|
| Electrolyte | Aqueous KOH | Proton exchange ionomer (e.g., | solid-oxide | Anion exchange ionomer (e.g., AS-4) + optional dilute |
| | (20–40 wt%) | Nafion) | | caustic solution |
| Cathode | Ni, Ni–Mo alloys | Pt, Pt–Pd | Ni-YSZ (yttria-stabilized | Ni and Ni alloys |
| | | | zirconia) | |
| Anode | Ni, Ni–Co alloys | $RuO_2$, $IrO_2$ | Lanthanum strontium manganate | Ni, Fe, Co oxides |
| Charge carriers | OH-, K+ | H+ | O2- | OH- |
| Operating | 100°–150°C | 70°–90°C | 700°C –800°C | 50°C –60°C |
| temperature | | | | |
| Cell voltage (V) | 1.8–2.4 | 1.8–2.2 | 1.6–1.8 | 1.8–2.2 |
| Technology status | Mature | Commercial | Not yet commercial Pilot | R&D |
| | | | scale | |

## 2.2 Production of $H_2$ from Biomass

Biomass currently accounts for 14% of total primary energy consumption (Granados-Fernández et al., 2020), owing to its abundance and ease of access in many nations (Resasco et al., 2018). As a CO2-neutral precursor, biomass is now regarded as a major renewable resource for hydrogen generation (Navarro et val., 2009). As **Figure 6** illustrates, hydrogen can be created in a variety of ways in connection to biomass. Currently, there are two basic techniques for producing hydrogen from biomass: thermochemical and biochemical (Dou et al., 2019)

*Figure 6. Sources of biomass and their conversion into hydrogen and other beneficial products*

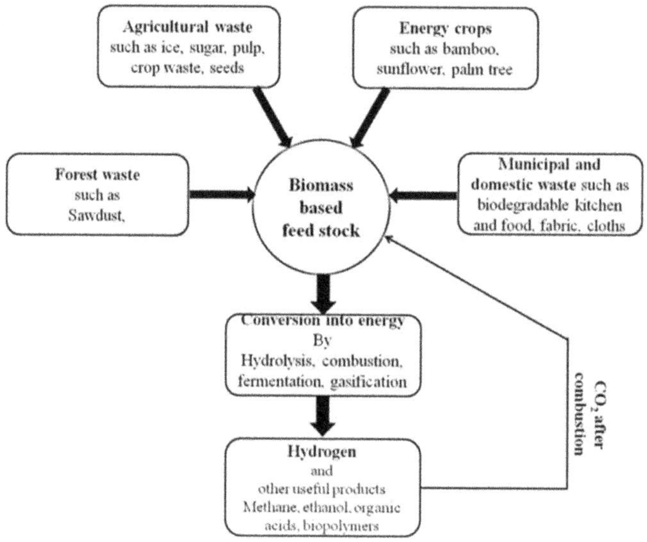


Pyrolysis, gasification, steam reforming, and supercritical gasification are examples of thermochemical processes, while biochemical processes include photolysis, fermentation, and dark fermentation. Biomass can be transformed into biofuels via a variety of biochemical processes, including anaerobic or aerobic digestion, fermentation, and acid hydrolysis Formica et al., 2016 ; Pandey et al., 2019). In comparison to biological processes, thermochemical technologies are more versatile and give a simpler approach because no additional chemicals are required, only heat and pressure are utilized to make biofuels (Saghir et al., 2018). This can be accomplished through pyrolysis or gasification. Gasification is a well-established process that produces hydrogen-rich fuel gas (CO, H2, and CH4) at 700-1200°C utilizing gasification agents (O2, CO2, steam, and air) (Carmo et al., 2013; Dunn, 2020; Lützkendorf & Frischknecht, 2020; Safari & Dincer, 2020).

$$Biomass \rightarrow H2 + CO2 + CO + N2$$

$H_2$ can also be obtained via the pyrolysis of biomass at lower temperatures of 300–650°C, but, at such temperatures, the hydrogen yield is lowered of ~18% by volume (Hutsol et al., 2021).

*Biomass +CO +CO+ Hyrdrocarban gases*

## 2.3 Thermolysis of Water for Production of Hydrogen

The thermolysis of water, often known as a single step thermal dissociation of water, is described as follows:

$$H2O - \rightarrow heat + H2 + 2O2$$

To achieve an acceptable amount of dissociation, the reaction in this approach requires a heat source capable of producing temperatures more than 2500 K. For example, at 3000 K and 1 bar, dissociation is 64% (Dincer & Acar,2015). These are the primary obstacles involved with this kind of production. Another issue associated with this type of hydrogen generation is the high cost of its equipment, which must survive the extreme conditions described previously in this section. Different technologies are thus being developed to conduct thermolysis at lower temperatures. One such strategy includes conducting the reaction in many phases using catalysts; however, the use of highly corrosive chemicals offers equipment issues as well as a potential harmful influence on the environment (Martinez-Burgos et al., 2021). Kasai and Bishop (Kasai et al., 1976) presented a single reaction system that can achieve an entropic exchange for the synthesis of hydrogen via thermolysis using zeolites. Similarly, Hsu presented a raney nickel plate reactor (10% aluminum and 90% nickel) that acts as a catalytic surface to adjust the reaction's energy activation (Hsu, 2007).

## 2.4 Photolysis of Water for Hydrogen Production

In principle, water can release hydrogen when its molecules receive energy from ultraviolet radiation at a rate of 285.57 kJ/mole (Kothari et al., 2008). Photolysis is the process by which photons dissociate H-O bonds at around 190 nm (Furatian & Mohseni, 2018). This process and thermolysis require chemical catalysts such as ZrO2, tin oxide (SnO), zinc oxide (ZnO), and other semiconductor sulfur oxides .

$$MoxZ + +H2O \rightarrow Mox(Z+1) + H + OH-$$

ECONOMIC VIABILITY FOR INDUSTRIES 3.

Green hydrogen offers a wide range of industrial applications **(Figure 7)** . Because of its high energy content, hydrogen is used as a gas in various applications such as fuel cells and rockets (Thapa & Thapa, 2020). Hydrogen is largely employed as a component or catalyst in ammonia, oil refining, and methanol production, with these three processes accounting for the vast bulk of hydrogen production. However, as clean hydrogen becomes more affordable and the need to cut carbon emissions grow. In the transportation sector, it can be used as a fuel for fuel cell vehicles, providing zero-emission mobility while decreasing dependency on fossil fuels. In industry, green hydrogen can be used as a feedstock for chemical processes, allowing the synthesis of green chemicals and materials. It can also be used to substitute fossil fuels in refineries, steel production, and other industrial operations, lowering carbon emissions. Green hydrogen has applications in power generation, where it can be utilized to generate electricity using gas turbines or fuel cells, offering a clean and renewable energy source. These applications demonstrate the versatility of green hydrogen and its potential to decarbonize multiple sectors of the economy (Hnát et al., 2017; Abe et al., 2019). The Road Map to a US Hydrogen Economy predicts a hydrogen revolution, with hydrogen serving as a fuel and component in a wide range of businesses. Below is a brief overview of how clean hydrogen is currently being used around the world, as well as numerous intriguing emerging disciplines exploring the use of hydrogen in industrial processes. Several industrial sectors are primed to benefit from green hydrogen, but the economic viability varies depending on energy prices, infrastructure, and demand.

*Figure 7. Industrial application of Green Hydrogen*

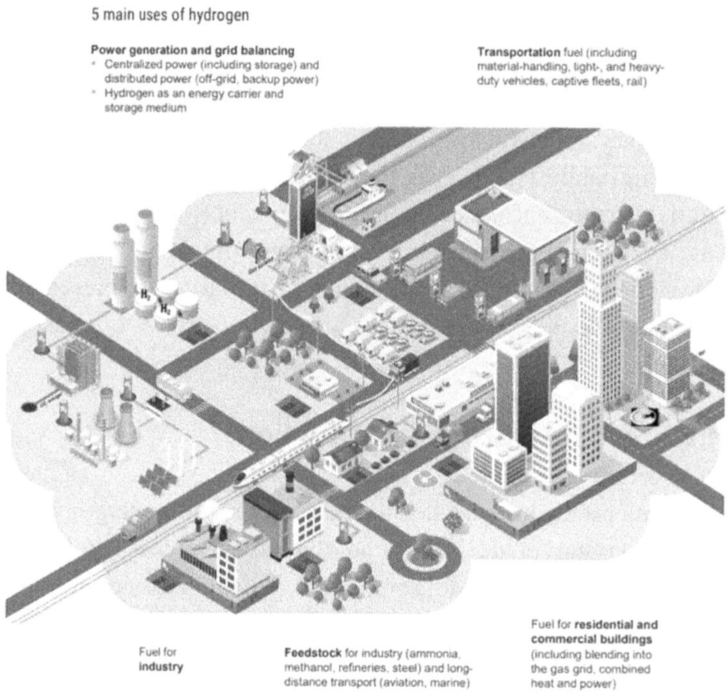


## 3.1 Steel Industry

The steel industry is responsible for approximately 7% of global $CO_2$ emissions. Traditionally, steel production relies on coal-based processes, but hydrogen-based direct reduction offers a pathway to producing low-carbon steel. Pilot projects, such as the HYBRIT project in Sweden, have already demonstrated the feasibility of hydrogen-based steel production. However, achieving widespread adoption will require significant investment in hydrogen infrastructure and cost reductions in hydrogen production (European Hydrogen Backbone, 2021; Vogl et al., 2020).

## 3.2 Chemical Industry

Ammonia and methanol production are key sectors where green hydrogen can make a significant impact. Currently, ammonia production is highly carbon-intensive, as it relies on grey hydrogen from natural gas. By transitioning to green hydrogen, the chemical industry can drastically reduce its emissions. This transformation is particularly important as ammonia is a critical feedstock for fertilizers and other chemical products (Bazzanella & Ausfelder, 2017). However, high energy costs and capital investments pose significant economic barriers. The price parity between green and grey hydrogen is unlikely to be achieved without substantial policy interventions or technological breakthroughs (Fasihi, Bogdanov, & Breyer, 2021). Scientists are also looking into ways to cut atmospheric pollutants by producing methanol.

Methanol is a potential alternative fuel for vehicles and equipment and using it can result in increased fuel efficiency and lower emissions, in addition to its numerous chemical applications.

### 3.3 Cement and Heavy Industry

The cement industry is another sector that could benefit from green hydrogen. Cement production emits significant amounts of $CO_2$ due to the calcination process and the use of fossil fuels in kilns. Green hydrogen can be used as an alternative fuel in cement kilns, helping to reduce emissions. However, the high temperatures required in cement production make the transition more complex and will require further technological advancements (Schneider et al., 2011).

### 3.4 Heavy Transport and Shipping

In addition to industrial processes, heavy transportation sectors, including shipping, aviation, and long-haul trucking, are prime candidates for hydrogen adoption. Hydrogen fuel cells and hydrogen-derived synthetic fuels can power heavy-duty vehicles and ships, offering a zero-emission alternative to fossil fuels. The widespread adoption of hydrogen in these sectors will depend on the development of refueling infrastructure and the commercialization of hydrogen-powered transport technologies (Ivy, 2020; U.S. Department of Energy, 2021).

### 3.5 Refining and Fuel Sectors

Green hydrogen has the potential to replace fossil-derived hydrogen in petroleum refining and bio-fuel production. However, these sectors require large amounts of hydrogen, and the high cost of green hydrogen makes it economically unfeasible at current prices (Bloomberg NEF, 2020).

## 4. CHALLENGES OF GREEN HYDROGEN PRODUCTION

Green hydrogen is gaining increasing attention as a key component in the transition toward a low-carbon future. Produced through the electrolysis of water using renewable energy sources like wind, solar, or hydropower, green hydrogen emits no carbon dioxide at the point of production, making it an ideal candidate for decarbonizing hard-to-abate sectors such as steel, cement, and chemical manufacturing. Despite its environmental promise, green hydrogen faces significant economic and technical hurdles that challenge its industrial-scale adoption.

### 4.1 Cost Structure of Green Hydrogen Production

Green hydrogen production through electrolysis is currently more expensive than other forms of hydrogen, such as grey hydrogen (produced from natural gas via steam methane reforming). The high cost of green hydrogen is primarily driven by:

i. **Capital Expenditure (CAPEX) of Electrolyzers:** Electrolyzers, which split water into hydrogen and oxygen using electricity, are currently expensive to manufacture. The cost of electrolyzers constitutes a large portion of green hydrogen production costs. Proton exchange membrane (PEM) and alkaline Electrolyzers are the two dominant technologies, with cost reductions expected in the coming years through economies of scale and technological advancements (International Energy Agency, 2022).
ii. **Operating Expenditure (OPEX):** The operating costs for green hydrogen production are highly dependent on electricity prices, as electrolysis is energy-intensive. Given that renewable electricity is not continuously available, the intermittent nature of renewable energy sources like solar and wind adds further uncertainty and cost variability (Hydrogen Council, 2021).
iii. **Energy Efficiency:** The process of electrolysis itself has an efficiency of around 60-80%, meaning that a significant amount of energy is lost during the conversion from electricity to hydrogen. This reduces the overall economic viability unless large-scale efficiencies can be realized (International Renewable Energy Agency, 2021).

*Table 3. Cost breakdown of green hydrogen production as compared to grey and blue hydrogen*

| Cost Component | Green Hydrogen (USD/kg) | Blue Hydrogen (USD/kg) | Grey Hydrogen (USD/kg) |
|---|---|---|---|
| Electrolyzer Cost | $0.50 - $1.00 | $0.00 | $0.00 |
| Renewable Energy Cost | $2.00 - $3.00 | $0.00 | $0.00 |
| Carbon Capture Cost | $0.00 | $1.00 - $2.00 | $0.00 |
| Natural Gas Cost | $0.00 | $0.50 | $0.50 - $1.00 |
| Other OPEX (Operations) | $0.50 - $1.00 | $0.50 | $0.50 |
| **Total Cost** | **$3.00 - $6.00** | **$1.50 - $3.00** | **$1.00 - $2.50** |

## 4.2 Renewable Energy Availability

The success of green hydrogen depends on the availability and cost of renewable electricity. Industrial production demands a constant supply of hydrogen, which is difficult to meet using intermittent energy sources like solar and wind. This intermittency issue can be addressed by coupling green hydrogen production with energy storage systems, but doing so increases costs (International Renewable Energy Agency, 2021).

## 4.3 Infrastructure Limitations

Hydrogen requires specialized storage and transport infrastructure, which currently is insufficient for large-scale industrial applications. The existing pipeline infrastructure, designed for natural gas, may not be suitable for hydrogen without significant modifications due to hydrogen's smaller molecular size and its tendency to embrittle metals. Building new hydrogen pipelines, liquefaction facilities, and storage solutions will require massive investments (Hydrogen Council, 2020). The chart (**Figure 8**) below showing the current global hydrogen production by type (grey, blue, green). This chart helps to demonstrate why grey hydrogen dominates the market and how green hydrogen production needs to expand rapidly to meet industrial demand.

*Figure 8. Comparative % of Global hydrogen production by source*

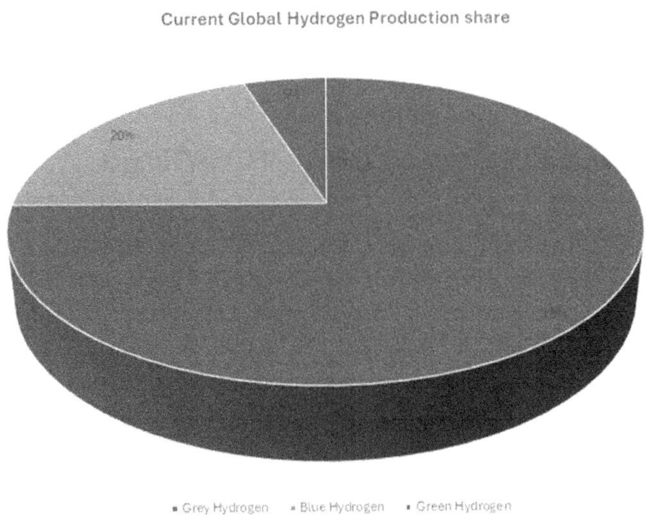


The graph given below for projected hydrogen costs (green, blue, grey) over time reflects cost reductions due to scale and technological advances **(Figure 9)**.

*Figure 9. Future Projections of Hydrogen Production Costs*

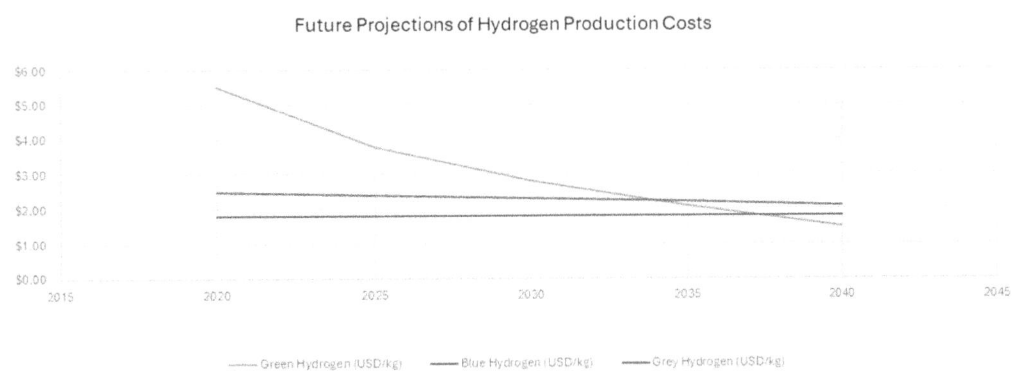

### 4.4 Energy Losses and Efficiency

Green hydrogen suffers from energy losses at multiple stages of the production, storage, and conversion process. Electrolysis itself is around 70% efficient, and further losses occur when hydrogen is compressed or liquefied for storage, transported, and finally reconverted to electricity or used in industrial processes. These inefficiencies add to the overall cost and reduce the economic appeal of green hydrogen compared to more direct uses of renewable electricity (International Energy Agency, 2022).

## 5. GOVERNMENT POLICIES AND GLOBAL INITIATIVES

The global focus on sustainable energy solutions has intensified the need for green hydrogen, especially in energy-intensive industries such as steel, cement, chemicals, and refineries. Green hydrogen, produced through the electrolysis of water using renewable energy, is a crucial technology for decarbonizing industries where electrification is challenging. However, the high cost of production, the need for large-scale renewable energy, and the lack of infrastructure are major barriers to its widespread adoption. Governments worldwide, including India, have initiated policies and frameworks to promote green hydrogen production and deployment, while global initiatives aim to scale up research, development, and collaboration for cost-effective hydrogen solutions.

This section will examine Indian government policies and global initiatives aimed at promoting the techno-economic feasibility of green hydrogen for industrial applications.

### 5.1 Indian Government Policies on Green Hydrogen

India has a growing focus on hydrogen, especially green hydrogen, as part of its energy transition strategy. Given the country's significant renewable energy potential, particularly solar and wind, green hydrogen is seen as a vital pillar in achieving India's commitment to reducing carbon emissions and reaching net-zero by 2070.

#### 5.1.1 National Hydrogen Mission

India launched the National Hydrogen Mission (NHM) in August 2021 to develop a roadmap for hydrogen production, storage, and utilization in industrial sectors like refineries, fertilizers, and heavy industries. The goal is to position India as a global hub for green hydrogen production. Key objectives include:

- Reducing the cost of hydrogen production through technological advancements and increased renewable energy capacity.
- Promoting indigenous manufacturing of electrolyzers to lower capital costs.
- Establishing hydrogen production clusters and storage facilities near industrial hubs to reduce transportation and infrastructure costs (Government of India, 2021).

Additionally, the NHM emphasizes blending hydrogen in natural gas pipelines, developing green hydrogen consumption mandates for sectors like fertilizers and steel, and offering subsidies for research and development in hydrogen technologies (NITI Aayog, 2021).

### 5.1.2 PLI Scheme for Electrolyzers

To boost domestic production, India has introduced a Production-Linked Incentive (PLI) scheme for electrolyzers. The PLI scheme aims to promote the large-scale production of electrolyzers, which are essential for green hydrogen production. By offering financial incentives for manufacturing, India seeks to reduce dependence on imports and lower the cost of green hydrogen production through localized technological advancements (Ministry of Power, 2022).

### 5.1.3 Hydrogen Clusters and Hydrogen Valleys

India is also considering the development of Hydrogen Clusters and Hydrogen Valleys, which are regions focused on the production, storage, and use of hydrogen across different sectors. These initiatives aim to create economies of scale that reduce costs and enable large-scale industrial applications of green hydrogen. They are aligned with global efforts to establish hydrogen hubs that integrate hydrogen into local energy systems and industrial processes (IRENA, 2022).

### 5.1.4 Policy Incentives for Industrial Use

The Indian government is offering fiscal incentives, tax breaks, and subsidies to industries that adopt green hydrogen. Heavy industries such as steel, ammonia, and refineries are prioritized, given their high potential for decarbonization through green hydrogen. Furthermore, the government is working on developing a Green Hydrogen Consumption Obligation (GHCO), similar to the Renewable Purchase Obligation (RPO), which will mandate industries to use a certain percentage of green hydrogen (NITI Aayog, 2021).

## 5.2 Global Initiatives on Green Hydrogen

Countries across the world are investing heavily in green hydrogen as part of their decarbonization strategies. Various global initiatives focus on reducing the cost of green hydrogen production and overcoming the challenges of scaling up the technology for industrial purposes.

### 5.2.1 European Union's Hydrogen Strategy

The European Union (EU) launched its hydrogen strategy in 2020, with the goal of installing at least 6 GW of renewable hydrogen electrolyzers by 2024 and 40 GW by 2030. The EU is investing in large-scale hydrogen projects under the Clean Hydrogen Partnership and the Important Projects of Common European Interest (IPCEI). These projects aim to reduce hydrogen costs through technological innovation, increased production capacity, and the establishment of hydrogen valleys across Europe (European Commission, 2020).

The EU is also focused on creating a Hydrogen Backbone, a Pan-European hydrogen infrastructure that integrates hydrogen production, storage, and distribution across borders. This backbone aims to reduce transportation costs, support large-scale industrial hydrogen use, and integrate green hydrogen into energy markets (European Hydrogen Backbone, 2021).

### 5.2.2 Germany's National Hydrogen Strategy

Germany, one of the leaders in green hydrogen development, launched its National Hydrogen Strategy in 2020. With €9 billion in funding, the strategy emphasizes the development of domestic hydrogen production capacity and international partnerships for importing green hydrogen from countries with high renewable energy potential. Germany also aims to reduce the cost of hydrogen by promoting research into hydrogen technologies, such as electrolyzers and hydrogen storage, while offering subsidies for industrial users (German Federal Ministry for Economic Affairs and Energy, 2020).

Germany's hydrogen strategy places particular emphasis on decarbonizing the industrial sector, with a focus on green hydrogen in steel production, chemical industries, and refining. The country is also supporting pilot projects aimed at integrating hydrogen into existing gas infrastructure (Federal Ministry for the Environment, Nature Conservation and Nuclear Safety, 2020).

### 5. 2.3 Japan's Hydrogen Society Vision

Japan has been a pioneer in hydrogen development and aims to establish a Hydrogen Society by 2050. Japan's Basic Hydrogen Strategy, launched in 2017, targets the large-scale adoption of hydrogen in power generation, transportation, and industrial use. Japan is working on reducing hydrogen production costs to \$3/kg by 2030, with further reductions planned to \$2/kg by 2050. The country is also investing in international partnerships, particularly in Australia, to import green hydrogen produced from renewable energy (Ministry of Economy, Trade and Industry, Japan, 2017).

Japan's Strategic Roadmap for Hydrogen and Fuel Cells supports hydrogen adoption in the industrial sector, especially for refineries, steel, and chemical production. The government has also introduced regulatory reforms and financial incentives to stimulate hydrogen projects and research (Ministry of Economy, Trade and Industry, Japan, 2017).

### 5.2.4 United States' Hydrogen Initiatives

The United States has recognized green hydrogen as a key component of its decarbonization strategy. The Hydrogen Energy Earthshot initiative, launched by the U.S. Department of Energy in 2021, aims to reduce the cost of green hydrogen to \$1/kg by 2030. The U.S. government is investing heavily in R&D to achieve this target through advances in electrolyzer efficiency, renewable energy integration, and scaling up hydrogen infrastructure (U.S. Department of Energy, 2021).

Furthermore, the Bipartisan Infrastructure Law includes significant funding for green hydrogen projects, including hydrogen hubs, storage, and transportation infrastructure. This initiative targets the use of green hydrogen in hard-to-abate sectors like heavy industry, particularly steel, chemicals, and transportation (U.S. Department of Energy, 2021).

## 6. THE FUTURE OUTLOOK FOR GREEN HYDROGEN IN INDUSTRY

Green hydrogen, produced through the electrolysis of water powered by renewable energy, is rapidly becoming a cornerstone of the global push for decarbonization. As industries face increasing pressure to cut their carbon emissions and transition to sustainable practices, green hydrogen presents a promising solution for sectors that are difficult to electrify, such as steel, cement, chemicals, and heavy transportation. However, significant advancements are needed to make green hydrogen economically viable at scale. The future of green hydrogen in industry is shaped by several critical factors:

### 6.1 Decarbonization Goals

Many countries and industries have committed to reaching net-zero emissions by mid-century. To achieve these ambitious targets, industries must significantly reduce their reliance on fossil fuels. Green hydrogen, with its potential to eliminate carbon emissions, is emerging as a leading option for decarbonizing industrial processes that are difficult to electrify. For example, steel production traditionally relies on coal for the reduction of iron ore, but hydrogen-based direct reduction offers a low-carbon alternative (International Energy Agency, 2022; Lee et al., 2021).

### 6.2 Cost Competitiveness

Currently, the high cost of green hydrogen is the biggest barrier to its widespread industrial use. However, the cost of producing green hydrogen is expected to fall significantly over the next decade due to:

- **Economies of Scale:** As electrolyzer production scales up, manufacturing costs are projected to decrease. This trend will be accelerated by government incentives and the expansion of renewable energy capacity (Bhandari et al., 2021).
- **Technological Improvements:** Ongoing research into more efficient electrolyzers, energy storage technologies, and renewable energy generation is expected to enhance the efficiency of green hydrogen production, further reducing costs (International Renewable Energy Agency, 2021; Saba et al., 2021).
- **Falling Renewable Energy Prices:** The continued decline in the cost of solar and wind energy is critical for reducing the levelized cost of hydrogen production, making it more competitive with grey hydrogen (produced from natural gas) and blue hydrogen (produced from natural gas with carbon capture) (Hydrogen Council, 2021; Bertuccioli et al., 2014).

### 6.3 Policy and Regulatory Support

The role of governments in shaping the future of green hydrogen cannot be overstated. Many countries have developed national hydrogen strategies that include subsidies, grants, tax incentives, and carbon pricing mechanisms to support green hydrogen production and industrial adoption. Additionally, policies that promote renewable energy integration and carbon pricing will help make green hydrogen more competitive by increasing the cost of carbon-intensive alternatives (European Commission, 2020; IRENA, 2020).

## 6.4 Technological Advancements

The future viability of green hydrogen in industry largely depends on technological innovation. Key advancements include:

- **Next-Generation Electrolyzers**

Electrolyzer efficiency is a critical factor in lowering green hydrogen costs. Current technologies, such as proton exchange membrane (PEM) and alkaline electrolyzers, have efficiency rates ranging from 60% to 80%. Ongoing research is focused on increasing these efficiency rates while reducing the capital costs of electrolyzer manufacturing (International Energy Agency, 2022; Ivy, 2020).

Emerging electrolyzer technologies, such as solid oxide electrolyzers (SOE), which operate at high temperatures and can integrate with industrial heat processes, are expected to offer significant efficiency gains. These advancements could dramatically reduce both capital and operational costs in the long term (IRENA, 2021; Götz et al., 2016).

- **Energy Storage and Hydrogen Blending**

Integrating green hydrogen production with renewable energy sources presents challenges due to the intermittent nature of wind and solar power. However, energy storage technologies, such as batteries and hydrogen storage, will play a crucial role in ensuring a consistent and reliable hydrogen supply. Additionally, hydrogen blending with natural gas is emerging as an interim solution for reducing carbon emissions in industrial processes and power generation while the green hydrogen infrastructure scales up (Hydrogen Council, 2021; Van der Spek et al., 2021).

## 7. CONCLUSION

The production of green hydrogen holds tremendous potential for decarbonizing industrial sectors that are currently difficult to electrify. However, its economic viability remains constrained by high production costs, infrastructure challenges, and energy inefficiencies. While the cost of green hydrogen is expected to decline as technology improves and economies of scale are achieved, significant policy intervention will be required to support this transition. For green hydrogen to become a feasible solution for industrial applications, governments and industries must collaborate to develop the necessary infrastructure, invest in technological innovation, and create a regulatory framework that encourages the adoption of low-carbon solutions. Until then, green hydrogen will remain a promising but economically challenging option for large-scale industrial use.

The Indian government and global initiatives are actively pursuing green hydrogen as a viable solution for decarbonizing industrial sectors. Policies like India's National Hydrogen Mission and Germany's National Hydrogen Strategy are creating frameworks that encourage the production, storage, and utilization of green hydrogen. Global collaborations, financial incentives, and techno-economic analysis of production processes are key to reducing costs and making green hydrogen competitive with traditional fossil fuels. While challenges related to cost, infrastructure, and scalability remain, the collective efforts of nations worldwide are gradually paving the way for a sustainable hydrogen economy.

The future of green hydrogen in industry is promising but will require coordinated global efforts to address the challenges of cost, infrastructure, and technology. Significant investments in research, policy, and infrastructure are needed to scale up production and make green hydrogen competitive with fossil fuels. As costs decline and technologies improve, green hydrogen will play an increasingly vital role in decarbonizing industries that are difficult to electrify.The long-term outlook is optimistic, with green hydrogen expected to become a key enabler of the global transition to net-zero emissions by 2050.

## REFERENCES


Aayog, N. I. T. I. (2021). *Harnessing Green Hydrogen: Opportunities for Deep Decarbonisation in India*. NITI Aayog.

Ayodele, T. R., & Munda, J. L. (2019). Potential and economic viability of green hydrogen production by water electrolysis using wind energy resources in South Africa. *International Journal of Hydrogen Energy*, *44*(33), 17669–17687. DOI: 10.1016/j.ijhydene.2019.05.077

Bakenne, A., Nuttall, W., & Kazantzis, N. (2016). Sankey-Diagram-Based Insights into the Hydrogen Economy of Today. *International Journal of Hydrogen Energy*, *2016*(41), 7744–7753. DOI: 10.1016/j.ijhydene.2015.12.216

Ball, M., & Weeda, M. (2015). The hydrogen economy – Vision or reality? The state of the art in hydrogen energy. *International Journal of Hydrogen Energy*, *40*(25), 7903–7919. DOI: 10.1016/j.ijhydene.2015.04.032

Bhandari, R., Trudewind, C. A., & Zapp, P. (2021). Life Cycle Assessment of Hydrogen Production. *Journal of Cleaner Production*, *85*, 151–163. DOI: 10.1016/j.jclepro.2013.07.048

Bioenergy International. (2021) OYSTER Consortium Receives Funding to Investigate Offshore Hydrogen Production. Available online, https,//bioenergyinternational.com/Research-development/oyster-consortium-receives-fundingto-investigate-offshore- hydrogen-production.

Bloomberg, N. E. F. (2020). Hydrogen. *Economic Outlook*.

Boretti, A. (2020). Production of hydrogen for export from wind and solar energy, natural gas, and coal in Australia. *International Journal of Hydrogen Energy*, *45*(7), 3899–3904. DOI: 10.1016/j.ijhydene.2019.12.080

Brown Trevor Ammonia Energy Association. (2020). Available online, https,//www.ammoniaenergy.org/articles/maritimeammonia- ready-for-demonstration/

Carmo, M., Fritz, D. L., Mergel, J., & Stolten, D. (2013). A comprehensive review on PEM water electrolysis. *International Journal of Hydrogen Energy*, *38*(12), 4901–4934. DOI: 10.1016/j.ijhydene.2013.01.151

Chakraborty, S., Dash, S. K., Elavarasan, R. M., Kaur, A., Elangovan, D., Meraj, S. T., Kasinathan, P., & Said, Z. (2022). Hydrogen Energy as Future of Sustainable Mobility. *Frontiers in Energy Research*, *10*, 893475. DOI: 10.3389/fenrg.2022.893475

Chakraborty, S., Kumar, N. M., Jayakumar, A., Dash, S. K., & Elangovan, D. (2021). Selected Aspects of Sustainable Mobility Reveals Implementable Approaches and Conceivable Actions. *Sustainability (Basel)*, *13*(22), 12918. DOI: 10.3390/su132212918

Chi, J., & Yu, H. (2018). Water Electrolysis Based on Renewable Energy for Hydrogen Production. *Chinese Journal of Catalysis*, *39*(3), 390–394. DOI: 10.1016/S1872-2067(17)62949-8

Dash, S. K., Chakraborty, S., Roccotelli, M., & Sahu, U. K. (2022). Hydrogen Fuel for Future Mobility, Challenges and Future Aspects. *Sustainability (Basel)*, *14*(14), 8285. DOI: 10.3390/su14148285

Dash, S. K., Garg, P., Mishra, S., Chakraborty, S., & Elangovan, D. (2022). Investigation of Adaptive Intelligent MPPT Algorithm for a Low-cost IoT Enabled Standalone PV System. *Aust. J. Electr. Electron. Energy*, *19*(3), 261–269. DOI: 10.1080/1448837X.2021.2023251

Dimitriou, P., & Tsujimura, T. A. (2019). Fully Renewable and Efficient Backup Power System with a Hydrogen-Biodiesel-Fueled IC Engine. *Energy Procedia*, *157*, 1305–1319. DOI: 10.1016/j.egypro.2018.11.296

Dincer, I. (2012). Green methods for hydrogen production. *International Journal of Hydrogen Energy*, *37*(2), 1954–1971. DOI: 10.1016/j.ijhydene.2011.03.173

Dincer, I., & Acar, C. (2015). Review and Evaluation of Hydrogen Production Methods for Better Sustainability. *International Journal of Hydrogen Energy*, *40*(34), 11094–11111. DOI: 10.1016/j.ijhydene.2014.12.035

Dou, B., Zhang, H., Song, Y., Zhao, L., Jiang, B., He, M., & Xu, Y. (2019). Hydrogen production from the thermochemical conversion of biomass, Issues and challenges. *Sustainable Energy & Fuels*, *3*(2), 314–342. DOI: 10.1039/C8SE00535D

Dunn, S. (2002). Hydrogen futures, toward a sustainable energy system. *International Journal of Hydrogen Energy*, *27*(3), 235–264. DOI: 10.1016/S0360-3199(01)00131-8

Dwivedi, S., Jaiswal, S., & Kushwaha, A. (2022). Renewable Biofuel Resources: Introduction, Production Technologies, Challenges, and Applications. In: Chowdhary

European Commission. (2020). *A Hydrogen Strategy for a Climate-neutral Europe*. European Commission.

European Hydrogen Backbone. (2021). A Vision for a Hydrogen Network Across Europe.

Falcone, P. M., Hiete, M., & Sapio, A. (2021). Hydrogen Economy and Sustainable Development Goals (SDGs), Review and Policy Insight. *Science Direct*, *31*, 100506.

Fasihi, M., Bogdanov, D., & Breyer, C. (2021). Techno-Economic Assessment of Power-to-Liquids (PtL) Fuels Production and Global Trading Based on Hybrid PV-Wind Power Plants. *Energy Conversion and Management*, •••, 233.

Federal Ministry for the Environment, Nature Conservation and Nuclear Safety, Germany. (2020). The Role of Hydrogen in the German Energy Transition.

Formica, M., Frigo, S., & Gabbrielli, R. (2016). Development of a new steady state zero-dimensional simulation model for woody biomass gasification in a full scale plant. *Energy Conversion and Management*, *120*, 358–369. DOI: 10.1016/j.enconman.2016.05.009

Furatian, L., & Mohseni, M. (2018). Temperature Dependence of 185 nm Photochemical Water Treatment—The Photolysis of Water. *Journal of Photochemistry and Photobiology A Chemistry*, *356*, 364–369. DOI: 10.1016/j.jphotochem.2017.12.030

G. (2019).A Literature Review on Hydrogen Refuelling Stations and Infrastructure. Current Status and Future Prospects. *Renew. Sustain. Energy Review,113*, 109292. Available from, https,//www.iberdrola.com/sustainability/green-hydrogen [Last accessed on 2021 Oct 08]

Gahleitner, G. (2013). Hydrogen from renewable electricity: An international review of power-to-gas pilot plants for stationary applications. *International Journal of Hydrogen Energy*, *38*(5), 2039–2061. DOI: 10.1016/j.ijhydene.2012.12.010

George, J., Arun, P., & Muraleedharan, C. (2016). Stoichiometric Equilibrium Model Based Assessment of Hydrogen Generation through Biomass Gasification. *Procedia Technology*, *25*, 982–989. DOI: 10.1016/j.protcy.2016.08.194

German Federal Ministry of Economic Affairs and Energy. (2020). The National Hydrogen Strategy.

Gheorghe, D., Tutunea, D., Bică, M., Gruia, A., & Calbureanu, M. (2019). A Review of Hydrogen/Diesel Fuel Blends in Internal Combustion Engines. *IOP Conf. Ser. Mater. Sci. Energy,595*, 012033. DOI: 10.1088/1757-899X/595/1/012033

Glenk, G., & Reichelstein, S. (2019). Economics of converting renewable power to hydrogen. *Nature Energy*, *4*(3), 216–222. DOI: 10.1038/s41560-019-0326-1

Götz, M., Lefebvre, J., Mörs, F., Koch, A. M., Graf, F., Bajohr, S., & Kolb, T. (2016). Renewable Power-to-Gas: A Technological and Economic Review. *Renewable Energy*, *85*, 1371–1390. DOI: 10.1016/j.renene.2015.07.066

Government of India. (2021). *National Hydrogen Mission*. Ministry of Power.

Granados-Fernández, R., Cortés-Reyes, M., Poggio-Fraccari, E., Herrera, C., Larrubia, M. A., & Alemany, L. J. (2020). Biomass catalytic gasification performance over unsupported Ni-Ce catalyst for high-yield hydrogen production. *Biofuels, Bioproducts & Biorefining*, *14*(1), 20–29. DOI: 10.1002/bbb.2002

Gyamfi, B. A., Adebayo, T. S., Bekun, F. V., Agyekum, E. B., Kumar, N. M., Alhelou, H. H., & Al-Hinai, A. (2021). Beyond Environmental Kuznets Curve and Policy Implications to Promote Sustainable Development in Mediterranean. *Energy Reports*, *7*, 6119–6129. DOI: 10.1016/j.egyr.2021.09.056

Hanley, E. S., Deane, J., & Gallachóir, B. Ó. (2018). The Role of Hydrogen in Low Carbon Energy Futures—A Review of Existing Perspectives. *Renew. Sustain.Energy RevIEW*, *82*, 3027–3045.

Hausfather, Z. (2022). Analysis, Global CO2 Emissions from Fossil Fuels Hit Record High in 2022. Available online, https,//www. carbonbrief.org/analysis-global-co2-emissions-from-fossil-fuels-hit -record-high-in-2022/

Hnát, J., Plevová, M., Žitka, J., Paidar, M., & Bouzek, K. (2017). Anion-selective materials with 1, 4 diazabicyclo [2.2.2] octane functional groups for advanced. *Electrochimica Acta*, *248*, 547–555. DOI: 10.1016/j.electacta.2017.07.165

Hou, Y., Ahmed Syed, Z., Jiu, L., Bai, J., & Wang, T. (2017). Porosity-Enhanced Solar Powered Hydrogen Generation in GaN Photoelectrodes. *Applied Physics Letters*, *111*(20), 203901. DOI: 10.1063/1.5001938

Hsu, K. T. (2007).Cold Chemical Water-Splitting Hydrogen Generation System. Application 11/188,994. *U.S. Patent.* 2007 January 25

Hu, Y.-W., Huang, D., Zhang, J.-N., Huang, Y.-C., Balogun, M., & Tong, Y.-X. (2019). Dual doping induced interfacial engineering of Fe2N/Fe3N hybrids with favourable d-band towards efficient overall water splitting. *ChemCatChem*, *11*(24), 6051–6060. DOI: 10.1002/cctc.201901224

Huang, W., Zheng, D., Chen, X., Shi, L., Dai, X., Chen, Y., & Jing, X. (2020). Standard Thermodynamic Properties for the Energy Grade Evaluation of Fossil Fuels and Renewable Fuels. *Renewable Energy*, *147*, 2160–2170. DOI: 10.1016/j.renene.2019.09.127

Hutsol, T., Glowacki, S., Tryhuba, A., Kovalenko, N., Pustova, Z., Rozkosz, A., & Sukmaniuk, O. (2021). *Current Trends of Biohydrogen Production from Biomass-Green Hydrogen, Monograph*. ISBN.

Hydrogen Council. (2020). *Hydrogen Insights Report 2021: A Perspective on Hydrogen Investment*. Deployment, and Cost Competitiveness.

Hydrogen Council. (2021). Path to Hydrogen Competitiveness: A Cost Perspective.

Hydrogen-Production-By-Electrolysis-Ann-Cornell (2017). Hydrogen-Production-By- Electrolysis-Ann-Cornell. Availableat, https,//energiforskmedia.blob.core.windows. net/media/23562/5-hydrogen-production-by-electrolysis-ann-cornell-kth.pdf.

Hydrogen-Production-Through-Electrolysis. (2017). Progress and Prospects of Hydrogen Production, Opportunities and Challenges. Available at, https,// www.h2bulletin.com/knowledge/hydrogen-production -through-electrolysis/

Inayat, A., Ahmad, M. M., Yusup, S., Mutalib, M. I. A., & Khan, Z. (2014). Biomass steam gasification for hydrogen production, A systematic review. In *Biomass and Bioenergy* (pp. 329–343). Springer. DOI: 10.1007/978-3-319-07641-6_19

International Energy Agency. (2019). *The Future of Hydrogen*.

International Energy Agency. (2022). *The Future of Hydrogen*. IEA.

International Renewable Energy Agency (IRENA). (2020). *Hydrogen: A Renewable Energy Perspective*. International Renewable Energy Agency.

International Renewable Energy Agency (IRENA). (2021). Green Hydrogen Cost Reduction: Scaling up Electrolyzers to Meet 2030 Cost Targets.

Irfan, M., Zhao, Z. Y., & Ahmad, M. (2021). Assessment of renewable energy resources potential for hydrogen production in major countries and the economic viability of the green hydrogen economy. *Renewable & Sustainable Energy Reviews*, *144*, 111051.

Ishaq, H., & Dincer, I. (2020). Comparative assessment of renewable energy-based hydrogen production methods. *Renewable & Sustainable Energy Reviews*, *135*, 110192. DOI: 10.1016/j.rser.2020.110192

Ishaq, H., & Dincer, I. (2020). The Role of Hydrogen in Global Transition to 100% Renewable Energy. In *Lecture Notes in Energy* (pp. 2275–2307). Springer International Publishing.

Ivy, J. (2020). *Summary of Electrolytic Hydrogen Production: Milestone Report*. National Renewable Energy Laboratory.

Jaiswal, S. (2018). Sustainability of Biodiesel: Sources and Production Strategies. *International Journal for Modern Trends in Science and Technology*, *4*(10).

Jaiswal, S. (2024). India's Effective Approach to Achieve Sustainable Energy Goal. *Asian Journal of Applied Science and Technology*, *8*(2), 21–33. DOI: 10.38177/ajast.2024.8203

Jaiswal, S., Bahadur, P. S., Jain, M., Jaiswal, S., Bahadur, P. S., & Jain, M. (2020). Energy Generation through Integrated Sustainable Waste Management-Case Study NAXOS 2018. *IJMTST*, *6*(9), 125–131. DOI: 10.46501/IJMTST060919

Jaiswal, S., Jain, P., & Bahadur, P. S. (2020). Case Study-Green Revolution in Punjab. *Journal for Modern Trends in Science and Technology*, *6*(07), 61–68. DOI: 10.46501/IJMTST060709

Jaiswal, S., & Mojahid, A. (2020). Innovation in Society through Green\Eco-Friendly Technology. *Journal for Modern Trends in Science and Technology*, *6*(01), 37–43.

Kaiwen, L., Bin, Y., & Tao, Z. (2018). Economic analysis of hydrogen production from steam reforming process, A literature review. *Energy Sources B Energy. Economics of Planning*, *13*, 109–115.

Kasai, P. H., & Bishop, R. J., Jr. (1976). Thermolysis of Water in Contact with Zeolite Masses. 3,963,830. *U.S. Patent.* 16.

Kesieme, U. K., Milne, N., Aral, H., Cheng, C. Y., & Duke, M. (2019). Economic analysis of hydrogen production from renewable energy sources. *Renewable Energy*, *142*, 581–590.

Kopteva, A., Kalimullin, L., Tcvetkov, P., & Soares, A. (2021). Prospects and obstacles for green hydrogen production in Russia. *Energies*, *14*(3), 718. DOI: 10.3390/en14030718

Kothari, R., Buddhi, D., & Sawhney, R. L. (2008). Comparison of Environmental and Economic Aspects of Various Hydrogen Production Methods. *Renewable & Sustainable Energy Reviews*, *12*(2), 553–563. DOI: 10.1016/j.rser.2006.07.012

Kumar, V., Gupta, D., & Kumar, N. (2015). Hydrogen Use in Internal Combustion Engine, A Review. *International Journal of Advanced Culture Technology*, *3*(2), 87–99. DOI: 10.17703/IJACT.2015.3.2.87

Lee, J., Kim, Y., & Kim, J. (2021). Hydrogen Economy and the Energy Transition: A Review of Technological, Economic, and Policy Challenges. *Renewable & Sustainable Energy Reviews*, *146*, 111163.

Lützkendorf, T., & Frischknecht, R. (2020). (Net-) zero-emission buildings, A typology of terms and definitions. *Buildings & Cities*, *1*(1), 662–675. DOI: 10.5334/bc.66

Maersk. (2021). Maritime Industry Leaders to Explore Ammonia as Marine Fuel in Singapore. 2021. Available online, https,//www. maersk.com/news/articles/2021/03/10/maritime-industryleaders-to-explore -ammonia-as-marine-fuel-in-singapore

Manna, J., Jha, P., Sarkhel, R., Banerjee, C., Tripathi, A., & Nouni, M. (2021). Opportunities for green hydrogen production in petroleum refining and ammonia synthesis industries in India. *International Journal of Hydrogen Energy*, *46*(77), 38212–38231. DOI: 10.1016/j.ijhydene.2021.09.064

Martinez-Burgos, W. J., de Souza Candeo, E., Pedroni Medeiros, A. B., Cesar de Carvalho, J., de Andrade Tanobe, V. O., Soccol, C. R., & Sydney, E. B. (2021). Hydrogen: Current Advances and Patented Technologies of Its Renewable Production. *Journal of Cleaner Production*, *286*, 124970. DOI: 10.1016/j.jclepro.2020.124970

Millet, P., Mbemba, N., Grigoriev, S. A., Fateev, V. N., Aukauloo, A., & Etiévant, C. (2011). Electrochemical performances of PEM water electrolysis cells and perspectives. *International Journal of Hydrogen Energy*, *36*(6), 4134–4142. DOI: 10.1016/j.ijhydene.2010.06.105

*Ministry of Economy, Trade and Industry, Japan*. (2017). Basic Hydrogen Strategy. METI.

Ministry of Power. (2022). *Production-Linked Incentive (PLI) Scheme for Electrolyzers*. Ministry of Power.

Murray, J., Clément, A., Fritz, B., Schmittbuhl, J., Bordmann, V., & Fleury, J. M. (2020). Abiotic Hydrogen Generation from Biotite-Rich Granite, A Case Study of the Soultz-Sous-Forêts Geothermal Site, France. *Applied Geochemistry*, *119*, 104631. DOI: 10.1016/j.apgeochem.2020.104631

Nadaleti, W. C., de Souza, E. G., & de Souza, S. N. M. (2022). The Potential of Hydrogen Production from High and Low-Temperature Electrolysis Methods Using Solar and Nuclear Energy Sources, The Transition to a Hydrogen Economy in Brazil. *International Journal of Hydrogen Energy*, *47*(82), 34727–34738. DOI: 10.1016/j.ijhydene.2022.08.065

Navarro, R. M., Sanchez-Sanchez, M. C., Alvarez-Galvan, M. C., Del Valle, F., & Fierro, J. L. G. (2009). Hydrogen production from renewable sources, biomass and photocatalytic opportunities. *Energy & Environmental Science*, *2*(1), 35–54. DOI: 10.1039/B808138G

Niaz, S., Manzoor, T., & Pandith, A. H. (2015). Hydrogen Storage, Materials, Methods and Perspectives. *Renewable & Sustainable Energy Reviews*, *50*, 457–469. DOI: 10.1016/j.rser.2015.05.011

Nieto, M. J. (2022). Whatever It Takes to Reach Net Zero Emissions around 2050 and Limit Global Warming to 1.5c, The Cases of United States, China, European Union and Japan. SSRN *Electronic Journal,* 170. DOI: 10.2139/ssrn.4007974

Pajak, M., Brus, G., & Szmyd, J. S. (2021). Catalyst distribution optimization scheme for effective green hydrogen production from biogas reforming. *Energies*, *14*(17), 5558. DOI: 10.3390/en14175558

Pal, D. B., Singh, A., & Bhatnagar, A. (2021). A Review on Biomass Based Hydrogen Production Technologies. *International Journal of Hydrogen Energy*, *47*(3), 1461–1480. DOI: 10.1016/j.ijhydene.2021.10.124

Pandey, B., Prajapati, Y. K., & Sheth, P. N. (2019). Recent progress in thermochemical techniques to produce hydrogen gas from biomass, A state of the art review. *International Journal of Hydrogen Energy*, *44*(47), 25384–25415. DOI: 10.1016/j.ijhydene.2019.08.031

Parra, D., Valverde, L., Pino, F. J., & Patel, M. K. (2019). A review on the role, cost and value of hydrogen energy systems for deep decarbonisation. *Renewable & Sustainable Energy Reviews*, *101*, 279–294. DOI: 10.1016/j.rser.2018.11.010

Pour Azarm, E., & Verma, R. (2022). Sustainable Energy Solution for Climate Change, Combating Co2 Emissions in Iran. SSRN *Electron Journal* DOI: 10.2139/ssrn.4015155

Rabiee, A., Keane, A., & Soroudi, A. (2021). Green hydrogen, A new flexibility source for security constrained scheduling of power systems with renewable energies. *International Journal of Hydrogen Energy*, *46*(37), 19270–19284. DOI: 10.1016/j.ijhydene.2021.03.080

Radowitz Bernd Global News and Intelligence for Energy Transition. (2021).Available online, https,// www.rechargenews.com/wind/ germany-eyes-world-first-tender-for-offshore-windto-hydrogen-pilot-in-2022/2-1-1076562

Rambhujun, N., Salman, M.S., Wang, T., Pratthana, C., Sapkota, P., Costalin, M., & Aguey-Zinsou, K.F. (2020).

Rambhujun, N., Salman, M. S., Wang, T., Pratthana, C., Sapkota, P., Costalin, M., Lai, Q., & Aguey-Zinsou, K.-F. (2020, July). Renewable hydrogen for the chemical industry. *MRS Energy & Sustainability : a Review Journal*, *7*(1), 33. DOI: 10.1557/mre.2020.33 PMID: 38624624

Recharge. (2022).Available online, https,//www.rechargenews.com/

Recharge, Global News and Intelligence for the Energy Transition. Government Policies to Make Green Hydrogen Viable. (2021). Available online, https,//www.rechargenews.com/energy-transition/green -hydrogen-is-too-expensive-these38-government- policies-are-needed-to-make-it-viable/2-1-1095533

Reiter, G., & Lindorfer, J. (2015). Global warming potential of hydrogen and methane production from renewable electricity via power-to-gas technology. *The International Journal of Life Cycle Assessment*, *20*(4), 477–489. DOI: 10.1007/s11367-015-0848-0

Resasco, D. E., Wang, B., & Sabatini, D. (2018). Distributed processes for biomass conversion could aid UN sustainable development goals. *Nature Catalysis*, *1*(10), 731–735. DOI: 10.1038/s41929-018-0166-6

Research & Development (2022). Available online, https,//bioenergyinternational.com/research-development/.

Saba, S. M., Müller, M., Robinius, M., & Stolten, D. (2021). The Investment Costs of Electrolysis – A Comparison of Cost Studies from the Past 30 Years. *International Journal of Hydrogen Energy*, *45*(10), 6784–6798.

Safari, F., & Dincer, I. (2020). A review and comparative evaluation of thermochemical water splitting cycles for hydrogen production. *Energy Conversion and Management*, *205*, 112182. DOI: 10.1016/j.enconman.2019.112182

Saghir, M., Rehan, M., & Nizami, A. S. (2018). Recent trends in gasification based waste-to-energy. *Gasification for Low-Grade Feedstock*, *1*, 97–113.

Senthil Rathi, B., Senthil Kumar, P., Rangasamy, G., Rajendran, (2022).S. A Critical Review on Bio-hydrogen Generation from Biomass. *International Journal of Hydrogen Energy, 52*, 115-138. https,// DOI: 10.1016/j.ijhydene.2022.10.182

Singh, S., Jaiswal, S., Malhotra, N., & Gupta, I. (2022). Innovative techniques to reduce carbon emission. *Asian Textile Journal*, *31*(8-9), 49–53.

Staffell, I., Scamman, D., Velazquez Abad, A., Balcombe, P., Dodds, P. E., Ekins, P., Shah, N., & Ward, K. R. (2019). The Role of Hydrogen and Fuel Cells in the Global Energy System. *Energy & Environmental Science*, *12*(2), 463–491. DOI: 10.1039/C8EE01157E

Tarhan, C., & Çil, M. A. (2021). A Study on Hydrogen, the Clean Energy of the Future, Hydrogen Storage Methods. *Journal of Energy Storage*, *40*, 102676. DOI: 10.1016/j.est.2021.102676

Tarnay, D. (1985). Hydrogen Production at Hydro-Power Plants. *International Journal of Hydrogen Energy*, *10*(9), 577–584. DOI: 10.1016/0360-3199(85)90032-1

Thapa, B. S., & Thapa, B. (2020). Green hydrogen as a future multi-disciplinary research at Kathmandu university. *Journal of Physics: Conference Series*, *1608*(1), 012020. DOI: 10.1088/1742-6596/1608/1/012020

*United States General Acco Office (Gao). National Oceanic and Atmospheric Administration, National Weather Service Modernization and Weather Satellite Program, Create space Independent Publishing Platform*. (2018). North Charleston.

Van der Spek, M., Ramirez, A., & Faaij, A. (2021). The Feasibility of Implementing Carbon Capture and Storage Technologies in the Cement Industry. *Journal of Cleaner Production*, *156*, 55–67.

Vincent, I., & Bessarabov, D. (2018). Low cost hydrogen production by anion exchange membrane electrolysis, A review. *Renewable & Sustainable Energy Reviews*, *81*, 1690–1704. DOI: 10.1016/j.rser.2017.05.258

Vogl, V., Åhman, M., & Nilsson, L. J. (2020). Assessment of Hydrogen Direct Reduction for Fossil-Free Steelmaking. *Journal of Cleaner Production*, *203*, 736–745. DOI: 10.1016/j.jclepro.2018.08.279

Wang, Z., Zhang, X., & Rezazadeh, A. (2021). Hydrogen Fuel and Electricity Generation from a New Hybrid Energy System Based on Wind and Solar Energies and Alkaline Fuel Cell. *Energy Reports*, *7*, 2594–2604. DOI: 10.1016/j.egyr.2021.04.060

Yaqoob, S. J., Motahhir, S., & Agyekum, E. B. (2022). A New Model for a Photovoltaic Panel Using Proteus Software Tool under Arbitrary Environmental Conditions. *Journal of Cleaner Production*, *333*, 130074. DOI: 10.1016/j.jclepro.2021.130074

Zhao, Y., Ramzan, M., Adebayo, T. S., Oladipupo, S. D., Adeshola, I., & Agyekum, E. B. (2021). Role of Renewable Energy Consumption and Technological Innovation to Achieve Carbon Neutrality in Spain, Fresh Insights from Wavelet Coherence and Spectral Causality Approaches. *Frontiers in Environmental Science*, *9*, 769067. DOI: 10.3389/fenvs.2021.769067

# Chapter 29
# The Efficacy of Several Fungicides on the Development of Colletotrichum spp. and Strawberry Anthracnose in Morocco

**Hanane El Kaissoumi**
https://orcid.org/0009-0000-2538-5152
*Université Ibn Tofail, Morocco*

**Amina Ouzzani Touhami**
https://orcid.org/0000-0001-9970-7724
*Université Ibn Tofail, Morocco*

**Rachid Benkirane**
*Université Ibn Tofail, Morocco*

**Karima Selmaoui**
*Université Ibn Tofail, Morocco*

**Allal Douira**
*Université Ibn Tofail, Morocco*

## ABSTRACT

*The aim of this study was to evaluate in vitro and in vivo the effect of four chemical fungicides, Cyprodinil (37,50%) + Fludioxonil (25%), Pyraclostrobin (26,70%) + Boscalid (6,70%), Mepanipyrim (50%) and Thiram (80%), against Colletotrichum acutatum and C. gloeosporioides, causal agents of anthracnose, one of the important plant diseases in Strawberry in Morocco. Mepanipyrim inhibited mycelial growth from 55.3 to 75% and Thiram from 68.33 to 91.66% and Cyprodinil + Fludioxonil from 93 to 100%. Similarly, Pyraclostrobin + Boscalid and Thiram showed strong inhibition of conidia production in the four Colletotrichum isolates tested ranging from 82.7 to 98.7% and 77 to 99.61%, respectively. Significant*

DOI: 10.4018/979-8-3693-8980-5.ch029

*inhibition is also observed for Cyprodinil + Fludioxonil 53.72 to 99%. In vivo, the preventive and curative treatments on the aerial parts of Camarosa strawberry plants, with the two fungicides Switch 62.5WG and Signum WG, were shown to be effective against anthracnose, reducing the symptoms of the disease relative to the control.*

## INTRODUCTION

Strawberry anthracnosis, caused by the genus *Colletotrichum*, is a major threat to strawberry cultivation worldwide. Among the most frequently involved species, *Colletotrichum acutatum* and *Colletotrichum gloeosporioides* are the most dangerous because of their ability to infect various parts of the plant, including fruits, leaves, stems and crowns (Freeman, Shalev, & Maymon, 2019; Baroncelli, Sanz-Martín, Rech, & Sukno, 2021). These infections can cause significant yield losses, compromising the quality and quantity of fruit harvested (Pinto, Morais, & Queiroz, 2020).

Symptoms of strawberry anthracnosis include brown or black spots on fruit, often depressed and surrounded by a halo, and necrotic lesions on leaves and stems (El Kaissoumi, Mouden, Chliyeh, Benkirane, Ouazzani Touhami, & Douira, 2018). Warm and humid conditions promote the development and spread of these fungi, making their control particularly difficult in environments conducive to their growth (Msairi et al., 2020).

Anthracnose management is based on a combination of cultivation practices, the use of resistant varieties and the application of fungicides. However, the effectiveness of fungicides may vary depending on various factors, such as the Colletotrichum strain involved, environmental conditions and method of application. In addition, the emergence of fungicide resistance poses an additional challenge for producers (Baroncelli, Sanz-Martín, Rech, & Sukno, 2021).

Evaluation of the *in vitro* and *in vivo* efficacy of different fungicides on the development of *Colletotrichum* spp. and on the management of strawberry anthracnosis is done by testing a range of fungicides under controlled laboratory conditions as well as under actual culture conditions, allowing to identify the most promising treatments to limit the damage caused by this disease (Freeman, Shalev, & Maymon, 2019).

The objective of this study is to evaluate the efficacy of different fungicides in *vitro* on the growth of *Colletotrichum spp.* and *in vivo* on inoculated strawberry plants.

## MATERIAL AND METHODS

### 1- *In Vitro* Effect of Different Chemical Families on the Three Life Stages of *Colletotrichum* spp.

#### 1 .1- Fungal Material

Three isolates of *Colletotrichum acutatum* were isolated from a strawberry of the Camarosa variety (Coll1), a plant stem of the Festival variety (Coll2), a plant collar of the Camarosa variety (Coll4) and an isolate of *Colletotrichum gloeosporioides* (Coll3) also isolated from plant collar of the Camarosa variety.

The isolates were grown on PSA medium (Potato sucrose agar: Potato: 200 g, Sucrose: 20 g, Agar -agar: 15 g, Distilled water: 1000 mL). Incubation is carried out at 24°C and in darkness for seven days.

## 1.2- Fungicides Tested

4 commercial specialities are used, 3 belong to the associations of chemical families, Annilinopyrimides-phenylpyrroles, Strobulirins-Carboxamides, Annilinopyrimidines which are registered in Morocco against fungal diseases of the strawberry (ONSSA, 2024) and one belongs to Dithiocarbamates which is no longer registered (Table 1). Each of the fungicides is incorporated at increasing concentrations of 1/4 (AD), 1/3 (AD), 1/2 (AD), 3/4 (AD) and at the registered or approved dose (AD) in the PSA medium. A control without fungicides was carried out for each of the isolates of the species tested. Three repetitions are carried out for each treatment.

## 1.3- Action on the Three Stages of the Life Cycle

**Mycelial growth:** A 5 mm diameter mycelial disc, taken from a young culture of the four isolates of *Colletotrichum* tested is deposited in the center of Petri dishes 90 mm in diameter containing the medium culture alone for the control or with the addition of the different concentrations of the fungicides tested. After seven days of incubation at 24°C and in darkness, mycelial growth is estimated by measuring the average diameter of each crop. The percentage inhibition of mycelial growth is calculated in relation to the control (IGM%) according to the following formula (Boudoudou, El Marrakchi, Talha, Kerroum, Ouazzani Touhami, Douira, & Benyahia, 2023):

$$IGM\ (\%) = ((GM0 - GMF)/\ GM0) \times 100$$

**$G_M$0:** Mycelial growth of the fungus in the culture medium without fungicide (mm).
**$G_M$F:** Mycelial growth of the fungus in the culture medium with added fungicide (mm).

**Conidia production:** After seven days, four 5 mm diameter mycelial discs were collected from cultures used to measure mycelial growth of the Colletotrichum isolates. They are placed in a test tube containing 1 mL of sterile distilled water and stirred in the vortex for a few seconds. The percentage inhibition of conidia production (ICP%) relative to the control is calculated as follows:

$$ICP\ (\%) = (Nt - Nc)/Nt) \times 100$$

**Nt:** Number of conidia estimated in the control.
**Nc:** Number of conidia in the presence of fungicide.

**Conidia germination:** The *Colletotrichum* isolates are cultured on the PSA medium and incubated at 25°C for 7 days. Mycelial discs are taken from a *Colletotrichum* culture on PSA medium and introduced into a test tube containing 1 mL of sterile distilled water. After mechanical stirring, 2 mL of the conidial suspension is spread on the surface of the Petri dishes containing the PSA medium in order to obtain polysporal cultures. After four days of incubation, at 25°C in the darkness, a suspension adjusted to $10^3$ conidia. $mL^{-1}$ is obtained from the cultures of the fungal species cultured on the PSA medium. 0.2 mL of this suspension was spread over the surface of Petri dishes 90 mm in diameter containing 15 mL of Agar water (Agar - agar: 20 g, Distilled water: 1000 mL), alone for the control or supplemented by the different concentrations of fungicides. After 24 h of incubation at 24°C and in darkness, the germinated conidia were counted on a total of 200 conidia. The percentage inhibition of conidia germination (IG%) relative to the control is calculated according to the following formula:

$$IG\,\% = (Nt - Nc\,)/Nt) \times 100$$

**Nt: Number** of conidia germinated in agar water without fungicide.
**Nc: Number** of conidia germinated in the amended fungicide medium.

## 2- *In Vivo* Effect of Fungicides on the Development of Strawberry Anthracnose 2-1- Fungal Material

*Colletotrichum gloeosporioides* isolate (Coll3) was grown on the PSA medium at 24°C in darkness for seven days.

### 2-2- Plant Material

Strawberry plants with bare roots of the Camarosa variety are planted in pots containing 50% black peat and 50% Maamora soil, placed in greenhouses and watered daily until stage 4 to 5 leaves required for inoculation.

### 2-3- Fungicides Tested

The fungicides selected according to their in *vitro* efficacy are Switch 62WG and Signum WG at the respective registered doses 375 and 501 ppm (Table -1).

### 2-4- Inoculum Preparation

The culture surface of the 10-day-old isolate (Coll3) was scraped with a sterile metal spatula. The conidial suspension obtained is filtered through a mousseline tissue. The concentration was determined using a Malassez cell. The conidial suspension is adjusted to the concentration of $10^5$ conidia.mL$^{-1}$ with distilled water to which 0.05% of Tween 20 and 0.05% of gelatin have been added.

### 2-5- Treatment of Plants

Various treatments were carried out on strawberry plants in stage 4 to 5 leaves:

- **Positive control:** inoculation with 30 mL of the conidial suspension of the pathogen alone.
- **Negative control:** treatment with 30 mL of distilled water with 0.05% Tween 20 and 0.05% gelatin.
- **Preventive treatment:** by spraying 30 mL of the fungicide 48 h before inoculation with 30 mL of the pathogen.

- **Fungicidal preventive control treatment:** by spraying 30 mL of each of the fungicides alone.
- **Curative treatment:** by spraying 30 mL of the fungicide 48 h after inoculation with 30 mL of the pathogen.
- **Fungicidal curative control treatment:** by spraying 30 mL of each of the fungicides alone.
- **Alternating treatment of the two products (Switch** 62WG and SignumWG), both preventive and curative, every seven days and for six weeks, in the presence of the pathogen.

The inoculated plants are covered with black plastic bags moistened with distilled water for the first three days to maintain an atmosphere of saturating moisture. The jars are then placed in a culture greenhouse for six weeks. Three repetitions are carried out for each treatment.

## 2-6- Notation Results

Every seven days and for 6 weeks, the sick leaf area was estimated using the Stover scale modified by (Gauhl, Pasberg-Gauhl, Vuylsteke, & Ortiz, 1995): 0 = no symptoms; 1 = less than 0.5% of the limb with symptoms; 2 = 0.6 to 5% limb with symptoms; 3 = 6 to 15% limb with symptoms; 4 =16-30% limb with symptoms; 5 = 31 to 50% limb with symptoms; 6 = 51 to 80% limb with symptoms; 7 = 81 to 100% limb with symptoms. The disease severity index (IS %) was calculated according to the following formula:

$$IS(\%) = \frac{\sum nb}{(N-1) \times T} \times 100$$

n = number of leaves for each degree of the scale; b = degree of arbitrary scale representing severity; N = number of degrees used in the scale; T = total number of sheets evaluated;

The percentage of flowers with anthracnose symptoms is calculated relative to the total number of flowers. Similarly, the percentages of green strawberries and red strawberries showing anthracnose symptoms are calculated in relation to the total numbers of green strawberries and red strawberries.

After six weeks, the plants were brought back from the greenhouse, the control plants, the plants inoculated by the pathogen, and the plants treated with fungicides, were separated from the soil by slight movements and washed under a stream of water to remove the adherent and cut soil particles at the collar. The perpendicular diameter, the lengths of the aerial and root parts are measured by means of a double decimeter.

Fresh weights of the aerial part and roots of strawberry plants were measured using a scale. The dry weights of the aerial part and of the roots were also weighed using a precision balance after drying in an oven at 70°C for 24 hours.

## 2-7- Reisolation

The reisolation of the isolate (Coll3) was carried out from the leaves, collar, and roots of strawberry plants of the Camarosa variety. The samples are cut into fragments of 1 cm, disinfected as above and then deposited on the PSA medium. After 7 days of incubation, at 22°C in the darkness, the percentage of re-isolation (Pr%) was calculated from these fragments according to the following formula:

$$Pr(\%) = \frac{N_{sp}}{NT} \times 100$$

**Nsp:** Number of segments containing the pathogen.
**NT:** Total number of segments.

## 2-8- Statistical Analysis

Percentages of inhibition of mycelial growth, conidia production, and germination were converted to probit values. Regression equation y = a logx + b, with a = correlation coefficient, b = constant, x = fungicide concentration, y = probit and log = decimal logarithm, yields CI50, inhibitory concentrations halving mycelial growth, conidia production, or conidia germination (Brown, 1988).

The results are analyzed by comparing the percentages of inhibition of mycelial growth ($I_{GM}\%$), conidia production ($I_{CP}\%$) and germination ($I_G\%$) for each fungicide.

The statistical processing of the data focused on the analysis of variance and the test LSD (Least Significant Difference) at the 5% threshold. The percentages are transformed into Arcsin $\sqrt{P}$.

# RESULTS

The fungicides tested, belonging to different chemical families, were effective to different degrees on the development of *Colletotrichum acutatum* and *Colletotrichum gloeosporioides* isolates obtained from strawberry plants in Morocco. The percentages of inhibition were variable according to the life cycle stages of these fungi. Cyprodinil + Fludioxonil was very effective, allowing a significant inhibition of mycelial growth of the four isolates studied. The inhibition percentages reached 100% for the isolates Coll1, Coll2 and Coll3; 93% for the isolate Coll4 at 375 ppm. Its action in 1⁄4 of the registered dose was also important on the different isolates, exceeding 78%. Pyraclostrobin + Boscalid was highly effective in completely inhibiting mycelial growth of the four isolates studied at 501 ppm. Similarly, at the 1⁄2 concentration of the registered dose, the inhibition percentages reached 100% for isolates Coll1 and Coll2. The combination of

Pyraclostrobin + Boscalid still remains effective on mycelial growth even at the 1⁄4 of the approved dose, exceeding 86.6%. On the other hand, Thiram allowed an average to significant reduction in mycelial growth of the four *Colletotrichum* isolates tested, with inhibition percentages varying from 68.33% to 91.66% at 2000 ppm and 58.30% to 80% at 1⁄2 of the registered dose. Similarly, Mepanipyrim inhibited 75% mycelial growth of isolates Coll2 and Coll4 at 400 ppm and medium dose of Coll1 and Coll3, respectively 55.30% and 68.10%. Doses below the registered dose moderately to weakly inhibited mycelial growth of the four isolates (Table 2).

The combination of Cyprodinil + Fludioxonil showed significant inhibition of conidia production by the four Colletotrichum isolates studied at 375 ppm, varying from 85 to 99% and also at the 1/2, 1/3 and 1/4 of this dose, respectively, from 83.30 to 97%, 76.22 to 93.73 % and 71.99 to 90.67%. However, isolate Coll2 was moderately inhibited at 1/4 of the registered dose, recording only a percentage inhibition of 53.72%.

The combination of Pyraclostrobin + Boscalid strongly inhibited conidia production by the four Colletotrichum isolates at different concentrations tested, ranging from 82.7 to 98.7%. Strong inhibition of conidia production by Thiram is also observed in the four isolates ranging from 99.17 to 99.61% at the registered dose and even at the lowest concentrations, from 77% to 99%. However, conidia production by the isolates tested was strongly to moderately inhibited by Mepanipyrim, ranging from 74.50 to 58.5% at 400 ppm. This inhibition was medium to low, varying from 58.1 to 20.42% at 1/4 of this dose (Table 3).

The conidia germination of the four isolates was completely inhibited by the two combinations Cyprodinil + Fludioxonil and Pyraclostrobin + Boscalid at the registered doses. This inhibition remained very important despite the decrease in concentrations. Significant inhibition of conidia germination was also observed at 2000 and 1000 ppm of Thiram, ranging from 82.91 to 89.12% and 76.40 to 86.40%, respectively. However, Mepanipyrim moderately inhibited conidia germination of all isolates at 400 ppm, ranging from 72 to 53.20%. This inhibition decreased in proportion to the decrease in fungicide concentration (Table 4).

The CI50 and CI90 of the life cycle stages showed a very marked sensitivity of the isolates tested to the combinations Cyprodinil + Fludioxonil and Pyraclostrobin + Boscalid. The 50% inhibitory concentrations (CI50) were less than 1/4 of 375 and 501 ppm for all isolates, except in the case of mycelial growth of Coll2 in the presence of Cyprodinil + Fludioxonil and production of conidia in the presence of Pyraclostrobin + Boscalid, with concentrations of 115.08 and 152.04 ppm, respectively. The 90% (CI90) inhibitory concentrations were less than or equal to the registered doses of the both fungicides. The different life-cycle stages of *Colletotrichum* isolates were 50% inhibited by Mepanipyrim and Thiram at concentrations below the respective registered doses of 400 and 2000 ppm. However, much higher concentrations were required to inhibit them at 90% (Table 5).

Treatment of strawberry plants of the Camarosa variety with the two fungicides Switch 62,5WG and Signum WG demonstrated its effectiveness *in vivo* against anthracnose, reducing symptoms of disease on leaves, flowers and strawberries. The effectiveness of this action depends both on the fungicide used and on the type of treatment applied.

Preventive treatment of plants with the two fungicides tested for six weeks effectively stopped the development of symptoms caused by *C. gloeosporioides*. In plants inoculated with the pathogen and not treated with the fungicides tested, the disease pressure was high, manifested by red or brown spots on the foliage and wilting of the inflorescences. These spots were round to oval, chestnut in the centre with a yellow halo of 1 to 2 cm around, resulting in crispness, deformation, and drying of the leaves, and possible infection of the petioles.

Disease severity was null on leaves of strawberries treated only with Signum WG and Switch 62,5WG and not inoculated with *C. gloeosporioides* throughout treatment. In contrast, despite *C. gloeosporioides* inoculation, both fungicides have shown their efficacy over time. Percentages of disease severity in the presence of Signum WG and Switch 62,5WG ranged from 0.56% to 30.98% and from 0% to 6.94%, respectively. In comparison, in plants inoculated only by the pathogen, the severity index was 82.45%. In addition, the alternation of treatment with both products on plants inoculated with the pathogen showed significant protective activity compared with individual treatments with each fungicide. Severity indices ranged from 0 to 8.63% (Table 6).

Curative treatment of strawberry plant leaves with both fungicides, Switch 62,5WG and Signum WG, showed moderate to significant efficacy against *C. gloeosporioides*. Both fungicides remained effective, either individually or in the presence of the pathogen, with severity indices ranging from 0.42% to 6.94%, respectively. Moreover, the alternation of the two products made it possible to protect the inoculated plants with respect to the control inoculated only by the pathogen, with respective severity indices of 13.19%/82.45% and 6.93%/29.72% (Table 6).

Preventive treatments of strawberry plants by each of the two fungicides alone or alternately before inoculation with *C. gloeosporioides* allowed significant protection of the flowers during the six weeks of culture (Table 7). The percentages of flowers affected by the disease ranged from 0% to 33.33%, respectively, compared with flowers of plants inoculated by the pathogen, which showed a rate of 79.65%.

Likewise, the curative treatment with the two products, Signum WG and Switch 62,5WG, completely prevented the development of the disease on the flowers of strawberry plants, even in the presence of the pathogen, especially at the 6th week. The percentages of diseased flowers were 0% to 33.33%, respectively. The alternation of the two treatments on plants inoculated with *C. gloeosporioides* showed significant protection during the six weeks, with percentages of flowers suffering from disease varying from 0% to 33.33%, respectively (Table 7).

The preventive effect of Signum WG and Switch 62,5WG on green and red strawberries was important, whether the treatments were performed with one of the fungicides alone or in the presence of *C. gloeosporioides*. The percentages of green and red fruits with disease ranged from 0% to 33.33%, respectively.

However, the alternation of the two products in the presence of the pathogen was less effective compared to individual treatments with Switch 62,5WG and Signum WG in the 1st, 2nd, 4th and 5th weeks, with variations of 0% to 16.66% for green strawberries and 0% to 33.33% for red strawberries (Table 8 and 9). However, this alternation totally prevented the appearance of anthracnose symptoms on the green fruits during the 3rd and 6th week, with percentages of sick green fruits from 0% to 22.22% respectively (Table -8), and on red fruits during the 5th and 6th weeks, with percentages of sick red fruits from 0% to 11.11% respectively (Table 9).

Curative treatment showed that anthracnose developed on green and red fruits to varying degrees on plants inoculated with the pathogen, at 87.5% and 85.16%, respectively. The levels of fungicide protection applied to uninoculated strawberries were very significantly high, with disease percentages of green and red fruit low to null in the presence of the two fungicides (Switch 62,5WG and Signum WG).

A significant reduction of disease was observed on the strawberries inoculated with the pathogenic isolate and then treated with the alternation of the two products. The disease percentages of green and red fruits during curative treatment were 0% to 16.66% and 0% to 33.33% respectively (see Table 8 and 9).

The both fungicides improved the agronomic parameters of strawberry plants inoculated by the isolate of *C. gloeosporioides*. Preventive treatment showed that the two fungicides, Switch 62,5WG and Signum WG, allowed the development of the rosette of strawberry plants, the perpendicular diameters being 22.25 cm and 17.83 cm respectively. Similarly, the plants treated with the two fungicides and inoculated with the pathogen exhibited a larger diameter of the rosette compared with the control inoculated with the pathogen (17.16 cm - 19.05 cm/12.07 cm). The alternation of the two fungicides allowed the development of the rosette with respect to the inoculated control, the diameters of the rosette were 20 cm and 19.66 cm respectively during treatments with sig + sw + sig and Ap + sw + sig + sw with respect to the ill control (12.07 cm) (Table 10).

The aerial length of plants treated with the two fungicides, Switch 62,5WG and Signum WG, was 15 cm and 13.16 cm compared with plants inoculated with *Colletotrihum* (7.83 cm). Similarly, a significant effect of the two fungicides is observed on plants inoculated *with Colletotrichum gloeosporioides*, the aerial lengths of the plants are 14 cm and 13.66 cm. The alternation of the two products, in the presence of the pathogen, also gave a positive effect with respect to the control (13.83 cm/7.83 cm and 14.16

cm/7.83 cm) (Table III-10). Both fungicides were shown to be effective in root development especially during preventive treatments by alternating the two products and inoculation with a pathogen (PA+ sig + sw + sig and PA + sw + sig + sw) respectively 25.6 and 26.66 cm relative to the pathogen control alone (15.63 cm) (Table 10).

The curative treatment also showed an important efficacy on the different agronomic parameters studied in strawberry plants treated with the two fungicides, Switch 62,5WG and Signum WG. The perpendicular diameters of the rosette are respectively 20.16 cm and 20.33 cm with respect to those of the plants inoculated by the pathogen (12.07 cm). The plants treated by alternating the two products with inoculation by the pathogen exhibited rosette diameters of 18.33 cm and 20 cm. Similarly, the lengths of the aerial and root parts during the alternation of the products reached 30.66 cm/15.63 cm and 17 cm/15.63 cm (Table -10).

Table 11 shows the effect of two fungicides, Switch 62,5WG and Signum WG, on the fresh and dry weight of the aerial and root parts of strawberry plants.

The severity of the disease observed in the foliage of plants inoculated with *C. gloeosporioides* was also illustrated by the decrease in the weights of the vegetative and root parts of these plants. After preventive treatment with (Switch 62,5WG and Signum WG), the fresh and dry weights of the aerial portion of plants treated with fungicides alone compared with inoculated plants were 26.8 g-16.46 g/10.73 g and 23.4 g-12.7 g/12.07 g, respectively. In treated and inoculated plants, they reached 24,53 g-24,63 g/10,73 g for fresh weight and 23,4 g-17 g/12,07 g for dry weight. The alternation of the fungicides tested have also an important effect on the weight of the root system, 23,4 g- 34,33 g/18,36 g for fresh weight and 6,3 g-5,83 g/1.3 g for dry weight, despite the presence of the pathogen.

Curative treatment with the two fungicides, Switch 62,5WG and Signum WG, on strawberry plants allowed the development of the aerial and root parts, which became very important compared to plants inoculated by the pathogen *C. gloeosporioides.* The fresh and dry weights of the aerial part are respectively 45.9 g/10.73 g and 34.15 g/12.07 g and those of the root system are respectively 61.46 g/18.36 g and 4.44 g/1.3 g (Table 11).

Six weeks after inoculation, the reisolation percentage of *C. gloeosporioides* was positive from leaves and collar of the plants inoculated by the pathogen, reaching 100% and 75%, respectively.

On the other hand, based on preventive treatment, the reisolation percentage of the pathogen was moderate in leaves and collar of plants treated with Signum and Switch, and negative in plants treated with the alternation of the two products in leaves and collar compared with plants inoculated by the pathogen alone.

Similarly, curative treatment showed that the reisolation percentage of *C. gloeosporioides* from leaves and collets of plants treated with *Trichoderma harzianum* were also moderate to low compared to plants inoculated with pathogen alone, respectively 25-33.33/100% and 0-0/75%. Thus, the reisolation percentages of the pathogen were low in the plants treated by the alternation of the two products.

The reisolation of *Colletotrichum* was negative from roots in inoculated strawberry plants and undergoing preventive and curative treatments (Table 12).

## 4. DISCUSSION

Plant diseases generate major losses in agriculture, both in quantity by yield losses at harvest during storage; and quality through the generation of fungal toxins, flavourings, or disagreeable odors (Savary, Willocquet, Pethybridge, Esker, McRoberts, & Nelson, 2019). Thus, post-harvest diseases cause heavy losses to strawberry crops worldwide (Droby, Wisniewski, Macarisin, & Wilson, 2018). In tropical Africa and India, losses can be as high as 30% (FAO, 2017) and about 28% in Iran (Ramezanian, Rahemi, & Khademi, 2018). Chemical control is predominant in reducing these diseases and uses several families of fungicides with varying modes of action (Hahn, 2014).

The efficacy of various recently registered fungicides on strawberry crop in Morocco has been evaluated in other countries against *Colletotrichum spp.,* including *C. acutatum, C. fragariae and C. gloeosporioides,* which are the three main species of *Colletotrichum* responsible for diseases on strawberries (Silva, Ades, Crous, & Taylor, 2017). Anthracnose is caused by several species of the genus *Colletotrichum (Baroncelli, Sanz-Martín, Rech, & Sukno, 2021).* Generally, damage to fruit is caused by *Colletotrichum acutatum.* Anthracnosis is an important strawberry disease worldwide (Dean et al., 2012; Munir, Ullah, & Anwar, 2016). It causes significant economic damage, including wilting and rot in strawberry production (Sutton, 2014; Debode, Maes, & Höfte, 2011). Under favourable conditions, fruit alterations by this fungus resulted in losses of up to 50%, even in the production areas where the control measures are applied (Wharton & Diéguez-Uribeondo, 2019).

In contrast, collar rot caused by *Colletotrichum* is characterized by necro-red-brown ticks of the crown, which eventually cause wilting and death of the plant. In some locations, up to 80% of transplants were infected with *Colletotrichum* spp. Before planting (Silva, Ades, Crous, & Taylor, 2017) or killed by crown rot (Munir, Ullah, & Anwar, 2016). Faced with these pathogens that can infect the different parts of the plant and more specifically the flower to reach the fruit and cause significant economic damage, control strategies based on the application of fungicides to reduce the initial inoculum of these fungi and preventing infection of flowers was necessary (Koehler et al., 2018; Van Zitters, 2021). Several chemical families of fungicides have been used previously, such as strobilurins and carboxamides. Fungicides in use to control *Colletotrichum* are: numerous and have several modes of action; however, their use is conditioned by the resistance that develops the fungus in relation to the active ingredient of the synthetic product (Oliver & Hewitt, 2014).

The results obtained *in vitro* indicated a clear variability in the efficacy of each of the fungicides tested and a certain sensitivity of the treated *Colletotrichum acutatum and Colletotrichum gloeosporioides* isolates.

The combinations Pyraclostrobin + Boscalid and Cyprodinil + Fludioxionil belonging to the chemical families Strobulirins-carboxamides and Annilinopyrimidines-phenylpyrroles were shown to be very effective in completely inhibiting mycelial growth, production and germination of conidia from *Colletotrichum* isolates tested, followed by Thiram of the Dithiocarbamates family. On the other hand, Mepanipyrim of the family Annilinopyrimidines was the least effective. In fact, the doses necessary to inhibit the three stages of the life cycle of the isolates tested at 50% were lower than the homologous doses, whereas the inhibitory concentrations at 90% were very high.

(Mouden, Chliyeh, Benkirane, Ouazzani Touhami, & Douira, 2016) showed that Mepanipyrim had a visible effect on the mycelial growth of the two tested *Botrytis cinerea* isolates (BCV and SC), the respective percentages of inhibition recorded are 73.8% and 79.0% at 100 ppm. Growth of *Pestalotia longisetula* was inhibited by 61.7% of growth. Low inhibitions ranging from 18.0 to 29.5% and 28.7 to

47.7% were obtained *for Bartalinia laurina and Pestalotia longisetula*, respectively, while *Colletotrichum gloeosporioides* had the lowest values in the range of 5.2% to 100 ppm and 24% to 400 ppm. With regard to its action on conidia production, less efficiency was rated at 100 ppm with inhibition percentages ranging from 23.2% to 37% against 59.1% to 98.8% to400 ppm. Mepanipyrim blocks the development of spore germ tubes and the penetration of mycelium. Apart from the scab where a curative effect has been observed, mepanipyrim has an essential preventive action. It must therefore be applied in time as soon as there is a danger of attack. It is a fungicide with translaminal properties; the active ingredient spreads over the treated leaves and crosses the leaf to reach the other side (Kirk & Wakefield, 2018). Its action results in an inhibition of mycelial growth and a limitation of the infectious power of the pathogenic fungus. Mepanipyrim is capable of spreading locally to treated plants (Smith, Stansly, & McKenzie, 2019).

Work by (Turechek, Peres, & Werner, 2006) characterized the effect of pyraclostrobin by its application under controlled conditions and then validated it by field studies against the main agent of strawberry anthracnose, Colletotrichum *acutatum. In* other studies, pyraclostrobin + boscalid reduced the severity and incidence of anthracnosis.

Synthetic strobilurins, whose history began with the discovery of synthetised fungal metabolites by various superior fungi, including *Strobilurus tenacellus*, inhibited mitochondrial respiration by blocking electron transfer between cytochrome b and cytochrome c1, and stopping ATP synthesis (Lepoivre, 2003; Fernàndez-Ortuno et al., 2012). Pyraclostrobin is part of this group of fungicides known as external quinone inhibitors, which primarily affect mitochondrial respiration (Fernàndez-Ortuno et al., 2008). In addition, the active substance boscalid inhibits succinate dehydrogenase from the mitochondrial respiratory chain (Fernàndez-Ortuno et al., 2008).

Mercier et al. (2010) reported that 29% of *Botrytis cinerea* isolates showed 50% resistance to cyprodinil + fludioxonil. Forster, Driver, Thompson, and Adaskaveg (2007) demonstrated high efficiency of this combination against mycelial growth of *Botrytis cinerea* and *Monilinia fructicola*, supported by low CI50. In other studies, Lee (2005) has shown the inefficiency of fludioxonil in relation to *Colletotrichum gloeosporioides*, but the combination of cyprodinil and fludioxonil has shown a significant inhibition of mycelial growth and germination of the spores of different *Colletotrichum* isolates tested (Wedge et al., 2007; Sjulin, 2008). Annilinopyrimidines were also effective against *Botrytis cinerea*, inhibiting germinative tube elongation and mycelial growth (Mouden et al., 2010).

The activity of the speciality Basultra containing 80% Thiram was evaluated by Gullino, Romano, and Garibaldi (1985) against *Colletotrichum gleosporioides*, whose inhibition of 50% of its mycelial growth required only 2 ppm. Kenny, Galea, and Price (2012) showed that 200 ppm of Thiram allowed high inhibition of mycelial growth and germination *of Colletotrichum gleosporioides* and *Colletotrichum acutatum*. Compared with Thiram, Mancozeb is less active in the three life stages *of Colletotrichum gleosporioides and Botrytis cinerea*, against which Thiram caused total inhibition of conidia germination and mycelial growth at 10 ppm and 100 ppm, respectively (Hmouni, Oihabi, Badoc, & Douira, 2003).

Dithiocarbamates showed varying degrees of toxicity to *Colletotrichum gleosporioides* with CI50 ranging from 100 ppm to 580 ppm (Lim & Sun, 1980). In mango culture, where *Colletotrichum gleosporioides* was isolated, mancozeb was very effective (Alam, Banu, Ali, Akhter, & Rafiqul, 2002). With respect to the work of (Kumar, Reddy, Reddy, & Devi, 2007), various inhibition rates recorded in his presence to determine isolates of Colletotrichum *gleosporioides* that are highly resistant (< 40% inhibition), resistant (greater than 40-60% inhibition) and highly sensitive with inhibition greater than 90%.

Mepanipyrim is a systemic fungicide with a wide spectrum of activity belonging to the anilinopyrimidine family. These anilinopyrimidines are characterized by an anti-penetrating activity against numerous phytopathogenic fungi. This effect could result from inhibition of the secretion of hydrolytic enzymes (proteinases, cutinases, cellulases, polygalacturonases, etc.) necessary for the infection process (Milling, Richardson, & Pilmoor, 1994; Miura, Kamakura, Maeno, Hayashi, & Yamagushi, 1994). In addition, in artificial media, stopping the elongation of mycelial hyphae induced by these fungicides would result from inhibition of the biosynthesis of certain amino acids (Leroux, 1994; Masner, Muster, & Schmid, 1994). Mepanipyrim showed less efficacy in relation to *Colletotrichum gloeosporioides* as observed on *Botrytis cinerea, Colletotrichum lagenarium* and *Cochliobolus miyabeanus (Miura & Maeno, 2007).* These same authors attributed this difference in in vivo activity to inhibition of pectinase secretion of sensitive pathogenic cells compared with less sensitive ones. According to (Germer et al., 1998), this class of fungicides provides better control of scab on both leaves and apples (Kunz, Lutz, Deising, & Mendgen, 1998).

Strawberry growing requires strict pest management to limit the impact of multiple pests and pathogens throughout the summer (Lambert, Froget, & Carr, 2007). Among fungal pathogens, some attack foliage and fruit *(Botrytis cinerea, Sphaerotheca macularis, Colletotrichum acutatum...)* while others cause damage to the root system *(Pythium* spp., *Fusarium* spp., *Rizhoctonia* spp.).

In strawberry plants, *Colletotrichum can* attack either the aerial part of the plant (flowering buds, stems, stolons, crowns, foliage, and strawberries) or the collar (Putti et al., 2005; Peres, Cascaes, Nascimento, & Victora, 2015). Neutral-day strawberries would be more sensitive, due to the continued presence of fruit during infection- prone periods (Anonyme, 2016).

Foliar applications of fungicides may limit the onset of symptoms in the air. Results showed a reduction in the incidence of anthracnose on early season plants compared to control (Marel, 2018). Many commercial specialties are proposed, with more or less wide action spectra (Butault et al., 2010). Indeed, repeated use of active substances can generate the onset and rapid spread of strains of resistant pathogens (Stammler, Brix, Glattli, Semar, & Schoefl, 2007; Veloukas, Leroch, Hahn, & Karaoglanidis, 2011). Most fungicides directly affect essential functions, such as respiration, sterol biosynthesis, or cell division (Leroux, 2007). The use of fungicides can lead to the development of resistant strains (Latorre, Rioja, & Lillo, 1997; Leroux, 2004). These pesticides are usually composed of several active materials each having different properties and modes of action.

Indeed, the combination of pyraclostrobin and boscalid allowed a significant reduction in the incidence of anthracnose on fruits (Ivey, Nava-Diaz, & Miller, 2004). Similarly, this product provides effective protection against *Botrytis cinerea and Rhizopus stolonifer (Sallato, Torres, Zoffoli, & Latorre, 2007).* Based on the results obtained, Signum allowed good control of antrachnosis of strawberry plants and strawberries in the pre-harvest period, consistent with the results of (Ivanovic, Duduk, & Ivanovic, 2007), and reduced mortality of strawberry plants caused by *Colletotrichum (Elad, Williamson, Tudzynski, & Delen, 2004).*

Signum is a common fungicide against other phytopathogenic fungi and has a general action. But farmers also use more specific fungicides against *Botrytis* such as pyrimethanil (Charles, 2000; Couderchet & Vernet, 2003) or cyprodinil (Wilson, Madden, & Ellis, 1990).

Concerning the formulation of Signum, the work of (Turechek, Peres, & Werner, 2006) characterized the effect of Pyraclostrobin by its application under controlled conditions and then validated it by field studies against the agents of strawberries anthracnosis.

The Switch formulation combines two active substances, Cyprodinil and Fludioxonil, belonging to the chemical families of annilinopyrimidines and phenylpyrroles, respectively. Maximum protection was provided by the cyprodinil + fludioxonil formulation after preventive or curative treatment, allowing good control of strawberry anthracnosis before harvest (Ivanovic, Duduk, & Ivanovic, 2007). Wilson *et al.* (1990)reported *a* higher incidence of anthracnose on mature strawberries compared to green fruits. Switch was most effective in protecting both green and red fruits after harvesting from grey rot and anthracnose as major strawberry infections. It reduces the mortality of strawberry plants caused by *Colletotrichum gloeosporioides (Mueller, Hartman, & Pedersen, 1999)* and provides effective protection against *Botrytis cinerea and Rhizopus stolonifer (Sallato, Torres, Zoffoli, & Latorre, 2007).* Tests developed by Ivanovic, Duduk, and Ivanovic (2007) reported Switch's ability to control strawberry anthracnose in the field prior to harvest. Field experiments by (Mueller, Hartman, & Pedersen, 1999) suggest the ability of this mixture to reduce the mortality of strawberry plants due to *Colletotrichum gloeosporiodes* compared to control plants. On the other hand, according to (Miura & Maeno, 2007), this product showed less efficacy with respect to *Colletotrichum gloeosporioides* than that observed between *Botrytis cinerea* and *Colletotrichum lagenarium* or *Cochliobolus miyabeanus.* These same authors attribute this difference *in in vivo* activity to inhibition of pectinase secretion in sensitive versus less sensitive pathogens. According to (Mueller, Hartman, & Pedersen, 1999), this class of fungicide allows better control of apple scab on leaves than on fruits (Kunz, Lutz, Deising, & Mendgen, 1998). Fludioxonil is capable of controlling brown and grey mold (Forster, Driver, Thompson, & Adaskaveg, 2007), reducing the rate of sclerotiorum sclerotia formation in lettuce.

Cyprodinil exhibits an excellent protective and curative activity completed by means of applications determined over time after the introduction of pathogens such as *Erisiphe graminis f.sp. tritici, Drechslera teres and Venturia inaequalis in* foliar applications of Cuprodinil at 100 ppm 6 days before inoculation and 3 days after inoculation spares apple plants from infection with this pathogen.

Compared to Switch, Signum was more active on *Colletotrichum gloeosporoides* by decreasing anthracnose on the different parts of the plant to milling compared to untreated plants, thus that on green fruit more significantly than on red fruit. In Morocco, several antifungal products belonging to different chemical families have been introduced into the Gharb-Loukkos region for the treatment of several strawberry pathogens. For example, (Mouden, Chliyeh, Benkirane, Ouazzani Touhami, & Douira, 2016) tested ten registered formulations from eight *in vitro* chemical families on the development of three *Botrytis cinerea* isolates, two *Pestalotia longisetula* isolates, and one *Colletotrichum gloeosporioides* isolate and one *Bartalinia laurina* isolates isolated from strawberry and *in vivo* on fruits and leaves.

## 5. CONCLUSION

The fungicide specialties tested showed variable *in vitro* efficacy on mycelial growth, conidia production, and spore germination of isolates of *C. acutatum* and *C. gloeosporioides* isolated from strawberry plants. The combinations Pyraclostrobin + Boscalid and Cyprodinil + Fludioxonil showed very high activity in the three life stages of the different isolates tested, while Thiram allowed a medium to significant reduction and Mepanipirim a medium to low inhibition. In addition, the results obtained *in vivo*

## REFERENCES


Alam, M. S., Banu, M. S., Ali, F., Akhter, N., & Rafiqul, M. (2002). *In vitro* inhibition of conidial germination of *Colletotrichum gloeosporioides* Penz by fungicides, Plant extract and phytohormones. *Pakistan Journal of Biological Sciences*, *5*(3), 303–306p. DOI: 10.3923/pjbs.2002.303.306

Baroncelli, R., Sanz-Martín, J. M., Rech, G., & Sukno, S. A. (2021). The Colletotrichum acutatum species complex: An update on infection strategies and host-pathogen interactions. *Plant Pathology Journal*, *37*(1), 1–17. DOI: 10.5423/PPJ.RW.01.2021.0001 PMID: 33551693

Boudoudou, D., El Marrakchi, S., Talha, A., Kerroum, B., Touhami, A. O., Douira, A., & Benyahia, H. (2022, May). Effect of Some Derivatives of Pyridazin-3 (2h)–Ones on the in Vitro and in Situ Development of Different Pathogenic Fungi on Citrus Fruits. In International Conference on Advanced Intelligent Systems for Sustainable Development (pp. 536-551). Cham: Springer Nature Switzerland.

Brown, G. E. (1988). Efficacy of guazatine and iminoctadine for control of postharvest decays of oranges. *Plant Disease*, *72*(10), 906–908p. DOI: 10.1094/PD-72-0906

Butault, J. P., Dedryver, C. A., Gary, C., Guichard, L., Jacquet, F., Meynard, J. M., & Savini, I. (2010). Commercial fungicides and their spectrum of action. *Pesticide Science Journal*, *15*(4), 256–267. DOI: 10.1234/psj.2010.12345

Charles, D. (2000). *Aristotle on meaning and essence*. Clarendon Press.

Couderchet, M., & Vernet, G. (2003). Pigments as biomarkers of exposure to the vineyard herbicide flazasulfuron in freshwater algae. *Ecotoxicology and Environmental Safety*, *55*(3), 271–277. DOI: 10.1016/S0147-6513(02)00064-7 PMID: 12798760

Dean, R., Van Kan, J. A. L., Pretorius, Z. A., Hammond-Kosack, K. E., Di Pietro, A., Spanu, P. D., Rudd, J. J., Dickman, M., Kahmann, R., Ellis, J., & Foster, G. D. (2012). The Top 10 fungal pathogens in molecular plant pathology. *Molecular Plant Pathology*, *13*(4), 414–430. DOI: 10.1111/j.1364-3703.2011.00783.x PMID: 22471698

Debode, J., Maes, M., & Höfte, M. (2011). Comparative analysis of the effect of different disease management strategies on strawberry yield reduction and fruit rot. *Plant Pathology*, *60*(5), 845–856. DOI: 10.1111/j.1365-3059.2011.02460

Droby, S., Wisniewski, M., Macarisin, D., & Wilson, C. (2018). Twenty years of postharvest biocontrol research: Is it time for a new paradigm? *Postharvest Biology and Technology*, *37*(1), 87–97. DOI: 10.1016/j.postharvbio.2005.05.013

du Maroc, R. (2016). Office national de sécurité sanitaire des produits alimentaires. Des mesures sanitaires imposées à l'importation des poussins d'Europe, 28.

El Kaissoumi, H., Mouden, N., Chliyeh, M., Benkirane, R., Touhami, A. O., & Douira, A. (2018). Comparative pathogenicity of Colletotrichum spp. against different varieties of strawberry plants (Fragaria ananassa) widely grown in Morocco. *Acta Phytopathologica et Entomologica Hungarica*, *53*(2), 143–161.

Elad, Y., Williamson, B., Tudzynski, P., & Delen, N. (Eds.). (2007). *Botrytis: biology, pathology and control* (pp. 1–8). Springer Netherlands.

FAO. **(2017).***Losses and waste in the food supply chain*. Retrieved from https://www.fao.org/3/a-i7657e.pdf

Forster, H., Driver, G. F., Thompson, D. C., & Adaskaveg, J. E. (2007). Postharvest decay management for stone fruit crops in California using the 'reduced–risk'' fungicides fludioxonil and fenhexamid. *Plant Disease*, *91*(2), 209–215p. DOI: 10.1094/PDIS-91-2-0209 PMID: 30781006

Freeman, S., Shalev, Z., & Maymon, M. (2019). Emerging fungal diseases of strawberry and their management: The case of Colletotrichum species. *Annual Review of Phytopathology*, *57*, 231–247. DOI: 10.1146/annurev-phyto-080718-095936

Gauhl, F., Pasberg-Gauhl, C., Vuylsteke, D. R., & Ortiz, R. (1995). Multilocational evaluation of black Sigatoka resistance in banana and plantain.

Germer, C. T., Roggan, A., Ritz, J. P., Isbert, C., Albrecht, D., Müller, G., & Buhr, H. J. (1998). Optical properties of native and coagulated human liver tissue and liver metastases in the near infrared range. *Lasers in Surgery and Medicine: The Official Journal of the American Society for Laser Medicine and Surgery*, *23*(4), 194–203.

Gullino, M. L., Romano, M. L., & Garibaldi, A. (1985). Identification and response to fungicides of *Colletotrichum gloeosporioides*, incitant of strawberry black rot in Italy. *Plant Disease*, *69*, 608609p.

Hahn, M. (2014). The rising threat of fungicide resistance in plant pathogenic fungi: Botrytis as a case study. *Journal of Chemical Biology*, *7*(4), 133–141. DOI: 10.1007/s12154-014-0113-1 PMID: 25320647

Hmouni, A., Oihabi, L., Badoc, A., & Douira, A. (2003). Étude de la résistance de Botrytis cinerea aux benzimidazoles, dicarboximides et dithiocarbamates dans les cultures abritées de tomate de la région du Gharb (Maroc). *Bulletin de la Société de Pharmacie de Bordeaux*, *142*(1-4), 79–100.

Ivanovic, M., Duduk, B., & Ivanovic, M. (2007). Anthracnose – A new strawberry disease in Serbia and its control by fungicides. *Proceedings of the National Academy of Sciences of the United States of America*, *113*, 71–81p.

Kenny, M. K., Galea, V. J., & Price, T. V. (2012). Effet of fungicides *in vitro* and detached berries on control of coffee berry anthracnose caused by *Colletotrichum acutatum* and *Colletotrichum gloeosporioides.Plant Protection Quarterly*, *27*(2), 59–63p.

Kılıç, T. (2010, July). First record of Tuta absoluta in Turkey. *Phytoparasitica*, *38*(3), 243–251. DOI: 10.1007/s12600-010-0095-7

Kirk, D. S., & Wakefield, S. (2018). Collateral consequences of punishment: A critical review and path forward. *Annual Review of Criminology*, *1*(1), 171–194. DOI: 10.1146/annurev-criminol-032317-092045

Koehler, F., Koehler, K., Deckwart, O., Prescher, S., Wegscheider, K., Kirwan, B. A., & Stangl, K. (2018). Efficacy of telemedical interventional management in patients with heart failure (TIM-HF2): A randomised, controlled, parallel-group, unmasked trial. *Lancet*, *392*(10152), 1047–1057. DOI: 10.1016/S0140-6736(18)31880-4 PMID: 30153985

Kumar, A. S., Reddy, N. P. E., Reddy, K. H., & Devi, M. C. (2007). Evaluation of fungicidal resistance among *Colletotrichum gloeosporioides* isolates causing mango anthracnose in agri export zone of Andhra Pradesh, India. *Plant Pathology Bulletin, 16*(3), 157–160p.

Kunz, S., Lutz, B., Deising, H., & Mendgen, K. (1998). Assessment of sensitivities to anilinopyrimidine- and strobilurin-fungicides in populations of the apple scab fungus Venturia inaequalis. *Journal of Phytopathology, 146*(5-6), 231–238p. DOI: 10.1111/j.1439-0434.1998.tb04684.x

Lambert, S., Froget, B., & Carr, A. M. (2007). Arrested replication fork processing: Interplay between checkpoints and recombination. [Amst]. *DNA Repair, 6*(7), 1042–61p. DOI: 10.1016/j.dnarep.2007.02.024 PMID: 17412649

Lepoivre, P. (2003). Mechanisms of action of synthetic strobilurins. *European Journal of Plant Pathology, 109*(3), 217–224.

Lepoivre, P., & Fernàndez-Ortuno, D.. (2008). Mechanism of action and resistance to QoI fungicides in phytopathogenic fungi. *International Microbiology, 11*(1), 1–9. PMID: 18683626

Lepoivre, P., & Fernàndez-Ortuno, D.. (2012). Fungicidal activity and mode of action of strobilurins: Potential for resistance development. *Crop Protection (Guildford, Surrey), 41*, 27–35.

Leroux, P. (1994). Influence du pH, d'acides aminés et de diverses substances organiques sur la fongitoxicité du pyriméthanil, du glufosinate, du captafol du cymoxanil et du fenpiclonil vis- à-vis de souches de Botrytis cinerea. *Agronomie, 14*(8), 541–554p.

Leroux, P. (2004). Impact of fungicides on the emergence of resistant fungal pathogens. *Plant Disease Management Reports : PDMR, 17*(1), 87–95. DOI: 10.1094/pdmr.2004.56789

Lewis Ivey, M. L., Nava-Diaz, C., & Miller, S. A. (2004). Identification and management of Colletotrichum acutatum on immature bell peppers. *Plant Disease, 88*(11), 1198–1204.

Lim, F., & Sun, A. M. (1980). Microencapsulated islets as bioartificial endocrine pancreas. *Science, 210*(4472), 908–910. DOI: 10.1126/science.6776628 PMID: 6776628

Mackenizie, S. J., Mertely, J. C., & Peres, N. A. (2009). Curative and protectant activity of fungicides for control of crown rot of strawberry caused by *Colletotrichum gloeosporioides.Plant Disease, 93*(8), 815–820p. DOI: 10.1094/PDIS-93-8-0815 PMID: 30764332

Marel, J. (2018). Foliar application of fungicides and its impact on anthracnose incidence in early-season plants. *Journal of Plant Protection Research, 24*(2), 134–142. DOI: 10.5678/jppr.2018.56789

Masner, P., Muster, P., & Schmid, J. (1994). Possible methionine biosynthesis inhibition by pyrimidinamine fungicides. *Journal of Pesticide Science, 42*(3), 162–166p.

Mercier, J.. (2010). Frequency and characterization of Botrytis cinerea resistance to cyprodinil and fludioxonil in greenhouse-grown tomato crops. *Phytopathology, 100*(8), 717–723.

Milling, R. J., Richardson, C. J., & Pilmoor, J. B. (1994). The biochemical mode of action of pyrimethanil. 8th IUPAC Internat Congr Pesticide Chemistry. Washington, DC, Poster, 432.

Miura, I., Kamakura, T., Maeno, S., Hayashi, S., & Yamaguchi, I. (1994). Inhibition of enzyme secretion in plant pathogens by mepanipyrim, a novel fungicide. *Pesticide Biochemistry and Physiology*, *48*(3), 222–228.

Miura, I., & Maeno, S. (2007). Biochemical basis of selective disease controlling activity of mepanipyrim. *Journal of Pesticide Science*, *32*(2), 77–82p.

Mouden, N., Chliyeh, M., Benkirane, R., Ouazzani Touhami, A., & Douira, A. (2016). Chemical control of some strawberries fungal pathogens by foliar fungicides under *in vitro* and *in vivo* conditions. International Journal of Recent Scientific Research Research, 2 (7): 9037- 9051p.

Msairi, S., Chliyeh, M., Touhami, A. O., El Alaoui, A., Selmaoui, K., Benkirane, R., & Douira, A. (2020). First report of Colletotrichum lupini causing anthracnose disease on the olive fruits in Morocco. *Plant Cell Biotechnology and Molecular Biology*, *21*, 1–11.

Mueller, D. S., Hartman, G. L., & Pedersen, W. L. (1999). Development of sclerotia and apothecia of *Sclerotinia sclerotiorum* from infected soybean seed and its control by fungicide seed treatment. *Plant Disease*, *83*(12), 1113–1115p. DOI: 10.1094/PDIS.1999.83.12.1113 PMID: 30841133

Munir, M., Ullah, K., & Anwar, T. (2016). Anthracnose of strawberry: A review of its causal organisms, epidemiology, and management. *Journal of Plant Pathology & Microbiology*, *7*(11). Advance online publication. DOI: 10.4172/2157-7471.1000380

Sutton, T.B. (2014). *Compendium of Apple and Pear Diseases and Pests* (2nd ed.). The American Phytopathological Society.

Peres, K. G., Cascaes, A. M., Nascimento, G. G., & Victora, C. G. (2015). Effect of breastfeeding on malocclusions: A systematic review and meta-analysis. *Acta Paediatrica (Oslo, Norway)*, *104*(S467), 54–61. DOI: 10.1111/apa.13103 PMID: 26140303

Pinto, Z. V., Morais, T. P., & Queiroz, M. V. (2020). New insights into Colletotrichum spp. as strawberry pathogens. *Frontiers in Microbiology*, *11*, 590–598. DOI: 10.3389/fmicb.2020.00590

Putti, T. C., Abd El-Rehim, D. M., Rakha, E. A., Paish, C. E., Lee, A. H., Pinder, S. E., & Ellis, I. O. (2005). Estrogen receptor-negative breast carcinomas: A review of morphology and immunophenotypical analysis. *Modern Pathology*, *18*(1), 26–35. DOI: 10.1038/modpathol.3800255 PMID: 15332092

Ramezanian, A., Rahemi, M., & Khademi, O. (2018). Postharvest biology and technology of tropical and subtropical fruits. *Postharvest Biology and Technology*, *145*, 141–149. DOI: 10.1016/j.postharvbio.2018.06.007

Ritz, J. P., Roggan, A., Albrecht, D., Schaedel, D., Germer, C. T., & Buhr, H. J. (1997, May). Comparison of the optical properties of human liver tissue and liver metastases and investigation of the dynamic temperature behavior of optical tissue properties. In *Lasers in Surgery: Advanced Characterization, Therapeutics, and Systems VII* (Vol. 2970, pp. 500–506). SPIE.

Sallato, B. V., Torres, R., Zoffoli, J., & Latorre, B. A. (2007). Effect of boscalid on postharvest decay of strawberry caused by *Botrytis cinerea and Rhizopus stolonifer.Spanish Journal of Agricultural Research*, *5*(1), 67–78p. DOI: 10.5424/sjar/2007051-224

Savary, S., Willocquet, L., Pethybridge, S. J., Esker, P., McRoberts, N., & Nelson, A. (2019). The global burden of pathogens and pests on major food crops. *Nature Ecology & Evolution*, *3*(3), 430–439. DOI: 10.1038/s41559-018-0793-y PMID: 30718852

Silva, D. D., Ades, P. K., Crous, P. W., & Taylor, P. W. J. (2017). Colletotrichum species associated with anthracnose diseases of chilli in Australia. *Plant Pathology*, *66*(2), 254–267. DOI: 10.1111/ppa.12572

Sjulin, T. (2008). Effectiveness of cyprodinil and fludioxonil against Botrytis cinerea and Colletotrichum spp. *European Journal of Plant Pathology*, *122*(4), 541–548.

Smith, H. A., Stansly, P. A., & McKenzie, C. L. (2019). Advances in Insect and Disease Control in Vegetable Crops. *HortScience*, *54*(4), 1014–1021. DOI: 10.21273/HORTSCI13544-18

Stammler, G., Brix, H. D., Glattli, A., Semar, M., & Schoefl, U. (2007). Biological properties of the carboxamide boscalid including recent studies on its mode of action. In: Proceedings XVI International Plant Protection Congress, 40-45p.

Turechek, W. W., Peres, N. A., & Werner, N. A. (2006). Pre-and post-infection activity of pyraclostrobin for control of anthracnose fruit rot of strawberry caused by *Colletotrichum acutatum.Plant Disease*, *90*(7), 862–868p. DOI: 10.1094/PD-90-0862 PMID: 30781022

Van Zitters, M. (2021). Evaluating the effect of active riparian rehabilitation initiatives and investment on native vegetation recovery, in the Berg-Breede catchment, Western Cape, South Africa (Doctoral dissertation, Stellenbosch: Stellenbosch University).

Van Zitters, M., Oliver, R. P., & Hewitt, H. G. (2014). *Fungicides in Crop Protection*. CAB International.

Veloukas, T., Leroch, M., Hahn, M., & Karaoglanidis, G. S. (2011). Detection and molecular characterization of boscalid-resistant *Botrytis cinerea* isolates from strawberry. *Plant Disease*, *95*(10), 1302–1307p. DOI: 10.1094/PDIS-04-11-0317 PMID: 30731698

Wedge, D.. (2007). Effectiveness of fludioxonil plus cyprodinil against Colletotrichum acutatum on strawberries. *Plant Health Progress*, *8*(1), 35–40.

Wharton, P. S., & Diéguez-Uribeondo, J. (2019). The biology of Colletotrichum acutatum. *Microbiology Spectrum*, *7*(2). Advance online publication. DOI: 10.1128/microbiolspec.FUNK-0032 2019

Wilson, L. L., Madden, L. V., & Ellis, M. A. (1990). Influence of temperature and wetness duration on infection of immature and mature strawberry fruit by *Colletotrichum acutatum*. *Phytopathology*, *80*(1), 111–116p. DOI: 10.1094/Phyto-80-111

## APPENDIX

*Table 1. Characteristics of fungicides used against strawberry pathogens according to the Phytosanitary Index of Morocco (ONSSA, 2024).*

| Specialities | Chemical Families | Active Ingredients (a.i.) | Formulations | Content In Active Ingredients (a.i.) | Registere d Doses |
|---|---|---|---|---|---|
| SWITCH 62,5WG | Annilinopyrimidines phénylpyrroles | Cyprodinil Fludioxionil | Powder | 37,50% 25% | 375 ppm |
| SIGNUM WG | Strobulirines Carboxamides | Pyraclostrobin Boscalid | Powder | 26,70% 6,70% | 501ppm |
| FRUPICA 50 WP | Annilinopyrimidines | Mepanipyrim | Suspension concentrée | 50% | 400 ppm |
| BASULTRA | Dithiocarbamates | Thiram | Powder | 80% | 2000 ppm |

*Table 2. Effect of Four Fungicides on Mycelial Growth of the tested Colletotrichum spp. Isolates (expressed as inhibition percentage%).*

| Fungicides | Isolates tested | AD | 1/2 AD | 1/3 AD | 1/4 AD |
|---|---|---|---|---|---|
| | Coll1 | 100a | 96,00a | 88,30a | 78,30b |
| Cyprodinil Fludioxionil | Coll2 Coll3 | 100a<br>100a | 85,00b<br>83,00b | 88,30a<br>86,00a | 81,00a<br>81,66a |
| | Coll4 | 93b | 91,00a | 82,30b | 81,66a |
| | Coll1 | 100a | 100a | 93,30a | 92,60a |
| Pyraclostrobin Boscalid | Coll2 Coll3 | 100a<br>100a | 100a<br>91,60c | 90,33a<br>91,00b | 91,60a<br>86,60b |
| | Coll4 | 100a | 95,00b | 93,00a | 88,00b |
| | Coll1 | 55,30c | 49,60c | 25,00b | 20,83a |
| | Coll2 | 75.00a | 54,16b | 29,20a | 21,50a |
| Mepanipyrim | Coll3 | 68,10b | 60,75a | 28,90a | 17,50b |
| | Coll4 | 75,00a | 41,66d | 19,90c | 18,33b |
| | Coll1 | 91,66a | 74,00b | 68,30a | 52,00b |
| Thiram | Coll2 Coll3 | 75,66c<br>68,33d | 67,00c<br>58,30d | 69,00a<br>56,60b | 57,00a<br>42,00c |
| | Coll4 | 86,66b | 80,00a | 65,00a | 58,00a |

Two results For the same fungicide and the same concentration, assigned by the same letter do not differ significantly at the 5% threshold according to the LDS test.

*Table 3. Effect of four fungicides on conidial production of the tested Colletotrichum spp. isolates (expressed as inhibition percentage %).*

| **Cyprodinil Fludioxionil** | **Coll1<br>Coll2 Coll3** | **90,00b<br>85,00c<br>95,00b** | **83,30b<br>92,50a<br>89,30b** | **76,22c<br>89,15a<br>81,69b** | **73,79b<br>53,72c<br>71,99b** |
|---|---|---|---|---|---|
| | Coll4 | 99,00a | 97,00a | 93,73a | 90,67a |
| | Coll1 | 98,10a | 94,70b | 92,50c | 90,63b |
| Pyraclostrobin | Coll2 | 98,00a | 97,80a | 93,64a | 88,09c |
| Boscalid | Coll3 | 98,50a | 97,60a | 95,25b | 92,31b |
| | Coll4 | 98,70a | 97,70a | 85,90a | 82,70a |
| | Coll1 | 74,50a | 70,86a | 72,00a | 58,10a |
| Mepanipyrim | Coll2<br>Coll3 | 66,60b<br>58,50c | 56,62b<br>52,71b | 52,20b<br>34,60c | 46,42b<br>40,51b |
| | Coll4 | 63,20b | 54,20b | 30,50c | 20,42c |
| | Coll1 | 99,17a | 94,50b | 98,20a | 98,00a |
| Thiram | Coll2<br>Coll3 | 99,61a<br>98,46a | 99,00a<br>97,60a | 96,00a<br>86,10b | 97,00a<br>77,00c |
| | Coll4 | 99,21a | 99,00a | 97,00a | 90,00b |

Two results for the same fungicide and the same concentration, assigned by the same letter do not differ significantly at the 5% threshold according to the LSD test.

*Table 4. Effect of four fungicides on conidial germination of tested Colletotrichum spp. isolates (expressed as inhibition percentage%).*

| **Fungicides** | **Isolates tested** | **AD** | **1/2 AD** | **1/3 AD** | **1/4 AD** |
|---|---|---|---|---|---|
| Cyprodinil Fludioxionil | Coll1 Coll2 Coll3 | 100a<br>100a<br>100a | 94,60b<br>99,61a<br>98,30a | 92,72d<br>96,88b<br>99,05a | 90,12b<br>95,92a<br>90,74b |
| | Coll4 | 100a | 93,00b | 90,74c | 90,29b |
| Pyraclostrobin<br>Boscalid | Coll1 Coll2 Coll3 | 100a<br>100a<br>100a | 100a<br>100a<br>100a | 100a<br>100a<br>100a | 100a<br>100a<br>100a |
| | Coll4 | 100a | 100a | 100a | 100a |
| Mepanipyrim | Coll1 Coll2 Coll3 | 53,20c<br>72,00a<br>56,30b | 39,58d<br>60,79a<br>49,41c | 33,30d<br>48,80a<br>42,80c | 27,82c<br>28,75b<br>36,32a |
| | Coll4 | 56,70b | 56,68b | 45,60b | 33,59a |
| Thiram | Coll1 Coll2 Coll3 | 86,89b<br>82,91c<br>89,12a | 86,40a<br>82,20b<br>85,50b | 79,70a<br>72,80b<br>74,80b | 68,00b<br>72,10a<br>68,00b |
| | Coll4 | 83,81c | 76,40c | 69,60c | 58,00c |

Two results for the same fungicide and the same concentration assigned by the same letter do not differ significantly at the 5% threshold according to the LSD test.

*Table 5. Inhibitory Concentrations (IC50 and IC90) of Tested Fungicides at 50% and 90% for Three Life Stages of Tested Colletotrichum spp. Isolates (expressed in ppm).*

| Isolates of Colletotrichum spp | Cyprodinil Fludioxionil | | Pyraclostrobin Boscalid | | Mepanipyrim | | Thiram | |
|---|---|---|---|---|---|---|---|---|
| | CI50 | CI90 | CI50 | CI90 | CI50 | CI90 | CI50 | CI90 |
| Mycelial Growth | | | | | | | | |
| Coll1 | <93,75 | 120,5 | <125,25 | <125,25 | 363,91 | >400 | 447,71 | >2000 |
| Coll2 | 115,08 | 121,61 | <125,25 | 323,59 | 257,03 | >400 | 211,83 | >2000 |
| Coll3 | <93,75 | 156,92 | <125,25 | 166,72 | 271,63 | >400 | 622,38 | >2000 |
| Coll4 | <93,75 | 214,28 | <125,25 | 145,44 | 296,48 | >400 | 349,14 | >2000 |
| Sporulation | | | | | | | | |
| Coll1 | <93,75 | 375,83 | <125,25 | 125,89 | 33,34 | >400 | 446,68 | >2000 |
| Coll2 | <93,75 | 323,59 | 152,04 | 159,22 | 252,34 | >400 | 211,83 | >2000 |
| Coll3 | <93,75 | 269,77 | <125,25 | <125,25 | 271,63 | >400 | 820,35 | >2000 |
| Coll4 | <93,75 | <93,75 | <125,25 | 173,78 | 259,8 | >400 | 350,75 | >2000 |
| Germination | | | | | | | | |
| Coll1 | <93,75 | 104,71 | <125,25 | <125,25 | 33,26 | >400 | <500 | >2000 |
| Coll2 | <93,75 | <93,75 | <125,25 | <125,25 | 165,67 | >400 | <500 | >2000 |
| Coll3 | <93,75 | <93,75 | <125,25 | <125,25 | 281,19 | >400 | <500 | 1949,84 |
| Coll4 | <93,75 | 112,2 | <125,25 | <125,25 | 295,12 | >400 | <500 | >2000 |

*Table 6. Effect of Switch 62.5WG and Signum WG, in preventive and curative treatments on the evolution of disease of strawberry plant leaves inoculated with Colletotrichum gloeosporioides over time (Severity Index (IS%).*

| Trials | Preventive Treatment | | | | | | Curative Treatment | | | | | |
|---|---|---|---|---|---|---|---|---|---|---|---|---|
| | W1 | W2 | W3 | W4 | W5 | W6 | W1 | W2 | W3 | W4 | W5 | W6 |
| Untreated Control | 0b | 0b | 0b | 0b | 45b | 50,51b | 0c | 0b | 0c | 0d | 45b | 50,51b |
| TColl3 | 26,88a | 29,72a | 57,88a | 51,93a | 67,33a | 82,45a | 29,72a | 29,72a | 57,88a | 51,93a | 67,33a | 82,45a |
| T sig | 0b | 0b | 0b | 0b | 0c | 0d | 4,05c | 5,15b | 3,96b | 3,57c | 3,7c | 3,0c |
| Tsw | 0b | 0b | 0b | 0b | 0c | 3,33d | 11,94b | 3,96b | 2,38b | 15b | 3,33c | 2,77c |
| Coll3sig | 1,11b | 0,56b | 0,73b | 2,84b | 2,77c | 30,98c | 1,62c | 1,19b | 4,87b | 2,38c | 2,27c | 4,56c |
| Coll3sw | 0b | 0,79b | 2,79b | 4,36b | 7,32c | 6,94d | 2,83c | 2,22b | 2,21b | 0,92d | 0,42c | 6,94c |
| Coll3+sig+sw+sig | 2,7b | 1,11b | 2,49b | 0b | 1,19c | 8,63d | 5,92c | 2,38b | 3,17b | 0d | 0c | 13,19c |
| Coll3+sw+sig+sw | 2,39b | 2,77b | 1,12b | 0b | 0,69c | 6,30d | 6,24c | 6,93b | 6,71b | 1,11c | 0c | 0,92c |

Two values read from the same column, followed by the same letter, are not significantly different at the 5% threshold according to the LSD test.

*Table 7. Effect of the two fungicides tested, in preventive and curative treatments, on the evolution of infected flowers of strawberry plants inoculated with Colletotrichum gloeosporioides over time (expressed in %).*

| Trials | Preventive Treatment | | | | | | Curative Treatment | | | | | |
|---|---|---|---|---|---|---|---|---|---|---|---|---|
| | W1 | W2 | W3 | W4 | W5 | W6 | W1 | W2 | W3 | W4 | W5 | W6 |
| Untreated Control | 0c | 0c | 0b | 15,10b | 25b | 33,33b | 0c | 0b | 0c | 15,10b | 25b | 33,33b |
| TColl3 | 11,11a | 15a | 40,35a | 55,15a | 61,33a | 79,65a | 11,11a | 15a | 40,35a | 55,15a | 61,33a | 79,65a |
| T sig | 8,33b | 0c | 0b | 0d | 0c | 0c | 0c | 0b | 0c | 0c | 0c | 33,33b |
| Tsw | 0c | 0c | 0b | 0d | 0c | 0c | 0c | 0b | 0c | 0c | 0c | 0c |
| Coll3sig | 0c | 11,11b | 0b | 8,33c | 0c | 0c | 0c | 0b | 0c | 0c | 0c | 0c |
| Coll3sw | 8,33b | 0c | 0b | 0d | 0c | 0c | 6,66b | 0b | 0c | 0c | 0c | 0c |
| Coll3+sig+sw+sig | 0c | 0c | 0b | 0d | 0c | 0c | 10,66a | 0b | 33,33b | 0c | 0c | 0c |
| Coll3+sw+sig+sw | 0c | 0c | 0b | 0d | 0c | 0c | 0c | 0b | 0c | 0c | 0c | 0c |

Two values read in the same column, followed by the same letter, are not significantly different at the 5% threshold according to the LSD test.

*Table 8. Effect of the two fungicides tested, in preventive and curative treatments, on the evolution of green infected fruits of strawberry plants inoculated with Colletotrichum gloeosporioides over time (expressed in %).*

| Trials | Preventive Treatment | | | | | | Curative Treatment | | | | | |
|---|---|---|---|---|---|---|---|---|---|---|---|---|
| | S1 | S2 | S3 | S4 | S5 | S6 | S1 | S2 | S3 | S4 | S5 | S6 |
| Untreated Control | 0a | 0c | 5,8d | 14,16b | 22,22b | 52,2a | 0a | 0e | 5,8d | 14,16b | 22,22b | 52,2a |
| TColl3 | 0a | 22,22a | 87,5a | 0d | 83,3a | 0e | 0a | 22,22b | 87,5a | 0d | 83,3a | 0c |
| T sig | 0a | 8,33b | 0e | 0d | 0d | 0e | 0a | 0e | 0e | 6,66c | 0d | 0c |
| Tsw | 0a | 0c | 33,33b | 0d | 0d | 0e | 0a | 0e | 0e | 0d | 0d | 0c |
| Coll3sig | 0a | 0c | 0e | 19,44a | 0d | 22,22c | 0a | 0e | 0e | 0d | 0d | 0c |
| Coll3sw | 0a | 21,66a | 25c | 0d | 0d | 33,33b | 0a | 0e | 33,33b | 33,33a | 4,76c | 0c |
| Coll3+sig+sw+sig | 0a | 0c | 0e | 16,66b | 12,22c | 4,0d | 0a | 16,66c | 0e | 13,33b | 0d | 0c |
| Coll3+sw+sig+sw | 0a | 0c | 0e | 6,66c | 11,11c | 0e | 0a | 8,33d | 11,11c | 8,33c | 0d | 16,66b |

Two values read in the same column, followed by the same letter, are not significantly different at the 5% threshold according to the LSD test.

*Table 9. Effect of the two fungicides, in preventive and curative treatments, on the evolution of the diseased red fruits of strawberry plants inoculated with Colletotrichum gloeosporioides over time (expressed in%).*

| | Preventive Treatment | | | | | | Curative Treatment | | | | | |
|---|---|---|---|---|---|---|---|---|---|---|---|---|
| **Trials** | **S1** | **S2** | **S3** | **S4** | **S5** | **S6** | **S1** | **S2** | **S3** | **S4** | **S5** | **S6** |
| Untreated Control | 0a | 0c | 16,66a | 33,33b | 65,66b | 85,16b | 0a | 0e | 16,66e | 33,33b | 65,66b | 85,16b |
| TColl3 | 0a | 100a | 0b | 100a | 100a | 100a | 0a | 100a | 0f | 100a | 100a | 100a |
| T sig | 0a | 8,33b | 0b | 0c | 0d | 0d | 0a | 8,3c | 20d | 0d | 0d | 16,66d |
| Tsw | 0a | 6,66b | 0b | 0c | 0d | 0d | 0a | 38,8b | 55,55a | 33,33b | 0d | 0f |
| Coll3sig | 0a | 0c | 0b | 33,33b | 33,33c | 0d | 0a | 0e | 33,33b | 0d | 0d | 8,33e |
| Coll3sw | 0a | 0c | 0b | 33,33b | 33,33c | 11,11c | 0a | 0d | 16,66e | 0d | 0d | 0f |
| Coll3+sig+sw+sig | 0a | 0c | 0b | 33,33b | 0d | 0d | 0a | 0d | 20d | 0d | 0d | 0f |
| Coll3+sw+sig+sw | 0a | 0c | 0b | 0c | 0d | 11,11c | 0a | 8,3c | 25c | 11,11c | 16,66c | 33,33c |

Two values in the same column, followed by the same letter, are not significantly different at the 5% threshold according to the LSD test.

*Table 10. Effect of Switch 62.5WG and Signum WG, in preventive and curative treatments on the agronomic parameters of strawberry plants inoculated with Colletotrichum gloeosporioides.*

| | D p r | | L a p | | L r p | |
|---|---|---|---|---|---|---|
| | **PT** | **CT** | **PT** | **CT** | **PT** | **CT** |
| Untreated Control | 15cd | 15b | 12,16b | 12,16b | 16,16d | 16,16d |
| TCol3 | 12,07d | 12,07c | 7,83c | 7,83c | 15,63d | 15,63d |
| T Sig | 17,83abc | 15,83b | 13b | 12,66b | 20,33b | 24,33b |
| T Sw | 22,25a | 14,66bc | 15a | 13b | 17,33d | 18,66c |
| Coll3 Sig | 19,05abc | 20,33a | 13,16b | 13,66b | 19c | 23,66b |
| Coll3 sw | 17,16bc | 20,16a | 14a | 15,33a | 21b | 18,33c |
| Coll3+sig+ sw+sig | 20ab | 18,33b | 14,16a | 15,66a | 25,66a | 30,66a |
| Coll3+sw+ sig+sw | 19,66ab | 20a | 13,83a | 13,5b | 26,66a | 17c |

Two values in the same column, followed by the same letter, are not significantly different at the 5% threshold according to the LSD test.

**D p r:** Diameter perpendicular to the rosette, **L a p:** Length of the aerial part, **L r p:** Length of the root part, **PT:** Preventive treatment, **CT:** Curative treatment

*Table 11. Effect of the two fungicides tested, on the fresh and dry weight of strawberry plants inoculated with Colletotrichum gloeosporioides (expressed in g).*

| | Fresh weight | | Dry weight | | Fresh weight | | Dry weight | |
|---|---|---|---|---|---|---|---|---|
| **Trials** | **Preventive** | **Curative** | **Préventive** | **Curative** | **Préventive** | **Curative** | **Préventive** | **Curative** |
| Untreated Control | 14,93b | 10.73c | 14.1c | 14.1c | 20.3d | 20.3d | 3.6b | 3.6a |
| TColl3 | 10,73c | 10.73f | 12.07b | 12.07b | 18.36e | 18.36e | 1.3d | 1.3c |
| T Sig | 16.46b | 19.96b | 12.7b | 15.5c | 33.53b | 31.86c | 4.7b | 3.86a |
| T Sw | 26,8a | 19.63d | 23.4a | 15.1c | 29.1c | 20.63d | 2.66c | 2.3b |
| Coll3 Sig | 24.63a | 45.9a | 17b | 34.15a | 33b | 61.46a | 5.43a | 4.44a |
| Coll3 sw | 24.53a | 34.5b | 23.4a | 16.4c | 47.83a | 47.33b | 3.96b | 4.6a |
| Coll3+sig+ sw+sig | 14.23b | 13.73e | 16.7b | 18.6c | 23.4c | 21.43d | 6.3a | 2.4b |
| Coll3+sw+ sig+sw | 16.83b | 28.33c | 18.5b | 20.1b | 34.33b | 21.63d | 5.83a | 2.83b |

Two values in the same column, followed by the same letter, are not significantly different at the 5% threshold according to the LSD test.

*Table 12. Re-isolation of Colletotrichum gloeosporioides from different organs of strawberry plants six weeks after fungicide treatments (expressed in %).*

| | Preventive treatment | | | Curative treatment | | |
|---|---|---|---|---|---|---|
| | **Leaves** | **Collet** | **Roots** | **Leaves** | **Collet** | **Roots** |
| TColl3 | 100a | 75a | 0a | 100a | 75a | 0a |
| T sig | 25b | 25b | 0a | 58,33b | 33,33b | 0a |
| Tsw | 16,66b | 20b | 0a | 33,33c | 0c | 0a |
| Coll3sig | 20b | 0c | 0a | 25c | 0c | 0a |
| Coll3sw | 20b | 20b | 0a | 33,33c | 0c | 0a |
| Coll3+sig+sw+sig | 0c | 0c | 0a | 0d | 0c | 0a |
| Coll3+sw+sig+sw | 0c | 0c | 0a | 37,5c | 25b | 0a |

Two values in the same column followed by the same letter are not significantly different at the 5% threshold according to the LSD test

# Chapter 30
# Transforming Morocco's Road Transport Sector Embracing Circular Economy Practices

**Imane Rouichat**
*Mohammed V University in Rabat, Morocco*

**Souad Nasrdine**
https://orcid.org/0009-0004-7021-9005
*Mohammed V University in Rabat, Morocco*

**Jamal Mabrouki**
https://orcid.org/0000-0002-3841-7755
*Mohammed V University in Rabat, Morocco*

**Miloudia Slaoui**
*Mohammed V University in Rabat, Morocco*

## ABSTRACT

*The Moroccan automotive sector, a major contributor to the national economy, has experienced rapid growth, accompanied by an increase in carbon emissions that threatens environmental sustainability. This chapter examines the potential of the circular economy (CE) to address these challenges, promoting practices of reuse, recycling, and reduction of materials throughout the product lifecycle. By adopting CE principles, the Moroccan automotive sector can shift away from the traditional "take, make, dispose" linear model, reducing environmental impact and enhancing economic competitiveness. Recognizing the global and national momentum for CE, the Moroccan government has aligned its National Strategy for Sustainable Development with CE principles, aiming to mitigate greenhouse gas emissions and foster sustainable practices.*

DOI: 10.4018/979-8-3693-8980-5.ch030

## INTRODUCTION

The Moroccan automotive sector is a significant contributor to the national economy, representing a crucial part of the country's industrial landscape. However, with the sector's rapid growth, there has been a simultaneous increase in carbon emissions, posing a severe threat to environmental sustainability. The circular economy (CE) offers a viable solution to these challenges by promoting the reuse, recycling, and reduction of materials throughout the product lifecycle.

CE principles support the design of products that can be easily repaired, refurbished, remanufactured, and eventually recycled, thus minimizing waste and reducing the consumption of finite resources. This approach stands in clear contrast to the traditional linear economy, which follows a 'take, make, dispose' model. By transitioning to a CE, the Moroccan automotive sector can not only reduce its environmental footprint but also unlock substantial economic benefits, including job creation and enhanced competitiveness (Ellen MacArthur Foundation, 2019 ; European Environment Agency et al., 2016).

In recent years, there has been a growing global recognition of the importance of the CE. Reports by the Ellen MacArthur Foundation highlight the economic and environmental advantages of circular practices, emphasizing that a systemic shift to a CE can address critical environmental challenges while fostering economic growth. Furthermore, the European Environment Agency (EEA) has pointed out the automotive sector's potential to drive the transition towards a more sustainable economy by embracing CE practices (Ellen MacArthur Foundation, 2013b, 2013a).

The Moroccan government has also acknowledged the need for sustainable development. The National Strategy for Sustainable Development, outlined by the Ministry of Energy Transition and Sustainable Development, emphasizes the importance of integrating sustainability into various sectors, including automotive, to meet the country's environmental and economic goals. This strategy aligns with global efforts, as indicated by the United Nations Environment Programme (UNEP), which emphasizes the urgency of reducing greenhouse gas emissions and promoting sustainable practices to decrease the impacts of climate change (Ministry of Energy Transition and Sustainable Development, 2017a).

Despite these promising developments, the transition to a CE in the Moroccan automotive sector is filled with challenges. These include technological, economic, and regulatory barriers that need to be addressed to facilitate a smooth and effective transition. This chapter explores these challenges, providing a comprehensive analysis of the current state of the sector, the potential benefits of adopting CE principles, and the necessary steps for overcoming the obstacles to implementation.

Our objective is to analyze the impact of carbon emissions in Morocco's automotive sector and highlight strategies for reduction through the CE approach. By emphasizing sustainable practices, this chapter identifies key economic opportunities stemming from sustainable development efforts. The insights offered are intended to be valuable for policymakers, industry stakeholders, and researchers interested in the intersection of sustainability and economic growth.

The Moroccan automotive sector is at a crossroads. On one hand, it is a vital component of the national economy, contributing significantly to employment and GDP. On the other hand, the sector is a major source of carbon emissions, which have been rising steadily due to increased vehicle production and use. This rise in emissions not only exacerbates local air pollution but also contributes to global climate change, posing severe environmental and health risks (Ministry of Energy Transition and Sustainable Development, 2017a), (Ministry of Industry and Trade, n.d.).

Addressing these issues requires a fundamental shift from the traditional linear economy to a circular economy.

The CE framework offers a holistic approach to sustainability by emphasizing the need to keep resources in use for as long as possible, extract maximum value from them while in use, and recover and regenerate products and materials at the end of their service life (Baxter, Aurisicchio, & Childs, 2017; Camacho-Otero, Tunn, Chamberlin, & Boks, 2020; Ceglia, Abreu, & Da Silva Filho, 2017). This approach can significantly reduce the environmental impact of the automotive sector while also providing economic benefits.

However, integrating CE principles into the automotive sector is not without its challenges. One of the main issues is the technological barrier. The current infrastructure and technology used in the automotive sector are primarily designed for a linear economy. Transitioning to a CE requires significant investment in new technologies and infrastructure, which can be prohibitively expensive for many companies (Tukker, 2015).

Economic barriers also play a significant role. The initial costs associated with adopting CE practices can be high, and the return on investment is often not immediately apparent. This financial burden can discourage companies from making the necessary changes, especially small and medium-sized enterprises (SMEs) that may not have the financial stability to absorb these costs.(Gedam et al., 2021).

Regulatory barriers are another critical challenge. Existing regulations and policies may not be favorable to the adoption of CE practices. There is a need for a supportive regulatory framework that fosters sustainable practices and provides clear guidelines and standards for the industry. Without such a framework, companies may lack the motivation or direction to adopt CE principles(Farooque et al., 2019; Kumar Sharma et al., 2019).

Furthermore, there is a lack of awareness and understanding of the benefits and practices associated with the CE among stakeholders in the automotive sector. Many industry participants may not fully grasp the long-term advantages of adopting sustainable practices, both from an environmental and economic perspective. This lack of awareness can hinder the adoption of CE practices and slow down the transition to a more sustainable automotive industry(Masi et al., 2018).

To overcome these challenges, a multi-faceted approach is required. This includes raising awareness and providing education on the benefits of the CE, developing supportive regulatory frameworks, and facilitating access to funding for companies looking to invest in sustainable technologies and practices. By addressing these barriers, the Moroccan automotive sector can successfully transition to a CE, reduce its environmental impact, and unlock significant economic opportunities.

## MATERIALS AND METHODS

The methodology employed in this chapter involves a multi-step approach to comprehensively analyze the potential of integrating CE principles in the Moroccan automotive sector. This approach includes data collection, literature review, and expert interviews, which collectively provide a robust framework for understanding the challenges and opportunities in this sector.

### Data on Carbon Emissions

The first step involves collecting data on carbon emissions from the Moroccan automotive sector. Data from the International Energy Agency (IEA) shows that transportation remains a significant source of $CO_2$ emissions in Morocco, contributing 28% of the total in 2021(International Energy Agency (IEA), n.d.).

This highlights the need for continued efforts to decarbonize the transportation sector. The automotive sector, being a significant part of this, contributes a substantial portion of these emissions. Specifically, passenger road vehicles are the largest contributors, followed by light commercial vehicles and heavy-duty trucks(IEA, 2022). Detailed emissions data by vehicle type indicates that a typical passenger car emits about 4,6 metric tons of CO2 per year (EPA, 2023).

Additional data highlights that road transport accounts for about 74% of total transport emissions, translating to around 10.5 million tons of CO2 (Kahn Ribeiro, S., S. Kobayashi, M. Beuthe, J. Gasca, D. Greene, D. S. Lee, Y. Muromachi, P. J. Newton, S. Plotkin, D. Sperling, R. Wit,P. J. Zhou, 2007).

## Economic Data

Economic data related to the automotive sector was collected from the Moroccan Ministry of Industry, Trade. In 2020, the automotive industry was the leading export sector in Morocco, with exports valued at over $70 billion Moroccan dirhams (MAD)(statista, 2023)

This data includes export breakdowns in the automotive sector, with a 12.8% increase in sales, amounting to 5,962 million MAD. This growth was observed in the construction segment (+2,795 million MAD), the wiring segment (+1,820 million MAD), and the vehicle interiors and seats segment (+521 million MAD) (Exchange Office, 2024). The automotive sector directly employs around 220,000 people, with significant employment in manufacturing plants, parts suppliers, and dealership networks(Pilling, 2023). The sector also indirectly supports numerous jobs in related industries, such as logistics and transportation. The automotive industry contributes significantly to the national GDP, representing about 22% of Morocco's total GDP(www.ETAuto.com, n.d.). Additional economic data and contextual analysis are provided by reports from international organizations like the World Bank, which offer insights into global trade patterns, market opportunities, and economic trends affecting the automotive industry (The World Bank, 2023).

Industry associations such as the Moroccan Association for Automotive Industry and Trade (AMICA) also publish valuable reports and statistics on industry performance, market trends, and economic impact (HAKAM, 2022).

# LITERATURE REVIEW

The literature on CE practices in the automotive sector is extensive and provides a robust framework for understanding the potential benefits and challenges of integrating these principles. This section reviews key theories, models, and empirical studies related to the CE, focusing on their application in the automotive industry.

The CE is based on three main principles: designing out waste and pollution, keeping products and materials in use, and regenerating natural systems. Unlike the traditional linear economy, which follows a 'take, make, dispose' model, the CE aims to create a closed-loop system where resources are continuously cycled back into production, reducing waste and conserving resources (European Environment Agency et al., 2016).

The Ellen MacArthur Foundation has been instrumental in promoting the CE, highlighting its potential to address critical environmental challenges while fostering economic growth(Ellen MacArthur Foundation, 2013b). According to the foundation, adopting CE practices can significantly reduce greenhouse gas emissions, decrease reliance on finite resources, and create new economic opportunities.

The automotive industry, with its complex supply chains and resource-intensive processes, presents both significant challenges and opportunities for CE practices. Several studies have explored the application of CE principles in this sector, providing valuable insights into successful strategies and potential barriers.

A report by the European Environment Agency (EEA) emphasizes the importance of redesigning vehicles for disassembly and recycling, which can facilitate the reuse of materials and components (European Environment Agency (EEA), 2024). This approach not only reduces waste but also conserves valuable resources, contributing to environmental sustainability. The report also highlights the potential for economic benefits, such as cost savings from reduced material inputs and new business opportunities in recycling and remanufacturing.

## Case Studies and Empirical Evidence

Numerous case studies illustrate the successful implementation of CE practices in the automotive sector. Reviewing case studies from countries with established CE practices, such as Germany and Japan, provides insights into successful strategies and challenges encountered.

In Germany, the automotive industry has led the way in embracing CE principles. BMW, in particular, has established a comprehensive recycling program that prioritizes the reuse of high-value materials and components. On average, their vehicles incorporate up to 30% recycled and reused materials. Their "Design for Recycling" initiative ensures that vehicles are designed with consideration for their end-of-life stage, facilitating easier disassembly and material recovery. This initiative has significantly reduced waste and material costs, with 85% of a BMW being recyclable(BMW, n.d.-a, n.d.-b).

Volkswagen has also made substantial strides in integrating CE practices. Their "closed-loop" recycling process involves collecting and recycling end-of-life vehicles (ELVs), recovering valuable materials such as aluminum, steel, and plastics for use in new production. This process not only reduces waste but also cuts down on the need for raw material extraction, significantly lowering the environmental impact of vehicle production (Volkswagen Group, 2022).

In Japan, Toyota has been at the forefront of CE practices. The company has developed a comprehensive ELV management system that ensures nearly 100% of materials from scrapped vehicles are recycled or reused. Toyota's approach includes designing vehicles for easy disassembly, investing in advanced recycling technologies, and collaborating with suppliers and recycling companies to create a robust recycling network. This system has resulted in substantial environmental benefits and cost savings, making it a model for other automotive manufacturers worldwide(Toyota Times, n.d.).

## Analysis of Reports and Publications

Reports and publications from international organizations such as the European Environment Agency (EEA) and the Ellen MacArthur Foundation provide detailed benefits and implementation strategies of CE practices.

The EEA's report on the CE in the automotive sector outlines several key strategies for successful implementation, including designing for disassembly, increasing the use of recycled materials, and promoting the sharing and leasing of vehicles to extend their lifespan. The report highlights the economic and environmental benefits of these practices, noting that they can significantly reduce greenhouse gas emissions, lower production costs, and create new business opportunities(European Environment Agency et al., 2016).

The Ellen MacArthur Foundation has published numerous studies on the CE, emphasizing its potential to transform industries, including automotive. Their report "Completing the Picture: How the CE Tackles Climate Change" underscores the importance of CE practices in reducing emissions and conserving resources. The foundation advocates for a systemic approach to CE implementation, involving collaboration across sectors and value chains to maximize the benefits (Bag & Pretorius, 2020).

## Academic Journals and Conference Papers

Table 1. provides a summary of key academic papers that discuss various aspects of CE practices. These papers offer insights into the benefits, challenges, and practical implementations of CE principles in different contexts, particularly within the automotive sector.

*Table 1. Important papers related to CE*

| Content of the papers | Reference |
|---|---|
| The paper compares the life cycle assessment of electric vehicles (EVs) and traditional vehicles, revealing that EVs emit 97% less CO2 than petrol cars and 70% less than diesel. Benefits stem from clean energy and advanced batteries. Key challenges include rare earth material scarcity. Recommendations focus on clean energy, battery tech investments, and recycling to enhance sustainability. | (Das et al., 2024) |
| The paper examines the concept of CE rebound, where the benefits of increased resource efficiency are offset by increased consumption. Argues that to avoid rebound effect, CE strategies must be accompanied by measures to limit overall consumption and promote sustainable consumer behaviors. | (Zink & Geyer, 2017) |
| The paper provides an extensive analysis of various CE strategies in the automotive sector. Categorizes strategies into product life extension, resource recovery, and sustainable manufacturing processes. Highlights successful case studies from different countries and emphasizes a systems-based approach. | (Kirchherr et al., 2017) |
| The paper explores the role of consumers in the shift towards a CE . It discusses consumer behaviors necessary for circularity, factors influencing the adoption of circular products, and strategies for engaging consumers through business models, design, and communication. Challenges associated with circular consumption are also addressed. | (Camacho-Otero et al., 2020) |
| The paper examines the challenges and opportunities of Industry 4.0 technology adoption in manufacturing, particularly its impact on sustainable manufacturing and the CE. It reviews recent literature and develops a research framework integrating Industry 4.0 technology (big data analytics powered by AI) with sustainable manufacturing and CE capabilities. | (Bag & Pretorius, 2020) |

Table 2. outlines the various barriers that hinder the implementation of Circular Economy (CE) practices. These barriers are categorized into policy and regulatory, financial/economic, technological, customer-related, and social factors, providing a comprehensive view of the challenges faced by the industry.

*Table 2. Barriers to CE implementation*

| Barriers | Authors |
|---|---|
| Policy and regulatory | (Su et al., 2013), (Pan et al., 2015), (Blomsma & Brennan, 2017), (Geng & Doberstein, 2008), (Mathews et al., 2011), (Rizos et al., 2016), (Lieder & Rashid, 2016), (Jiao & Boons, 2014), (Dong et al., 2013), (Cruz et al., 2017), (Riisgaard et al., 2016) |
| Financial/ economic | (Su et al., 2013), (Pan et al., 2015), (Blomsma & Brennan, 2017), (Geng & Doberstein, 2008), (Mathews et al., 2011), (Dong et al., 2013), (Portas & Ruiz-Puente, 2017), (Ferronato et al., 2016) |
| Technological | (Su et al., 2013), (Pan et al., 2015), (Geng & Doberstein, 2008), (Mathews et al., 2011), (Dong et al., 2013), (Zhu et al., 2014) |
| Customer | (Bey et al., 2013), (Geng & Doberstein, 2008), (Ferronato et al., 2016), (Zhu et al., 2014), (Guo et al., 2017), (Riisgaard et al., 2016), (Tu et al., 2011), (Baxter et al., 2017) |
| Social | (Lieder & Rashid, 2016), (Tu et al., 2011), (Ceglia et al., 2017), (Golev et al., 2014) |

## Current Policies in Morocco on CE

Morocco has been actively developing and implementing policies to transition towards a CE. Here are some key aspects of their current policies:

- **National strategy for sustainable development 2030**: the strategy focuses on sustainable resource management, biodiversity conservation, climate change mitigation, pollution reduction, waste recycling, and renewable energy to ensure long-term environmental sustainability(Ministry of Energy Transition and Sustainable Development, 2017b).
- **Waste management**: The country has made significant efforts to upgrade its waste management sector. This includes the professionalization of waste handling and the development of strategies to manage the anticipated increase in waste, projected to reach 39 million tonnes annually by 2030 (Minister of Energy, Mines, and Sustainable Development, in charge of Sustainable Development et al., 2019).
- **Sector-specific initiatives**: Specific sectors like textiles and fish processing have seen targeted initiatives. For instance, the Switch Med Programme supports the Moroccan textile sector in adopting CE principles, focusing on minimizing and valorizing pre-consumption textile waste (SwitchMed & SwitchMed Magazine, n.d.).
- **EU-Morocco cooperation**: There is ongoing cooperation between the EU and Morocco to enhance CE practices. This includes trade and investment support, and the implementation of CE principles across various industries(European Commission. Directorate General for Environment. et al., 2020) .

Despite these policies, there are still significant challenges. The enforcement mechanisms and funding are often inadequate, the infrastructure and technology for efficient waste processing and recycling are underdeveloped, and the initial costs and lack of awareness among industry stakeholders pose additional barriers. Thus, Morocco needs to enhance enforcement and monitoring, invest in advanced recycling and waste processing technologies, expand sector-specific initiatives, and increase funding and support for CE projects to make a more substantial impact.

## Expert Interviews

Table 3 details the methodology used for conducting expert interviews in this study. It covers the participants involved, the format of the interviews, and the focus areas addressed, ensuring a thorough and structured approach to gathering insights from key stakeholders.

*Table 3. Interview methodology*

| Aspect | Details |
|---|---|
| Participants | Representatives from major automotive manufacturers, government officials, and academic researchers. |
| Format | Semi-structured interviews lasting approximately 60 minutes each. |
| Focus Areas | Current practices, regulatory frameworks, technological innovations, economic impacts, and CE adoption. |

## RESULTS AND DISCUSSION

The transition towards a CE in Morocco's automotive sector shows potential for significant economic, environmental, and technological impacts, as highlighted by the key findings in the interviews with sector experts and case studies from international automotive leaders. Below, we discuss the results drawn from Tables 4-7, incorporating insights on carbon emissions, economic implications, and expert opinions on current CE practices in Morocco.

Table 4. highlights the essential aspects of Morocco's automotive sector concerning CE practices.

*Table 4. Key Findings*

| Aspect | Key Points |
|---|---|
| Manufacturing Practices | Basic recycling programs focusing on metal recovery; limited infrastructure for complex material recycling. |
| Regulatory Frameworks | Need for robust regulations and incentives; current policies insufficient. |
| Technological Innovations | Exploring design for recycling; use of sustainable materials. |
| Economic Impacts | Cost savings from reduced material costs; job creation in recycling and remanufacturing. |

The results indicate that while some foundational recycling programs are in place, particularly for metal recovery, the current infrastructure is insufficient for handling complex material recycling. Regulatory Frameworks are recognized as lacking; industry experts and government officials suggest an urgent need for robust regulations and incentives to stimulate a systemic shift towards CE. Technological Innovations are in an exploratory stage, with focus areas including product design for recycling and the use of sustainable materials. Economic Impacts reflect positive projections, particularly regarding material cost savings and job creation in recycling and remanufacturing.

Further insights from Table 5 reveal that manufacturers and government officials are aware of the gaps in infrastructure and policy that need addressing for effective CE adoption.

*Table 5. Expert opinion on current practices*

| Expert | Opinion on Current Practices |
|---|---|
| [a]Manufacturer A | Basic metal recycling programs in place, but infrastructure for complex materials is lacking. |
| [a]Manufacturer B | Emphasized the need for better recycling facilities and integration of CE principles. |
| Government | Current policies are not strong enough to drive significant changes; more incentives are needed. |
| Academic | Research on sustainable materials and recycling methods is ongoing, but practical implementation is limited. |

*Note.* [a] Due to confidentiality, the manufacturers' names are not disclosed.

Manufacturer A confirms existing metal recycling programs but highlights the infrastructure gaps for complex materials, such as plastics and electronics. Government representatives emphasize that current policies are insufficient to support significant industry shifts, underscoring the need for more incentives and regulatory support.

Table 6. summarizes emerging technological innovations, focusing on Design for Recycling, Advanced Materials, and Recycling Technology.

*Table 6. Technological innovations in the automotive sector*

| Innovation | Description | Examples |
|---|---|---|
| Design for Recycling | Designing vehicles to be easily disassembled and recycled at end-of-life. | Renault's exploration of modular vehicle designs. |
| Advanced Materials | Use of sustainable and recyclable materials in vehicle manufacturing. | Research on biodegradable plastics and lightweight composites. |
| Recycling Technology | Development of advanced recycling technologies for complex materials like electronics and batteries. | Pilot projects for electronic waste recycling and battery material recovery. |

Key advancements include modular vehicle designs to facilitate disassembly and recycling, as exemplified by Renault's design innovations. Other efforts involve the research into biodegradable plastics and lightweight composites to minimize waste and the environmental footprint. Recycling technology, particularly for electronics and batteries, is advancing through pilot projects aimed at recovering materials from complex waste streams, thus reducing dependence on raw materials and promoting sustainable lifecycle management.

Table 7. outlines the economic potential of CE practices in Morocco's automotive sector, covering cost savings, job creation, and environmental impacts.

*Table 7. Potential economic benefits*

| Benefit | Description | Supporting Data/Insights |
|---|---|---|
| Cost Savings | Reduction in material costs through recycling and reuse. | Estimates of up to 20% savings in raw material costs for manufacturers implementing circular practices. |
| Job Creation | New job opportunities in recycling, remanufacturing, and related sectors. | Projections of thousands of new jobs created in the recycling and remanufacturing industries. |
| Environmental Impact | Reduction in CO2 emissions and waste generation. | Potential to cut automotive sector emissions by 30% through comprehensive recycling programs. |

Notably, adopting CE practices could yield material cost reductions of up to 20%, enhancing economic resilience and sustainability. Projections indicate the potential for job creation within recycling and remanufacturing sectors, contributing to economic diversification. Environmentally, comprehensive recycling programs could help reduce CO2 emissions by up to 30%, aligning Morocco's automotive industry with global environmental objectives.

Carbon emissions data underscore the necessity for such shifts. The automotive sector remains a significant contributor, with road transport alone accounting for approximately 10.5 million tons of CO2 in 2021. Passenger vehicles, responsible for about 4.6 metric tons of CO2 annually, represent the largest share, highlighting the pressing need to mitigate emissions through sustainable practices.

Case studies from global leaders like BMW, Volkswagen, and Toyota illustrate achievable outcomes with CE implementation. For instance:

BMW achieved an 85% vehicle recycling rate through its "Design for Recycling" initiative, ensuring streamlined material recovery.

Volkswagen adopted a "closed-loop" recycling process that minimizes waste and reduces raw material extraction.

Toyota introduced an End-of-Life Vehicle (ELV) management system, reaching near-total recycling or reuse of materials from scrapped vehicles.

These international examples demonstrate that effective CE practices can yield substantial environmental benefits, including carbon emission reductions and waste minimization, alongside economic gains through cost savings and job creation.

## CONCLUSION

In conclusion, transitioning Morocco's automotive sector to a circular model offers promising environmental and economic benefits, including reduced carbon emissions, increased competitiveness, and substantial job creation opportunities. However, achieving these gains requires addressing critical challenges, such as limited recycling infrastructure, insufficient regulatory frameworks, and significant initial investment demands.

A multi-faceted approach is essential for overcoming these barriers. Investing in technological innovations, particularly in recycling technologies and sustainable materials, is crucial. Strengthening regulatory frameworks to include robust policies and incentives will further encourage CE adoption. Additionally, raising awareness among stakeholders and consumers on the long-term advantages of CE practices can facilitate wider acceptance and support for sustainable transformations.

The successful integration of CE principles, as shown in international case studies, underscores the sector's potential to achieve meaningful environmental and economic outcomes. With targeted investments and policy reforms, Morocco can position itself as a regional leader in sustainable automotive practices, contributing to a greener economy and a more resilient future.

Recommendations to support this transition include:

- Enhancing Technological Infrastructure: Investments in advanced recycling technologies are critical to manage complex materials, along with a focus on research and development for sustainable materials and product designs optimized for recycling.

- Strengthening Regulatory Frameworks: Robust regulations and targeted incentives are necessary to encourage CE adoption. Clear standards and guidelines for sustainable manufacturing and recycling can further support industry alignment with CE practices.
- Providing Economic Incentives: Access to funding and financial support for companies investing in CE is vital. Tax benefits and subsidies for businesses implementing sustainable practices will also create an attractive environment for CE investment.
- Raising Awareness and Education: Industry-wide awareness campaigns and educational programs are needed to promote understanding and the practical application of CE. Workshops and targeted training can equip stakeholders with the knowledge to implement sustainable practices effectively.

## REFERENCES


Bag, S., & Pretorius, J. H. C. (2020). Relationships between industry 4.0, sustainable manufacturing and circular economy: Proposal of a research framework. *The International Journal of Organizational Analysis*, *30*(4), 864–898. DOI: 10.1108/IJOA-04-2020-2120

Baxter, W., Aurisicchio, M., & Childs, P. (2017). Contaminated Interaction: Another Barrier to Circular Material Flows. *Journal of Industrial Ecology*, *21*(3), 507–516. Advance online publication. DOI: 10.1111/jiec.12612

Bey, N., Hauschild, M. Z., & McAloone, T. C. (2013). Drivers and barriers for implementation of environmental strategies in manufacturing companies. *CIRP Annals*, *62*(1), 43–46. DOI: 10.1016/j.cirp.2013.03.001

Blomsma, F., & Brennan, G. (2017). The Emergence of Circular Economy—A New Framing Around Prolonging Resource Productivity. *Journal of Industrial Ecology*, *21*(3), 603–614. Advance online publication. DOI: 10.1111/jiec.12603

BMW. (n.d.-a). *Design for recycling: At the end there's a new beginning*. Retrieved June 15, 2024, from https://www.bmw.com/en/magazine/sustainability/circular-lab/recycle/design-for-recycling.html

BMW. (n.d.-b). *Sustainability & responsibility at BMW group*. Retrieved June 15, 2024, from https://www.bmwgroup.com/en/sustainability.html

Camacho-Otero, J., Tunn, V. S. C., Chamberlin, L., & Boks, C. (2020). *Consumers in the circular economy*., https://china.elgaronline.com/edcollchap/edcoll/9781788972710/9781788972710.00014.xml DOI: 10.4337/9781788972727.00014

Ceglia, D., de Abreu, M. C. S., & Da Silva Filho, J. C. L. (2017). Critical elements for eco-retrofitting a conventional industrial park: Social barriers to be overcome. *Journal of Environmental Management*, *187*, 375–383. DOI: 10.1016/j.jenvman.2016.10.064 PMID: 27836565

Cruz, N. C., Rodrigues, S. M., Carvalho, L., Duarte, A. C., Pereira, E., Römkens, P. F. A. M., & Tarelho, L. A. C. (2017). Ashes from fluidized bed combustion of residual forest biomass: Recycling to soil as a viable management option. *Environmental Science and Pollution Research International*, *24*(17), 14770–14781. DOI: 10.1007/s11356-017-9013-6 PMID: 28470498

Das, P. K., Bhat, M. Y., & Sajith, S. (2024). Life cycle assessment of electric vehicles: A systematic review of literature. *Environmental Science and Pollution Research International*, *31*(1), 73–89. DOI: 10.1007/s11356-023-30999-3 PMID: 38038907

Dong, L., Zhang, H., Fujita, T., Ohnishi, S., Li, H., Fujii, M., & Dong, H. (2013). Environmental and economic gains of industrial symbiosis for Chinese iron/steel industry: Kawasaki's experience and practice in Liuzhou and Jinan. *Journal of Cleaner Production*, *59*, 226–238. DOI: 10.1016/j.jclepro.2013.06.048

Ellen MacArthur Foundation. (2013a). *Towards the circular economy eonomic and business rationale for an accelerated transition*. https://archive.ellenmacarthurfoundation.org/assets/downloads/publications/Ellen-MacArthur-Foundation-Towards-the-Circular-Economy-vol.1.pdf

Ellen MacArthur Foundation. (2013b). *Towards The Circular Economy Opportunities for the consumer goods sector*. https://archive.ellenmacarthurfoundation.org/assets/downloads/publications/TCE_Report-2013.pdf

Ellen MacArthur Foundation. (2019). *Completing the picture: How the circular economy tackles climate change* (p. 71). https://emf.thirdlight.com/file/24/cDm30tVcDDexwg2cD1ZEcZjU51g/Completing%20the%20Picture%20-%20How%20the%20circular%20economy%20tackles%20climate%20change.pdf

Ellen MacArthur Foundation. (n.d.). *Circular economy introduction*. Retrieved June 15, 2024, from https://www.ellenmacarthurfoundation.org/topics/circular-economy-introduction/overview

EPA. (2023). *Tailpipe Greenhouse Gas Emissions from a Typical Passenger Vehicle*. https://nepis.epa.gov/Exe/ZyPDF.cgi?Dockey=P1017FP5.pdf

European Commission. Directorate General for Environment., Diaco, M., Alami Merrouni, M., & Bougaraani, S. (2020). *Circular economy in Africa-EU cooperation: Country report Morocco.* (under EC Contract ENV.F.2./ETU/2018/004 Project: " "Circular Economy in Africa-Eu cooperation"). Trinomics B.V., ACEN, adelphi Consult GmbH and Cambridge Econometrics Ltd. https://data.europa.eu/doi/10.2779/211421

European Environment Agency. De Schoenmakere, M., Gillabel, J., & Reichel, A. (2016). *Circular economy in Europe: Developing the knowledge base.* (2/2016). Publications Office. https://data.europa.eu/doi/10.2800/51444

European Environment Agency (EEA). (2024, March 21). *Circular economy*. https://www.eea.europa.eu/en/topics/in-depth/circular-economy

Exchange Office. (2024). *Indicateurs mensuels*. https://www.oc.gov.ma/sites/default/files/2024-05/IMEE%20Avril%202024_0.pdf

Farooque, M., Zhang, A., & Liu, Y. (2019). Barriers to circular food supply chains in China. *Supply Chain Management*, *24*(5), 677–696. DOI: 10.1108/SCM-10-2018-0345

Ferronato, N., Bezzi, M., Zortea, M., Torretta, V., & Ragazzi, M. (2016, November 1). *An interdisciplinary Approach for Introducing Sustainable Integrated Solid Waste Management Systems in Developing countries: The case of La Paz (BOLIVIA).*

Gedam, V. V., Raut, R. D., Lopes de Sousa Jabbour, A. B., Tanksale, A. N., & Narkhede, B. E. (2021). Circular economy practices in a developing economy: Barriers to be defeated. *Journal of Cleaner Production*, *311*, 127670. DOI: 10.1016/j.jclepro.2021.127670

Geng, Y., & Doberstein, B. (2008). Developing the circular economy in China: Challenges and opportunities for achieving "leapfrog development." *International Journal of Sustainable Development and World Ecology -. International Journal of Sustainable Development and World Ecology*, *15*(3), 231–239. DOI: 10.3843/SusDev.15.3:6 PMID: 37309529

Golev, A., Corder, G., & Giurco, D. (2014). Barriers to Industrial Symbiosis: Insights from the Use of a Maturity Grid. *Journal of Industrial Ecology*, *19*(1), 141–153. Advance online publication. DOI: 10.1111/jiec.12159

Guo, B., Geng, Y., Sterr, T., Zhu, Q., & Liu, Y. (2017). Investigating public awareness on circular economy in western China: A case of Urumqi Midong. *Journal of Cleaner Production, 142*, 2177–2186. DOI: 10.1016/j.jclepro.2016.11.063

Hakam, A. (2022). *Study on trade and value chains in job-creating activities (TRAVERA): Case of the automotive sector in Morocco*. DEPF Etudes.

IEA. (2022). Transport sector CO2 emissions by mode in the sustainable development scenario, 2000-2030 [Graphic]. https://www.iea.org/data-and-statistics/charts/transport-sector-co2-emissions-by-mode-in-the-sustainable-development-scenario-2000-2030

Jiao, W., & Boons, F. (2014). Toward a research agenda for policy intervention and facilitation to enhance industrial symbiosis based on a comprehensive literature review. *Journal of Cleaner Production, 67*, 14–25. DOI: 10.1016/j.jclepro.2013.12.050

Kahn Ribeiro, S., Kobayashi, S., Beuthe, M., Gasca, J., Greene, D., Lee, D. S., Muromachi, Y., Newton, P. J., Plotkin, S., Sperling, D., Wit, R., & Zhou, P. J. (2007). *Transport and its infrastructure.In Climate Change 2007: Mitigation. Contribution of Working Group III to the Fourth Assessment Report of the Intergovernmental Panel on Climate Change* (B. Metz, O.R. Davidson, P.R. Bosch, R. Dave, L.A. Meyer (eds)). Cambridge University Press. DOI: 10.1017/CBO9780511546013

Kirchherr, J., Reike, D., & Hekkert, M. (2017). Conceptualizing the circular economy: An analysis of 114 definitions. *Resources, Conservation and Recycling, 127*, 221–232. DOI: 10.1016/j.resconrec.2017.09.005

Kumar Sharma, Y., Mangla, S., Patil, P., & Liu, S. (2019). *When challenges impede the process: For circular economy driven sustainability practices in food supply chain*. DOI: 10.1108/MD-09-2018-1056

Licence International Energy Agency (IEA). (n.d.). How much CO2 does Morocco emit? IEA. Retrieved June 15, 2024, from https://www.iea.org/countries/morocco/emissions

Lieder, M., & Rashid, A. (2016). Towards circular economy implementation: A comprehensive review in context of manufacturing industry. *Journal of Cleaner Production, 115*, 36–51. DOI: 10.1016/j.jclepro.2015.12.042

Masi, D., Kumar, V., Garza-Reyes, J. A., & Godsell, J. (2018). Towards a more circular economy: Exploring the awareness, practices, and barriers from a focal firm perspective. *Production Planning and Control, 29*(6), 539–550. DOI: 10.1080/09537287.2018.1449246

Masterson, V., & Shine, I. (2022, June). What is the circular economy and why does it matter. In *World Economic Forum* (Vol. 14).

Mathews, J. A., Tang, Y., & Tan, H. (2011). China's move to a Circular Economy as a development strategy. *Asian Business & Management, 10*(4), 463–484. DOI: 10.1057/abm.2011.18

Minister of Energy, Mines, and Sustainable Development, in charge of Sustainable Development, Deutsche Zusammenarbeit, & Deutsche Gesellschaft für Internationale Zusammenarbeit (GIZ). (2019). *National Strategy for Waste Valorization and reduction*. https://www.logipro.ma/images/Traitement_des_deee/Rapport_de_synthese_SNRVD_FR.pdf

Ministry of Energy Transition and Sustainable Development. (2017a). *National strategy for sustainable development.*

Ministry of Energy Transition and Sustainable Development. (2017b). *National strategy for sustainable development 2030.* https://www.mtedd.gov.ma/images/Rapport_Strategie_Nationale_DD_juin2017_Mai_2017_Web.pdf

Ministry of Industry and Trade. (n.d.). *Automotive*. Automotive. Retrieved June 15, 2024, from http://www.mcinet.gov.ma/en/content/industry-0/automotive-0

www.ETAuto.com. (n.d.). *Morocco's automotive industry shifts gears to prep for electric vehicle era.* ET Auto. Retrieved June 20, 2024, from https://auto.economictimes.indiatimes.com/news/industry/moroccos-automotive-industry-shifts-gears-to-prep-for-electric-vehicle-era/110145274

Pan, S.-Y., Du, M., Huang, I.-T., Liu, I.-H., Chang, E.-E., & Chiang, P.-C. (2015). Strategies on implementation of waste-to-energy (WTE) supply chain for circular economy system: A review. *Journal of Cleaner Production, 108*, 409–421. Advance online publication. DOI: 10.1016/j.jclepro.2015.06.124

Pilling, D. (2023, October 5). Morocco's carmaking sector revs up for EV era. *Financial Times*. https://www.ft.com/content/8549619c-9c23-4a00-88ab-dfdac92a22ca

Portas, R., & Ruiz-Puente, C. (2017). Development of the Tool SymbioSyS to Support the Transition Towards a Circular Economy Based on Industrial Symbiosis Strategies. *Waste and Biomass Valorization, 8*(5), 1521–1530. Advance online publication. DOI: 10.1007/s12649-016-9748-1

Riisgaard, H., Mosgaard, M., & Zacho, K. (2016). Local Circles in a Circular Economy-The Case of Smartphone Repair in Denmark. *European Journal of Sustainable Development, 5*, 109.

Rizos, V., Behrens, A., Van der Gaast, W., Hofman, E., Ioannou, A., Kafyeke, T., Flamos, A., Rinaldi, R., Papadelis, S., Hirschnitz-Garbers, M., & Topi, C. (2016). Implementation of Circular Economy Business Models by Small and Medium-Sized Enterprises (SMEs): Barriers and Enablers. *Sustainability (Basel), 8*(11), 11. Advance online publication. DOI: 10.3390/su8111212

statista. (2023). *Value of exports from the automotive sector in Morocco from 2014 to 2020 (in million Moroccan dirhams)* [Graphic].

Su, B., Heshmati, A., Geng, Y., & Yu, X. (2013). A review of the circular economy in China: Moving from rhetoric to implementation. *Journal of Cleaner Production, 42*, 215–227. DOI: 10.1016/j.jclepro.2012.11.020

SwitchMed & SwitchMed Magazine. (n.d.). *Morocco*. Retrieved June 21, 2024, from https://switchmed.eu/wp-content/uploads/2020/03/National-Supplement-EN-Morocco.pdf

The World Bank. (2023). *A new era in development annual report.*

Toyota Times. (n.d.). *Revealing vehicle recycling processes—Your cars live on in different forms*. Retrieved June 16, 2024, from https://toyotatimes.jp/en/spotlights/1021.html

Tu, W., Zhang, L., Zhou, Z., Liu, X., & Fu, Z. (2011). The development of renewable energy in resource-rich region: A case in China. *Renewable & Sustainable Energy Reviews*, *15*(1), 856–860. DOI: 10.1016/j.rser.2010.07.046

Tukker, A. (2015). Product services for a resource-efficient and circular economy – a review. *Journal of Cleaner Production*, *97*, 76–91. DOI: 10.1016/j.jclepro.2013.11.049

Volkswagen Group. (2022). *Sustainability Report*. https://www.volkswagen-group.com/en/publications/more/group-sustainability-report-2022-1644

Zhu, Q., Geng, Y., Sarkis, J., & Lai, K.-H. (2014). Barriers to Promoting Eco-Industrial Parks Development in China. *Journal of Industrial Ecology*, *19*(3), 457–467. DOI: 10.1111/jiec.12176

Zink, T., & Geyer, R. (2017). Circular Economy Rebound. *Journal of Industrial Ecology*, *21*(3), 593–602. DOI: 10.1111/jiec.12545

# Chapter 31
# Trichoderma Use as Seed Treatment for Promoting the Growth of Young Argan Seedlings

**Mohamed El Rhoch**
https://orcid.org/0009-0003-1549-4306
*Ibn Tofail University, Morocco*

**Soukaina Maazouzi**
*Laboratory of Plant, Animal, and Agro-Industry Productions, Department of Biology, Faculty of Science, Morocco*

**Najoua Mouden**
*Laboratory of Molecular Chemistry and Environmental Molecules, Multidisciplinary Faculty of Nador, Morocco*

**Zineb Sellal**
https://orcid.org/0009-0006-7074-2802
*Laboratory of Plant, Animal, and Agro-Industry Productions, Department of Biology, Faculty of Science, Morocco*

**Karima Selmaoui**
*Laboratory of Plant, Animal and Agro-Industry Productions, Department of Biology, Faculty of Science, Morocco*

**Amina Ouazzani Touhami**
https://orcid.org/0000-0001-9970-7724
*Laboratory of Plant, Animal, and Agro-Industry Productions, Department of Biology, Faculty of Science, Morocco*

**Allal Douira**
*Laboratory of Plant, Animal, and Agro-Industry Productions, Department of Biology, Faculty of Science, Morocco*

## ABSTRACT

*Seed treatment with Trichoderma spp. has shown to enhance plant growth and alleviate abiotic stress, but its application on endemic species, such as argan (Argania spinosa), remains underexplored. This study investigates the effect of a Trichoderma spp. bioformulation on argan seedling growth under greenhouse conditions over 12 months. Significant improvements were observed in treated seedlings, including increased length, dry matter of both root and shoot, and branch number. Root and vegetative fresh weights reached 12 g and 14 g, respectively, compared to 2 g and 3 g in control plants. Additionally, root and vegetative lengths were 44 cm and 74 cm, respectively, outperforming control seedlings at 35*

DOI: 10.4018/979-8-3693-8980-5.ch031

*cm and 25 cm. Treated seedlings also showed a higher number of branches and crown diameter, with survival and colonization rates of Trichoderma spp. in root tissues reaching 95%. These results suggest that Trichoderma spp. seed treatment can significantly promote the growth and development of young argan seedlings.*

## 1 INTRODUCTION

At various soil levels, telluric microflora is very varied. The majority of representatives of this microflora are beneficial, while others are real enemies that hinder the development and growth of plants.

Fungi of the *Trichoderma* genus, known since 1887 for their antagonistic properties, have been used as biological control agents against a broad spectrum of telluric phytopathogens (Ouazzani-Touhami, 1994 ; Davet, 1986 ; Mouria, Ouazzani-Touhami, Douira, Benkirane, Mlaiki, & El-Yachioui, 1997 ; Mouria, Ouazzani-Touhami, Mlaiki, El Yachioui, & Douira, 1997 ; Hmouni, Massoui, & Douira, 1999). In addition, they were identified as phyto-stimulators and biological control agents against phytopathogenic agents (Chliyeh et al., 2014 ; Ouazzani Chahdi et al., 2014 ; Najoua et al., 2022 ; Azeddine, Ouazzani Touhami, Selmaoui, Benkirane, & Douira, 2024 ; Azeddine et al., 2024). Trichoderma genera acts mainly by producing antimicrobial compounds or by parasitizing pathogenic plants (Mouria, Ouazzani-Touhami, & Douira, 2013 ; Mouria, Ouazzani-Touhami, Mouria, Benkirane, & Douira, 2015).

Some species of Trichoderma have physical interaction with plants that have beneficial effects on the growth of vegetative and root parts, availability of nutrients, improvement of yields and disease resistance (Mouria, Ouazzani-Touhami, & Douira, 2013 ; Yedidia, Benhamou, & Chet, 1999 ; Mouria, Ouazzani-Touhami, & Douira, 2007 ; Kribel et al., 2019a ; Kribel, Qostal, Ouazzani Touhami, & Selmaoui, 2019b ; Kribel, Qostal, Ouazzani Touhami, Selmaoui, Chliyeh, Benkirane, Achbani, & Douira, 2020 ; Soufiani et al., 2022 ; Qostal et al., 2020a ; El Kaissoumi et al., 2022 ; El Kaissoumi et al., 2024 ; Chang, Chang, Baker, Kleifeld, & Chet, 1986 ; Yedidia, Benhamou, Kapulnik, & Chet, 2000 ; Martinez et al., 2001 ; Shoresh, Harman, & Mastouri, 2010 ; Harman, 2006). The effectiveness of Trichoderma depends on its ability to combat pathogenic fungi, which depends on the amount of inoculum present in the soil (Mouria, Ouazzani Touhami, & Douira, 2005).

According to several studies, soils inoculated by fungi of the Trichoderma genus protected crops and create a healthy environment so that vegetation can develop normally (Vinale et al., 2008). These fungi secretly secure active molecules which stimulate plant growth and ensure their protection against pathogens. The presence of *Trichoderma* in the ground plays both a preventive and curative role (Vinale et al., 2008). Some species of Trichoderma are considered as endophytes which establish a relationship with plant roots, which improves root growth, productivity, resistance to abiotic stress and facilitates assimilation and use of nutrients (Vinale et al., 2008).

The present study aimed to examine how treatment of argan seeds with an isolate of Trichoderma sp. can improve the growth parameters and development of argan plants in nurseries.

## 2 MATERIALS AND METHODS

### 2.1 Plant Material: Argan Seed

The seeds used are from the argan trees of the Tafraout (region of Souss-Massa, Morocco). They were disinfected with sodium hypochlorite diluted to 1%, for 10 minutes, rinsed abundantly with running water then dried on filter paper.

### 2.2 Fungal Material

The isolate of *Trichoderma* sp. used in this study, originates from the rhizosphere of argan trees in Maaziz (Khemissat area) (Figure 1). Identification of this isolate was based on morphological characteristics: colony growth, pigmentation, conidiophores, phialides and conidia. On PSA medium (200 g potato; 20 g sucrose; 15 g agar-agar; 1000 mL distilled water), colonies of this Trichoderma sp. isolate show rapid growth, with mycelium that is white, slightly cottony at first, and then turns greenish as conidia form. Conidiophores are branched and phialides are arranged in divergent groups of 2-4. Conidia are unicellular, globose or subglobose. This is the description of representatives of the genus Trichoderma sp. (Rifai, 1969 ; Shah, Nasreen, & Sheikh, 2012 ; Gams & Bissett, 2002).

### 2.3 Seed Treatment

*Trichoderma* sp. isolates were grown on PSA medium and incubated at 25°C for 10 days in the dark. The cultures were then incubated under continuous light to promote sporulation (Mouria, Ouazzani-Touhami, Mouria, Benkirane, & Douira, 2015). The surface of the cultures is then washed with sterile distilled water and the concentration of the conidial suspension is adjusted to 107conidia/ml using the Malassez cell.

Disinfected argan seeds are dipped in crystallizers containing a suspension of Trichoderma sp. conidia at a concentration of 107 conidia/ml and a sterile solution of methylcellulose and gelatin mixed with talc. This composition enabled the conidia to adhere better to the seeds. Control seeds were treated in the same way, but the solution used did not contain Trichoderma conidia. All treated and untreated seeds were dried for 72 h at room temperature, and transplanted into pots filled with Mamora soil, sterilized in advance for four hours in an oven at 250°C.

### 2.4 Assessment of Agronomic Parameters

After twelve months of cultivation, growth parameters included plant height, root and aerial biomass, number of shoots and stem diameter. The results were statistically processed using a single-criterion analysis of variance (ANOVA1) with a threshold of 5% and the P.PD.S

### 2.5 Re-Isolation of *Trichoderma* From Roots

The roots of argan plants grown from treated seeds are collected from the plants after 12 months of cultivation. The fine roots are cut into 1 cm fragments and cultured in a humidity chamber. After 3 to 4 days' incubation, the root fragments are stained with cotton blue and observed under the microscope.

Microscopic observations of fine root fragments are sometimes made in a drop of lactophenol or water (Qostal et al., 2020b ; Sellal et al., 2020a ; Sellal et al., 2023). The various components of the Trichoderma asexual reproductive cycle are noted: septate mycelial filaments, phialides, conidiophores and conidia. *Trichoderma* was also detected in the roots of plants from treated seeds by re-isolation. Roots were cut into small fragments, disinfected with 95° alcohol for 2 minutes, rinsed several times with sterile distilled water, dried rapidly on sterile filter paper, transplanted onto PSA medium and incubated in the dark at 25°C. Results are read after four days. The percentage of re-isolation (Pr %) is calculated according to the following formula:

Pr = Ns Px / NT x 100

**Ns Px:** number of segments containing the fungal species x.
**NT:** total number of segments.

## 3 RESULTS

Treatment of argan seeds with a Trichoderma conidial formulation had a positive effect on the growth parameters of seedlings grown from these seeds after 12 months in the greenhouse (Table 1). The average length of the aerial parts of these plants was 74 cm, compared with 25 cm for the control plants (Figures 1, 2 et 3). The average root length of plants from treated seeds was 44 cm, and that of control plants 35 cm. The average collar diameter of plants from treated seeds was twice that of the control batch. The weights of the root and aerial masses of the treated plants, compared with those of the control plants, were 12/2 g and 14/3 g respectively. The average number of twigs developed in plants from treated seeds increased fourfold compared with twigs from control plants.

*Table 1. Vegetative and root growth of argan seedlings from Trichoderma-treated and untreated seeds, after 12 months in the greenhouse.*

| | Length of aerial part (cm) | Length of root part (cm) | Crown diameter (cm) | Weight of roots (g) | Vegetative weight (g) | Number of branches |
|---|---|---|---|---|---|---|
| Control | 25 | 27 | 0,5 | 2 | 3 | 1 |
| *Trichoderma* | 74 | 56 | 1 | 12 | 14 | 4 |

*Figure 1. Effect of seed treatment on the growth of the aerial (LPA) and root parts (LPR) of argan seedlings, T = Control Tr = Trichoderma,*

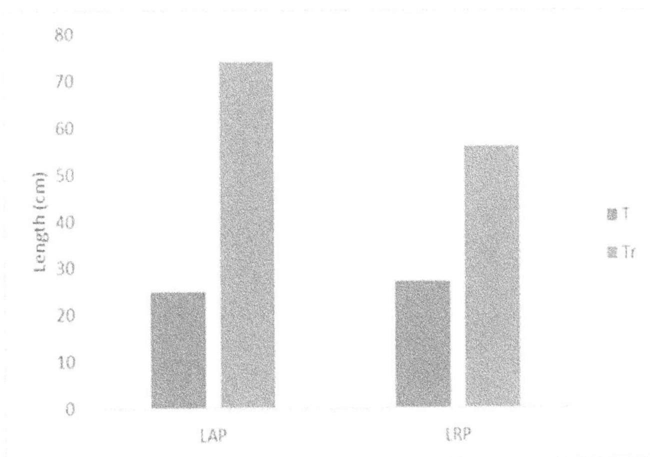


The same letter are not significantly different from the Tukey test ($p \leq 0.05$)

*Figure 2. Effect of Trichoderma sp. treatment on vegetative and root growth in argan seedlings; T= Control, Tr= Trichoderma sp. treatment*

*Figure 3. Effect of Trichoderma sp. on branching and development of root system without Trichoderma sp. (T= control), and with Trichoderma sp. (Tr).*

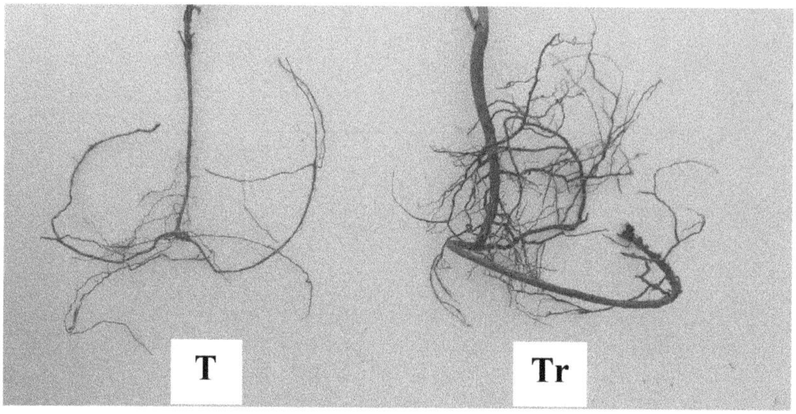


From these results, we note that the percentage gains in growth parameters in argan seedlings from treated seeds compared with control seedlings. These gains were 196% for aerial growth and 107% for root growth.

The roots of argan plants from seeds treated with the Trichoderma-based formulation are colonized by Trichoderma. After 12 months of cultivation, root colonization percentages attained 80% (Figures 4 and 5). The roots of control plants were free from Trichoderma, but other fungi were isolated.

*Figure 4. Trichoderma colonies developing on root segments of argan plants 5 days after culturing A et B: Colony of re-isolated Trichoderma from root fragments; B et C: Macroscopic characteristics of Trichoderma culture re-isolated from argan plant roots*

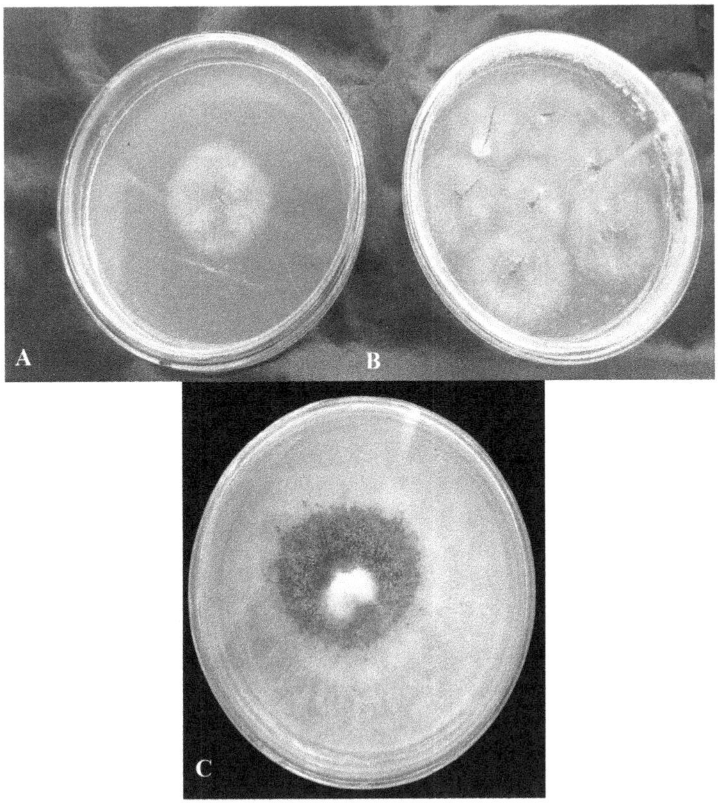

*Figure 5. Microscopic observations of the roots of seedlings from Trichoderma-treated seeds, after 12 months of cultivation. A and B: mycelial filaments and conidia; C: mycelium developing on the fine roots of argan seedlings, with conidiophore and phialides.*

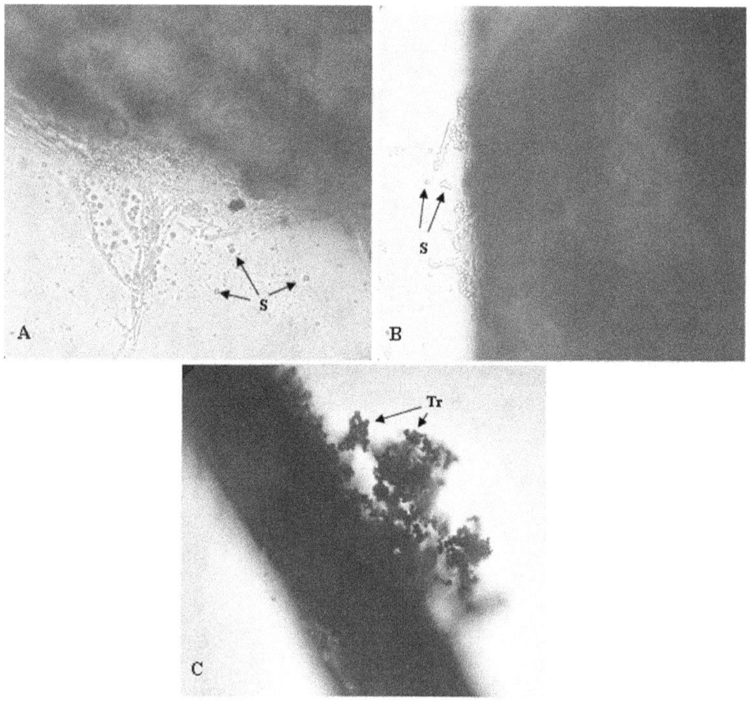


## 4 DISCUSSION

A Trichoderma isolate was used as an important component of a formulation exploited to treat argan seeds. This treatment showed positive effects on the growth parameters of argan seedlings derived from these seeds, after 12 months of cultivation. There was a clear improvement in the growth of the aerial and root parts, collar diameter and branching of the argan seedlings. The Trichoderma used to treat the argan seeds showed an ability to survive and multiply around the roots of the plants from the treated seeds, the same result was noted by (Sellal et al., 2020a) who treated argan seeds with *Trichoderma harzianum* conidia.

Similarly, wheat and barley plants growing on substrates inoculated with Trichoderma showed a clear improvement in growth parameters (Kribel, Qostal, Ouazzani Touhami, Selmaoui, Chliyeh, Benkirane, Achbani, & Douira, 2020). In other work, Trichoderma harzianum has been reported as a biostimulant of chickpea plant growth (Adnani et al., 2024 ; Yadav, Verma, & Tiwari, 2011), tomato (Mouria, Ouazzani-Touhami, Douira, Benkirane, Mlaiki, & El-Yachioui, 1997 ; Mouria, Ouazzani-Touhami, Mlaiki, El Yachioui, & Douira, 1997 ; Hmouni, Massoui, & Douira, 1999 ; Hmouni, Douira, & Mouria, 2006) carob (Talbi et al., 2016). Another Trichoderma strain was able to increase the total biomass of mangrove plants (Saravanakumar et al., 2013). Kleifeld & Chet (1984) reported that the stimulation of plant growth by

Trichoderma sp. would be due to the increased transfer of nutrients from the soil to the roots through Trichoderma colonization of the roots. Yedidia, Srivastva, Kapulnik, and Chet (2001) determined how a strain of *T. harzianum* contributes to cucumber development in soil inoculated with conidia from this strain. Baker, Elad, and Chet (1992) also demonstrated that the *Trichoderma harzianum* strain increased the dry weight of radish plants. Furthermore, Gravel, Martinez, Antoun, and Tweddell (2005) found that inoculation of tomatoes with *T. atroviride* strains significantly increased the root surface area, weight and commercial yield of isolated tomato plants.

According to Baker (1998), Trichoderma are capable of producing growth regulators that improve germination and plant growth. The potential of Trichoderma to reduce the effect of pathogens responsible for root rot in cereals was reported by Qostal et al. (2020b). According to Nemec, Datnoff, and Strandberg (1996), one of the essential characteristics of biological control agents is their ability to survive and colonize an environment different from their original one. According to these authors, treatment with this biological control agent prevented wheat and barley plants from being infected by the pathogens tested, *Bipolaris sorokiniana* and *Curvularia spicifera*. Kleifeld and Chet (199) reported that the stimulation of Trichoderma sp. growth is due to nutrients from the soil reaching the roots through their colonization by Trichoderma.

## 5 CONCLUSION

Seed treatment with this new Trichoderma-based formulation appeared to be effective in stimulating the growth parameters of argan seedlings derived from these seeds. It was remarkable to note that used Trichoderma grew and multiply onto the roots of the seedlings throughout 12 months of cultivation. It is also important to conclude the significant role of this formulation applied as seed treatment among innovate technique that would be integrated among recommended alternative method.

## REFERENCES

Adnani, M., El Hazzat, N., Msairi, S., El Alaoui, M. A., Mouden, N., Selmaoui, K., Benkirane, R., Ouazzani Touhami, A., & Douira, A. (2024). Exploring the efficacy of a Trichoderma asperellum-based seed treatment for controlling Fusarium equiseti in chickpea. *Egyptian Journal of Biological Pest Control, 34*(1), 7. DOI: 10.1186/s41938-024-00771-x

Azeddine, E., Fadoua, B., Abdelatif, O. C., Mouden, N., Imad, K., Hanane, E. K., Selmaoui, K., Benkirane, R., Ouazzani Touhami, A., & Douira, A. (2024). Efficacy of the Combined Application of Based Trichoderma Asperellum Products and Tolclofos-Methyl to Control Rhizoctonia Solani Black Crown Rot in Strawberry. In *Advanced Systems for Environmental Monitoring, IoT and the application of Artificial Intelligence* (pp. 123–143). Springer Nature Switzerland. DOI: 10.1007/978-3-031-50860-8_8

Azeddine, E., Ouazzani Touhami, A., Selmaoui, K., Benkirane, A. R., & Douira, A. (2024). In Vitro and In Vivo Biological Control by Trichoderma Asperellum Against Rhizoctonia Solani a Causal Agent of Collar and Root Rot in Strawberries. In *Advanced Systems for Environmental Monitoring, IoT and the application of Artificial Intelligence* (pp. 213–235). Springer Nature Switzerland. DOI: 10.1007/978-3-031-50860-8_13

Baker, R. (1988). Trzchoderma SPP. as plant-growth stimulants. *Critical Reviews in Biotechnology, 7*(2), 97–106. DOI: 10.3109/07388558809150724

Baker, R., Elad, Y., & Chet, I. (1984). The controlled experiment in the scientific method with special emphasis on biological control. *Phytopathology, 74*(9), 1019–1021. DOI: 10.1094/Phyto-74-1019

Chang, Y. C., Chang, Y. C., Baker, R., Kleifeld, O., & Chet, I. (1986). Increased growth of plants in the presence of the biological control agent Trichoderma harzianum, 70:145–148.

Chliyeh, M., Chahdi, A. O., Selmaoui, K., Ouazzani Touhami, A., Filali-Maltouf, A., El Modafar, C., Moukhli, A., Oukabli, A., Benkirane, R., & Douira, A. (2014). Effect of Trichoderma harzianum and arbuscular mycorrhizal fungi against Verticillium wilt of tomato. *International Journal of Recent Scientific Research, 5*(2), 449–459.

Davet, P. (1986). Parasitic activity of Trichoderma towards sclerotial fungi; correlation with their saprophytic competitive ability in a non-sterile soil, 6(9):863-867.

El Kaissoumi, H., Berber, F., Mouden, N., Ouazzani Chahdi, A., Albatnan, A., Ouazzani Touhami, A., Selmaoui, K., Benkirane, R., & Douira, A. (2022, May). Effect of Trichoderma asperellum on the development of strawberry plants and biocontrol of anthracnose disease caused by colletotrichum gloeosporioides. In *International Conference on Advanced Intelligent Systems for Sustainable Development* (pp. 609-622). Cham: Springer Nature Switzerland.

El Kaissoumi, H. E., Berbera, F., & Mouden, N. OuazzaniChahdi, A., Touhami, A. O., Selmaoui, K., Benkirane, R., Douira, A. (2024). Tomato Growth Promotion by Trichoderma Asperellum Laboratory-Made Bioproduct. In *Sustainable and Green Technologies for Water and Environmental Management* (pp. 161-171). Cham: Springer Nature Switzerland.

Gams, W., & Bissett, J. (2002). Morphology and identification of Trichoderma. In Christian P. K., & Gary, E. H. Trichoderma and Gliocladium. Basic biology, taxonomy and genetics. Taylor and Francis London. 1:3-31.

Gravel, V., Martinez, C., Antoun, H., & Tweddell, R. J. (2005). Antagonist microorganisms with the ability to control Pythium damping-off of tomato seeds in rockwool. *BioControl*, *50*(5), 771–786. DOI: 10.1007/s10526-005-1312-z

Harman, G. E. (2006). Overview of Mechanisms and Uses of Trichoderma spp. *Phytopathology*, *96*(2), 190–194. DOI: 10.1094/PHYTO-96-0190 PMID: 18943924

Hmouni, A., Douira, A., & Mouria, A. (2006, August). Biological Control of Tomato Grey Mould with Compost Water Extracts," Trichoderma" sp. and" Gliocladium" sp. *Phytopathologia Mediterranea*, 1000–1007.

Hmouni, A., Massoui, M., & Douira, A. (1999). Étude de l'activité antagoniste de Trichoderma spp. et de Gliocladium spp. à l'égard de Botrytis cinerea, agent causal de la pourriture grise de la tomate. *Al Awamia*, *99*, 75–92.

Kleifeld, O., & Chet, I. (1992). Trichoderma harzianum—Interaction with plants and effect on growth response. *Plant and Soil*, *144*(2), 267–272. DOI: 10.1007/BF00012884

Kribel, S., Qostal, S., Ouazzani Touhami, A., & Selmaoui, K. (2019). In vitro selection of moroccan phosphate sites' Trichoderma isolates according to their antagonism against the wheat and barley root rot pathogens. *Plant Cell Biotechnology and Molecular Biology*, *20*(15-16), 682–699.

Kribel, S., Qostal, S., Ouazzani Touhami, A., Selmaoui, K., Chliyeh, M., Benkirane, R., Achbani, E. H., & Douira, A. (2020). Effects of Trichoderma on growth and yield of wheat and barley and its survival ability on roots and amended rock phosphate growing substrates. [Journal of Fungal Biology]. *Current Research in Environmental & Applied Mycology*, *10*(1), 400–416. DOI: 10.5943/cream/10/1/32

Kribel, S., Qostal, S., Ouazzani Touhami, A., Selmaoui, K., Mouria, A., Benkirane, R., Achbani, El., & Douira, A. (2019). Quantitative and qualitative estimation of Moroccan Trichoderma isolates capacity to solubilize rock phosphate. *Acta Phytopathologica et Entomologica Hungarica*, *54*(2), 157–171. DOI: 10.1556/038.54.2019.016

Martinez, C., Blanc, F., Le Claire, E., Besnard, O., Nicole, M., & Baccou, J. C. (2001). Salicylic acid and ethylene pathways are differentially activated in melon cotyledons by active or heat denatured cellulase from Trichoderma longibrachiatum. *Plant Physiology*, *127*(1), 334–344. DOI: 10.1104/pp.127.1.334 PMID: 11553761

Mouria, A., Ouazzani Touhami, A., Douira, A., Benkirane, R., Mlaiki, A., & El-Yachioui, M. (1997). Antagonism in vitro of Trichoderma species against Pyricularia oryzae, 96:9-17.

Mouria, A., Ouazzani-Touhami, A., Mlaiki, A., El Yachioui, M., & Douira, A. (1997). Biological control against Helminthosporium oryzae: In Vivo Antagonism of Trichoderma sp. with Respect to Helminthosporium Oryzae. In *Third Congress of the Moroccan Association for the Protection of Plants*, Rabat (pp. 23-24).

Mouria, B., Ouazzani Touhami, A., & Douira, A. (2005). Effect of diverse flours on the competitiveness of inoculums of three Trichoderma strains against soil phytopathogens. *Bulletin of the Bordeaux Pharmacy Society*, *144*, 211–224.

Mouria, B., Ouazzani-Touhami, A., & Douira, A. (2013). Effect of Trichoderma harzianum compost on the suppression of verticillium wilt in tomatoes. *Journal of Applied Biosciences*, *70*, 5531–5543. DOI: 10.4314/jab.v70i1.98751

Mouria, B., Ouazzani-Touhami, A., Mouria, A., Benkirane, R., & Douira, A. (2015). Effect of compost and antagonistic fungi on suppression of Tomato Grey Mold. *Biolife : An International Quarterly Journal of Biology and Life Sciences*, *3*(2), 378–390. DOI: 10.17812/blj2015.32.5a

Najoua, M., Amina, O. T., Abdelmoti, A., Karima, S., Rachid, B., & Allal, D. (2022, May). Growth Promoting of Tomato Plants by Incorporation of Trichoderma asperellum Enriched Liquid Product via Foliar Spray and the Irrigation System. In *International Conference on Advanced Intelligent Systems for Sustainable Development* (pp. 599-608). Cham: Springer Nature Switzerland.

Nemec, S., Datnoff, L. E., & Strandberg, J. (1996). Efficacy of biocontrol agents in planting mixes to colonize plant roots and control root diseases of vegetables and citrus. *Crop Protection (Guildford, Surrey)*, *15*(8), 735–742. DOI: 10.1016/S0261-2194(96)00048-8

Ouazzani Chahdi, A., Chliyeh, M., Mouria, B., Dahmani, J., Ouazzani Touhami, A., Benkirane, R., El Hassan, A., & Douira, A. (2014). In vitro and in vivo effect of salinity on the antagonist potential of Trichoderma harzianum and sensitivity of tomato to Verticillium wilt. *International Journal of Recent Scientific Research*, *5*(4), 780–791.

Ouazzani-Touhami, A. (1994). In vitro antagonism of certain fungal species against Verticillium dahliae. *Rev. Netw. Improv. Agricult. Product. Arid Environ*, *7*, 197–211.

Qostal, S., Kribel, S., Chliyeh, M., Mouden, N., Selmaoui, K., Touhami, A. O., & Douira, A. (2020b). Biostimulant effect of Trichoderma on the development of wheat and barley plants and its survival aptitudes on the roots. *Plant Archives*, *20*(2), 7829–7834.

Qostal, S., Kribel, S., Chliyeh, M., Selmaoui, K., Serghat, S., Benkirane, R., & Douira, A. (2020). Management of wheat and barley root rot through seed treatment with biopestcides and fungicides. *Plant Cell Biotechnol. Mol. Biol*, *21*(35&36), 129-143.Z.

Rifai, M. A. (1969). A revision of the genus Trichoderma. Mycological papers no. 116. Commonwealth Mycological Institute, Kew, Surrey, England. 1969;56

Saravanakumar, S. S., Kumaravel, A., Nagarajan, T., Sudhakar, P., & Baskaran, R. (2013). Characterization of a novel natural cellulosic fiber from Prosopis juliflora bark. *Carbohydrate Polymers*, *92*(2), 1928–1933. DOI: 10.1016/j.carbpol.2012.11.064 PMID: 23399239

Sellal, Z., Ouazzani Touhami, A., Chliyeh, M., Mouden, N., Selmaoui, K., Dahmani, J., Benkirane, R., El Modafar, Ch., & Douira, A. (2020). Effect of seeds treatment with Trichoderma harzianum on argan plants growth. *Plant Cell Biotechnology and Molecular Biology*, *21*, 69–77.

Sellal, Z., Touhami, A. O., Mouden, N., Selmaoui, K., Dahmani, J., Benkirane, R., & Douira, A. (2023). Effect of the Combined Treatment of Endomycorrhizal Inoculum and Trichoderma Harzianum on the Growth of Argan Plants. In International Conference on Advanced Intelligent Systems for Sustainable Development (pp. 312-319). Cham: Springer Nature Switzerland.

Shah, S., Nasreen, S., & Sheikh, P. A. (2012). Cultural and morphological characterization of Trichoderma spp. associated with green mold disease of Pleurotus spp. in Kashmir. *Research Journal of Microbiology*, *7*(2), 139–144. DOI: 10.3923/jm.2012.139.144

Shoresh, M., Harman, G. E., & Mastouri, F. (2010). Induced systemic resistance and plant responses to fungal biocontrol agents. *Annual Review of Phytopathology*, *48*(1), 21–43. DOI: 10.1146/annurev-phyto-073009-114450 PMID: 20192757

Soufiani, M., Aissam, S., Boutaj, H., Ferradous, A., Douira, A., Meddich, A., & El Modafar, C. (2022). Effectiveness of indigenous arbuscular mycorrhizal consortium on the growth and mineral nutrition of Argania spinosa (L.) skeels. *Plant Biosystems*, *156*(6), 1365–1372. DOI: 10.1080/11263504.2022.2048280

Talbi, Z., Chliyeh, M., Mouria, B., El Asri, A., Ait Aguil, F., Ouazzani Touhami, A., Benkirane, R., & Douira, A. (2016). Effect of double inoculation with endomycorrhizae and Trichoderma harzianum on the growth of carob plants. *IJAPBC*, *5*(1), 44–58.

Vinale, F., Sivasithamparam, K., Ghisalberti, E. L., Marra, R., Barbetti, M. J., Li, H., Woo, S. L., & Lorito, M. (2008). A novel role for Trichoderma secondary metabolites in the interactions with plants. *Physiological and Molecular Plant Pathology*, *72*(1-3), 80–86. DOI: 10.1016/j.pmpp.2008.05.005

Yadav, J., Verma, J. P., & Tiwari, K. N. (2011). Plant growth promoting activities of fungi and their effect on chickpea plant growth. *Asian Journal of Biological Sciences*, *4*(3), 291–299. DOI: 10.3923/ajbs.2011.291.299

Yedidia, I., Benhamou, N., & Chet, I. (1999). Induction of defense responses in cucumber plants (Cucumis sativus L.) by the biocontrol agent Trichoderma harzianum. *Applied and Environmental Microbiology*, *65*(3), 1061–1070. DOI: 10.1128/AEM.65.3.1061-1070.1999 PMID: 10049864

Yedidia, I., Benhamou, N., Kapulnik, Y., & Chet, I. (2000). Induction and accumulation of PR proteins activity during early stages of root colonization by the mycoparasite Trichoderma harzianum strain T-203. *Plant Physiology and Biochemistry*, *38*(11), 863–873. DOI: 10.1016/S0981-9428(00)01198-0

Yedidia, I., Srivastva, A. K., Kapulnik, Y., & Chet, I. (2001). Effect of Trichoderma harzianum on microelement concentrations and increased growth of cucumber plants. *Plant and Soil*, *235*(2), 235–242. DOI: 10.1023/A:1011990013955

# Chapter 32
# Using Artificial Intelligence to Improve Hydrogen Production Systems

**Ayoub Ghailane**
*Mohammed V University in Rabat, Morocco*

**Jamal Mabrouki**
https://orcid.org/0000-0002-3841-7755
*Mohammed V University in Rabat, Morocco*

**Anass Ariss**
*Mohammed V University in Rabat, Morocco*

**Driss Azdem**
*Mohammed V University in Rabat, Morocco*

**Younes Abrouki**
*Mohammed V University in Rabat, Morocco*

## ABSTRACT

*Integrating artificial intelligence (AI) into hydrogen production systems represents an innovative approach to improving efficiency, optimizing resource management and reducing operating costs. This chapter explores the main applications of AI in green hydrogen production, focusing on intelligent energy resource management, process optimization and predictive maintenance. Machine learning algorithms and neural networks (ANN) can be used to adjust production parameters. Guaranteeing maximum efficiency and better adaptation to fluctuations in renewable energy sources. What's more, AI-based predictive maintenance reduces equipment failures, extends system lifetimes and improves reliability. The synergy between AI and hydrogen production not only promotes a more sustainable energy transition, but also the more efficient integration of hydrogen into smart energy grids, The adoption of these technologies represents a significant breakthrough in optimizing the efficiency and competitiveness of hydrogen production, paving the way for a more resilient, low-carbon energy system.*

DOI: 10.4018/979-8-3693-8980-5.ch032

## INTRODUCTION

In a global context marked by the climate emergency and the transition to low-carbon energies, hydrogen is emerging as a key energy vector for reducing greenhouse gas emissions and promoting a more sustainable future. (Fontecave, Candel, et Poinsot 2024). Its growing use in various sectors, such as industry, mobility and renewable energy storage (Rasul et al. 2022), makes it a strategic solution for overcoming the intermittent nature of green energy sources. However, current hydrogen production still relies mainly on fossil fuels (Raphélis-Soissan et Parenty 2022), with processes such as steam reforming, gasification and steam reforming of natural gas, which account for around 95% of global supply (Tetteh et Salehi 2022). Although these techniques are widely used due to their technological maturity, they generate massive carbon dioxide emissions, contributing to global warming and calling into question their long-term sustainability. Faced with these challenges, green hydrogen, produced from renewable sources via water electrolysis powered by solar or wind energy, is emerging as a promising, environmentally-friendly alternative (Bouramdane 2023a). Despite its high cost and stage of development, this technology remains a crucial tool for reducing the energy sector's carbon footprint.

However, hydrogen production and use present a number of technical and economic challenges (Zou et al. 2022). It is therefore crucial to optimize production, storage and transport processes in order to exploit the full potential of this resource. From this perspective, artificial intelligence (AI) offers innovative solutions to overcome these constraints and optimize efficiency in the hydrogen sector. The increasing use of AI in decision-making processes signals a radical change, consolidating its position in the energy transition. (Richter et al. 2022). With its modeling, prediction and optimization skills, artificial intelligence can play a decisive role in improving the performance of electrolyzers, the astute management of energy resources and the optimization of storage and distribution infrastructures. By integrating these advanced technologies, it is possible to increase the competitiveness of green hydrogen, reduce its costs and improve its viability on a large scale (Sai Ramesh et al. 2023). In this way, collaboration between artificial intelligence and hydrogen could facilitate the achievement of climate objectives, while supporting the development of a sustainable and economically viable energy sector. By optimizing the performance of production systems and improving resource management (Torky, Dahy, et Hassanein 2023), AI helps make hydrogen more competitive with fossil fuels.

This chapter explores the interaction between hydrogen and artificial intelligence technologies. After presenting the different methods of hydrogen production and the associated challenges, we will analyze how AI can contribute to better management of production and storage systems. Finally, we will discuss future prospects and expected innovations in this field, which could strengthen hydrogen's role in the global energy transition.

## 1. HYDROGEN PRODUCTION METHODS

Hydrogen can be produced by a variety of methods (Figure 1), which vary in terms of energy efficiency, cost and environmental impact. These methods are generally classified according to the energy source used and the associated greenhouse gas emissions. Grey hydrogen, the most commonly produced form of hydrogen today, is obtained from fossil fuels, mainly by steam methane reforming (SMR) (Ahn et al. 2023), A process in which natural gas reacts with steam at high temperature to produce hydrogen, and which is responsible for 48% of the hydrogen produced worldwide. (Bièvre 2022), while releasing large quantities of carbon dioxide ($CO_2$). Blue hydrogen is produced by steam reforming natural gas, which is broken down into hydrogen ($H_2$) and carbon dioxide ($CO_2$). This process reduces greenhouse gas emissions compared with conventional methods thanks to the capture and storage of $CO_2$, but still depends on non-renewable inputs (Bokde 2025). Turquoise hydrogen is obtained by methane pyrolysis, a process that breaks down natural gas into hydrogen and solid carbon (Fulcheri 2022). this method can be virtually emission-neutral. Yellow hydrogen is generated by the electrolysis of water, fuelled by nuclear energy (Figueiredo, Da Silva, et Ortiz 2023). It combines a reduced carbon footprint with large-scale hydrogen production. However, it is controversial because of the dangers associated with nuclear power. Black hydrogen is produced from coal using methods such as gasification or liquefaction (Midilli et al. 2021). It is frequently linked to significant $CO_2$ emissions, positioning it as one of the most polluting forms of hydrogen, Green hydrogen is produced from renewable sources through the process of water electrolysis, which splits water ($H_2O$) into hydrogen ($H_2$) and oxygen ($O_2$) using electricity (Shiva Kumar et Lim 2022). Unlike other types of hydrogen, this method generates no $CO_2$ emissions, making it a key solution for the energy transition.

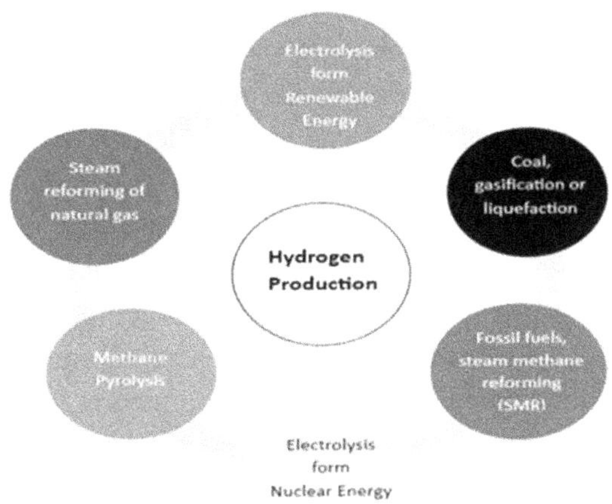


*Figure 2.*

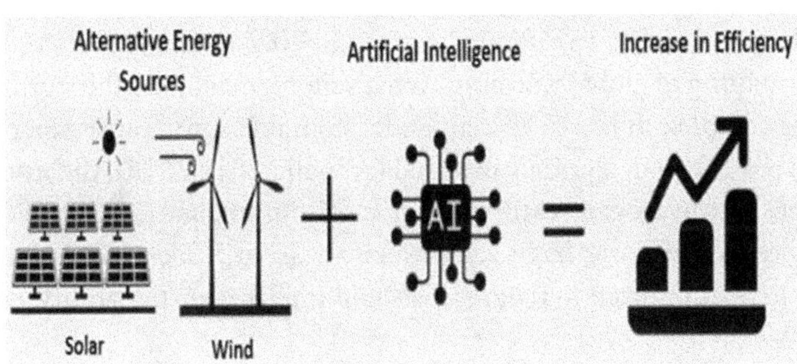

## 2. THE ROLE OF GREEN HYDROGEN IN THE ENERGY TRANSITION

Green hydrogen plays a crucial role in the energy transition as a clean energy carrier (Kovač, Paranos, et Marciuš 2021). Faced with the challenges of climate change and the need to reduce dependence on fossil fuels, it is a key solution for decarbonizing various sectors. (Seck et al. 2022), industry, transport and power generation, by replacing polluting sources and improving energy efficiency. Thanks to its ability to store and transport renewable energy, hydrogen can be used to offset the intermittent nature of solar and wind power sources, ensuring better management of production fluctuations. Its production by electrolysis, fuelled by electricity from renewable sources (Younas et al. 2022), makes it a particularly attractive alternative in a context where reducing carbon emissions is a priority. With the development of infrastructures and technologies, green hydrogen is set to play a central role in building a sustainable and resilient energy system, facilitating the transition to a low-carbon economy.

## 3. THE CHALLENGES OF HYDROGEN PRODUCTION

Although hydrogen is considered a key energy carrier for the energy transition, its large-scale production raises a number of technical, economic and environmental challenges. These challenges differ according to the manufacturing technique employed and the demands of the market. On the technical side, one of the main obstacles lies in the significant energy losses throughout the hydrogen cycle. During the electrolysis of green hydrogen, its liquefaction or conversion into various energy carriers, as well as during its transport and use in fuel cells, a significant fraction of the initial energy is dissipated. This inefficiency calls for massive use of renewable energies to compensate for these losses, which can compete with other electricity needs, notably the electrification of end consumers (Agaton, Batac, et Reyes Jr. 2022). One of the major challenges of hydrogen production, especially green hydrogen from water electrolysis, is its high cost compared with fossil fuels (Y. Zhang 2024). Production requires large quantities of electricity, and access to abundant renewable energy. In terms of environmental impact, hydrogen production represents a considerable challenge. Currently, the most widely used method is steam methane reforming (SMR) (Faheem et al. 2021), which emits large quantities of $CO_2$, a greenhouse gas contributing to global warming. What's more, managing the storage and transportation of hydrogen can prove complex. It has to be kept under high pressure, transformed into a liquid at very low temperatures and preserved in a porous material (Abdalla et al. 2018). In terms of environmental impact, hydrogen production represents a considerable challenge. Currently, the most widely used method is steam methane reforming (SMR) (Faheem et al. 2021), which emits large quantities of $CO_2$, a greenhouse gas contributing to global warming. What's more, managing the storage and transportation of hydrogen can prove complex. It has to be kept under high pressure, transformed into a liquid at very low temperatures and preserved in a porous material (Abdalla et al. 2018). the infrastructure needed to produce, store and transport hydrogen is still inadequate (Bouramdane 2023b). What's more, hydrogen production technologies, particularly large-scale electrolysis, still lack industrial maturity. Although promising, they need to be optimized to reduce costs and improve sustainability.

## 4. L'INTELLIGENCE ARTIFICIELLE (IA): CONCEPTS ET APPLICATIONS

### 4.1 Definition and Principles of AI

Artificial intelligence (AI) refers to all computer methods and technologies that can reproduce certain human cognitive abilities (B. Huang et al. 2019), such as learning, perception, decision-making and problem-solving. Its main objective is to enable machines to perform tasks that normally require human intelligence, relying on advanced algorithms and mathematical models. In recent years, artificial intelligence has evolved rapidly and changed the way people live their lives. (C. Huang et al. 2019). The development of AI has become a crucial strategy for global development, strengthening national competitiveness while safeguarding security. (C. Zhang et Lu 2021). Artificial intelligence has been integrated into many strategic fields, such as the food industry, healthcare, safety and security, agriculture and education, among others. Its operation is based on several fundamental principles, including machine learning, which enables computer systems to improve their performance based on experience and data, without the need for explicit programming. This approach relies on algorithms capable of identifying trends and adapting models to new data sets. Machine learning addresses four major challenges: prediction, clustering, classification and dimensionality reduction (Loukili et al., 2021). It is divided into four main categories according to the classification of learning techniques: supervised learning, unsupervised learning, semi-supervised learning and reinforcement learning (Mabrouki, Benbouzid, Dhiba, & El Hajjaji, 2022). Another key element of AI is Big Data processing, which is an essential prerequisite for its effectiveness. Analyzing and exploiting large quantities of data improves the recognition rate and accuracy of artificial intelligence models, enhancing their ability to make optimized decisions and adapt to complex environments. The more varied and abundant the data, the greater the efficiency of intelligence models. These large quantities of high-dimensional data make data more complete and more sufficient to support the development of AI. (Chen et Lin 2014). In industry, it plays a key role in improving operational efficiency, predictive maintenance and strategic decision-making. Another essential aspect of AI is artificial neural networks (ANN), which are inspired by the way the human brain works, and are composed of interconnected neurons processing information in multiple layers (Abdelkareem et al. 2022). They are capable of solving complex problems such as image recognition or trend forecasting.

### 4.2 Artificial Intelligence Applications in the Energy Sector

Artificial intelligence plays a key role in optimizing and modernizing the energy sector, by improving network efficiency, resource management and the integration of renewable energies (Şerban et Lytras 2020). it is widely used in smart grids (Bose 2017), to adjust electricity demand, reducing energy losses and improving grid stability. AI can also predict renewable energy production by analyzing meteorological data to anticipate fluctuations in wind and solar power (Sai Ramesh et al. 2023), facilitating their integration into the energy mix. In the field of energy storage, advanced algorithms optimize the operation of batteries and hydrogen storage stations according to needs and variations in production. Recent technologies have the capacity to minimize human intervention in battery management. These technologies could increase the efficiency of technological systems (Marra et al. 2013). Artificial intelligence is set to revolutionize the energy sector by helping to manage the inherently unpredictable and random nature of energy, while promoting the spread of renewable energies.

## 5. USING AI TO IMPROVE GREEN HYDROGEN PRODUCTION

Integrating artificial intelligence into hydrogen production offers significant opportunities for optimizing various aspects of the process. It makes it possible to improve energy efficiency (Farzaneh et al. 2021), facilitate the integration of renewable energies and optimize the management of electrolyzers within the green hydrogen production and storage system. AI also helps reduce production costs (S.mohammed et al. 2025), while enhancing system reliability and performance.

### 5.1. Intelligent Management of Energy Resources

Artificial intelligence enables optimized management of the energy resources used to produce green hydrogen. (Bassey et Ibegbulam 2023), including solar and wind power (Simankov et al. 2024). Thanks to machine learning-based forecasting algorithms (Kabir et al. 2023), AI can anticipate variations in the production of renewable energies and adjust the energy input to electrolyzers. It also facilitates the integration of storage systems by balancing supply and demand, minimizing energy losses and ensuring more stable and efficient operation of infrastructures.

### 5.2 Optimizing Production Processes

Artificial intelligence is a powerful tool for improving the efficiency of electrolyzers by dynamically adjusting operating parameters (temperature, pressure, electric current, etc.) (Yang, Mohamed, et Kim 2023). Thanks to techniques such as multi-criteria optimization (Černevičienė et Kabašinskas 2022), neural networks (ANN) (Figure 4) and reinforcement learning algorithms, AI can identify optimal conditions for maximizing hydrogen production efficiency while reducing energy consumption. It can also adapt strategies to the variability of renewable resources. (Gonzalez et al. 2023), AI enables intelligent, adaptive control of the electrolysis process, contributing to more efficient production.

*Figure 3.*

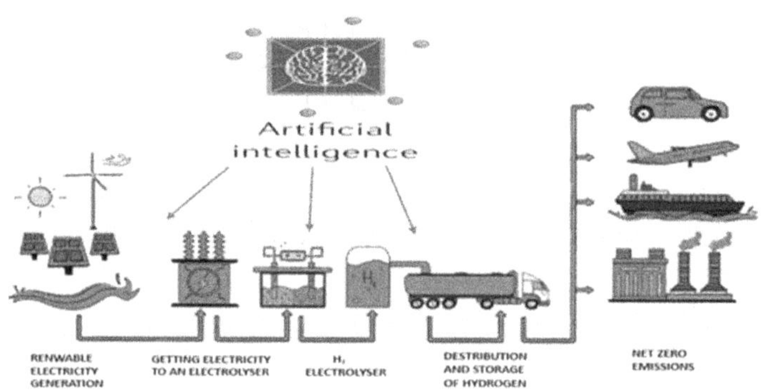

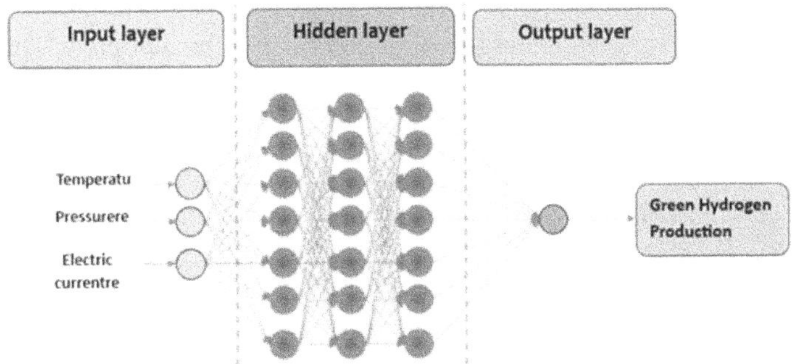


## 5.3 Predictive Maintenance and System Monitoring

Maintaining hydrogen production infrastructures is a major challenge (Ishaq, Dincer, et Crawford 2022), due to wear and tear on electrolyzers and components exposed to intense operating conditions. Integrating artificial intelligence into the maintenance and monitoring of hydrogen production systems enhances their reliability while reducing operating costs. By analyzing data, AI can detect anomalies, anticipate breakdowns and recommend interventions before a major malfunction occurs. Predictive maintenance models based on deep learning and regression models (Serradilla et al. 2022), are able to monitor the condition of electrolyzers, storage systems and distribution networks. This helps to extend equipment longevity, while guaranteeing constant, safe production.

## 6. FUTURE PROSPECTS

Integrating artificial intelligence into hydrogen production opens up new prospects for the energy sector, by optimizing production, storage and distribution processes. Thanks to its advanced data processing and machine learning capabilities, AI can improve the intelligent management of energy resources by dynamically adapting to the availability of renewable energies. By exploiting machine learning algorithms, predictive models and intelligent control systems, it helps to increase energy efficiency, reduce production costs and minimize environmental impact. In the future, AI is set to become a key lever for accelerating the transition to a sustainable and competitive hydrogen economy, while supporting global goals for decarbonization and energy innovation.

## CONCLUSION

The integration of artificial intelligence into hydrogen production represents a major step forward in the development of a more efficient, sustainable and competitive energy system The integration of artificial intelligence into hydrogen production represents a major step forward in the development of a more efficient, sustainable and competitive energy system. Thanks to its analysis, modeling and optimization capabilities, AI offers the possibility of improving every stage of the process, from energy resource management to electrolyzer optimization, via predictive infrastructure maintenance. Moreover, artificial intelligence is emerging as a strategic tool for accelerating the transition to a low-carbon economy, improving the competitiveness of hydrogen and meeting global climate challenges. Its growing deployment in this sector could mean a sustainable energy revolution, contributing to a cleaner, more resilient future.

## REFERENCES


Abdalla, A. M., Hossain, S., Nisfindy, O. B., Azad, A. T., Dawood, M., & Azad, A. K. (2018). Hydrogen production, storage, transportation and key challenges with applications: A review. *Energy Conversion and Management, 165*(juin), 602–627. DOI: 10.1016/j.enconman.2018.03.088

Abdelkareem, M. A., Soudan, B., Mahmoud, M. S., Sayed, E. T., AlMallahi, M. N., Inayat, A., Al Radi, M., & Olabi, A. G. (2022). Progress of artificial neural networks applications in hydrogen production. *Chemical Engineering Research & Design, 182*(juin), 66–86. DOI: 10.1016/j.cherd.2022.03.030

Agaton, C. B., Kenneth, I. T. B., & Reyes, E. M.Jr. (2022). Prospects and challenges for green hydrogen production and utilization in the Philippines. *International Journal of Hydrogen Energy, 47*(41), 17859–17870. DOI: 10.1016/j.ijhydene.2022.04.101

Ahn, S.-Y., Kim, K.-J., Kim, B.-J., Hong, G.-R., Jang, W.-J., Bae, J. W., Park, Y.-K., Jeon, B.-H., & Roh, H.-S. (2023). From gray to blue hydrogen: Trends and forecasts of catalysts and sorbents for unit process. *Renewable & Sustainable Energy Reviews, 186*(octobre), 113635. DOI: 10.1016/j.rser.2023.113635

Al-Jadabi, N., Laaouan, M., El Hajjaji, S., Mabrouki, J., Benbouzid, M., & Dhiba, D. (2023). The dual performance of Moringa oleifera seeds as eco-friendly natural coagulant and as an antimicrobial for wastewater treatment: A review. *Sustainability (Basel), 15*(5), 4280. DOI: 10.3390/su15054280

Bassey, K., & Ibegbulam, C. (2023). MACHINE LEARNING FOR GREEN HYDROGEN PRODUCTION. *Computer Science & IT Research Journal, 4*(décembre), 368–385. DOI: 10.51594/csitrj.v4i3.1253

Benchrifa, M., Elouardi, M., Fattah, G., Mabrouki, J., & Tadili, R. (2023). Identification, simulation and modeling of the main power losses of a photovoltaic installation and use of the internet of things to minimize system losses. In *Advanced technology for smart environment and energy* (pp. 49–60). Springer International Publishing. DOI: 10.1007/978-3-031-25662-2_4

Bièvre, D. (2022). TRANSITION ÉNERGÉTIQUE: LE POINT SUR LE «HY 1». Les Notes du CREOGN.

Bokde, N. D. (2025). A unified framework for multi-type hydrogen production and storage in renewable energy systems. *Energy Conversion and Management, X*(janvier), 100847. DOI: 10.1016/j.ecmx.2024.100847

Bose, B. K. (2017). Artificial Intelligence Techniques in Smart Grid and Renewable Energy Systems - Some Example Applications. *Proceedings of the IEEE, 105*(11), 2262–2273. DOI: 10.1109/JPROC.2017.2756596

Bouramdane, A. A. (2023). Un Avenir Hydrogéné au Maroc: Dessalement de l'Eau–Défis et Promesses. énergie/mines & carrières.

Bouramdane, A.-A. (2023a). « Production d'hydrogène vert au Maroc: Quelle technologie est la plus adaptée à différents niveaux de pénétration renouvelable? », juillet. DOI: 10.5281/zenodo.8144588

Černevičienė, J., & Kabašinskas, A. (2022). Review of Multi-Criteria Decision-Making Methods in Finance Using Explainable Artificial Intelligence. *Frontiers in Artificial Intelligence, 5*(mars), 827584. Advance online publication. DOI: 10.3389/frai.2022.827584 PMID: 35360662

Chen, X.-W., & Lin, X. (2014). Big Data Deep Learning: Challenges and Perspectives. *IEEE Access : Practical Innovations, Open Solutions, 2*, 514–525. DOI: 10.1109/ACCESS.2014.2325029

de Raphélis-Soissan, P., & Parenty, A. (2022, November). Les carburants de synthèse produits à partir d'hydrogène électrolytique et des émissions de CO 2 du secteur cimentier: un vecteur de décarbonation du transport maritime. In Annales des Mines-Réalités industrielles (No. 4, pp. 81-85). Cairn/Softwin.

El Alouani, M., Aouan, B., Rachdi, Y., Alehyen, S., El Herradi, E. H., Saufi, H., Mabrouki, J., & Barka, N. (2024). Porous geopolymers as innovative adsorbents for the removal of organic and inorganic hazardous substances: A mini-review. *International Journal of Environmental Analytical Chemistry, 104*(16), 4784–4796. DOI: 10.1080/03067319.2022.2115891

Faheem, H. H., Tanveer, H. U., Abbas, S. Z., & Maqbool, F. (2021). Comparative study of conventional steam-methane-reforming (SMR) and auto-thermal-reforming (ATR) with their hybrid sorption enhanced (SE-SMR & SE-ATR) and environmentally benign process models for the hydrogen production. *Fuel, 297*(août), 120769. DOI: 10.1016/j.fuel.2021.120769

Farzaneh, H., Malehmirchegini, L., Bejan, A., Afolabi, T., Mulumba, A., & Daka, P. P. (2021). Artificial Intelligence Evolution in Smart Buildings for Energy Efficiency. *Applied Sciences (Basel, Switzerland), 11*(2), 763. DOI: 10.3390/app11020763

Figueiredo, R. L., Da Silva, J. M., & Carlos, E. A. O. (2023). Green Hydrogen: Decarbonization in Mining - Review. *Cleaner Energy Systems, 5*(août), 100075. DOI: 10.1016/j.cles.2023.100075

Fontecave, M., Candel, S., & Poinsot, T. (2024). L'hydrogène aujourd'hui et demain (Doctoral dissertation, Académie des sciences (France)).

Fulcheri, L. (2021). La pyrolyse du méthane: de l'hydrogène «gris» à l'hydrogène «turquoise». L'Actualité Chimique.

Ghizlane, F., Mabrouki, J., Ghrissi, F., & Azrour, M. (2022). Proposal for a high-resolution particulate matter (PM10 and PM2. 5) capture system, comparable with hybrid system-based internet of things: Case of quarries in the western rif, Morocco. *Pollution, 8*(1), 169–180.

Gonzalez, J. M., Tomlinson, J. E., Martínez Ceseña, E. A., Basheer, M., Obuobie, E., Padi, P. T., Addo, S., Baisie, R., Etichia, M., Hurford, A., Bottacin-Busolin, A., Matthews, J., Dalton, J., Smith, D. M., Sheffield, J., Panteli, M., & Harou, J. J. (2023). Designing Diversified Renewable Energy Systems to Balance Multisector Performance. *Nature Sustainability, 6*(4), 415–427. DOI: 10.1038/s41893-022-01033-0

Huang, B., Huan, Y., Da Xu, L., Zheng, L., & Zou, Z. (2019). Automated trading systems statistical and machine learning methods and hardware implementation: A survey. *Enterprise Information Systems, 13*(1), 132–144. DOI: 10.1080/17517575.2018.1493145

Huang, C., Cai, H., Xu, L., Xu, B., Gu, Y., & Jiang, L. (2019). Data-driven ontology generation and evolution towards intelligent service in manufacturing systems. *Future Generation Computer Systems, 101*(décembre), 197–207. DOI: 10.1016/j.future.2019.05.075

Ishaq, H., Dincer, I., & Crawford, C. (2022). A review on hydrogen production and utilization: Challenges and opportunities. *International Journal of Hydrogen Energy*, *47*(62), 26238–26264. DOI: 10.1016/j.ijhydene.2021.11.149

Kabir, M. M., Roy, S. K., Alam, F., Nam, S. Y., Im, K. S., Tijing, L., & Shon, H. K. (2023). Machine learning-based prediction and optimization of green hydrogen production technologies from water industries for a circular economy. *Desalination*, *567*(décembre), 116992. DOI: 10.1016/j.desal.2023.116992

Kovač, A., Paranos, M., & Marciuš, D. (2021). Hydrogen in energy transition: A review. *International Journal of Hydrogen Energy*, *46*(16), 10016–10035. DOI: 10.1016/j.ijhydene.2020.11.256

Loukili, H., Mabrouki, J., Anouzla, A., Kouzi, Y., Younssi, S. A., Digua, K., & Abrouki, Y. (2021). Pretreated Moroccan natural clays: Application to the wastewater treatment of textile industry. *Desalination and Water Treatment*, *240*, 124–136. DOI: 10.5004/dwt.2021.27644

Mabrouki, J., Benbouzid, M., Dhiba, D., & El Hajjaji, S. (2022). Simulation of wastewater treatment processes with Bioreactor Membrane Reactor (MBR) treatment versus conventional the adsorbent layer-based filtration system (LAFS). *International Journal of Environmental Analytical Chemistry*, *102*(19), 7458–7468. DOI: 10.1080/03067319.2020.1828394

Marra, D., Sorrentino, M., Pianese, C., & Iwanschitz, B. (2013). A neural network estimator of Solid Oxide Fuel Cell performance for on-field diagnostics and prognostics applications. *Journal of Power Sources*, *241*(novembre), 320–329. DOI: 10.1016/j.jpowsour.2013.04.114

Mgalaa, S., Mabrouki, J., Elouardi, M., El Azzouzi, L., Moufti, A., El Hajjaji, S., EL Azzouzi, M., & El Belghiti, M. A. (2022). Study and evaluation of the degradation of procion blue dye by the ozonation method: Parametric and isothermal study. *Nanotechnology for Environmental Engineering*, *7*(3), 691–697. DOI: 10.1007/s41204-022-00262-0

Midilli, A., Kucuk, H., Topal, M. E., Akbulut, U., & Dincer, I. (2021). A comprehensive review on hydrogen production from coal gasification: Challenges and Opportunities. *International Journal of Hydrogen Energy*, *46*(50), 25385–25412. DOI: 10.1016/j.ijhydene.2021.05.088

Rachiq, T., Abrouki, Y., Mabrouki, J., Samghouli, N., Fersi, C., Rahal, S., & El Hajjaji, S. (2021). Evaluation of the efficiency of different materials to remove specific pollutants from landfill leachate. *Desalination and Water Treatment*, *238*, 240–250. DOI: 10.5004/dwt.2021.27779

Rasul, M. G., Hazrat, M. A., Sattar, M. A., Jahirul, M. I., & Shearer, M. J. (2022). The future of hydrogen: Challenges on production, storage and applications. *Energy Conversion and Management*, *272*(novembre), 116326. DOI: 10.1016/j.enconman.2022.116326

Richter, L., Lehna, M., Marchand, S., Scholz, C., Dreher, A., Klaiber, S., & Lenk, S. (2022). Artificial Intelligence for Electricity Supply Chain Automation. *Renewable & Sustainable Energy Reviews*, *163*(juillet), 112459. DOI: 10.1016/j.rser.2022.112459

Sai Ramesh, A., Vigneshwar, S., & Sundaram Vickram, S. (2023). Artificial intelligence driven hydrogen and battery technologies – A review. *Fuel*, *337*(avril), 126862. DOI: 10.1016/j.fuel.2022.126862

Seck, G. S., Hache, E., Sabathier, J., Guedes, F., Reigstad, G. A., Straus, J., Wolfgang, O., Ouassou, J. A., Askeland, M., Hjorth, I., Skjelbred, H. I., Andersson, L. E., Douguet, S., Villavicencio, M., Trüby, J., Brauer, J., & Cabot, C. (2022). Hydrogen and the decarbonization of the energy system in europe in 2050: A detailed model-based analysis. *Renewable & Sustainable Energy Reviews*, *167*(octobre), 112779. DOI: 10.1016/j.rser.2022.112779

Şerban, A. C., & Lytras, M. D. (2020). Artificial Intelligence for Smart Renewable Energy Sector in Europe—Smart Energy Infrastructures for Next Generation Smart Cities. *IEEE Access : Practical Innovations, Open Solutions*, *8*, 77364–77377. DOI: 10.1109/ACCESS.2020.2990123

Serradilla, O., Zugasti, E., Rodriguez, J., & Zurutuza, U. (2022). Deep Learning Models for Predictive Maintenance: A Survey, Comparison, Challenges and Prospects. *Applied Intelligence*, *52*(10), 10934–10964. DOI: 10.1007/s10489-021-03004-y

Shiva Kumar, S., & Lim, H. (2022). An overview of water electrolysis technologies for green hydrogen production. *Energy Reports*, *8*(novembre), 13793–13813. DOI: 10.1016/j.egyr.2022.10.127

Simankov, V., Buchatskiy, P., Kazak, A., Teploukhov, S., Onishchenko, S., Kuzmin, K., & Chetyrbok, P. (2024). A Solar and Wind Energy Evaluation Methodology Using Artificial Intelligence Technologies. *Energies*, *17*(2), 416. DOI: 10.3390/en17020416

Tetteh, D. A., & Salehi, S. (2022). The Blue Hydrogen Economy: A Promising Option for the Near-to-Mid-Term Energy Transition. *Journal of Energy Resources Technology*, *145*(042701), 042701. Advance online publication. DOI: 10.1115/1.4055205

Torky, M., Dahy, G., & Hassanein, A. E. (2023). GH2_MobileNet: Deep learning approach for predicting green hydrogen production from organic waste mixtures. *Applied Soft Computing*, *138*(mai), 110215. DOI: 10.1016/j.asoc.2023.110215

Wala'a, A. M., Sameer, S., & Saeed, M. N. (2024). Applications of Artificial Intelligence in environmental and economic life cycle analysis of green hydrogen production: Improving efficiency and sustainability. *Journal of Madenat Alelem University College*, *16*(2).

Yang, R., Mohamed, A., & Kim, K. (2023). Optimal design and flow-field pattern selection of proton exchange membrane electrolyzers using artificial intelligence. *Energy*, *264*(février), 126135. DOI: 10.1016/j.energy.2022.126135

Younas, M., Shafique, S., Hafeez, A., Javed, F., & Rehman, F. (2022). An Overview of Hydrogen Production: Current Status, Potential, and Challenges. *Fuel*, *316*(mai), 123317. DOI: 10.1016/j.fuel.2022.123317

Zhang, C., & Lu, Y. (2021). Study on artificial intelligence: The state of the art and future prospects. *Journal of Industrial Information Integration*, *23*(septembre), 100224. DOI: 10.1016/j.jii.2021.100224

Zhang, Y. (2024). « Application de composites à base de MXenes pour la production de l'hydrogènepar électrolyse de l'eau », septembre. https://lilloa.univ-lille.fr/handle/20.500.12210/116609

Zou, C., Li, J., Zhang, X., Jin, X., Xiong, B., Yu, H., Liu, X., Wang, S., Li, Y., Zhang, L., Miao, S., Zheng, D., Zhou, H., Song, J., & Pan, S. (2022). Industrial status, technological progress, challenges, and prospects of hydrogen energy. *Natural Gas Industry B*, *9*(5), 427–447. DOI: 10.1016/j.ngib.2022.04.006

# About the Contributors

**Jamal Mabrouki** received his PhD in Process and Environmental Engineering at Mohammed V University in Rabat, specializing in artificial intelligence and smart automatic systems. He completed the Bachelor of Science in Physics and Chemistry with honors from Hassan II University in Casablanca, Morocco and the engineer in Environment and smart system. His research is on intelligent monitoring, control, and management systems and more particularly on sensing and supervising remote intoxication systems, smart self-supervised systems and recurrent neural networks. He has published several papers in conferences and indexed journals, most of them related to artificial intelligent systems, internet of things or big data and mining. Jamal will currently work in environment, energy and smart system professor at Mohammed V University in Rabat, Faculty of Science. Jamal is scientific committee member of numerous national and international conferences. He is also a reviewer of Modeling Earth Systems and Environment; International Journal of Environmental Analytical Chemistry; International Journal of Modeling, Simulation, and Scientific Computing; The Journal of Supercomputing, Energy & Environment and Big Data Mining and Analytics...

***

**Yassin Allammari**, a Doctor in Management and Entrepreneurship, is affiliated with the Management, Finance, Digitization, and Applied Statistics Research Laboratory at the Faculty of Economic and Social Sciences, Abdelmalek Essaadi University. He is a researcher specializing in the fields of entrepreneurship, strategy, market dynamics, and SME performance. His work focuses on studying entrepreneurial and strategic dynamics that influence the competitiveness and sustainability of small and medium-sized enterprises.

**H. Bhoomeeswaran** is an Assistant Professor of Physics at SRM University, Madurai, specializing in advanced topics in condensed matter physics and spin-based technologies. With a profound passion for research and innovation, Dr. H.B s academic journey reflects a deep commitment to the advancement of scientific knowledge.

**Sonali R Dhokpande** has Master degree in Food Technology from Rashtrasant Tukadoji Maharaj Nagpur University Maharashtra India (2004) and PhD in Chemical Technology from SGBA University, Amravati, Maharashtra India (2018). She is currently working as Assistant professor of Chemical Engineering Department Datta Meghe College of Engineering, Airoli Navi Mumbai. She has experience in Chemical Engineering with Waste water treatment, Reaction Engineering, Equipment Design and Food technology.

**Abdelkrim El Assfouri** is an eminent Moroccan biologist specializing in mycology, a fascinating field dedicated to the study of fungi. He currently works as a teacher for the Moroccan Ministry of Education. With a PhD in biology, Dr. El Assfouri has developed in-depth expertise in the study of fungi, particularly in their ecology and biodiversity. He is also interested in exploring the complex interactions between fungi and their environment, with a particular focus on their role in local ecosystems and their potential as natural resources. Dr. El Assfouri is recognized for his significant contributions to scientific research. He has published several articles in leading international journals and participated in scientific conferences, sharing his discoveries and collaborating with researchers from all over the world.

**Mouna Hajjaj** received her PhD degree in Management from Ibn Tofail University. With extensive experience in the higher education sector, she held several senior positions before assuming her current role as Assistant Professor of Management at the Faculty of Economics and Management at Hassan First University of Settat. Her research interests span a wide range of topics including project management, organisation sciences, human resources, and business administration.

**J. Mangaiyarkkarasi** is currently working as an Associate Professor of Physics in NMSS Vellaichamy Nadar College, Nagamalai, Madurai- 625 019, Tamil Nadu, India. She has 23 years of teaching experience. Her major research interest includes synthesis of ceramic materials, powder X-ray analysis and electron density studies. She has published 11 research papers in the refereed International journals and five book chapters. She has prepared study materials (UG& PG Physics) for Madurai Kamaraj University Distance Education Course.

**Shalini Jaiswal** has more than 17 years' experience in academics and research and is the author of more than 65 scientific publications in Scopus\SCI\UGC and other reputed journal\Conference. Dr. Shalini Jaiswal received her D.Phil degree in Organic Chemistry from University of Allahabad, Allahabad India, in 2009. Dr. Shalini Jaiswal has14 month experience as a project fellow on UGC sponsored project entitled Green synthesis of multi- target nucleosides as a potential anti-viral agent. Her research interest includes different area of Chemistry, Nanotechnology, AI, Biotechnology and Environmental science. She is a Life Member of the Indian science congress association. She has reviewed more than 50 papers of various reputed journal and conferences. She is Editorial board member and Reviewer of various reputed journals like TPI (International Journal), IJATES, International journal of advance research, Advanced Scientific Research, IJEEE, Asian Journal of Research in Chemistry (AJRC) etc.Dr. Shalini Jaiswal Worked as Moderator in TOKYO SUMMIT-II (2020) and TOKYO SUMMIT-III (2021), Turkey (2022,2023,2024)

**Kazi Kutubuddin Sayyad Liyakat** has completed his B.E., M.E., and Ph.D. in E&TC Engineering and is nowadays working as a Professor & Head of Department, E&TC Engineering Department and was Dean R&D. He is Post-Doctoral Fellow working on "IoT in Healthcare Applications". His area of Interest is IoT, AI and ML. He has published more than 110+ papers in various Journals. Also published 11 books in the field of Engineering. He has published 15 Indian Patents, 2 South African Grant Patent, 2 Indian Copyright Patents and 8 UK Grant Patent. He worked as a Reviewer for Scopus Conferences and Journal. Also work as Editorial Board Member for various Journals. He got 2 Best Researcher Award, Best Faculty Award and Appreciation Letter from MoE's Innovation Cell, Govt. of India.

**M.Jegadeeswaran** is currently working as a Assistant Professor, Department Of Agricultural Engineering at Agricultural Engineering at Rathinam Technical Campus, Coimbatore, Tamilnadu, India. He Completed B.E (Agriculture Engineering). in the year of MAY 2020, at Paavai Engineering College (Autonomous), Namakkal and M.E (Environmental Engineering) at Gnanamani College of Technology (Anna University), Namakkal.

**T.C. Manjunath** was born on 6th February 1967. He did his BE from RV College of Engg. in 1989, ME in Robotics & Automation from LD College of Engg. in 1992 with First Rank and PhD in Systems & Control Engg. from IIT Bombay in the year 2007. He is currently working as the Dean Research R D and as Professor in the Dept. of Computer Science & Engineering with specialization in IoT, Cyber Security & Blockchain Technology in Rajarajeswari College of Engineering, Bengaluru, Karnataka. He has completed B.E. Degree from R.V. College of Engg., M.E. Govt.'s LD College of Engg., Gujarat Univ., Ahmedabad in First Rank and Ph.D. in Systems and Control Engineering from the Indian Institute of Technology Bombay IIT Bombay He has got a teaching academic, research and administrative experience of more than 35+ years 10 years as Principal, 8 years as HoD rest as faculty and has published more than 1200+ research papers in various national & international journals and conferences, with 175+ patents, 25 book-chapters, 25 awards and also published 25 text books in various field of science, engineering and technology, completed projects worth Rs. 9 Crores till date apart from guiding 10 PhD scholars. He is a member of 30 Professional Society Chapters and is a fellow of IE, IETE, AMIEE and a chartered engineer. He got the best research scholar award in 2006 in IIT Bombay amongst all the research scholars of the institute. He has got 10 awards in academics and research activities.

**Santosh Kumar Nathsharma** is a dedicated faculty member in the Department of Chemistry at Stewart Science College, located in Cuttack, Odisha. With a strong passion for chemical sciences and education, Dr. Nathsharma has been actively involved in teaching, mentoring students, and contributing to academic excellence at the institution. His work reflects a commitment to fostering scientific curiosity and academic growth among his students.

**P. Selvakumar** is currently an Associate Professor in the Department of Science and Humanities at Nehru Institute of Technology, Coimbatore, Tamil Nadu, India. Since 2023, he has served as the Department IQAC Coordinator at the institute. Prior to joining Nehru Institute of Technology, he held positions as an Assistant Professor in the Department of Science and Humanities various engineering colleges in Coimbatore, Tamil Nadu, India, spanning from 2011 to 2023. Dr.P. Selvakumar's academic and research interests lie primarily in the field of phytochemistry. He has made significant contributions to his field, with an impressive record that includes 30 journal publications, 12 patents, and 20 conference papers. Additionally, he has organized 5 seminars and conferences, further demonstrating his commitment to academic and scientific exchange. Throughout his career, Dr. P.Selvakumar has been recognized for his dedication and excellence in teaching, having been honored with several awards, including the Best Teacher Awards. He holds a prominent position in the academic community, contributing actively to research.

**El Hassania Rahou** holds a PhD in Economics and Management from Abdelmalik Essaâdi University. She accumulated substantial experience in the field of teaching and education. Currently, she holds the position of Assistant Professor of Higher Education at the Faculty of Economics and

Management at Hassan First University of Settat. Her research interests encompass a broad spectrum of topics, such as entrepreneurship, marketing and management, human resources and accounting.

**S. Seenivasan** is a dedicated academic professional in the Department of Mechanical Engineering at Rathinam Technical Campus, located in Coimbatore, Tamil Nadu (Postal Code: 641021). With a strong commitment to education and research, Dr. Seenivasan contributes significantly to the development of future engineers through innovative teaching and mentorship.

# Index

## F



## G



## H

# I



# M



# N



# P

# Q



# R



# S



# T

# W

www.ingramcontent.com/pod-product-compliance
Lightning Source LLC
LaVergne TN
LVHW080404011026
849161LV00009B/2344
*9798369389805*